Green Energy and Technology

AF183393

Climate change, environmental impact and the limited natural resources urge scientific research and novel technical solutions. The monograph series Green Energy and Technology serves as a publishing platform for scientific and technological approaches to "green"—i.e. environmentally friendly and sustainable—technologies. While a focus lies on energy and power supply, it also covers "green" solutions in industrial engineering and engineering design. Green Energy and Technology addresses researchers, advanced students, technical consultants as well as decision makers in industries and politics. Hence, the level of presentation spans from instructional to highly technical.

Indexed in Scopus.

Indexed in Ei Compendex.

Adel Mellit · Lassaad Sbita · Karim Kemih ·
Malek Ghanes
Editors

Proceedings of the 2nd International Conference on Green Energy Conversion System

ICGECS 2023, 29 September–1 October 2023, Djerba, Tunisia

Springer

Editors
Adel Mellit
Faculty of Sciences and Technology
Jijel University
Jijel, Algeria

Karim Kemih
Faculty of Sciences and Technology
Jijel University
Jijel, Algeria

Lassaad Sbita
LR18ES34-ENIG Research Laboratory,
National Engineering School of Gabes
University of Gabes
Gabes, Tunisia

Malek Ghanes
École Centrale Nantes, CNRS, LS2N
Nantes Université
Nantes, France

ISSN 1865-3529 ISSN 1865-3537 (electronic)
Green Energy and Technology
ISBN 978-981-97-6150-0 ISBN 978-981-97-6148-7 (eBook)
https://doi.org/10.1007/978-981-97-6148-7

© The Editor(s) (if applicable) and The Author(s), under exclusive license to Springer Nature Singapore Pte Ltd. 2024

This work is subject to copyright. All rights are solely and exclusively licensed by the Publisher, whether the whole or part of the material is concerned, specifically the rights of translation, reprinting, reuse of illustrations, recitation, broadcasting, reproduction on microfilms or in any other physical way, and transmission or information storage and retrieval, electronic adaptation, computer software, or by similar or dissimilar methodology now known or hereafter developed.
The use of general descriptive names, registered names, trademarks, service marks, etc. in this publication does not imply, even in the absence of a specific statement, that such names are exempt from the relevant protective laws and regulations and therefore free for general use.
The publisher, the authors and the editors are safe to assume that the advice and information in this book are believed to be true and accurate at the date of publication. Neither the publisher nor the authors or the editors give a warranty, expressed or implied, with respect to the material contained herein or for any errors or omissions that may have been made. The publisher remains neutral with regard to jurisdictional claims in published maps and institutional affiliations.

This Springer imprint is published by the registered company Springer Nature Singapore Pte Ltd.
The registered company address is: 152 Beach Road, #21-01/04 Gateway East, Singapore 189721, Singapore

If disposing of this product, please recycle the paper.

Contents

Atomic Hydrogen Adsorption on Two-Dimensional AlN: DFT Study

Amina Boudiaf, Malika Gallouze, and Abdelhafid Kellou

Abstract Based on Density Functional Theory (DFT) calculations, we investigate the structural and energetics properties of the atomic hydrogen adsorption on two-dimensional (2D) hexagonal AlN. The obtained results show that hydrogen in hollow site induces volume expansion (~ 1.12%) of AlN monolayer. Whereas volume contraction is found for hydrogen in bridge, top Al and top N adsorption sites, due to the slight decrease of the lattice parameters (~ 0.32–0.8%). Hydrogen atom prefers to form a bond with nitrogen than aluminum in AlN monolayer with bond length H-N of 1.05 Å.

Keywords AlN-2D · Structural and energetic properties · Hydrogen adsorption · DFT

1 Introduction

Recently, Two-dimensional (2D) materials have attracted great interest in the field of materials science due to their unique physicochemical properties and rare physical phenomena that occur when charge and heat transport are confined to a plane [1, 2]. Following the discovery of graphene [3], various 2D nanomaterials are also being investigated for their unique electrical, optical, mechanical, magnetic and thermal properties [4, 5]. Monolayers that have characteristics better than or comparable to graphene can be used in wide range of applications [6, 7]. Other interesting monolayer materials include group III-nitride monolayers, such as Aluminum Nitride (AlN), which can be used in a variety of applications [2, 8]. AlN monolayers have been the subject of extensive experimental and theoretical investigations. Most of the physical properties of 2D monolayers can be modified using a variety of methods, including doping, strain, confinement in heterostructures, and planar curvature tuning.

A. Boudiaf (✉) · M. Gallouze · A. Kellou
Theoretical Physics and Didactics Laboratory, Faculty of Physics, USTHB, BP 32 El Alia, Bab Ezzouar, Algiers, Algeria
e-mail: aboudiaf@usthb.dz

© The Author(s), under exclusive license to Springer Nature Singapore Pte Ltd. 2024
A. Mellit et al. (eds.), *Proceedings of the 2nd International Conference on Green Energy Conversion System*, Green Energy and Technology,
https://doi.org/10.1007/978-981-97-6148-7_1

Hydrogen fuel cells are emerging as one of the possible sources of clean energy, with water being the only product of the reaction. Currently, hydrogen is stored by cryogenic compression, liquefaction, or absorption/adsorption on metals or compounds [5, 9]. DFT has been used for many studies on hydrogen storage to investigate the adsorption of atoms on different surfaces [2, 5, 10, 11].

This work aims to the atomic hydrogen adsorption study on two-dimensional hexagonal AlN using the density functional theory method (DFT).

2 Computational Details

Our calculations are performed using the Density Functional Theory (DFT) as implemented in the pseudo-potential plane wave self-consistent field package (Quantum Espresso) [12]. We have used a plane-wave basis set for the expansion of the single particle Kohn–Sham wave functions. The many-body problem of interacting electrons and nuclei is mapped to a series of one-electron equations, the so-called Kohn and Sham (KS) equations [13].

The exchange and correlation interactions are described by the generalized gradient approximation of Perdew et al. (GGA-PBE) [14], and ultra-soft pseudo-potentials (USPPs) of Vanderbilt [15] are used with non-linear core corrections. In the first irreducible Brillouin zone (BZ), the slab calculations are conducted using a $4 \times 4 \times 1$ *k*-points mesh along the three basic axes in reciprocal space, according to the Monkhorst–Pack scheme. The wave functions are expanded with plane wave basis with 544 eV of cutoff energy. All structures are allowed to fully relax using the Broydene-Fletcher-Goldforbe-Shanno (BFGS) scheme until the total energy converged to less than 10^{-5} Ry/atom. The maximum force converged to less than 0.004 Ry/bohr and the maximum displacement is 0.002 Å.

The hexagonal AlN system is described using the supercell approach, as given in Fig. 1. A monolayer AlN with 8 atoms (4 atoms of Al and 4 atoms of N) is separated by 30 Å from its image along the *z*-axis. The highly symmetric top, bridge and hollow sites of hydrogen atomic adsorption on AlN monolayer are represented in Fig. 2.

The adsorption energy of the hydrogen atom on hexagonal AlN monolayer is defined as:

$$E_{ads} = E(AlNH) - E(AlN) - E(H) \quad (1)$$

Fig. 1 Structure of hexagonal AlN monolayer ($2 \times 2 \times 1$)

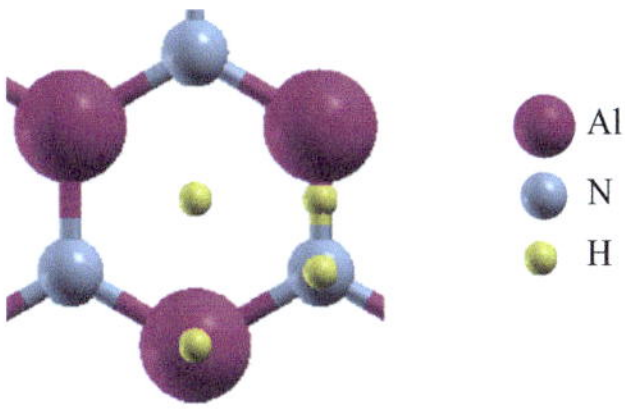

Fig. 2 The adsorption sites for a hydrogen atom on AlN monolayer. H in the center of the hexagon: Hollow site. H in the middle of Al-N band: Bridge site. H above Al or N atom: top Al or top N site

where E(AlNH), E(AlN) and E(H) are the calculated total energies of AlNH, AlN monolayer and the isolated hydrogen atom respectively.

3 Results and Discussion

3.1 Structural and Energetic Properties of AlN Monolayer

This section is devoted to the study of structural and energetic properties of hexagonal AlN. Firstly, for Al-fcc, our calculated lattice parameter (a_0), bulk modulus (B) and cohesive energy (E_{coh}) are 4.02 Å, 78 GPa and – 3.26 eV respectively. For N_2 molecule, the calculated bond length ($d_{N\text{-}N}$) and binding energy (E_b) are 1.10 Å and – 9.76 eV. Our results are in good agreement with the experimental ones [16, 17], as well as with those predicted theoretically [18]. This fact reflects that the pseudopotential method is quite appropriate for these elements and can describe the electronic structure of these materials.

Based on density functional theory (DFT) calculations, it is proposed that hexagonal AlN with a graphene-like structure is the most stable two dimensional structure [8]. The structure of monolayer AlN is optimized using a periodically repeating supercell (2 × 2 × 1—Fig. 1) with a 2D hexagonal lattice and a spacing of 30 Å between AlN planes along the *z*-axis. After relaxation, no buckling height is obtained (planar structure), the lattice constant size of the monolayer AlN is 3.12 Å, and the bond length is 1.79 Å, which are slightly larger and smaller than the bulk AlN with wurtzite structure value of 3.11 and 1.90 Å [19] respectively. This is due to lattice distortion. These results are consistent with previous studies [20, 21]. The calculated formation energy of AlN monolayer is – 7.95 eV, this negative value means that the process of creating AlN system is exothermic.

3.2 Hydrogen Adsorption on AlN Monolayer

In order to study the hydrogen adsorption on hexagonal AlN monolayer, hydrogen atom occupies the adsorption sites. Namely, the top site: H is placed above Al or N atom, the bridge site: H is placed in the middle of Al-N band and the hollow site: H occupies the center of the hexagon, as shown in Fig. 2. Table 1 collects our calculated equilibrium lattice parameter (a_0), interatomic distance ($d_{H\text{-}Al/H\text{-}N/Al\text{-}N}$) and adsorption energy (E_{ads}). Structures of monolayer AlN with adsorption sites of hydrogen before and after relaxation are also given. After relaxation, the equilibrium lattice parameter of AlN slightly increases compared to that before relaxation, in the case of H in hollow site. This fact means that hydrogen in hollow site induces volume expansion of AlN monolayer (by about 1.12%). Whereas volume contraction is found for the remaining adsorption sites, due to the slight decrease of lattice parameters (by about 0.32–0.8%). It is shown from this table that interatomic distances between H, Al and N atoms increase for all the considered adsorption sites.

According to the adsorption energy definition (Eq. 1), the larger E_{ads}, the stronger the interaction between hydrogen atom and the AlN layer. A negative value of E_{ads} indicates that the adsorption is endothermic. From Table 1, it is clearly seen that, after relaxation, the most stable configurations adsorption sites are hydrogen in bridge and top N sites, which correspond to the lowest calculated E_{ads} (− 2.03 eV). It can be concluded that hydrogen atom prefers to form a bond with nitrogen atom than aluminum in AlN monolayer with bond length H-N value of 1.05 Å.

4 Conclusions

The DFT based pseudo-potential method is applied to investigate the structural and energetic properties of hydrogen atomic adsorption on two-dimensional AlN monolayer. Our results point out that no buckling height is found in a perfect AlN monolayer, which results in the fact that AlN is a planar structure.

When hydrogen atom is placed in the hollow adsorption site, it induces a volume expansion of the AlN monolayer by about 1.12%. Whereas a volume contraction of AlN is found in the case of hydrogen occupying the bridge, top Al and top N adsorption sites, which is due to the slight decrease of the lattice parameters a_0 (about 0.32–0.80%).

As a result, hydrogen atom prefers the bridge and the top N adsorption sites, according to the calculated adsorption energies. This fact means that hydrogen forms a bond with nitrogen than aluminum atom in AlN monolayer, with bond length H-N of 1.05 Å.

Table 1 Calculated lattice parameter a_0 (Å), interatomic distance $d_{H\text{-}Al/H\text{-}N/Al\text{-}N}$ (Å) and adsorption energy E_{ads} (eV) for atomic hydrogen adsorption sites on AlN monolayer, before and after relaxation

			a_0		E_{ads}
Before relaxation	Hollow site		6.24	$d_{Al\text{-}N}$ = 1.80 $d_{H\text{-}N}$ = 1.80 $d_{H\text{-}Al}$ = 1.80	–
	Top view 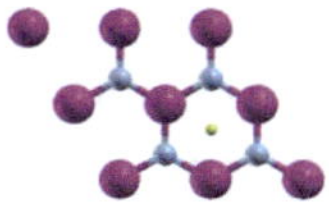	Side view 			
After relaxation	Top view 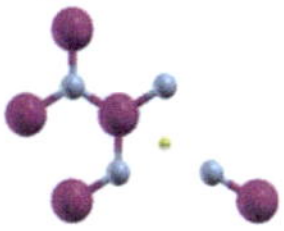	Side view 	6.31	$d_{Al\text{-}N}$ = 1.83 $d_{H\text{-}N}$ = 1.84 $d_{H\text{-}Al}$ = 1.82	0.32
Before relaxation	Bridge site		6.24	$d_{Al\text{-}N}$ = 1.80 $d_{H\text{-}N}$ = 0.90 $d_{H\text{-}Al}$ = 0.90	–
	Top view 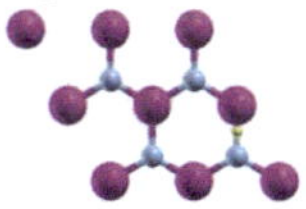	Side view 			
After relaxation	Top view 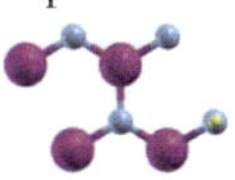	Side view 	6.19	$d_{Al\text{-}N}$ = 1.91	– 2.03
				$d_{H\text{-}N}$ = 1.05	
				$d_{H\text{-}Al}$ = 2.48	
Before relaxation	Top Al site		6.24	$d_{Al\text{-}N}$ = 1.80 $d_{H\text{-}N}$ = 2.59 $d_{H\text{-}Al}$ = 1.87	–
	Top view 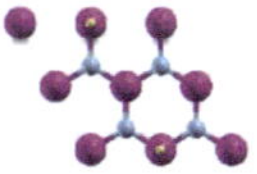	Side view 			
After relaxation	Top view 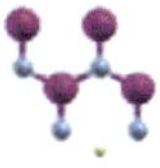	Side view 	6.22	$d_{Al\text{-}N}$ = 1.89 $d_{H\text{-}N}$ = 2.67 $d_{H\text{-}Al}$ = 1.70	– 1.34
Before relaxation	Top N site		6.24	$d_{Al\text{-}N}$ = 1.80 $d_{H\text{-}N}$ = 1.87 $d_{H\text{-}Al}$ = 2.59	–
	Top view 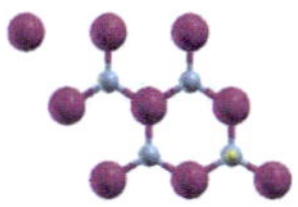	Side view 			
After relaxation	Top view 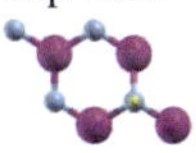	Side view 	6.19	$d_{Al\text{-}N}$ = 1.91	– 2.03
				$d_{H\text{-}N}$ = 1.05	
				$d_{H\text{-}Al}$ = 2.48	

References

1. Abdullah NR, Abdullah BJ, Tang C, Gudmundsson V (2021) Interactions between dopant atoms. Mater Sci Semicond Process 135:106073
2. Beshkova M, Yakimova R (2020) Properties and potential applications of two-dimensional AlN. Vacuum 176:109231
3. Novoselov KS, Geim AK, Morozov SV, Jiang D, Zhang Y, Dubonos SV, Grigorieva IV, Firsov AA (2004) Electric field effect in atomically thin carbon films. Science 22:2–6
4. Zollner K, Cummings AW, Roche S, Fabian J (2021) Graphene on two-dimensional hexagonal BN, AlN, and GaN: electronic, spin-orbit, and spin relaxation properties. Phys Rev B 103(7):075129
5. Villagracia AR, Ong HL, David M, Arboleda N Jr (2019) First principles investigation on H2 adsorption on the pristine 2-dimensional hexagonal aluminum. Earth Environ Sci 268:012135
6. Khan K, Tareen AK, Aslam M, Wang R, Zhang Y, Mahmood A, Ouyang Z, Zhang H, Guo Z (2020) Recent developments in emerging two-dimensional materials and their applications. J Mater Chem C 8:387–440
7. Ma Q, Ren G, Xu K, Ou JZ (2021) Tunable optical properties of 2D materials and their applications. Adv Opt Mater 9:2001313
8. Ben J, Liu X, Wang C, Zhang Y, Shi Z, Jia Y, Zhang S, Zhang H, Yu W, Li D, Sun X (2021) 2D III-nitride materials: properties, growth, and applications. Adv Mater 33:2006761
9. Eftekhari A, Fang B (2017) Electrochemical hydrogen storage: opportunities for fuel storage, batteries, fuel cells, and supercapacitors. Int J Hydrogen Energy 42:25143–25165
10. Irvin K, Rojas M, Rey A, Villagracia C, Lorenzo J, David M, Arboleda NB (2018) Ca and K decorated germanene as hydrogen storage: an ab initio study. Int J Hydrogen Energy 43:4393–4400
11. Wang Y, Song N, Song X, Zhang T, Yang D (2018) A first-principles study of gas adsorption on monolayer AlN sheet. Vaccum 147:18–23
12. Giannozzi P, Baroni S, Bonini N et al (2009) QUANTUM ESPRESSO: a modular and open-source software project for quantum simulations of materials. J Phys Condens Matter 21
13. Kohn W, Sham LJ (1965) Self-consistent equations including exchange and correlation effects. Phys Rev 140
14. Perdew JP, Burke K, Ernzerhof M (1996) Gradient approximation made simple. Phys Rev 77:3865–3868
15. Vanderbilt D (1990) Soft self-consistent pseudopotentials in a generalized eigenvalue formalism. Phys Rev B 41:7892–7895
16. Kittel C (1996) Introduction to solid state physics, 7th edn
17. Wang K, Reeber RR, Wang KAI, Reeber R (2000) The perfect crystal, thermal vacancies and the thermal expansion coefficient of aluminium expansion coefficient of aluminium. Philos Mag A 80:1629–1643
18. Berrached I, Rabahi L, Gallouze M, Kellou A (2019) J Magn Magn Mater Nitriding Effect Struct Stab Magn Proper FeAl Alloys DFT Study 480:79–86
19. Morales FM, Manuel JM, Garcia R, Reuters B, Kalisch K, Vescan A (2013) Evaluation of interpolations of InN, AlN and GaN lattice and elastic constants for their ternary and quaternary alloys. J Phys D Appl Phys 46:245502
20. Santos RB, Mota FDB, Rivelino R, Kakanakova-gueorguie A (2016) Van der Waals stacks of few-layer h-AlN with graphene: an ab initio study of structural, interaction and electronic properties. Nanotechnology 27:145601
21. Ghorbanzadeh M, Fereidoon A, Hamed A (2017) Interlayer interaction and mechanical properties in multi-layer graphene, Boron-Nitride, Aluminium-Nitride and Gallium-Nitride graphene-like structure: a quantum-mechanical DFT study. Superlattices Microstruct 112:30–45

Site Preference of Hydrogen Atom in Mg_3La Alloy: A DFT Study

N. Mosteghanemi, L. Rabahi, L. Rouaiguia, and A. Kellou

Abstract The Pseudo-Potential method (PP) based on Density Functional Theory (DFT), was applied to investigate the site preference of hydrogen atom in the Mg_3La alloy, which exhibits Fcc structure (space group 225). The relative stability of hydrogen atom at various interstitial sites (Wyckoff positions: 24d, 24e, 32f and 48g) was determined, with a particular attention to the hydrogenation effects on the stability and structural properties of the Mg_3La alloy. The results were discussed in the light of the available theoretical and experimental results.

Keywords DFT calculations · Hydrogen insertion · Relative stability · Structural properties

1 Introduction

Magnesium-based alloys have attracted significant attention as hydrogen storage materials in recent years, because of their high hydrogen storage capacity (up to 7.6 wt% in the form of MgH_2). However, the high thermodynamic stability and sluggish kinetics of Mg [1–3] limit the application of Mg-based alloys. Many efforts have been proposed for kinetic and thermodynamic tuning of Mg-based alloys such as nanostructuring [4], nanocatalyst [5] and alloying [6].

Magnesium-rich rare earth (RE), one of the several magnesium-based alloys owns considerable hydrogen storage capacity [7]. The specificity of Mg-RE systems is that both Mg and RE could react with H and form hydrides with large negative enthalpy of formation [7]. The Mg_3La is an illustration of this class of compounds, to which very little research has been done to study the effect of hydrogen [8, 9] on stability of

N. Mosteghanemi (✉) · L. Rabahi · L. Rouaiguia · A. Kellou
Theoretical Physics and Didactics Laboratory, Faculty of Physics USTHB, BP 32, Bab-Ezzouar, Algiers, Algeria
e-mail: nmosteghanemi@usthb.dz

L. Rabahi
Research Center in Industrial Technologies (CRTI), BP 64, Cheraga, Algiers, Algeria

© The Author(s), under exclusive license to Springer Nature Singapore Pte Ltd. 2024

A. Mellit et al. (eds.), *Proceedings of the 2nd International Conference on Green Energy Conversion System*, Green Energy and Technology,
https://doi.org/10.1007/978-981-97-6148-7_2

this compound. Mg_3La was found to be stable within the DO_3 (space group number 225) structure [10, 11], at ambient conditions.

In this paper, a first-principal plane-wave pseudopotential method based on density functional theory is used to investigate the site preference of hydrogen atom in the Mg_3La with Fcc structure (space group 225) and the relative stability of hydrogen atom at various interstitial sites (Wyckoff positions: 24d, 24e, 32f and 48g).

2 Computational Details

All the calculations of this study were performed using the density functional theory (DFT) employing pseudo-potential methods (PP) as implemented in the pseudo-potential plane wave self-consistent field package (Quantum Espresso) [12]. The so-called Khon Sham (KS) equations [13, 14] were used to map the many-body issue of interacting electrons and nuclei to a set of one electron equations. The exchange and correlation interactions are described by the generalized gradient approximation of Perdew et al. (GGA-PBE) [15] and ultra-soft pseudo-potentials (USPPs) of Vanderbilt are used [16]. A kinetic energy cut-off of 500 eV and a well converged value of the k-point mesh over the Brillouin zone were considered. Namely $8 \times 8 \times 8$ for the cubic structure. All structures were allowed to fully relax using BroydenFletcher-Goldforb-Shanno (BFGS) scheme [17] until the total energy has converged to less than 105 eV/atom. The maximum force has converged to lower than 0.004 eV/Å.

3 Results and Discussion

3.1 *Structural Properties of Mg_3La*

A series of total energy versus total volume calculations, for the Mg_3La compound in the cubic structure (space group 225), were performed in order to determine the static structural parameters at 0 K. The corresponding values were fitted with the universal Birch-Murnaghan equation of state [18]. The obtained lattice parameters (a), bulk modulus (B) and its pressure derivative (B′) and formation enthalpies (ΔH_f) are presented in Table 1. The available experimental and theoretical values are also given.

Our calculated results are in good agreement with the experimental and theoretical values. The bulk modulus of Mg_3La is around 37 GPa, this means that our system is rigid. All B' values obtained fall between 3 and 4 indicating a good fit.

In order to determine the phase stability of Mg_3La compound, the formation enthalpy has been calculated using the following equation:

Table 1 The optimized lattice constant (a_0), bulk modulus (B), first pressure derivative of bulk modulus (B′) and formation enthalpy ΔH_f of Mg_3La

Crystal structures		a_0 (Å)	B (GPa)	B′	ΔH_f (eV/at)
FCC (Fm-3m) (Space group 225)	This work	7.502	36.9	3.51	− 0.133
	Exp.	7.494 [19]	–	–	− 0.134 [21]
	Th.	7.539 [20]	39.6 [20]	–	− 0.139 [22]
Mg_3LaH_6-H (24d)	This work	7.801	46.9	3.84	− 0.028
Mg_3LaH_6-H (24e)	This work	8.298	36.6	3.53	0.091
Mg_3LaH_8-H (32f)	This work	9.478	27.1	3.42	0.138
Mg_3LaH_{12}-H (48g)	This work	9.732	17.5	3.62	0.556

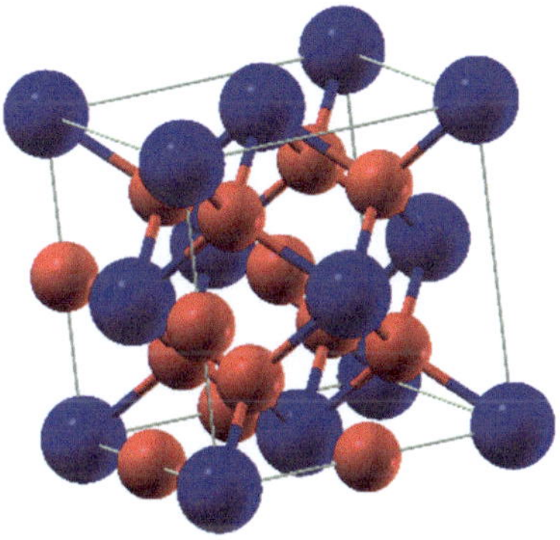

Fig. 1 Crystal structure of Mg_3La (space group 225)

$$\Delta H_f = \frac{1}{x+y}\left[E_{solid}^{Mg_3La} - xE_{solid}^{Mg} - yE_{solid}^{La}\right] \quad (1)$$

where x and y refer respectively to the number of Mg and La atoms in the unit cell, whereas $E_{solid}^{Mg_3La}$, E_{solid}^{Mg} and E_{solid}^{La} are respectively the total energies of Mg_3La, Mg and La in their corresponding ground-state phases. It is important to note that stable structures have the lowest heat of formation values. Also, negative values of this parameter mean that the reaction (formation) is exothermic and can take place without any external energy supply [23]. The Mg_3La is thermodynamically stable in the cubic structure at 0 K, as already found in previous works [21, 22]. The crystal structure of the studied system is given by Fig. 1.

3.2 Hydrogen Effects on Thermodynamic Stability of Mg_3La

In order to study the hydrogen insertion in the Mg_3La alloy, the relative stability of atomic hydrogen at various interstitial sites has been investigated. We have considered Mg_3LaH_6 (6H in 24d and 24e), Mg_3LaH_8 (8H in 32f) and Mg_3LaH_{12} (12H

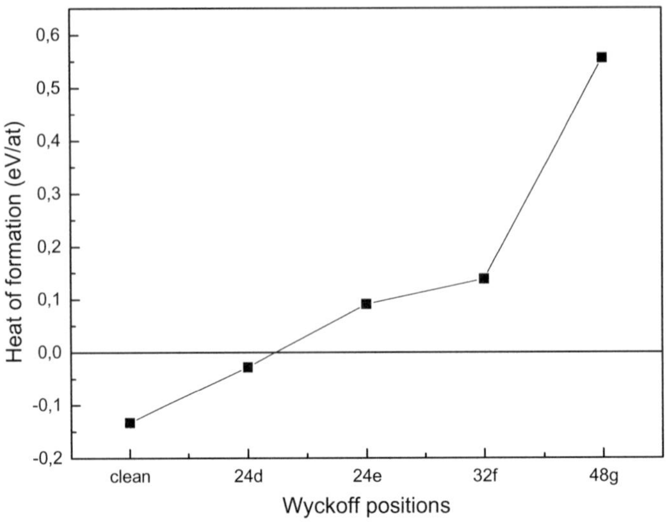

Fig. 2 The heat of formation as a function of Wyckoff positions of hydrogen atoms: Mg_3LaH_6 (24d), Mg_3LaH_6 (24e), Mg_3LaH_8 (32f) and Mg_3LaH_{12} (48g)

in 48g) hydrides. All the sites were fully occupied by H atoms, so the symmetry is preserved. Lattice parameters (a), bulk modulus (B) and its pressure derivative (B′) and formation enthalpies (ΔH_f) of these hydrides are collected in Table 1. For better discussion of the relative stability of Mg_3LaH_n (n = 6, 8 and 12) hydrides, the heat of formation as function of H atoms in different Wyckoff positions are represented in Fig. 1. The general trend is that the cell parameters increase (which means volume) by increasing the H amount. In term of stability, the Mg_3LaH_6 (H in 24d) forms favorable hydride because of the negative value of its heat of formation as it is shown in Fig. 2 and Table 1.

Whereas the other hydrides are not thermodynamically stable (positive heat of formation values). It is also interesting to note that H atom would prefer the 24d site characterized by the lowest heat of formation − 0.028 eV/at compared to the other hydrides.

Ouyang et al. [24] have synthetized Mg_3LaH_n (n = 3, 6) hydrides (partially hydrogenated) and Mg_3LaH_9 hydride which is totally hydrogenated. But, to our knowledge, no studies have carried out on the structural properties of Mg_3LaH_n (n = 6, 8 and 12) so no available results for comparison.

4 Conclusion

The purpose of this study is to use the DFT pseudo-potential method to investigate the site preference of hydrogen atoms in Mg_3La alloy. The main results can be summarized as follows.

The Mg_3La is more stable in the fcc phase (space group 225), and the predicted cell parameters are in good agreement with the available experimental and other theoretical results at 0 K.

The insertion of hydrogen atoms increases volume of the unit cell.

Hydrogen prefers 24d site with 6H and Mg_3LaH_6 is the resultant hydride with lowest heat of formation − 0.028 eV/at.

Mg_3LaH_6 is energetically the most stable and hydride can be formed without any external energy supply.

References

1. Zhu M, Lu Y, Ouyang L, Wang H (2013) Thermodynamic tuning of mg-based hydrogen storage alloys: a review. Materials 6:4654–4674. https://doi.org/10.3390/ma6104654
2. Shao H, He L, Lin H, Li HW (2018) Progress and trends in magnesium-based materials for energy-storage research: a review. Energ Technol 6:445–458. https://doi.org/10.1002/ente.201700401
3. Luo Q, Li J, Li B, Liu B, Shao H, Li Q (2019) Kinetics in Mg-based hydrogen storage materials: enhancement and mechanism. J Magn Alloys 7:58–71. https://doi.org/10.1016/j.jma.2018.12.001
4. Rafatnejad M, Raygan S, Sefidmooy AM (2020) Investigation of dehydrogenation performance and air stability of MgH_2–PMMA nanostructured composite prepared by direct high-energy ball-milling. Mater Renew Sustain Energy 9:1–12. https://doi.org/10.1007/s40243-020-00174-6
5. Zhong HC, Huang YS, Du ZY, Lin HJ, Lu XJ, Cao CY et al (2020) Enhanced hydrogen Ab/De-sorption of Mg(Zn) solid solution alloy catalyzed by YH_2/Y_2O_3 nanocomposite. Int J Hydrogen Energy 45:27404–27412. https://doi.org/10.1016/j.ijhydene.2020.07.021
6. Yang T, Wang P, Xia C, Li Q, Liang C, Zhang Y (2019) Characterization of microstructure, hydrogen storage kinetics and thermodynamics of a melt-spun $Mg_{86}Y_{10}Ni_4$ alloy. Int J Hydrogen Energy 44:6728–6737. https://doi.org/10.1016/j.ijhydene.2019.01.148
7. Ouyang LZ, Dong HW, Zhu M (2007) Mg_3Mm compound based hydrogen storage materials. J Alloys Compd 446–447:124–128. https://doi.org/10.1016/j.jallcom.2007.02.160
8. Karonik VV, Kazakov DN, Andrievskii RA, Bogachkova OP (1984) Phase-equilibria in the Mg–Mg_3La–Mg_2NiLE–Mg_2Ni system and reaction of its alloys with hydrogen. Inorg Mater 20:207–211
9. Zhang QA, Liu YJ, Si TZ (2006) Crystal structures and hydrogen-induced decomposition of $La(Mg_{1-x}Al_x)_3$ alloys. J Alloys Compd 417:100–103. https://doi.org/10.1016/j.jallcom.2005.09.027
10. Zhou W, Jin S, Dai W, Lyon JT, Lu C (2021) Theoretical study on the structural evolution and hydrogen storage in NbH n (n = 2–15) clusters. Int J Hydrogen Energy 46:17246–17252
11. Zaluska A, Zaluski L, Ström-Olsen JO (2001) Structure, catalysis and atomic reactions on the nano-scale: a systematic approach to metal hydrides for hydrogen storage. Appl Phys A Mater Sci Process 72:157–165. https://doi.org/10.1007/s003390100783

12. Giannozzi P, Baroni S, Bonini N, Calandra M, Car R, Cavazzoni C et al (2009) Quantum ESPRESSO: a modular and open-source software project for quantum simulations of materials. J Phys: Condens Matter 21:395502. https://doi.org/10.1088/0953-8984/21/39/395502
13. Kulikov NI (1982) Self-consistent band structure calculations and optical properties of the La-H system. J Less-Common Metals 88:307–316
14. Hohenberg P, Kohn W (1964) Inhomogeneous electron gas. Phys Rev 136:B864–B871. https://doi.org/10.1103/PhysRev.136.B864
15. Perdew JP, Burke K, Ernzerhof M (1996) Generalized gradient approximation made simple. Phys Rev Lett 77:3865–3868. https://doi.org/10.1103/PhysRevLett.77.3865
16. Vanderbilt D. Rapid communications soft self-consistent pseudopotentials in a generalized eigenvalue formalism, vol 41 (n.d.)
17. Fischer TH, Almliw J (1992) General methods for geometry and wave function optimization. J Phys Chem 96:9768–9774
18. Birch F (1947) Finite elastic strain of cubic crystals. Phys Rev 71:809–824
19. Wang YF, Zhang WB, Wang ZZ, Deng YH, Yu N, Tang BY et al (2007) First-principles study of structural stabilities and electronic characteristics of Mg–La intermetallic compounds. Comput Mater Sci 41:78–85. https://doi.org/10.1016/j.commatsci.2007.03.003
20. Ganeshan S, Shang SL, Zhang H, Wang Y, Mantina M, Liu ZK (2009) Elastic constants of binary Mg compounds from first-principles calculations. Intermetallics (Barking) 17:313–318. https://doi.org/10.1016/j.intermet.2008.11.005
21. NayebHashemi A, Clark J, Nayeb-Hashemi BA (1981) Magnetic susceptibility, electrical resistivity, and neutron spectroscopy of some cubic cerium intermetallics, vol 23. North-Holland Physics Publishing
22. Zhang H, Shang S, Saal JE, Saengdeejing A, Wang Y, Chen LQ et al (2009) Enthalpies of formation of magnesium compounds from first-principles calculations. Intermetallics (Barking) 17:878–885. https://doi.org/10.1016/j.intermet.2009.03.017
23. Rabahi L, Gallouze M, Grosdidier T, Bradai D, Kellou A (2017) Energetics of atomic hydrogen absorption in C15-Fe2Zr Laves phases with ternary additions: a DFT study. Int J Hydrogen Energy 42:2157–2166. https://doi.org/10.1016/j.ijhydene.2016.11.131
24. Ouyang LZ, Huang JM, Fang CJ, Wang H, Liu JW, Zhang QA et al (2013) The high capacity and controllable hydrolysis rate of Mg3La hydride. J Alloys Compd 580. https://doi.org/10.1016/j.jallcom.2013.03.153

Experimental and Digital Simulation Investigations of Harmonics Injection by CFLs into a LV Network

Mohamed Hajjej and Lassaad Sbita

Abstract Today, the demand for electrical appliances, such as power converters, economic lamps, LEDs, speed drives, and regulators, is increasing in recent years, but despite their robustness and good dynamic responses, they inject harmonic currents into the network. This article presents in-depth studies on the effect of the use of Compact fluorescent lamps CFL on a Low Voltage (LV) local area network, as well as an interest in modeling this type called non-linear charge under PSIM and MATLAB Software.

Keywords Compact fluorescent lamps CFLSs · Power quality · Power factor · Total harmonic distortion THD · Non-linear loads

1 Introduction

Lighting is considered one of the most important stresses on energy systems. It is estimated by [1, 2] that light consumes 25–35% of the world's energy. The use of compact fluorescent lamps (CFLs) instead of incandescent lamps (IL) can reduce approximately 80% of the total lighting load as they have high lighting efficiency. However, these non-linear charges inject harmonic currents into the power grid because of the electronic ballast that makes them. This is why the EMC has set limits, to limit the effect of these loads on the equipment connected to a public network [3].

Experimental and digital Modeling attempts were administered to define an accurate model for the General Electric lamps to be used later in digital simulation or dedicated software. So, a new simulation model for a General Electric lamp is presented and used in this paper. And in this study takes into account the current and voltage

M. Hajjej (✉) · L. Sbita
esearch Laboratory of Process, Energies, Environment and Electrical Systems, National Engineering School of Gabes, University of Gabes, Zrig Eddakhlania 6029, Tunisia
e-mail: mohamedmongihajjej@gmail.com

L. Sbita
e-mail: lassaad.sbita@enig.rnu.tn

© The Author(s), under exclusive license to Springer Nature Singapore Pte Ltd. 2024

A. Mellit et al. (eds.), *Proceedings of the 2nd International Conference on Green Energy Conversion System*, Green Energy and Technology,
https://doi.org/10.1007/978-981-97-6148-7_3

distortion up to ring 21, the voltage and current values RMS, active Power P, reactive power Q, and the apparent power S are also involved in the correct quantification of the observed phenomenon, it should be noted that all these quantities have been calculated by IEEE standard 1459 [4, 5]. These quantities will be used to determine the effects of these lamps on the distribution network and will be used in other studies to minimize the flow of currents into the distribution network and to compensate for the reactive energy decided.

This document is organized as follows: the first Sects. 2, present the global model and the real model obtained from the studied lamp. In addition, we present an experimental test on a single lamp with all associated values. Section 3 presents the equivalent model of the same studied load, after that Sect. 4 presents the effect of the massive use of these lamps on a three-phase distribution network. Finally, Sects. 5 present, a summary and a conclusion for the full work with the future work to be made to resolve these problems.

2 Full CFL Modeling and Harmonics Generated by One CFL

The circuit consists of four sections: an AC bridge rectifier, an EMI filter, a half-bridge inverter, and the CFL tube. The CFL circuit has been designed for a nominal mains voltage of 230 Vrms, 50–60 Hz. According to [6] the AC mains voltage is rectified by four Diodes (D1–D4) and smoothed by the buffer capacitor to get a DC supply voltage for the half-bridge inverter. An Electromagnetic Interference (EMI) filter minimizes the disturbance towards the mains (because of the rapid changes of di/dt and dv/dt at high switching frequency). At last, the half-bridge inverter is used to convert the DC voltage to AC voltage at a high switching frequency suitable to drive the fluorescent lamp. Figure 1a shows the block diagram of the circuit, and Fig. 1b presents the studied lamps' electronic ballast.

The type of the studied lamp is General Electric FLE20 20W 230 V/50–60 Hz. And to achieve this objective, we modeled the electronic ballast of the lamp, under PSIM software as shown in Fig. 2.

The full circuit model for this lamp is shown in Fig. 2. The tube of the lamp, is replaced by an equivalent resistance, is that by referring to the various previous research [7, 8]. and based on this research, the approximate value of this resistance is 100 Ω. More else, the same half-bridge inverter is used to convert the DC voltage to AC voltage at a very High switching frequency suitable to drive the fluorescent lamp here to drive the equivalent resistance, this is why the results remain non-linear at the level of the source. The part list of key components is summarized in Table 1.

Figure 3a, b shows respectively the simulated voltage and current waveforms absorbed by one CFL model lamp supplied by 230 V–50 Hz line, and their FFT components. The harmonic and power quantities are summarized in Table 2.

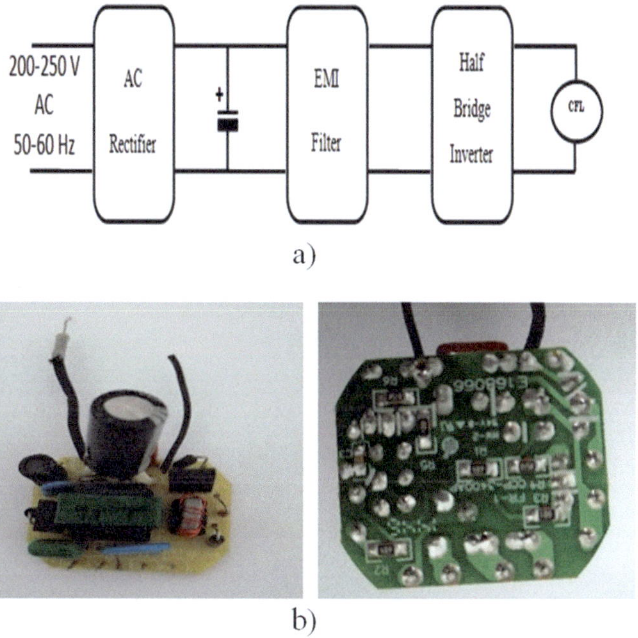

Fig. 1 **a** A schematic diagram of a CFL lamp. **b** Photos of electronic ballast of the investigated GE FLE20 20W CFL

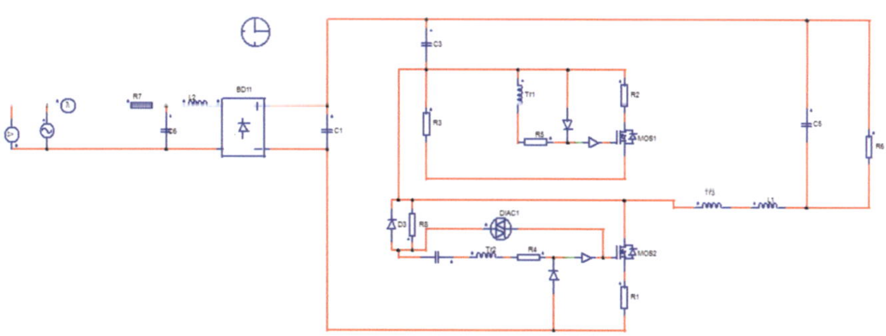

Fig. 2 Simulation model of 20 W GE CFL lamp in PSIM

This test is carried out by a three-phase analyzer from the manufacturer Chauvin Ar-noux. Wish is configured in two-phase mode. The simulated results for one single model CFL showing the voltage and current waveforms are already given in Fig. 4. As for experimental measurements in Fig. 4, the simulated waveforms given in Fig. 3 show that the current reproduces the same shapes. This implies that this model (Fig. 2) can be used as a reference for the following research (for example compensation of the reactive energy decided, implantation of filtering devices).

Table 1 The full circuit Lamp model components values

Component	Value	Component	Value
C1	100 nF	R1	4.78 Ω
C2	0.1 nF	R2, R3	680 kΩ
C3	2.2 μF	R4, R5	10 Ω
C4	47 nF	R6, R7	510 Ω
C5	3.2 nF	L1	1mH
C6	4.7 μF	L2	1.4 mH
V	230 V	f	50 Hz

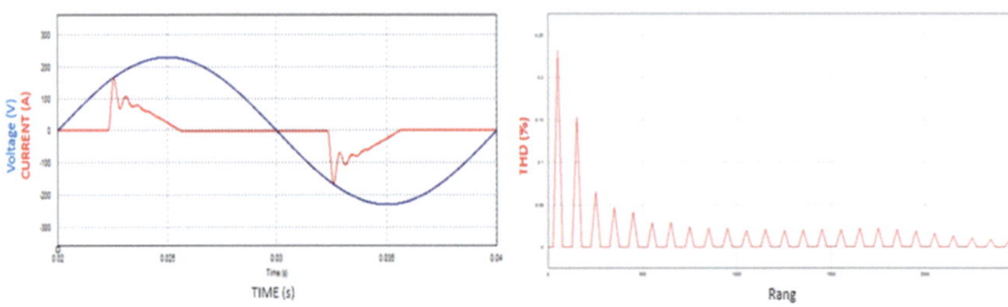

a) b)

Fig. 3 CFL model Simulation results and FFT components. **a** Simulated current (mA) and voltage (V) waveforms for one CFL (voltage in blue, current in red * 250) in PSIM, **b** FFT components of one CFL model

Table 2 Summary of harmonic and power quantities of one CLF model

P	PF	I1	I_RMS
18.52W	0.691	0.16 A	0.14 A
S (VA)	V_THD	I_THD	V_RMS
25.38	2.07	89.83	220 V

Fig. 4 Experimental current (in blue) and voltage (in yellow) waveforms for one GE CFL. Components of one CFL model

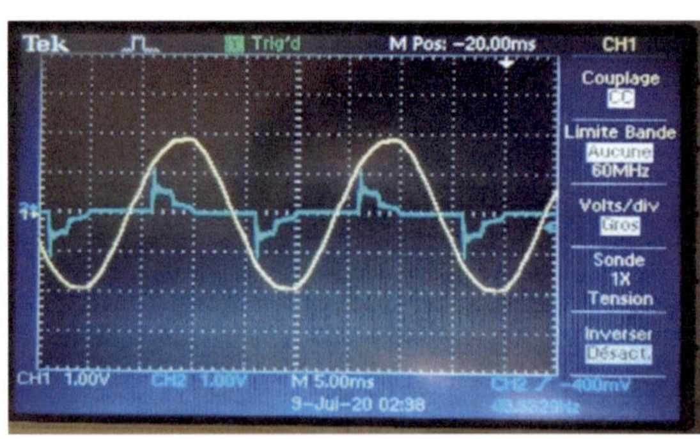

3 Equivalent Model of One CFL Lamp

A traditional lamp for residential use with a power lower than 30 W is equipped with an electronic ballast composed of four blocks as it is represented in Sect. 3. However, for economic reasons (valid only in simulation). Previous research [9–12] led to the representation of such a lamp by an equivalent model composed of a simple rectifier with a capacitive filter in parallel at its output, and an equivalent resistive load (Req). Figure 5a shows the equivalent model of a GE lamp tested in the laboratory such that the input impedance is represented by an RLC filter. The circuit is supplied by a 230 V–50 Hz alternating voltage source. According to [13] several tests were carried out to determine the approximation of the value of the equivalent resistance. In our case, the value found is around 2400 Ω. Figure 5, c show, respectively, the current curves of the model and that of the equivalent model of the studied lamp, and their FFT components.

The equivalent model takes the same characteristics (amplitudes and shape) as the real model, from which this model is validated and can be used in future research (ex: elimination of harmonics, reactive compensation, filtering).

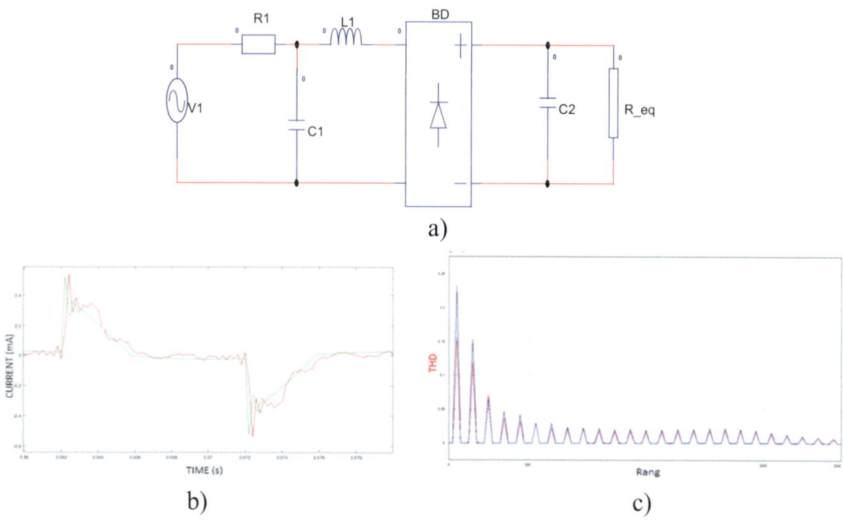

Fig. 5 **a** Equivalent model of one CFL lamp in PSIM software. **b** Simulated current (mA) of the full model and the equivalent model waveforms for one CFL (full model current in blue, equivalent model current in red). **c** FFT components of one CFL model (in blue) and the equivalent model (in red)

4 Simulated and Experimental Results for Large Number of Equivalents Model of CFLS

Figure 6a shows the overall scheme proposed in the experimental simulation for a large number of models of GE lamps on a three-phase network system, to quantify the effect of the use of these non-linear loads on the distribution network. This figure contains 75 lamps model distributed over 3 AC phases (25 lamps in parallel for each phase). A Power Analyzer (Chauvin Arnoux C.A_8336) is used to obtain data in terms of electrical parameters for any point, and this device not only shows data on the spot but can also store data digitally which can later be retrieved using a computer and software provided by the Power analyzer manufacturer (named Pat2). Figure 6b, c show, respectively, the simulated and current waveforms absorbed by the simulated equivalent CFL model lamp supplied by three lines of source voltage (230 V_50 Hz), and their FFT components.

NOTE: It should be noted that the same theoretical measurement parameters carried out for a single General Eclectic lamp were applied for this new arrangement. The harmonics and power quantities are well presented, Table 3 summarizes the power parameter for this new arrangement.

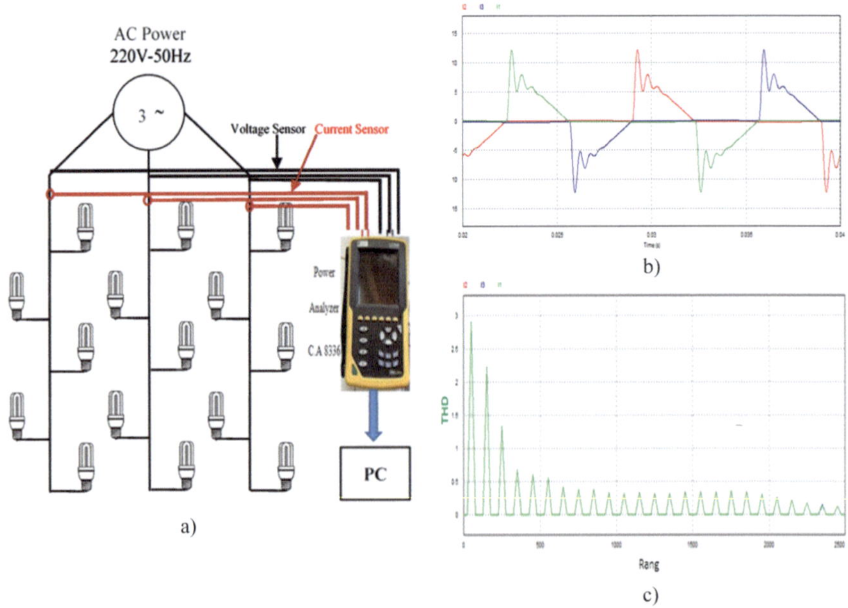

Fig. 6 **a** Proposed scheme of CFL equivalent model distribution system used in the simulation. **b** Simulated current waveforms of 75 CFL model lamps in three sources. **c** FFT components of 75 CFL model supplied by 3 lines source

Table 3 Power quantities

P	PF	I1	S (VA)	I_RMS	F
268.25 W	0.849	2.85 A	668.8	3.04 A	50 Hz

5 Conclusion and Future Research

In this work, experimental work is carried out to investigate and measure the power factor, and the harmonic performance of using CFLs on the distribution network, after we have presented the model of an economic lamp of the General Electric type under PSIM software to study the effect of the different factors involved. We have also determined the equivalent model of such a system called an economical lamp, finally, a section leads to research work in progress, to bring together the different effects, such as the solution is implementing a parallel Active Power Filter APF based on types of control different from the existence.

References

1. Shrivastava A, Singh B (2010) PFC Cuk converter based electronic ballast for an 18 W compact fluorescent lamp. In: 2010 international conference on industrial and information systems (ICIIS), pp 393–397
2. Radakovic Z, Topalis FV, Kostic M (2005) The voltage distortion in low-voltage networks caused by compact fluorescent lamps with electronic gear. Electr Power Syst Res 73(2):129–136
3. Chapman D (2002) Harmonics causes and effects. Copper Development Association
4. IEEE standard definitions for the measurement of electric power quantities under sinusoidal, nonsinusoidal, balanced, or unbalanced conditions. IEEE Std. 1459-2010 (Revision of IEEE Std. 1459-2000), pp 1–40, 19 Mar 2010
5. Mohamed H, Mohamed N, Lassaad S (2021) Experimental investigations of a CFLs currents harmonics injection into an electrical network grid. In: 2021 4th international symposium on advanced electrical and communication technologies (ISAECT), pp 01–05. https://doi.org/10.1109/ISAECT53699.2021.9668434
6. Shafi MA, McMahon RA (2006) Implementation and verification of CFL models for use in performance analysis in high frequency electronic ballasts. In: 3rd IET international conference on power electronics, machines and drives, pp 561–565
7. Gelvez JC, Ramirez A (2017) Measurement-based characterization of residential lighting devices: CFL and LED. North Am Power Symp (NAPS) 2017:1–6. https://doi.org/10.1109/NAPS.2017.8107190
8. Henao-Muñoz AC, Herrera-Murcia JG, Saavedra-Montes YAJ (2018) Experimental characterization of compact fluorescent lamps for harmonic analysis of power distribution systems. TecnoLógicas 21(42):79–94
9. Nohra AFH, Kanaan HY, Al-Haddad K (2012) A study on the impact of a massive integration of compact fluorescent lamps on power quality in distribution power systems. In: 2012 international conference on renewable energies for developing countries (REDEC), Beyroute 28–29 Nov 2012, pp 1–6
10. Cunill-Sola J, Sainz JJML (2011) Low-watt compact fluorescent lamps with electronic ballasts. In: Powereng'11, Malaga, Spain, May 2011, pp 1–5

11. Mohamed K, Shareef H, Mohamed A (2011) Analysis of harmonic emission from dimmable compact fluorescent lamps. In: Proceedings of the 2011 international conference on electrical engineering and informatics, pp 1–5. https://doi.org/10.1109/ICEEI.2011.6021744
12. Jiang C, Salles D, Xu W, Freitas W (2012) Assessing the collective harmonics impact of modern residential loads-part application. IEEE Trans Power Deliv 27(4):1947–1995
13. Molina J, Sainz L (2014) Model of electronic ballast compact fluorescent lamps. IEEE Trans Power Delivery 29(3):1363–1371. https://doi.org/10.1109/TPWRD.2013.2284095

New Hysteresis Control Block for Shunt Active Power Filter for Harmonics Compensation Generated by CFLs with PQ Theory

Mohamed Hajjej, Lassaad Sbita, Abdelali El Aroudi, and Kuntal Mandal

Abstract This paper proposes a New Modulation Hysteresis Control block for a three phase four wire Shunt Active Power Filter (SAPF) for currents harmonics compensation generated by Compact Fluorescents Lamps (CFLs). This study employs the PQ power control strategy. The performance of the SAPF is evaluated under both simple hysteresis controller and under the new modulation hysteresis controller conditions using the Matlab/Simulink platform. The findings simulation results demonstrate that the proposed new technique can improve the power quality of the main, and the total harmonic distortion (THD) optimization, which also confirms the effectiveness of the approach.

Keywords Power quality · Harmonic compensation · Hysteresis controller · PQ theory · Shunt active power filter

1 Introduction

In power system, the power quality issues are more significant factor due to the massive increase of semiconductor devices and nonlinear loads in various applications in the recent years. Harmonic currents are a major concern among these

M. Hajjej (✉) · L. Sbita
Research Laboratory of Process, Energies, Environment and Electrical Systems, National Engineering School of Gabes, University of Gabes, Zrig Eddakhlania 6029, Tunisia
e-mail: mohamedmongihajjej@gmail.com

L. Sbita
e-mail: lassaad.sbita@enig.rnu.tn

A. El Aroudi · K. Mandal
Department of Electronics, Electrical Engineering and Automatic Control, Universitat Rovira i Virgili, Tarragona, Spain
e-mail: abdelali.elaroudi@urv.cat

K. Mandal
e-mail: kuntal.mandal@urv.cat

© The Author(s), under exclusive license to Springer Nature Singapore Pte Ltd. 2024
A. Mellit et al. (eds.), *Proceedings of the 2nd International Conference on Green Energy Conversion System*, Green Energy and Technology,
https://doi.org/10.1007/978-981-97-6148-7_4

issues. Shunt active power filters (SAPFs) are commonly used for managing current harmonics, reactive power compensation, and neutral current compensation in distribution systems. These systems are typically connected at the point where the nonlinear load and utility grid meet which is known as Point of Common Coupling (PCC). They are more effective in mitigating current harmonics than passive power filters (PPFs). But these passive filters give a fixed compensation, resonance effect, sizing, etc. [1]. Currently, the focus of the researchers and developers is on both design and control part of SAPFs for three phases four wires nonlinear loads. Different SAPFs topologies have been developed to address these issues [2, 3]. As some researchers focus on the design part of the SAPFs, other researchers focus on their control part [4]. This paper utilizes PQ theory for reference currents calculation and hysteresis current controllers for pulse generation.

This paper is organized as follows: Sect. 2 describes the power and the control part of the used three-phase four-wire Shunt Active Power Filter SAPF. The summary of the gate controller and optimization techniques are presented in Sect. 3, and the simulation results of harmonic mitigation are discussed in detail in Sect. 4. Finally, Sect. 5 concludes this work.

2 Three Phase Four Wire Shunt Active Power Filter with Four Legs and a Split Capacitor Topology

As illustrated in Fig. 1 the fundamental process of the SAPF, wherein a non-linear load is being powered by a three-line source, and the SAPF is connected at the point of common coupling (PCC) to inject compensating current $i_{f_{(a,b,c)}}$ into the PCC. This compensating current is generated by taking the harmonic currents present in the load current $i_{L_{(a,b,c)}}$, but shifted by 180° in phase [5, 6], effectively cancelling out the harmonic current in the system, and making the source current $i_{S_{(a,b,c)}}$ free from harmonics. The non-linear loads currents are sensed to find out harmonic contents, which are then used to calculate the compensate currents $i_{c_{(a,b,c)}}*$ to control the switched Power Device ($F_{(a,b,c)}$ and $\overline{F}_{(a,b,c)}$) of the SAPF.

2.1 Mathematical Fundamentals of the Harmonic Detection Technique

The load current is influenced by the line impedance connected to the power sources during steady-state conditions, and the equations for calculating the load current can be derived as follows

$$i_{L_{(a,b,c)}}(t) = i_{p_{(a,b,c)}}(t) + i_{q_{(a,b,c)}}(t) + i_{h_{(a,b,c)}}(t)$$
$$i_{p_{(a,b,c)}}(t) = i_1 \cos(\varphi_1),\ i_{q_{(a,b,c)}}(t) = i_1 \sin(\varphi_1)$$

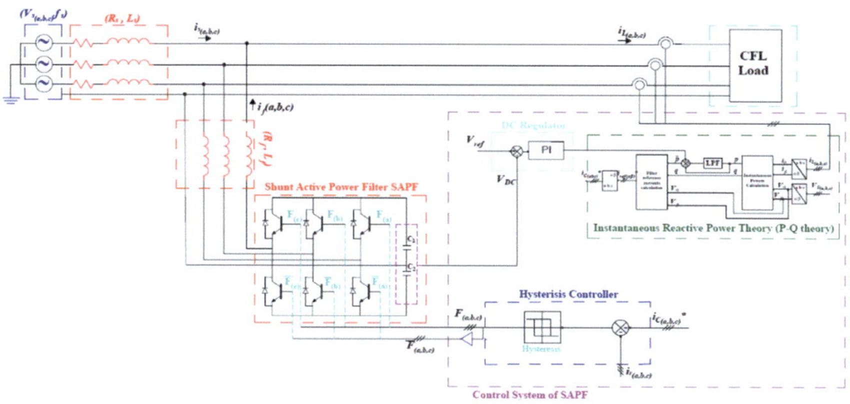

Fig. 1 Circuit diagram of a three-phases four-wire shunt active power filter with SRF and hysteresis control

$$i_{h_{(a,b,c)}}(t) = i_3\sqrt{2}\sin\left(3\omega t + \varphi_3\frac{\pi}{180}\right) + \sum_{h=5}^{49} i_h\sqrt{2}\sin\left(h\omega t + \varphi_h\frac{\pi}{180}\right) \quad (1)$$

It is clear from Eq. (1) that the variables $i_{p_{(a,b,c)}}(t)$, $i_{q_{(a,b,c)}}(t)$ and $i_{h_{(a,b,c)}}(t)$ represent the fundamental active current, fundamental reactive current, and the harmonic current of the Three-Phases Four-wires load respectively. To compensate the harmonic and reactive currents, the active current can be isolated from the equation. Thus, the required compensation currents (2) that the active power filter must generate are those that need to be injected.

$$i_{compensation}(t) = i_{f_{(a,b,c)}}(t) = -i_{q_{(a,b,c)}}(t) + i_{h_{(a,b,c)}}(t) \quad (2)$$

2.2 Review of the Instantaneous Reactive Power Theory (PQ Theory)

In 1983, Akagi et al. [7] introduced the "Generalized Theory of the Instantaneous Reactive Power in Three-Phase Circuits," commonly known as the instantaneous power theory or PQ theory. This theory is based on instantaneous values in three-phase power systems with or without neutral wire, and can be applied to steady-state or transitory operations with various voltage and current waveforms. The PQ theory involves transforming the three-phase voltages and currents in the (*a*-*b*-*c*) coordinates to the (α-β-0) coordinates using the Clarke transformation, and then calculating the PQ theory instantaneous power components.

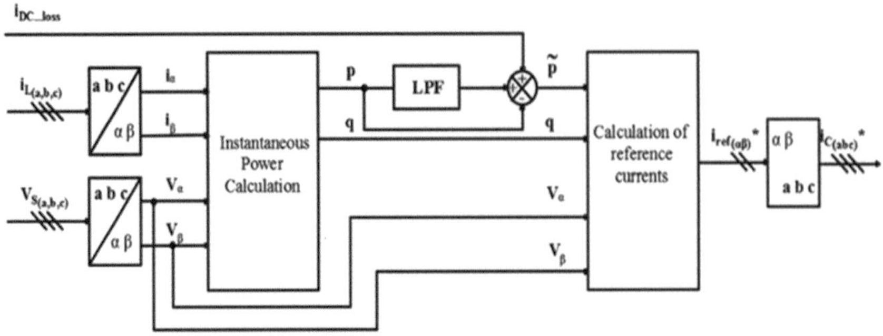

Fig. 2 Block diagram of the PQ theory

The PQ theory's advantage is that it separates the zero sequence components which do not contribute to the alpha and beta components. Therefore, it can be neglected in a balanced system. Figure 2 represented the block diagram of the proposed methodology.

2.3 General Electric CFLs Loads Presentation

In this section, the equivalent model of the studied nonlinear load is considered. To validate the studied lamp, several simulation and identification research works are carried out in order. The equivalent model of one 20 W General Electric brand compact fluorescent lamp powered by 230 V line source, and the different parameters (C_1, C_2, R_1, R_{eq}, and L_1 of this model are presented in [8].

3 Summary of the Gate Controller and Optimization Techniques

3.1 Conventional Fixed Hysteresis Band

The conventional hysteresis fixed band technique employs two bands, namely the upper band and the lower band represented by $\Delta+$ and $\Delta-$, respectively. The modulated current is confined between these two bands. When the error current ε_a approaches the upper band $\Delta+$, the upper IGBT switch F_a, is switched off while the lower switch $\overline{F_a}$ is switched on, and vice versa for the lower band $\Delta-$. This ensures that the compensating current $i_{c_{(a,b,c)}}*$ tracks the reference current $i_{r_{(a,b,c)}}$, this achieving robustness and good dynamic performance compared to other compensation methods. Figure 3a, b illustrate the block diagram of the conventional hysteresis control and the waveform of the control signals for a single phase (phase a). It is to

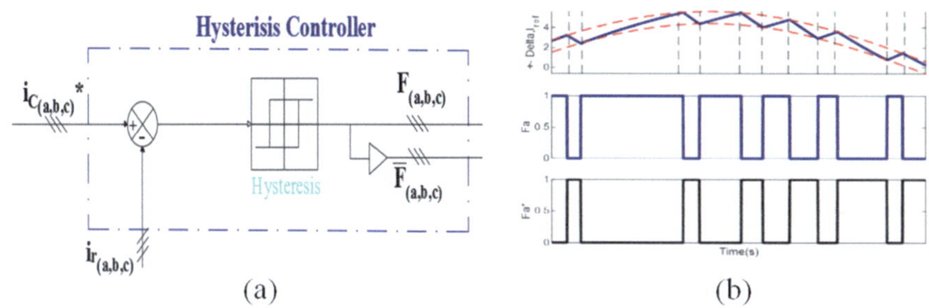

Fig. 3 Block diagram of the constant hysteresis modulation and their principle of generating control signals

be noted that the lower side of the inverter is controlled by the same signal but with a NOT logic gate.

3.2 New Modulation Hysteresis Control Technique

As previously mentioned, the inverter's IGBT switches are controlled using a hysteresis band to generate gate pulses, by comparing the actual currents $i_{c_{(a,b,c)}}*$ with the reference currents $i_{r_{(a,b,c)}}$. Figure 4 illustrates this process. In order to enhance and ensure the aforementioned results and to mitigate Total Harmonic Distortion (THD) issues, even at high or low power levels, a new technique proposed by Bodetto et al. [9] is implemented in this study. This technique involves adding an extra forward loop in the control stage to adapt the hysteresis window Δ, utilizing the forwarded current reference of Compact Fluorescent Lamps (CFLs).

This technique is implemented in order to avoid the harmonic distortion close to the shifted zero-crossing of the CFLs current $i_{L_{(a,b,c)}}(t_1)$ and $i_{L_{(a,b,c)}}(t_2)$ for all CFLs

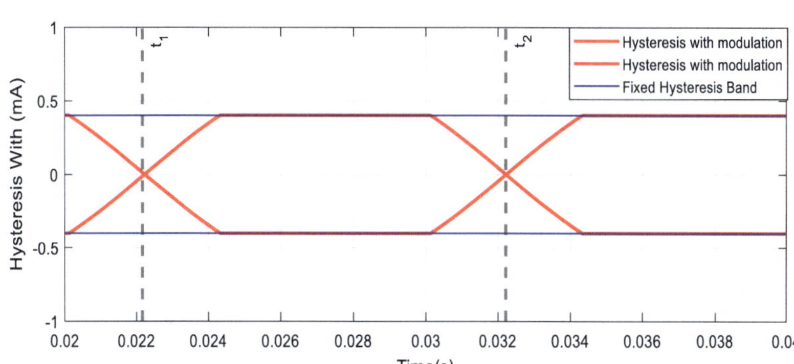

Fig. 4 The evolution of the new modulated hysteresis window Δ, in terms of the fixed hysteresis block

current levels. Where t_1 and t_2 represent the shifted zero crossing aera respectively, equal to $\frac{\pi}{2} + t1$ and $\pi + t1$ for all CFLs current levels. The new strategy does not require a complex system like vector PWM. Simple hysteresis is sufficient to control the switching device triggers.

4 Results and Discussions

The studied system is simulated using Matlab/Simulink software, the load is constituted of 75 equivalent model lamps distributed over the 3 AC phases (25 equivalent model lamps in parallel for each phase). A three phase four wire SAPF with three legs and the ford leg connected to a split capacitor (2C topologies) is connected in parallel with the load. Table 1 summarizes the simulation parameters. The SAPF is connected to PCC at time instant t = 0.02 s. Such that the founded results in this section are carried out only in phase a, knowing that the other two phases b and c are phase-shifted by 120° and − 120° with respect to phase a.

The simulated results voltage (three phases) and currents waveforms before compensation are presented in Fig. 5a, b respectively. Figure 6a, b shows the source currents waveforms after compensation with the fixed hysteresis band, and with the modulation hysteresis controller.

The source current waveforms lustered in Fig. 5b for phase a (and other two phases) is non-sinusoidal, and shifted to the voltage current, the THD = 90% before compensation. After connected the active filter with the fixed hysteresis controller as

Table 1 System parameters

Parameters	Value
Source voltage and frequency (V_S, F_S)	230 V, 50 Hz
Source impedance (R_S, L_S)	2 mΩ, 1 mH
Filter inductance (R_f, L_f)	1 mΩ, 0.1 μH
DC link capacitor (C_1, C_2)	2200 μF
Reference voltage V_{ref}	600 V

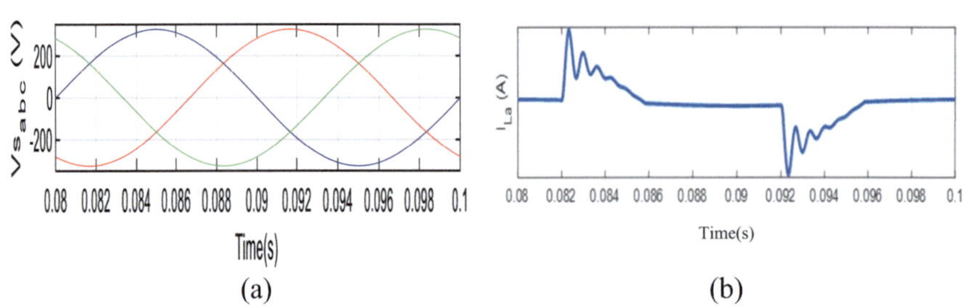

Fig. 5 Source voltage and current waveforms before compensation (THD_i = 90% from [8])

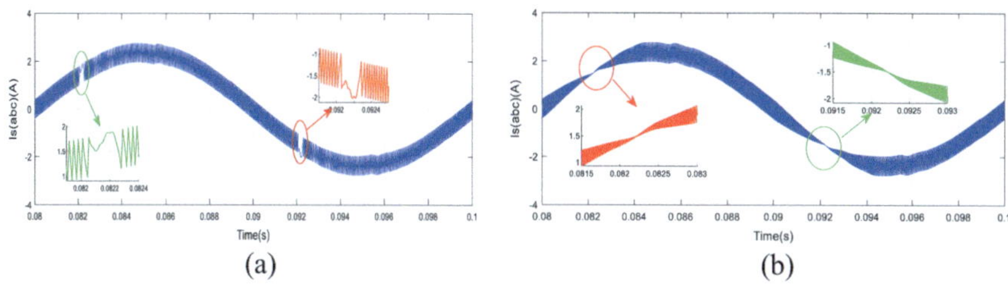

Fig. 6 Source current waveforms after compensation: **a** with the hysteresis controller, **b** with the modulation hysteresis controller

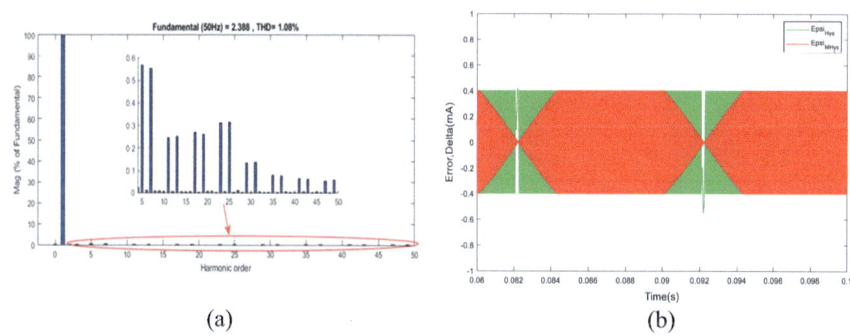

Fig. 7 FFT components of the load currents after compensation (THD = 1.8%), Error Signal after compensation with the hysteresis and the modulated hysteresis controller

mentioned in Fig. 6a, the source current becomes sinusoidal, and the THD reduced to 3%. Consequently, the PQ theory with a simple fixed hysteresis controller is capable of satisfying the IEEE-519 standard limits. In order to more improvement of the power quality [near the zero crossing aera, and the shifted zero crossing aera (t1 and t2)], the new technique was applied in this new arrangement and we can see in Fig. 6b that the power quality (the source current waveforms) is greatly improved and as a result the THD is reduced to 1.08% as presented in Fig. 7a, this also detailed and presented in Fig. 7b for both cases in the same figure.

5 Conclusion

This paper presented a three phase four wire SAPF for harmonic reduction and reactive power compensation generated by General Electric GE Compact Fluorescents Lamps with PQ and hysteresis controller. Active filters are a reliable option for compensating harmonics. The construction of passive filters is simple, but they are used to filter only a band of harmonics. In contrast, SAPFs are capable of compensating for most harmonic frequencies. Harmonics are reduced significantly to less than 5% as prescribed by IEEE-519 standards.

Acknowledgements We are most thankful for the Research Laboratory: Process, Energetic, environment and electrical systems which Tunisian higher ministry and scientific research code is LR18ES34. That has provided financial and technical supports for the larger project development from which this paper grew.

References

1. Akagi H (2005) Active harmonic filters. Proc IEEE 93(12):2128–2141
2. Nagi Reddy B et al (2019) Performance and dynamic analysis of single switch AC–DC buck-boost buck converter. Int J Innov Technol Explor Eng 8(4):307–313
3. Ramachandra Y et al (2019) Analysis of recent developments in brushless DC motors controlling techniques. Int J Innov Technol Explor Eng 8(5):522–526
4. Singh B et al (1998) Harmonic elimination, reactive power compensation and load balancing in three-phase, four wire electric distribution systems supplying non-linear loads. Int J Elect Power Syst Res 44(2):93–100
5. Wu F, Sun B, Zhao K, Sun L (2013) Analysis and solution of current zero-crossing distortion with unipolar hysteresis current control in grid-connected inverter. IEEE Trans Ind Electron 60(10):4450–4457
6. Dahono PA (2009) New hysteresis current controller for single-phase full-bridge inverters. IET Power Electron 2(5):585–594
7. Akagi H, Kanazawa Y, Nabae A (1983) Generalized theory of the instantaneous reactive power in three-phase circuits, IPEC'83. In: International power electronics conference, Tokyo, Japan, pp 1375–1386
8. Mohamed H, Mohamed N, Lassaad S (2021) Experimental investigations of a CFLs currents harmonics injection into an electrical network grid. In: 2021 4th international symposium on advanced electrical and communication technologies (ISAECT), pp 01–05. https://doi.org/10.1109/ISAECT53699.2021.9668434
9. Bodetto M, El Aroudi A, Cid-Pastor A, Calvente J, Martínez-Salamero L (2016) Design of AC–DC PFC high-order converters with regulated output current for low-power applications. IEEE Trans Power Electron 31(3):2012–2025

A Comparative Study of Performed SV PWM Three Phase Inverters Strategies Drive

Nadia Gasmi and Mouna Ben Hamed

Abstract Sinusoidal pulse width modulation (SPWM) and space vector pulse width modulation (SVPWM) are the most popular modulation strategies for multi-level inverters. This paper provides the theoretical analysis and simulation results of SPWM and SVPWM for three level inverters. This paper focuses on implementation of Space Vector Pulse Width Modulation (SVPWM) Technique for Three level Neutral Point Clamped (NPC) Multilevel Inverter and compare it with Sinusoidal Pulse Width Modulation Technique (SPWM). Simulation results were shown and compared in the view of THD for both the techniques. The result shows that the SVPWM is better for three level NPC inverter in THD point of view.

Keywords Three level neutral point clamped (NPC) multilevel inverter · SVPWM · SPWM · THD · NPC

1 Introduction

Multilevel inverters are widely used in high-power high-voltage applications due. There are four types of MLIs are present namely, Diode Clamped, Flying Capacitor, cascaded [1] and Modular multilevel Inverters. We have two most popular modulation schemes namely SPWM and SVPWM are available [2].

The SVPWM has been widely used in 3-phase 3-level inverter NPC control system because; it has a higher utility efficiency of DC-side voltage. Although the SVPWM has many advantages, it is difficult to implement. The most difficult factor is calculating the duty cycles for each power switch, as well as determining the vector sector and pulse sequence in each switching cycle. In this article, the three-tiered SVPWM and SPWM methods are explained in detail and compared using THD analysis [3].

N. Gasmi (✉) · M. B. Hamed
PEESE (Processes, Energetics, Environments and Electrical Systems) Research Laboratory LR18ES34, National Engineering School of Gabes University, Zrig Eddakhlania, Tunisia
e-mail: gasminadia809@hotmail.com

© The Author(s), under exclusive license to Springer Nature Singapore Pte Ltd. 2024

A. Mellit et al. (eds.), *Proceedings of the 2nd International Conference on Green Energy Conversion System*, Green Energy and Technology,
https://doi.org/10.1007/978-981-97-6148-7_5

In this paper we were implemented the SVPWM scheme for three level NPC inverters. The simulation results of SVPWM scheme employed circuit are compared with the SPWM scheme employed circuit in THD perspective, and it shows that the SVPWM is better compared to the SPWM scheme employed circuit in terms of THD.

2 Three Level NPC Inverter

The three level NPC topology consist of 12 switches including 6 diodes. It is necessary to connect two capacitors in series both charged with V_{dc} [4]. Each phase led contains 4 series switches with two diodes clamped in series. The main function of the diodes is to clamp the upper switches at higher potential to the zero dc-link point. A particular switching pattern will give the output of three level NPC inverter. The circuit diagram shown in Fig. 6 gives idea about the arrangement of the power circuit required for the three level NPC inverter. In this circuit, the diodes are arranged in such a way that it yields different stages of voltages with respect to neutral point N. The splitting of the voltage level is considered according capacitors C1 and C2 arranged in series with each other having neutral point N in between them. The division of voltages are obtained as $V_{dc}/2$, 0, $-V_{dc}/2$. So, it is named as three level topology [4].

In order to deduce the knowledge model of the inverter, we introduce the connection function Sis of the switch which describes the state of every switch $(1 = closed, 0 = opened)$. In this function '*i*', is the number of the of the commutation cell, $i \in \{1,2,3\}$, '*s*', is the number of the semi-conductor. The output voltages of the inverter relatively to the middle point M with using the connection functions of the half-arm are defined as follows [7]:

$$\begin{cases} B_{k1} = \overline{B_{k4}} \\ B_{k2} = \overline{B_{k3}} \end{cases} \tag{1}$$

For the leg k, we define a half leg connection function where $k = 1, 2, 3$: where $b = 1$ refers to the lower half leg and $b = 0$ to the upper half leg

$$\begin{cases} F_{k1}^{b} = F_{k1}.F_{k2} \\ F_{k0}^{b} = F_{k3}.F_{k4} \end{cases} \tag{2}$$

Connect functions for switches in parallel are defined as follows [7]:

$$\begin{cases} F_{k7} = F_{k1}F_{k2}(1 - F_{k3}) \\ F_{k8} = F_{k4}F_{k5}(1 - F_{k6}) \end{cases} \tag{3}$$

Introducing these functions into the system (3) results in the following system:

$$\begin{cases} V_{AM} = F_{11}F_{12}U_{c1} - F_{13}F_{14}U_{c2} \\ = (F_{11}F_{12} - F_{13}F_{14})U_c \\ V_{BM} = F_{21}F_{22}U_{c1} - F_{23}F_{24}U_{c2} \\ = (F_{21}F_{22} - F_{23}F_{24})U_c \\ V_{CM} = F_{31}F_{32}U_{c1} - F_{33}F_{34}U_{c2} \\ = (F_{31}F_{32} - F_{33}F_{34})U_c \end{cases} \quad (4)$$

In matrix form (4) is written

$$\begin{cases} V_{AM} = F_{11}^b U_{c1} - F_{10}^b U_{c2} \\ V_{BM} = F_{21}^b U_{c1} - F_{20}^b U_{c2} \\ V_{CM} = F_{31}^b U_{c1} - F_{30}^b U_{c2} \end{cases} \quad (5)$$

The potentials of nodes A, B and C of Three phase three-level inverter relatively to the middle point M in the case $U_{c1} = U_{c2} = U_c$ are given by the following system [7]:

$$\begin{bmatrix} V_{AM} \\ V_{BM} \\ V_{CM} \end{bmatrix} = \begin{bmatrix} F_{11}^b - F_{10}^b \\ F_{21}^b - F_{20}^b \\ F_{31}^b - F_{30}^b \end{bmatrix} U_c \quad (6)$$

Compound voltages are given by:

$$\begin{bmatrix} U_{AB} \\ U_{BC} \\ U_{CA} \end{bmatrix} = \begin{bmatrix} 1 & -1 & 0 \\ 0 & 1 & -1 \\ -1 & 0 & 1 \end{bmatrix} \left(\begin{bmatrix} F_{11}^b \\ F_{21}^b \\ F_{31}^b \end{bmatrix} U_{c1} - \begin{bmatrix} F_{10}^b \\ F_{20}^b \\ F_{30}^b \end{bmatrix} U_{c2} \right) \quad (7)$$

The simple voltages at the boundaries of the load are given by the following system:

$$\begin{bmatrix} V_A \\ V_B \\ V_C \end{bmatrix} = \frac{1}{3} \begin{bmatrix} 2 & -1 & -1 \\ -1 & 2 & -1 \\ -1 & -1 & 2 \end{bmatrix} \begin{bmatrix} F_{11}^b - F_{10}^b \\ F_{21}^b - F_{20}^b \\ F_{31}^b - F_{30}^b \end{bmatrix} U_c \quad (8)$$

3 Control Technique for 3-Level

Pulse width modulation technique is used for inverter output voltage control and if we switch device in particular fashion harmonics can be eliminated [6]. In Pulse Width Modulation Technique, the switching signal [5] is generated by comparing fundamental reference signal with carrier wave of required switching frequency. The frequency of reference signal determines the frequency of fundamental output wave and its amplitude controls the modulation index. There are various PWM scheme available for controlling multi-level inverter such as

- Sinusoidal Pulse Width Modulation (SPWM)
- Third Harmonic Injection Pulse Width Modulation (THI-PWM)
- Space Vector Pulse Width Modulation (SVPWM)
- Selective Harmonic Elimination Pulse Width Modulation (SHE-PWM).

3.1 The Basic SVPWM Principle of the 3-Phase 3-Level Inverter

There are various steps for implementing the three levels SVPWM inverter. They are

- No of switching states
- No of voltage vectors and corresponding voltages.
- Sector identification.
- Determining the region in the sector.
- Calculating the active vectors switching time periods.
- Generation of gating signals.

The total number of switching states in an "N" level inverter is "N3". So, the total number of switching states in a "3" level inverter is "33" that is 27 switching states. 24 states are active states and 3 zero states [7].

In this topology there are mainly two pairs of switches and also two diodes are used in a three-level diode clamped multi-level inverter. The DC bus voltage is divided into three different voltage levels with the help of two series connections of DC capacitors that is to say C1 and C2 as shown in Fig. 1. The voltage across each capacitor is Vdc/2 that is to say voltage across capacitor C1 and C2 is equals to Vdc/2 [8]. There are mainly twelve active combinations were taken and using these switching states twelve active voltage vectors are produced. There are nonzero voltage vectors and are from V1 to V12. Figure 1 shows the Three Level Diode Clamped Multilevel Inverter (MLI) [7].

There are 27 statuses when controlling the power switches for each leg; each status can be mapped to the vector diagram of α-β coordinate plane. The 27 vectors can form 18 sectors, as shown in Fig. 2 [6, 9].

$$\begin{pmatrix} V_\alpha \\ V_\beta \end{pmatrix} = \frac{2}{3}\begin{pmatrix} 1 & \cos\left(\frac{2\pi}{3}\right) & \cos\left(\frac{2\pi}{3}\right) \\ 0 & \sin\left(\frac{2\pi}{3}\right) & -\sin\left(\frac{2\pi}{3}\right) \end{pmatrix}\begin{pmatrix} V_a \\ V_b \\ V_c \end{pmatrix} \tag{9}$$

$$\begin{pmatrix} V_\alpha \\ V_\beta \end{pmatrix} = \frac{2}{3}\begin{pmatrix} 1 & -\frac{1}{2} & -\frac{1}{2} \\ 0 & \frac{\sqrt{3}}{2} & -\frac{\sqrt{3}}{2} \end{pmatrix}\begin{pmatrix} V_a \\ V_b \\ V_c \end{pmatrix} \tag{10}$$

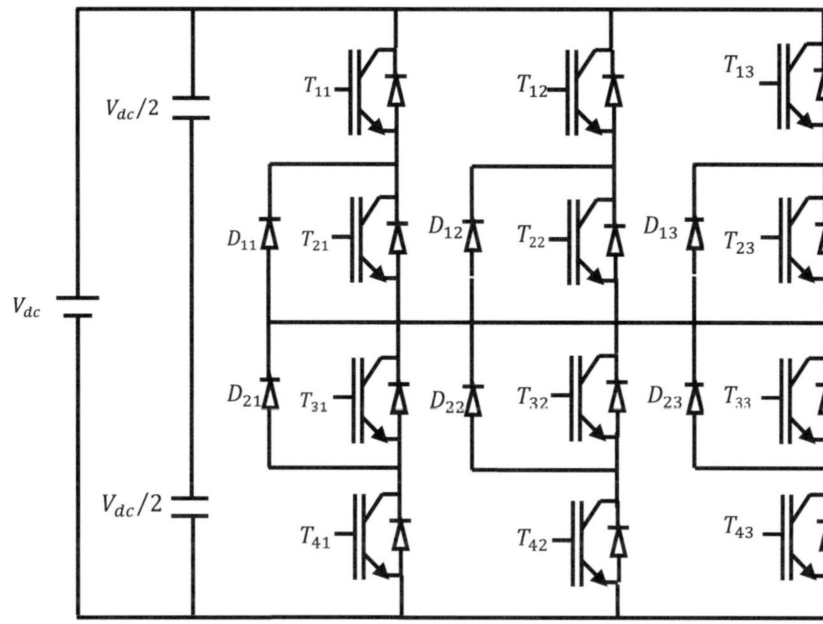

Fig. 1 Three level NPC inverter [5]

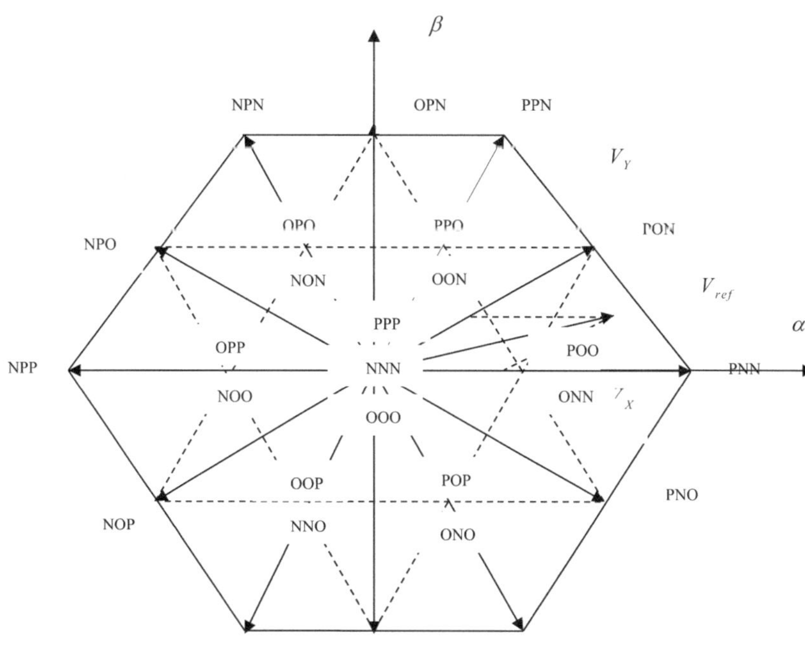

Fig. 2 The 3-phase 3-level inverter SVPWM vector diagram [9]

Assume a reference vector V_{ref}. According to the theory of the SVPWM, we must find out two nearest vectors V_X, V_Y and one zero vector V_Z in Fig. 2 to form the vector V_{ref}. Figure 2 shows the relationship between V_{ref} and V_X, V_Y, V_Z. So, the vector PNN (V_X), PNN (V_Y), and NNN (V_Z) can be selected to form V_{ref}. If the dwell time of V_X, V_Y, V_Z in a specified interval T_S are T_X,T_Y,T_Z. respectively, the following functions must be satisfied [2, 7]:

$$V_X T_X + V_Y T_Y + V_Z T_Z = V_{ref} T_S \tag{11}$$

$$T_X + T_Y + T_Z = T_S \tag{12}$$

(a) **Finding the sector in the hexagon**

Alfa (α) is calculated and then the sector, in which the ref vector V* is located, is determined as [7];

If α is in the range of $0° \leq \alpha < 60°$, **then** vector V* lies in Sector a,
If α is in the range of $60° \leq \alpha < 120°$, **then** vector V* lies in Sector b,
If α is in the range of $120° \leq \alpha < 180°$, **then** vector V* lies in Sector c,
If α is in the range of $180° \leq \alpha < 240°$, **then** vector V* lies in Sector d,
If α is in the range of $240° \leq \alpha < 300°$, **then** vector V* lies in Sector e,
If α is in the range of $300° \leq \alpha < 360°$, **then** vector V* lies in Sector f.

(b) **Finding the respective region in the sector**

From Fig. 3b m_1 and m_2 in sector 'a'can be calculated as

$$m_1 = d - c = m_n \cos\alpha - \left(\frac{2}{\sqrt{3}} m_n \sin\alpha\right)\cos\left(\frac{\pi}{3}\right) \tag{13}$$

$$m_1 = m_n\left(\cos\alpha - \frac{\sin\alpha}{\sqrt{3}}\right) = m_n \frac{2}{\sqrt{3}} \sin\left(\frac{\pi}{3} - \alpha\right) \tag{14}$$

$$m_2 = a = \frac{b}{\sin\left(\frac{\pi}{3}\right)} = \frac{2}{\sqrt{3}} b = \frac{2}{\sqrt{3}} m_n \sin\alpha \tag{15}$$

$$m_n = \frac{V_{ref}}{\frac{2U_{dc}}{3}} \tag{16}$$

And by,

If the value of m_1, m_2 and $(m_1 + m_2) < 0.5$, **then** V^* lies in ***Reg 1***,
If the valve of $m_1 > 0.5$, then V^* lies in ***Reg 2***,
If the value of $m_2 > 0.5$, **then** V* lies in **Reg 3**,
If the value of m_1 and $m_2 < 0.5$ and $(m_1 + m_2) > 0.5$, **then** V* lies in **Reg 4**.

(c) **Calculating the respective switching times for T_a,T_b,T_c**

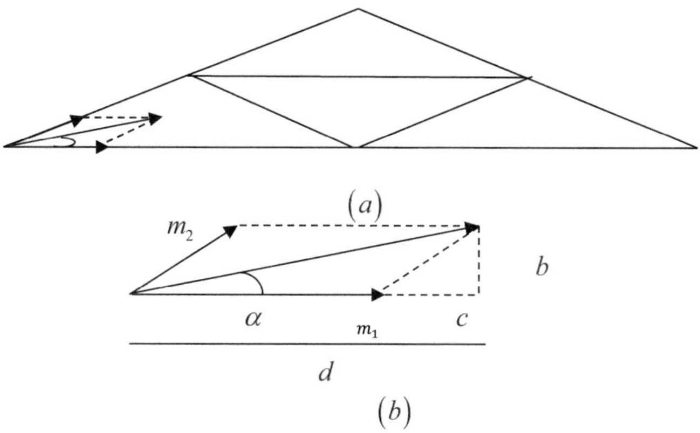

Fig. 3 Space vector diagram for $\boldsymbol{m}_1$ and $\boldsymbol{m}_2$ in sector 'a' [9]

Table 1 Switching times for $\mathbf{T_a,T_b,T_c}$ [8]

T_a	$2k\sin\left(\frac{\pi}{3}-\theta\right)$	$T_s-2k\sin\theta$
T_b	$T_s-2k\sin\left(\frac{\pi}{3}+\theta\right)$	$2k\sin\left(\frac{\pi}{3}+\theta\right)-T_s$
T_c	$2k\sin\theta$	$T_s-2k\sin\left(\frac{\pi}{3}-\theta\right)$
T_a	$2k\sin\theta-T_s$	$2T_s-2k\sin\left(\frac{\pi}{3}+\theta\right)$
T_b	$2k\sin\left(\frac{\pi}{3}-\theta\right)$	$2T_s-2k\sin\left(\frac{\pi}{3}+\theta\right)$
T_c	$2T_s-2k\sin\left(\frac{\pi}{3}+\theta\right)$	$2k\sin\left(\frac{\pi}{3}+\theta\right)-T_s$

T_a, T_b, T_c switching times of Sector 'a' is given in Table 1.

(d) **Finding the switching states based on above step**

At any time by taking the switching transition of only one switch; the switching transition orders are shown below and they are obtained for every region in Sector 'a' if all switching transition states in every region are used. Therefore, switching states for Sector 'a' given below [10]

Reg 1: *1-1-1, 0-1-1, 00-1, 000, 100, 110, 111*
Reg 2: *0-1-1, 1-1-1, 10-1, 100*
Reg 3: *0-1-1, 11-1, 10-1, 100, 110*
Reg 4: *00-1, 10-1, 11-1, 110.*

3.2 The Basic SPWM Principle of the 3-Phase 3-Level Inverter

The sinusoidal PWM technique is a modulation technique which is one of the most common and popular approach for generating a sinusoidal wave. In this method the

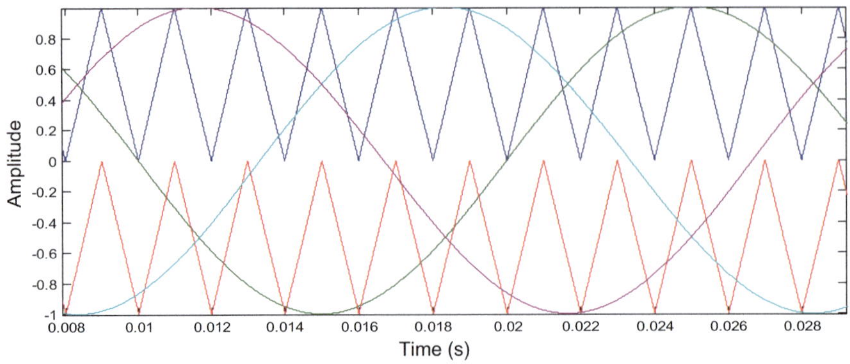

Fig. 4 Comparison of sine and triangular wave for three level SPWM [6]

pulse widths are determine by comparing a triangle wave which is also known as carrier wave with a sinusoidal wave of desired fundamental frequency. Further these pulses are the controlling factor to control the state of switches used in inverter. Specifically, the pulse width of SPWM is modulated such as to make the average output voltage waveform corresponding to a sinusoidal wave [9].

Comparison of fundamental sine wave and high frequency triangular wave is as shown in figure (Fig. 4). The compared triangular wave is called as carrier waves and the desired sine wave is termed as reference wave. Both these waves are compared in a comparator and the output of the comparator is given to the switching devices like TCRs, MOSFETs, IGBTs or any other switching devices [2, 11].

The comparator output is high when the magnitude of the sine wave is higher than the magnitude of the triangular wave. As SPWM strategy with a two carrier, this strategy is characterized by the modulation index m, and the modulation ratio r are defined as follows [1, 2]:

$$\mathrm{m}m = \frac{f_m}{f_{pm}} \tag{17}$$

$$= \frac{f_{pm}}{f_m} r = \frac{V_m}{U_{pm}} \tag{18}$$

4 Simulation Results Comparison of SPWM and SVPWM

Figure shows the waveforms and the harmonic spectra of the voltage obtained at the output of the three-level inverter for the two strategies (SPWM, SVPWM). The results between the two (SPWM-SVPWM) control strategies are compared and summarized in Table 2. The simulation and the analysis of the curves deliver very similar results

Table 2 Comparisons between SPWM and SVPWM for the three-level NPC inverter

Type	*Va*	*Vb*	*Vc*
SPWM	*191.88*	*180.19*	*167.66*
SVPWM	*71.68*	*78.35*	*78.34*
	UAB	*UBC*	*UCA*
SPWM	*192.83*	*169.29*	*177.39*
SVPWM	*74.42*	*79.82*	*74.40*

for both control techniques, with a slight advantage for the control system with vector PWM. This is verified by the FFT analysis results, where it can be seen that the Total Harmonic Distortions (THD) of the SVPWM technique are reduced compared with SPWM for all parameters (inverter voltage and the stator currents) (Figs. 5, 6, 7, 8, 9, 10, 11, 12, 13, 14, 15, 16, 17, 18, 19, 20, 21, 22, 23 and 24).

The table points out a remarkable difference between the distortions introduced by the two modulation techniques. In fact, the Total Harmonic Distortion (THD %) produced by the SVPWM is equal to 71.68, while that introduced by the SPWM is equal to 191.88.

The modeling of the inverter of three level has been detailed. The comparative study addresses two control strategies, which are slightly different for the two types of control.

It is verified by the FFT analysis The THD spectra showed that SVPWM offers a better choice. The system studied in this paper is part of a complete system that can be of interest in a wind power generation system.

the rotor is powered by a three phase, three level inverter controlled primarily by SPWM, and then by SVPWM. SVPWM command of the inverter is much better than the SPWM technique (oscillations in this case are more significant).

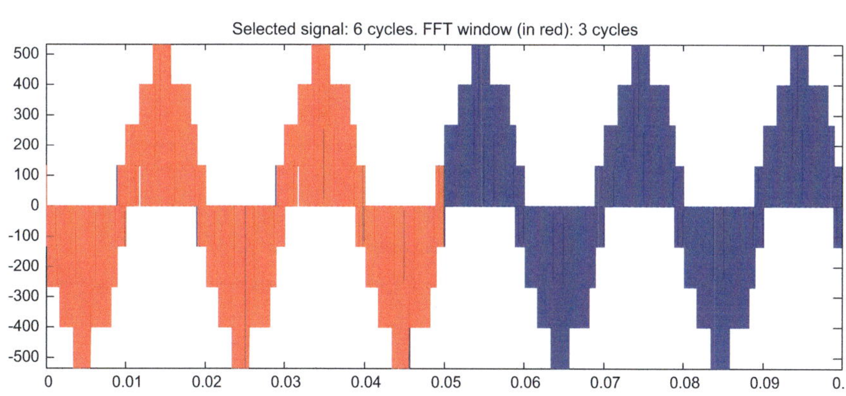

Fig. 5 FFT analysis window for line voltage ***Va*** in a simple three level NPC inverter with SPWM scheme

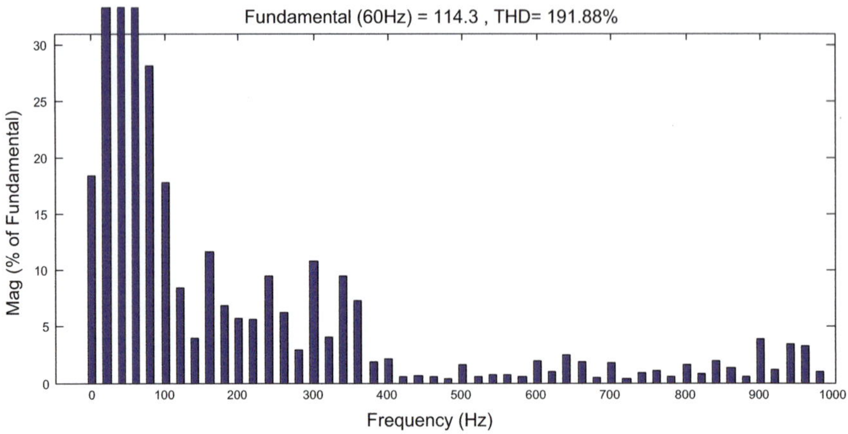

Fig. 6 FFT analysis at the 50 Hz of fundamental frequency for the line voltage ***Va*** and calculation of THD for three level NPC inverter with SPWM scheme

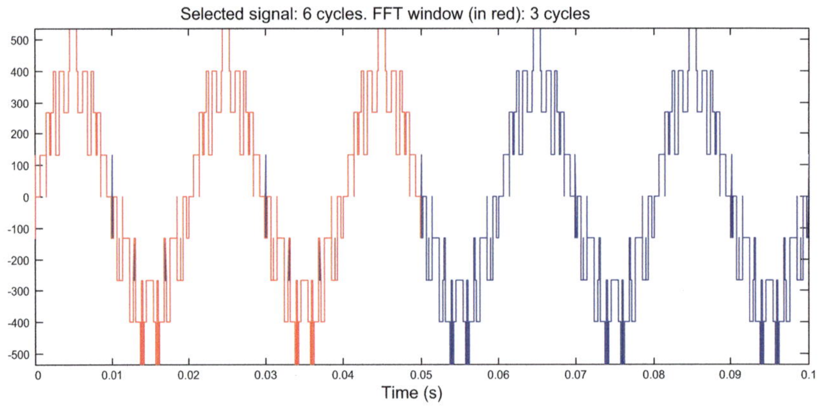

Fig. 7 FFT analysis window for line voltage *Va* in a simple three level NPC inverter with SVPWM scheme

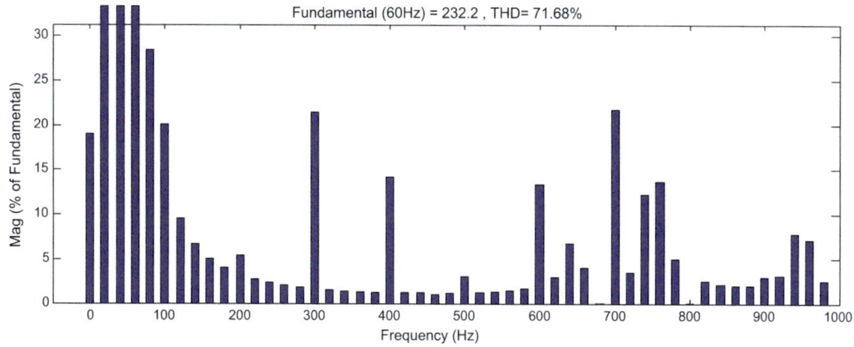

Fig. 8 FFT analysis at the 50 Hz of fundamental frequency for the line voltage *Va* and calculation of THD for three level NPC inverter with SVPWM scheme

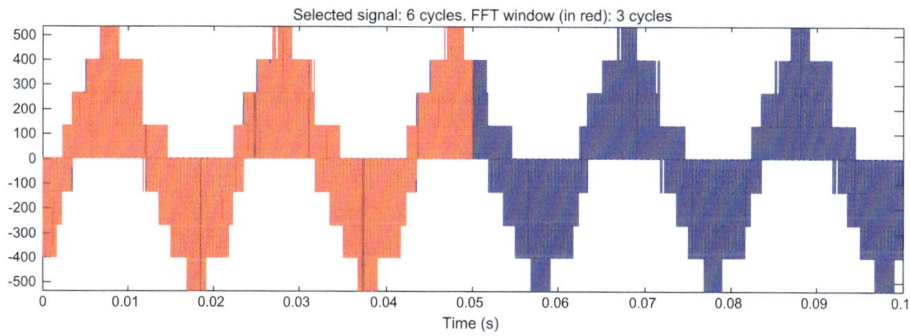

Fig. 9 FFT analysis window for line voltage *Vb* in a simple three level NPC inverter with SPWM scheme

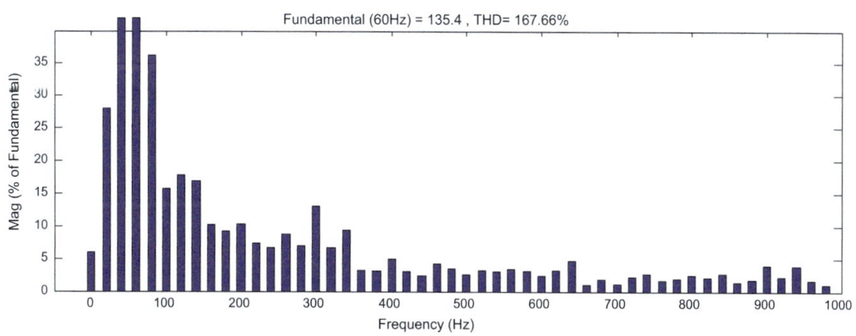

Fig. 10 FFT analysis at the 50 Hz of fundamental frequency for the line voltage *Vc* and calculation of THD for three level NPC inverter with SPWM scheme

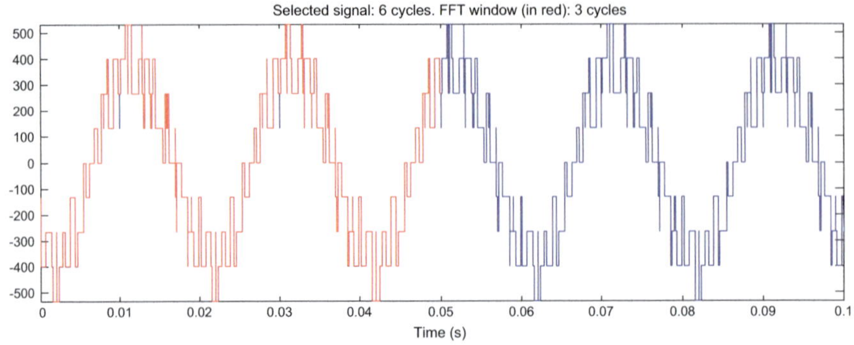

Fig. 11 FFT analysis window for line voltage *Vc* in a simple three level NPC inverter with SVPWM scheme

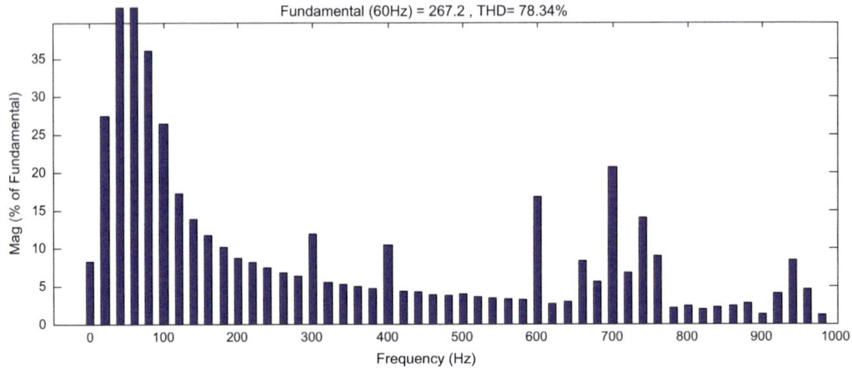

Fig. 12 FFT analysis at the 50 Hz of fundamental frequency for the line voltage *Vc* and calculation of THD for three level NPC inverter with SVPWM scheme

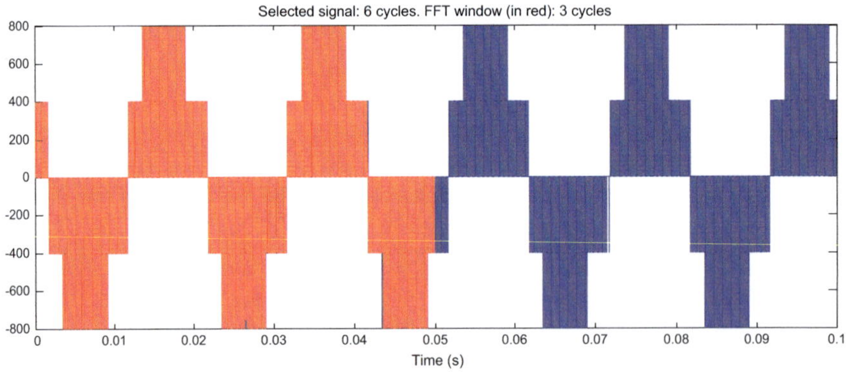

Fig. 13 FFT analysis window for line voltage *UAB* in a simple three level NPC inverter with SPWM scheme

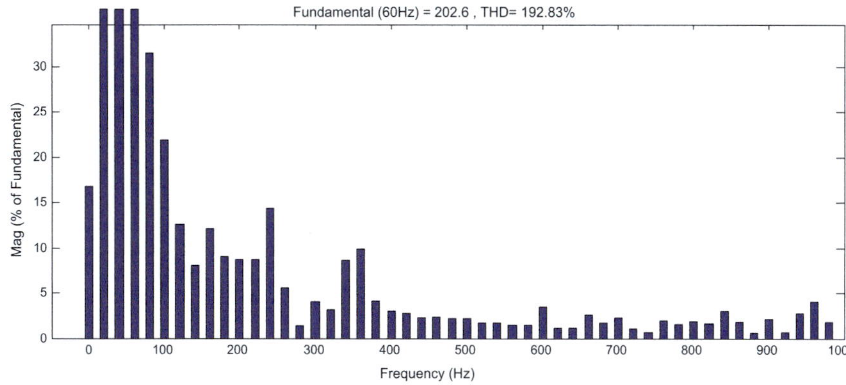

Fig. 14 FFT analysis at the 50 Hz of fundamental frequency for the line voltage *UAB* and calculation of THD for three level NPC inverter with SPWM scheme

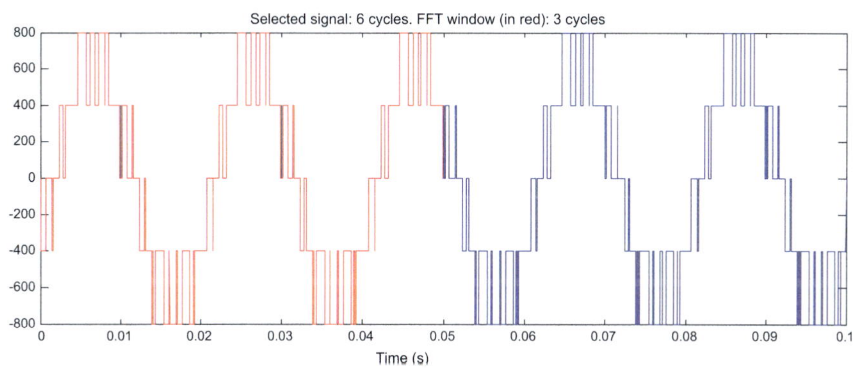

Fig. 15 FFT analysis window for line voltage *UAB* in a simple three level NPC inverter with SVPWM scheme

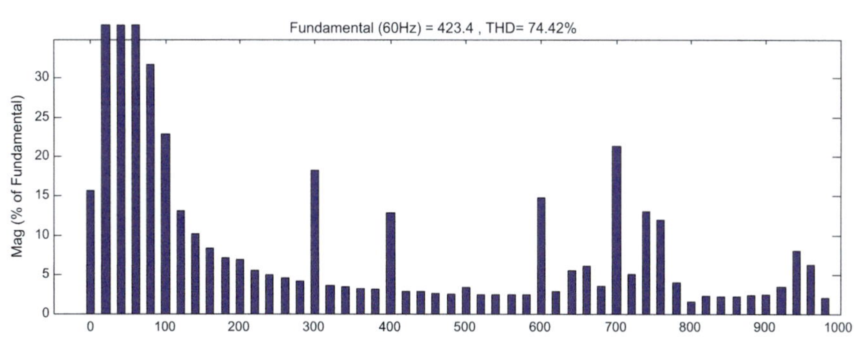

Fig. 16 FFT analysis at the 50 Hz of fundamental frequency for the line voltage *UAB* and calculation of THD for three level NPC inverter with SVPWM scheme

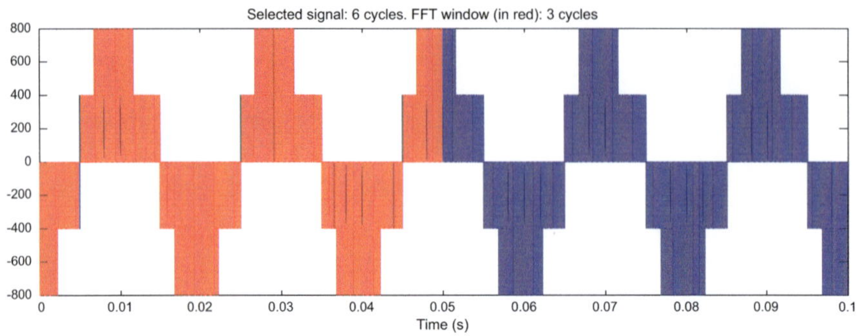

Fig. 17 FFT analysis window for line voltage *UBC* in a simple three level NPC inverter with SPWM scheme

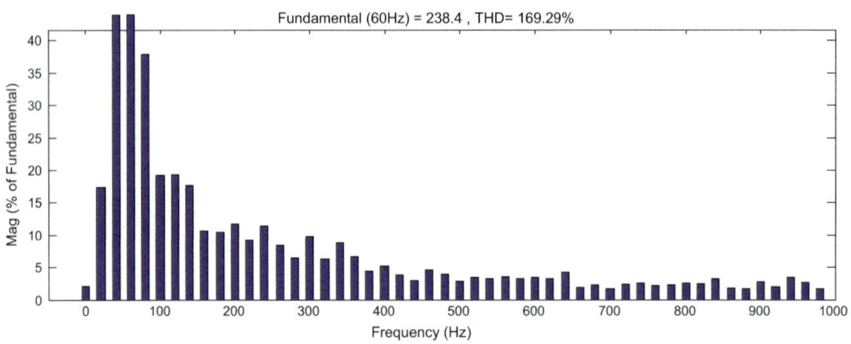

Fig. 18 FFT analysis at the 50 Hz of fundamental frequency for the line voltage *UBC* and calculation of THD for three level NPC inverter with SPWM scheme

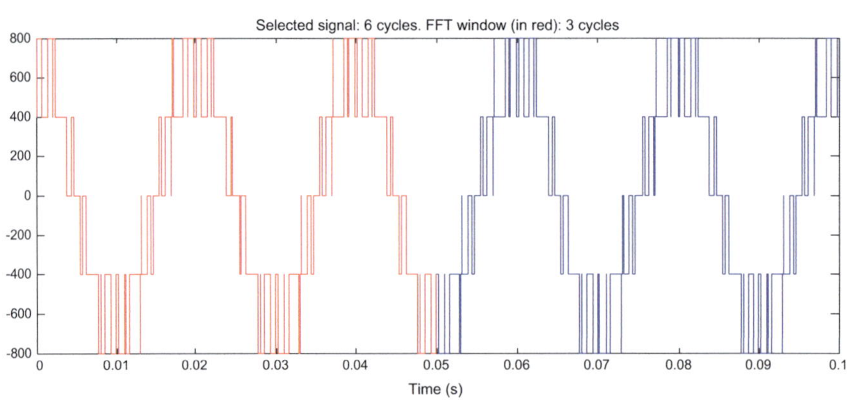

Fig. 19 FFT analysis window for line voltage *UBC* in a simple three level NPC inverter with SVPWM scheme

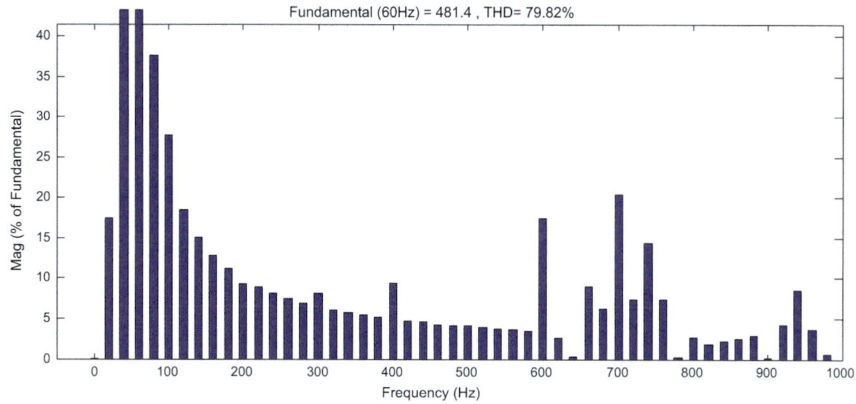

Fig. 20 FFT analysis at the 50 Hz of fundamental frequency for the line voltage *UBC* and calculation of THD for three level NPC inverter with SVPWM scheme

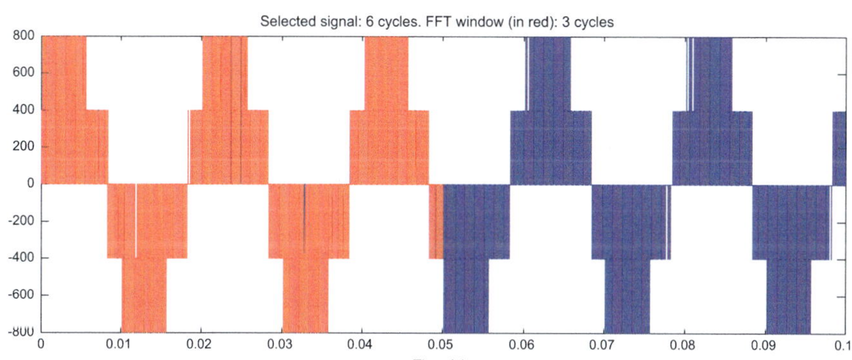

Fig. 21 FFT analysis window for line voltage *UCA* in a simple three level NPC inverter with SPWM scheme

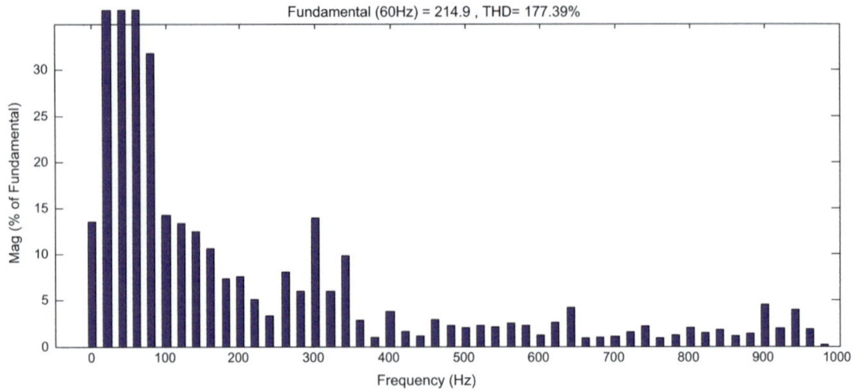

Fig. 22 FFT analysis at the 50 Hz of fundamental frequency for the line voltage *UCA* and calculation of THD for three level NPC inverter with SPWM scheme

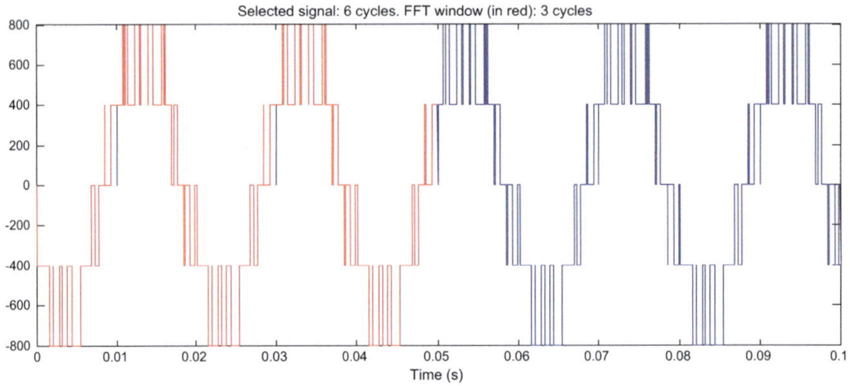

Fig. 23 FFT analysis window for line voltage *UCA* in a simple three level NPC inverter with SVPWM scheme

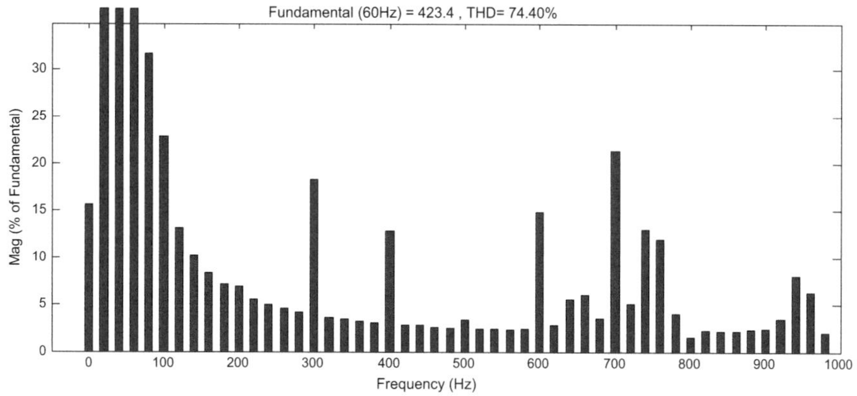

Fig. 24 FFT analysis at the 50 Hz of fundamental frequency for the line voltage *UCA* and calculation of THD for three level NPC inverter with SVPWM scheme

5 Conclusion

In this paper, SVPWM technique is used to reduce the harmonics. In first comparative analysis of Space Vector PWM with conventional SPWM for a three level Inverter is carried out. SPWM inverter voltage THD values are greater than SVPWM. SPWM and SVPWM are implemented in MATLAB/SIMULINK software and its performance is compared with conventional PWM techniques. The comparative study addresses two control strategies, which are slightly different for the two types of control. It is verified by the FFT analysis The THD spectra showed that SVPWM offers a better choice.

References

1. Al-Odienat A, Al-Momani M, Al-Maitah K, Al-Mbaideen A. The dynamic response of SPWM vs. SVPWM synchronverter
2. Bandaru U, Priyadarshini YSI, Sathyanarayana MS (2021) Space vector PWM algorithms for three-level inverter. Adv Aspects Eng Res 16:30–51
3. Chaurasia MK, Gidwani L. Simulation analysis of SVPWM and THIPWM for induction machine drive without filter
4. Lamreoua A, Benslimane A, Bouchnaif J, El Ouariachi M (2021) An improved sinusoidal (PWM) and vector (SVPWM) current control for a three-phase photovoltaic inverter connected to a non-linear load. In: Proceedings of the 2nd international conference on electronic engineering and renewable energy systems: ICEERE 2020, 13–15 Apr 2020, Saidia, Morocco. Springer, Singapore, pp 481–494
5. Deng J (2023) Pulse-width voltage modulation in the python-based open-source simulator motulator: a realization of SVPWM, harmonic analyzer and over-modulation on two and three-level inverters, and a solution for unbalanced neutral point voltage on three-level inverters

6. Hassanabad AH, Sarsabahi Y, Bohre AK, Ngwenyama MK (2022) Design and simulation of space vector pulse width modulation inverter based transformer-less in small wind turbines. World Acad J Eng Sci 9(2):1–8
7. Amor YA, Kheldoun A, Metidji B, Hamoudi F, Merazka A, Lazoueche Y (2021) Design and implementation of three-level T-type inverter based on simplified SVPWM using cost-effective STM32F4 board. Int J Digital Signals Smart Syst 5(1):20–39
8. Alzuabidi OHAA (2022) Study and implementation sinusoidal PWM inverter fed 3-phase induction motor. Int J Nonlin Anal Appl 13(1):3293–3303
9. Kadum AA (2020) PWM control techniques for three phase three level inverter drives. TELKOMNIKA (Telecommun Comput Electron Control) 18(1):519–529
10. Zhang G (2023) Development status of space vector pulse width modulation (SVPWM) control method for multi-level inverter. J Phys Conf Ser 2479(1):012053
11. Islam MT, Ayon SI (2020) Performance analysis of three-phase inverter for minimizing total harmonic distortion using space vector pulse width modulation technique. In: 2020 23rd international conference on computer and information technology (ICCIT). IEEE, pp 1–4

Thermodynamic and Environmental Comparison of Various Recuperated Cycles for Gas Turbine Applications

Faiza Brahimi and Messaouda Guemmadi

Abstract The considerable amount of energy contained in the exhaust gases of gas turbines represents a valuable potential that cannot be overlooked. The objective of this study is to identify the cycle that contributes the most to improving the performance of the simple gas turbine cycle (Joule-Brayton cycle) and reducing CO_2 emissions to the environment through the recovery of exhaust gases. Many studies have focused on methods for better exploiting this potential. In this context, three different cycles with exhaust gas recovery were analyzed in this work: the cycle with conventional recovery, the cycle with alternative recovery, and the cycle with staged recovery. The aim of this study is to evaluate the benefits of the best cycle with exhaust gas recovery based on energy performance and environmental impact. The energy analysis shows a significant improvement in energy efficiency compared to that of the cycle without recovery. To better understand the behavior of these technologies and achieve their best integration, we analyze the influence of certain key parameters such as compression ratio, ambient temperature, and heat exchanger effectiveness. The results of the environmental analysis show that the amount of CO_2 emissions in the cycles with recovery is significantly reduced.

Keywords Gas turbine · Joule Brayton cycle · Exhaust gas recovery · Gas turbine improvement

---

F. Brahimi (✉) · M. Guemmadi
Département de Génie Mécanique, Faculté de Technologie, Université M'hamed Bougara Boumerdes, Boumerdes, Algérie
e-mail: f.brahimi@univ-boumerdes.dz

F. Brahimi
Laboratoire Dynamique des Moteurs et Vibro-Acoustique, Université M'hamed Bougara Boumerdes, Boumerdes, Algérie

M. Guemmadi
Laboratoire d'Energétique Mécanique et Matériaux, Université de Tizi-Ouzou, Tizi-Ouzou, Algérie

© The Author(s), under exclusive license to Springer Nature Singapore Pte Ltd. 2024

A. Mellit et al. (eds.), *Proceedings of the 2nd International Conference on Green Energy Conversion System*, Green Energy and Technology,
https://doi.org/10.1007/978-981-97-6148-7_6

1 Introduction

With the increase in population worldwide and the development of countries, the demand for energy continues to grow [1, 2]. To meet the growing energy needs, production must be increased by implementing new installations or improving existing ones. The optimization of energy systems such as gas turbines represents a continuous technological challenge, both for economic and environmental reasons. Gas turbines, with their diverse applications such as power plants, oil and gas fields, aircraft engines, etc., have undergone remarkable development to improve their energy performance, reduce their fuel consumption, and their pollutant emissions.

An alternative for improving gas turbines is the recovery of their exhaust gases, which represent heat lost to the atmosphere. This alternative has become an attractive research topic [3–5]. Over the past decade, research on gas turbines with heat recovery has made rapid progress, and several technologies have been developed [6], such as the combined cycle gas turbine and steam turbine using water or other alternative fluids [7, 8]. In addition, the organic Rankine cycle (ORC) has been widely used to generate energy from low and medium temperature heat sources [9, 10]. The air Joule and supercritical CO_2 Joule cycles have been suggested for recovering residual heat from gas turbines [11]. Several studies on the recovery of exhaust gases from gas turbines have been conducted to demonstrate the potential of different recovery cycles [12–15].

In light of this research, recovery cycles offer the opportunity to provide an interesting improvement in the performance of the simple gas turbine cycle (Joule-Brayton cycle). The main objective of this study is to search for thermodynamic processes that can be applied to improve the performance of existing gas turbines by recovering lost heat via different recovery cycles. The specificity of this study is to find the appropriate cycle for recovering lost heat without the requirement of a steam-water cycle and that is suitable for sites with water scarcity, based on energy performance and CO_2 emissions.

Therefore, four configurations are studied: the simple Joule-Brayton cycle, the cycle with conventional recuperation, the cycle with alternative recuperation, and the cycle with simultaneous (or staged) recuperation. The modeling results of all cycles are presented, and the energy efficiencies of the systems are determined and compared to select the best recuperation cycle. Optimal key parameters of the cycle, which provide maximum performance to the system, are obtained through sensitivity analysis. The effects of pressure ratio, ambient temperature, and heat exchanger efficiency on performance are highlighted.

In addition to the energy analysis of the cycles, an environmental analysis is conducted. The contribution of the proposed cycles to the reduction of CO_2 emissions is emphasized to focus on clean power plants.

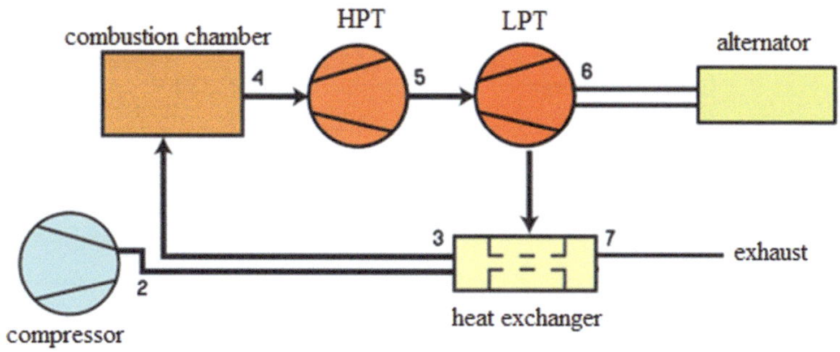

Fig. 1 Gas turbine configuration with conventional recuperation

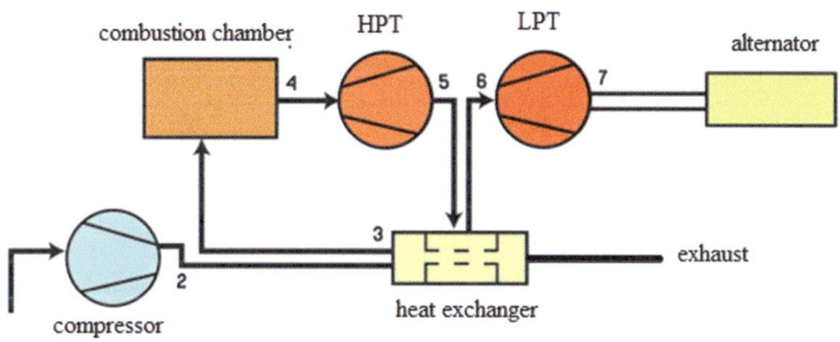

Fig. 2 Gas turbine configuration with alternative recuperation

2 Cycles Description

Figure 1 shows a simple gas turbine consisting of a compressor, a combustion chamber, and a turbine. The proposed installations are presented in Figs. 1, 2 and 3, and refer to cycles with conventional, alternative, and staged recoveries. These regenerative cycles involve recovering heat from the exhaust gases and using it to increase the temperature of the air supplied by the compressor before it is sent to the combustion chamber. They differ in the location and number of heat exchangers that ensure heat recovery.

3 Energetic Analysis

Figures 4, 5 and 6 show the thermodynamic process on the temperature-entropy (T, s) diagram for the three recovery cycles.

Equations 1, 2 and 3 represent the energy balance relationships applied for each component of the different cycles.

Compressor:

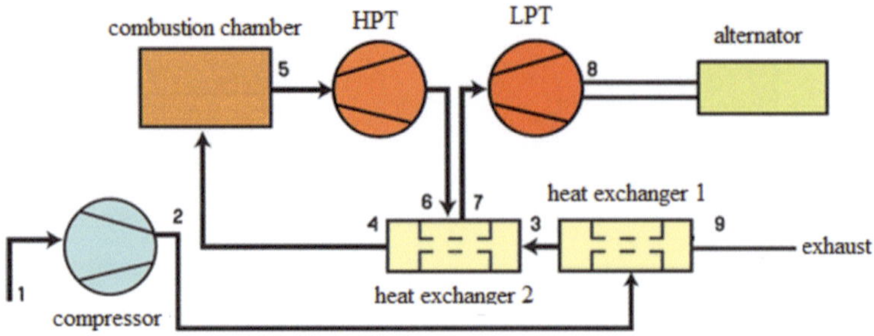

Fig. 3 Gas turbine configuration with staged recuperation

Fig. 4 Conventionnel recuperation cycle

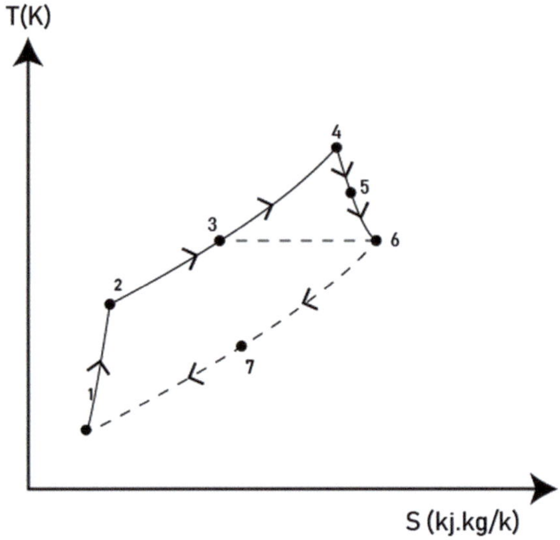

$$P_C = \frac{1}{\eta_{is,c}}\left[\dot{m}_{air} Cp_{air} T_1 \left(\pi_{air}^{\frac{\gamma_{air}-1}{\gamma_{air}}} - 1\right)\right] \tag{1}$$

Combustion chamber:

$$P_{cc} = \dot{m}_c . PCI . \eta_{cc} \tag{2}$$

High pressure turbine:

$$P_{THP} = \eta_{is,T} . m_g C_{P g} T_{eTHP} \left[1 - \left(\frac{1}{\pi_{air}}\right)^{\frac{\gamma_g - 1}{\gamma_g}}\right] \tag{3}$$

Low pressure turbine:

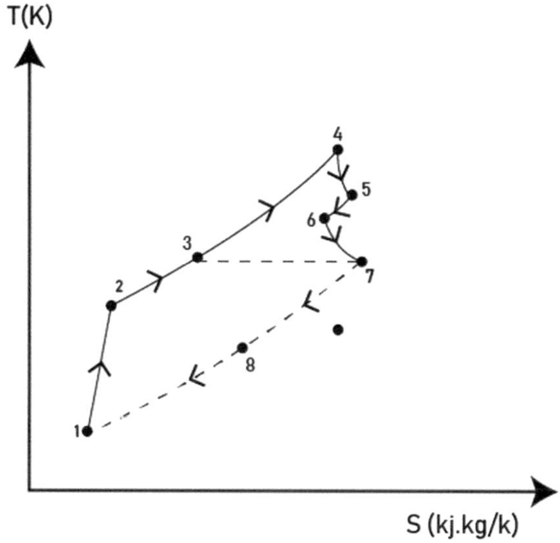

Fig. 5 Alternative recuperation cycle

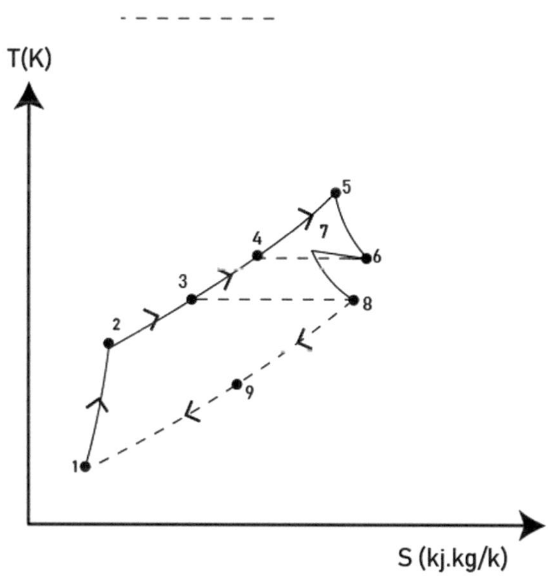

Fig. 6 Staged recuperation cycle

$$P_{THP} = \eta_{is,T}.m_g C_{Pg} T_{eTHP}\left[1-\left(\frac{1}{\pi_{air}}\right)^{\frac{\gamma_g-1}{\gamma_g}}\right] \quad (4)$$

Pc, Pcc, P_{HPT}*, and* P_{BPT}*:* *power of the compressor, combustion chamber, high-pressure turbine, and low-pressure turbine, respectively* [W]

m_{air} *and* m_g:	*mass flow rates of air and fuel, respectively* [kg/s]
Cp_{air} *and* Cp_g:	*Specific heat at constant pressure of air and exhaust gases, respectively* [J/kg K]
Π_{air}:	*Pressure ratio*
$\eta_{is,c}$ and $\eta_{is,T}$:	Isentropic efficiencies of the compressor and turbines, respectively
η_{cc}:	Combustion chamber efficiency

Based on the first law of thermodynamics, the energy efficiency is defined by Eq. (5).

$$\eta_{th} = \frac{(P_{THP} + P_{TB}) - Pc}{Pcc} \tag{5}$$

4 Environmental Impact Analysis

Power plants that produce electricity or drive machines are considered a significant source of CO_2. Due to the consumption of fossil fuels, they emit a large amount of CO_2. Recently, technological development has focused on ways to have clean and safe power plants. The purpose of this section is first to highlight the contribution of recovery cycles in preserving the Earth's ecosystem by minimizing pollutants emissions, and then to select the most environmentally friendly cycle. The amount of CO_2 emissions reduction, which is one of the decisive parameters, is evaluated by Eqs. (6) and (7).

$$M_{CO_2} = \alpha_{CO_2}.P_{cycle} \tag{6}$$

$$\alpha_{CO_2} = HR_{Ng}.EF_{Ng} \tag{7}$$

5 Results and Discussions

Based on the energy model presented above, a computer program is developed in MATLAB 9.2 to model the four cycles, evaluate their thermodynamic performance, and perform a parametric analysis. Energy efficiency is used to assess and compare the performance of the cycles. The results of the thermodynamic analysis are shown in Fig. 7. They indicate that for high-pressure turbine inlet temperatures (T_{eHPT}) below 1200 K, the cycle with alternative recuperation provides the maximum energy efficiency.

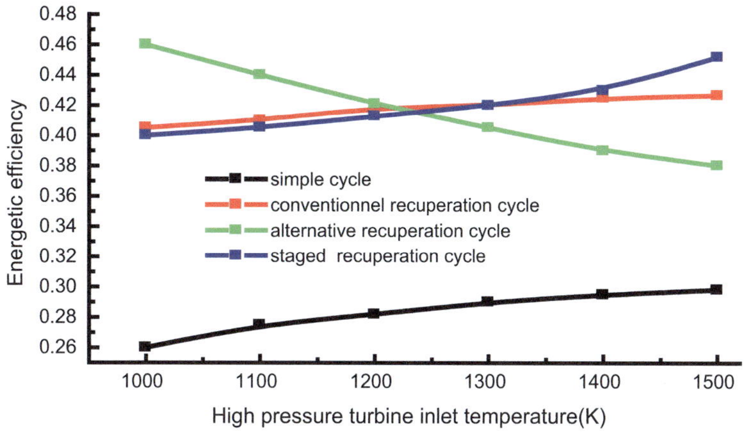

Fig. 7 Comparison of thermal efficiencies of the four cycles for different high-pressure turbine inlet temperatures

For a T_{eHPT} of 1200 K, the performances of the three recuperated cycles are superior to that of the simple cycle but very close to each other. The three cycles add around 24 points to the energy efficiency of the simple cycle. For a T_{eHPT} of 1500, we observe that the difference in efficiencies is greater. The energy efficiency of the simple cycle is improved by about 16 points with conventional recuperation, about 8 points with alternative recuperation, and 18 points with staged recuperation. It is then evident that at high temperatures, the cycle with staged recuperation performs better. A sensitivity analysis is performed to evaluate the effects of various parameters such as pressure ratio, ambient temperature, and heat exchanger efficiency on the performance of the different cycles considered.

5.1 Pressure Ratio Effect

In Fig. 8a, b, which show the evolution of thermal efficiency with pressure ratio for different T_{eHPT}, we observe that for the three recuperated cycles, the efficiencies increase with increasing T_{eHPT} over different ranges of the pressure ratio. As T_{eHPT} increases, the range of the compression ratio widens. The same figures reveal that by increasing the pressure ratio, the efficiency passes through a maximum value for each recuperation cycle. For different T_{eHPT} (1200 and 1500 K), we observe that for low pressure ratios, the cycle with conventional recuperation is the most efficient. For a pressure ratio of 6, its efficiency can exceed 36%. For moderate pressure ratios, it is observed that the cycle with staged recuperation is more efficient than the cycle with alternative recuperation, before being surpassed by the latter for higher pressure ratios.

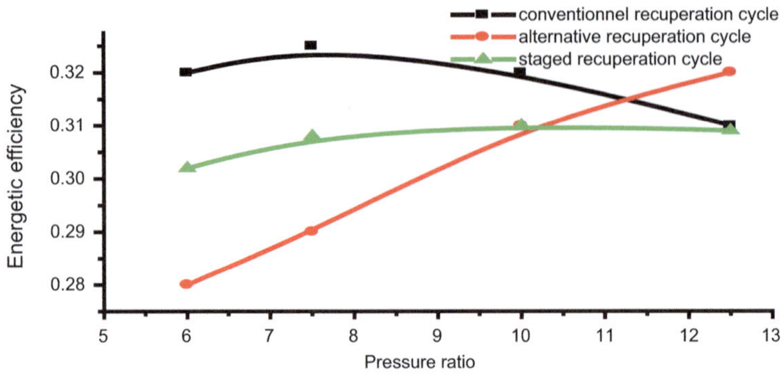

(a)

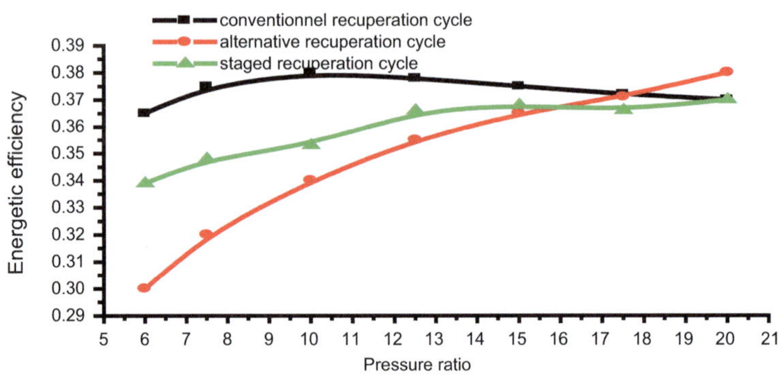

(b)

Fig. 8 Thermal efficiencies obtained with recuperated cycles as a function of ambient temperature for T_{eHPT} **a** 1200 K and **b** 1500 K

5.2 Ambient Temperature Effect

Gas turbines can be used in harsh weather conditions. Far from ISO standards, ambient temperature varies considerably throughout the year and can often reach over 50 °C, leading to a significant decrease in performance. Figure 9a, b show the degree of sensitivity of the thermal efficiency of the different cycles as a function of ambient temperature ranging from 273 to 323 K.

For a low T_{eHPT} (1200 K) and high ambient temperatures, it is shown that the cycle with alternative recuperation would have a thermodynamic advantage over the other cycles. However, for low temperatures, the performances of the three cycles are very close. For higher T_{eHPT} (1500 K), the performance of the cycle with staged recuperation is judged to be better regardless of the ambient temperature.

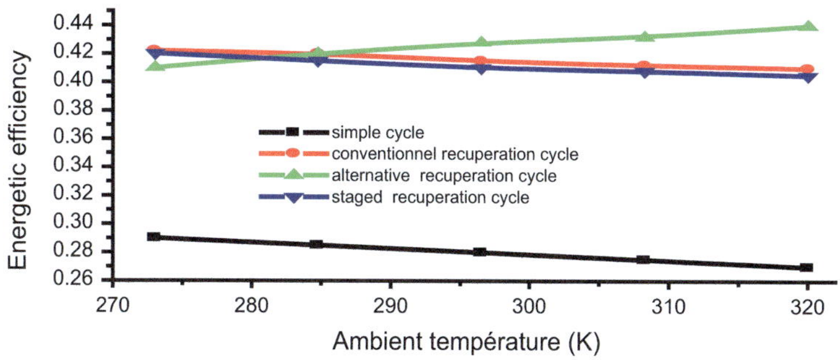

(a)

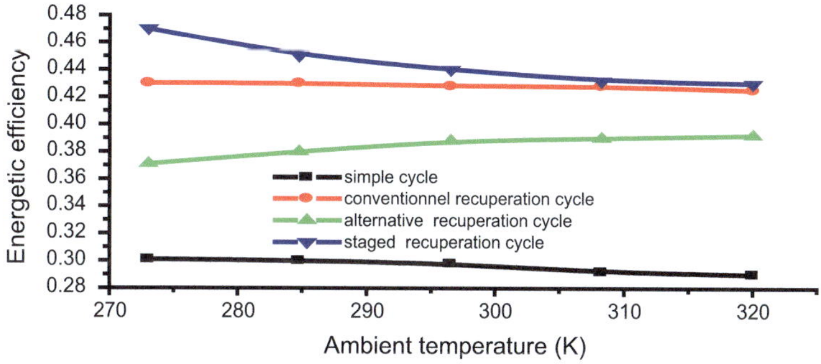

(b)

Fig. 9 Thermal efficiencies obtained with recuperated cycles compared to that of the simple cycle as a function of ambient temperature for T_{eHPT} **a** 1200 K and **b** 1500 K

5.3 *Exchanger Efficiency Effect*

Figure 10a, b show the evolution of thermal efficiency as a function of heat exchanger efficiency for different T_{eHPT}. This figure shows that regardless of the type of recuperation, increasing the efficiency of the heat exchangers leads to a proportional increase in thermal efficiency. With a T_{eHPT} of 1200 K, the performances of the three cycles with recuperation are close for low efficiencies, but for higher efficiencies, these performances diverge and the cycle with alternative recuperation shows a rapid increase compared to the other cycles. With a T_{eHPT} of 1500 K and low efficiencies, the performances of the cycles with alternative and staged recuperation are close and better than those of the cycle with conventional recuperation. But for higher efficiencies, the cycle with staged recuperation is judged to be the most efficient.

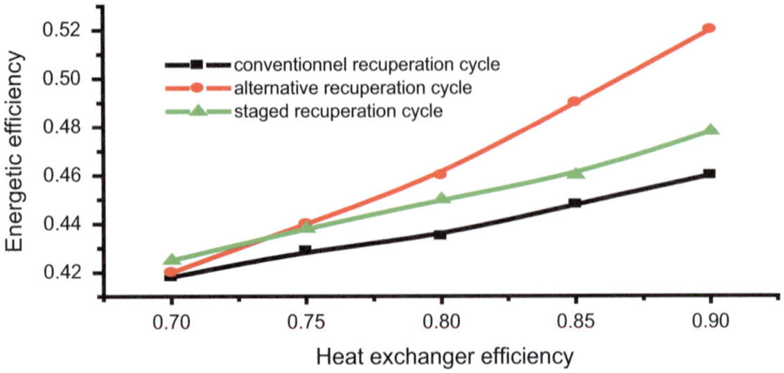

(a)

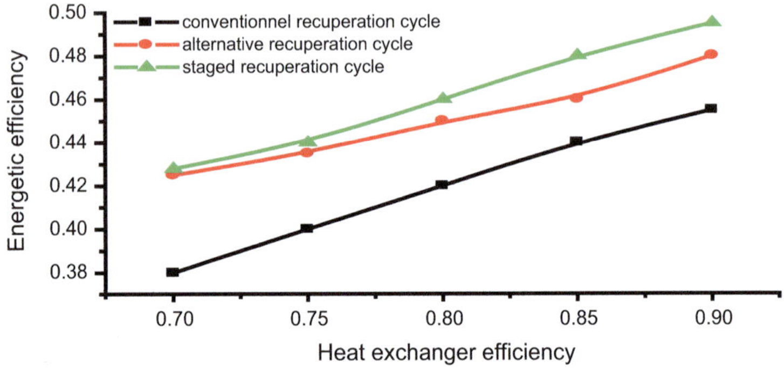

(b)

Fig. 10 Thermal efficiencies obtained with recuperated cycles as a function of heat exchanger efficiency for T_{eHPT} **a** 1200 K and **b** 1500 K

6 Environmental Analysis

This preliminary section provides interesting information on the potential of recuperated cycles to reduce CO_2 emissions. The environmental benefits of recuperated cycles, compared to the simple Joule-Brayton cycle which has the highest emissions, are presented in Fig. 11. It is observed that the lowest value of CO_2 emissions is obtained by adopting the cycle with staged recuperation. The figure also shows that the amount of CO_2 emissions in each cycle depends on T_{eHPT}. Increasing this temperature generates more CO_2 emissions for all cycles.

For low T_{eHPT}, the cycles with conventional, alternative, and staged recuperation can provide a reduction of 5%, 10%, and 36%, respectively, compared to the simple cycle. But for high T_{eHPT}, the reductions are more significant, exceeding 17%, 35%, and 47%, respectively. Based on these results, we can conclude that recuperated cycles, which allow for a reduction in CO_2 emissions, are considered environmentally

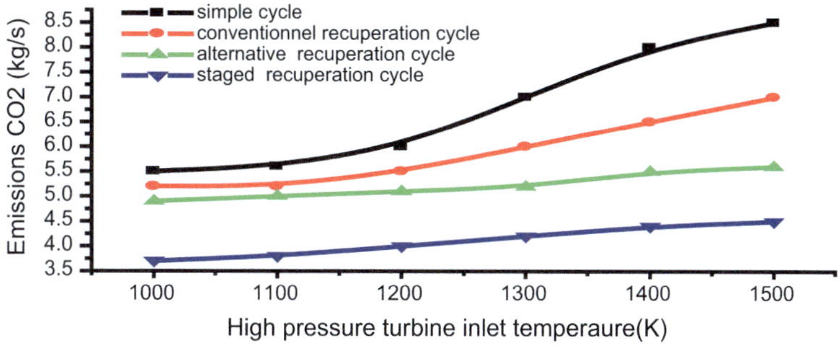

Fig. 11 Evolution of CO_2 emissions for the four cycles as a function of temperature

friendly cycles, and the cycle with staged recuperation is the most favorable regardless of T_{eHPT}.

7 Conclusion

This study illustrates the potential of recuperated cycles to improve the performance of simple cycle gas turbines. Unlike the traditional combined cycle, these alternatives do not require bulky steam equipment. These cycles are suitable when water supply is limited or with weight and physical space restrictions.

We determined the performance of each cycle, and the amount of CO_2 emissions was evaluated as the primary indicator of environmental performance. The comparative analysis, based on energy and environmental efficiency, of the different cycles allowed us to determine the range of each configuration's use, to achieve the best possible efficiency and the lowest CO_2 emissions.

Through a parametric analysis, the effects of certain parameters on the performance of these systems were highlighted. According to this analysis, pressure ratio, ambient temperature, and heat exchanger efficiency are identified as factors affecting the performance of the different recuperated cycles.

References

1. Fengrui S, Yuedong Y, Mingqiang C, Xiangfang L, Lin Z, Ye M (2017) Performance analysis of superheated steam injection for heavy oil recovery and modelling of well bore heat efficiency. Energy 125:795–804
2. Pastor-Martinez E, Rubio-Maya C, Ambriz-Díaz VM, Belman-Flores JM, Pacheco-Ibarra JJ (2018) Energetic and exergetic performance comparison of different polygeneration arrangements utilizing geothermal energy in cascade. Energy Convers Manage 168(15):252–269
3. He MG, Zhang XX, Zeng K, Gao KA (2011) A combined thermodynamic cycle used for waste heat recovery of internal combustion engine. Energy 36:6821–6829

4. Zhu S, Deng K, Qu S (2013) Energy and exergy analyses of a bottoming Rankine cycle for engine exhaust heat recovery. Energy 58:448–457
5. Zhang L, Pan Z, Zhang Z, Shang L, Wen J, Chen S (2018) Thermodynamic and economic analysis between organic Rankine cycle and Kalina cycle for waste heat recovery from steam-assisted gravity drainage process in oilfield. J Energy Resour Technol 140(12):122005. https://doi.org/10.1115/1.4041093
6. Saidur R, Rezaei M, Muzammil WK (2012) Technologies to recover exhaust heat from internal combustion engines. Renew Sust Energy Rev 16:5649–5659
7. Cho SK (2015) Investigation of the bottoming cycle for high efficiency combined cycle gas turbine system with supercritical carbon dioxide power cycle. In Proceedings of ASME turbo expo 2015, turbine technical conference and exposition GT 15–19 June 2015, Montréal, Canada
8. Campos Rodriguez CE, Escobar Palacio JC, Venturini OJ, Silva Lora EE, Cobas VM, Marques dos Santos D, Lofrano Dotto FR, Gialluca V (2013) Exergetic and economic comparison of ORC and Kalina cycle for low temperature enhanced geothermal system in Brazil. Appl Therm Eng 52(1):109–119
9. Hugo DP (2020) Working fluid dependence on source temperature for organic Rankine cycles. J Energy Resour Technol 142(1):012103. https://doi.org/10.1115/1.4045383
10. Mokheimer EMA (2021) Energy and exergy performance comparative analysis of a solar-driven organic Rankine cycle using different organic fluids. J Energy Resour Technol 143(10):102107. https://doi.org/10.1115/1.4050343
11. Brahimi F, Madani B, Ghemmadi M (2022) Comparative thermodynamic environmental and economic analyses of combined cycles using air and supercritical CO_2 in the bottoming cycles for power generation by gas turbine waste heat recovery. Energies 15:9066. https://doi.org/10.3390/en15239066
12. Salpingidou C, Vlahostergios Z, Misirlis D, Donnerhack S, Flouros M, Goulas A, Yakinthos K (2017) Thermodynamic analysis of recuperative gas turbines and aero engines. Appl Therm Eng 124:250–260. https://doi.org/10.1016/j.applthermaleng.2017.05.169
13. Li Y, Zhang G, Wang L, Yang Y (2020) Part-load performance analysis of a combined cycle with intermediate recuperated gas turbine. Energy Convers Manage 205:112346. https://doi.org/10.1016/j.enconman.2019.112346
14. Salpingidoua C, Misirlis D, Vlahostergiosa Z, Donnerhackb S, Flourosb M, Goulasa A, Yakinthosa K (2016) Investigation of the performance of different recuperative cycles for gas turbines/aero engine applications. Chem Eng Trans 52. https://doi.org/10.3303/CET1652086
15. Salpingidou C, Vlahostergios Z, Misirlis D, Donnerhack S, Flouros M, Goulas A, Yakinthos K (2017) Thermodynamic analysis of recuperative gas turbines and aero engines. Appl Therm Eng. https://doi.org/10.1016/j.applthermaleng

Application of Artificial Intelligence for the Supervision of a Multi-source System

Chafiaa Serir, Samia Bensmail, Katia Tadjine, and Djamila Rekioua

Abstract Renewable energy sources offer unlimited potential for clean electricity production and their usage is rapidly growing. Hybrid energy systems incorporate various sources, requiring efficient control and supervision mechanisms. In this regard, this study proposes an artificial intelligence-based energy management system, utilizing fuzzy logic control, for a hybrid wind/photovoltaic system with battery storage. The proposed approach adjusts the system output to supply the load demand, offering benefits of precise control of the battery's state of charge (SOC) across five states. This is a marked improvement over traditional management systems that rely on two SOC states, namely SOCmin-SOCmax.

Keywords Artificial intelligence · Fuzzy logic controller · Wind · Photovoltaic

1 Introduction

The utilization of renewable and decarbonized energy sources has proven to have significant positive economic [1, 2] and environmental impacts [3, 4]. Hybrid systems, such as those combining photovoltaic and wind sources, have garnered considerable attention due to their numerous benefits. These systems are capable of generating constant power regardless of weather and load fluctuations by intelligently drawing from different sources to meet energy demand and maintain battery charge [5]. However, the intermittent nature of wind and photovoltaic power generation can make it challenging to operate a sustainable energy system continuously.

C. Serir (✉) · K. Tadjine · D. Rekioua
Laboratoire de Technologie Industrielle et de l'Information (LTII), Faculté de Technologie, Université de Bejaia, 06000 Bejaia, Algeria
e-mail: chafiaa.serir@univ-bejaia.dz

S. Bensmail
Laboratoire de Technologie Industrielle et de l'Information (LTII), Department of Electrical Engineering, University of Bouira, Bouira, Algeria

© The Author(s), under exclusive license to Springer Nature Singapore Pte Ltd. 2024

A. Mellit et al. (eds.), *Proceedings of the 2nd International Conference on Green Energy Conversion System*, Green Energy and Technology,
https://doi.org/10.1007/978-981-97-6148-7_7

Energy storage is therefore vital to balance the supply and demand of electricity by storing excess energy and releasing it when required.

The development of environmentally friendly and safe hybrid systems is thus imperative [6, 7]. Effective energy management systems play a crucial role in balancing the supply and demand of energy and reducing peak demands during unforeseen periods. Numerous types of energy management controls (EMCs) have been explored in various applications, such as telecommunications [8], water pumping [9], smart grids [10], and electrification [11]. This paper focuses on the study of a hybrid system (photovoltaic-wind) with battery storage to power a load electrification system, utilizing power optimization and intelligent control. The optimization strategy for the PV system is performed using two MPPT methods: Pertub & Observ (P&O) [12–19] and Fuzzy Logic Controller (FLC) [20, 21] For wind turbines, two MPPT methods have been studied: optimal torque control (OTC) [22] and fuzzy logic controller (FLC).The implementation of smart management utilizing artificial intelligence (AI) based on a fuzzy logic controller (FLC) is more robust compared to conventional techniques based on logical states. It can determine the operational variables of the hybrid system based on weather conditions, enabling the system to efficiently and quickly determine the power output of each source. Moreover, it keeps the battery charged, preventing power failures and prolonging battery lifespan. The model is simulated using Matlab/Simulink.

2 MPPT Algorithms

2.1 *Photovoltaic System Optimization*

P&O technique. Because of its simplicity, this is a widely used technique. The panel voltage is intentionally disrupted, and the panel power is observed in relation to the previous power. In other words: If the disturbance with a voltage step increases the power variation of the panel, the next disturbance will be in the same direction. If the power variation decreases, a new perturbation step is made in the opposite direction. The benefits of this approach can be summarized as follows: the technique is relatively simple, and easy to implement [23–25].

FLC technique. A fuzzy logic controller (FLC) is introduced to find the operating point corresponding to the maximum power for various solar irradiances and temperatures. The inputs of the FLC in this case are the power variation (ΔP_{pv}) and the voltage variation (ΔV_{pv}). The output is a variation of the reference voltage ($V_{pv,ref}$). It is relatively easy to establish rules to reach to the maximal point. These rules are affected by ΔP_{pv} and ΔV_{pv}. According to Table 1, the voltage (Vpv, ref) should, however, decrease if the power (Ppv) does as well. Table 1 shows the controller's fuzzy rules. The inference interface is implemented using Mamdani's method [26].

Table 1 Fuzzy decision rules

	ΔP_{pv}						
ΔV_{pv}	BN	MN	SN	Z	SP	MP	BP
BN	BP	BP	MP	Z	MN	BN	BN
MN	BP	MP	SP	Z	SN	MN	BN
SN	MP	SP	SP	Z	SN	SN	MN
Z	BN	MN	SN	Z	SP	MP	BP
SP	MN	SN	SN	Z	SP	SP	MP
MP	BN	MN	SN	Z	SP	MP	BP
BP	BN	BN	MN	Z	MP	BP	BP

2.2 Wind System Optimization

OTC technique. In this method, the torque of the generator is adjusted to its optimum at different wind speeds to obtain the maximum value of the power coefficient and thus the maximum energy efficiency [22]. Torque control allows a fast approach to the optimum with simple measurements, but this control requires knowledge of the maximum power curve of the wind turbine.

FLC technique. The fuzzy sets and the membership functions must be defined first, followed by the fuzzy input variables.

$$\begin{cases} \Delta P_t = P_t(k) - P_t(k-1) \\ \Delta \Omega_t = \Omega_t(k) - \Omega_t(k-1) \\ \Omega_t\_ref(k) = \Omega_t(k-1) + \Delta\Omega_t\_ref(k) \end{cases} \tag{1}$$

2.3 Simulation Result

From the results of the optimization of the Siemens 110 photovoltaic panel simulated under standard conditions (STC) illustrated in the figure (see Fig. 1), it can be clearly seen that the rhythm obtained with the fuzzy controller (FLC) is accurate, stable and faster than that obtained with (P&O).

We simulated the 1 kW wind turbine with a selected wind speed profile (see Fig. 2a). Despite the abrupt variations in wind speed, we can clearly see that the power obtained with (FLC) follows the profile accurately and with a better response time than (OTC) (see Fig. 2b).

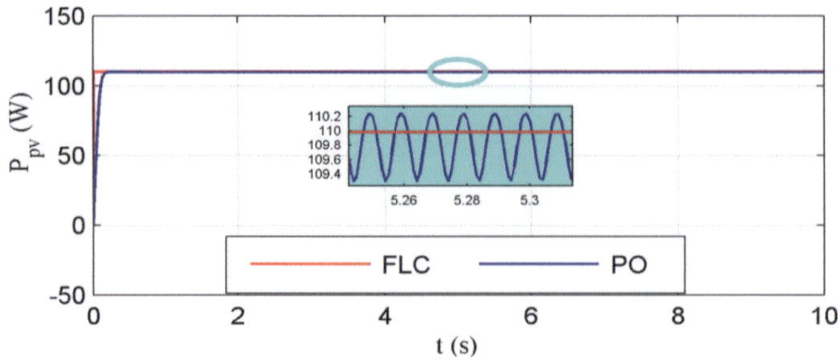

Fig. 1 Photovoltaic power

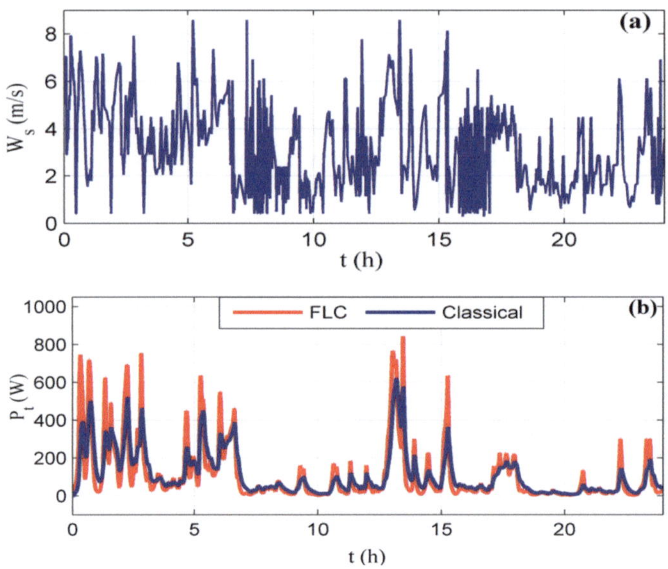

Fig. 2 **a** Wind speed profile, **b** turbine power

3 Smart Energy Management

3.1 Structure of the Global Controler System

To control the flow of energy between the various system components, a management system based on artificial intelligence (FLC) has been applied. Then the battery's charge/discharge process is controlled with greater precision, as the battery's SOC will be monitored in five states, improving protection against deep discharges and overloads and, consequently, battery life. The structure of the overall system studied with the intelligent management block is shown in the following figure (see Fig. 3).

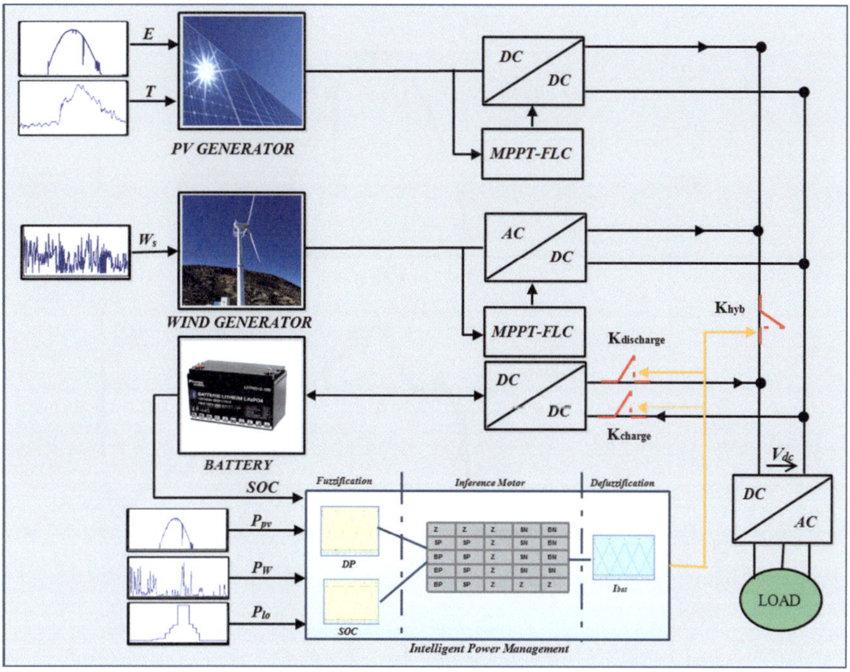

Fig. 3 Description of the studied system

3.2 System Simulation Results

The simulated system consists of a photovoltaic generator with 18 modules (110 Wc), a 1 kW wind turbine and 12 batteries (12 V–100 Ah). Measured profiles of solar irradiation, ambient temperature and wind speed, so the load profile, were used in the simulation (see Figs. 4a–c and 2a).

These figures show demonstrate the effective management of the overall system's operation (see Fig. 5) show that successful coordination of hybrid power and battery discharge perform great together to power the load, and that the system continuously protects the batteries between 30 and 90% of their capacity.

This proves the effectiveness of each decision rule in the fuzzy logic controller with five classes corresponding to variations in meteorological conditions. The SOC variation respects the limits imposed (see Fig. 6). Note also that the bus voltage DC is kept constant and follows its reference value (see Fig. 7).

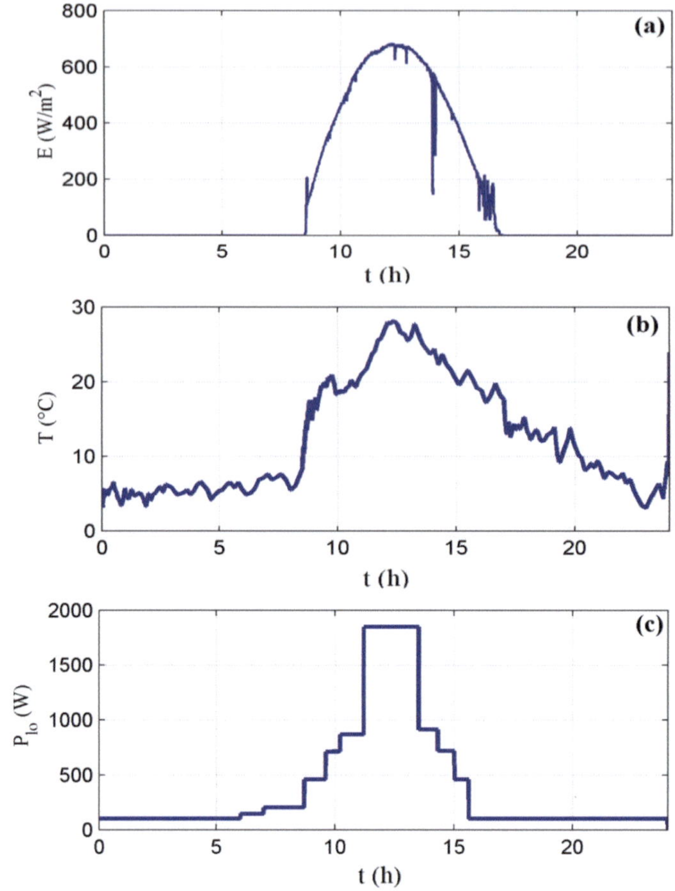

Fig. 4 The different profiles: **a** solar irradiance, **b** temperature, **c** load power

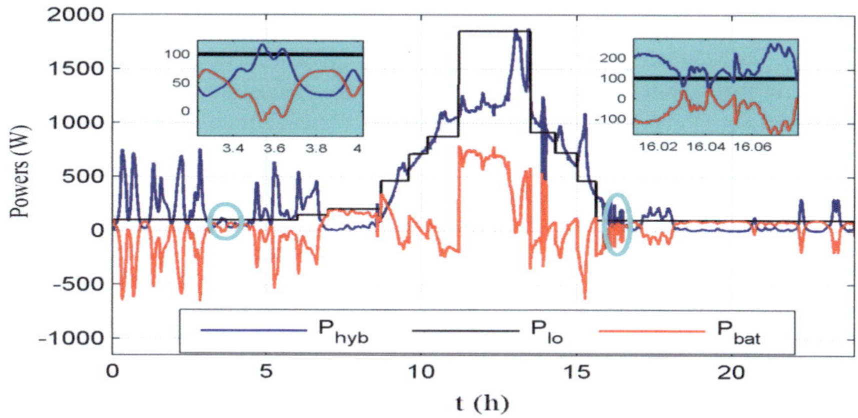

Fig. 5 Powers: hybrid, load and battery

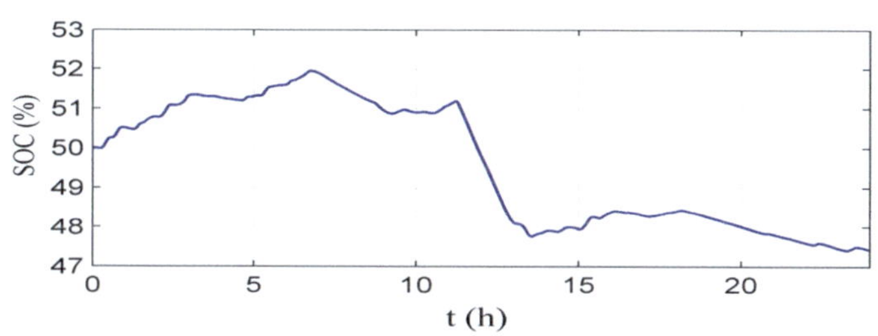

Fig. 6 Battery state of charge

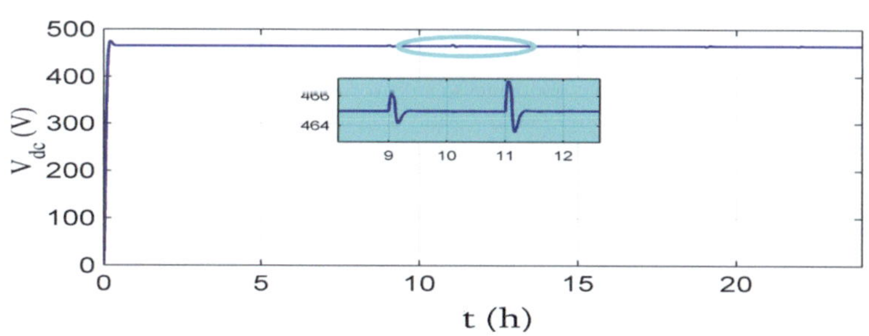

Fig. 7 DC bus voltage V_{dc}

4 Conclusion

The objective of this research was to optimize the power output, and to develop an energy management system for this purpose. Two maximum power point tracking approaches were implemented to maximize energy production from solar and wind power systems. The results showed that the FLC was more efficient in terms of stability, and accuracy. Furthermore, a smart management supervisor was developed to control the overall energy system. The findings demonstrated the robustness of the supervisor in making decisions between the different components of the system and improving the battery's lifespan by avoiding sulfation phenomena.

References

1. Rekioua D, Matagne E (2012) Optimization of photovoltaic power systems: modelization, simulation and control. series: green energy and technology. Springer
2. Busu M (2020) Analyzing the impact of the renewable energy sources on economic growth at the EU level usingan ARDL model. Mathematics 8:1367
3. Zheng Y, Weng Q (2020) Modeling the effect of green roof systems and photovoltaic panels for building energysavings to mitigate climate change. Remote Sens 12:2402
4. Pande J, Nasikkar P, Kotecha P, Varadarajan V (2021) A review of maximum power point tracking algorithms for wind energy conversion systems. J Mar Sci Eng 9:1187
5. Rekioua D, Bensmail S, Bettar N (2014) Development of hybrid photovoltaic-fuel cell system for stand-alone application. Int J Hydrogen Energy 39(3):1604–1611
6. Mohammedi A, Rekioua D, Rekioua T, Bacha S (2016) Valve regulated lead acid battery behavior in a renewable energy system under an ideal Mediterranean climate. Int J Hydrogen Energy 41(45):20928–20938
7. Tadjine K, Rekioua D, Belaid S, Rekioua T, Logerais PO (2022) Design, modeling and optimization of hybrid photovoltaic/wind turbine system with battery storage: application to water pumping. 09(3):655–667. ISSN 0360-3199
8. Sekar K, Duraisamy V (2015) Efficient energy management system for integratedrenewable power generation systems. J Sci Ind Res 74:3259
9. Serir C, Rekioua D, Mezzai N, Bacha S (2016) Supervisor control and optimization of multi-sources pumping system with battery storage. Int J Hydrogen Energy 41(45):20974–20986
10. Elhawary ME (2014) The smart grid-state-of-the art and future trends. Electr Power Compon Syst 42(3–4):239–250
11. Serir C, Rekioua D, Bensmail S, Belkaid A, Tadjine K, Hadji S, Colak I (2022) Electrification of a load by a hybrid photovoltaic-wind system with battery storage. IEEE
12. Rekioua D (2020) Power electronics in hybrid renewable energies systems. Green Energy Technol 39(77)
13. Belkaid A, Colak I, Kayisli K (2020) A novel approach of perturb and observe technique adapted to rapid change of environmental conditions and load. Electr Power Compon Syst 48(4–5):375–387
14. Achour A, Rekioua D, Mohammedi A, Mokrani Z, Rekioua T, Bacha S (2016) Application of direct torque control to a photovoltaic pumping system with sliding-mode control optimization. Electr Power Compon Syst 44(2):172–184
15. Mohanty S, Subudhi B, Ray PK (2016) A new MPPT design using grey wolf optimization technique for photovoltaic system under partial shading conditions. IEEE Trans Sustain Energy 7(1):181–188
16. Chao KH, Rizal MN (2021) A hybrid MPPT controller based on the genetic algorithm and ant colony optimization for photovoltaic systems under partially shaded conditions. Energies 14:2902
17. Meghni B, Ouada M, Saad S (2020) A novel improved variable-step-size P&O MPPT method and effective supervisory controller to extend optimal energy management in hybrid wind turbine. Electr Eng 102:763–778

18. Alik R, Jusoh A (2018) An enhanced P&O checking algorithm MPPT for high tracking efficiency of partially shade PV module. Sol Energy 163:570–580
19. Shadoul M, Yousef H, Al Abri A, Al-Hinai A (2021) Adaptive fuzzy approximation control of PV grid-connected inverters. Energies 14:942
20. Meskani A, Haddi A (2019) Modeling and simulation of an intelligent hybrid energy source based on solar energy and battery. Energy Procedia 162:97–106
21. Hassani H, Zaouche F, Rekioua D, Belaid S, Bacha B (2020) Feasibility of a standalone photovoltaic/battery system with hydrogen production. J Energy Storage 31:101–644
22. Rekioua D (2020) Power management and supervision of hybrid renewable energy systems, power management and supervision of hybrid renewable energy systems. In: Hybrid renewable energy systems. green energy and technology. Springer, Cham. ISBN: 978-3-030-34021-6
23. Mebarki N, Rekioua T, Mokrani Z, Rekioua D (2015) Supervisor control for stand-alone photovoltaic/hydrogen/battery bank system to supply energy to an electric vehicle. Int J Hydrogen Energy 40(39):13777–13788
24. Serir C, Rekioua D (2015) Control of photovoltaic water pumping system. J Electr Eng 15(2):339–344
25. Rekioua D (2018) Energy management for PV installations. Adv Renew Energies Power Technol 1:349–369
26. Serir C, Bensmail S, Tadjine K, Rekioua D (2023) fuzzy control of an electromechanical system powered by renewable energy. In: III international science conference "electrical and power engineering and electromechanics (EPEE-2023)", 15 June 2023 Odesa, Ukraine 621.3, 16–17

Renewable Hybrid System with Battery Storage for Water Pumping for Small-Scale Area

Katia Tadjine, Djamila Rekioua, Pierre-Olivier Logerais, Toufik Rekioua, and Chafiaa Serir

Abstract The paper is devoted to a study on design, optimization and experimental study of a renewable hybrid system (photovoltaic/wind turbine) with battery storage. The design of the studied system use the total incident energy approaches to determinate the size of photovoltaic and wind generators. The simulation study has been done under Matlab/Simulink. An application to pumping water is done for four various profiles measured at Bejaia area. To control the water pumping hybrid system, a power management control has been applied to coordinate the different sources. The simulation results obtained are shown and discussed to demonstrate the applicability of the suggested system. The overall hybrid pumping system with batteries has been then investigated by an experimental study, where tests have been made during different days and under various total dynamic heads (TDH) to show the different performances as flow rate and pump efficiency variations of the proposed system. A comparison of simulation and experimental results are presented and analyzed to conclude on the feasibility of the studied system for water pumping needs applied in a small-scale area of Bejaia a Mediterranean region in Algeria.

Keywords Photovoltaic panels · Wind turbine · Storage · Water pumping

1 Introduction

Intermittent nature of wind and photovoltaic energy production complicates the electrical system. By storing surplus energy and producing it when it is needed, energy storage is required to balance the demand and supply of power. As a result, a hybrid

K. Tadjine · D. Rekioua (✉) · T. Rekioua · C. Serir
Laboratoire de Technologie Industrielle et de l'Information (LTII), Faculté de Technologie, Université de Bejaia, 06000 Bejaia, Algeria
e-mail: djamila.ziani@univ-bejaia.dz

P.-O. Logerais
IUT de Sénart-Fontainebleau (UPEC) France, Université Paris-Est Créteil-CERTES, Lieusaint, France

© The Author(s), under exclusive license to Springer Nature Singapore Pte Ltd. 2024
A. Mellit et al. (eds.), *Proceedings of the 2nd International Conference on Green Energy Conversion System*, Green Energy and Technology,
https://doi.org/10.1007/978-981-97-6148-7_8

system is created that is both safer and environmentally friendly [1, 2]. When there are different sources in an electrical system, it is necessary to control the different sources. Some of power management control them are synthesis of the most significant supervisory controls [3] and others are focused on a very precise method and a well defined system [4–28]. Some EMs methods use "if-else" statements in the decision algorithms [10, 12, 15]. Others use more intelligent and predictive methods [13, 18, 21, 23, 28].

In this work, it is presented a to a study on design, optimization and experimental study of a renewable hybrid system (photovoltaic/wind turbine). Batteries are inserted to ensure the pump demand during the periods when there is neither solar irradiance nor high wind speeds. Measurements of wind speed, temperature, and solar radiation profiles were made for four days profiles using a data acquisition system. The proposed sizing method used the annual monthly average of the total incident [1]. The simulation study has been done under Matlab/simulink. An application to pumping water is done for four various profiles measured at Bejaia area. To control the water pumping hybrid system, a power management control has been applied to coordinate the different sources and to satisfy pump needs level and to protect the batteries used. The simulation results obtained are shown and discussed to demonstrate the applicability of the suggested system. The overall hybrid pumping system with batteries has been then investigated by an experimental study, where tests have been made during different days and under various total dynamic heads (TDH) to show the different performances as flow rate and pump efficiency variations of the proposed system. The obtained simulation results are very close to experimental ones whatever profile days variations and total dynamic heads values. This confirms the mathematical model proposed and applied to a renewable hybrid system with battery storage for water pumping applied in a small-scale area of Bejaia in Algeria.

2 Sizing Hybrid Pumping Photovoltaic/Wind Turbine with Batteries Storage

2.1 Geographical Area Location

The region of Bejaia, with its geographical coordinates (Latitude: 36.75, Longitude: 5.06667 36° 45′ 0″ North, 5° 4′ 0″ East), has a considerable potential of solar and wind energies (see Fig. 1). The solar irradiance, the wind speeds and the temperature measured using the software Homer pro are given in Fig. 2.

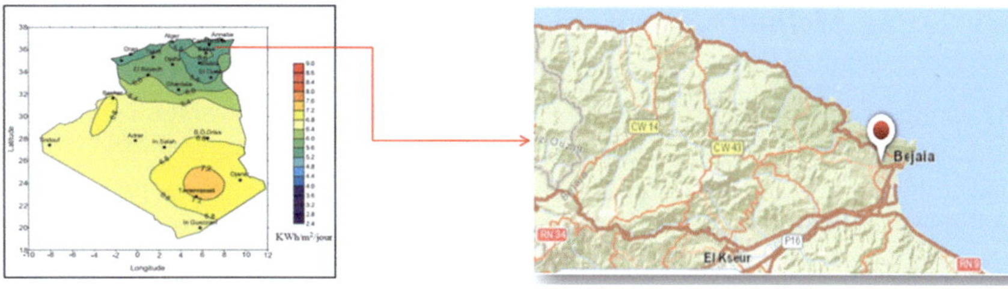

Fig. 1 Geographical position of Bejaia area

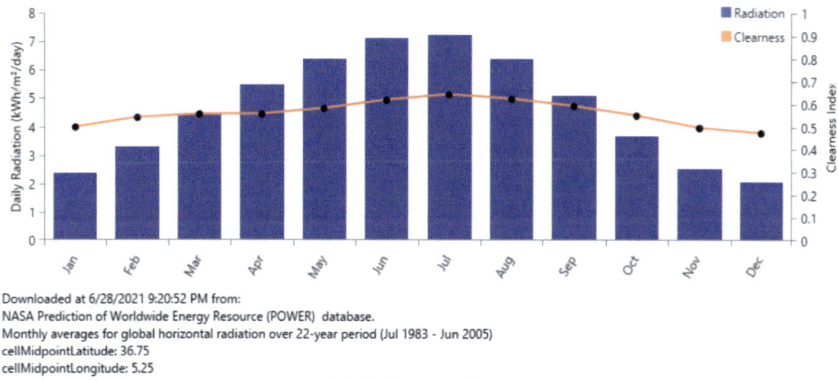

(a) Solar irradiance

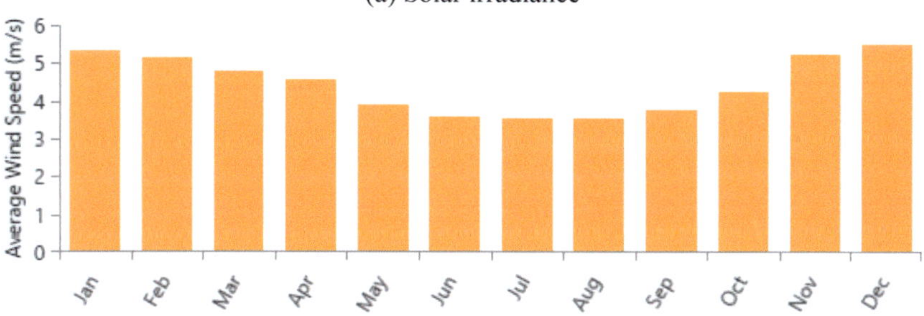

(b)Wind speeds

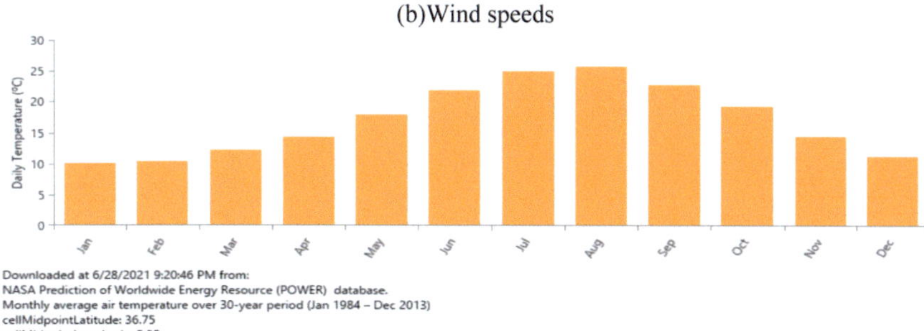

(c)Temperatures

Fig. 2 Weather conditions at Bejaia area using homer pro

2.2 Sizing Method

The sizing of the studied system is based on annual monthly average of the total incident energy method [1]. It determinates the components number of PV panels, wind turbine and batteries. After calculations, the different results are presented in the following figures (see Figs. 3, 4, 5 and 6).

The number of PV panels and wind turbine are determined (see Fig. 6). The average load energy required is 231.09 kWh, so, the combination of six (06 panels) and one (01) wind turbine is the closest to the required load energy (244.40 kWh).

The calculation of battery capacity gives 02 batteries of (100 Ah, 12 V).

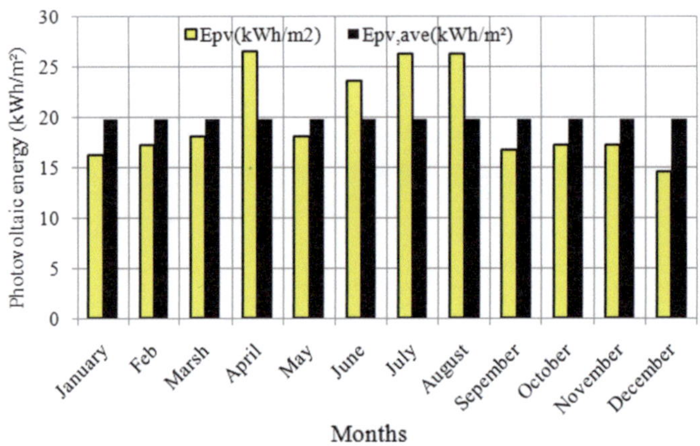

Fig. 3 Photovoltaic and its average energy along a year

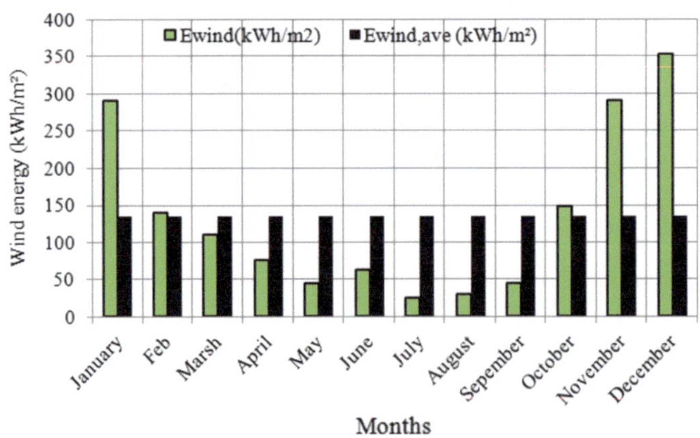

Fig. 4 Wind and its average energy along a year

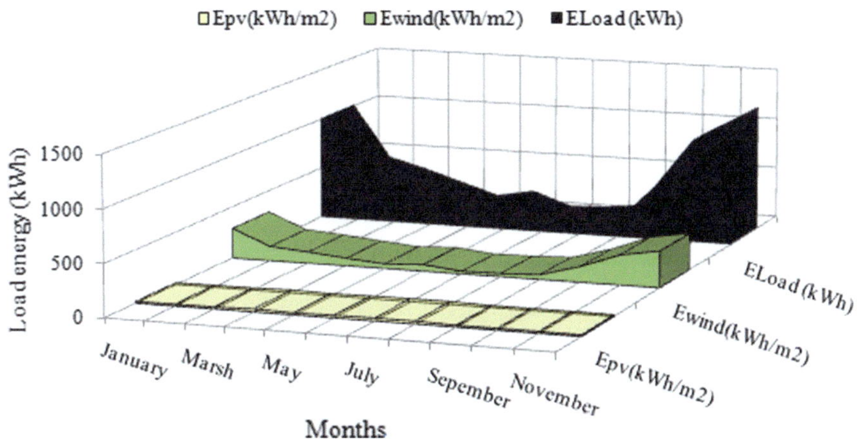

Fig. 5 Photovoltaic, wind an load energy during a year

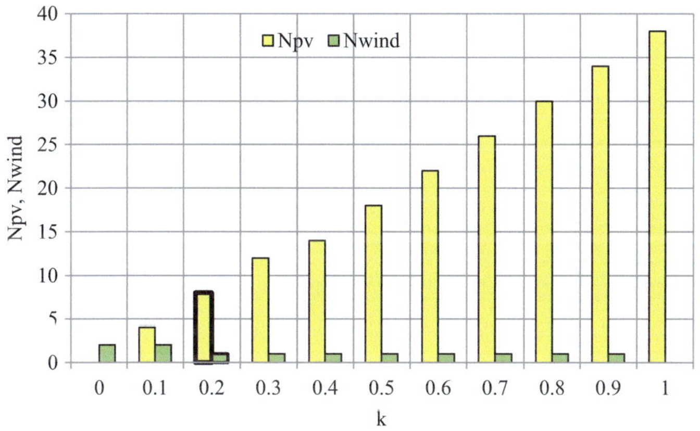

Fig. 6 Number of panels and wind turbine

3 Experimental Study of Hybrid Pumping System (PV/ Wind Turbine) with Batteries

3.1 Description of the System

The studied hybrid pumping system (photovoltaic/wind turbine) includes, a photovoltaic sub-system with a charge controller that allows to control the state of charge of the storage batteries at each time. The PV system consists of 06 solar panels of 175 Wp for each module, they are connected in three parallel strings, each string consists of six panels in series with a total power of 1050 W, A wind power system

Fig. 7 Schematic diagram of the water pumping system

associated with a load shedding system, this system consists of a wind turbine of 900 W that converts the kinetic energy of the wind into mechanical energy that flows into a synchronous generator with permanent magnets associated with a PD3 diode rectifier. Both energy sources are connected to a DC bus. The storage is provided by 02 batteries in serial (100 Ah, 12 V), which maintain the voltage of the 24 V DC bus. The load to be powered is an AC motor-pump unit connected through a single-phase inverter of 900W power. Figure 7 shows the block diagram of the overall hybrid pumping system with batteries.

3.2 Moto-Pump Modeling

3.2.1 Induction Motor Model

Its stator voltage equations in a stationary reference frame can be represented by [10]:

$$
\begin{aligned}
V_{s\alpha} &= R_{s\alpha} I_{s\alpha} + \frac{d}{dt}\Phi_{s\alpha} \\
V_{s\beta} &= R_{s\beta} I_{s\beta} + \frac{d}{dt}\Phi_{s\beta} \\
R_{r\alpha} I_{r\alpha} &+ \frac{d}{dt}\Phi_{r\alpha} + \omega\Phi_{r\beta} = 0 \\
R_{r\beta} I_{r\beta} &+ \frac{d}{dt}\Phi_{r\beta} + \omega\Phi_{r\alpha} = 0
\end{aligned}
\tag{1}
$$

With:

$$
\begin{aligned}
\Phi_{s\alpha} &= L_{s\alpha}.I_{s\alpha} + M_{sr\alpha}.I_{r\alpha} \\
\Phi_{s\beta} &= L_{s\beta}.I_{s\alpha} + M_{sr\beta}.I_{r\beta} \\
\Phi_{r\alpha} &= L_{r\alpha}.I_{r\alpha} + M_{sr\alpha}.I_{s\alpha} \\
\Phi_{r\beta} &= L_{r\beta}.I_{r\alpha} + M_{sr\beta}.I_{s\beta}
\end{aligned}
\tag{2}
$$

The electromagnetic torque can be written as:

$$
T_{em} = p.M_{sr\alpha}(I_{s\beta}.I_{r\alpha} - I_{s\alpha}.I_{r\beta}) \tag{3}
$$

where $V_{s\alpha}$, $V_{s\beta}$ are α-β stator voltages; $I_{s\alpha}$, $I_{s\beta}$, $I_{r\alpha}$, $I_{r\beta}$ are α-β stator and rotor currents; $\Phi_{s\alpha}$, $\Phi_{s\beta}$, $\Phi_{r\alpha}$, $\Phi_{r\beta}$ are α-β stator and rotor fluxes, $R_{s\alpha}$, $R_{s\beta}$, R_r are α-β stator and rotor resistances; L_s, L_r are stator and rotor self-inductances; M_{sr} is the mutual inductance; and p the pole pairs number.

3.2.2 Centrifugal Pump Modeling

The load torque of the centrifugal pump is proportional to the square of the rotor speed [1, 10, 18]:

$$
T_r = k_c.\Omega^2 \tag{4}
$$

where k_c is the proportionality constant and Ω is the angular velocity.

The TDH-Q characteristic is given as [1, 18]:

$$
TDH = EH + k.Q^2 \tag{5}
$$

where TDH, Q, TSH, and k are respectively the total dynamic head, the flow rate, elevation head and the constant depending on conduit diameter.

From a characteristic curve (P, TDH, Q) generated for a rotation speed N of the impeller pump, the characteristic (P_1, TDH_1, Q_1) can be obtained for any rotation speed N_1 [1, 18]. The centrifugal pumps confirm the similarity laws.

$$\begin{cases} Q_1 = Q.\left(\frac{N_1}{N}\right) \\ TDH_1 = TDH.\left(\frac{N_1}{3}\right)^2 \\ P_1 = P.\left(\frac{N_1}{N}\right)^2 \end{cases} \quad (6)$$

3.3 Experimental Results Obtained with the Hybrid Pumping Water System

The first test is devoted to a height of 3.5 m which can be used to supply water to the first level of a house for example. The second test will be assigned to a higher height of 11.0 m which can represents to the supplying of water to the upper floor of a house. During the two tests, the pumped water volume is about 500 L. Different performances are calculated such as flow rate and efficiencies (see Figs. 8 and 9). From the figures presented above, it could be noticed that the performances of the system does not depend on weather conditions. Indeed, whatever the values of the irradiance and wind speeds velocity, the performances of the system kept constant. In particular, the flow rate and the efficiency of the motor pump unit, which take slightly higher values when decreasing in height and remain almost constant throughout the day. Because the supply voltage is imposed by the batteries that constitute the input of the single-phase inverter.

4 Conclusion

A study on feasibility of a renewable system with battery storage for water pumping applied in a small-scale area of Bejaia are described in this work. A correct and precise sizing has been made to determinate the optimal configuration of the studied system. Also, a control approach has been applied in simulation study. It has been established that the different sources have been managed in an ideal manner to supplythe load demand regardless of weather variations. An application has been made for four various profile days and under different total dynamic heads. Obtained experimental results have been compared to simulation ones to validate the proposed mathematical model. Therefore, the study will be a useful reference point when making investment decisions in renewable hybrid water pumping systems for water domestic needs and irrigation in the Mediterranean region, particularly in the Bejaia area.

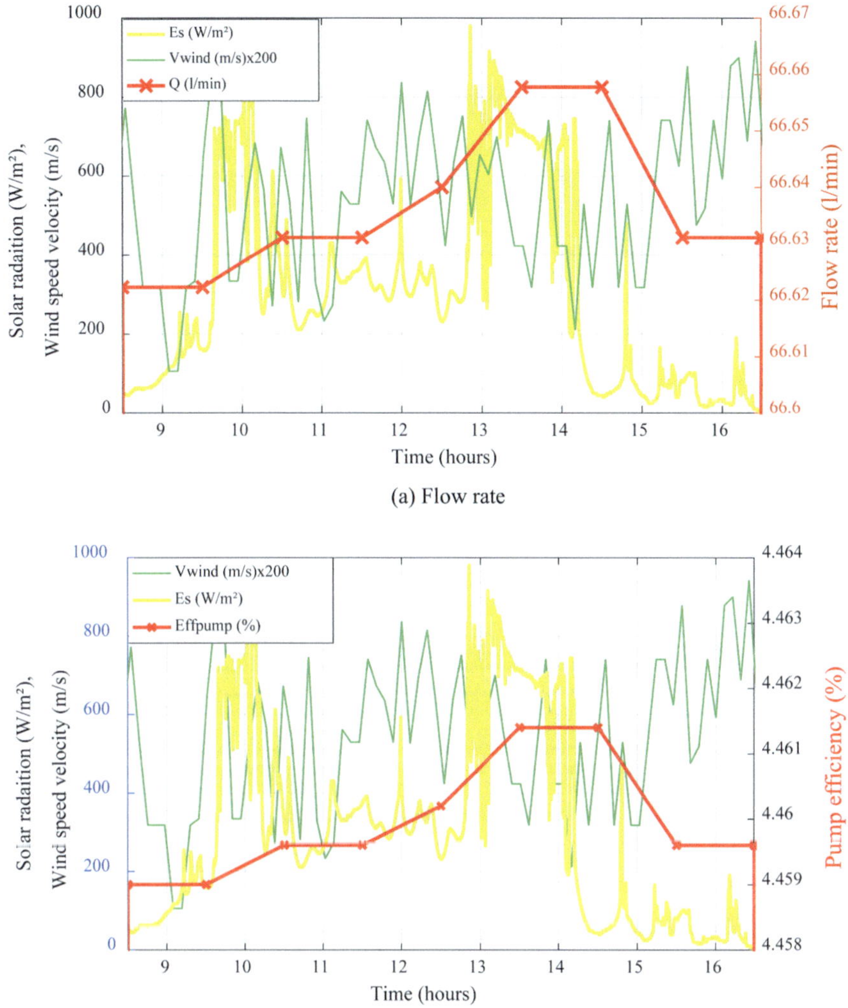

(a) Flow rate

(b) Pump efficiency

Fig. 8 Obtained performances during a day profile at TDH = 3.5 m

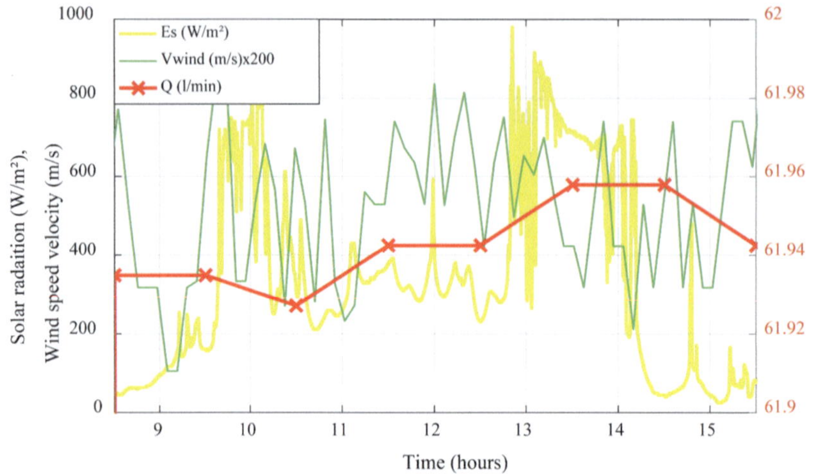

(a) Flow rate

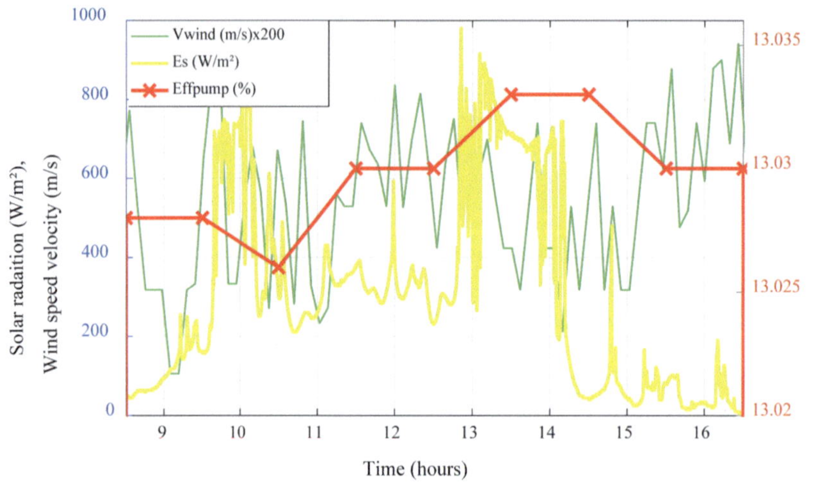

(b) Pump efficiency

Fig. 9 Obtained performances during a day profile at TDH = 11 m

References

1. Rekioua D, Matagne E (2012) Optimization of photovoltaic power systems: modelization, simulation and control, series: green energy and technology. Springer
2. Rekioua D, Bensmail S, Bettar N (2014) Development of hybrid photovoltaic-fuel cell system for stand-alone application International. J Hydro Energy 39(3):1604–1611
3. Aissou R, Rekioua T, Rekioua D, Tounzi A (2016) Robust nonlinear predictive control of permanent magnet synchronous generator turbine using Dspace hardware. Int J Hydrogen Energy 41(45):21047–21056

4. Mebarki N, Rekioua T, Mokrani Z, Rekioua D (2015) Supervisor control for stand-alone photovoltaic/hydrogen/ battery bank system to supply energy to an electric vehicle. Int J Hydrogen Energy 40(39):13777–13788
5. Dash V, Bajpai P (2015) Power management control strategy for a stand-alone solar photovoltaic-fuel cell-battery hybrid system. Sustain Energy Technol Assess 9:68–80
6. Hassani H, Zaouche F, Rekioua D, Belaid S, Rekioua T, Bacha S (2020) Feasibility of a standalone photovoltaic/battery system with hydrogen production. J Energy Storage 31:101644
7. Rekioua T, Rekioua D (2003) Direct torque control strategy of permanent magnet synchronous machines. In: 2003 IEEE bologna power tech conference proceedings, vol 2, 6 pp
8. Mohammedi A, Rekioua D, Rekioua T, Bacha S (2016) Valve regulated lead acid battery behavior in a renewable energy system under an ideal Mediterranean climate. Int J Hydrogen Energy 41(45):20928–20938
9. Rekioua D (2020) Power electronics in hybrid renewable energies systems, green energy and technology, 39–77. Springer
10. Idjdarene K, Rekioua D, Rekioua T, Tounzi A (2011) Wind energy conversion system associated to a flywheel energy storage system. Analog Integr Circ Sig Process 69(1):67–73
11. Bukar AL, Tan CW, Lau KY (2019) Optimal sizing of an autonomous photovoltaic/wind/ battery/diesel generator microgrid using grasshopper optimization algorithm Sol Energy 188:685–696
12. Zhang W, Maleki A (2022) Modeling and optimization of a stand-alone desalination plant powered by solar/wind energies based on back-up systems using a hybrid algorithm. Energy 254:124341
13. Khan AA, Minai AF, Pachauri RK, Malik H (2022) Optimal sizing control, and management strategies for hybrid renewable energy systems: a comprehensive review. Energies 15(17):6249
14. Kamal MM, Ashraf I, Fernandez E (2022) Planning and optimization of microgrid for rural electrification with integration of renewable energy resources. J Energy Storage 52:104782
15. Lalouni S, Rekioua D, Idjdarene K, Tounzi AM (2014) An improved MPPT algorithm for wind energy conversion system. J Electr Syst 10(4):484–494
16. Masih A, Verma HK (2022) Optimum sizing and simulation of hybrid renewable energy system for remote area. Energy Environ 33(5):933–951
17. Abid MZ, Yousif M, Ullah S, Hassan M (2022) Design, sizing and economic feasibility of a hybrid PV/diesel/battery based water pumping system for farmland. Int J Green Energy 19(6):614–637
18. Obaid W, Hamid A-K, Ghenai C (2021) Hybrid water pumping system design: a case study in Dubai, United Arab Emirates. Case Stud Therm Eng 26:101121
19. Angadi S, Yaragatti UR, Suresh Y, Raju AB (2021) An effective standalone hybrid wind-photovoltaic water pumping system with reduced power converter count. Int Trans Electr Energy Syst 31(12):e13140
20. Serir C, Rekioua D (2019) Control of photovoltaic water pumping system. J Electr Eng 15(2):339–344
21. Mokrani Z, Rekioua D, Mebarki N, Rekioua T, Bacha S (2016) Energy management of battery-PEM Fuel cells Hybrid energy storage system for electric vehicle. In: 2016 International renewable and sustainable energy conference (IRSEC). Marrakech, Morocco, pp 985–990
22. Tamalouzt S, Benyahia N, Rekioua T, Rekioua D and Abdessemed R (2015) Wind turbine-DFIG/photovoltaic/fuel cell hybrid power sources system associated with hydrogen storage energy for micro-grid applications, In: 2015 3rd International renewable and sustainable energy conference (IRSEC). Marrakech, Morocco, pp 1–6
23. Rekioua D (2020) Hybrid renewable energy systems overview. In: Hybrid renewable energy systems: optimization and power management control. Springer, pp 1–37
24. Aissou R, Rekioua T, Rekioua D, Tounzi A (2016) Application of nonlinear predictive control for charging the battery using wind energy with permanent magnet synchronous generator. Int J Hydrogen Energy 41(45):20964–2097
25. Rekioua D (2023) Energy storage systems for photovoltaic and wind systems: a review. Energies 16:3893

26. Belaid S, Rekioua D, Oubelaid A, Ziane D, Rekioua T (2022) A power management control and optimization of a wind turbine with battery storage system, J Energy Storage 45:103613
27. Ipsakis D, Voutetakis S, Seferlis P, Stergiopoulos F, Elmasides C (2009) Power management strategies for a stand-alone power system using renewable energy sources and hydrogen storage. Int J Hydrogen Energy 34(16):7081–7095
28. Zhang Z, Lin Z, Zhao Z, Zhou J, Qu C (2023) MPC implementation for optimal active power control of large-scale wind-PV-battery hybrid system. In: 42nd Chinese control conference (CCC). Tianjin, China, pp 2765–2770

DSP Real-Time Implementation of an Optimized Steady State DFIM Control for Electric Vehicles

Mouna Zerzeri and Adel Khedher

Abstract The choice of traction motor for the propulsion system is crucial in their design. In order to contribute to the actual research applied to electric vehicle powertrains field, this paper focuses on Doubly-Fed Induction Machines (DFIM) in traction systems. The theoretical study examines the DFIM's steady-state features and performances based on the d-q reference model. Hence, an optimized control for an accurate rotor voltage amplitude and angle adjusting in steady state is proposed. The obtained reference voltages will be injected into a Voltage Source Inverters (VSI) in both stator and rotor sides, which are controlled through a Space Vector Pulse Width Modulation (SVPWM) technique. The overall of optimized control scheme has been implemented in real-time on a Digital Signal Processor (DSP) and has been experimentally tested on a laboratory prototype designed around two variable speed drives that feed a DFIM. The obtained results show that by properly determining the rotor voltage, the DFIM can operate optimally in a wide range of speed variations.

Keywords Electric vehicle powertrain · DFIM · Steady-state · Optimized control · Experimental tests · Digital signal processor

1 Introduction

The increasing awareness of the negative impact of fossil fuel emissions on the environment, including air pollution and climate change due to the emission of harmful pollutants from traditional fuel-powered vehicles, has led to a push for cleaner transportation options, including electric vehicles (EVs) [1–5]. With the development of

M. Zerzeri (✉) · A. Khedher
Université de Sousse, Ecole Nationale d'Ingénieurs de Sousse, LATIS—Laboratory of Advanced Technology and Intelligent Systems, 4023 Sousse, Tunisie
e-mail: mounazer@hotmail.fr

M. Zerzeri
Université de Sousse, Institut Supérieur de Transport et de la Logistique de Sousse, LATIS—Laboratory of Advanced Technology and Intelligent Systems, 4023 Sousse, Tunisie

© The Author(s), under exclusive license to Springer Nature Singapore Pte Ltd. 2024
A. Mellit et al. (eds.), *Proceedings of the 2nd International Conference on Green Energy Conversion System*, Green Energy and Technology,
https://doi.org/10.1007/978-981-97-6148-7_9

battery technology and electric motor technology, EVs have become more practical and efficient, and has led to lower costs and longer ranges.

Traction motors play a crucial role in EVs by converting electrical energy into mechanical energy to propel the wheels. Therefore, the efficiency and performance of the traction motor significantly affect the overall performance of the EV. In the EV industry, various types of traction motors are utilized, and the selection of the motor type is based on several factors, such as the vehicle's power and torque requirements, the desired efficiency, and cost considerations. Academic research has extensively explored this topic, analyzing, and comparing various motor technologies [4–7], as shown in Fig. 1.

These studies have highlighted the unique features of Doubly Fed Induction Motors (DFIMs), such as their ability to operate at maximum torque (up to twice the rated speed) and generate twice the rated power, making them a highly promising option for EV applications [1–5].

The main objective and novelty of this work is to use the DFIM in variable speed traction systems, with the aim of achieving maximum torque and high efficiency. To achieve this, it is necessary to have a precise understanding of the steady-state operation of the drive. However, literature reviews on this topic often focus on dynamic control, rather than steady-state operation. Many previous approaches are based on the classical steady-state equivalent circuit of the DFIM, which may not be accurate [8, 9]. This paper proposes an optimized control based on determining the optimal values of amplitude and angle of the rotor voltage. This study develops a model of the DFIM in the *d-q* reference frame and uses this model to analyze the characteristics of the DFIM under steady-state operation. The simulation results of the DFIM revealed three modes of operation: synchronous, super-synchronous and sub-synchronous speeds, and experimental tests are carried out to validate the theoretical approach. The tests involve implementing the proposed optimized control in real-time on a Digital Signal Processor (DSP) and testing this technique on a laboratory prototype of two inverters feeding the DFIM.

The remainder of this paper is organized as follows: Section 2 focuses on modeling the various components of the electric traction system. In Sect. 3, the proposed optimized control in steady state is presented given the optimal values of amplitude

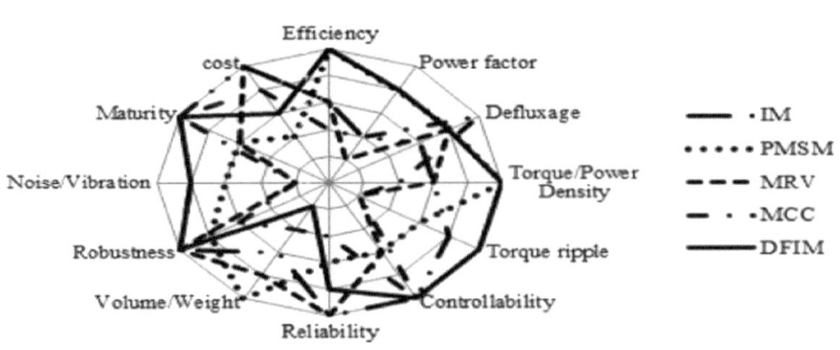

Fig. 1 Comparison of electric motors for traction application

and the angle of the rotor voltage. Experimental results are outlined in Sect. 4. Finally, the paper concludes with some remarks and conclusions.

2 Modeling of Proposed Electric Vehicle Architecture

The proposed EV architecture uses one mechanical output, a DFIM, two inverters, a motor controller and an energy storage device as shown in Fig. 2.

2.1 Battery Pack Modeling

The battery used in this study can be modeled by the following Eq. (1):

$$V_{dc} = E_0 - \frac{r_0}{1 + r_0 c_0 p} I_b \tag{1}$$

where E_0, r_0 and c_0 are constants given by the constructor.

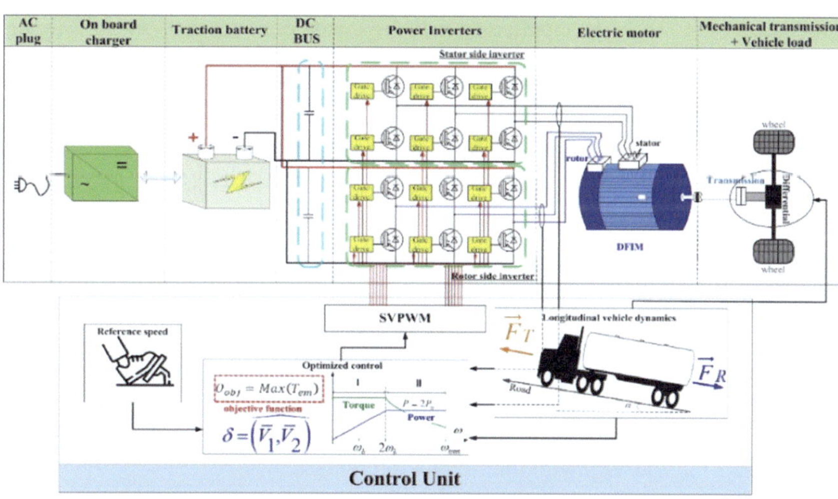

Fig. 2 Representative diagram of the electric vehicle drive train

2.2 Traction Motor Modeling

The traction motor can be modeled by the complex equations in the d-q reference as follows:

$$\begin{cases} \overline{V}_1 = \left(R_s + L_s\left[j\omega_s + \frac{d}{dt}\right]\right)\overline{I}_1 + M\left[j\omega_s + \frac{d}{dt}\right]\overline{I}_2 \\ \overline{V}_2 = M\left[j\omega_r + \frac{d}{dt}\right]\overline{I}_1 + \left(R_r + L_r\left[j\omega_r + \frac{d}{dt}\right]\right)\overline{I}_2 \end{cases} \tag{2}$$

where R_s and R_r are respectively the stator and rotor resistances, L_s and L_r are respectively the leakage inductances of the stator and the rotor, ω_s and ω_r are respectively the stator and rotor angular velocity. $\overline{V}_1$, $\overline{V}_2$, $\overline{I}_1$, and $\overline{I}_2$ are respectively the stator and rotor voltage as well as the stator and rotor current formulated as follows:

$$\begin{cases} \overline{V}_1 = V_{sd} + jV_{sq}\,, & \overline{V}_2 = V_{rd} + jV_{rq} \\ \overline{I}_1 = I_{sd} + jI_{sq}\,, & \overline{I}_2 = I_{rd} + jI_{rq} \end{cases}$$

2.3 Inverter Modeling

In this study, the DFIM is supplied by two VSI each consisting of six IGBTs and six antiparallel diodes. S_{ai}, S_{bi}, S_{ci} are the logical signals switches respectively TAH_i, TBH_i, TCH_i. The voltage expressions V_{an_i}, V_{bn_i}, V_{cn_i} can be represented by:

$$\begin{cases} V_{an_i} = \dfrac{V_{dc}}{3}(2S_{ai} - S_b - S_c) \\ V_{bn_i} = \dfrac{V_{dc}}{3}(2S_{bi} - S_a - S_c) \\ V_{cn_i} = \dfrac{V_{dc}}{3}(2S_{ci} - S_a - S_b) \end{cases} \tag{3}$$

with $i = \{s, r\}$. The generation of the command signals and the accurate switching times for the inverter's switches is realized using the PWM technique. This technique ensures the synchronization of the frequency and amplitude of the fundamental output voltages (V_{A_i}, V_{B_i} and V_{C_i}) with their target references.

2.4 Vehicle Dynamics Modeling

Environment modeling

The dynamics of the vehicle considers the longitudinal direction, and it is determined by adding up all the total forces applied to the vehicle during a traction operation.

One can find the basic equation that describes the longitudinal dynamics of a vehicle on the road in the following expression:

$$m\frac{dV}{dt} = F_T - F_R \tag{4}$$

where m represents the mass of the vehicle, V represents the velocity of the vehicle, and $\frac{dV}{dt}$ represents the velocity change rate. F_T is the traction force, and F_R is the total resistive force, which are defined as follows:

$$\begin{cases} F_T = \mu_{tire} mg \\ F_R = f_1(V - V_{wind})^2 + f_2 \sin\alpha + f_3 \cos\alpha \end{cases} \tag{5}$$

μ_{tire} is the coefficient of friction between the tire and the road, g is the acceleration due to gravity, f_1, f_2 and f_3 are constants that depend on the vehicle parameters, V_{wind} is the wind speed, and α is the road inclination relative to the horizontal.

Mechanical coupling modeling

To model the mechanical coupling between the wheels and the motor, we introduce a gear that allows the transmission of the motor torque to the driving wheels via a gear ratio. In this study, we assume that the gear transmits a lossless gear ratio, which implies a unitary transmission efficiency. The longitudinal speed of the vehicle is approximated by:

$$V = \frac{r_{wheel}}{N_{red}} \Omega_m \tag{6}$$

where N_{red} is the gear ratio, r_{wheel} is the wheel radius, and Ω_m is the speed of the motor. The transmission of the motor torque to the driving wheels, $T_{l_{wheel}}$ is defined by:

$$T_{l_{wheel}} = r_{wheel} F_R \tag{7}$$

We deduce from the expression (7), the load torque applied to the DFIM, T_l, in the case of the vehicle is driven on a flat road and the wind speed is negligible:

$$T_l = \frac{T_{l_{wheel}}}{N_{red}} = \frac{r_{wheel}}{N_{red}} \left(\frac{1}{2} \frac{r_{wheel}}{N_{red}} f_1 \Omega_m^2 + f_2 \cos\alpha \right) \tag{8}$$

3 The Proposed Optimized Control in the Steady State

The control voltages are defined by the complex form as follows:

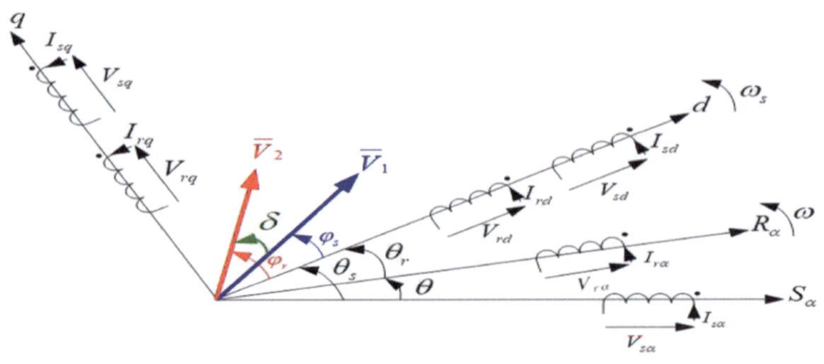

Fig. 3 Different spatial displacement angles

$$\begin{cases} \overline{V}_1 = |\overline{V}_1| = \sqrt{2}V_s \\ \overline{V}_2 = \sqrt{2}V_r e^{j\delta} = \sqrt{2}kV_s e^{j\delta} \end{cases} \tag{9}$$

where $k = \frac{V_r}{V_s}$, and $\delta = \widehat{(\overline{V_1}, \overline{V_2})}$ as shown in the schematic representation of Fig. 3.

The analysis of steady-state performance of the powertrain relies on the definition of three optimization parameters: the ratio k, the angle δ, the slip $s = \frac{\omega_r}{\omega_s}$.

Therefore, the rotor and stator currents of the DFIM can be written in the steady state as follows:

$$\begin{cases} \overline{I}_1 = \frac{1}{\overline{\Delta}}\left(\overline{Z}_r\overline{V}_1 - \overline{Z}_m\frac{\overline{V}_2}{s}\right) \\ \overline{I}_2 = \frac{1}{\overline{\Delta}}\left(-\overline{Z}_m\overline{V}_1 + \overline{Z}_s\frac{\overline{V}_2}{s}\right) \end{cases} \tag{10}$$

With:

$$\begin{cases} \overline{Z}_r = \frac{R_r}{s} + jL_r\omega \\ \overline{Z}_s = R_s + jL_s\omega \\ \overline{Z}_m = jM\omega \\ \overline{\Delta} = \overline{Z}_s\overline{Z}_r - \overline{Z}_m^2 \end{cases} \tag{11}$$

By substituting Eqs. (9) and (10), $\overline{I}_1$ and $\overline{I}_2$ can be expressed in terms of δ, k and s as follows:

$$\begin{cases} \bar{I}_1 = \sqrt{2}\frac{V_s}{|\bar{\Delta}|^2}\left(\bar{Z}_r\bar{\Delta}^* - \bar{Z}_m\bar{\Delta}^*\frac{k}{s}\, e^{j\delta}\right) \\ \bar{I}_2 = \sqrt{2}\frac{V_s}{|\bar{\Delta}|^2}\left(-\bar{Z}_m\bar{\Delta}^* + \bar{Z}_s\bar{\Delta}^*\frac{k}{s}\, e^{j\delta}\right) \end{cases} \tag{12}$$

From the optimization parameters, we can express the electromagnetic torque T_{em} as the sum of three components, T_s, T_r and T_{sr}, formulated as follows:

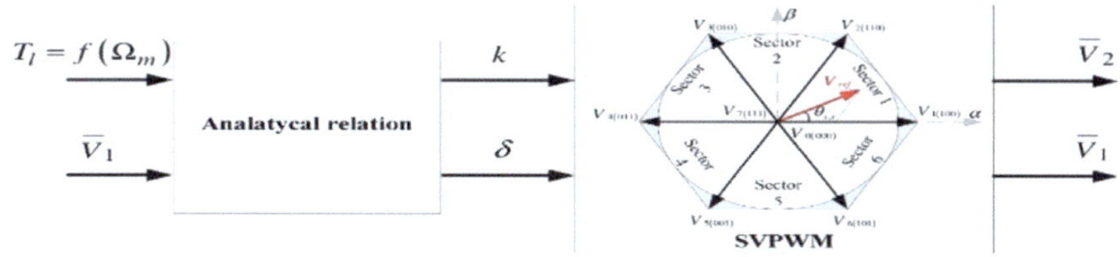

Fig. 4 Proposed optimized control law

$$\begin{cases} T_{sr} = 3\dfrac{P}{\omega}\dfrac{kV_s^2}{s}\dfrac{X_m}{\Delta^2}\left[\left(R_sX_r - \dfrac{R_r}{s}X_s\right)\cos\delta - \left(R_s\dfrac{R_r}{s} + \sigma X_rX_s\right)\sin\delta\right] \\ T_s = 3\dfrac{P}{\omega}\left(\dfrac{X_m}{\Delta}\right)^2 V_s^2\dfrac{R_r}{s} \\ T_r = -3\dfrac{P}{\omega}\left(\dfrac{X_m}{\Delta}\right)^2\dfrac{V_r^2}{s^2}R_s \end{cases} \tag{13}$$

Retaining the optimal torque capability is a very important task in EV applications. The dynamic performance of the machine is affected by the inverter's maximum output current, and voltage applied to it. In this context, the proposed optimized control in this paper is implemented based on analytical resolution, which determines the appropriate k and δ to achieve the optimal torque at any operating mode. This allows the extension of the maximum torque operating range of DFIM until maximum speed, i.e., up to twice the nominal speed. For the optimal torque operation, T_{em} value should be maximized under the voltage and current ratings of the PWM inverter. The maximum inverter voltage applied to the machine is limited by the DC-link voltage and the PWM strategy.

An SVPWM strategy is adopted in this paper for the VSI control, whose eight combinations of voltage vectors $\vec{V}_{i(i-[0,\ldots,7])}$ are possible to describe the states of IGBTs. The SVPWM algorithm is performed by rotating a reference vector $\vec{V}_{ref}$ around the state diagram, which is composed of six basic nonzero vectors forming a hexagon of origin coincides with the zero vectors $\left(\vec{V}_0, \vec{V}_7\right)$. Figure 4 illustrates the proposed optimized control law based on analytical approach.

4 Experimental Results

To show the effectiveness of the proposed DFIM in electric traction system, an experimental test bench equipped with two variable speed drives were designed. These latter are built using a diode rectifier and two-level voltage inverters with 1200 V/30 A IGBTs and their associated control board as well as DFIM speed, stator, and rotor acquisition board. It consists of a DSP Board *eZdsp F28335* from Spectrum Digital. Its operating clock frequency is 150 MHz. The development tool

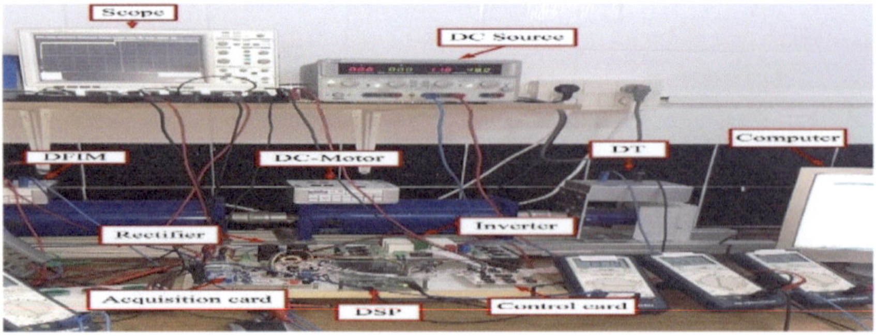

Fig. 5 Photo of the developed experimental test bench of the variable speed drives feeding a DFIM

Table 1 Nominal values and parameters of the DFIM used in experimental test

Nominal values		Parameters	
P_n	0.5 kW	R_s	15.26 Ω
V_{sn}	230 V	R_r	8.17 Ω
I_{sn}	1.8 A	L_s	0.426 H
I_{rn}	3.54 A	L_r	0.175 H
f	50 Hz	M	0.24 H
N_n	1440 rpm	J	0.0178 kg m^2
P	2	f	0.00539 N m red^{-1} s

used for the DSP is the Texas Instruments Code Composer Studio *v.3.3*. The two inverters are controlled by the SVPWM technique with the switching frequency is set to $f_{sw} = 5$ kHz. The proposed algorithm which includes the modulation and commutation schemes was implemented on TMS320F28335 DSP. A photo of the realized prototype is illustrated in Fig. 5.

The parameters of the used DFIM in experimental test are summarized in Table 1.

The DFIM cannot operate directly in dual feed mode, it must be in short-circuited rotor to ensure the asynchronous operation mode. Therefore, an electrical circuit has been carried out to guarantee this operation. We considered a short-circuited rotor with no load and then at $t = 10$ s, we supply the rotor. The stator reference voltage is characterized by a frequency equal to 50 Hz and RMS value equal to 230 V. Based on the study of DFIM in steady state, we have highlighted two types of rotor voltage injection in order to highlight sub- and super-synchronous speed for the DFIM motor operating mode. We illustrate by Fig. 6 the torque versus slip characteristic for $k = 0.08$ and for two load angles δ:

- $\delta_1 = 180°$ to highlight the super-synchronous motor operating mode.
- $\delta_2 = 300°$ to highlight the motor operating mode in sub-synchronous.

For a motor mode, we can observe from Fig. 6 that when $\delta_2 \in [-90°, 90°]$, the torque-speed characteristic of DFIM is shifted to sub-synchronous operation.

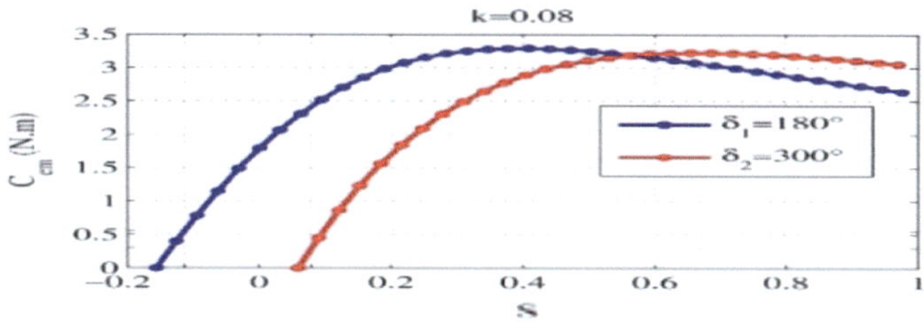

Fig. 6 Electromagnetic torque of DFIM versus slip s for $k = 0.08$

Whereas, when $\delta_1 \in [90°, 270°]$, this characteristic is shifted for super-synchronous operation.

The proposed optimized control algorithm is tested on the developed experimental prototype. In Fig. 7a, we present the SVPWM pulse signals that control the different power switches of the same inverter leg for a switching frequency of 5 kHz. Furthermore, Fig. 7b shows the programmed dead time that prevents the short circuit between two complementary transistors. This dead time is programmed to 2μ.

Experimental results for these two scenarios are given in Figs. 8 and 9.

These figures show the rotational speed, the inverter line-to-line voltage U_{rab} on the rotor side, the rotor, and the stator currents respectively. The experimental results show that a judicious choice of the optimization parameters allows extracting

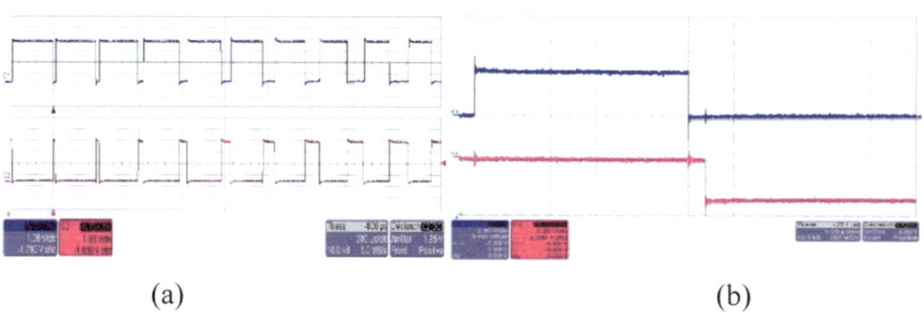

Fig. 7 Control signals

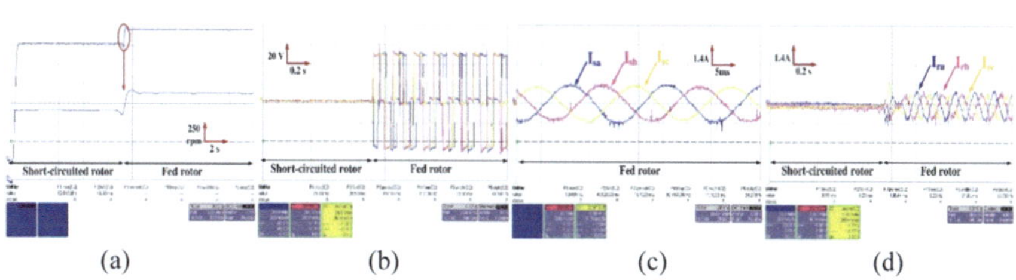

Fig. 8 Experimental results for $k = 0.08$ and $\delta = 180°$: **a** N (rpm), **b** U_{rab} (V), **c** I_s (A), **d** I_r (A)

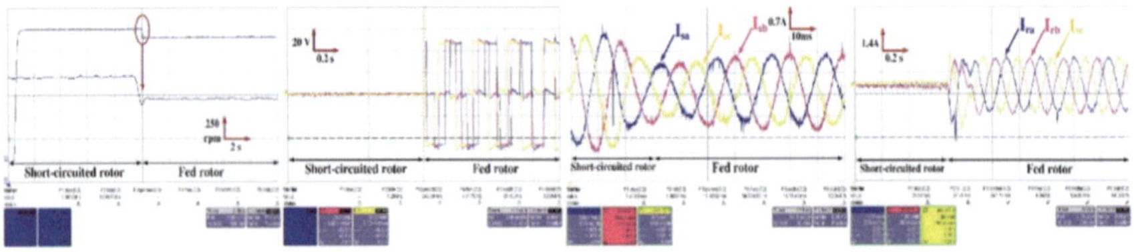

(a) (b) (c) (d)

Fig. 9 Experimental results for $k = 0.08$ and $\delta = 300°$: **a** N (rpm), **b** U_{rab} (V), **c** I_s (A), **d** I_r (A)

a maximum of the electromagnetic torque over a wide speed range covering the sub- and super-synchronous region to meet the mechanical demand of the electric vehicle.

5 Conclusion

In this paper, an optimized control strategy was proposed with the aim to fully explore the torque and speed capabilities of the DFIM applied in an electric vehicle system. The novelty in our design resides in investigating the DFIM's steady-state features and performances based on the *d-q* reference model, which are a function of the injected rotor voltage. This latter depends on three parameters: k, δ and s. By varying k and δ of the rotor injected voltage, the DFIM mechanical characteristics are shifted from the super-synchronous to the sub-synchronous speed range to generate the mechanical power. It was found that summing-up the synchronous torque with the asynchronous torque makes it possible to increase the DFIM electromagnetic torque in wide speed range and to improve the overload capability. This is a valuable and attractive feature of EV applications. Experimental tests are carried out to show that the speed exceeds the nominal one and operates in a motor mode when varying the load angle δ and the slip s, which validates the theoretical approach of the DFIM integration into traction system.

References

1. Zerzeri M, Jallali F, Khedher A (2017) Steady-state analysis of a variable speed traction system using DFI-motor. In: 18th international conference on sciences and techniques of automatic control and computer engineering (STA). IEEE, Monastir, pp 198–204
2. Zerzeri M, Jallali F, Khedher A (2019) A robust nonlinear control based on SMC approach for doubly-fed induction motor drives used in electric vehicles. In: 1st international conference on signal, control and communication (SCC). IEEE, Hammamet, pp 138–144
3. Zerzeri M, Jallali F, Khedher A (2019) Robust FOC analysis of a DFIM using an SMFO: application to electric vehicles. In: International conference on signal, control and communication (SCC). IEEE, Hammamet, pp 250–255

4. Zerzeri M, Khedher A (2021) Optimal speed–torque control of doubly-fed induction motors: analytical and graphical analysis. Comput Electr Eng 93:107258
5. Zerzeri M, Khedher A (2021) A sensorless mixed DFIM control strategy based on fuzzy-PI speed controller and current sliding mode controller for electric vehicles. Int J Model Ident Control 38(3–4):271–281
6. Cao W, Bukhari AAS, Aarniovuori L (2019) Review of electrical motor drives for electric vehicle applications. Mehran Univ Res J Eng Technol 38(3):525–540
7. Phatak JP, Venkatesha L, Raviprasad CS (2019) An alternative procedure for selection of induction motor for electric vehicle application. J Comput Theor Nanosci 16(8):3351–3358
8. Yu K, Tang P (2018) Novel equivalent circuit model and theoretical analysis of doubly fed machine. IEEE Trans Energy Convers 34(2):1073–1081
9. Ge J, Xu W, Liu Y, Xiong F (2022) Novel equivalent circuit model applicable to all operation modes for brushless doubly fed induction machines. IEEE Trans Industr Electron 69(12):12540–12550

Wind Turbine Emulator Using Three-Level Inverter-Fed Induction Motor Drive Controlled Through DTC Strategy

Salma Jnayah, Mouna Zerzeri, Intissar Moussa, and Adel Khedher

Abstract Induction motors (IM) are widely applied in manifold industrial applications, notably in the renewable energy sector, owing to their low maintenance requirements, economical cost, and high efficiency. This article outlines a Wind Turbine Emulation (WTE) system using a squirrel cage asynchronous machine under Direct Torque Control (DTC) approach. The proposed DTC strategy is based on a three-level Neutral Point Clamped (NPC) inverter fed the IM to proficiently imitate a small-scale wind-turbine behavior in a laboratory setting by regulating the waveform of the speed. The suggested control system reacts effectively as a wind turbine emulator in research laboratory by controlling the speed waveform with minimum torque and flux ripples. The simulation findings, carried out using Matlab/SIMULINK environment, confirmed the suitability of the DTC in WTE application.

Keywords IM · Wind turbine emulator · DTC · Three-level inverter · Torque ripples

Nomenclatures

$i_{s\alpha}, i_{s\beta}$	Stator currents in (α, β) frame
$\varphi_{s\alpha}, \varphi_{s\beta}$	Stator fluxes in (α, β) frame
L_s, L_r, M_{sr}	Stator, rotor and mutual inductances
R_s, R_r	Stator and rotor resistances
$V_{s\alpha}, V_{s\beta}$	Stator voltages in (α, β) frame
ω	Electric speed
ρ	Density of the air
β, λ	Pitch angle and tip speed ratio
V_v	Wind speed

S. Jnayah (✉) · M. Zerzeri · I. Moussa · A. Khedher
Université de Sousse, Ecole Nationale d'Ingénieurs de Sousse, LATIS- Laboratory of Advanced Technology and Intelligent Systems, 4023 Sousse, Tunisie
e-mail: salmajnayah@gmail.com

© The Author(s), under exclusive license to Springer Nature Singapore Pte Ltd. 2024

A. Mellit et al. (eds.), *Proceedings of the 2nd International Conference on Green Energy Conversion System*, Green Energy and Technology,
https://doi.org/10.1007/978-981-97-6148-7_10

C_p	Power coefficient
R	Blade radius

1 Introduction

The rapid industrialization in the past few decades and the proliferation of electrical domestic appliances have resulted in large global needs for electrical energy. In response to this growing demand, Renewable Energy Sources (RES) have become increasingly important in recent years as consumers around the world seek to move towards more sustainable sources of energy, which are eco-friendly and Cost-effective [1]. Among several green energy sources, wind energy has emerged as a highly economical choice for electricity production, and it is deemed as the most auspicious form of clean resources thanks to the technological progress and government investment incentives [2–4]. Nevertheless, given the challenges of conducting research studies on-site and in order to simulate the control strategies of Wind Energy Conversion Systems (WECS), the requirement for modeling a setup that imitate the Wind Turbine (WT) properties is highly recommended [5, 6]. Hence, a Wind Turbine Emulator (WTE) serves as an affordable platform to test and validate WECS based on power electronic converters. Furthermore, it is essential to conduct rigorous laboratory testing on upcoming innovations, while considering the potential impact of wind patterns that may still be ambiguous or conjectural. As a result, numerous methodologies for designing WTE have been examined and developed in the literature. In [7], the modeling of a WTE has been categorized into three distinct groups based on its utility. These groups comprise of: (i) an emulator that possesses a general structure, (ii) an emulator that displays dynamic behavior, and (iii) an emulator without dynamic behavior. The authors in [8] designed the WTE in a manner to take into account the effect of a tower shadow and inertia compensation at different wind speeds. Reference [9] introduced a digital signal controller to the WTE that is both cost-effective and highly efficient. The experimental results prove that the WTE performs satisfactorily despite encountering wind turbulence and the tower shadow effect.

The process of constructing the WTE is built on connecting the generator shaft to a motor known as a prime mover, which is controlled through a variable speed drive. Various research works have analyzed three main categories of prime movers namely: Induction Motor (IM), DC motor (DCM) and Permanent Magnet Synchronous Motor (PMSM) [10–12]. However, DC motors necessitate regular maintenance, and they are also costly and bulky. Besides, IMs are designed with a simple construction, require minimal maintenance, and have better overload performances when compared to PMSMs. Therefore, emulators using IM are viable alternatives that allow researchers to evaluate wind turbines performances without the need for physical testing. The key challenge in wind emulation is to choose the best control method that can replicate the static and dynamic performances of a given wind turbine. Hence, numerous control strategies have been developed and investigated. In case of IMs, Field-Oriented Control is proposed to improve the WTE dynamics [13]. However, this approach is renewed by its sensitivity to parameter variations since it is based on PI controllers. Likewise, Sliding Mode Control (SMC) has been designed to manage the WTE to track the reference values and get the desired output. Nerveless, it cannot ensure the convergence of the system states within a finite time and may result in chattering phenomenon that can affect the system reliability. As a promising alternative, Direct Torque Control (DTC) has been proposed since it is the most prevalent and widespread approach for controlling WTE that produces the fastest torque response [14]. It is also distinguished by its structural simplicity and resilience against parameter uncertainties. However, hysteresis controllers used in DTC strategy engender variable switching frequency which can cause undesirable torque and flux ripples especially at low speeds. Therefore, this paper exhibits the challenge of developing a laboratory variable speed wind turbine emulator to mimic the functioning of a small-scale wind turbine based on three-phase IM fed with three-level inverter and controlled through DTC strategy. Using three-level inverter in the closed loop suggested system will enhance its reliability and improve its dynamic response since low THD and less torque and flux ripples will be guaranteed.

This paper is structured as follows: Sect. 2 deals with the mathematical model and the DTC control algorithm of the three-phase IM fed with three level inverter. Section 3 details the WTE general scheme responsible for emulating the variable reference speed profile. Section 4 presents the entire system description of the WTE controller using DTC strategy. In Sect. 5, simulation results are conducted to show the effectiveness of the proposed system under different wind speed profiles. Finally, the main conclusions are given in Sect. 6.

2 DTC of IM Fed with Three Level Inverter

2.1 IM Control Model

To ensure the control of IM through DTC strategy, it is necessary to present its state model in the Concordia stationary reference frame (α, β) as given below:

$$\left[\dot{X}\right] = [A][X] + [B][U] \tag{1}$$

where [X], [A], [B], and [U] denote respectively the state vector, the state matrix, the control matrix and the control vector. Thus, the system describing the input/output state model of the IM can be written in Eq. (2):

$$\begin{bmatrix} \dot{i}_{s\alpha} \\ \dot{i}_{s\beta} \\ \dot{\varphi}_{s\alpha} \\ \dot{\varphi}_{s\beta} \end{bmatrix} = \begin{bmatrix} -\left(\frac{R_s}{\sigma L_s} + \frac{R_r}{\sigma L_r}\right) & -\omega & \frac{R_r}{\sigma L_r L_s} & \frac{\omega}{\sigma L_s} \\ \omega & -\left(\frac{R_s}{\sigma L_s} + \frac{R_r}{\sigma L_r}\right) & \frac{-\omega}{\sigma L_r L_s} & \frac{R_r}{\sigma L_r L_s} \\ -R_s & 0 & 0 & 0 \\ 0 & -R_s & 0 & 0 \end{bmatrix} * \begin{bmatrix} i_{s\alpha} \\ i_{s\beta} \\ \varphi_{s\alpha} \\ \varphi_{s\beta} \end{bmatrix} + \begin{bmatrix} \frac{1}{\sigma L_s} & 0 \\ 0 & \frac{1}{\sigma L_s} \\ 1 & 0 \\ 0 & 1 \end{bmatrix} * \begin{bmatrix} V_{s\alpha} \\ V_{s\beta} \end{bmatrix} \quad (2)$$

With $\sigma = 1 - \left(M_{sr}^2/L_sL_r\right)$ represents the Blondel coefficient.

The mechanical equation of the IM is established as follows:

$$J\,\dot{\Omega}_{IM} = T_e - f\,\Omega_{IM} - T_l \quad (3)$$

where T_e is the electromagnetic torque, T_l is the load torque, f and J are the viscous friction coefficient and the inertia moment, respectively. Ω_{IM} is the mechanical speed.

2.2 *DTC Applied to Three-Level NPC Inverter*

In this work, we suggested to develop the direct torque control strategy using a three-level NPC inverter to address the drawbacks of the conventional DTC. The proposed approach focuses on reducing torque, flux, and current ripples in order to enhance the system overall performances [15]. The aim of this control technique is to choose the suitable voltage vector to regulate the amplitudes of flux and torque, while keeping them within their hysteresis bands. Hence, this method relies on identifying the appropriate sequences to be applied to the three-level inverter power switches. By using hysteresis controllers, the estimated flux and torque are compared to their desired values as depicted in Fig. 1.

In the reference frame (α, β), the stator flux magnitude is determined by:

$$\varphi_s = \sqrt{\varphi_{s\alpha}^2 + \varphi_{s\beta}^2} \quad (4)$$

where the stator flux can be estimated as:

$$\varphi_{s\alpha} = \int (V_{s\alpha} - R_s.i_{s\alpha})\,dt \quad \text{and} \quad \varphi_{s\beta} = \int \left(V_{s\beta} - R_s.i_{s\beta}\right)dt \quad (5)$$

The equation used to estimate the electromagnetic torque takes into account the estimated flux magnitude and the measured stator currents as given below:

$$T_e = 1.5 \times p\left(\varphi_{s\alpha} \times i_{s\beta} - \varphi_{s\beta} \times i_{s\alpha}\right) \quad (6)$$

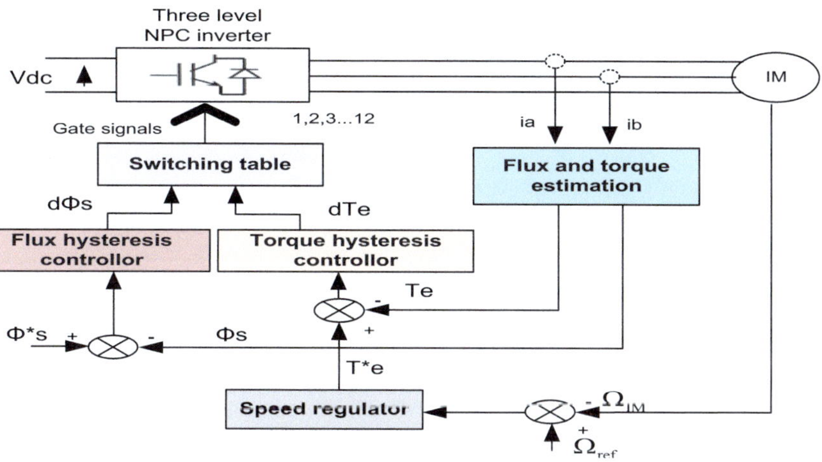

Fig. 1 DTC scheme using three level inverter

Stator voltages $V_{s\alpha}$ and $V_{s\beta}$ are derived from the Eqs. (7) and (8).

$$V_s = V_{s\alpha} + jV_{s\beta} \tag{7}$$

$$\begin{cases} V_{s\alpha} = 1/\sqrt{6} * V_{dc} \times (2K_a - K_b - K_c) \\ V_{s\beta} = 1/\sqrt{2} * V_{dc} \times (K_b - K_c) \end{cases} \tag{8}$$

where $K_a = 0.5 * (S_{AH1} + S_{AH2} - 1)$; $K_b = 0.5 * (S_{BH1} + S_{BH2} - 1)$; $K_c = 0{,}5 * (S_{CH1} - S_{CH2} - 1)$

SAHi are the upper switches of the first leg of the 3L inverter (Fig. 2).
SBHi are the upper switches of the second leg of the 3L inverter.
SCHi are the upper switches of the third leg of the 3L inverter.

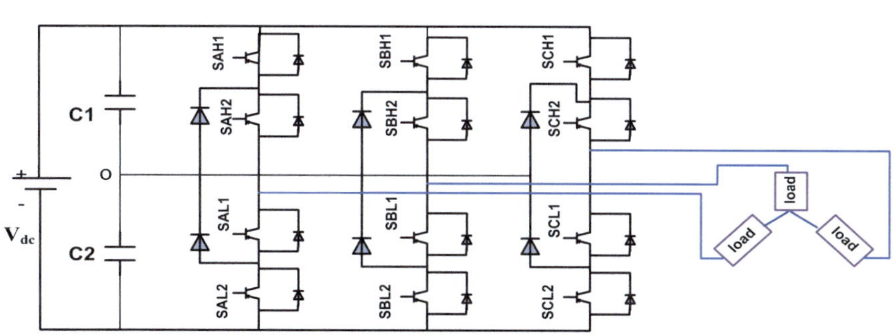

Fig. 2 3 level NPC inverter

The three-level NPC inverter's voltage vectors can be chosen using a switching table that takes into account the stator flux position, which expression is given in Eq. (9), and the output states of the two flux and torque hysteresis controllers. This table is detailed in [16].

$$\theta_s = \arctan\left(\varphi_{s\beta}/\varphi_{s\alpha}\right) \tag{9}$$

3 WTE Design

The Block diagram of the wind power system based on the WTE outlined in this paper is presented in Fig. 3 [17]. It intends to reproduce the mechanical behavior of the IM similar to an aerodynamic turbine. The emulation principle entails the calculation of the reference wind speed and comparing it to the measured motor speed, which is fed by three level inverter and controlled through DTC strategy.

The equation describing the wind turbine mechanical power can be stated as:

$$P_t = 0.5 * \rho * C_p(\lambda, \beta) * AV_v^3 \tag{10}$$

in which C_p is given by the following expression:

$$C_p(\lambda, \beta) = (C_1 - C_2(\beta - C_3)) \sin\left(\frac{C_4(\lambda + C_5)}{C_6 - C_7(\beta - C_8)}\right) - C_9(\lambda - C_{10})(\beta - C_{11}) \tag{11}$$

and the tip speed ratio λ can be declared as:

$$\lambda = \Omega_t R / V_v \tag{12}$$

Then, the wind turbine mechanical torque can be expressed as:

$$T_t = P_t / \Omega_t = 0.5 * \left(\rho * C_p(\lambda, \beta) * ARV_v^2\right)/\lambda \tag{13}$$

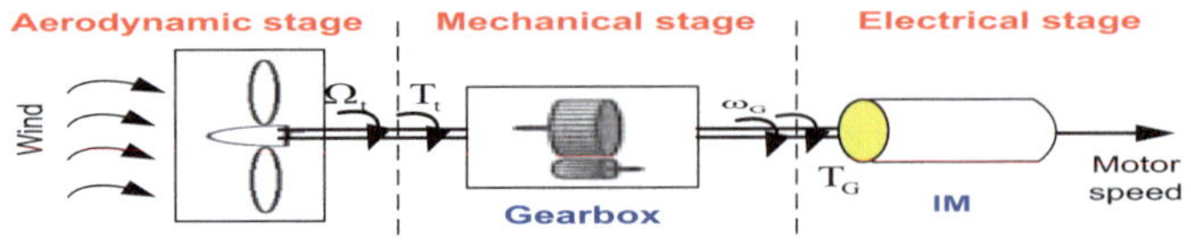

Fig. 3 Energy conversion process for wind turbine

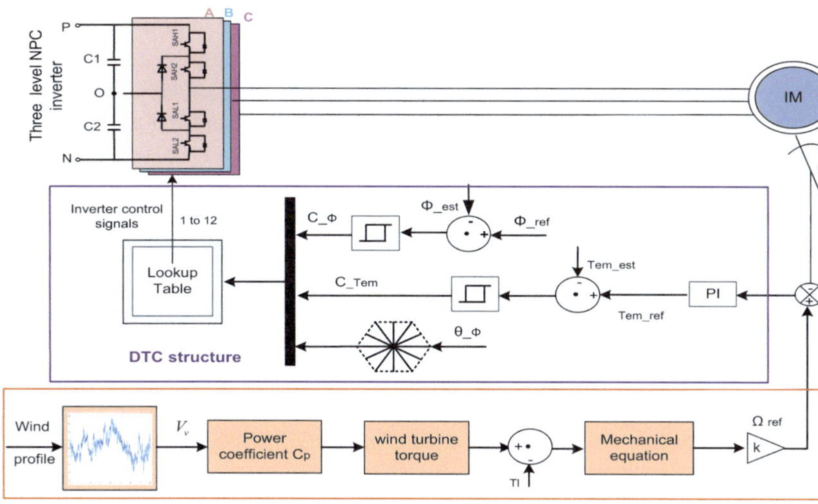

Fig. 4 Scheme of the WTE based DTC controller

The power turbine coefficients that depend on the turbine geometry are given in [18]:

$C_1 = 0.5$; $C_2 = 0.00167$; C_3, C_8 *and* $C_{11} = 2$; $C_4 = 3.14$; $C_5 = 0.1$; $C_6 = 18.5$; $C_7 = 0.3$; $C_9 = 0.00184$ and $C_{10} = 3$.

The wind turbine parameters used for software simulations are:

Rotor radius = 3.61 m; Maximum power coefficient = 0.52; Gearbox ratio = 13; Blade number = 3; Air density = 1.225 kg/m^3.

4 Whole System Description of the WTE Controller

The designed system input in this paper is the wind profile where the output is the IM speed [19]. The entire WTE control scheme based on DTC strategy and including the wind profile, the turbine and the IM fed by three level NPC inverter is described in Fig. 4. The IM parameters used in this work is presented in the Apendix.

5 Simulation Results

Based on the block diagram shown in Fig. 4, the simulation results using Matlab/SIMULINK have been accomplished to evaluate the suggested control efficiency. Hence, simulation findings were performed using a three level NPC inverter to feed a 1.5 kW three-phase IM. The emulator receives the reference speed from the wind turbine model, and then the principal role of the proposed control scheme is to facilitate the reliable imitation of the emulator dynamic features. The advocated emulator, developed in this paper, is controlled through DTC strategy. Intending to confirm the accuracy of the reference tracking capabilities and the proficiency of the controller behavior, the software test envisages to analyze how well the system functions when subjected to a variable wind speed profile that follows three different shapes such as sinusoidal shape, random shape and Mexican-hat shape. Results shown in Fig. 5a–d display the speed waveforms, the electromagnetic torque, the stator current waveforms and the stator flux magnitude for the first wind profile which is reported in [17, 19]. Likewise, Fig. 6a–d present the WTE performances under DTC with a random wind profile. The dynamic response of the proposed WTE based on the Mexican-hat wind speed is depicted in Fig. 7a–d. Based on the obtained simulation results, the proposed WTE demonstrates its capability to generate and imitate the characteristics of a realistic wind turbine. The acquired waveforms in Fig. 5a indicate that the prime motor speed faithfully tracks the reference speed trajectory of the wind turbine with almost neglected error. It is also noted from Fig. 5b that the electromagnetic torque generated by the three phase IM controlled with DTC strategy tracks its reference value accurately. The shape of the stator currents presented in Fig. 5c remains sinusoidal and varies in accordance with the speed and torque fluctuations. Concerning the module of the stator fluxes, depicted in Fig. 5d, it is featured by a very quick response with almost minimal fluctuations around the reference value despite the electromagnetic torque variable curve. Hence, the decoupling between torque and flux is granted under DTC control strategy. Therefore, there is clear evidence that the system accurately replicates the theoretical torque-speed properties of the wind turbine.

The WTE performances under stochastic change of wind speed, which is depicted in Fig. 6a, are revealed in Fig. 6. It is apparent from Fig. 6b that he motor is effectively

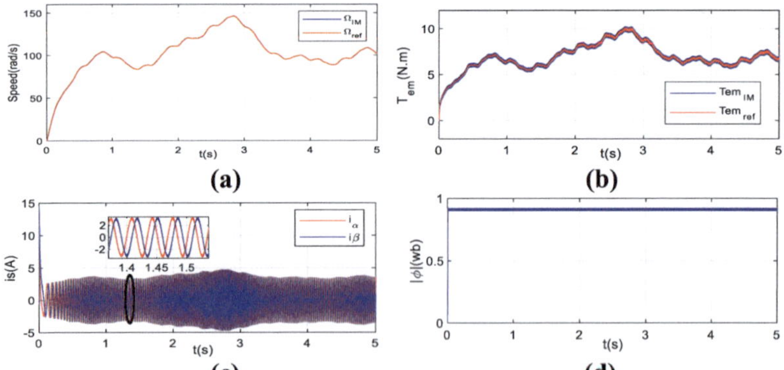

Fig. 5 Simulation results related to the proposed WTE based on IM fed with three level inverter and controlled through DTC When the wind speedfollows a sinusoidal shape: **a** WT and IM speeds; **b** WT and IM torques; **c** IM stator currents; **d** flux magnitude

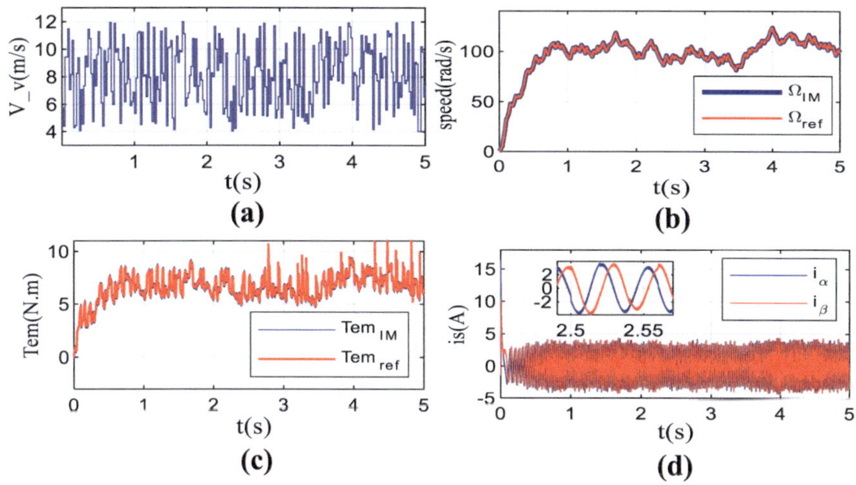

Fig. 6 WTE response for a random wind profile based on DTC strategy: **a** turbine and IM speeds; **b** IM and reference electromagnetic torque; **c** IM stator currents

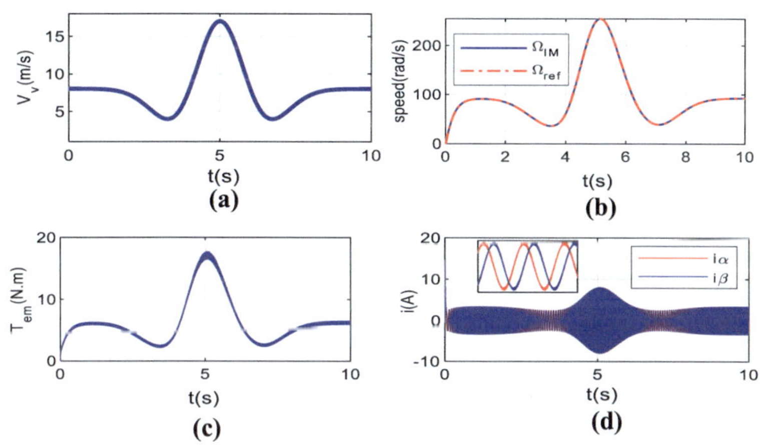

Fig. 7 WTE dynamic response for a Mexican-hat wind profile: **a** wind speed; **b** turbine and motor speed responses; **b** IM torque; IM currents in (α, β) frame

tracking the desired speed pattern and exhibiting a strong correlation with the desired performance, which implies that the system is operating as intended. Furthermore, the IM electromagnetic torque, presented in Fig. 6c is closely following the reference value, indicating that the motor is functioning as envisaged, and responding appropriately to the control strategy. As shown in Fig. 6d, the stator currents maintain a sinusoidal shape and fluctuate in response to the speed and torque variations. The good performance of the IM in terms of speed and torque reference tracking under random wind conditions would indicate that the controlled system through

DTC approach is designed and optimized to cope with the variability of the wind resource, ensuring that the WTE plant operates reliably and efficiently even under challenging weather conditions.

The analysis of the WTE performances illustrated in Fig. 7 under a Mexican-hat wind profile would assess the ability of the controlled system to mimic the dynamic behavior of wind turbine characteristics in a research laboratory since the IM generated variables track almost faithfully their reference values.

6 Conclusion

This research paper introduces a proposal to enhance the performance of the WTE system, which aims to replicate the dynamic characteristics of a realistic wind turbine in a laboratory setting. The proposed design relies on a three phase IM acting as the primary driver, fed with three level NPC inverter and controlled through the DTC strategy. The reliability and effectiveness of the entire closed loop control scheme using three different wind shapes is proved by simulations since the generated torque and speed track thoroughly the reference trajectories despite speed profile variations. Thus, the overall good performance of the DTC based emulator is s guaranteed.

Appendix

The parameters of the three-phase induction motor used for simulations are expressed in SI units.

Rated power	1.5 kW
Rated speed	1435 tr/min
Rated frequency	50 Hz
Rated current	5.5/3.2 A
Number of pole pairs	2
Stator resistance	5.72 Ω

Rotor resistance	4.28 Ω
Stator inductance	0.464 H
Rotor inductance	0.464 H
Mutual inductance	0.44 H
Moment of inertia	0.0049 kg m^2
Viscous friction coefficient	0.002

References

1. Zhang Z, Zhao Y, Qiao W, Qu L (2015) A discrete-time direct torque control for direct-drive PMSG-based wind energy conversion systems. IEEE Trans Ind Appl 51(4):3504–3514
2. Sahri Y, Tamalouzt S, Belaid SL, Bajaj M, Belkhier Y, Singh AR, Kamel S (2023) Effectiveness analysis of twelve sectors of DTC based on a newly modified switching table implemented on a wind turbine DFIG system under variable wind velocity. Ain Shams Eng J: 102221
3. Benbouhenni H (2022) Amelioration effectiveness of torque and rotor flux control applied to the asynchronous generator for dual-rotor wind turbine using neural third-order sliding mode approaches. Int J Eng 35(3):517–530
4. Eial Awwad A (2022) Dynamic performance enhancement of a direct-driven PMSG-based wind turbine using a 12-sectors DTC. World Electr Vehic J 13(7):123
5. Moussa I, Khedher A (2020) A theoretical and experimental study of a laboratory wind turbine emulator using DC-motor controlled by an FPGA-based approach. Electr Power Compon Syst 48(4–5):399–409
6. Behera PK, Mendi B, Sarangi SK, Pattnaik M (2021) Robust wind turbine emulator design using sliding mode controller. Renew Energy Focus 36:79–88
7. Diop AD, Nichita C, Belhache JJ, Dakyo B, Ceanga E (2000) Error evaluation for models of real time wind turbine simulators. Wind Eng 24(3):203–221
8. Munteanu I, Bratcu AI, Bacha S, Roye D, Guiraud J (2010) Hardware-in-the-loop-based simulator for a class of variable-speed wind energy conversion systems: design and performance assessment. IEEE Trans Energy Convers 25(2):564–576
9. Neammanee B, Sirisumrannukul S, Chatratana S (2007) Development of a wind turbine simulator for wind generator testing. Int Energy J 8(1)
10. Andrzej J, Janusz B (2018) Laboratory setup with squirrel-cage motors for wind turbine emulation. In: 2018 applications of electromagnetics in modern techniques and medicine (PTZE). IEEE, pp 1–4
11. Sirouni Y, El Hani S, Naseri N, Aghmadi A, El Harouri K (2018) Design and control of a small scale wind turbine emulator with a DC motor. In: 2018 6th international renewable and sustainable energy conference (IRSEC), pp 1–6
12. Zeinoddini-Meymand H, Kamel S, Khan B (2022) Design and implementation of a novel intelligent strategy for the permanent magnet synchronous motor emulation. Complexity 2022
13. Moussa I, Zerzeri M, Khedher A (2022) Robust WTE sensorless design based on soft-VSI structure for three-phase IM drive using sliding mode observer controller. In: International conference on electrical sciences and technologies in Maghreb (CISTEM), pp 1–7
14. Arafa OM, Abdallah ME, Aziz GA (2018) Realisation and HIL testing of wind turbine emulator based on DTC squirrel cage induction motor drive. Int J Industr Electron Drives 4(3):155–168
15. Jnayah S, Khedher A (2022) Sensorless DTC-SVM applied to an induction motor controlled by a three-level inverter using SMSFO. In: COMPEL—the international journal for computation and mathematics in electrical and electronic engineering
16. Jnayah S, Khedher A (2019) DTC of induction motor drives fed by two and three-level inverter: modeling and simulation. In: 19th international conference on sciences and techniques of automatic control and computer engineering (STA), pp 376–381
17. Moussa I, Khedher A (2022) Wind turbine emulator controlled through field programmable gate array: modeling, analysis, and design. In: Recent advances in renewable energy technologies. Academic Press, pp 251–283
18. Moussa I, Khedher A (2022) Software in-the-loop simulation of an advanced SVM technique for 2ϕ-inverter control Fed a TPIM as wind turbine emulator. Electronics 11(2):187
19. Zerzeri M, Moussa I, Khedher A (2022) Sliding mode control for wind turbine emulator based on advanced space vector modulation technique for two-phase induction motor drive. IETE J Res: 1–14

Experimental and Numerical Investigation on Heat Transfer in Helically Coiled Tube Heat Exchanger Under Constant Wall Heat Flux

Sami Missaoui, Romdhane Ben Slama, and Bechir Chaouachi

Abstract Waste heat recovery from helically coiled tube heat exchanger of a household refrigerator system is attractive due to its free energy consumption for residential and commercial hot water production. In this chapter, we present an experimental and numerical investigation on a modified household refrigerator for hot water production useful for domestic purpose. This device helps undergraduate mechanical engineering students in understanding the basic natural heat transfer phenomenon between the helical condenser coil and the water in the tank. From the results, after 300 min of heating, the maximum water temperature achieved in the cylindrical tank is 51 °C. Therefore, this modified refrigerator can be utilized in hotels, industry and domestic purposes which need hot water.

Keywords Waste heat recovery · Hot water production · Experiment · CFD · Constant wall heat flux

1 Introduction

Energy used for domestic hot water production in the commercial building sector is the fourth largest energy user after lighting and air conditioning. The statistics confirmed that the water heating accounts for a large proportion of energy consumption. Almost 75% of the population use electric water heaters, 22% use wood, oil or gas water heaters and 3% use solar thermal systems. Besides this huge amount of energy consumption, water heating is also a major source of carbon dioxide emissions. The use of solar energy, geothermal energy, air and other green energy has been an effective approach to tackle the problem of high energy consumption in recent years, and it is an effective way to solve the problem of high energy consumption. With its availability and environment-friendliness, a coupled unit of a domestic

S. Missaoui (✉) · R. B. Slama · B. Chaouachi
Laboratory of Energy, Water, Environment and Processes (LR18ES35), National Engineering School of Gabes (ENIG), University of Gabes, Zrig Eddakhlania, Tunisia
e-mail: Missaouisami1988@Gmail.com; Sami.Missaoui@Ensit.rnu.tn

© The Author(s), under exclusive license to Springer Nature Singapore Pte Ltd. 2024
A. Mellit et al. (eds.), *Proceedings of the 2nd International Conference on Green Energy Conversion System*, Green Energy and Technology,
https://doi.org/10.1007/978-981-97-6148-7_11

refrigerator with water heaters is a promising energy source and offers an economical solution by reducing the energy consumed for water heaters and the carbon dioxide emissions. In this context, Dai et al. [1] investigated the heat transfer in helical coil heat exchangers in a household heat pump water heater. The results indicated that the condenser coil design has a direct effect on the thermal performance of the coupled machines. Missaoui et al. [2] performed a numerical and experimental investigation on heat transfer in helical coil heat exchangers for domestic refrigerator and water heating. The results indicated that the helical coil with variable pitch improve the heat transfer coefficient and the performance of the system than the normal helical condenser coil. Ye et al. [3] performed a numerical investigation on the performance of the heat pump water heater with wrap-around condenser coil. The results indicated that the heat transfer coefficient and the performance of the heat pump are improved by using the helical condenser with variable pitch coil. Slama [4] conducted an experimental investigation on the refrigerator coupled to a water heater and heating floor with the aim to sage energy and to reduce the CO_2 emissions. The results indicated that the water temperature in the cylindrical tank reaches the maximum value equal to 60 °C. Missaoui et al. [5] conducted a numerical analysis of the heat pump water heater with immersed helically coiled tubes. The results indicated that the helical coil pitch presents a direct effect on the velocity and water profiles during the heating process. Missaoui et al. [6] proposed a numerical model containing both domestic refrigerator and water heating unit. The results indicated that the hot water distribution in the cylindrical water tank has a direct effect on the performance of the coupled system. Based on the above literature review, there is a few works integrated the refrigeration machines with hot water production unit. According to those researches, the investigation concerning the heat transfer in helically coiled tube heat exchanger immersed in the tank and their impact on the performance of the refrigeration machines for domestic hot water production in residential and commercial buildings are limited. The aim of this study is to investigate the heat transfer between the helical coil tube type heat exchanger and the water inside the tank, the velocity and the temperature of the water during heating process. Therefore, this study is the first numerical and experimental investigation of this new coupled unit.

2 System Description

The experiments are performed in a setup, which is schematically shown in Fig. 1. It contains a compressor, an evaporator, an expansion valve and a spiral condenser coil which is totally immersed in the storage tank.

Figure 2 shows the schematic diagram and the geometric details of the immersed helically coiled tube in the water tank. In the experiment, the condenser coil with 13 m length is well insulated with 50 mm thickness of glass wool in order to minimize the heat losses to the surrounding ambient.

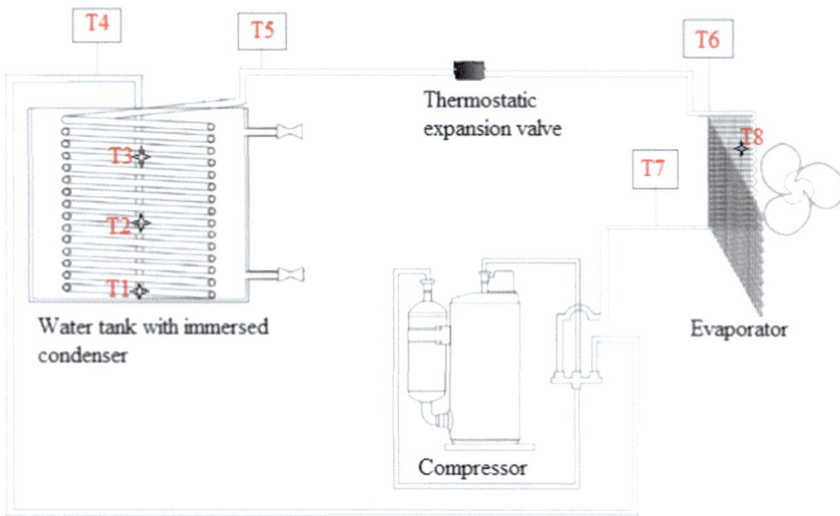

Fig. 1 Schematic of the experimental setup

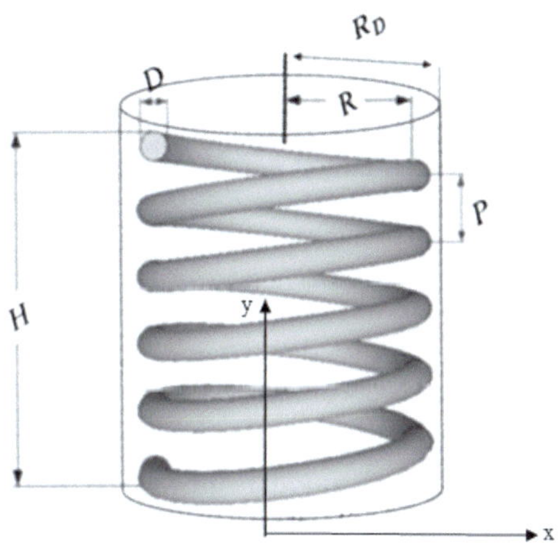

Symbols	Value (m)
H	0.38
P	0.03
D	0.006
R	0.16
R_D	0.19

Fig. 2 Manufacture process of the helically coiled tube heat exchanger

3 Calculation Methods

Heat water storage efficiency is calculated from the useful energy Q_u over the absorber energy Q_a as follows:

$$\eta = \frac{Q_u}{Q_a} \quad (1)$$

The useful energy Q_u is calculated from the substraction of the loss energy from the absorber energy as follows:

$$Q_{useful} = Q_{abs} - Q_{loss} \tag{2}$$

The absorber energy can be also calculated from the compressor power Pc and the performance coefficient COP during the function operating time of the compressor (t) as follows:

$$Q_{abs} = P_{comp} \cdot COP.t \tag{3}$$

The loss energy is calculated from the thermal loss ratio U, the water temperature increases from initial state to final state ΔT and the operating time of the compressor (t) as follows:

$$Q_{loss} = U \cdot \Delta T \cdot t \tag{4}$$

The thermal loss ratio is written as follows:

$$U = \frac{\rho \cdot Cp \cdot V}{\Delta t} \cdot \ln\left(\frac{T_i - T_{am}}{T_f - T_{am}}\right) \tag{5}$$

where V is the water capacity, ρ is the density, Cp is the specific heat, Δt is the overall time of heating, T_{am} is the ambient temperature and T_i and T_f are the initial and final water temperature in the tank, respectively.

4 Numerical Model

The mathematical model is based on solving the continuity, momentum and energy equations with a finite volume discretization method using the software "ANSYS Fluent".

4.1 *Boundary Conditions*

In the present study, the water tank with immersed condenser coil is simplified to a two dimensional axisymmetric model. R134a is a working fluid, which is assumed with a constant specific heat, constant thermal conductivity, constant dynamic viscosity and a constant thermal expansion coefficient. These properties are determined by using Engineering Equation Solver (EES) software based on the experimental test performed. The water fluid density is treated with Boussinesq approximation. With this approximation, the density is solved as a constant value during the whole heating process. The constant heat flux equal to 1450 W/m^2 which is produced along the surface of the helical coil heat exchanger is used as a heat source in order to heat the cold water from the initial temperature equal to T = 20 °C.

4.2 Meshing

As shown in Fig. 3, the mesh is generated by ICEM-CFD software package. For calculations accuracy, a fine grid-size is created near the surfaces of the condenser coil and a coarse one is conducted away from the coil.

5 Model Validation

An experimental setup of a household refrigerator system with integrated domestic hot water production unit is built for the purpose of the model validation. According to the obtained results, it has been noted that there is a good agreement between the experimental and the calculated results (Fig. 4). Therefore, the proposed model of the integrated system is accuracy for further control and optimization application.

Fig. 3 Geometry and mesh of the storage tank with immersed helical coil heat exchanger

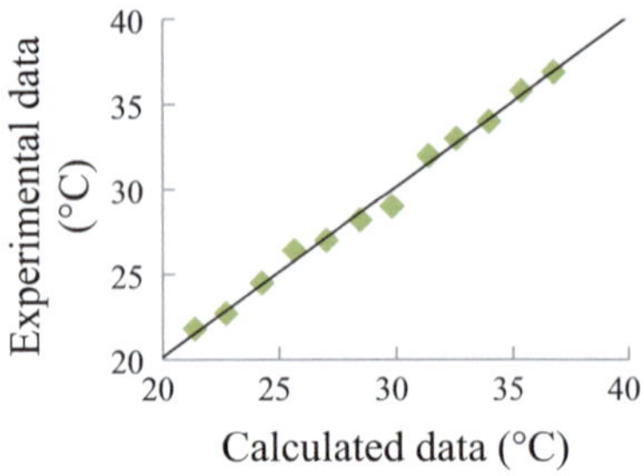

Fig. 4 Experimental measurements versus simulated water temperature at t = 180 min

6 Results and Discussion

6.1 Temperature

Figure 5 shows the experimentally obtained results of the water temperature and the evaporator temperature during the heating time. From these results, it has been observed that the water temperature increases with the increase of the heating time. While the evaporator temperature varies between T = − 13 °C and T = − 17 °C, the evaporator part of this coupled unit is not affected by the increase of the water temperature in the storage tank.

Figure 6 shows the water temperature distribution in the tank under different heating time defined respectively by t = 60 min, t = 180 min, t = 240 min and t = 300 min. According to the obtained results, it has been noted that the thermal stratification gradually builds up as the hot water flows upwards and cold water flows downwords along the vertical direction of the cylindrical tank. Also, it can be seen from the CFD results, during the heating process, the water at a higher temperature flows upwards and a higher thermal stratification builds at a lower part of the tank (from 0.1 m down to the tank bottom).

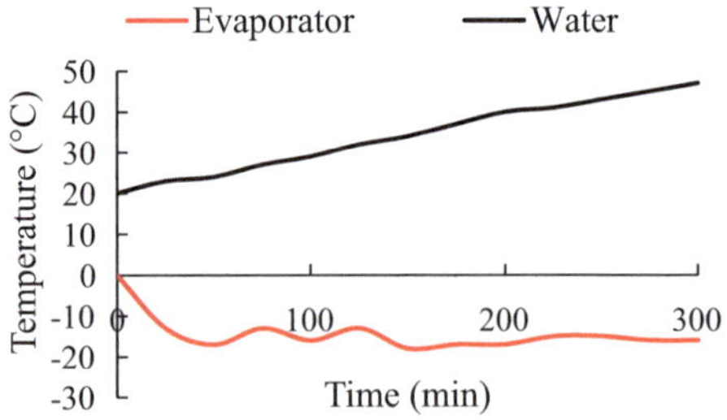

Fig. 5 Variation of temperature during the heating process

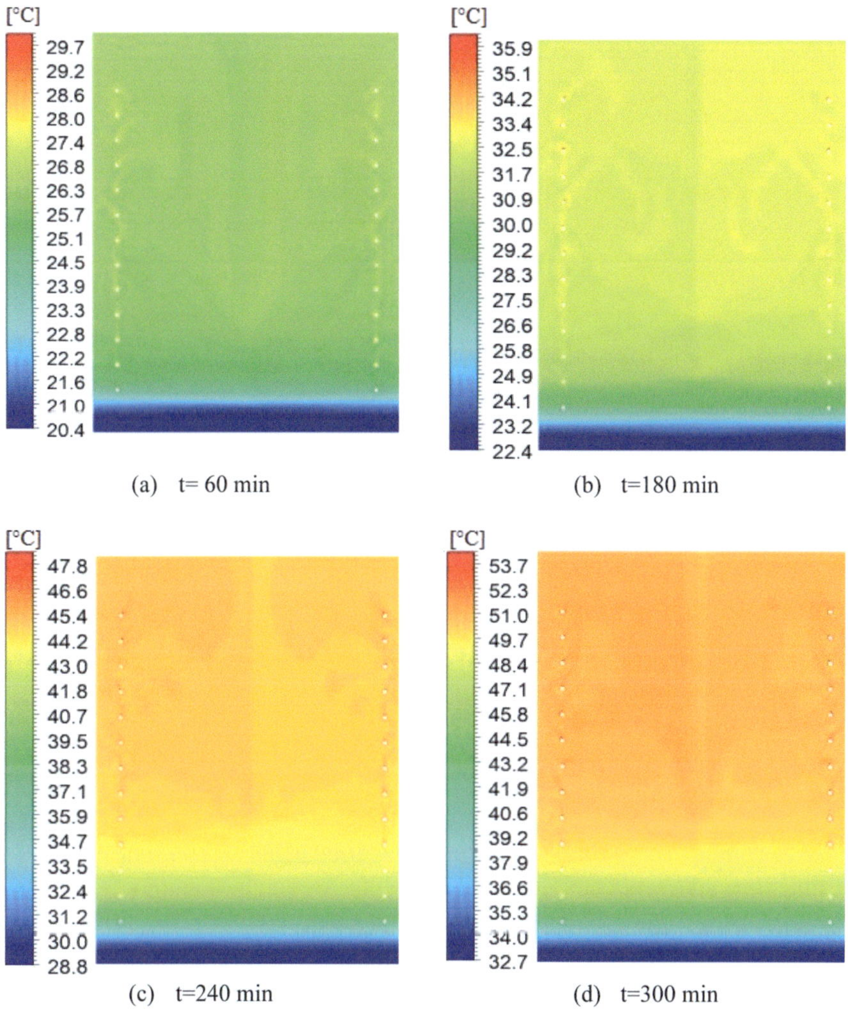

Fig. 6 Water temperature profile at different heating times

6.2 *Efficiency*

Figure 7 shows the variation of the efficiency value of the new coupled system with the heating time. From these results, it can be seen that with the increase of heating time, the performance decreases. This is because the water temperature in the tank is heated fully without used for domestic purposes or changed. In fact, in order to keep the efficiency of this coupled system higher then we have to use another storage tank for hot water storage.

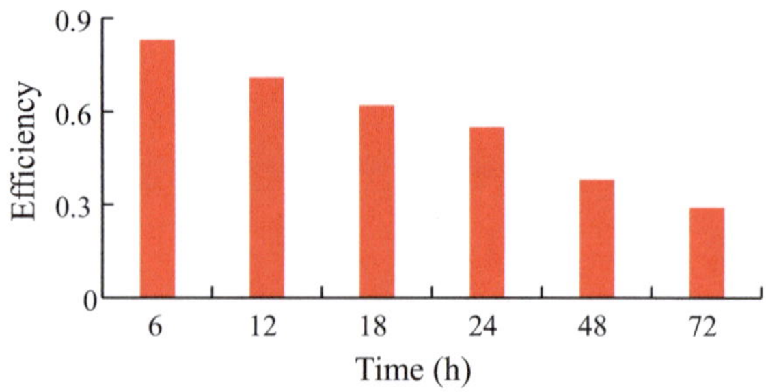

Fig. 7 Thermal efficiency of the domestic refrigerator for hot water production

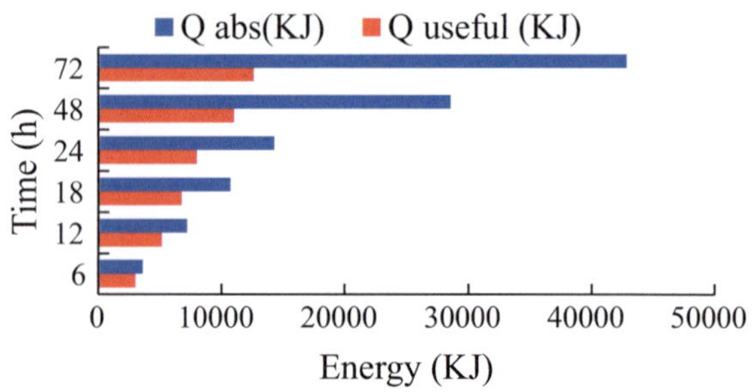

Fig. 8 Energy distribution in the cylindrical tank with immersed helical condenser coil

6.3 *Heat Energy*

Figure 8 shows the variation of the heat energy released into a cylindrical water tank and the condenser's heat load energy recovered for water heating during the heating time. According to the obtained results, it has been noted that with increasing the heating time the energy are improved. During the overall heating process, the heat absorbed by the water still higher than the heat energy recovered for water heating. This is meanly due to the increase of the temperature surface of the condenser coil as a result of the high temperature of the water inside the storage tank without using it or changing it.

6.4 *Velocity*

Figure 9 shows the water velocity distribution in the cylindrical tank under different heating times defined respectively by t = 60 min, t = 180 min, t = 240 min and t = 300 min. Due to the relative high water temperature in the upper part of the tank, there is an upward flow with a vertical fluid velocity up to 0.013 m/s. The high water velocity is due to the important heat transfer from the helical coil in the middle part. At t = 240 min, there is an upward flow of 0.009 m/s in the upper part of the tank meaning that the relatively warmer water rises up. At t = 300 min, the water velocity increases in the upper part. This can be explained by the increase of the

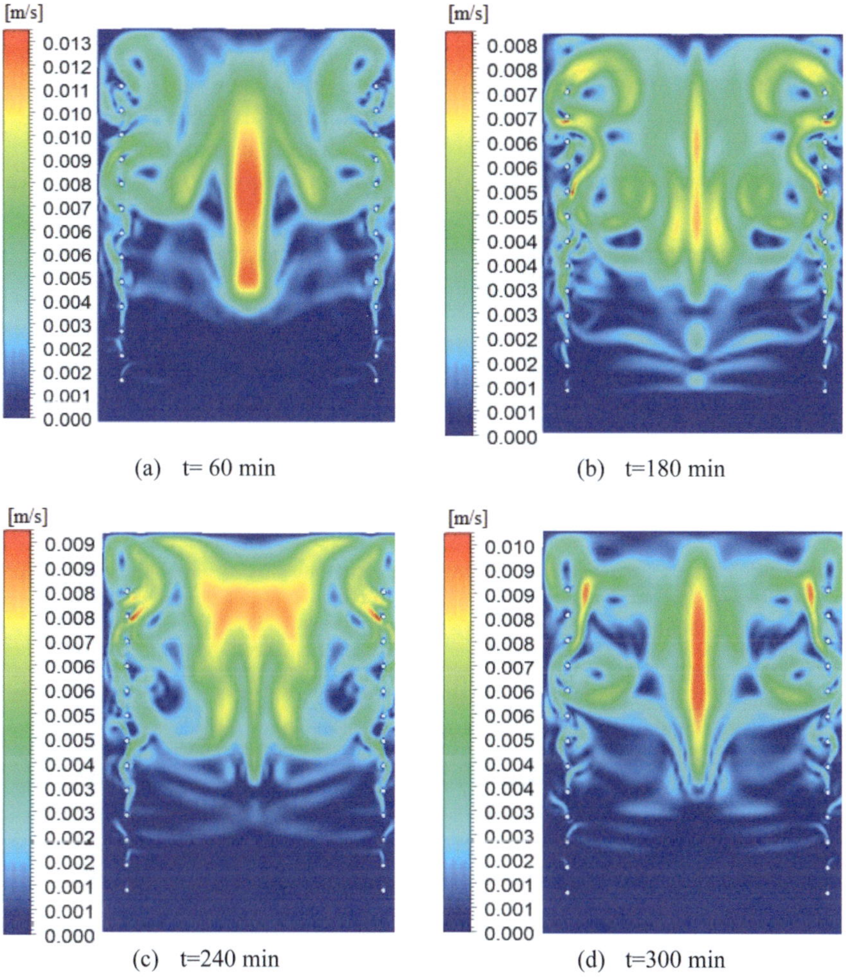

Fig. 9 Water velocity profile at different heating time

water temperature in the upper part and the decrease of the thermal stratification of the storage tank.

7 Concluding Remarks

The present chapter addresses the characterization of the natural heat transfer between the refrigerant R134a flowing inside the helical condenser coil and the water following outside of the surface for an imposed helical coil heat flux. Looking towards the numerical and experimental results, it is concluded that:

- The numerical results agree very well with the experimental results.
- The maximum water temperature achieved in the cylindrical tank after 300 min of heating is 51 °C. With continuation heating, this value is better.
- The waste heat recovery from the helical coil heat exchanger of a domestic refrigerator can be utilized for industrial and domestic purposes.
- This multipurpose machine contributes for the protection of global environment.

Acknowledgements The authors are grateful to the members of the Laboratory of Energy, Water, Environment, and Processes (LR18ES35) for their financial support.

References

1. Dai N, Li S (2018) Simulation and performance analysis on condenser coil in household heat pump water heater. Sustain Cities Soc 36:176–184
2. Missaoui S, Driss Z, Slama RB, Chaouachi B (2022) Experimental and numerical analysis of a helical coil heat exchanger for domestic refrigerator and water heating. Int J Refrig 133:276–288
3. Ye Q, Li S (2019) Investigation on the performance and optimization of heat pump water heater with wrap-around condenser coil. Int J Heat Mass Transf 143:118556
4. Slama RB (2013) Refrigerator coupling to a water-heater and heating floor to save energy and to reduce carbon emissions. Comput Water Energy Environ Eng 2:21–29
5. Missaoui S, Driss Z, Slama RB, Chaouachi B (2021) Numerical analysis of the heat pump water heater with immersed helically coiled tubes. J Energy Storage 39:102547
6. Missaoui S, Driss Z, Slama RB, Chaouachi B (2021) Experimentally validated model of a domestic refrigerator with an immersed condenser coil for water heating. Int J Air-Condition Refrig 29(3):2150022

CFD Simulation of Heat Transfer in Helically Coiled Tube Heat Exchanger for Water Heating

Sami Missaoui, Romdhane Ben Slama, and Bechir Chaouachi

Abstract In order to improve the heat transfer in helically coiled tube heat exchanger immersed in the water tank, we have developed a new coupled model containing both domestic refrigerator and water heater. In the numerical simulation, the three dimensional storage tank with immersed helical coil heat exchanger was simplified as an axisymmetric two dimensional model in order to reduce the calculated capacity and a total of mesh volumes. According to the obtained results, this model is able to predict the effect of condenser coil designs on the heating process of the domestic refrigerator coupled with water heating system. The results indicated that the heating characteristics in the variable pitch coil were improved in comparison with that of normal coil.

Keywords Computational fluid dynamics (CFD) · Hot water production · Energy saving · Heat transfer · Helical condenser coil design · Structure optimization

1 Introduction

Helically coiled tube are widely used in a wide variety of engineering areas, such as chemical industries, air-conditioning engineering, heating, heat recovery systems and medical equipment.

Therefore, the development of a helical coil tube heat exchanger could aid in the achievement of energy conservation, emission reduction and renewable energy goals. Several versions of heat pump water heaters were produced. But only a handful of them addressed the helical condenser coil immersed in the water tank. Mahdi et al. [1] conducted an experimental study on the melting behavior of a phase change material in a conical coil latent heat thermal energy storage unit. The results indicated that the melting rate in the conical coil LHSU is enhanced by 22% in comparison with that of

S. Missaoui (✉) · R. B. Slama · B. Chaouachi
Laboratory of Energy, Water, Environment and Processes (LR18ES35), National Engineering School of Gabes (ENIG), University of Gabes, Gabes, Tunisia
e-mail: Missaouisami1988@Gmail.com; Sami.Missaoui@Ensit.rnu.tn

© The Author(s), under exclusive license to Springer Nature Singapore Pte Ltd. 2024
A. Mellit et al. (eds.), *Proceedings of the 2nd International Conference on Green Energy Conversion System*, Green Energy and Technology,
https://doi.org/10.1007/978-981-97-6148-7_12

normal coil LHSU. Missaoui et al. [2] presented a numerical model to investigate the thermal performance of the domestic refrigerator for water heating. The results indicated that the hot water in the tank presents a direct effect on the performance of the system. Dai et al. [3] carried out a numerical investigation on heat pump water heater (HPWH). The results indicated that the heat transfer coefficient of variable diameter coil was improved in comparison with that of constant diameter coil. Missaoui et al. [4] conducted a numerical analysis on heat pump water heater with immersed helically coiled tubes. The results indicated that the helical coil pitch presents a direct effect on the velocity and water profiles during the heating process. Li et al. [5] carried out a numerical investigation on the performance of heat pump water heater (HPWH) with wrap-around condenser coil. The results indicated that the performance of the system was improved by using the helically coiled tube heat exchanger with variable pitch coil. Missaoui et al. [6] concluded from thermal modeling tests that employing a variable-diameter coil is superior to using a constant-diameter coil in terms of the heat transfer coefficient and water temperature distribution. Li et al. [7] used numerical methods to investigate the effect of the coil structure designs on the thermal performances of the heat pump water heater. Missaoui et al. [8] also verified the relationship between thermal performances and geometric parameters of the helical coil. They indicated that the coil geometry presented a direct effect on the heating process. Missaoui et al. [9] proposed a numerical model containing both domestic refrigerator and water heating unit. The results indicated that the hot water distribution in the cylindrical water tank has a direct effect on the performance of the coupled system. Missaoui et al. [10] presented numerically investigation on the effects of the geometrical parameters such as tube diameter, coil pitch and number of turns on both thermal stratification in the tank and natural convection heat transfer along the helical condenser coil immersed in the water tank. They indicated that the thermal characteristics are affected when varying the geometrical parameters of the helical condenser coil. In order to improve the heat transfer in helically coiled tube heat exchanger, the computational Fluid Dynamics (CFD) software has become an important method of current research. Therefore, the goal of this paper is to analyze the influence of the condenser coil designs on the heating performance of a domestic refrigerator for hot water production.

2 Cases Study

In order to analyze and compare the helical condenser coil designs on the heating effect of the water in the storage tank, the condenser coil designs are considered in this section. Particularly, variable pitch coil and normal helical condenser coil are studied. The structure of these two coil designs are presented in Fig. 1.

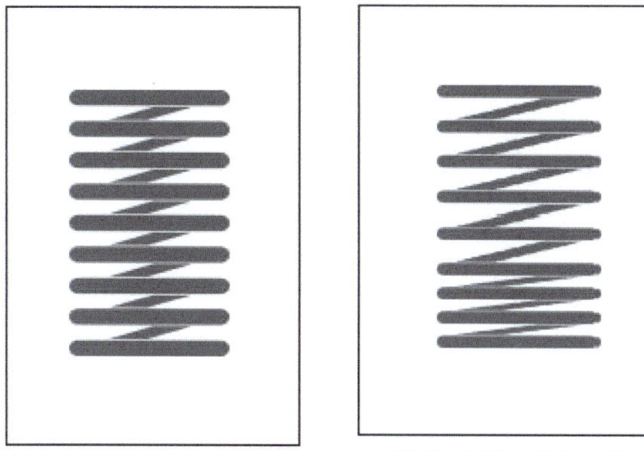

(a) Normal condenser coil (b) Variable pitch coil

Fig. 1 The structure parameters of these two coil designs

3 System Description

Figure 2 shows the helically coiled tube heat exchanger immersed in the cylindrical water tank. The condenser is made up of smooth copper tube with inner and outer diameter defined respectively by di = 0.004 m and do = 0.006 m. The coupled system containing both domestic refrigerator and water heaters have been given in more detail in our previous published paper [11]. As shown in Fig. 2, the water tank is made of plastic and is covered with a layer of stainless steel foil. Also, it is insulated with 5 cm glass wool.

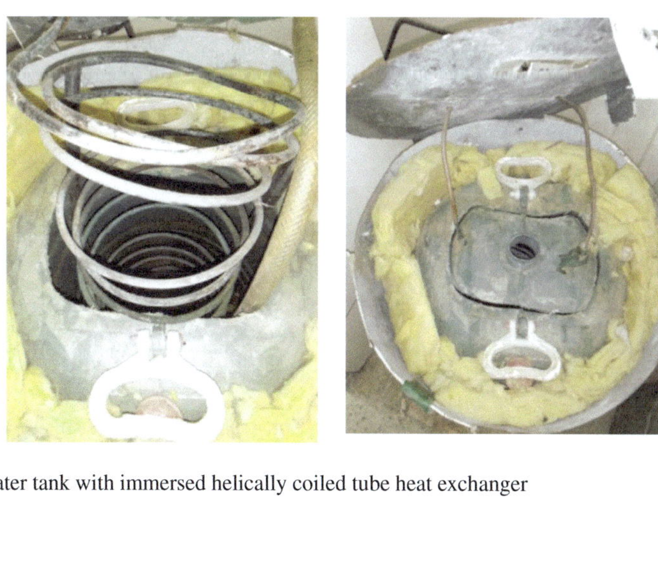

Fig. 2 Water tank with immersed helically coiled tube heat exchanger

4 Model Development

The computational fluid dynamics (CFD) code was used to create a dynamic model of a domestic refrigerator with an immersed condenser coil.

4.1 Condenser Model

The three dimensional of condenser coil is modeled as many small circles in symmetry plane. Depending on the heat transferred between the condenser coil surface and the water inside the storage tank, the heat transferred through the condenser coil is determined according to this equation:

$$Q_{con} = m_r \cdot \left(h_{con,r,i} - h_{con,r,o}\right) \quad (1)$$

The heat gained by the water is written as follow:

$$Q_w = C_{p,w} \cdot m_w \cdot \left(T_{w,o} - T_{w,i}\right) \quad (2)$$

The total heat transfer is written as follow:

$$Q_{con} = Q_w = U_c \cdot A_c \cdot \left(T_{c,r} - T_w\right) \quad (3)$$

The heat flux between the water and the condenser coil can be determined by the following equation:

$$q = U_c \cdot \Delta T_{c,w} \quad (4)$$

The overall heat transfer coefficient Uc can be calculated as the following relation:

$$U_c = \left(\frac{d_{c,o}}{\alpha_{c,r} d_{c,i}} + \frac{1}{\alpha_o} + \frac{d_{c,o}}{2\lambda_c} \ln \frac{d_{c,o}}{d_{c,i}}\right)^{-1} \quad (5)$$

4.2 Water Tank Model

The water tank is built based on the following assumptions to simplify the CFD model:

The tank is assumed as adiabatic.
The water flow is laminar.
Natural convection is modeled with Boussinesq approximation.

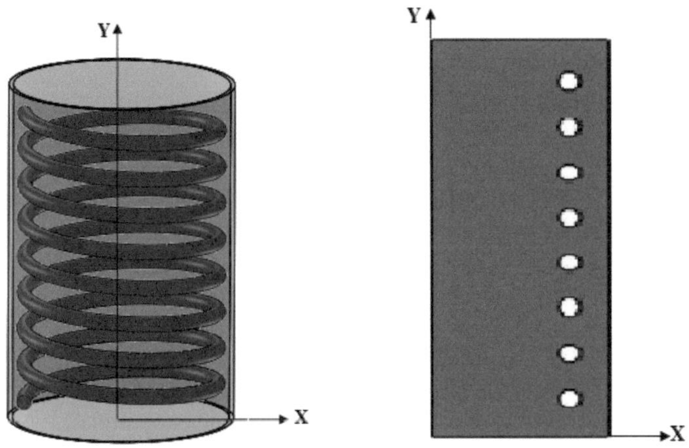

(a) 3D model of water tank (b) 2D model of the water tank

Fig. 3 Simplified model of the tank with helically coiled tube heat exchanger

The three-dimensional cylindrical water tank is simplified to an axisymmetric two-dimensional model as shown in Fig. 3.

5 Numerical Model

The numerical model is built by utilizing the program "ANSYS Fluent" to solve the Navier–Stokes equations using a finite volume discretization method.

5.1 Boundary Conditions

As shown in Table 1, the experimental results are used to establish the heat flow of the condenser coil, which is then used as the initial boundary condition in the CFD model.

Table 1 Main physical properties of the water in CFD simulation [12]

Water property	Symbol	Values	Units	Method
Density	ρ	994	kg/m^3	Boussinesq
Specific heat	Cp	4183	J/(kg K)	Constant
Thermal conductivity	λ	0.6107	W/(m K)	Constant
Dynamic viscosity	μ	0.0007196	kg/(m.s)	Constant
Thermal expansion coefficient	β	0.000347	1/K	Constant

5.2 Meshing

The grids were fine-tuned around the condenser coil to improve calculation accuracy, while the grids become further loosened away from the coil as shown in Fig. 4.

For calculation accuracy, three computational meshes of 4162, 24,202 and 122,005 are solved by fixing the temporal discretization to one second. Case with 122,005 cells performs no better than the case 24,202 cells. According to the obtained results (Fig. 5), there is no reason to increase the cells beyond 24,202 and this case is used for all further investigation.

Fig. 4 Structured mesh in variable pitch coil

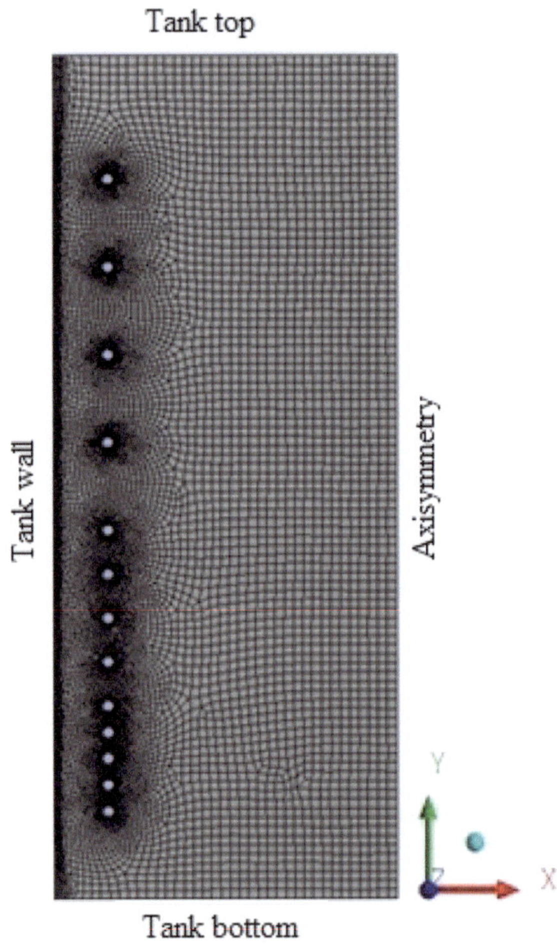

Fig. 5 Mesh independence test

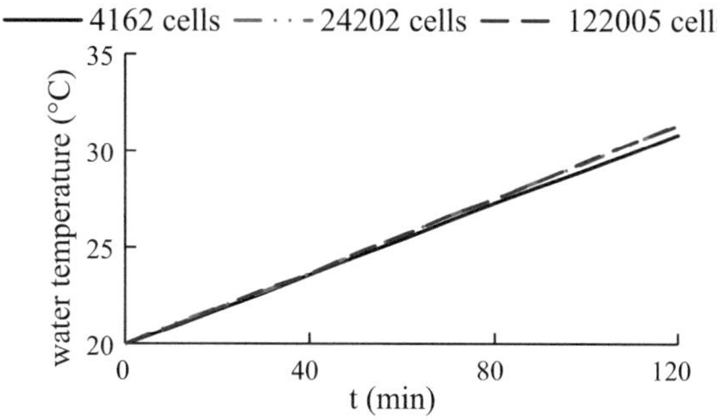

6 Mathematical Formulation

In their two-dimensional axisymmetric model, the Navier–Stokes and energy equations were assumed to govern the fluid flow and heat transfer phenomena inside the storage tank.

The continuity equation is written as follows:

$$\frac{\partial u_x}{\partial x} + \frac{\partial u_y}{\partial y} = 0 \tag{6}$$

The momentum equations are written as follows:

$$\frac{\partial u_x}{\partial t} + u_x \frac{\partial u_x}{\partial x} + u_y \frac{\partial u_x}{\partial y} = -\frac{1}{\rho_{w0}} \frac{\partial P}{\partial x} + \nu_w \left[\frac{\partial^2 u_x}{\partial x^2} + \frac{\partial^2 u_x}{\partial y^2} \right] \tag{7}$$

$$\frac{\partial u_y}{\partial t} + u_x \frac{\partial u_y}{\partial x} + u_y \frac{\partial u_y}{\partial y} = -\frac{1}{\rho_{w0}} \frac{\partial P}{\partial y} + \nu_w \left[\frac{\partial^2 u_y}{\partial x^2} + \frac{\partial^2 u_y}{\partial y^2} \right] - g\beta(T_w - T_{w0}) \tag{8}$$

The energy equation is written as follows:

$$\frac{\partial T}{\partial t} + u_x \frac{\partial T}{\partial x} + u_y \frac{\partial T}{\partial y} = \lambda_w \left[\frac{\partial^2 T}{\partial x^2} + \frac{\partial^2 T}{\partial y^2} \right] \tag{9}$$

7 Model Validation

An own experimental facility was used to obtain the experimental data in order to validate the model results as presented in Fig. 6. According to these results, it has been noted that the numerical results of the new proposed model show good agreement with the experimental results.

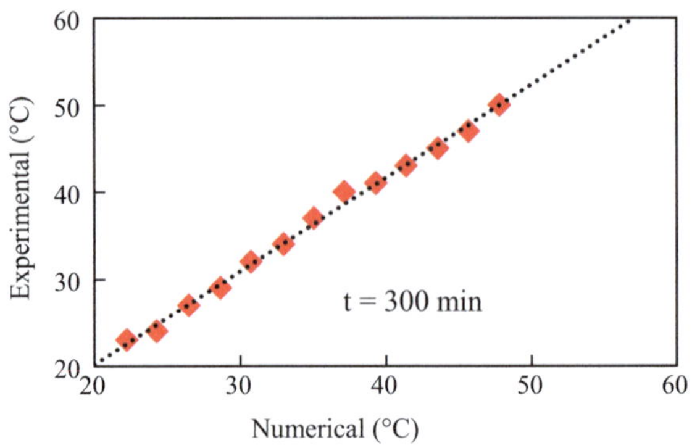

Fig. 6 Comparison of the numerical results with the experimental data

8 Numerical Results

In this work, we are interested to study the heat transfer, water temperature and water velocity distribution in the cylindrical water tank with immersed helical condenser coil. The numerical investigation was performed concurrently for variable pitch coil and normal coil to compare their thermal performance.

8.1 Velocity Field

Water velocity distribution in the storage tank for normal helically coiled tube heat exchanger and variable pitch coil during heating time equal to t = 300 min were demonstrated in Fig. 7. According to these results, it has been noted that the water velocity is on its maximum for variable pitch coil. From the results, variable pitch coil provides the highest velocity profiles in comparison to the normal coil design. Obviously, with reducing the coil pitch in the lower part of the water tank, the water velocity of lower part is larger than that of upper part, so water in the bottom tank is quickly heated indicating that the effect of helical coil design on thermal characteristics of domestic refrigerator for residential hot water production.

8.2 Water Temperature Distribution

Figure 8 shows the temperature distribution in the storage tank with immersed variable pitch coil and normal coil at t = 300 min. According to these results, a domestic refrigerator with variable pitch coil presents a temperature value superior to the normal helically coiled tube heat exchanger. Clearly, with reducing the coil pitch in the lower part of the storage tank, the temperature stratification in the tank is reduced which results in the mixing process between different temperature layers.

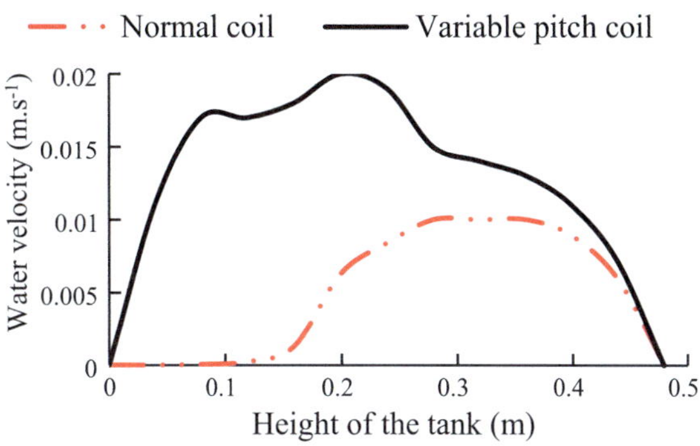

Fig. 7 Water velocity distribution in the storage tank under two coil structure at t = 300 min

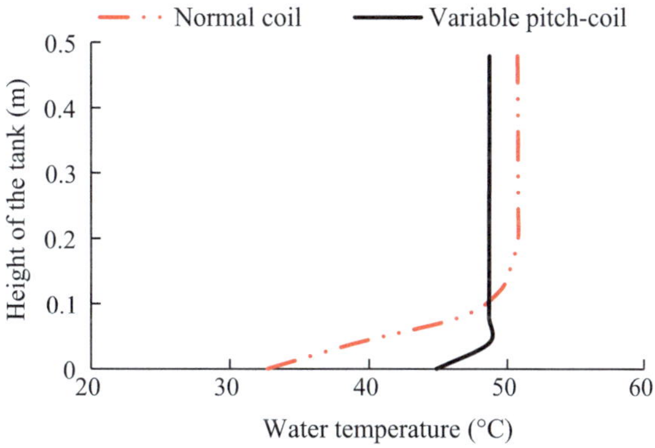

Fig. 8 Water temperature distribution

8.3 Overall Heat Transfer Coefficient

Figure 9 shows the variation of the heat transfer for normal and variable pitch coil at t = 300 min. According to these results, it is clear that the overall heat transfer coefficient decreases with increasing heating time. The distribution of the heat transfer coefficient shows the improvement of the heat transfer when using the helically coiled tube heat exchanger with variable pitch coil for domestic hot water production.

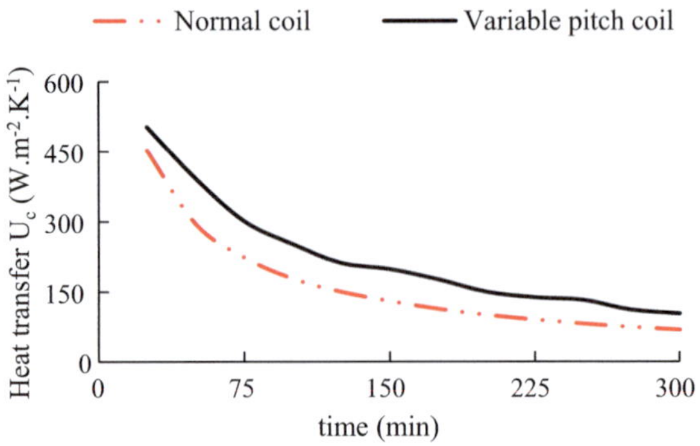

Fig. 9 Heat transfer coefficient of normal and variable pitch coil versus heating time

9 Conclusion

In this paper, a two-dimensional model of the cylindrical water tank with immersed helically coiled tube heat exchanger is established to analysis the influence of condenser coil configuration on the heating performance of the coupled refrigeration machines for hot water production. According to the numerical results, it has been confirmed that the water velocity, the water temperature and the heat transfer coefficient in the cylindrical water tank with immersed helically coiled tube is improved by appropriately optimizing the condenser coil designs.

Acknowledgements The authors are grateful to the members of the Laboratory of Energy, Water, Environment, and Processes (LR18ES35) for their financial support.

References

1. Mahdi MS, Mahood HB, Campbell AN, Khadom AA (2019) Experimental study on the melting behavior of a phase change material in a conical coil latent heat thermal energy storage unit. Appl Thermal Eng 175:114684
2. Missaoui S, Driss Z, Ben Slama R, Chaouachi B (2021) Experimentally validated model of a domestic refrigerator with an immersed condenser coil for water heating. Int J Air-Conditioning Refrig 29(3):2150022
3. Dai N, Li S (2018) Simulation and performance analysis on condenser coil in household heat pump water heater. Sustain Cities Soc 36:176–184
4. Missaoui S, Driss Z, Slama RB, Chaouachi B (2021) Numerical analysis of the heat pump water heater with immersed helically coiled tubes. J Energy Storage 39:102547
5. Dai N, Li S, Ye Q (2019) Performance analysis on the charging and discharging process of a household heat pump water heater. Int J Refrig 98:266–273
6. Missaoui S, Driss Z, Ben Slama R, Chaouachi B (2022) Experimental and numerical analysis of a helical coil heat exchanger for domestic refrigerator and water heating. Int J Refrig 133:276–288
7. Ye Q, Li S (2019) Investigation on the performance and optimization of heat pump water heater with wrap-around condenser coil. Int J Heat Mass Transf 143:118556

8. Missaoui S, Driss Z, Ben Slama R, Chaouachi B (2022) Effects of pipe turns on vertical helically coiled tube heat exchangers for water heating in a household refrigerator. Int J Air-Conditioning Refrig 30(1):1–13
9. Missaoui S, Ben Slama R, Chaouachi B (2022) Theoretical analysis on a household heat pump water heater with immersed condenser coil. Adv Model Thermodynamic Syst 236–252
10. Missaoui S, Driss Z, Ben Slama R, Chaouachi B (2023) Effects of helical condenser coil designs on the heating process of the domestic refrigerator for water production: a numerical study. Numer Heat Transf Part A Appl. https://doi.org/10.1080/10407782.2023.2186550
11. Missaoui S, Driss Z, Slama RB, Chaouachi B (2023) Numerical and experimental investigation on thermal characteristics of a helical condenser coil under different geometry design. Euro-Mediterr J Environ Integr 8:203–218
12. Missaoui S, Driss Z, Ben Slama R, Chaouachi B (2022) Experimental and numerical investigation of the heat transfer in helically coiled tubes heat exchanger and design optimization. In: The 10th international workshop on simulation for energy, sustainable development and environment, January 2022

Cost Optimization Based on a Linear Programming Approach for a Hybrid Renewable Energy System

Ramia Ouederni, Bechir Bouaziz, and Faouzi Bacha

Abstract The objectives of the studies presented here are to develop an efficient method adapted to the energy production of an isolated hybrid power system located on the island of Djerba, Tunisia. To this end, we have proposed two optimization methods, based respectively on linear programming (LP) and software (HOMER), with the aim of minimizing the energy cost of the hybrid system while respecting constraints linked to all the elements of the system. We therefore compared the two optimization strategies. The proposed linear program yields an electricity cost of 0.206$/kWh, lower than the HOMER cost of 0.308$/kWh. This system would therefore have advantages in terms of energy independence in particular.

Keywords Hybrid renewable energy systems · Optimal cost · MATLAB linear programming

1 Introduction

The ever-increasing demand for energy, and the consequences this may have in the medium term, has led to a growing debate about the planet's energy future. In fact, the national energy balance reveals the preponderance of the building sector in the country's energy consumption. Under these conditions, the ability of fossil fuels to meet energy needs can only be assured for a few decades yet. Fossil fuels may offer an alternative in the short/medium term, but they also raise real ecological problems.

R. Ouederni · B. Bouaziz (✉) · F. Bacha
Computer Lab for Industrial Systems, National Institute of Applied Sciences and Technology, Centre Urbain Nord, BP 676, 1080 Tunis Cedex, Tunisia
e-mail: bechir.bouaziz@isimg.tn

R. Ouederni
e-mail: Ramia.ouederni@isimg.tn

F. Bacha
e-mail: faouzi.bacha@esstt.rnu.tn

© The Author(s), under exclusive license to Springer Nature Singapore Pte Ltd. 2024
A. Mellit et al. (eds.), *Proceedings of the 2nd International Conference on Green Energy Conversion System*, Green Energy and Technology,
https://doi.org/10.1007/978-981-97-6148-7_13

There are 21 uninhabited islands in Tunisia. They are far from each other. Lack of electricity and the high cost of traditional energy production are the main obstacles to setting up a network in isolated areas. These islands must be able to count on a constant and regular supply of electricity from renewable sources, notably photovoltaic (PV) and wind power (WT). Developing these sources could generate significant social and economic benefits for the islands. According to the National Institute of Meteorologic, Tunisia has the largest solar deposits in the world. Average annual sunshine exceeds 3000 h, the highest in the Mediterranean basin at 1800/m^2/year. The wind potential is also significant, with annual wind speeds of 6 m/s [1]. However, their random and irregular nature means that they cannot guarantee a continuous supply.

Consequently, the use of hydroelectric storage and a battery system linked to a photovoltaic/wind power system is an attractive solution for providing reliable, low-cost electricity to consumers on isolated islands. However, this solution relies on the optimized design of the combination and sizing of the renewable energy generation systems coupled with the storage systems. The research community has recently developed various calculation and optimization methods to assess the profitability and feasibility of these solutions [2, 3]. Some research [4, 5] has developed different methodologies to determine the optimal size of renewable energy generation systems combined with different storage systems.

The optimization problem can be solved using two methods: the deterministic method and the stochastic method. In the deterministic method, all the variables and parameters of the microgrid are known precisely. This method is impractical for optimizing microgrids, as there will be unpredictable variations, i.e., variation in solar radiation or load. Research work [6] exploits a large dataset to deal with the uncertain nature of the deterministic method. The problems associated with this deterministic approach are generally solved using mixed integer linear programming [7], also using hybrid optimization of multiple energy resources (HOMER) software [8].

In this article, the optimization of the energy cost of renewable hybrid systems is considered using linear programming implemented in MATLAB and Homer software.

Both methods can be applied to different locations and energy scenarios. To illustrate our point, the example of a wind/PV/tidal/hydraulic/battery hybrid energy system has been provided.

In the first part of this article, we describe a general description of the proposed hybrid renewable energy system. In the second part, we show the method of optimization by linear programming. Finally, the results of the simulation are presented by the two methods of linear programming and the result obtained by the Homer software.

2 Homer Software Presentation

HOMER is a powerful tool for HRES system planning and design, enabling ideal component dimensions to be defined using of a techno-economic analysis. Numerous resources such as wind turbines, photovoltaic generators, fuel cells, small hydroelectric power plants, biomass, converters, batteries, and conventional generators can be integrated into HOMER. HOMER software also supports isolated hybrid renewable energy system (HRES).

Figure 1 illustrates the typical structure of the proposed HRES system designed in HOMER, comprising photovoltaic panels, wind turbines, tidal turbines, and hydraulic generators.

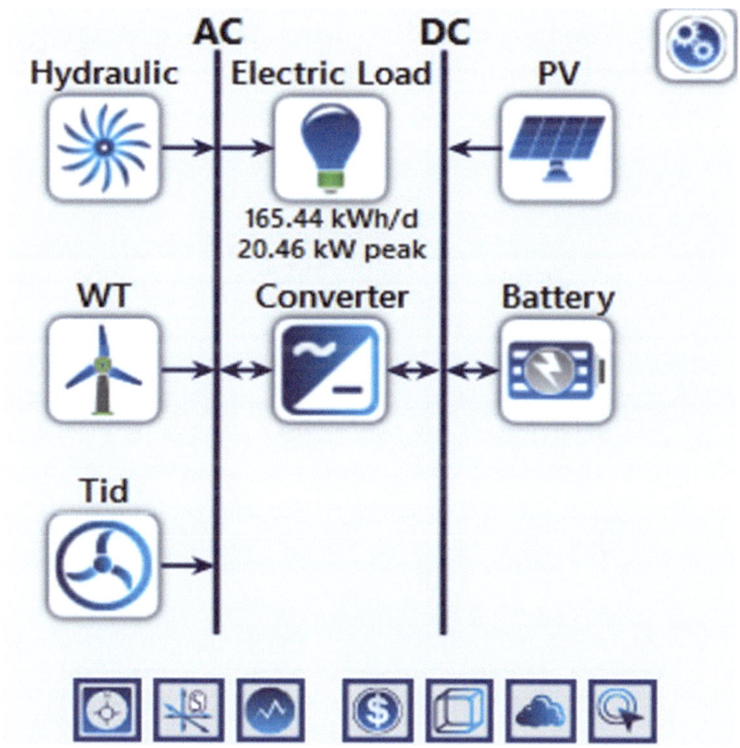

Fig. 1 Typical homer hybrid renewable energy system

2.1 System Overview

The proposed system is a renewable energy-based solution for the island of Djerba in Tunisia, shown in Fig. 2. It is composed of a photovoltaic unit, wind turbines, tidal turbines, hydroelectric as renewable generation sources. Table 1 depicts the data of the selected island. Table 2 lists the parameters for the different costs of each component of the configuration of the proposed HRES. By using the HOMER software, a summary of the main costs configurations of such as, the price of capital, replacement, operation and maintenance costs (O&M) of energy for the different sources, are provided. The meteorological data and the load of the studied island for the year 2020 are presented and collected by the National Aeronautics and Space Administration (NASA) power.

Table 3 shows the meteorological data required for each of the sources used in our system. For the proposed communities, the monthly load corresponding to residential demand is calculated. Consumption is divided into two periods: winter and summer.

Fig. 2 Location of the selected site

Table 1 Data of the selected site

Site	Latitude (°)	Longitude (°)	Altitude above sea level
Island of Djerba	33.83	11	147 km^2

Table 2 Different costs of each component of the hybrid renewable energy system

System component	Capital cost ($)	Replacement cost ($)	O&M cost ($/year)
PV System: (Trina Solar 330TSM)	25,000	25,000	10
Wind Turbine: (Generic 10 kW)	18,000	18,000	500
Tidal Turbine: (Generic Hydrokinetic 40 kW)	14,000	14,000	2700
Hydraulic: (5 kW Generic)	40,000	20,000	1200

Table 3 Monthly meteorological data for each source of the hybrid electric system

Months	Irradiation (kWh/m^2/day)	Wind speed (m/s)	Tidal speed (m/s)	Water flow (l/min)
Jan	3.02	6.5	1	2.3
Feb	3.98	6.34	0.5	4.6
Mar	5.1	6.02	0.3	5
Apr	6.27	6.01	0.5	4.5
May	6.88	5.88	0.5	3.7
Jun	7.43	5.61	0.4	3
Jul	7.62	5	0.6	3.2
Aug	6.96	4.83	0.7	2.8
Sep	5.54	5.22	0.9	2.5
Oct	4.16	5.18	1	3
Nov	3.16	6.02	1.2	4.7
Dec	2.69	6.67	1.4	5

In summer, this load is high due to weather conditions. In addition, winter loads are lower. Figure 3 shows the HOMER profiles for daily and seasonal profiles, with an average power of 6.9 kW.

3 Energy System Model

The proposed HRES configuration is an aggregation of the four renewable energy sources. The developed model of each power plant is studied and detailed in [8]. The output of a photovoltaic power plant is expressed by Eq. (1). Where η_{pv} is the generator efficiency, A_{pv} is the total area of the photovoltaic generator (m^2), and G_t is the total irradiance (kWh/m^2).

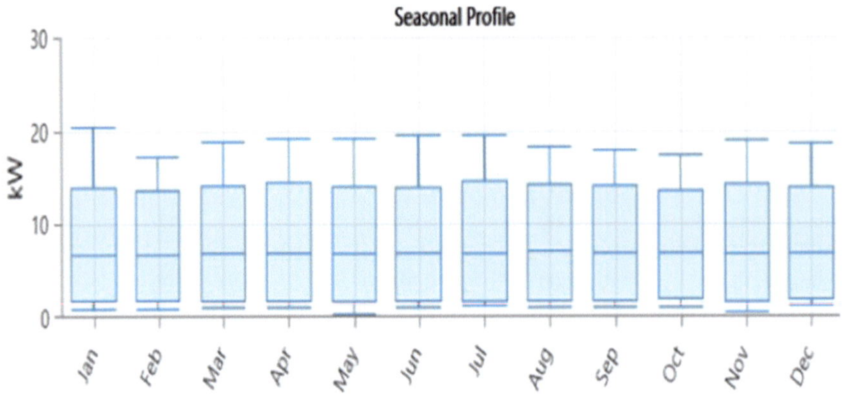

Fig. 3 Monthly load profile

$$P_{pv} = \eta_{pv} A_{pv} G_t \tag{1}$$

The wind generator can recover only some of that wind power, representing the power generated given by Eq. (2) and well detailed in [8]. Where: S_{wt} is the area traversed by the wind (m^2), ρ_{wt} is the air density (1.225 kg/m^3) and v_{wt} is the wind speed (m/s). C_{pw} is the efficiency factor.

$$P_{wt} = \frac{1}{2} \rho_{wt} S_{wt} C_{pw} v_{wt}^3 \tag{2}$$

Most of the operating principles of the tidal turbines are based upon wind turbines, as they work similarly. The power available can be calculated using Eq. (3) can detailed in [8]. Where: S_t is the turbine area (m^2), ρ_t is the density of water (1000 kg/m^3), v_t is the velocity of water (m/s) and C_{pt} is the power coefficient.

$$P_t = \frac{1}{2} \rho_t S_t C_{pt} v_t^3 \tag{3}$$

The generic 5 kW hydraulic pump component recently included in the HOMER component library has been used to model Eq. (4) in the HRES configuration. Where: P_h is power capacity, ρ_h is water density (1000 kg/m^3), g is the due acceleration at gravity (9.8 m/s^2), Q is the tip rate, and H is the effective gauge height through the turbine (m).

$$P_h = \rho g H Q \tag{4}$$

4 An Algorithm for Formulating the Optimization Problem

In this section, we will describe the different elements that allow us to solve the cost-minimization problem of the studied hybrid system. In the present work, we are interested in cost minimization through a linear algorithm that takes into consideration factors such as investment, maintenance, and energy costs, whose lifetimes are shown in Table 2. Its objective function is given by the following formula:

$$Min\, C_{Total} = \sum_{t=1}^{T} (C_{hs}(t)) \tag{5}$$

With C_{hs} from Eq. (5) is as follows:

$$C_{hs}(t) = \sum_{t=1}^{T} \left(c_{pv} C_{upv} P_{pv} + c_{wt} C_{uwt} P_{wt} + c_t C_{ut} P_t + c_h C_{uh} P_h\right) \tag{6}$$

The current unit costs of power generation being, respectively, C_{upv}, C_{uwt}, C_{ut}, and C_{uh}, for solar, wind turbine, tidal, and hydro power plants; the costs of electrical production of the hybrid system during the period T of the system for which it is intended, c_{pv}, c_{wt}, c_t, and c_h correspond to the decision variables for the choice of the installation.

4.1 Constraints

Equality and inequality constraints are used in optimized hybrid system planning in different contexts. For this problem, we are interested in the constraints of two orders, presented below.

In addition, the energy production constraint by the hybrid system must meet the demand at any time in the following way:

$$\sum_{t=1}^{T} P_{hs}(t) C_{hs} \geq P_d(t) \tag{7}$$

The total power generated must be within the range of the minimum and maximum powers that can be generated:

$$P_{min} \leq \sum_{t=1}^{T} P_{hs}(t) C_{hs} \leq P_{max} \tag{8}$$

With P_{hs}, which corresponds to the combination of the powers of each source used in our system, and C_{hs} is the total value of the costs of each source used in the system, P_d is the demand power.

4.2 Hybrid System Limits

The variable limits of each source in the hybrid system are shown in Eq. (9).

$$\begin{cases} 0 < c_{pv} \le N_{pv} \\ 0 < c_{wt} \le N_{wt} \\ 0 < c_t \le N_t \\ 0 < c_h \le N_h \end{cases} \tag{9}$$

4.3 Proposed Solution

Assuming the irradiance G, v_{wt} is the wind speed, v_t is the tidal velocity, and the water flow is Q where all are constant during the period T, the problem was viewed as a linear optimization function subjected to linear inequality constraint. The above-proposed scheduling of the hybrid system required all the variable limits given in the equation. That is shown by the upper and lower limits of various components of the hybrid system, this is necessary for the optimal scheduling of the hybrid system. This linear optimization problem is mainly solved by the function "linprog" defined in MATLAB software and formulated by Eq. (10).

$$\left[x_{fval}\right] = linprog\left(f, A, b, A_{eq}, b_{eq}, l_b, u_b\right) \tag{10}$$

where the optimized value is kept in the variables f_{val} and the other variable are defined as follows: f is an objective function, A is a matrix k by n, b is the vector of length k, A_{eq} is a matrix that summarizes all the equality constraints, b_{eq} is the vector of length m, u_b is a matrix of upper bound applied to the variables, and l_b is a matrix of lower bound applied to the variables.

5 Result and Discussion

The simulation results were obtained both from the proposed algorithm (implemented with MATLAB) and from the implementation performed with Homer software. In reality, the unit cost of electricity estimated by the linear algorithm was 0.17$/kWh,

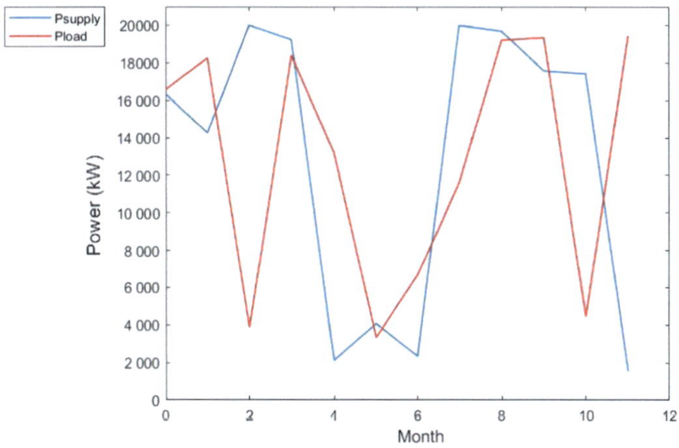

Fig. 4 Results of LP approach of power supply demand compared to power load demand

0.002$/kWh, and 0.0075$/kWh, 0.000383$/kWh for solar, wind, hydro, and tidal power, respectively. Thereafter, the other three sources are added to offset the load in the event that the contribution from the tidal turbine is not sufficient to meet the demand.

Figure 4, on the other hand, shows the energy demand versus the energy supply. The energy supplied in blue is almost always above the demand which is in red with a few exceptions. This explains the reliability of the proposed solution. The few cases, months 1, 4, 6, and 11 precisely, where the required loads were not met, can be attributed to the unavailability of enough primary resources needed for energy generation.

Figure 5 on the cost of electricity shows the same trend. This cost appears minimal at the outset and then increases after a while, remaining at a high level from the second to the fifth month. The pattern is directly correlated with the previous graph on the contribution of individual sources.

Figure 6 shows the simulation results for our proposed system. They show that the cost obtained for the economic parameters used to evaluate the proposed system is low.

Figure 7 shows the energy produced from each source using Homer software. The maximum amount of energy is consumed in December. The least energy supplied is observed during February.

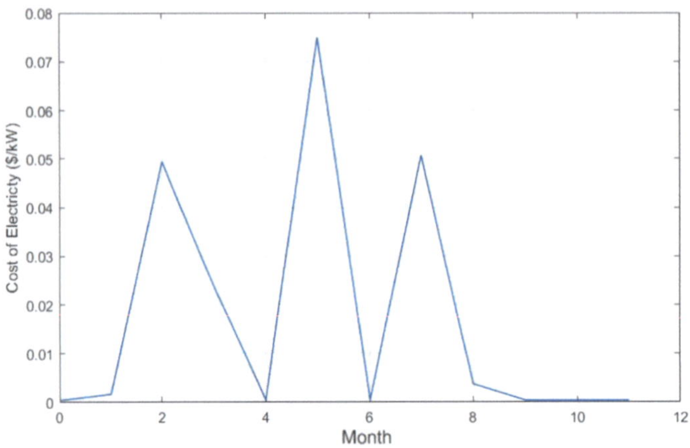

Fig. 5 Cost of electricity for hybrid renewable energy system

Cost			
NPC ($)	COE ($)	Operating cost ($/yr)	Initial capital ($)
$240,282	$0.308	$8,389	$131,838

Fig. 6 The economic parameters

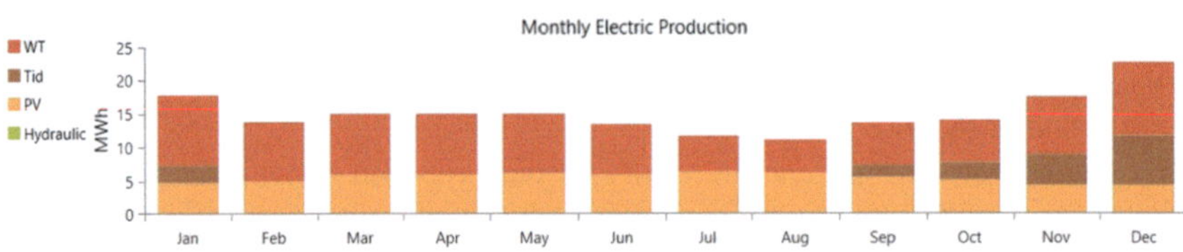

Fig. 7 Distribution of power supply by Homer software

6 Conclusion

This article briefly presents a comparative cost analysis of the optimization of a hybrid energy system including solar, wind, tidal, and hydroelectric power plants using a linear algorithm (LP) and the optimization software HOMER. These methods were presented and tested on the same data, and the results were compared. It emerged that the choice of different energy sources was considerably more efficient and dynamic with the linear algorithm than with the HOMER software. However, the linear program (LP) method reduced the cost of electricity (0.206$/kWh) compared to the HOMER method (0.308$/kWh), resulting in a faster payback period, which is an attractive incentive for investors. Finally, these two methods were compared in terms of their ability to meet load requirements. Hybrid systems are very useful for reducing the cost of electricity while keeping profit at an acceptable level.

References

1. National Institute of Meteorology-Tunisia. Available online: http://www.meteo.tn. Accessed on 22 June 2022
2. Minayégnan T, Oladapo O, Sako K, Marvillet C, Marvillet C (2021) Analysis of the technico-economic viability of an electric micro-grid with renewable electricity production sources in Elokato-Bingerville, Cote d'Ivoire. Smart Grid Renew Energy 12:183–202
3. Behabtu HA, Messagie M, Coosemans T, Berecibar M, Anlay Fante K, Kebede AA, Mierlo JV (2020) A review of energy storage technologies, application potentials in renewable energy sources grid integration. Sustainability 12:10511
4. Charabi Y, Abdul-Wahab S (2020) The optimal sizing and performance assessment of a hybrid renewable energy system for a mini-gird in an exclave territory. AIMS Energy 8:669–685
5. Diab AAZ, Sultan HM, Kuznetsov ON (2020) Optimal sizing of hybrid solar/wind/hydroelectric pumped storage energy system in Egypt based on different meta-heuristic techniques. Environ Sci Pollut Res 27:32318–32340
6. Li H, Wang X, Li F, Wang Y, Yu X (2018) A robust day-ahead electricity market clearing model considering wind power penetration. Energies 11:1772
7. Garip S, Ozdemir S (2022) Optimization of PV and battery energy storage size in grid-connected microgrid. Appl Sci 12:8247
8. Ouederni R, Bouaziz B, Bacha F (2022) Modeling and cost optimization of an islanded virtual power plant: case study of Tunisia. Turk J Electr Power Energy Syst 2(2):168–179

Power Comparative Analysis of Floating and Conventional Solar Photovoltaic System

Chayma Saad, Mohamed Naoui, and Aymen Flah

Abstract Photovoltaic (PV) solar power systems are gaining ground worldwide and represent a promising alternative to renewable energy generation. However, many studies have highlighted some disadvantages associated with the installation and operation of conventional solar power plants. Therefore, floating photovoltaic systems (FPV) have emerged as a new solar energy concept to reduce the negative impact on the environment caused by traditional photovoltaic system configurations. The FPV market has grown at an impressive rate over the past decade, with global installed capacity doubling year-on-year. This growth has been made possible due to the numerous advantages that FPV systems have over conventional systems, primarily related to land use and energy efficiency. The purpose of the study was to compare the FPV system with the conventional PV system (CPV). This research uses the MATLAB Simulink platform to simulate the entire block, and adapts the common photovoltaic power generation model to the floating solar photovoltaic power generation.

Keywords Floating photovoltaic system (FPV) · Conventional photovoltaic system (CPV) · Comparative analysis

---

C. Saad (✉) · M. Naoui
Processes, Energy, Environment and Electrical Systems, National Engineering School of Gabes, University of Gabes, Gabes, Tunisia
e-mail: Saadchaima96@gmail.com

M. Naoui
e-mail: Mednaouing@yahoo.com

A. Flah
The Private Higher School of Applied Sciences and Technologies of Gabes, University of Gabes, Gabes, Tunisia
e-mail: Flahaymening@gmail.com

© The Author(s), under exclusive license to Springer Nature Singapore Pte Ltd. 2024
A. Mellit et al. (eds.), *Proceedings of the 2nd International Conference on Green Energy Conversion System*, Green Energy and Technology,
https://doi.org/10.1007/978-981-97-6148-7_14

1 Introduction

Recently, photovoltaic (PV), which rely on solar panels to generate electricity, have become one of the most popular forms of electricity generation in the world [1]. However, multi-megawatt power generation requires vast tracts of land, limiting the use the land for other objectives, this makes it difficult for countries with limited lands to install more solar panels. To cope with this difficult situation, floating photovoltaic plants has been developed to overcome these challenges [2].

Floating solar, also known as floating photovoltaics (FPV) or floating photovoltaics is an environmentally friendly way of producing electricity that merges marine and renewable energy technologies. The first patents for this type of technology were registered in 2007 [3]. Since then, floating solar systems have been mainly installed in countries such as China, Japan and the United Kingdom, and are rapidly gaining popularity in the United States, especially in California and New Jersey. Because of the abundance of water spaces, FPV's installed capacity in recent years, is reaching 2.6 GW worldwide until mid-2020 [4]. This means FPV could play a key role in the transition to renewable energy in several countries.

Over the past several years, there have been numerous studies conducted on FPV technology and its effectiveness. Some of these studies have specifically delved into the design and operation of certain FPV systems, while others have focused on the feasibility of their implementation [5, 6]. Additionally, research has been conducted to evaluate the environmental benefits of the FPV [7]. Furthermore, there have been studies examining the electrical performance of FPV systems and comparing them to traditional PV systems [8, 9]. Given that FPV systems represents a cutting-edge area of research in the field of clean, renewable energy, it is crucial to conduct further research in order to ensure that countries remain up-to-date with the latest developments in this field.

2 Floating Photovoltaic System (FPVs)

The floating photovoltaic (FPV) is any type of solar cell array that floats on a body of water. The solar panels must be attached to a floating structure that keeps them above the water. Electricity produced is transmitted through underwater cables to a land-based power substation. Floating photovoltaic arrays has many similarities to traditional photovoltaic arrays, but there are also some differences, especially in terms of anchoring, floating systems, and energy dissipation of the array [10, 11].

There are a multitude of benefits that can be obtained by installing the floating photovoltaic system. These FPV are built on water and this assists in saving valuable land area for purposes like farming. The performance of floating photovoltaic modules is superior to that of shore-based systems, because in terrestrial systems, the high temperatures reduce both the performance and the lifespan of solar photovoltaic modules [7]. The implementation of floating structures on bodies of water has

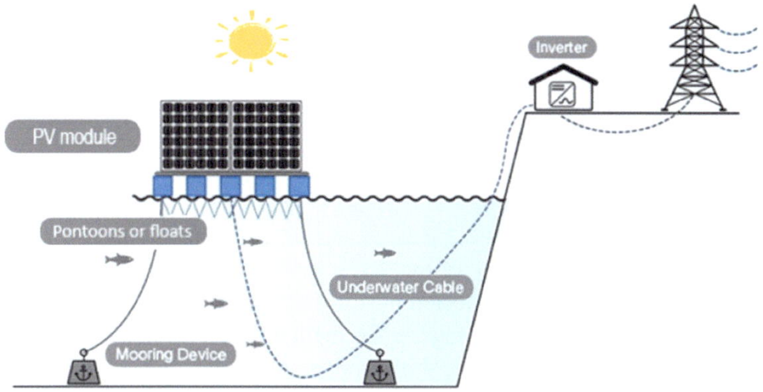

Fig. 1 Components of the FSPV system

proven to be an effective measure against water evaporation [12]. The water surface is relatively open, which can effectively prevent shade from trees, mountains, etc.

The FSPV comprises several key components, and depending on the design and technology used, the specific components may vary. However, basic parts typically include modules (PV), inverters, underwater cables, pontoons or floats and mooring and anchoring devices to position the entire system in a specific location (see Fig. 1).

Photovoltaic (PV) Module

A module consists of photovoltaic cells that produce direct current when they perceive light. The type of module used for FPV installations are practically the same module used in terrestrial PV as the crystalline type module, with a lower water permeability that protects the panels from internal corrosion [13].

Inverters

Inverters are used to convert the DC power generated by a photovoltaic system into AC form that can be used by an AC-based systems. In an FPV system, the location of the inverter can be determined based on the distance of the system from shore or the conditions of the sea area where the FPV is deployed.

Floating Structure: Pontoons or Floats

The floating system consists of a floating structure on which the PV system is installed to stay afloat. Floats are generally made of high-density polyethylene (HDPE) or medium density polyethylene (MDEP) [13].

Mooring Device

Since FPV systems are installed on bodies of water, fluctuations in water levels caused by monsoons, wind speed, or increases/decreases in water volume can cause problems for plants. To avoid this, the FPV system is anchored by a mooring system.

Under Water Cables

Electricity generated by photovoltaic systems is usually transmitted to substations via underwater cables. These cables can be extended underwater or above floaters.

3 Modelling of the PV Panel

Photovoltaic arrays are composed of photovoltaic cell modules connected in series or in parallel. Panels are also connected in series and parallel to ensure the required voltage, current and power. To represent the mathematical model of the PV array, the PV cell is modeled using the one diode model, as shown in Fig. 2.

For a PV array having number of parallel strings (N_p) and each string has number of series panels (N_s), the current and the voltage equations is presented in Eqs. (1) and (2):

$$I = I_{ph}N_p - I_sN_p\left(\exp\left(\frac{\left(V + I.R_s\frac{N_s}{N_p}\right)}{nN_s\frac{KTn}{q}}\right)\right) - \frac{\left(V + I.R_s\frac{N_s}{N_p}\right)}{R_{sh}\frac{N_s}{N_p}} \tag{1}$$

$$V = \frac{N_s.(K.t.n)}{q}ln\left(\frac{\left(N_p.I_{ph} - I\right) + N_p.I_s}{N_p.I_s}\right) - \frac{N_s}{N_p}.I.R_s \tag{2}$$

Furthermore, the mathematical equation for calculating power generation from the PV panel is given by Eq. (3):

$$P_{PV} = P_{STC} \times \left(1 + k_p(T - 25)\right) \times F_g \times F_{dirt} \times F_{cable} \times F_{inv} \times F_{mm} \tag{3}$$

where $P_{arraySTC}$ is the total array power (kWp); k_p is the temperature coefficient; F_g is the irradiation derating factor; F_{dirt} is the dirt derating factor; F_{cable} is the losses from cable; F_{inv} is the inverter derate factor and F_{mm} is the module mismatch derating factor.

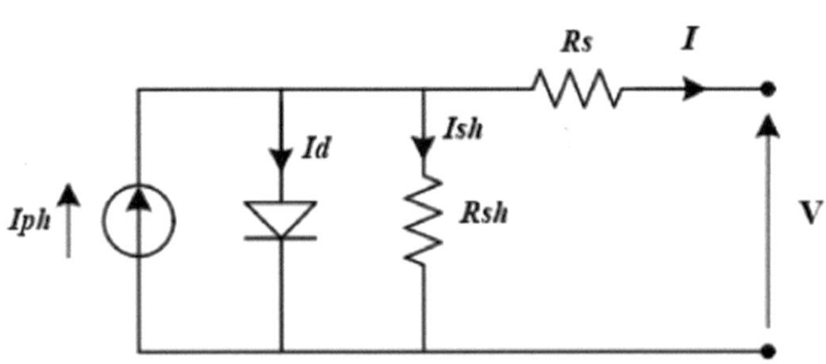

Fig. 2 One diode model of a photovoltaic cell

Fig. 3 Satellite view of suggested locations. *Source* Google

4 Results and Discussion

In this section, a power comparative analysis between the FPV and the CPV systems is presented. Two solar panels with identical characteristics are installed on site e.g., Gabes in Tunisia. One installed on a floating structure in the sea and the other on land (Fig. 3). MATLAB/Simulink was used to model and simulate the two systems.

Subsequently, a time interval was chosen (at midday) to see the difference between the temperature of CPV and FPV.

In the graph (Fig. 4), we can clearly see that the temperature outside the FPV system is lower than that of a land solar PV system. The reduction in temperature helps improve the efficiency of solar photovoltaic cells.

The simulations made make it possible to obtain the I–V and P–V curves of the PV module for different irradiation and temperatures (Fig. 5).

The comparison results of the two types of PV are presented in Fig. 6 at 13 h:

Figure 6 shows the power graphs generated by both types of photovoltaic installations. It can be seen that FPV has a higher total energy gain than traditional PVs where in approximately 1.4 s; the value in the FPV system corresponds to 167.7 W, while in the other system, it is 149.6 W. This difference is due to the lower temperature

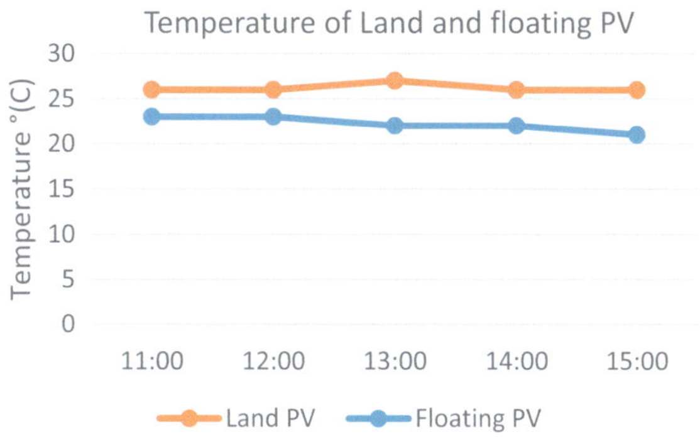

Fig. 4 Temperature differences in the two types of PV

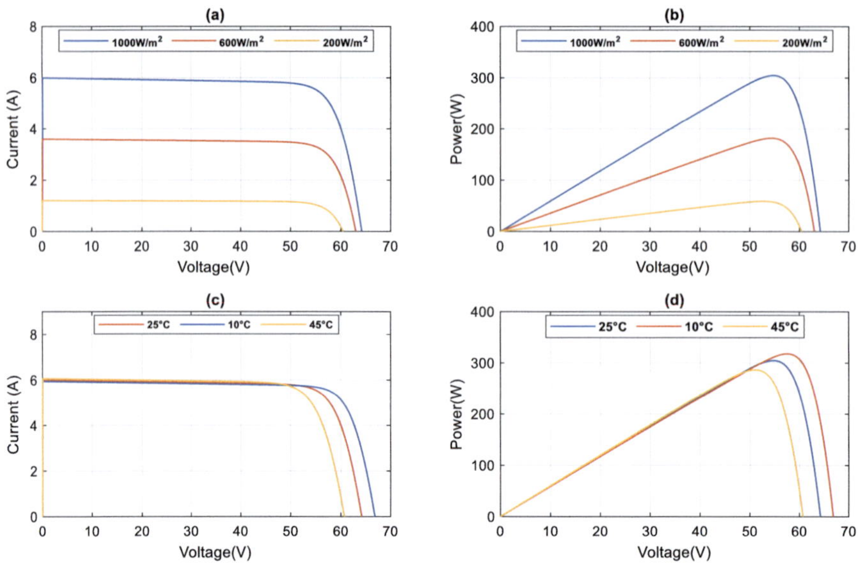

Fig. 5 **a** I-V characteristic for T = 25 °C, **b** P–V characteristic for T = 25 °C, **c** I–V characteristic for G = 1000 W/m^2, **d** P–V characteristic for G = 1000 W/m^2

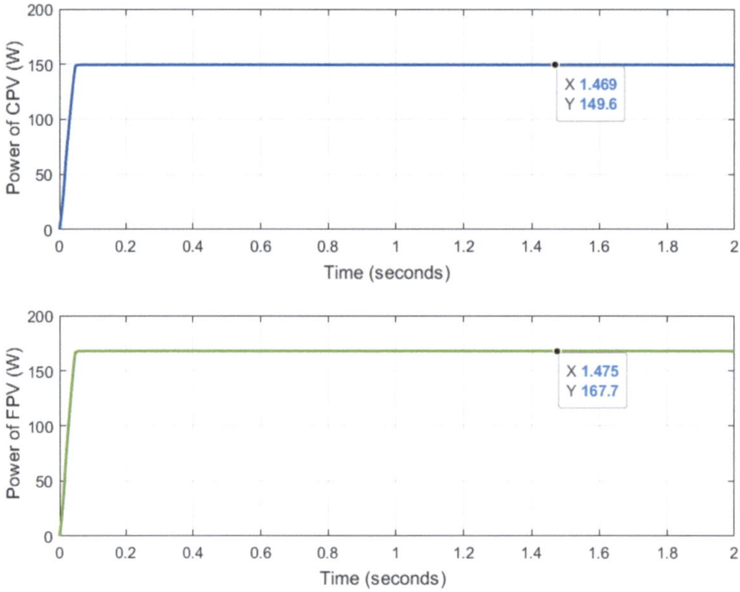

Fig. 6 Output power of conventional PV and floating PV system

profile, lesser shadow area and better cooling possibility on FPV system compared to conventional system.

5 Conclusion

An FPV system is a solar photovoltaic application in which photovoltaic modules are designed and installed to float on a body of water. Today, FPV systems represent a new way of utilizing solar energy systems that overcome the limitations of CPV systems such as the need for land.

This study describes and compares both of conventional PV and floating PV systems in the MATLAB-Simulink environment. The simulation results show that the production capacity of the FPV system is higher than that of the CPV system due to the natural cooling effects of water, which reduce thermal stress on the panels and minimize efficiency losses. Moreover, floating systems alleviate land use problems, making them a viable option in areas where available land is limited.

References

1. Choi SM, Kwon HG, Kim T, Moon HK, Cho HH (2022) Active cooling of photovoltaic (PV) cell by acoustic excitation in single-dimpled internal channel. Appl Energy 309:118466
2. Ateş AM (2022) Unlocking the floating photovoltaic potential of Türkiye's hydroelectric power plants. Renewable Energy 199:1495–1509
3. Wikipedia. https://en.wikipedia.org/wiki/Floating_solar. Last Accessed 21 May 2023
4. Micheli L, Talavera DL, Tina GM, Almonacid F, Fernández EF (2022) Techno-economic potential and perspectives of floating photovoltaics in Europe. Sol Energy 243:203–214
5. Ghigo A, Faraggiana E, Sirigu M, Mattiazzo G, Bracco G (2022) Design and analysis of a floating photovoltaic system for offshore installation: the case study of Lampedusa. Energies 15(23):8804
6. Refaai MRA, Dhanesh L, Ganthia BP, Mohanty M, Subbiah R, Anbese EM (2022) Design and Implementation of a floating PV model to analyse the power generation. Int J Photoenergy
7. Pouran HM, Lopes MPC, Nogueira T, Branco DAC, Sheng Y (2022) Environmental and technical impacts of floating photovoltaic plants as an emerging clean energy technology. Iscience 25(11)
8. Semeskandeh S, Hojjat M, Hosseini Abardeh M (2022) Techno–economic–environmental comparison of floating photovoltaic plant with conventional solar photovoltaic plant in northern Iran. Clean Energy 6(2):353–361
9. Makhija SP, Dubey SP, Bansal RC, Jena PK (2021) Techno-environ-economical analysis of floating PV/on-ground PV/grid extension systems for electrification of a remote area in India. Technol Econ Smart Grids Sustain Energy 6:1–10
10. Kumar M, Niyaz HM, Gupta R (2021) Challenges and opportunities towards the development of floating photovoltaic systems. Sol Energy Mater Sol Cells 233:111408
11. Miah MAR, Rahman SR, Kabir R (2021) Techno-economic analysis of floating solar PV integrating with hydropower plant in Bangladesh. In: 2021 IEEE green technologies conference (GreenTech). IEEE, pp 30–36
12. Vourdoubas J (2023) Assessment of the potential of installing floating photovoltaic systems in existing water reservoirs in Greece. Euro J Energy Res 3(1):17–23
13. Essak L, Ghosh A (2022) Floating photovoltaics: a review. Clean Technol 4(3):752–769

A Comparative Analysis of PID and SMC Control Topology for the Two Phases Interleaved DC DC Converter

Gaied Hajer, Naoui Mohamed, and Flah Aymen

Abstract The topology of a bidirectional DC-DC converter is typically crucial in systems that require the bidirectional exchange of electricity. It is frequently used in composite energy storage and microgrids. Electric cars, systems, etc. The architecture is interleaved to increase power handling capability and decrease input side ripple current. By allowing current sharing between several phases, interleaving improves the converter's power density by allowing for a smaller filter inductor and input and output capacitors. An N-phase non-isolated bidirectional DC-DC converter's design and generalized phase shift specifications are presented in this publication. In MATLAB/Simulink, a two-phase non-isolated interleaved bidirectional DC-DC converter's closed-loop simulation is performed while taking into account both the boost and buck modes of operation. The outcomes demonstrate equitable current sharing between phases, which lowers input current ripple, reduces the size of filter elements, and enhances converter thermal distribution. When integrating with non-isolated DC-DC converters in power systems, control strategies like proportional integral derivative (PID) and slide mode control (SMC) are also examined in terms of the parameters settling issue, response time, and complexity.

Keywords Non-isolated converter · Interleaved bidirectional buck-boost · Control techniques

1 Introduction

Interlaced DC-DC converters are commonly used in applications requiring efficient power conversion, such as switching power systems, solar converters, high-power telecommunications power supplies, as well as electric propulsion applications in hybrid and electric vehicles.

G. Hajer (✉) · N. Mohamed · F. Aymen
Processes, Energy, Environment and Electrical Systems (Code: LR18ES34), National Engineering School of Gabes, University of Gabes, Gabes, Tunisia
e-mail: hajergaied.enig@gmail.com

© The Author(s), under exclusive license to Springer Nature Singapore Pte Ltd. 2024

A. Mellit et al. (eds.), *Proceedings of the 2nd International Conference on Green Energy Conversion System*, Green Energy and Technology,
https://doi.org/10.1007/978-981-97-6148-7_15

The state of the art on interlaced DC-DC converters is constantly evolving, with active research aimed at improving the performance, efficiency and reliability of these converters. Recent advances are: Advanced topologies, Control and modulation, thermal management, component integration and Design optimization. These advances in interlaced DC-DC converters aim to improve their energy efficiency power density, reliability and electromagnetic compatibility, thus opening up new application opportunities in the fields of renewable energy, electric vehicles, telecommunications and more.

In this study, we combined a control strategy designed for the proposed topologies of the DC-DC interleaved converter and their interplay between two control techniques include proportional integral derivative (PID), slide mode control (SMC) [1, 2].

The use of several interlaced conversion stages also allows the power to be distributed between the different stages, thus reducing the heat load on each individual stage. This improves reliability and allows better management of heat dissipation.

2 Converter Description and Operation

2.1 Two-Phase Interleaved Bidirectional Converter Control

The elevating converter, which is popular because to its ease of use and great conversion range, is one of the converters most frequently employed in the process of charging/unloading a battery [3, 4]. By keeping the voltage levels of the battery constant and regulating the disturbances brought on by the connection and disconnection of loads and energy sources, this type of converter enables the adaptation of battery voltage levels to DC bus voltage level [5] (Fig. 1).

Each phase inductance value is given by Eq. 4:

$$L_1 = L_2 \cdots = L_n = \frac{v_{bat} D}{\Delta I_l F_{sw}} \tag{1}$$

DC link capacitance value is given by Eq. 5:

$$c_1 = \frac{D v_{dc}}{R_{dc} \Delta v_{dc} F_{sw}} \tag{2}$$

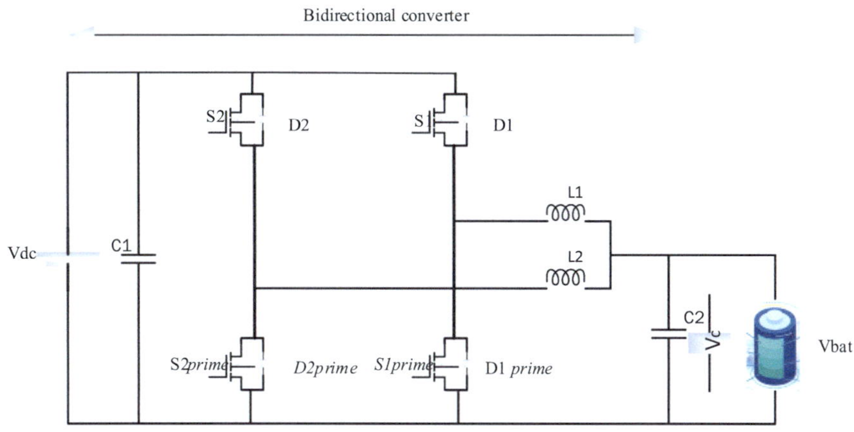

Fig. 1 A bidirectional interleaved DC DC converter- two phase

2.2 *Interleaved Bidirectional Converter Control and Design PID Control*

Controlling the DC-DC converter is one of the many applications for PID, which is a well-established and efficient control technology (see Fig. 2). The control and efficiency in renewable energy applications are being improved through the hybridization of PID with contemporary control techniques [6].

The PI corrector is defined by the following equation:

Or K_{pv} is the proportional gain and K_{iv} is the integration constant.

$$C_{gain} = K_{pv} + \frac{K_{iv}}{p} \tag{3}$$

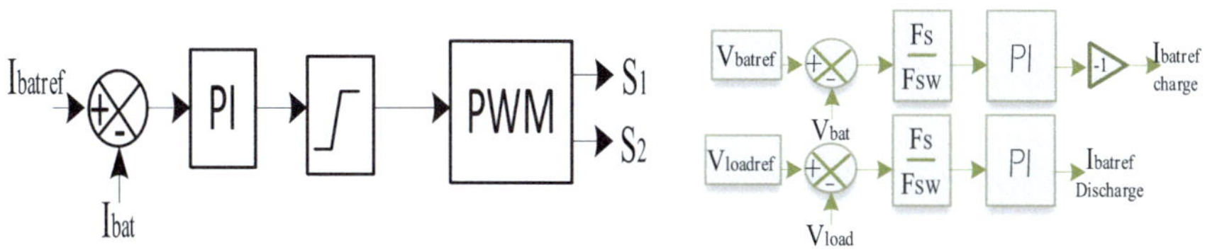

Fig. 2 A PID controller

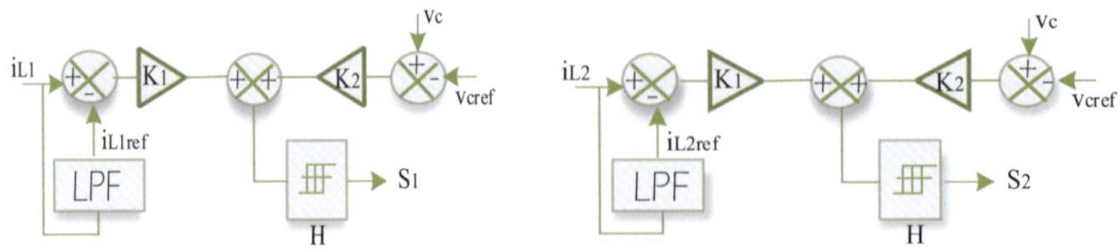

Fig. 3 A SMC controller

2.2.1 SMC Control

For optimal performance under a variety of load scenarios, the SMC control technique is combined with the DC-DC converter design in Fig. 3. The SMC receives the feedback and executes the iterations for switching the DC-DC converter [7, 8].

In this control scheme of converter, the sliding surface is chosen as a linear combination of both input inductor current I_l, and output capacitor voltage v_c of each buck boost DC DC converter as follows:

$$S(I_l, v_c) = k_1 e_1 + k_2 e_2 \tag{4}$$

where the coefficients K_1 and K_2 are proper gain, while e_1 and e_2 are the feedback errors of the state variabes I_l and v_c respectively has given by Eq:

$$e_1 = I_{lref} - I_l \tag{5}$$

$$e_2 = v_{cref} - v_c \tag{6}$$

where v_{cref} is the reference voltage of the output capacitor of each buck boost converter, while I_{lref} is the reference waveform of the input inductor current.

3 Comparative Performance Analysis

To validate the interlaced bidirectional converter functionality, as well as the comparison between the two control techniques analyzed in the previous sections, the measurements were performed using a battery-linked DC source.

The tests were configured with a fixed DC bus voltage $v_{Dc} = 48$ V and a battery voltage $v_{bat} = 24$ V.

The simulation is performed for the comparison between a PID control and an SMC control of an interleaved DC DC converter in two phases and three phases during the discharge/ load mode operation.

To simulate the converter during boost (discharge mode), low voltage side 24 V, 50 Ah the battery is considered to be powered by direct current 48 V bus voltage,

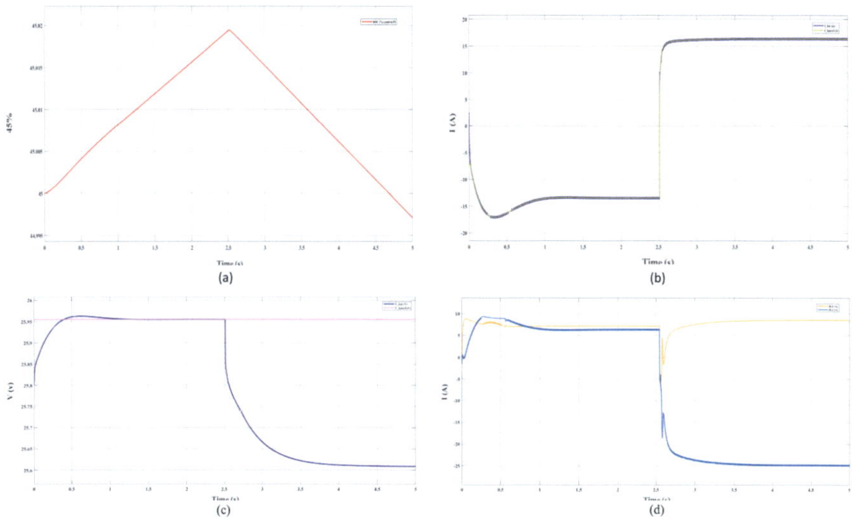

Fig. 4 Simulation results of two-phase interleaved converter during bock boost mode with control PID, **a** %SOC, **b** I batterie, **c** voltage of batterie, **d** inductor current

a resistive charge of 4.4 Ω is connected as part of this operation. Similarly, having the responsibility for operation, a 24 V battery, 50 Ah is connected to the DC bus replacement resistive charge.

It can be seen that the battery current with the PID control correctly follows the reference current. The voltage controller is capable of maintaining a constant output of 25 V during the transient charge of the battery and reducing the discharge as shown in Fig. 4b–d as well as the inductor current is the inverse of the battery current for the charging mode but for the unloading mode it is equal to i_{L1}. On the contrary for the SMC command, we can see the reverse battery current for each current of the inductor is to reduce the value without needing reference current control to beat and keep the voltage stable less losses and disturbances as shown in the Fig. 5b–d.

The PID control aims for precise regulation and rapid response, while the SMC control is designed to react quickly to disturbances and uncertainties under complex dynamic conditions.

In the SMC control for an interlaced DC-DC converter, the control signals of the conversion cells are modified to guide the system on the sliding surface. This allows rapid response to input voltage variations, disturbances or load variation.

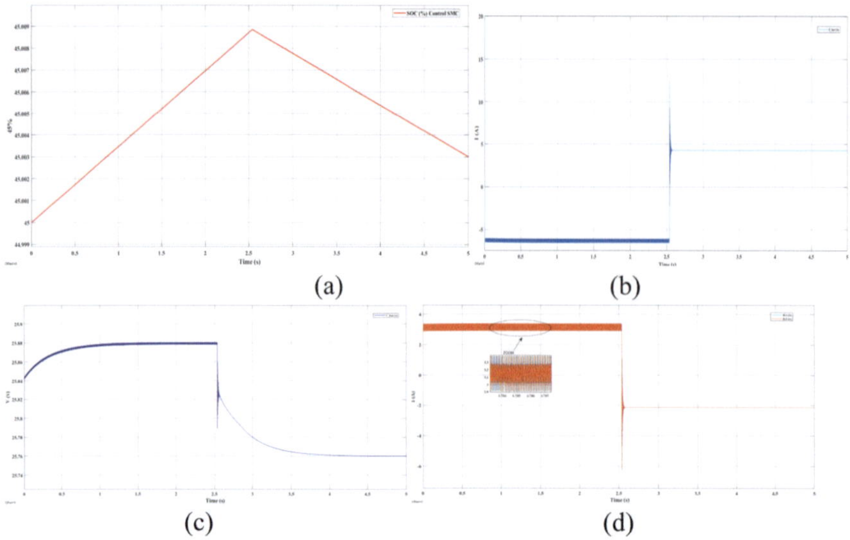

Fig. 5 Simulation results of two-phase interleaved converter during bock boost mode with control SMC **a** %SOC, **b** I batterie, **c** V batterie, **d** inductor current

4 Conclusion

This article described the modelling and control of a bidirectional DC-DC modulation converter. Two control topologies were discussed. The research also discusses the stability of the control methods for the two-phase non isolated DC-DC converter topologies.

The results of the simulation of SMC proposed for this type of converter shows the correct and reliable operation within its limits of the converter and its charge and discharge control of the battery without need of PID control for the battery current.

References

1. Liao H, Chen YT, Chen L, Chen JF (2022) Development of a bidirectional DC–DC converter with rapid energy bidirectional transition technology. Energies 15(13):4583
2. Mumtaz F, Yahaya NZ, Meraj ST, Singh B, Kannan R, Ibrahim O (2021) Review on non-isolated DC-DC converters and their control techniques for renewable energy applications. Ain Shams Eng J 12(4):3747–3763
3. Syah MN, Firmansyah E, Utomo DR (2022) Interleaved bidirectional DC-DC converter operation strategies and problem challenges: an overview. In: 2022 IEEE international conference in power engineering application (ICPEA). IEEE, pp 1–6
4. Rajesh P, Shajin FH, Kommula BN (2022) An efficient integration and control approach to increase the conversion efficiency of high-current low-voltage DC/DC converter. Energy Syst 13(4):939–958

5. Bastos MES, Nascimento SO, Torres VC, Almeida MS, Rendón MA, Rodrigues MC, Almeida PS, Oliveira JG (2022) Control strategy for an interleaved bidirectional DC–DC converter applied to battery management in a hybrid aircraft propulsion system. J Control Autom Electr Syst 33(3):965–973
6. Choudhury S, Khandelwal N (2023) A novel weighted superposition attraction algorithm-based optimization approach for state of charge and power management of an islanded system with battery and supercapacitor-based hybrid energy storage system. IETE J Res 69(2):825–838
7. Nicola M, Nicola CI (2022) Improved performance in the control of DC-DC three-phase power electronic converter using fractional-order SMC and synergetic controllers and RL-TD3 agent. Fractal Fractional 6(12):729
8. Gil-González W, Riffo S, Montoya OD, Restrepo C, Hernández JC (2023) Adaptive voltage control for second-order DC-DC converters supplying an unknown constant power load: a generalized PBC plus damping injection design. IEEE Access

Sliding-Mode and Lyapunov Function Based Control for a DC Microgrid with Renewable Generation a Solid Oxide Fuel Cell and Battery Storage

Mohamed Taieb Krakdia, Habib Kraiem, and Lassaad Sbita

Abstract In this paper, we propose a sliding mode and Lyapunov Function based control for a DC microgrid. The microgrid involving renewable distributed generators (PV system and wind turbine), energy storage system (Battery), a solid oxide fuel cell a dump load (aqua_electrolyzer) connected to the same dc bus and an inverter connected to LCL filter which is connected to an AC load Fig. 1. The behavior of the whole system is represented by mathematical models derived from the different sources of the microgrid the dump load and the storage system are presented in Sect. 2. The decentralized control of the microgrid components are depicted with synoptic figures presented in Sect. 2. The energy management algorithm is adopted to ensure the power balance under variable power generation. Simulation results demonstrate that the proposed control method has acceptable performance, including robust tracking and a velvety transition under generation scenario and load power variation in the large range.

Keywords Variable speed wind turbine · Photovoltaic (PV) · Battery · Solid oxide fuel cell · Dump load (aqua_electrolyzer) · Second order Sliding-mode controller · Sliding-mode controller · Lyapunov function-based controller · Energy management algorithm · AC linear load

M. T. Krakdia · L. Sbita
Research Unit of Photovoltaic, Wind and Geothermal Systems SPEG/UR11ES82, National Engineering School of Gabes, University of Gabes, Gabes, Tunisia
e-mail: kmronin@gmail.com

L. Sbita
e-mail: lassaad.sbita@gmail.com

H. Kraiem (✉)
Department of Electrical Engineering, College of Engineering, Northern Border University, Arar, Saudi Arabia
e-mail: kmronin@gmail.com

© The Author(s), under exclusive license to Springer Nature Singapore Pte Ltd. 2024
A. Mellit et al. (eds.), *Proceedings of the 2nd International Conference on Green Energy Conversion System*, Green Energy and Technology,
https://doi.org/10.1007/978-981-97-6148-7_16

1 Introduction

In [1] A Lyapunov based nonlinear controller is offered to stabilize the common DC bus. But, the system level controller to realize power balance is not presented in the paper. In [2], authors present a dc load supplied by a wind–PV hybrid system. But, battery is directly connected to dc bus and its control for charging/discharging is not cited. Adaptive Droop control proposed in [3] has its limitations to regulate the DC bus voltage when the microgrid contain multiple renewable sources such as PV system, wind system and HESS with different state of charges (SOCs) are connected together. In [4], backstepping control for DC microgrid was proposed, but the result of the connection of various renewable sources together to the common DC bus is not studied. In [5], the proposed work contains renewable sources based hybrid system for a green house, but is only for single phase. In [6] the proposed idea of author to control the DC–DC converter connected between the battery and the DC-link to maintain V_{dc} constant by controlling the charging/discharging of the battery. Dump load is absorbing the excess power (when state of charge of battery is above 0. 8) and unexpectedly the power demanded by the load becomes superior to the power generated by the source. The situation presented here the switches Q_1 and Q_2 both are still in off position so the battery can't supply the demand of load then the controller is in failure [7]. Authors proposed a PID controller that generated the reference current for the battery, a hysteresis band approach is adapted to switch either Q_1 or Q_2 of dc–dc converter. In this paper modified controller is developed as shown in Fig. 3. In [8] the storage system and the dump load are not considered this cause the unreliability of the system (Figs. 1, 2, 3, 4, 5, 6, 7, 8 and 9).

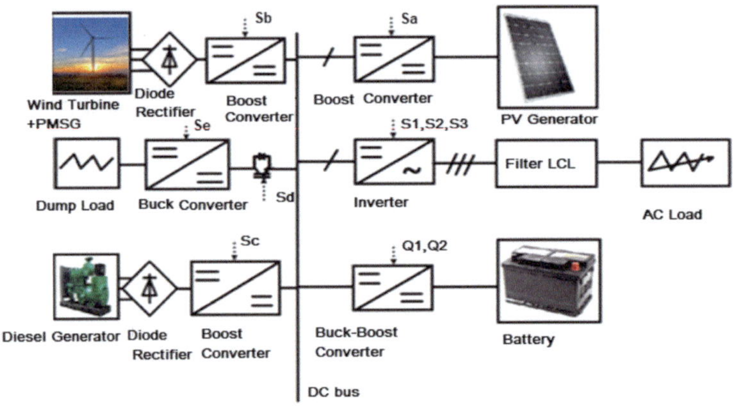

Fig. 1 Proposed microgrid

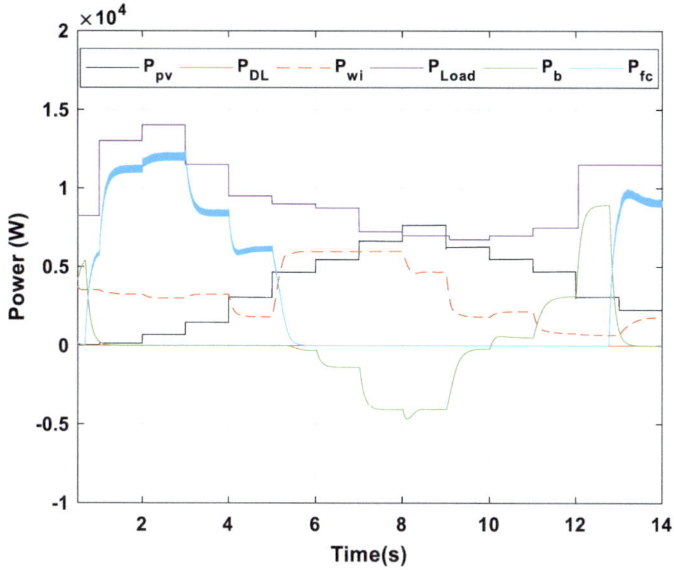

Fig. 2 Various powers (case-1): P_{fc}: Solid oxide fuel cell power, P_{Load}: AC variable linear load demand power, P_b: Battery power, P_{pv}: Pv system power, P_{wi}: wind generation system, P_{DL}: Dump load system power

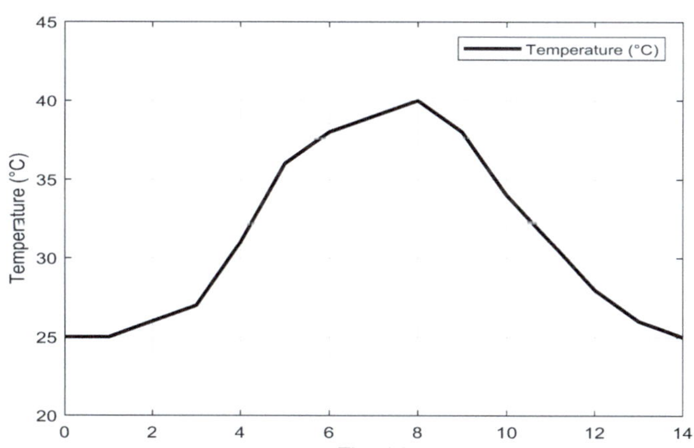

Fig. 3 The variation of the temperature (case-1)

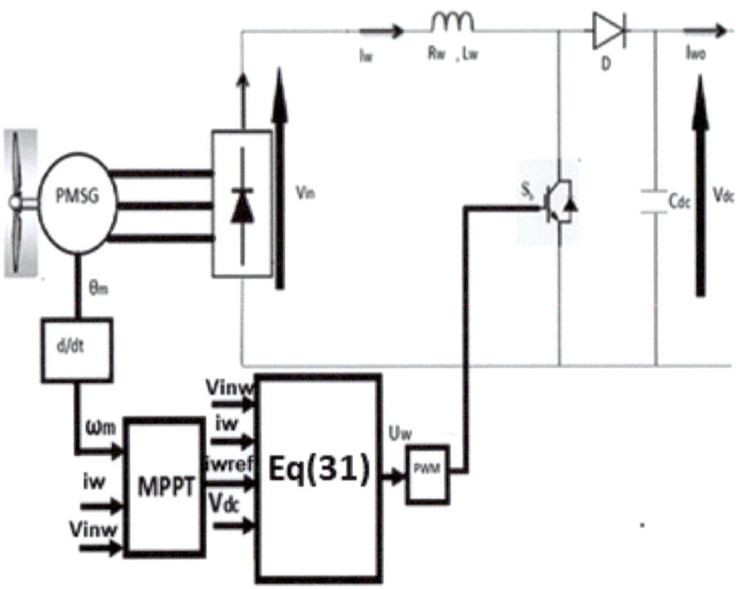

Fig. 4 Control scheme of the DC–DC Boost converter of the wind generation system

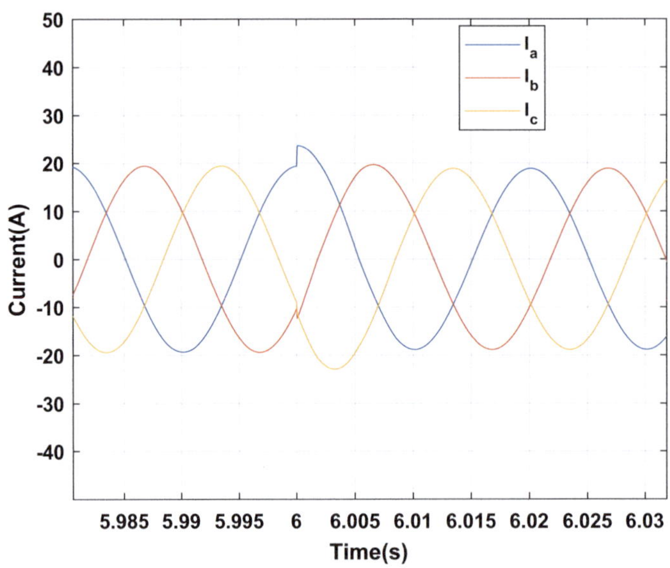

Fig. 5 The instantaneous current (case-1)

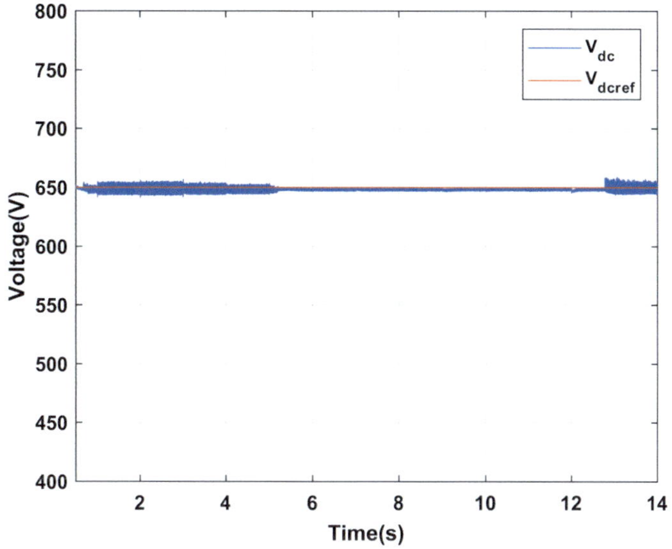

Fig. 6 DC bus voltage (case-1)

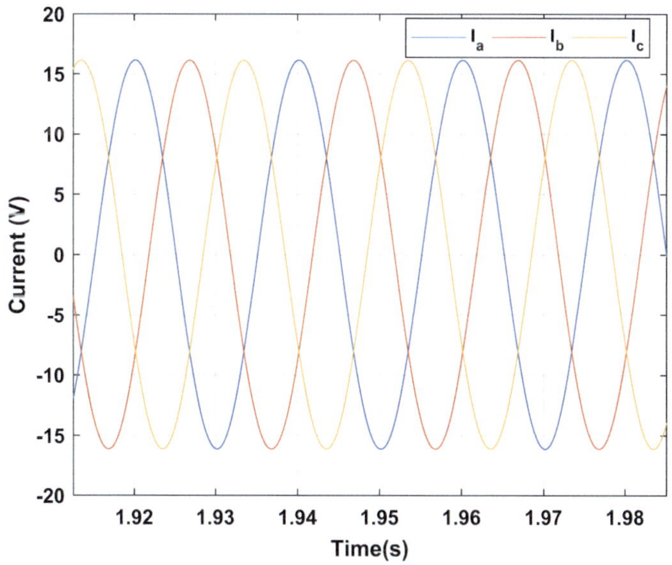

Fig. 7 The instantaneous current (case-2)

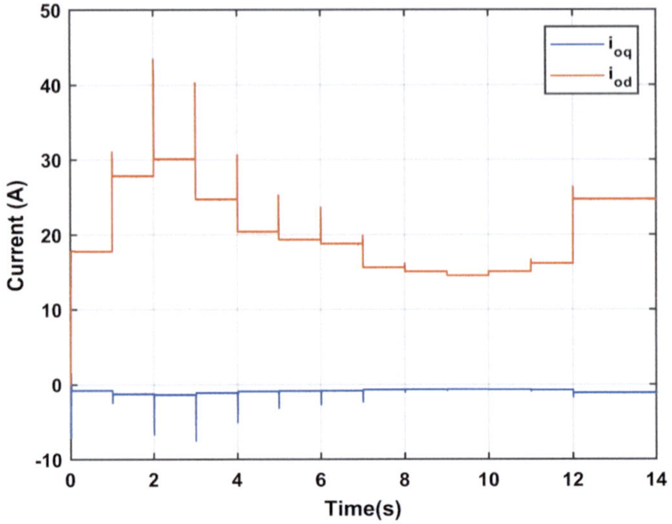

Fig. 8 Currents i_{od} _ i_{oq} (case-3)

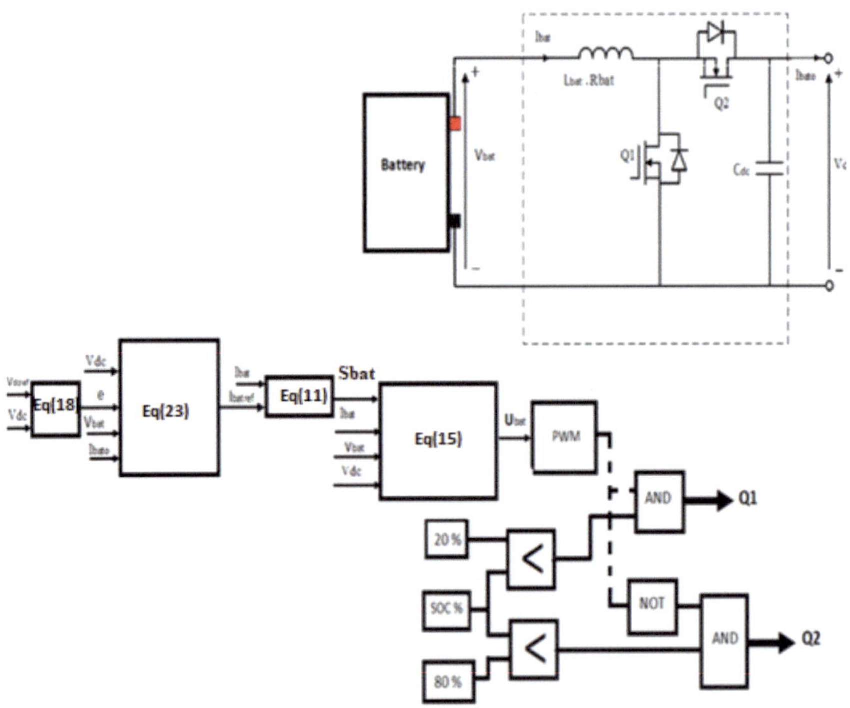

Fig. 9 Control scheme of the DC–DC buck-boost converter of the battery

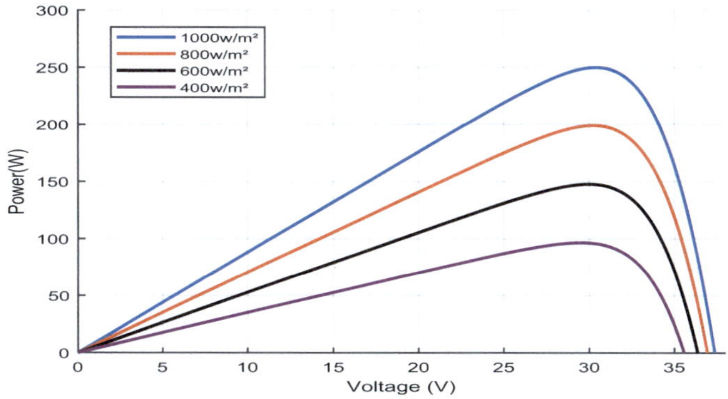

Fig. 10 Power-voltage curves of the module RENESOLA 250W curves at 25 °C for different Irradiance Levels

2 Modeling of DC Microgrid Components

2.1 Modelling PV Energy System in DC Microgrid

The module used in this paper is RENESOLA 250 W monocristalin type, Figs. 10 and 19 represent the simulation result of the module for a variable Irradiance. The PV system composed of 4 arrays are connected in parallel, each array consists of 8 modules connected in series. The PV system can be modelled as following:

$$\begin{cases} \frac{dx_0}{dt} = \frac{1}{C_{pv}} i_{pv} - \frac{1}{C_{pv}} x_1 \\ \frac{dx_1}{dt} = \frac{1}{L} x_0 - \frac{(1-u_{pv})}{L} V_{dc} \end{cases} \tag{1}$$

where

$$x = [x_0 x_1]^T = \left[V_{pv} I_L\right]^T. \tag{2}$$

2.2 Control of PV Generator: Sliding Mode with Slotine Surface Controller

The tracking error is:

$$e_{pv} = V_{pv} - V_{pvref} \tag{3}$$

The chosen sliding surface is:

$$S_{pv} = \left(\frac{d}{dt} + \lambda\right) e_{pv} \tag{4}$$

Then the sliding surface can be expressed as:

$$S_{pv} = \dot{e}_{pv} + \lambda e_{pv} \tag{5}$$

We replace the derivative of the error from (1) and (3) we get:

$$S_{pv} = \frac{1}{C_{pv}} i_{pv} - \frac{1}{C_{pv}} I_L + \lambda e_{pv} - \dot{V}_{pvref} \tag{6}$$

The derivate of the sliding surface becomes:

$$\dot{S}_{pv} = \frac{1}{C_{pv}} \dot{i}_{pv} - \frac{1}{C_{pv}} \dot{I}_L + \lambda \dot{e}_{pv} - \ddot{V}_{pvref} = -ksign(S_{pv}) \tag{7}$$

We replace the derivative of I_L from Eq. (2) we get:

$$u_{pv} = 1 + \frac{C_{pv}L}{V_{dc}} \left(-\frac{1}{C_{pv}} \dot{i}_{pv} + \frac{1}{LC_{pv}} V_{pv} - \lambda \dot{e}_{pv} + \ddot{V}_{pvref} - ksign(S_{pv}) \right) \tag{8}$$

where u_{pv} is the control low of the PV system.

The role of the MPPT algorithm is to calculate the optimal output voltage of the photovoltaic panel called V_{pvref} allowing the PV system to operate at its MPP. For that the well-known algorithm Incremental Conductance algorithm is used.

2.3 Modelling of the Battery System in DC Microgrid

The battery [9–12] connected to the DC bus, with a buck-boost converter, composed of the inductor, L_{bat} with its equivalent series resistance R_{bat}, output filter capacitor C_{dc} and two IGBT switches Q_1 and Q_2. The converter is utilized to regulated the DC bus voltage. The average state model of the buck-boost converter can be described by

$$\frac{dI_{bat}}{dt} = \frac{V_{bat}}{L_{bat}} - \frac{R_{bat}I_{bat}}{L_{bat}} - (u_{bat})\frac{V_{dc}}{L_{bat}} \tag{9}$$

$$\frac{dV_{dc}}{dt} = (u_{bat})\frac{I_{bat}}{C_{dc}} - \frac{I_{bato}}{C_{dc}} \tag{10}$$

where V_{dc}, I_{bat}, I_{bato}, C_{dc} and, u_{bat} are average state values Battery output voltage, Battery current, the output current of the converter, the output capacitor and control signal.

2.4 The Control of the Battery Current

Sliding mode control and super twisting algorithm control

The sliding surface chosen is

$$S_{bat} = I_{bat} - I_{batref}. \tag{11}$$

where I_{batref} the Battery reference current. By integrating (10) and (11) the derivative of S_{bat} is given by:

$$\dot{S_{bat}} = \frac{V_{bat}}{L_{bat}} - \frac{R_{bat}}{L_{bat}} I_{bat} - (u_{bat}) \frac{V_{dc}}{L_{bat}} - \dot{I}_{batref} \tag{12}$$

Then the equivalent term can be deduced as:

$$u_{eq} = \frac{V_{bat}}{V_{dc}} - \frac{R_{bat}}{V_{dc}} I_{bat} \tag{13}$$

Then the super twisting term is:

$$u_{st} = \alpha |S_{bat}|^{\frac{1}{2}} sign(S_{bat}) - \lambda \int sign(S_{bat}) \tag{14}$$

Then the global control of the battery is equal to:

$$u_{bat} = u_{eq} + u_{st} = \frac{V_{bat}}{V_{dc}} - \frac{R_{bat}}{V_{dc}} I_{bat} - \alpha |S_{bat}|^{\frac{1}{2}} sign(S_{bat}) - \lambda \int sign(S_{bat}) \tag{15}$$

The adaptive gains of the algorithm determined as following:

$$\dot{\alpha} = \begin{cases} \omega \sqrt{\frac{\varphi}{2}} & sign(|S_{bat}| - \gamma) if \alpha_a > \alpha_l \\ \mu if \alpha_a \leq \alpha_l \end{cases} \tag{16}$$

$$\lambda = \tau \alpha \tag{17}$$

where ω, λ, γ, τ, ω, μ, φ, α_a et α_l are positive constants.

2.5 The Control of the DC Bus Voltage: A Lyapunov Function for the Voltage Control

A Lyapunov function-based controller is proposed. The error term e is the difference between referential DC voltage and the real DC voltage, it can be expressed as:

$$e(t) = V_{dcref} - V_{dc} \tag{18}$$

The first Lyapunov candidate V is chosen as follows:

$$V = \frac{1}{2}C_{dc}e(t)^2 + \frac{1}{2}K_{Ib}(\int_{t_0}^{t} e(t)dt)^2 \tag{19}$$

where K_{Ib} is the integral gain of the DC bus voltage controller. After replacing the derivative of V_{dc} from (10) the derivative of V becomes:

$$\dot{V} = -C_{dc}e(t)\left((u_{bat})\frac{I_{bat}}{C_{dc}} - \frac{I_{bato}}{C_{dc}}\right) + K_{Ib}e(t)\int_{t_0}^{t} e(t)dt \tag{20}$$

In the voltage control process, it can be assumed that the battery current is controlled by the sliding-mode controller because the dynamic of the current controller is much faster than that of the voltage controller Figs. 27 and 28. In addition, R_{bat} very small, therefore (15) can be simplified as [13]:

$$u_{bat} \approx \frac{V_{bat}}{V_{dc}} \tag{21}$$

To ensure the convergence of the voltage control error to 0, the derivative of V should be negative this yield to:

$$-C_{dc}\left((u_{bat})\frac{I_{bat}}{C_{dc}} - \frac{I_{bato}}{C_{dc}}\right) + K_{Ib}\int_{t_0}^{t} e(t)dt = -K_p e(t) \tag{22}$$

where, K_p is a proportional gain positive constant. By substituting the value of u_{bat} from (21) in (22), I_{batref} can be defined as:

$$I_{batref} = (I_{bato} + K_{Ib}\int_{t_0}^{t} e(t)dt + K_p e(t))\frac{V_{dc}}{V_{bat}} \tag{23}$$

By integrating (20) to (23), and taking in consideration (21) the derivative of V becomes:

$$\dot{V} = -K_p e(t)^2 \leq 0 \tag{24}$$

Thus Lyapunov stability condition can be always realized, and the convergence of the DC bus voltage to its constant reference under all condition [13].

2.6 Modelling of Wind Energy System in DC Microgrid

The wind turbine proposed and simulated in this paper [14]: The DC–DC boost converter consists of the input inductor L_w with series resistance R_w, Diode D, output filter capacitor ($C_{dc} = C_w$) and IGBT switch S_b. The average state model of the boost converter can be described using the following differential equations:

$$\frac{dV_{dc}}{dt} = -\frac{i_{wo}}{C_{dc}} + (1 - u_w)\frac{i_w}{C_{dc}} \tag{25}$$

$$\frac{dx_3}{dt} = \frac{V_{inw}}{L_w} - \frac{R_w}{L_w} i_w - (1 - u_w)\frac{V_{dc}}{L_w} \tag{26}$$

where, x_3, x_2, i_{wo}, and u_w are average state values of wind current i_w, wind output voltage V_{dc}, the output current of the converter and control signal.

2.7 Control of Wind Energy System Current: Sliding Mode Control

The chosen surface S_w is:

$$S_w = i_{wref} - i_w \tag{27}$$

The derivative of S_w is:

$$S_w \cdot = \frac{di_{wref}}{dt} - \frac{di_w}{dt} \tag{28}$$

To guarantee the existence of sliding-mode surface, the condition below should be satisfied:

$$\dot{S_w} \cdot S_w < 0 \tag{29}$$

To ensure i_w converges to i_{wref} the derivative of S_w is set as:

$$\dot{S_w} = -K_{w1}S_w - K_{w2}\,sat(S_w) \tag{30}$$

where K_{w1} and K_{w2} are positive constants [13].

The control law u_w can be obtained by taking in view (26) and (30) and (28) and taking in consideration the change rate of the reference i_{wref} is neglected as follows:

$$u_w = 1 - \frac{V_{inw}}{V_{dc}} + \frac{R_w}{V_{dc}}i_w + \frac{K_{w1}S_wL_w}{V_{dc}} + \frac{K_{w2}sat(S_w)L_w}{V_{dc}} \tag{31}$$

The reference current is determined by the MPPT.

2.8 Modelling of Solid Oxide Fuel Cell System in DC Microgrid

The Solid Oxid Fuel Cell presented in this simulation [16, 17] is connected to a DC–DC boost converter as depicted in Fig. 14. The average state model of the boost converter can be described by:

$$\frac{dI_{SOFC}}{dt} = \frac{V_{inSOFC}}{L_{SOFC}} - \frac{R_{SOFC}}{L_{FC}}I_{SOFC} - (1 - u_{SOFC})\frac{V_{dc}}{L_{SOFC}} \tag{32}$$

$$\frac{dV_{dc}}{dt} = -\frac{I_{SOFCo}}{C_{dc}} + (1 - u_{SOFC})\frac{I_{SOFC}}{C_{dc}} \tag{33}$$

2.9 The Control of Solid Oxide Fuel Cell Current

The sliding surface is chosen as:

$$S_{SOFC} = I_{SOFCref} - I_{SOFC} \tag{34}$$

In the same manner of the wind energy system.

The control law u_{SOFC} can be obtained in the same manner of calculation of u_w as follows:

$$u_{SOFC} = 1 - \frac{V_{inSOFC}}{V_{dc}} + \frac{R_{SOFC}}{V_{dc}}I_{SOFC} + \frac{K_{SOFC1}.S_{SOFC}.L_{SOFC}}{V_{dc}} + \frac{K_{SOFC2}.sat(S_{SOFC}).L_{SOFC}}{V_{dc}} \tag{35}$$

where, K_{SOFC1}, K_{SOFC2} are positive constants.

2.10 The Control of the DC Bus Voltage: The Control of the DC Bus Voltage

The error term e_1 is the difference between referential DC voltage and the real DC voltage. Then it can be expressed as:

$$e_1 = V_{dcref} - V_{dc} \quad (36)$$

In the same manner of the battery system $I_{SOFCref}$ can be expressed:

$$I_{SOFCref} = (I_{SOFCo} + K_{SOFCI}.\int_{t_0}^{t} e_1(t)dt + K_{pSOFC}e(t))\frac{V_{dc}}{V_{inSOFC}} \quad (37)$$

where, K_{pSOFC} and K_{SOFCI} are positive gains constants.

2.11 The Load Side VSI Control

The specified root-mean-square (RMS) value of the output phase voltage and the load voltage frequency are 220 V and 50 Hz.

The equation of the LCL filter in the (dq) frame:

$$\frac{di_{ld}}{dt} = \frac{1}{L_f}(v_{id} - v_{od}) + \omega i_{lq} \quad (38)$$

$$\frac{di_{lq}}{dt} = \frac{1}{L_f}\left(v_{iq} - v_{oq}\right) - \omega i_{ld} \quad (39)$$

$$\frac{dv_{od}}{dt} = \frac{1}{C_f}(i_{ld} - i_{od}) + \omega v_{oq} \quad (40)$$

$$\frac{dv_{oq}}{dt} = \frac{1}{C_f}\left(i_{lq} - i_{oq}\right) - \omega v_{od} \quad (41)$$

$$\varepsilon_{d1} = i_{ldref} - i_{ld} \quad (42)$$

The sliding surface chosen here is $S_{dinv1} = \varepsilon_{d1}$.

The derivative of the sliding surface is:

$$\dot{S}_{dinv1} = \frac{di_{ldref}}{dt} - \frac{di_{ld}}{dt} \tag{43}$$

To guarantee the existence of sliding-mode surface, the condition (44) presented below should be satisfied:

$$\dot{S}_{dinv1} . S_{dinv1} < 0 \tag{44}$$

To ensure i_{ld} converges to i_{ldref}, the derivative of, S_{dinv1} is set as:

$$\dot{S}_{dinv1} = -K_1 S_{dinv1} - K_2\, sat(S_{dinv1}) \tag{45}$$

where, K_1, K_2 are positive constants.

The control law v_{id} can be obtained by taking in view (45) and (38) and (43) and taking in consideration the change rate of the reference i_{ldref} is neglected as follows:

$$v_{id} = v_{od} + L_f\left(-\omega i_{lq} + K_1 S_{dinv1} + K_2 sat(S_{dinv1})\right) \tag{46}$$

The sliding surface chosen here is $S_{qinv1} = \varepsilon_{q1}$.

In the same manner of calculation of v_{id} The control law v_{iq} can be expressed as:

$$v_{iq} = v_{oq} + L_f\left(\omega i_{ld} + K_3 S_{qinv1} + K_4 sat\left(S_{qinv1}\right)\right) \tag{47}$$

where, K_1, K_2 are positive constants.

To determine the reference current i_{lqref} and i_{ldref} can be calculated in the same manner of v_{iq}:

$$i_{ldref} = i_{od} + C_f\left(-\omega v_{oq} + K_5 S_{dinv2} + K_6 sat(S_{dinv2})\right) \tag{48}$$

where, K_5, K_6 are positive constants and S_{dinv2} the chosen sliding surface.

$$i_{lqref} = i_{oq} + C_f\left(\omega v_{od} + K_7 S_{qinv2} + K_8 sat\left(S_{qinv2}\right)\right) \tag{49}$$

where, K_7, K_8 are positive constants. and S_{qinv2} the chosen sliding surface.

2.12 Modelling of Dump Load System [18–20] in the Proposed Microgrid

The components of Buck converter are listed as follows: C_{buck} is the capacitor, I_{Lbuck} is the inductor, V_{dc} is the voltage source, switching devices are Q and D, and the electrolyser with its voltage V_e equal to the voltage of the capacitor C_{buck}, V_c.

The state space equation of buck converter is:

$$\frac{dI_{Lbuck}}{dt} = u_{buck}\frac{V_{dc}}{L_{buck}} - \frac{V_c}{L_{buck}} \tag{50}$$

$$\frac{dV_{Cbuck}}{dt} = \frac{I_{Lbuck}}{C_{buck}} - \frac{I_{obuck}}{C_{buck}} \tag{51}$$

The sliding surface is chosen as:

$$S_{buck} = I_{Lbuckref} - I_{Lbuck} \tag{52}$$

The switching function u_{buck} can be obtained in the same manner of calculation of u_w as follows:

$$u_{buck} = \frac{V_c}{V_{dc}} + K_{buck1}.S_{buck}. + K_{buck2}.sat(S_{buck}).\frac{L_{buck}}{V_{dc}}. \tag{53}$$

where K_{buck1}, K_{buck2} are positive constants [13].

2.13 *The Control of the Buck Converter* $\mathbf{V_e}$ *Voltage*

The error term e_b is the difference between referential V_{eref} voltage and the real V_e voltage. Then it can be expressed as:

$$e_b = V_{eref} - V_e \tag{54}$$

The reference current $I_{Lbuckref}$ can be calculated as follow in same manner of I_{DGref}:

$$I_{Lbuckref} = I_{obuck} + K_{Ibuck}.\left(\int_{t0}^{t} e_b(t)dt\right) + K_{pbuck}.e_b(t) \tag{55}$$

where K_{pbuck} is a constant positive proportional gain and K_{Ibuck} is a constant positive integral gain (Figs. 11, 12, 13, 14, 15, 16, 17, 18, 19, 20, 21, 22, 23, 24, 25, 26, 27 and 28).

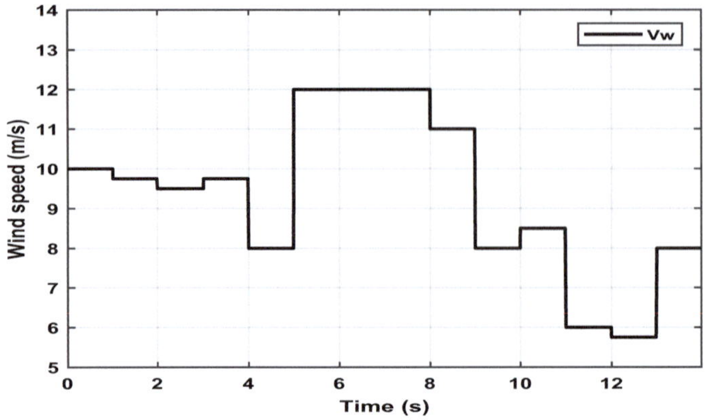

Fig. 11 Wind speed (case-1)

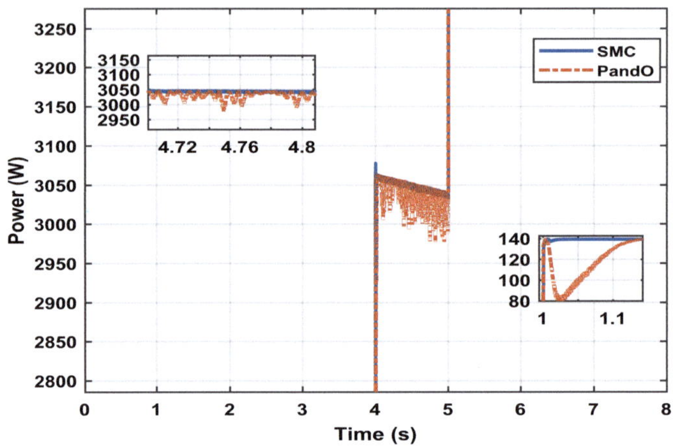

Fig. 12 Pv power (case-1)

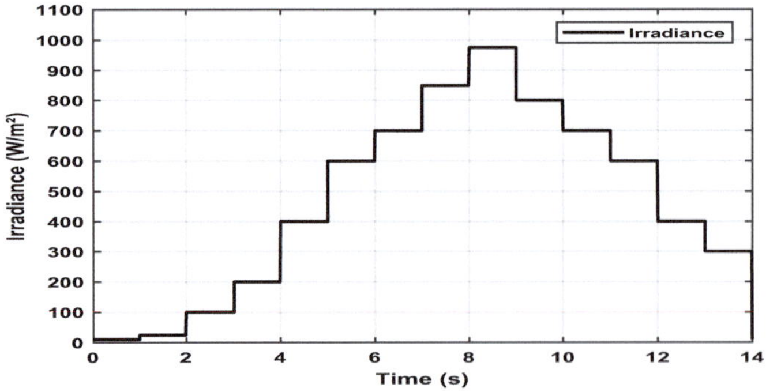

Fig. 13 The irradiation (case-1)

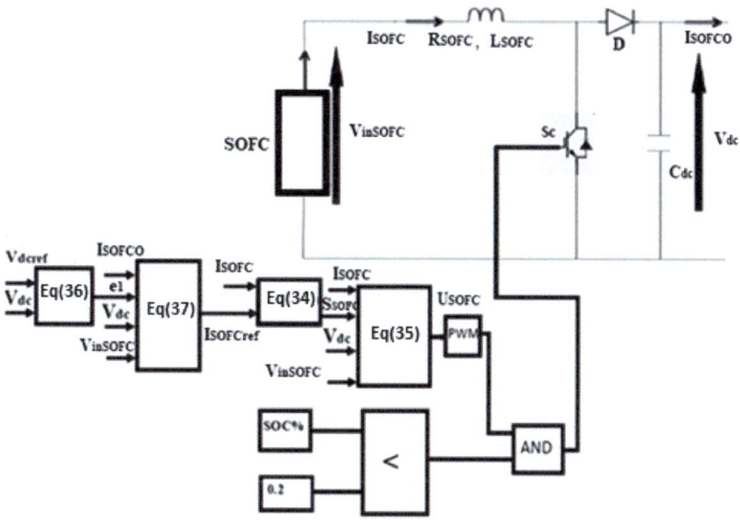

Fig. 14 Control scheme of the DC–DC Boost converter of solid oxide fuel cell system

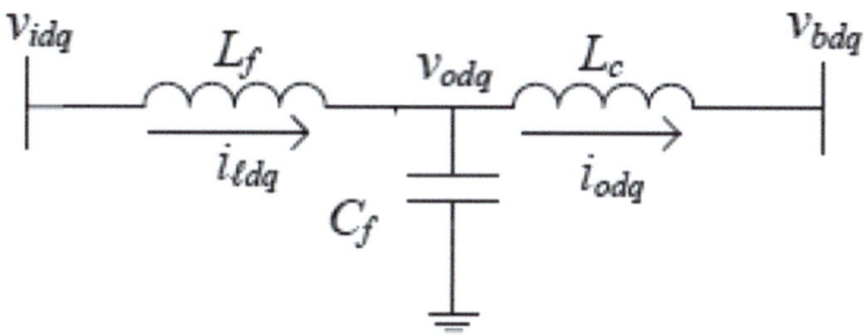

Fig. 15 LC L Filter

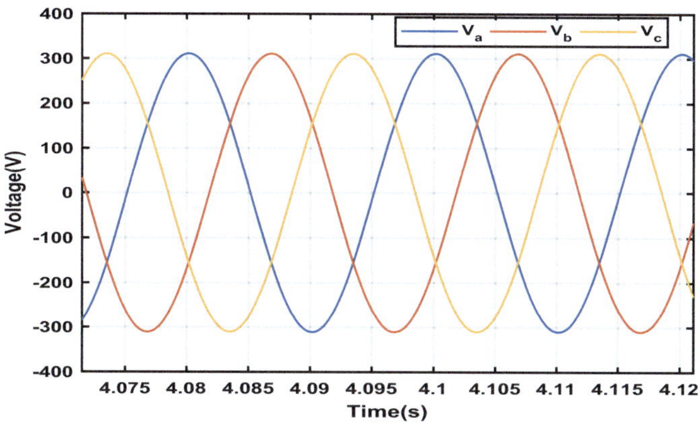

Fig. 16 The instantaneous current (case-3)

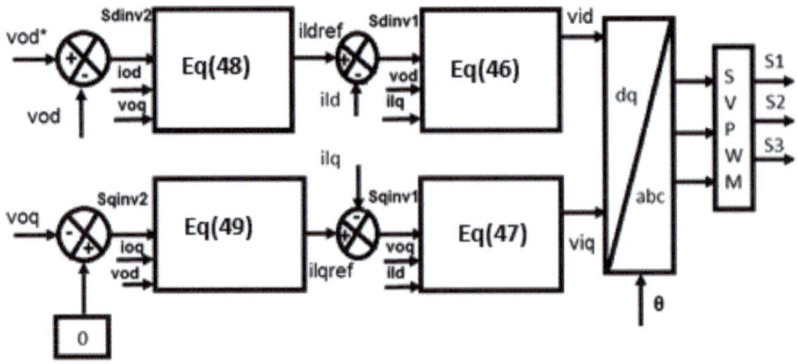

Fig. 17 The load side VSI controller

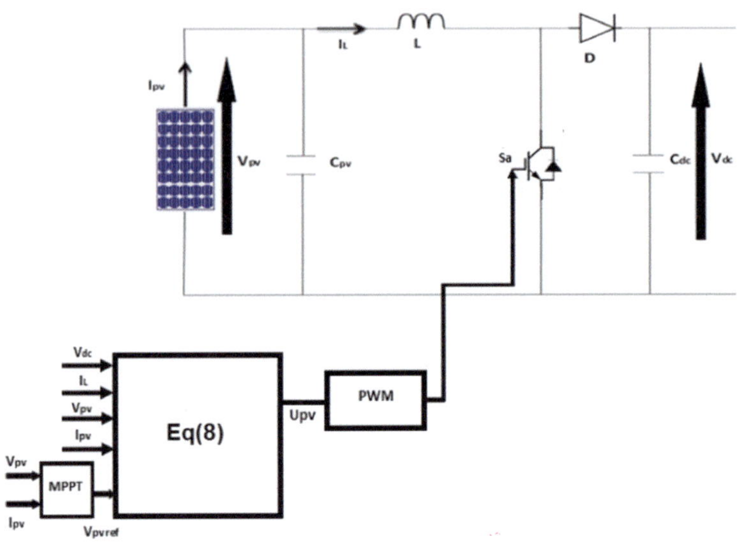

Fig. 18 Control scheme of the DC–DC boost converter of the Pv system

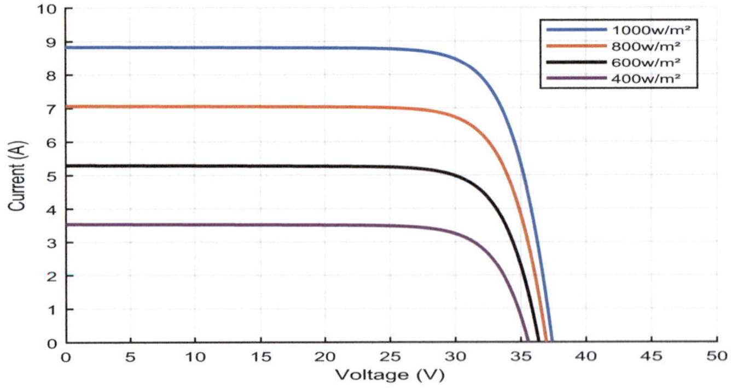

Fig. 19 Current–voltage curves of the Pv module at 25 °C

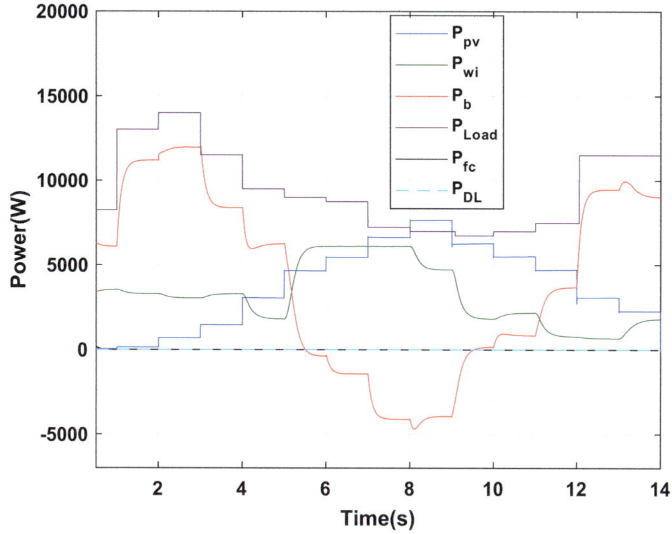

Fig. 20 Various powers (case-3): P_{fc}: Solid oxide fuel cell power, P_{Load}: AC variable linear load demand power, P_b: Battery power, P_{pv}: Pv system power, P_{wi}: wind generation system, P_{DL}: Dump load system power

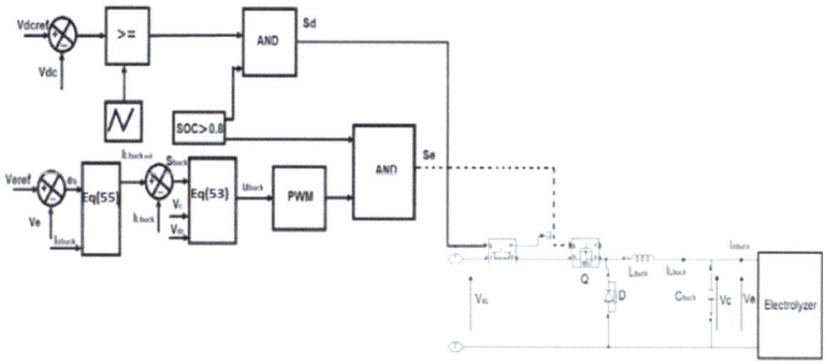

Fig. 21 Buck converter controller

3 Results and Discussion

3.1 Case-1: Consideration of Meteorological Changes

The proposed system is tested under the variation of the load Fig. 28, the variation of the temperature Fig. 3, the variation of the irradiation Fig. 13 and the variation of the wind speed Fig. 11, different powers are illustrated in Fig. 2, the battery and the solide oxide fuel cell maintain the power balance between the production and the demand of the load. The solid oxide fuel cell starts supplying the load at

Fig. 22 Energy management of the proposed microgrid

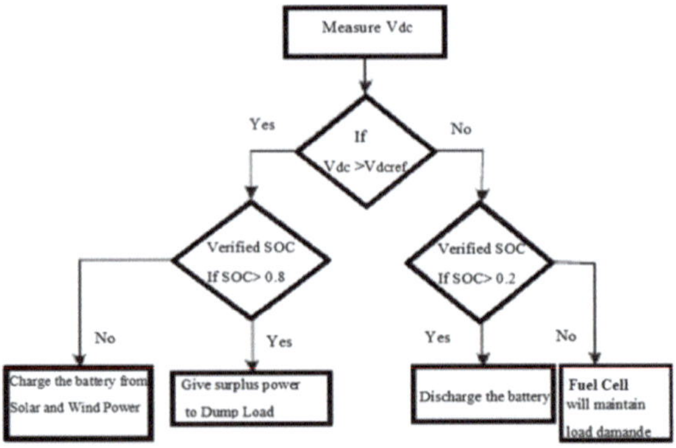

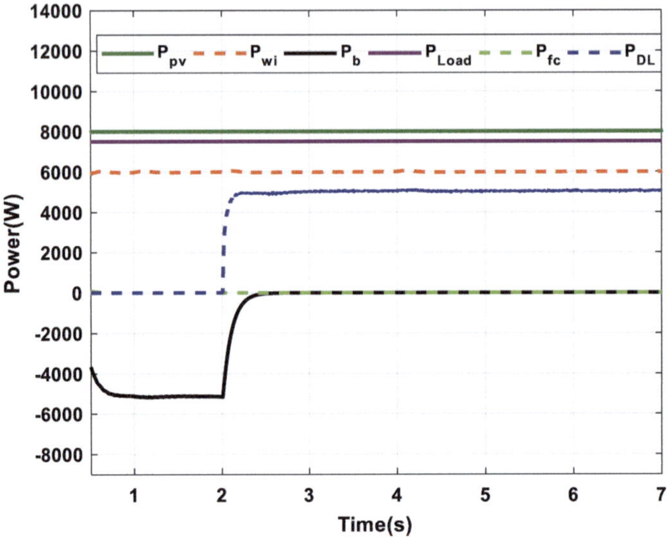

Fig. 23 Various powers (case-2): P_{fc}: Solid oxide fuel cell power, P_{Load}: AC variable linear load demand power, P_b: Battery power, P_{pv}: Pv system power, P_{wi}: wind generation system, P_{DL}: Dump load system power

the beginning of the operating cycle (at t = 1 s), when the SOC = 0. 2. From Fig. 2, it is shown that the battery response is very fast and can meet the charging demand quickly. In this process, the SOC of the battery reaches 0. 2 at t = 1 s the battery continues to momentarily supply power since Solid oxide fuel cell cannot act instantaneously due to its slow dynamics. Since the dynamics of Solid oxide fuel cell is slow, it starts delivering the power at t = 1 s and delivers the required power approximately at t = 1. 4 s as shown in Fig. 2. From Fig. 12, the slope can be observed in P_{pv} power due to temperature variations. From Fig. 12, Sliding mode controller provides a photovoltaique power with less steady state error and lower values of initial transients [15]. The DC bus voltage is shown in Fig. 6, it is kept

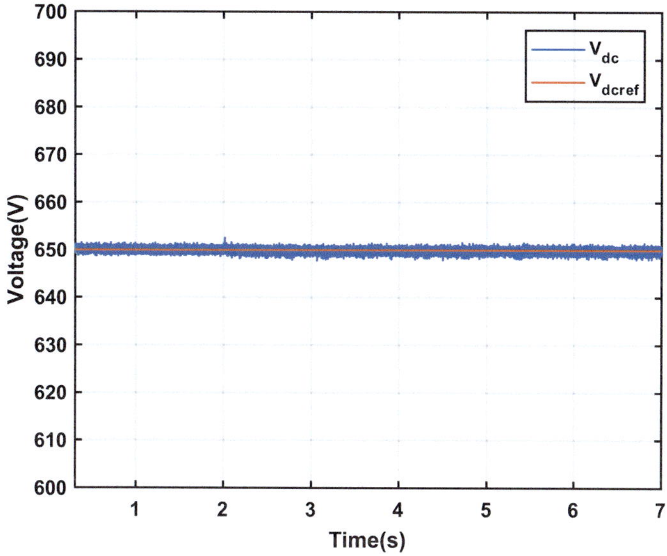

Fig. 24 DC bus voltage (case-2)

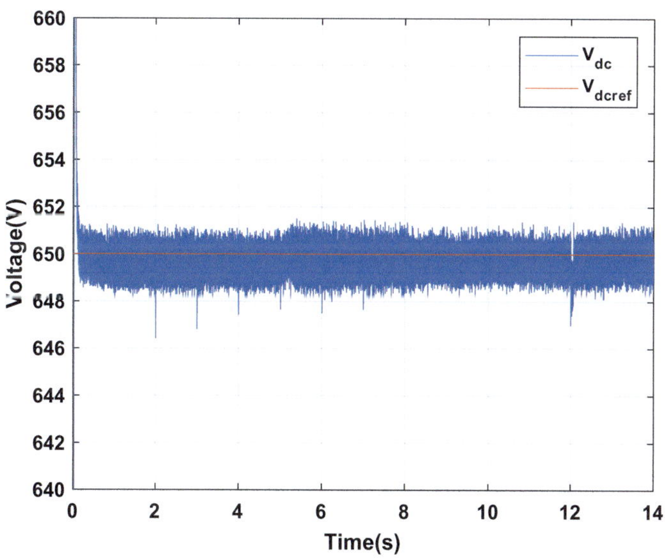

Fig. 25 DC bus voltage (case-3)

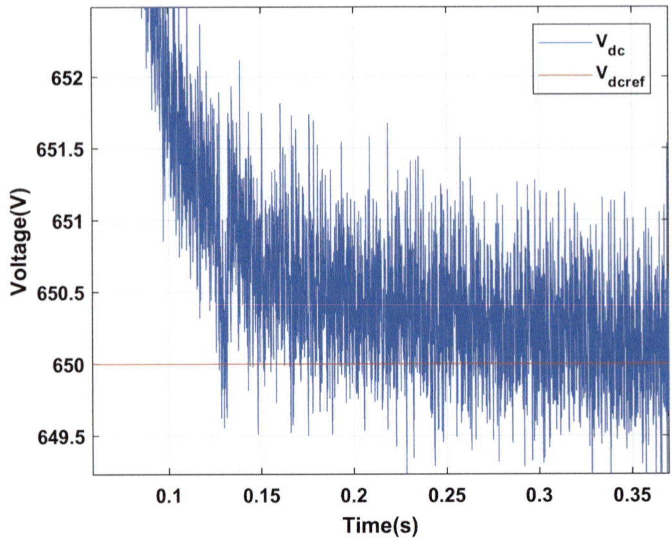

Fig. 26 Justification (case-2)

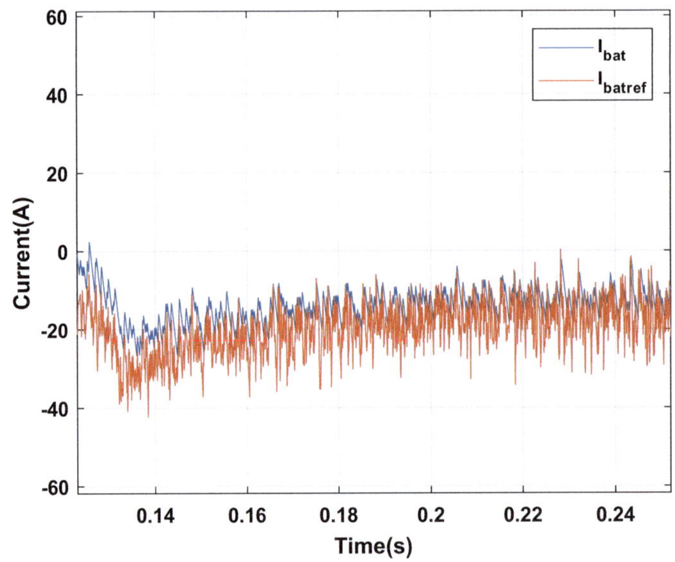

Fig. 27 Justification (case-2)

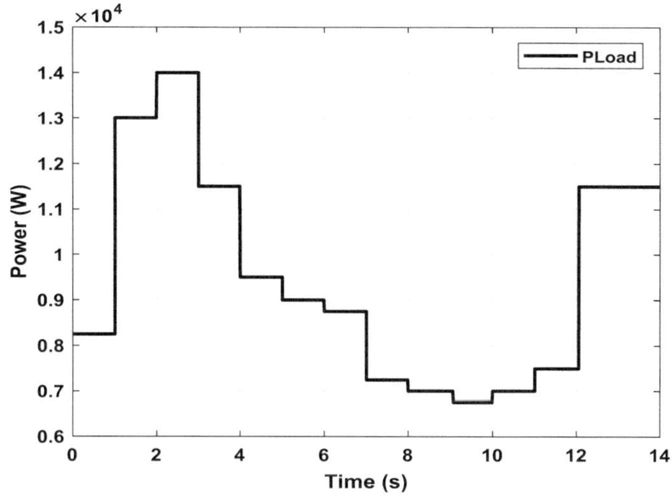

Fig. 28 The variation of the load (case-1)

constant around a value of 650 V. The instantaneous current is illustrated in Fig. 5, it can be seen that the wave form seems acceptable for the mentioned variations in case-1. The over shoot of the AC current is caused by the sudden variation of the load Fig. 5 the AC current stabilizes quickly and the current overshoot is not high.

3.2 Case- 2: Stand-Alone System Operation with Battery and Dump Load

In this case, battery SOC is acted near about 0. 8 and presuming that excess power is offered from the sources to charge the battery. In case-2 the power requested by the load is 7500 W, the power $P_{pv} = 8000$ W and the power $P_{wi} = 6000$ W at temperature T = 25 °C. As soon as the battery SOC becomes greater than 0. 8, according to the energy management algorithm Fig. 22 and the control of the buck converter of the dump load Fig. 21, the dump load absorbs the excess energy. From Fig. 23, we notice that when the SOC becomes greater than 0. 8 at t = 2. 0 s, the dump load takes over and the battery stops charging, which makes its power zero. Figure 24 illustrates the response of the DC bus voltage V_{dc} when the SOC becomes greater than 0. 8 at t = 2. 0 s, which shows that the DC bus voltage stabilizes quickly and the voltage overshoot is very small compared to PI controller [9]. Figures 26 and 27 demonstrate the dynamic of the current controller is much faster than that of the voltage controller. From Fig. 7, it can be seen the instantaneous current has an acceptable waveform.

3.3 Case-3: Variation of Powers SOC = 60%

The proposed system is tested under the variation of the load and the following meteorological variations like Case-1. Case-3 represents the response of the system without intervention of the Solid oxide fuel cell or the Dump Load since the SOC remains greater than 0. 2 and less than 0. 8. It is identical to case-1 from the point of view of the interaction of the powers, P_{pv}, P_{wi}, P_{Load} Fig. 20. The DC bus voltage remain constant around 650 V, Fig. 25. From Fig. 16, it can be seen that the current wave form seems acceptable for the mentioned variations in case-3. The current i_{oq} remain near by zero and i_{od} follow the demande of the load Figs. 8 and 15.

4 Conclusion

Stability analysis is carried out using composite Lyapunov function to ensure the stability of the DC microgrid, a sliding mode controller is proposed to control the current of each element of the microgrid a Lyapunov function controller is used to control the DC bus voltage. the performance of the controllers is satisfactory under steady state as well as dynamic conditions, the effectiveness and robustness of the proposed controllers are verified by simulation.

References

1. Iovine A, Carrizosa MJ, Damm G, Alou P (2018) Nonlinear control for dc microgrids enabling efficient renewable power integration and ancillary services for ac grids. IEEE Trans Power Syst 34(6):5136–5146
2. Dali M, Belhadj J, Roboam X, Blaquiere JM (2007) Control and energy management of a wind-photovoltaic hybrid system. In: European conference on power electronics and applications
3. Lu X, Guerrero JM, Sun K, Vasquez JC et al (2014) An improved droop control method for dc microgrids based on low bandwidth communication with dc bus voltage restoration and enhanced current sharing accuracy. IEEE Trans Power Electron 29(4):1800–1812
4. Roy TK, Mahmud MA, Oo AMT, Haque ME, Muttaqi KM, Mendis N (2018) Nonlinear adaptive backstepping controller design for islanded dc microgrids. IEEE Trans Ind Appl 54(3):2857–2873
5. Caliskan H, Dincer I, Hepbasli A (2012) Energy, exergy and sustainability analyses of hybrid renewable energy based hydrogen and electricity production and storage systems: modeling and case study. Appl Therm Eng (in press)
6. Bhende CN, Mishra S, Malla SG (2011) Permanent magnet synchronous generator-based standalone wind energy supply system. IEEE Trans Sustain Energy 2(4):361–73
7. Malla SG, Bhende CN (2014) Voltage control of stand-alone wind and solar energy system. School of Electrical Sciences, Indian Institute of Technology Bhubaneswar, Bhubaneswar 751013, India
8. Sait HH, Daniel SA, Babu PM (2008) Analysis and control of an autonomous hybrid wind-driven PM alternator and photovoltaic array without battery storage. In: TENCON 2008—IEEE region 10 conference

9. Omar N, Monem MA, Firouz Y, Salminen J, Smekens J, Hegazy O, Gaulous H, Mulder G, Van den Bossche P, Coosemans T, Van Mierlo J (2014) Lithium iron phosphate based battery—assessment of the aging parameters and development of cycle life model. Appl Energy 113:1575–1585
10. Saw LH, Somasundaram K, Ye Y, Tay AA (2014) Electro-thermal analysis of lithium iron phosphate battery for electric vehicles. J Power Sources 249:231–238
11. Tremblay O, Dessaint L-A (2009) Experimental validation of a battery dynamic model for EV applications. World Electric Veh J 3(May):13–16
12. Zhu C, Li X, Song L, Xiang L (2013) Development of a theoretically based thermal model for lithium ion battery pack. J Power Sources 223:155–164
13. Song Z, Hou J, Hofmann H, Li J, Ouyang M (2017) Sliding-mode and Lyapunov function-based control for battery/supercapacitor hybrid energy storage system used in electric vehicles. Energy. https://doi.org/10.1016/j.energy.2017.01.098
14. Haque ME, Negnevitsky M, Muttaqi KM (2010) A novel control strategy for a variable-speed wind turbine with a permanent-magnet synchronous generator. IEEE Trans Ind Appl 46(1):331–339
15. Farhat M, Barambones O, Sbita L (2017) A new maximum power point method based on a sliding mode approach for solar energy harvesting. Appl Energy 185:1185–1198
16. Padulles J, Ault GW, McDonald JR (2000) An integrated SOFC plant dynamic model for power systems simulation. J Power Sources 86:495–500
17. Zhu Y, Tomsovic K (2002) Development of models for analyzing the load-following performance of microturbines and fuel cells. Elect Power Syst Res 62:1–11
18. Dieguez PM, Ursua A, Sanchis P, Sopena C, Guelbenzu E, Gandia LM (2008) Thermal performance of commercial alkaline water electrolyzer: experimental study and mathematical modeling. Int J Hydrogen Energy 33:7338–7354
19. Gandia LM (2008) Thermal performance of commercial alkaline water electrolyzer: experimental study and mathematical modeling. Int J Hydrogen Energy 33:7338–7354
20. Ulleberg O (2003) Modeling of advanced alkaline electrolyzers: a system simulation approach. Int J Hydrogen Energy 28:21–33

Boost DC–DC Converter with MPPT for PV Application

Gasmi Nadia and Ben Hamed Mouna

Abstract The DC/DC converters are widely used in photovoltaic generation systems as interfaces between the photovoltaic panels and the load, enabling the maximum power point (MPP) follow-up. To extract the maximum power, it is necessary to adjust the load to match the current and voltage of the solar panel. The converter must be designed to be directly connected to the photovoltaic panel and perform operations to search for the maximum power point (MPPT). DC/DC converters, together with maximum power point tracking systems (MPPT), are used to avoid these losses.

Keywords DC–DC converters · Photovoltaic sources · PWM · Buck–boost · Boost converters · V–I characteristics · Maximum power point tracker

1 Introduction

Solar photovoltaic (PV) power generation is the process of converting light energy from the sun into electricity using solar modules/panels which is also known as the photovoltaic effect. A large number of PV panels are combined into arrays, installed in an optimal configuration to harness light energy from the sun and convert it in to DC current, which is then converted to alternating current (AC) using an inverter before it can be directly used or fed into the electrical grid. PV systems configurations are:

- Grid-connected-most common, easier to design are typically less expensive.
- Off-grid (stand-alone) configurations are expensive as they rely on batteries.

Now, we all know that a solar cell is basically just a PN junction diode with light shining on it when a photon from sunlight, with higher energy than the bandgap of the solar cell, strikes an electron from the p side, it is excited from the valence

G. Nadia (✉) · B. H. Mouna
PEESE (Processes, Energetics, Environments & Electrical Systems) Research Laboratory LR18ES34, National Engineering School of Gabes University, Gabes, Tunisia
e-mail: gasminadia809@hotmail.com

© The Author(s), under exclusive license to Springer Nature Singapore Pte Ltd. 2024

A. Mellit et al. (eds.), *Proceedings of the 2nd International Conference on Green Energy Conversion System*, Green Energy and Technology,
https://doi.org/10.1007/978-981-97-6148-7_17

band to the conduction band. We now have a hole in the valence band. The hole freely wanders around the valence band because it's a majority carrier waiting for an electron recombine with it. If the exited electron manages to reach the depletion region, it will be swept out by the electric field in the depletion region and end up at the n side as a majority carrier. The electron diffuses to the anode. This electron travels though the external circuit, powers the output de vice, and returns to the p side. Then, it recombines with the hole that was left in the valence band earlier. This cycle repeats itself and now we the hole that was left in the valence band earlier. This cycle repeats itself and now we have a current flow from the n side to the p side of the PN junction. We call that light generated current, which is opposite to the flow direction of a PN junction under forward bias. However, because of this flow of electrons, a voltage drop will be developed at the resistor. A positive anode develops on the left, while a negative cathode develops on the right. This voltage drops effectively forward biases the solar cell like how it does in a normal PN junction. This creates a current which we call dark current, flowing from the p side to the n side due to diffusion of electrons like in a normal PN junction. But in order for a solar cell to work for us, the light generated current will be much higher than the dark current. The two opposing lights generated current and dark current result in a net current flowing from the N side to the P side. This net current is the current that you will measure flowing through the wire and load.

1.1 Modeling of PV Cell

Solar cell is a P–N junction fabricated in a thin wafer or layer of semiconductor. The electromagnetic radiation of solar energy can be directly converted electricity through photovoltaic effect. Being exposed to the sunlight, photons with energy greater than the band-gap energy of the semiconductor are absorbed and create some electron–hole pairs proportional to the incident irradiation. Under the influence of the internal electric fields of the P–N junction, these carriers are swept apart and create a photocurrent, which is directly proportional to solar insolation. PV system naturally exhibits a nonlinear I–V and P–V characteristics, which vary with the radiant intensity and cell temperature. The equivalent circuit of a solar photovoltaic (PV) cell is given in Fig. 1. It includes a diode, a shunt resistance and a series resistance [1].

The voltage–current characteristic equation of a solar cell is given as;

$$I = I_{ph} - I_0\left[\exp\left(\frac{q(V + I.R_s)}{n.K_I N_s T_c}\right) - 1\right] - I_{sh} \tag{1}$$

$$I_{sh} = \frac{V + I.R_s}{R_s h} \tag{2}$$

In this equation I is the load current, I_{ph} is a photocurrent at a given irradiance at given T, I_s is the cell/diode saturation of dark current, V is the terminal voltage of

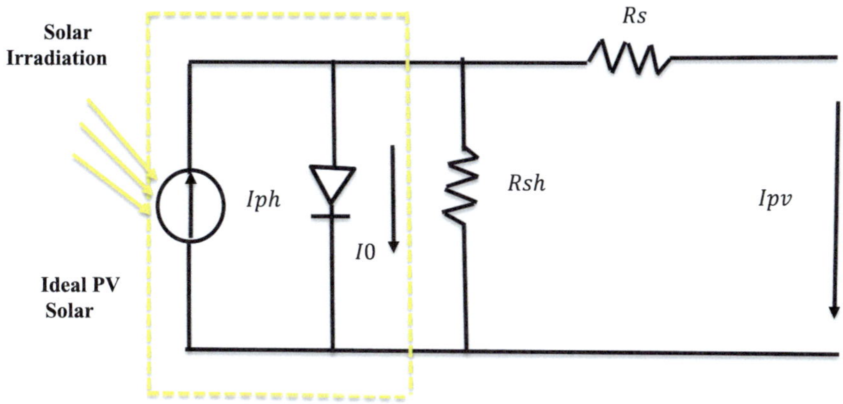

Fig. 1 Equivalent circuit of photovoltaic cell

the cell, $q\left(= 1.6 \times 10^{-19}\ \text{C}\right)$ is charge of an electron, $K_I\left(= 1.38 \times 10^{-23}\ \text{J/K}\right)$ is a Boltzmann's constant, T is the cell's working temperature, n is an diode ideal factor, RSH is a shunt resistance, and R_s is a series resistance. The photocurrent mainly depends on the solar insolation and cell's working temperature, which is described as [2];

$$I_{ph} = [I_{sc} + K_I(T - 298)]\frac{G}{1000} \tag{3}$$

where I_{sc} is the cells short-circuit current at a 25 °C and 1 kW/m^2, K_I is the cell's short-circuit current temperature coefficient, T_{Ref} is the cell's reference temperature, and G is the solar insolation in kW/m^2. On the other hand, the cell's saturation current varies with the cell temperature, which is described as.

$$I_{S0} = I_{rs}\left(\frac{T}{T_{Ref}}\right)^3 \exp\left[\frac{qE_{g0}\left(\frac{1}{T_{Ref}} - \frac{1}{T}\right)}{nK_I}\right] \tag{4}$$

$$I_{rs} = \frac{I_{sc}}{e^{\left(\frac{qV_{oc}}{nN_sK_IT}\right)} - 1} \tag{5}$$

where I_{rs} is the cell's reverse saturation current at a reference temperature and a solar radiation E_{g0} is the band-gap energy of the semiconductor used in the cell. The ideal factor K_I is dependent on PV technology [3]. With the help of above equation and description, we designed a Simulink module for PV module.

A practical solar cell comprises of parasitic resistances. Series resistance due to the combined resistances of contacts, metal grids and P and N layers. Shunt resistance due to leakage current through the P–N junction.

1.1.1 Short Circuit Current (I_{SC})

Maximum current obtained from a solar cell when the terminals are short circuited I_{sc} depends on solar irradiance level, area of the solar cell and characteristics of the material. I_{sc} the varies linearly with irradiance level variation of I_{sc} with temperature can be neglected.

1.1.2 Open Circuit Current (V_{OC})

Maximum voltage across the solar cell when the terminals are open circuited. V_{oc} depends on the quality of the material. V_{oc} is a strong function of temperature. It varies slightly with irradiance. Open circuit voltage is given as

$$V_{OC} = V_T Ln\left\{\left(\frac{I_{SC}}{I_0}\right) + 1\right\} \tag{6}$$

V_T voltage equivalent of temperature.

1.1.3 Fill Factor

It is the ratio of maximum power generated by a solar cell **to** the product of open circuit voltage and short array circuit current.

$$FF = \frac{V_{mp} I_{mp}}{V_{OC} I_{SC}} \tag{7}$$

Fill factor value gives an idea of the quality of the array and the closer the fill factor is to 1, the more the array can provide. Typical values are between 0.7 and 0.8.

1.1.4 Conversion Efficiency

The efficiency of a photovoltaic array is the ratio of the maximum electrical power that the array can produce to the amount of solar irradiance hitting the array

$$\eta = \frac{V_m I_{mp}}{P_{sun}} \tag{8}$$

The efficiently of a typical solar array is normally low at around 10–12%, depending on the type of cells being used.

2 DC–DC Converter

DC–DC converters are power electronic circuits that convert a DC source from one voltage level to another. It is used to step up (boost converter) or step down (Buck converter) the DC voltage to get the desired voltage.

A buck converter (step-down converter) is a DC–DC power converter which steps down voltage (while stepping up current) from its input (supply) to its output (load). It is a class of switched-mode power supply, typically containing at least two semiconductors (a diode and a transistor) and at least one energy storage component. Voltage ripple, filters made of capacitors, are normally added to the converter output (load-side filer) and input (supply-side filter). In this work, we choose to use a DC–DC buck converter as shown in Fig. 2 to achieve load adaptation in order to make PC Module work at its maximum power point. We consider this converter is more adequate for PV-MPPT control, because it allows us to control the input voltage V_{in}, which is equal to the PC Module voltage V_{pv} [4].

2.1 The Concept of DC–DC Converter

As previously mentioned, Boost converters are used to raise up the output voltage to a higher value by transferring the accumulated in the inductor energy which is converted into an output voltage. Figure 2 illustrates conventional boost converter having main inductor element L, chopper switch S, bypass diode D (to prevent back capacitor discharge) and capacitance C. The function of L is to transfer energy to the load during the off-state of the chopper switch and to maintain continuous current mode, while the function of C is to energize the load during the on-state of the chopper switch, and to reduce output voltage ripples [5].

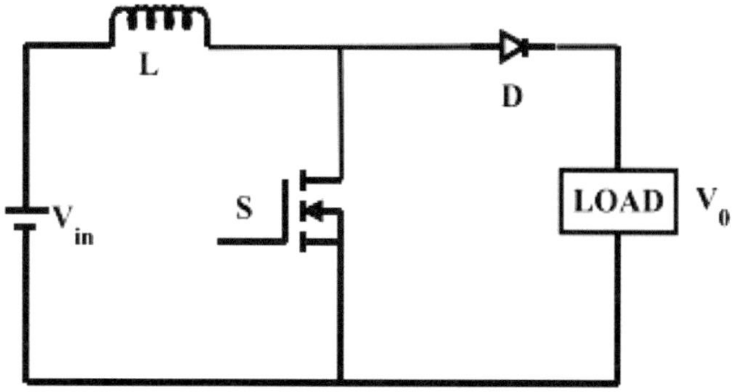

Fig. 2 DC–DC converter

2.2 Mathematical Model of DC–DC Converter

Taking into account the principal chopper time-diagram shown on Fig. 3, displaying the inductance voltage and current waveforms during two transistor operation modes. The inductance voltage during the on-state interval with $0 \leq t \leq DT$, assuming a linear current change due to large value of the inductance, is given by the Eq. (9) [1],

$$V_L(on) = V_{pv} = \mathrm{L}\frac{di_L(on)}{dt(on)} \cong \mathrm{L}\frac{\Delta i_{L\,on}}{\Delta T_{on}} = \mathrm{L}\frac{\Delta i_{L\,on}}{DT} \tag{9}$$

where T is the chopper switching period, D is the duty cycle, i_L is the inductor current, and i_{PV} is the solar panel current. For this circuit $i_L = i_{PV}$.

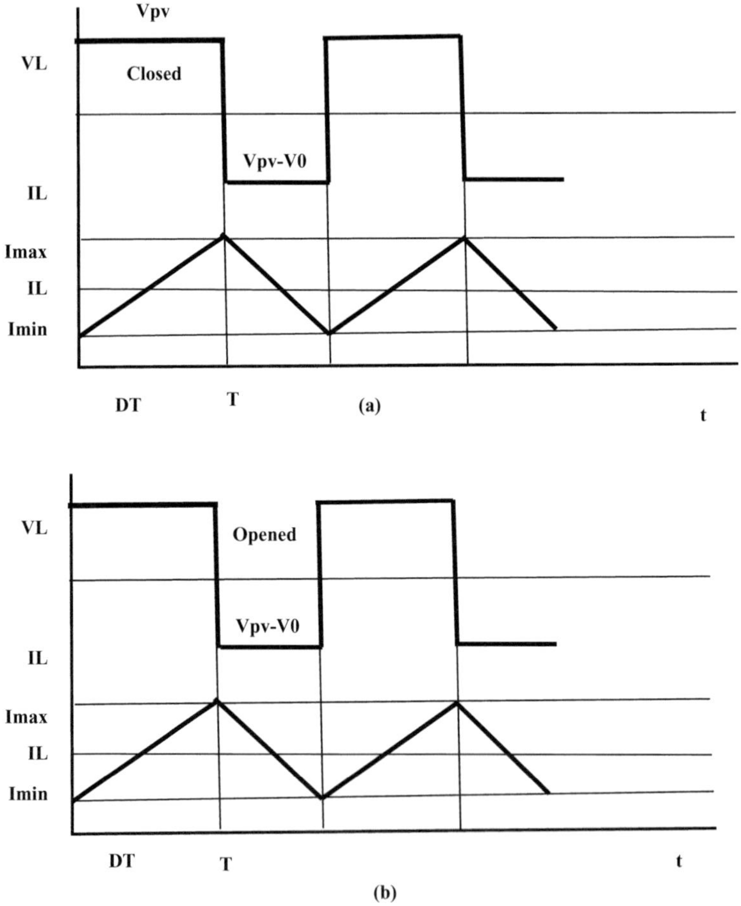

Fig. 3 Chopper waveforms. **a** During on-state; **b** during off-state

The inductance voltage during the off-state interval with $DT \leq t \leq (1-D)T$ is given by (10).

$$V_L(off) = V_{PV} - V_0 = \mathrm{L}\frac{di_L(off)}{dt}(off) \cong \mathrm{L}\frac{\Delta i_{L\,off}}{\Delta T_{on\,off}} = \mathrm{L}\frac{\Delta i_{L\,off}}{(1-D)T} \tag{10}$$

Considering the steady state conditions, it is observed that the net change in the inductor currents over one period is zero. This behavior arises due to the circuit's high inductive nature, causing the currents to exhibit a linear change. Consequently, combining (9) and (10) yields:

$$\Delta i_{L\,on} + \Delta i_{L\,off} = \frac{V_{PV}}{\mathrm{L}}DT + \frac{V_{PV} - V_0}{\mathrm{L}}(1-D)T = 0V_0 = \frac{V_{PV}}{1-D};\ and\ I_0 = \frac{V_0}{R_L} \tag{11}$$

where V_O, I_O are the load voltage and current respectively, and RL is the load resistance.

The critical inductance L_{cr} needed to realize continuous current mode is given by (12).

$$L_{cr} = \frac{(1-D)}{2f}R_L \tag{12}$$

where f is the chopping frequency. The required capacitance needed to realize minimum voltage ripples is given by the (13).

$$C - \frac{D}{\left(\frac{\Delta V_0}{V_0}\right)f.R_L} \tag{13}$$

where $\left(\frac{\Delta V_0}{V_0}\right)$ is the output voltage ripples.

The dissipated power of the circuit can be divided into:

(1) Inductor conduction losses due to internal inductor resistance r_{L1} can be expressed as in (14).

$$P_{rL1} = r_{L1}I_{Lrms}^2 = r_{L1}\left(I_{L1}^2 + \left(\frac{\Delta I_{L1}^2}{2\sqrt{3}}\right)\right) \tag{14}$$

where $\Delta I_{L1} = \frac{V_{PV}D}{L_{cr}f}$ presents the current change at the end of the on-state mode. $I_{L1} = \frac{V_{PV}}{R_L(1-D)^2}$—the average inductor current over complete switching period.

(2) The switch (Transistor) conduction loss due to internal bulk resistance r_{DS} during on-state mode is stated in (15).

$$P_{rDS} = r_{DS} I_{DS}^2 = r_{DS} \frac{D}{3} \left(4I_{min}^2 + I_{max}^2 - 2I_{min}.I_{max}\right) \quad (15)$$

where the minimum and maximum current can be stated as follow:

$$I_{min} = \frac{V_{PV}}{R_L(1-D)^2} - \frac{V_{PV}.D}{2L_{cr}f}; \quad I_{max} = \frac{V_{PV}}{R_L(1-D)^2} + \frac{V_{PV}.D}{2L_{cr}f} \quad (16)$$

(3) The diode conduction loss due to internal bulk resistance r_f during on-state interval is stated as in (17)

$$P_{rf} = r_f.I_{DM}^2 = r_f.\frac{(1-D)}{3} \left(4I_{min}^2 + I_{max}^2 - 2I_{min}.I_{max}\right) \quad (17)$$

(4) The capacitance conduction loss due to internal resistance r_{c1} among the operation period can be expressed as in (18):

$$P_{rC1} = r_{c1} I_{C1}^2 = r_{C1}\left[D\left(I_0^2 - I_{max}^2\right) + I_{max}^2\right] \quad (18)$$

Combining (14) to (18), the total chopper loss during one operation period is stated in (19)

$$P_{Loss} = P_{rL1} + P_{rDS} + P_{rf} + P_{rC1} \quad (19)$$

This paper primarily focuses on the behaviors of the boost chopper, with particular attention to the transistor switch, which acts as the power driver for the circuit. As a result, the following description will focus solely on transistor losses P_{rDS}.

3 Maximum Power Point Tracking (MPPT) of Solar PV System

To operate the photovoltaic modules in a manner that extracts the maximum power from the system. This can increase the tracking efficiency. The MPPT can be achieved by modifying the output voltage and current. MPPT is most effective under cold weather, cloudy or hazy days. A solar PV power system that delivers power to a load consist of the following main components: solar module, batteries and converter circuitry and MPPT controller. Generally, the MPPT controller is installed in between PV system and load [6] (Fig. 4).

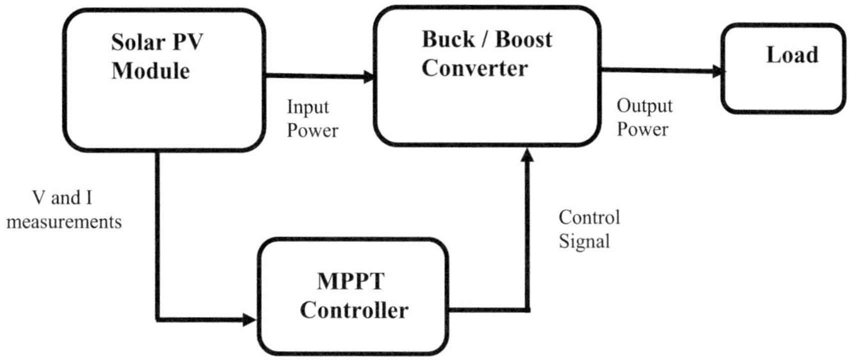

Fig. 4 Conventional solar PV system delivering power to load

- **Perturb and Observe technique (Hill Climbing)**

In this work we use Perturb and Observe technique (P&O). The concept behind the P&O method is to modify the operating voltage or current of the PV panel until the maximum power is obtained from it. The controller operates by continuously monitoring the solar array voltage. Hill Climbing method can fail under rapid environment change conditions. The P&O algorithm is a relatively simple yet powerful method for MPPT. The algorithm Follow an iterative approach to MPPT. A flowchart of the method can be seen in Fig. 5 [7].

4 Experimental Results

4.1 Effect of Variation of Irradiation

In Figs. 6 and 7 we can see the effect of change in solar irradiation on PV characteristics. From Fig. 6 we observe that as we increase the solar irradiation short circuit current increases. Variation in Solar irradiation effects mostly on current, as we can observe from Fig. 6 as we increase solar irradiation from 200 to 1000 W/m^2 current increases from 2 to 8 A approximately but effect of variation of solar irradiation on voltage is very less. Figure 7 shows the effect of variation of solar irradiation on P–V characteristics. As solar irradiation increases, power generated also increases. Increase in power is mainly due to increment in current.

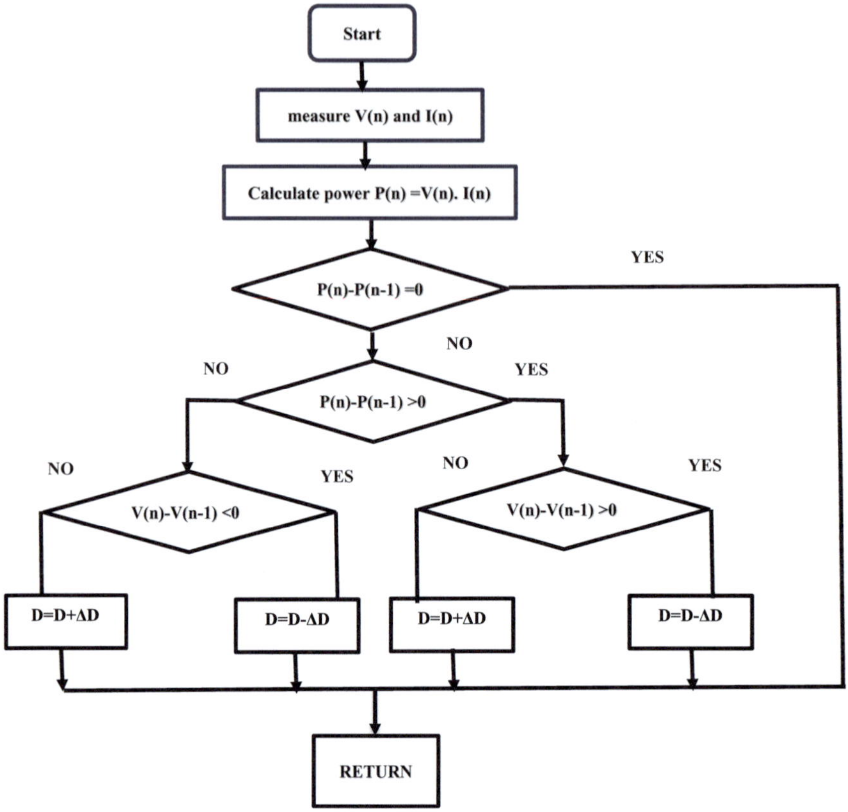

Fig. 5 Perturb and observe technique (Hill Climbing)

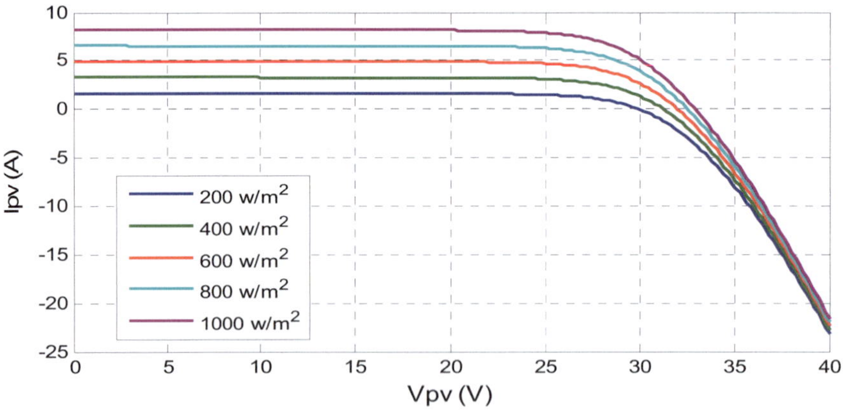

Fig. 6 *Ipv–Vpv* PV module curves for temperature T = 25 °C, and for different irradiance values

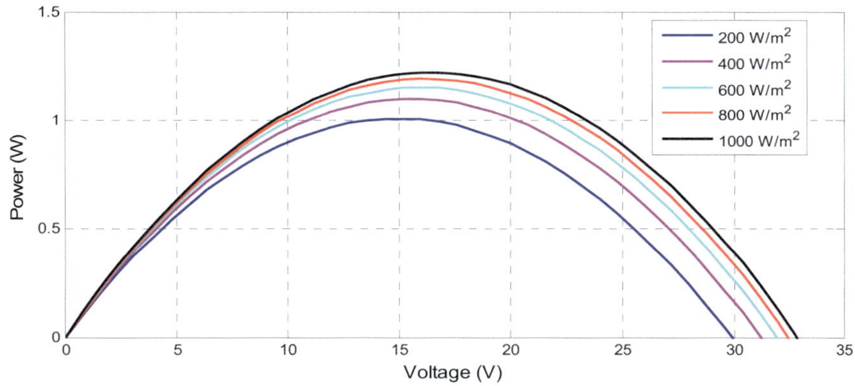

Fig. 7 *Ppv–Vpv* PV module curves for temperature T = 25 °C, and for different irradiance values

4.2 *Effect of Variation of Temperature*

The outcome of variation of temperature on I–V characteristics is shown in Fig. 8. From Fig. 8 we can see the variation of temperature mostly effects voltage, as we increase the temperature voltage decreases but current remains almost unaltered. Figure 9 shows effect of temperature variation on the P–V characteristics. As temperature increases power generated decreases, because on increment of temperature voltage decreases [8].

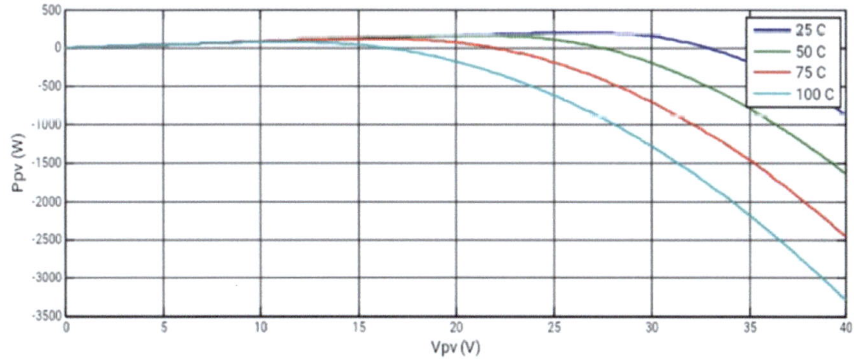

Fig. 8 Effect of variation of irradiation on I–V characteristics

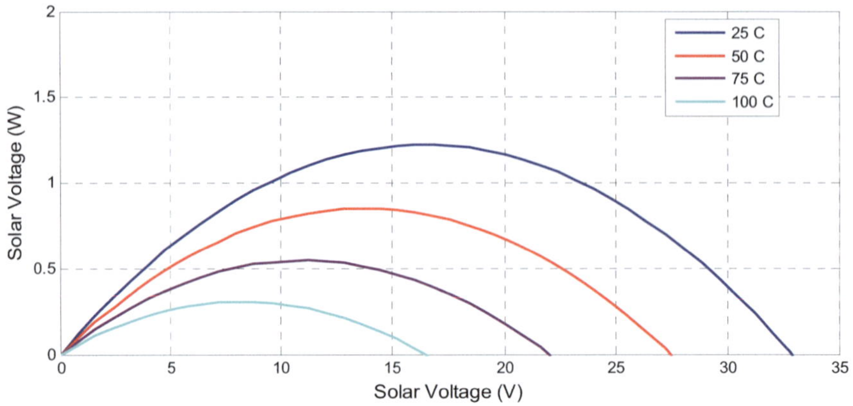

Fig. 9 Effect of variation of temperature on P–V characteristics

4.3 *Outputs After MPPT*

Output power and output voltage after maximum power point tracing are manifested in Figs. 10 and 11 respectively. As we observe from Fig. 10 maximum power is achieved at voltage 14 V; from Fig. 10 we can see we are able to track the output voltage where we can get the maximum power which is approximately 14 V. From Fig. 11 we can see the maximum power which is approximately 350 watts can be tracked.

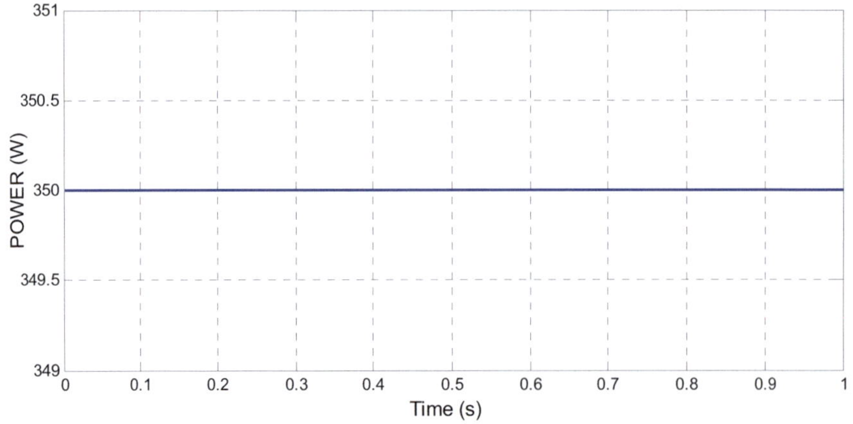

Fig. 10 Output power of PV module after MPPT

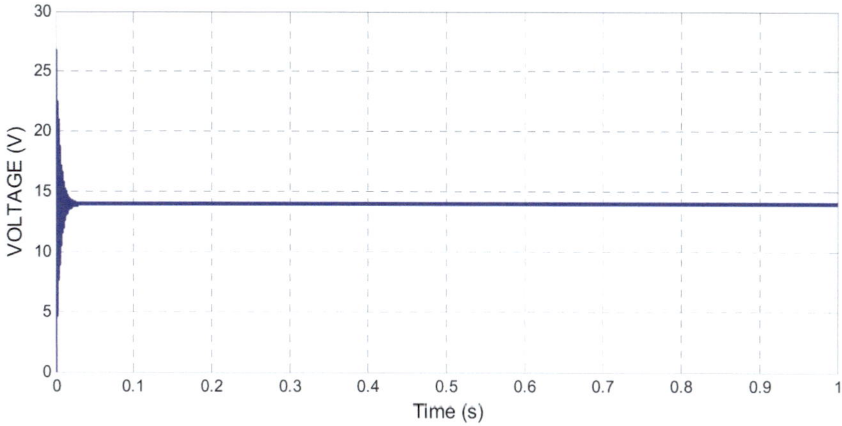

Fig. 11 Output voltage of PV module after MPPT

5 Conclusion

The buck–boost DC/DC converter topology is the only one which allows for the tracking of the PV module maximum power point regardless of temperature, irradiance and connected load. Connecting the buck–boost DC/DC converter to the output panel in a photovoltaic facility is considered a good practice to improve performance.

In this section we present the validation of our PV conversion control, as shown in Fig. 4. To verify the effectiveness of the adopted solution, we performed a set of simulation tests. Firstly, the control validation was performed with different values of irradiance falling within this interval: [200, 400, 600, 800, 1000 W/m^2]. Secondly, the validation was conducted with different temperature which include 25, 50, 75 and 100 °C. This practice ensures the PV panel maximum power point tracking for any solar irradiation, cell temperature and load conditions, which could lead to higher system efficiency for the facility.

References

1. Abdel-Rahim O, Wang H (2020) A new high gain DC-DC converter with model-predictive-control based MPPT technique for photovoltaic systems. CPSS Trans Power Electron Appl 5(2):191–200
2. Raghavendra KVG, Zeb K, Muthusamy A, Krishna TNV, Kumar SVP, Kim DH, Kim MS, Cho H-G, Kim HJ (2019) A comprehensive review of DC–DC converter topologies and modulation strategies with recent advances in solar photovoltaic systems. Electronics 9(1):31
3. Radhika S, Margaret V (2021, Feb) A review on dc-dc converters with photovoltaic system in dc micro grid. J Phys Conf Ser 1804(1):012155
4. Ba A, Ehssein CO, Mahmoud MEMOM, Hamdoun O, Elhassen A (2018) Comparative study of different DC/DC power converter for optimal PV system using MPPT (P&O) method. Appl Solar Energy 54:235–245

5. Rao C, Hajjiah A, El-Meligy MA, Sharaf M, Soliman AT, Mohamed MA (2021) A novel high-gain soft-switching DC-DC converter with improved P&O MPPT for photovoltaic applications. IEEE Access 9:58790–58806
6. Raj A, Praveen RP (2022) Highly efficient DC-DC boost converter implemented with improved MPPT algorithm for utility level photovoltaic applications. Ain Shams Eng J 13(3):101617
7. Nayak B, Mohapatra A, Mohanty KB (2017) Selection criteria of dc-dc converter and control variable for MPPT of PV system utilized in heating and cooking applications. Cogent Eng 4(1):1363357
8. Li S, Liao H, Yuan H, Ai Q, Chen K (2017) A MPPT strategy with variable weather parameters through analyzing the effect of the DC/DC converter to the MPP of PV system. Sol Energy 144:175–184

Random Sinusoidal PWM for Three Phase Inverter Using ATMega2560

Aymen Chaaira, Rabiaa Gamoudi, Habib Kraiem, and Lassaad Sbita

Abstract The Random sinusoidal pulse width modulation (RSPWM) is a non-conventional spread spectrum technique. It is proposed for the suppression of EMI and the improvement of power quality in a three-phase power inverter. This paper propose a digital modulation technique. It is implemented through a compact low-cost microcontroller ATMEGA2560 by developing suitable C coding for producing the required switching pulses for the inverter. A comparative analysis is also presented between the proposed random sinusoidal PWM modulation RSPWM and the conventional SPWM. It is shown that the implementation of RSPWM results in reduced EMI, a decreased Total Harmonic Distortion THD, and the noise effect is also improved significantly.

Keywords ATMEGA2560 · EMI · RSPWM · Three phase power inverter · THD

1 Introduction

Power inverter are commonly employed in applications such as variable speed drives (ASDs), uninterruptible power supplies (UPS), induction heating, and high voltage direct current transmission system. They rely heavily on PWM techniques for operation. In fact, PWM approaches have several advantages such as very low power

A. Chaaira (✉) · R. Gamoudi · L. Sbita
Processes, Energy, Environment and Electrical Systems Laboratory, University of Gabes, Gabes, Tunisia
e-mail: aymen.chaaira@gabes.r-iset.tn

L. Sbita
e-mail: lassaad.sbita@enig.rnu.tn

H. Kraiem
Department of Electrical Engineering, College of Engineering, Northern Border University, Arar 73222, Saudi Arabia
e-mail: ALhabeeb.Kareem@nbu.edu.sa

© The Author(s), under exclusive license to Springer Nature Singapore Pte Ltd. 2024

A. Mellit et al. (eds.), *Proceedings of the 2nd International Conference on Green Energy Conversion System*, Green Energy and Technology,
https://doi.org/10.1007/978-981-97-6148-7_18

loss, less noise emission, and willingly adjusted output voltage. They are key strategies for improving the performance of inverter drive systems. The PWM inverter, in general, operates at a constant switching frequency. However, they are source of harmonic pollution which must be minimized for a proper function. Recently, novel PWM approaches for noise reduction, known as RPWM methods, have been intensively researched [1–7]. Instead of a single set switching frequency, they employ broadband switching frequencies to disperse the noise spectrum. RPWM systems use distinct carrier switching frequencies at each switching instant to spread the voltage, current, and noise spectrum. Because of its simple methodology, RPWM techniques are gaining popularity as good solutions for noise reduction [1–5].

PWM approaches are divided into two categories: analog and digital. Analog approaches and analog controllers might involve drawbacks [8]:

- They are noise sensitive,
- They are impacted by temperature changes or component fluctuation,
- They are fixed and less adaptable than digital solutions,
- They require a large number of linked discrete components, which raises inventory costs and the manufacturing process's complexity, and
- They also need a large number of power supply for the multiple individual components.

As a result, digital PWM methods are preferable. They are more adaptable and less susceptible to ambient noise. DSPs, microcontrollers, microprocessors, FPGA, and ASICs can be used to develop digital PWM approaches. The current project makes use of an ATMEGA2560 microcontroller. It provides a more advantageous option for digital PWM approaches than other solutions due to its easy computational characteristics, and it is a cost-effective option.

In this paper, efforts are also taken to measure the essential power quality parameters such as Total Harmonic Distortion (THD) and welch power spectral density (PSD) to measure the harmonic spectrum using RSPWM modulation technique in three-phase inverter.

2 Proposed Control Method

The switching frequency and the used pulse width modulation (PWM) technology determine the harmonic distortion in the current. Due to practical constraints, the switching frequency cannot be raised beyond a particular range. Several efforts have been made to decrease distortion at a particular switching frequency by the use of a well-designed PWM approach. Specially developed PWM approaches for high power applications with low inverter switching frequency have been reported. Several modulating functions and frequency modulation of carrier have been researched at switching frequencies (say a few kilohertz) significantly higher than the maximum fundamental frequency (say 50 or 60 Hz).

Random sinusoidal pulse width modulation (RSPWM) is a switching method that is used for spreading voltage and current harmonics throughout a broad range of frequencies. The spectrum of electromagnetic and acoustic noise may be expanded across a broad frequency band by utilizing a random switching frequency. Thus wideband noise is obtained which is considerably ameliorated and less disturbing than narrowband noise. The suggested RSPWM approach reduces audible switching noise and spreads voltage and current harmonics across a broad region. As a result, RSPWM has gained popularity as an effective method for reducing noise generated in inverter drive circuits [1–6].

3 Random Sinusoidal PWM Generation

To randomize the switching frequency, a random function should be utilized.

A good random generator must be able to [2]:

- Create random numbers over a broad range,
- Generate random numbers with uniform distribution,
- Prevent periodicity.

Linear Congruential Generator (LCG) is a popular and most used method to generate random number. It utilizes a linear model to generate a random number defined as follows [9–11]:

$$X_i = \mathrm{mod}\,(aX_{i-1} + c, m)\ where \begin{cases} \mathrm{mod} : \textit{remainder of the Euclidean division}; \\ X_n,\ \mathrm{a, c, m} : \textit{integer value}; \\ \mathrm{m} : \textit{modulus}; \\ \mathrm{a} : \textit{multiplier}(0 < a < m); \\ \mathrm{c} : \textit{increment factor}(0 \le c < m); \\ \mathrm{X_0} : \textit{initial value}(\textit{seed})(0 \le X_0 < m); \end{cases}$$

where m, a, and c parameters should keep conditions:

$$\begin{cases} \gcd(m, c) = 1,\ (\gcd : \textit{greatest common divisor}) \\ \mathrm{mod}\,(a - 1, P) = 0\, \textit{where}\ P : \textit{all prime factors of}\ m, \\ \mathrm{mod}\,(a - 1, 4) = 0\, \textit{if}\ \ \mathrm{mod}\ (m, 4) = 0. \end{cases}$$

We choose $m = 2^{31}$, $a = 1103515245$, $X_0 = c = 12345$.

The RSPWM is programmed in c code and loaded to ATMEGA2560 microcontroller.

Then it was simulated under ISIS of Proteus to generate the control signals to gate the inverter's IGBTs as shown in Fig. 1. Figure 2 show the block diagram of implementation of RSPWM. After that, the simulated data is exported to Simulink / MATLAB.

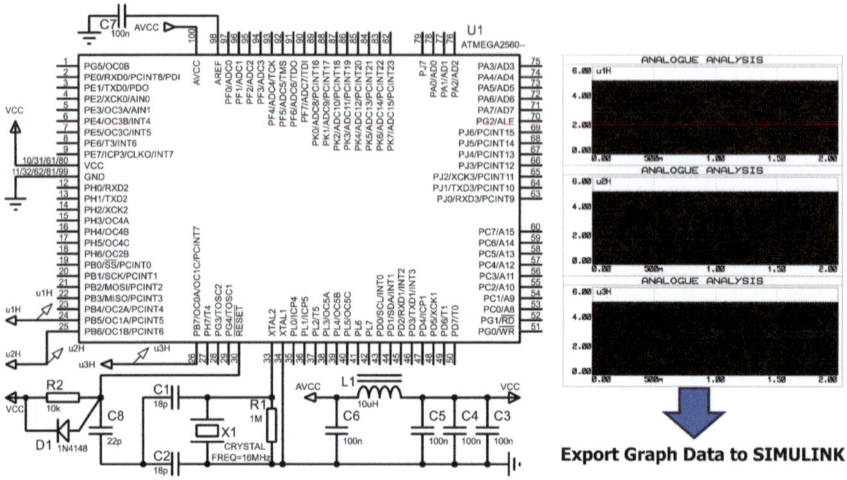

Fig. 1 ISIS simulation of ATMEGA2560 and data aquisition

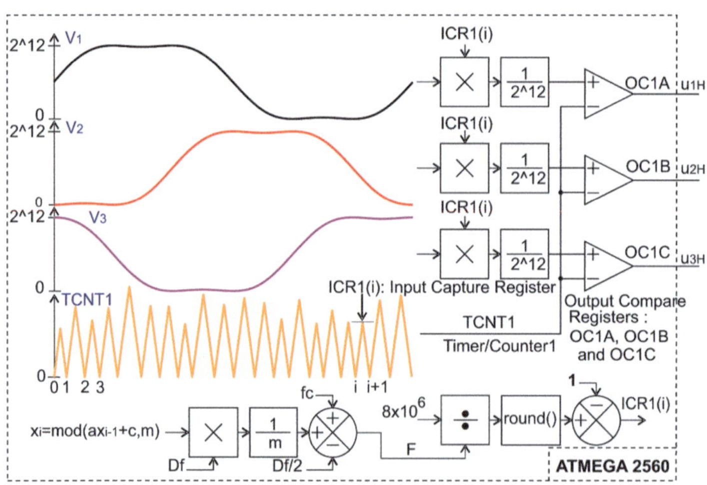

Fig. 2 Block diagram of implementation of proposed RSPWM

4 Proposed Simulation and Analysis

The RSPWM system that was implemented comprises of an ARDUINO board with the control program, an inverter, and a 4 kW three-phase asynchronous motor. The suggested RSPWM algorithm is run in the ATMEGA2560 controller, and the pulse signals to gate the inverter are generated through the ARDUINO board's output port. When measuring overall noise, including mechanical noise created by motor drives, the system is typically run at rated load and speed. However, in this study, we are only interested in the electromagnetic switching noise produced by the inverter switching mechanism. As a result, the simulation was carried out under no load conditions.

To confirm the effectiveness of harmonics dispersion, a three-phase inverter using the RSPWM modulation is built in MATLAB/Simulink and relevant simulations are conducted with the dc-link voltage Vdc = 566 V, the carrier frequency $f_c = 5$ kHz, Df = 3 kHz and the fundamental frequency $f_0 = 50$ Hz. The Simulink model is shown in Fig. 3.

Comparing the results obtained by conventional SPWM and RSPWM techniques Figs. 4, 5 and 6, it can be seen that, upon implementing RSPWM modulation technique, the THD of the AC source current is nearly equal to 5.47%, the limit specified by IEEE Std. 519 being 5% [12]. Even though the THD of the AC source current is slightly increased, the PSD has drastically decreased. Current harmonics move to higher frequencies as the carrier frequency increases, and harmonic magnitudes decrease.

From the simulation results Fig. 6, it is observed that PSD level is − 36 dB/Hz.

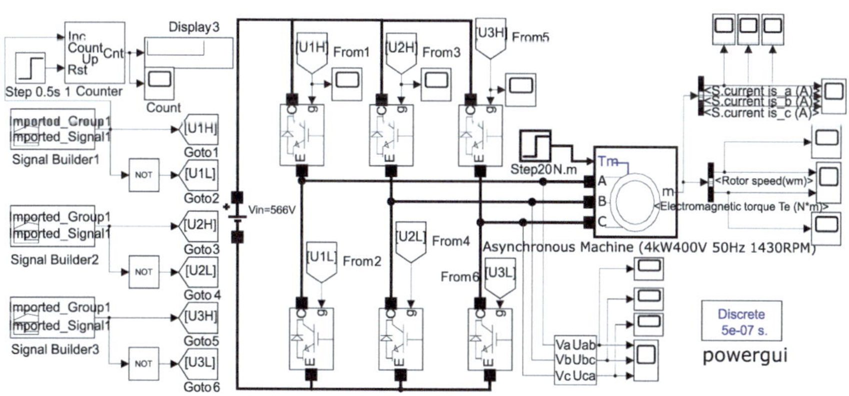

Fig. 3 Simulink model

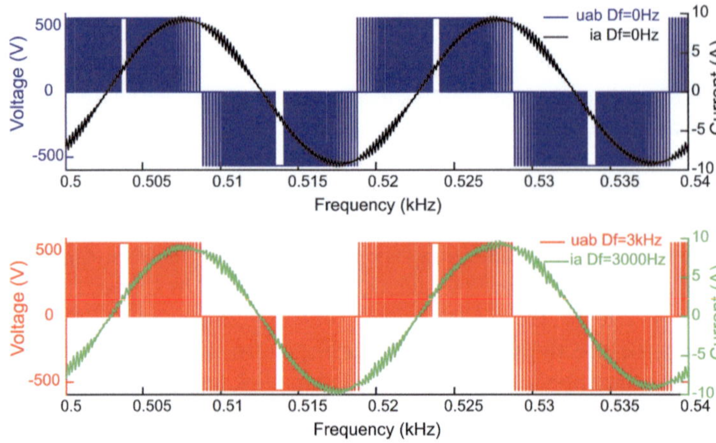

Fig. 4 Output voltage and current signal of the inverter under the SPWM (on the top) and RSPWM

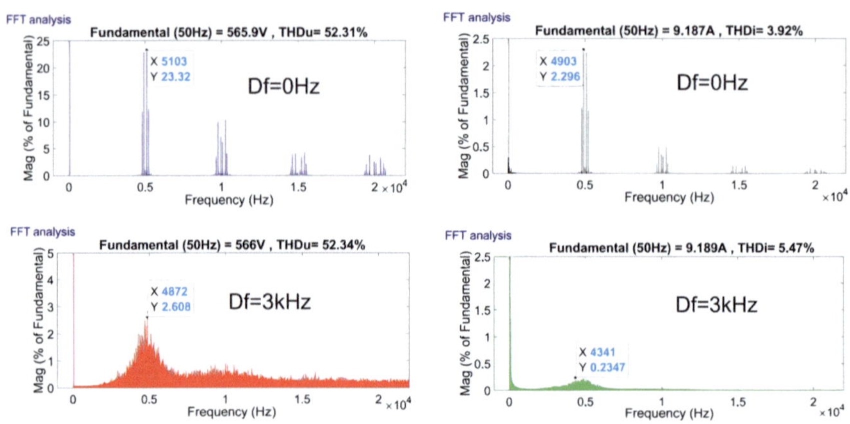

Fig. 5 FFT analysis (powergui)

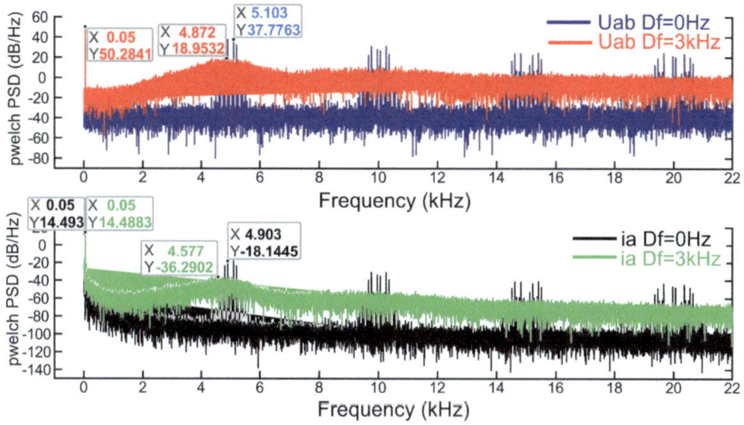

Fig. 6 The inverter output voltage and current PSD

5 Conclusion

The RSPWM technique is presented in this paper to preserve the advantageous characteristics of spread spectrum modulation, which is immune in nature and has a broad range of EMI spectrum. The amplitude of PSD and THD values are lowered by applying random frequency modulation techniques and are within the limitations of the IEEE Std. 519. The RSPWM modulation method offers the advantage of effective noise removal. It has a wide spectrum spread and is technologically robust.

References

1. Mohan Das R, Chandira Sekaran E (2019) A cloud system model employing random space vector pulse width modulation for noise reduction in VSI fed induction motor. Cluster Comput 22:347–360. https://doi.org/10.1007/s10586-018-1956-y
2. Kim K-S, Jung Y-G, Lim Y-C (2009) A new hybrid random PWM scheme. IEEE Trans Power Electron 24:192–200. https://doi.org/10.1109/TPEL.2008.2006613.(LCG)
3. Pindoriya RM, Rajpurohit BS, Kumar R (2020) A novel application of harmonics spread spectrum technique for acoustic noise and vibration reduction of PMSM drive. IEEE Access 8:103273–103284. https://doi.org/10.1109/ACCESS.2020.2999336
4. Jadeja R, Ved AD, Chauhan SK, Trivedi T (2020) A random carrier frequency PWM technique with a narrowband for a grid-connected solar inverter. Electr Eng 102:1755–1767. https://doi.org/10.1007/s00202-020-00989-6
5. Li G, Fu Z, Wang Y (2021) Electromagnetic vibration and noise suppression of induction motor based on RPWM selective spectrum shaping. IEEE Access 9:54509–54517. https://doi.org/10.1109/ACCESS.2021.3064848
6. Xu D, Yang M, Chen Y, Long J, Xu D (2022) EMI-regulated SiC-based motor drives with Markov chain pseudo-random frequency space vector pulse width modulation strategy. IEEE Trans Energy Convers 37:2348–2358. https://doi.org/10.1109/TEC.2022.3179138

7. Boopathi R, Jayanthi R, Ansari MMT (2020) Power quality improvement in wind energy conversion system using hybrid SVPWM inverter control technique for THD reduction. Int J Dyn Control 8:592–603. https://doi.org/10.1007/s40435-019-00556-3
8. Soomro J, Memon TD, Shah MA (2016) Design and analysis of single phase voltage source inverter using unipolar and bipolar pulse width modulation techniques. In: 2016 International conference on advances in electrical, electronic and systems engineering (ICAEES). IEEE, Putrajaya, Malaysia, pp 277–282. https://doi.org/10.1109/ICAEES.2016.7888052
9. Imamah I (2018) One time password (OTP) based on advanced encrypted standard (AES) and linear congruential generator (LCG). In: 2018 Electrical power, electronics, communications, controls and informatics seminar (EECCIS). IEEE, Batu, East Java, Indonesia, pp 391–394. https://doi.org/10.1109/EECCIS.2018.8692931
10. Entacher K (1998) Bad subsequences of well-known linear congruential pseudorandom number generators. ACM Trans Model Comput Simul 8:61–70. https://doi.org/10.1145/272991.273009
11. Hellekalek P (1998) Good random number generators are (not so) easy to find. Math Comput Simul 46:485–505. https://doi.org/10.1016/S0378-4754(98)00078-0
12. IEEE standard for harmonic control in electric power systems. IEEE. https://doi.org/10.1109/IEEESTD.2022.9848440

Thermal Behavior of Photovoltaic Modules Under Dirty and Different Shading Profiles

Nouamane Kellil, Catalina Rus Casas, and Adel Mellit

Abstract The energy efficiency of Photovoltaic (PV) systems is strongly linked to both manufacturing factors, such as the nature of the PV cells, and environmental factors (irradiance and temperature). However, during its operation, the PV system is subject to several types of faults that can affect its optimal operation. The soiling and shadowing, which create an inhomogeneous distribution of the incident irradiance on the PV cell or panel, is one of the most commonly encountered PV faults that leads to an overall degradation of the energy produced by this panel. The main objective of this paper is to study the thermal impact of the shadowing on the PV panels. For this, two types of shading have been created artificially: the first is the covering (total and partial) of the panel, and the second is the shading using a nearby opaque object. Thermal images were taken using the Fluke Ti25 IR thermal camera. The obtained results show the generation of hotspots on the shaded panel and a remarkable decrease in power produced.

Keywords PV panel · Shadow · PV faults · Soiling · Thermal images

1 Introduction

The demand for energy is still constantly increasing due to population growth and the improvement of living conditions. The depletion and non-renewability of certain fossil resources have made energy supply a strategic problem for some countries. The environmental aspect is considered as a major constraint of the production and

N. Kellil
Unité de Développement des Equipements Solaire/EPST CDER, RN 11, BP 386, 42415 Bou Ismail, Tipaza, Algeria

C. R. Casas
University of Jaen, Campus Las Lagunillas s/n, 23071 Jaén, Spain

A. Mellit (✉)
University of Jijel, 18000 Jijel, Algeria
e-mail: adelmellit2013@gmail.com

© The Author(s), under exclusive license to Springer Nature Singapore Pte Ltd. 2024
A. Mellit et al. (eds.), *Proceedings of the 2nd International Conference on Green Energy Conversion System*, Green Energy and Technology,
https://doi.org/10.1007/978-981-97-6148-7_19

exploitation of fossil energy, in particular the emission of greenhouse gases, which, according to the COP 27 agreement, must decrease by 45% by 2030 to limit global warming to 1.5 °C [1].

Renewable energy has been resorted to as an alternative to fossil energies, as they are the ideal solution for clean, safe, and sustainable energy. Solar energy is harnessed using various technologies to convert sunlight into electricity or heat for various applications. Photovoltaic (PV) cells, made of semiconductor materials such as silicon, generate a direct current, converting sunlight directly into electricity. PV modules, made up of multiple solar cells connected together, are the building blocks of solar PV systems. They are installed on rooftops, ground-mounted structures, or integrated into building materials. However, significant external factors may have a detrimental impact on the performance of PV systems, reducing their energy output and overall efficiency, such as shading and soiling. Because solar cells are connected in series, shading of one cell can affect the entire string. This can cause a significant decrease in overall power [2]. Furthermore, in a series-connected PV array, the current produced by each cell is determined by the weakest cell. Shading can cause mismatch losses, where shaded cells produce less current, reducing the overall output of the entire array.

The accumulation of pollutants on the surface of solar modules, such as dust, dirt, pollen, and bird droppings, reduces the amount of sunlight reaching the PV cells [3, 4]. Different studies have made it possible to develop a mathematical relationship to measure the density of dust accumulation on the active surface of the PV module based on the performance measurements of the PV module [5].

The main objective of this work is to inspect the thermal behaviour of the partially shaded and dusty PV modules using Infra-Red Thermographic Imaging (IRTI). This method is widely used for the detection and diagnosis of underperforming PV systems [6, 7].

2 Materials and Methodology

The investigation of the thermal behavior of partially shaded and dirty Poli-PV modules was carried out using Infra-Red thermographic imaging (IRTI) using Fluke Ti25 camera. The PV generator tested in this work is installed on the roof of the laboratory of the University of Jaen, located in the south of Spain. The PV generator consists of 13 Poly-Si PV modules connected in series (see Fig. 1). Table 1 illustrates the characteristics of PV modules. The overall size of the PV generator is estimated at 2.97 kWp.

A series of current measurements were performed over the whole PV generator, using the PVPM I–V tracer. Different profiles of partial shading have been created by covering different numbers of PV cells.

The images were taken using the Fluke Ti25 IR thermal camera (see Fig. 2). The general guidelines for outdoor IR thermography of PV modules and plants such as

Fig. 1 PV generator on-grid (2.97 kWp)

Table 1 Different configurations

PV panel type	AXITEC (AC-270P/156-60S)
Peak power	270 Wp (± 4 Wp)
V_{max}	31.12 V
I_{max}	8.71 A
V_{oc}	38.21 V
I_{sc}	9.25 A

sun irradiance rate, angle of view and other setting on an IR thermal camera have been considered as described in the IEC TS 62446-3:2017 [8].

We started measuring the thermal behavior of the PV modules in different configurations of shaded and dirty PV modules.

Fig. 2 IR thermal camera Fluke Ti25

3 Results and Analysis

In order to explain the impact of dirty and showing on the PV generator electric output current, an I–V characteristic curve has been traced for healthy, shaded and dirty PV modules. These measurements have been taken in $T_{amb} = 32$ °C and an average solar irradiance of 820 W/m^2 (see Fig. 3).

From Fig. 4, it can be clearly observed that a shift of around 10% of the V_{oc} for the partially shaded and dirty PV panels compared to the V_{oc} of the healthy case. In addition, a drop of almost 30% in I_{sc} is observed only for the complete shaded and half shaded PV panel case.

Figure 4, illustrates the behaviour of the characteristic curves I–V of the whole PV generator with one partial or complete shaded PV panel connected in series, which is not distorted despite the drop in P_{max}.

3.1 Photovoltaic Modules Under Partial Shading

Under sunlight irradiation, the PV panel, partially shaded and connected to the grid, exhibits hotspots. These hotspots are clearly depicted using the IRTI, as illustrated in Fig. 4. These hotspots are generally caused by the low current generated by the shaded PV cells, which behave like a resistance load.

One of the most important causes of energy production losses within PV generators is the partial shading of PV modules, as shown in Fig. 3. I_{sc} is less than the string current, which can cause PV string power dissipation when it is electrically connected to the other unshaded PV modules.

Figure 5 shows a false reflected shadow profile of the nearby object which can possibly lead to false measurements.

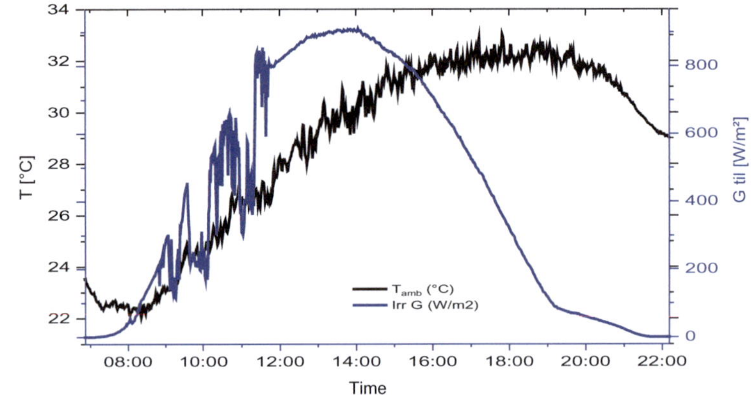

Fig. 3 Meteoritical conditions 07/27/2023

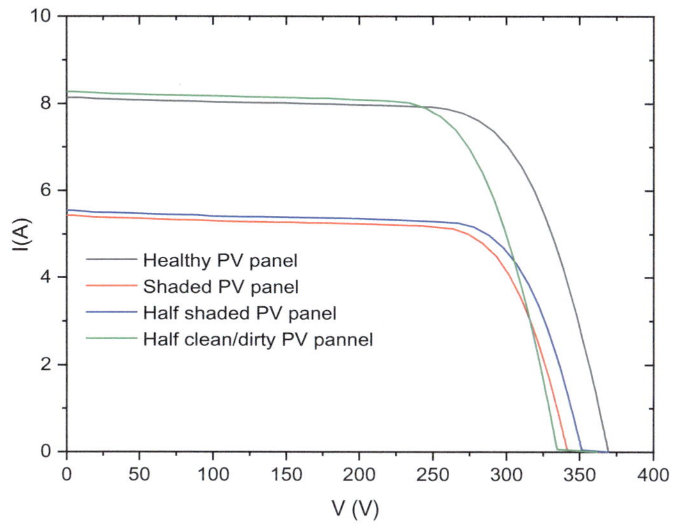

Fig. 4 Experimental I–V characteristic curves of 2.97 kWp PV generator in different conditions

3.2 Dirty Photovoltaic Modules

The accumulation of dust or dirt on the active surface of a PV module poses a significant issue, as materials causing light obstruction act as external resistances, thereby reducing the performance of the PV system [9, 10]. Can cause effects similar to those caused by shading. This type of PV fault is frequent in rooftop PV installations or even in deserts. This type of PV fault is temporary; the PV module should only be cleaned. From Fig. 6 the IRTI clearly shows the generation of hotspots.

Using electric I–V characterization, a comparative analysis between the cleaned and dirty PV (modules was performed under the same experimental conditions) have been performed. We have analyzed the different electric parameters of the PV module (P_{max}, V_{oc}, I_{max}, V_{max}) of both cleaned and dirty PV modules.

Figure 7 shows the difference between the electric performances of the two PV modules. Two PV modules (clean and dirty) were connected to an electric load. The deviation ΔP_{max} caused by the presence of dirt on the surface of the PV module increased considerably as the size of the PV generator increased (Fig. 8).

4 Conclusion and Perspectives

In this study, the thermal and electrical behaviors of partially shaded and dirty poly crystalline PV modules were investigated. These faults, linked to the environmental conditions, are common in most PV installations. They can decrease the performance of PV modules, causing their rapid degradation, potentially leading to a risk of fire.

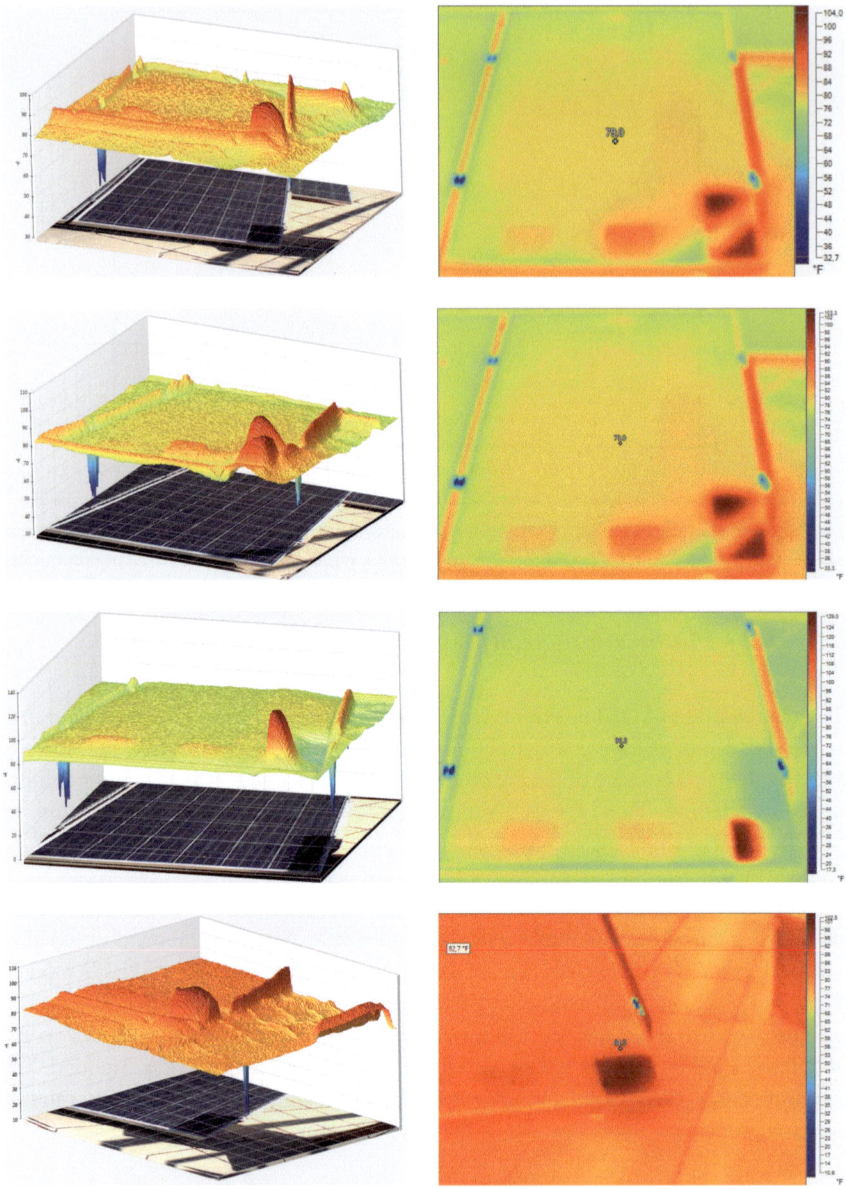

Fig. 5 IRTI of partial shaded PV panel

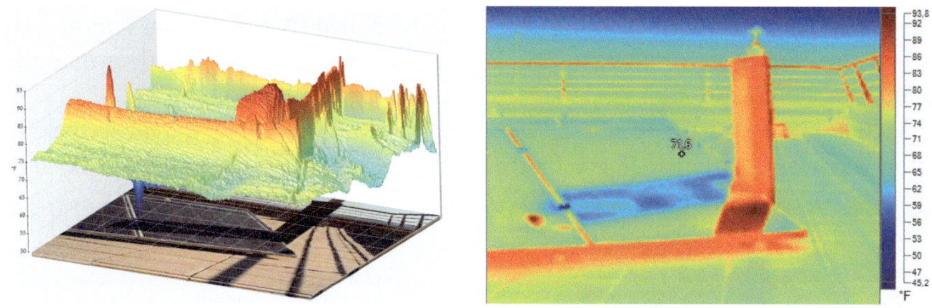

Fig. 6 False shading reflected on the PV panel due to the presence of nearby object

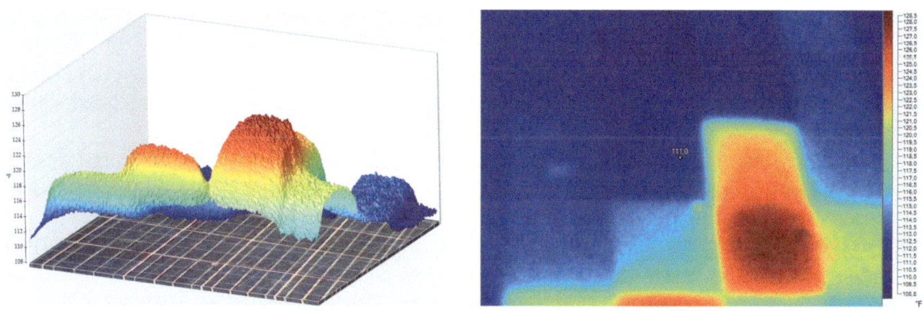

Fig. 7 Hot spots generated by dirt on the active surface of the PV panel

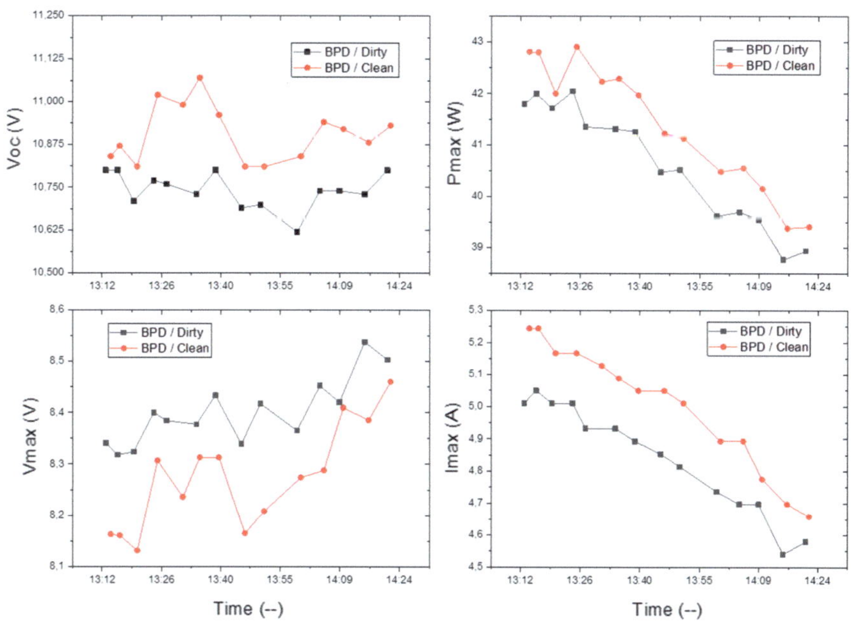

Fig. 8 Comparative representation of the electric parameters of the clean and dirty PV modules

This study demonstrates that an isolated, partially shaded PV module can significantly reduce the output performance of the PV generator, with a greater effect than this due to the accumulation of dirt on the active surface of the PV modules.

In our future work, we aim to develop fault detection and diagnosis models using Machine Learning (ML) methods, specifically Convolutional Neural Networks (CNN) and You Only Look Once (YOLO) for real-time PV fault detection.

References

1. Farajzadeh Z, Ghorbanian E, Tarazkar MH (2023) The impact of climate change on economic growth: evidence from a panel of Asian countries. Environ Dev 47:100898
2. Mellit A, Kalogirou S. Artificial intelligence and internet of things to improve efficacy of diagnosis and remote sensing of solar photovoltaic systems: challenges, recommendations and future directions. Renew Sustain Energy Rev 143:110889
3. Álvarez-Tey G, Clavijo-Blanco JA, Gil-García Á, Jiménez-Castañeda R, García-López C (2019) Electrical and thermal behaviour of crystalline photovoltaic solar modules in shading conditions. Appl Sci 9(15):3038
4. Pavan AM, Mellit A, De Pieri D (2011) The effect of soiling on energy production for large-scale photovoltaic plants. Solar Energy 85(5):1128–1136
5. Hosseini A, Mirhosseini M, Dashti R (2023) Analytical study of the effects of dust on photovoltaic module performance in Tehran, capital of Iran. J Taiwan Inst Chem Eng 148:104752
6. Buerhop C, Bommes L, Schlipf J, Pickel T, Fladung A, Peters IM (2022) Infrared imaging of photovoltaic modules: a review of the state of the art and future challenges facing gigawatt photovoltaic power stations. Prog Energy 4(4):042010
7. Schirripa Spagnolo G, Del Vecchio P, Makary G, Papalillo D, Martocchia A (2012) A review of IR thermography applied to PV systems. In: 2012 11th international conference on environment and electrical engineering, May 2012. IEEE, pp 879–884
8. (2017) IEC TS 62446-3-photovoltaic (PV) systems-requirements for testing documentation and maintenance. Part 3: Photovoltaic modules and plants—outdoor infrared thermography. IEC Geneva Switzerland
9. Sulaiman SA, Singh AK, Mokhtar MMM, Bou-Rabee MA (2014) Influence of dirt accumulation on performance of PV panels. Energy Procedia 50:50–56
10. Kazem HA, Chaichan MT, Al-Waeli AHA, Sopian K (2022) Effect of dust and cleaning methods on mono and polycrystalline solar photovoltaic performance: an indoor experimental study. Sol Energy 236:626–643

Deep Learning-Based Optimization of Energy Efficiency in D2D Communication for Hospital Logistics

Anouar Ben Abdennour, Mohamed Ouwais Kabaou, and Belgacem Chibani Rhimi

Abstract In this paper, we address the crucial issue of energy consumption minimization in device-to-device (D2D) communications in the healthcare logistics domain. The growing demand for real-time communication and data exchange in healthcare environments has led to an increase in the use of D2D technologies, but the associated energy consumption can have a long-term impact on the sustainability of these systems. To tackle this challenge, we propose exploring the use of neural networks to predict the energy consumption of different D2D communication protocols and optimize the parameters to minimize this consumption. The objective of this article is to provide a thorough study on the use of neural networks for energy consumption minimization in D2D communications for healthcare logistics. We will use simulations to evaluate the results of this approach, and we will discuss the implications of our conclusions for the design of sustainable communication systems in healthcare environments.

Keywords Mobile communication D2D · Energy · Neural network · Hospital logistic

1 Introduction

The use of D2D (Device-to-Device) communication in hospital logistics is a developing field that promises to greatly improve logistics operations in hospitals. D2D communication is a form of wireless communication that allows devices such

A. Ben Abdennour (✉) · M. O. Kabaou · B. C. Rhimi
National School of Engineering of Gabes, R.L MACS-ENIG, University of Gabes, Gabes, Tunisia
e-mail: ben.abdennour.anouar@gmail.com

M. O. Kabaou
e-mail: ouwayess@yahoo.fr

B. C. Rhimi
e-mail: bouahmed17@gmail.com

© The Author(s), under exclusive license to Springer Nature Singapore Pte Ltd. 2024

A. Mellit et al. (eds.), *Proceedings of the 2nd International Conference on Green Energy Conversion System*, Green Energy and Technology,
https://doi.org/10.1007/978-981-97-6148-7_20

as smartphones, tablets, and medical devices to communicate directly with each other without going through a central network. This technology is very useful for complex and constantly evolving environments like hospitals, where information must be quickly and effectively shared between different departments and healthcare professionals [1].

The use of D2D communication in hospital logistics can be used to improve patient tracking processes, inventory management, and medical equipment management. For example, medical devices could communicate directly with each other to track patient health information, while doctors' and nurses' smartphones could be used to share patient information and real-time inventory data [2].

The use of D2D communication in hospital logistics can also enable increased collaboration between different hospital departments, leading to improved patient care quality. Furthermore, by enabling more efficient resource management, D2D communication can also contribute to cost reduction for hospitals, which is critical in an industry that is constantly under pressure to reduce costs and improve efficiency [3-5].

Energy consumption can be a major issue in the field of hospital logistics, particularly when it comes to D2D communication. The medical devices and smartphones used for D2D communication can consume a large amount of energy to operate continuously, leading to increased electricity costs for hospitals. As a result, it is important to optimize energy consumption of devices to reduce costs and minimize environmental impact [6-8].

Energy optimization can be achieved by using technologies such as energy consumption management in devices, use of longer-lasting batteries, and data processing optimization to reduce energy consumption [9].

The goal of this article is to propose a new approach that aims to reduce energy consumption of connected objects and improve the efficiency and lifespan of the network.

2 Problem Formulation and System Model

2.1 *Problem Formulation*

Energy consumption can be a major issue in D2D communication in healthcare logistics. Mobile devices such as D2D communication terminals are often used for extended periods, which can result in high battery consumption and reduced battery life. This can compromise the quality of patient care, as terminals may shut down or become unavailable when needed [10, 11].

To minimize the impact of energy consumption on D2D communication, it may be necessary to optimize communication protocols to minimize listening time and number of transmissions, use larger capacity batteries, and implement power management strategies to conserve energy. It may also be possible to use deep learning techniques to minimize energy consumption by taking into account contextual parameters such as signal quality and transmission distance [12, 13].

How can the energy consumption of devices used for D2D communication in healthcare logistics be reduced while ensuring the quality of communication and maximizing battery life?

2.2 System Model

A neural network can be used to minimize energy consumption in D2D communication by using a neural network model to predict energy consumption based on different factors such as the distance between devices, transmission power, and transmission frequency. The model can be trained using historical data on energy consumption and related factors. Once trained, the model can be used to predict energy consumption for different configurations and optimize parameters by minimizing energy consumption [14-16].

For example, if the model predicts that energy consumption will increase when a certain distance is exceeded, the network can be used to determine the optimal distance between devices to minimize energy consumption. Similarly, the model can be used to optimize transmission power and frequency based on distance and other factors.

A neural network was used to optimize energy consumption of communication devices by following these steps:

1. Data Collection: We collected data on energy consumption and related factors for different D2D communication configurations in hospital logistics. The data may include the distance between devices, transmission power, transmission frequency, device type, etc.
2. Data Preparation: We prepared the data by normalizing and dividing it into training, validation and test sets.
3. Model Definition: We defined the neural network model by choosing the number of hidden layers, the number of neurons per layer, and the activation function. The inputs are the features associated with the communication configuration, and the output would be the predicted energy consumption.
4. Model Training: We trained the model using the training data and minimizing the error using a cost function such as means squared error.
5. Model Evaluation: We evaluated the model using the validation data to verify its accuracy and its ability to generalize to new data.
6. Parameter Optimization: We optimized the model parameters using methods such as gradient descent to minimize energy consumption.
7. Final Evaluation: We evaluated the final model using the test data to measure its accuracy and its ability to predict energy consumption based on related factors.

This model is used to predict energy consumption for different D2D communication configurations in hospital logistics and to determine the optimal configuration to minimize energy consumption.

3 Neural Network

A neural network is a signal processing model that is inspired by the human brain. It consists of several layers of interconnected neurons that work together to perform a specific task, such as image recognition, data classification, time series prediction, etc.

Each neuron in the network receives inputs, which can be numerical values or features extracted from input data, and produces an output based on these inputs. The weights associated with these inputs are optimized during the learning process to minimize a cost function that measures the error between the predicted outputs and the desired outputs.

Neural networks can be trained from scratch or pre-trained on large amounts of data for a specific task, then adapted to new data sets using deep learning methods [17].

3.1 Concept of Neural Network

The concept of neural network is explored through the most classical architecture: The Multi-Layer Perceptron (MLP). Although many other neural network architectures exist, they all share some of the basic principles defined in this section. The most fundamental element of a neural network is referred to as a neuron, and is represented by the following diagram [18].

A neuron is defined by an input vector x of dimension n (3 in Fig. 1), weights $\omega_1, \ldots, \omega_n$, *a* bias b, and an activation function σ. The activation function receives an affine combination of the inputs:

$$z = \sum_i (\omega_i x_i) + b$$

Given an input x, the output y of the neuron is calculated as

$$y = \sigma\left(\sum_i (\omega_i x_i) + b\right)$$

Some commonly used activation functions σ includes the logistic function:

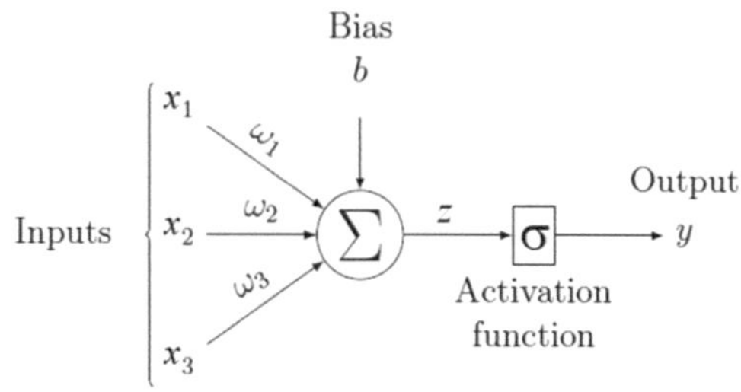

Fig. 1 Concept of neural network

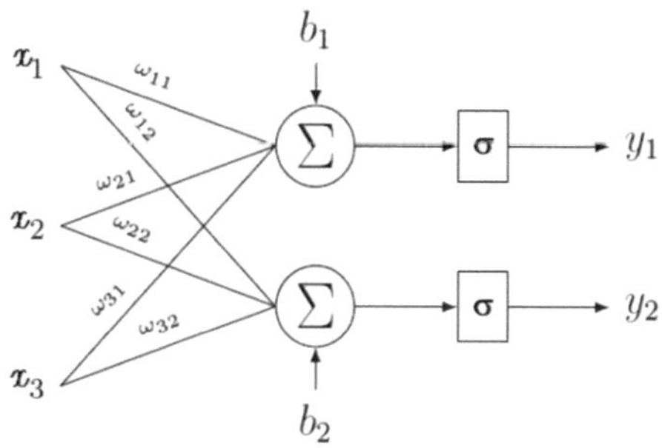

Fig. 2 Layers of neurons

$$y = \frac{1}{1 + e^{-z}}$$

the hyperbolic tangent function:

$$y = \tanh(z)$$

and the RELU (Rectified Linear Unit):

$$y = \max(0, z)$$

A layer of neurons is a group of neurons with the same inputs. Each neuron has its own weights and bias but is connected to the same input vector as the other neurons in the layer. A layer has as many outputs as it has neurons, so in Fig. 2, a layer with two neurons (and two outputs) is shown with a three-dimensional input vector [18–20].

3.2 *Multi-Layer Perceptron (MLP)*

A Multi-Layer Perceptron (MLP) is made up of several layers stacked together. The outputs of one layer serve as inputs for the next layer. In order to perform a regression with M outputs, the final layer should have M neurons. The dimension of the last layer is determined by the data, and all the other layers in the network are referred

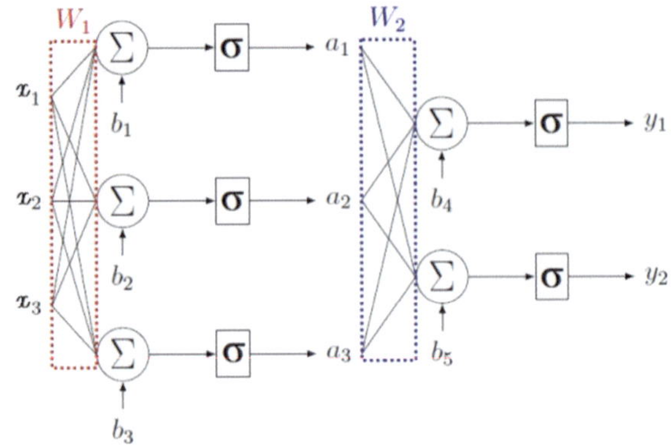

Fig. 3 Multi-layer perceptron (MLP)

to as hidden layers. Figure 3 illustrates an MLP with three inputs, two outputs, and a single hidden layer with three neurons [18, 19].

A multi-layer perceptron with three inputs, a single hidden layer, and two outputs can approximate any function in a closed and bounded space $\mathbb{R}^n$. This is due to neural networks being referred to as universal approximates, meaning that they have the ability to reach any desired precision with single layer neural network architecture, as formalized by the universal approximation theorem [19].

3.3 *Training a Neural Network*

Training a neural network, like other models, involves finding a set of parameters that minimize errors on the training set and ensure good generalization. In a neural network, the parameters that can be adjusted are the weights and biases of each neuron. The optimization of these weights is based on gradient descent, a technique that iteratively modifies the weights based on the observed error for each training data, as illustrated in Fig. 4 [18–20].

The training of a neural network involves comparing the network's output o_i, for each input value x_i, to the "true" value y_i. Based on the resulting errors, ϵ_i, the weights of the network are adjusted using a gradient algorithm, H, with the goal of minimizing a chosen loss function [18].

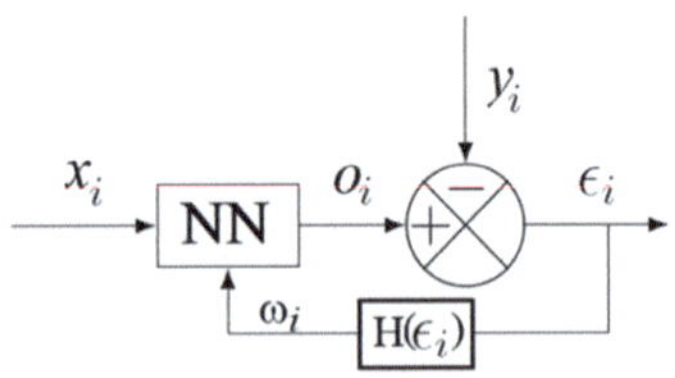

Fig. 4 Training a neural network

3.4 Neural Network Model Used

The neural network model structure used is:

Input Layers

- Device Distance (in meters)
- Transmission Power (in dBm)
- Transmission Frequency (in GHz)

Hidden Layers

One or multiple hidden layers can be used to capture complex relationships between inputs and outputs. Each of these layers can contain a certain number of neurons, and an activation function such as the sigmoid or rectified linear function can be used.

Output Layers: Total Energy Consumption (in Joules)

The model training will involve minimizing the error between predictions of the total energy consumption and the actual values, using a cost function such as mean squared error. Model parameter optimization can be performed using methods such as gradient descent to minimize energy consumption.

4 Proposed Model Implementation

The simulation parameters in the proposed neural network model to optimize energy consumption in D2D communication may include:

- Input data: This can include information such as device-to-device distance, transmission power, transmission frequency, etc.
- Network structure: This includes the number of hidden layers, number of neurons per layer, activation function, etc.
- Cost function: This is a function such as mean squared error to measure the error between predictions and true values.
- Optimization algorithms: These can include gradient descent to minimize energy consumption.
- Validation and test data: These are used to evaluate the model's accuracy and its ability to generalize to new data.
- Hyperparameters: This includes learning rate, number of epochs, number of hidden layers, etc.
- Evaluation metrics: This includes accuracy, cost, F1 metric, etc. to evaluate the model's performance.

5 Proposed Scenario and Simulation Results

The Proposed Scenario and Simulation Results is a critical aspect of evaluating the performance of a model. In this section, we outline the scenario that was considered for our simulation, the simulation results obtained and the analysis of the results. The simulation results will give us insights into the effectiveness of the proposed model in optimizing energy consumption in a D2D communication scenario.

5.1 Proposed Scenario

A scenario for minimizing energy consumption in D2D communications in the field of hospital logistics was proposed:

A hospital uses D2D communication devices to track and manage mobile medical equipment such as supply carts, gurneys, and automated external defibrillators. The devices use power management algorithms to optimize energy consumption by adjusting the transmission power based on the distance between devices. When a device is not in use, it is placed in sleep or hibernation mode to conserve energy.

Additionally, the hospital uses energy-efficient communication protocols to establish D2D communications. Location and tracking data are transmitted efficiently without compromising communication quality. Wireless communication networks are also designed to be low-energy consuming to minimize environmental impact.

By using these energy minimization techniques, the hospital can optimize logistics operations while saving energy and minimizing environmental impact. The battery life of the devices is also maximized, allowing for optimal use of D2D communication technology in the field of hospital logistics.

5.2 Simulation Results

Here is a simulation of a scenario for minimizing energy consumption in D2D communications in hospital logistics:

Suppose a hospital uses D2D communication devices to track supply carts.

The devices are equipped with power management algorithms to optimize energy consumption by adjusting the transmission power based on the distance between the devices. When a cart is not in motion, the device goes into standby mode to save energy.

- Simulation 1: No Energy-Saving Techniques Used. The D2D communication device constantly emits at full power, even when the cart is stationary. After 8 h of use, the device's battery is almost empty and needs to be recharged.
- Simulation 2: Energy-Saving Techniques Used. The D2D communication device uses the power management algorithm to adjust the transmission power based on

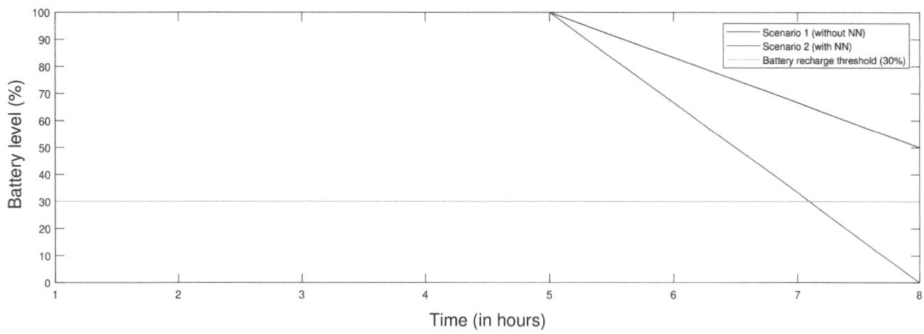

Fig. 5 Comparison of battery life with and without (NN)

the distance between the devices. When the cart is not in motion, the device is put into standby mode to save energy. After 8 h of use, the device's battery is only half empty and can be used for several more hours before needing to be recharged.

The simulation shows a significant difference in battery life when using energy consumption minimization techniques (with NN) compared to not using these techniques (without NN). When using the minimization techniques, the battery can be used for several hours before needing to be recharged, while when not using these techniques, the battery needs to be recharged after only 8 h of use. Figure 5 shows that the use of energy consumption minimization techniques can significantly extend the battery life of D2D communication devices.

5.3 Comparison of Energy Consumption with and Without Neural Network

Let's look at a hypothetical example comparing energy consumption with and without a neural network:

1. Scenario without a neural network:
   - Configuration:

     Devices located at an average distance of 10 m
     Transmission power of 1 W
     Transmission frequency of 2.4 GHz

   - Results: The total energy consumption is 50 Wh for 100 D2D transmissions.
2. Scenario with a neural network:
   - Configuration:

     Same configuration as the previous scenario

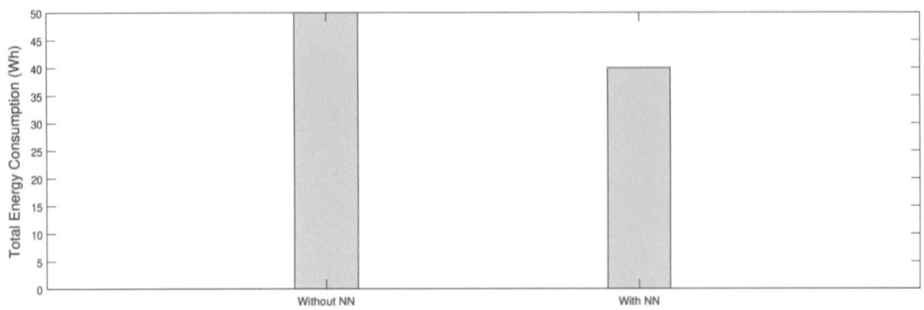

Fig. 6 Energy consumption comparison with and without neural network

- Neural network model:

  3 hidden layers
  10 neurons per layer
  ReLU activation function

- Training:

  Energy consumption data was collected for different D2D communication configurations
  The data was normalized and split into training, validation, and test sets
  The model was trained to minimize energy consumption using a cost function such as mean squared error.

- Results: The total energy consumption is 40 Wh for 100 D2D transmissions, a 20% reduction compared to the scenario without a neural network as shown in Fig. 6.

In this example, it can be seen that the use of a neural network can reduce energy consumption for D2D communication. Specific values depend on the chosen model and the data used for training, but this type of analysis can provide an indication of the potential effectiveness of using neural networks in this field.

6 Discussion

During the discussion of the simulation results, it is important to highlight key points such as differences between the simulated scenarios, observed trends, and relevant data points. It may be useful to compare the simulation results with available real-world data to verify the accuracy of the simulation.

It is also important to discuss the implications of the results for hospital logistics and determine the next steps to take to improve D2D communication systems used in this field. The simulation shows that minimizing energy consumption significantly increases battery life, it may be wise to continue efforts to optimize the D2D communication systems used in hospital logistics.

Finally, it is important to discuss the limitations and uncertainties associated with the simulation. The simulation does not take into account all the factors that affect energy consumption in real life, such as interference and signal quality variations. Therefore, it is important to take these limitations into account when interpreting the simulation results.

7 Conclusion

In conclusion, minimizing energy consumption in D2D communications in the healthcare logistics field is crucial for saving energy, reducing environmental impact, and improving healthcare quality. Techniques such as device hibernation, power emission regulation, power management algorithm usage, use of energy-efficient communication protocols, and deployment of low-energy wireless communication networks can aid in achieving this goal. Additionally, the use of neural networks can also be beneficial for minimizing energy consumption by using deep learning methods to adjust energy levels based on actual communication needs. It is therefore important to continue research to improve these techniques and implement them for optimal use of D2D communication technology in the healthcare logistics field.

References

1. Gismalla MSM et al (2022) Survey on device to device (D2D) communication for 5GB/6G networks: concept, applications, challenges, and future directions
2. Chakraborty C et al (2020) A comprehensive review on device-to-device communication paradigm: trends, challenges and applications
3. Albowarab MH et al (2018) Review on device-to-device communication in cellular based network systems
4. Singh R et al (2022) Smart device for effective communication in the healthcare system
5. Khujamatov K et al (2022) Device-to-device and millimeter waves communication for 5G healthcare informatics
6. Nauman A et al (2021) Reinforcement learning-enabled intelligent device-to-device (I-D2D) communication in narrowband internet of things (NB-IoT)
7. Fleischmann M et al (2000) A characterisation of logistics networks for product recovery
8. Cao J et al (2023) Energy efficiency resource allocation for cell-edge users with social-aware based grouping D2D
9. Jessica W et al (2018) Energy efficiency in logistics: an interactive approach to capacity utilization
10. Li R et al (2022) Research on automation control of university logistics management system based on wireless communication network
11. Zeadally S et al (2011) Energy-efficient networking: past, present, and future
12. Asadi A et al (2014) A survey on device-to-device communication in cellular networks

13. Chen X et al (2015) Energy-efficient resource allocation for D2D communication underlaying cellular networks
14. Wang X et al (2020) Energy-efficient power allocation and Q-learning-based relay selection for relay-aided D2D communication
15. Seoyoung Y et al (2022) Deep reinforcement learning based resource allocation for D2D communications underlay cellular networks
16. Zhi Y et al (2022) Deep reinforcement learning-based resource allocation for D2D communications in heterogeneous cellular networks
17. Bishop CM (1995) Neural networks for pattern recognition. Oxford University Press
18. Goodfellow I, Bengio Y, Courville A (2016) Deep learning. MIT Press
19. Haykin S (2009) Neural networks and learning machines, 3rd edn. Pearson Education
20. Alpaydin E (2010) Introduction to machine learning, 2nd edn. MIT Press

Design of a Smart Plug for Energy Management

Wided Saadi and Lassaad Sbita

Abstract With technological advancement and based on the principle of Communication of objects, oriented applications towards intelligent energy management are in line to be the simplest, most effective, while allowing to provide comfort to the human and minimize energy consumption. The use of certain standards communications protocols facilitates the deployment of wireless network. This article consists to design a «Smart Plug» able to receive and execute commands on various devices of the installation (lighting, heating, activation of rolling shutters) located on the same network.

Keywords Smart plug · Arduino · Isis · Programming language

1 Introduction

The development of digital electronics is booming [1]. To be convinced of this, just look around us at the explosion of electronic devices (Smartphones, computers, alarm systems, etc.). Following this technological advancement, objects can connect with each other and exchange data. This therefore makes it possible to have a connected and accessible world and to improve lifestyles.

Indeed, we are able to pilot and control our home at our fingertips, start or stop any equipment with a simple click, receive the status of our installation from anywhere in the world, manage the energy consumption of electrical appliances in an autonomous way. This new strategy affects several sectors in particular the home automation sector as well as the industrial sector and promises a bright and radiant future in terms of security, energy management and comfort for humans [2].

The objective of this work is to develop an example of a "Smart Plug" [3], assisted by a PC or a Smartphone, capable of receiving orders and executing them on various devices of the installation on the same network [4–7].

W. Saadi (✉) · L. Sbita
National Engineering School of Gabes, Zrig, 6029 Gabes, Tunisia
e-mail: saadi.wided@yahoo.fr

© The Author(s), under exclusive license to Springer Nature Singapore Pte Ltd. 2024
A. Mellit et al. (eds.), *Proceedings of the 2nd International Conference on Green Energy Conversion System*, Green Energy and Technology,
https://doi.org/10.1007/978-981-97-6148-7_21

This smart plug makes supervising and control easy and flexible during remote control [8]. It also measures power consumption. This is an interesting indicator for achieving energy saving.

This article is organized as follows: In the first part, we will make a conceptual study of our adopted application. The second part relates to the presentation of the hardware and software environments that we used to simplify the control and show the interest of the application as well as the simulation result obtained under ISIS. Finally, the conclusion is presented in the last section.

2 Study and Electronic Design

The main aim is to realize an intelligent energy optimization system [9]. To achieve our goal, we went through the following steps:

- Study and choice of components
- System design
- Programming and testing

2.1 General Structure

Our objective is the control of the smart socket, the acquisition of the various parameters such as the current, the power and the energy consumed. To do this, our system is made up of various components such as an Arduino card, XBee module, LCD display, ACS712 current sensor, relay card, electronic components, etc.

Figure 1 presents a block diagram of the solution we aim to achieve.

Figure 2 describes the different blocks of the system that we aim to achieve.

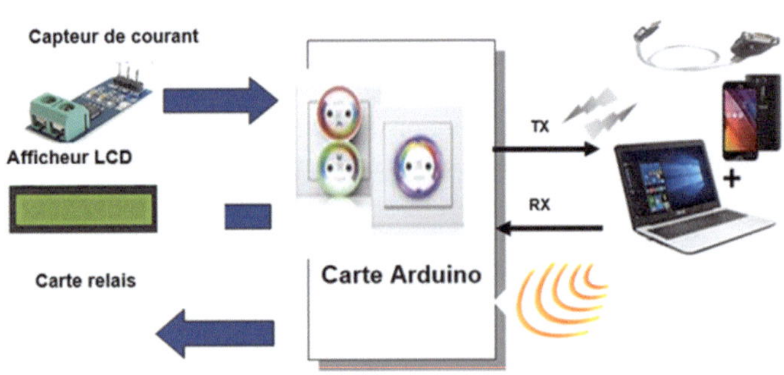

Fig. 1 Schéma synoptique de la solution proposée

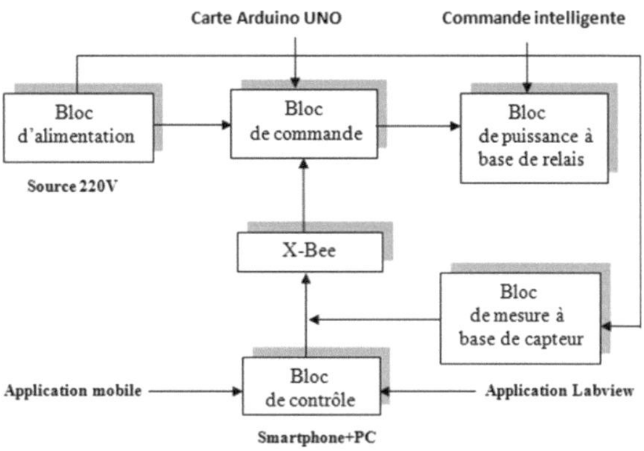

Fig. 2 Descriptive diagram of system's blocks

2.2 *Operating of the System*

In the acquisition card, the operation of the system is as follows:

After plugging such a device into a smart outlet, the current sensor starts measuring current consumption. Then, it calculates the power and energy consumed in a well-determined time. Then, it sends the measured values to the interface created by Labview on a PC through an X-Bee module and/or to the smart phone using a mobile application connecting to a PC via an internal connection. The person in charge of controlling energy consumption must register his own detail s in a database through a registration form, then he must identify his own login and password in order to access the supervision interface and control of the consumption of the connected devices so that it can be able to remotely control the equipment connected to it.

3 Simulation of the Electronic Card

Before moving on to the practical implementation, we used an ISIS which makes it possible to draw electronic diagrams, to simulate them and to produce the corresponding printed circuit. The latter allowed us to virtually simulate our system (Fig. 3).

This CAD has the possibility of carrying even hexadecimal codes for the realizations which contain programmable components or "Arduino" programmable cards as in our realization.

The simulation of our device under ISIS proves the success, this CAD even has a Virtual Terminal that can be used as an X-Bee (a USB terminal).

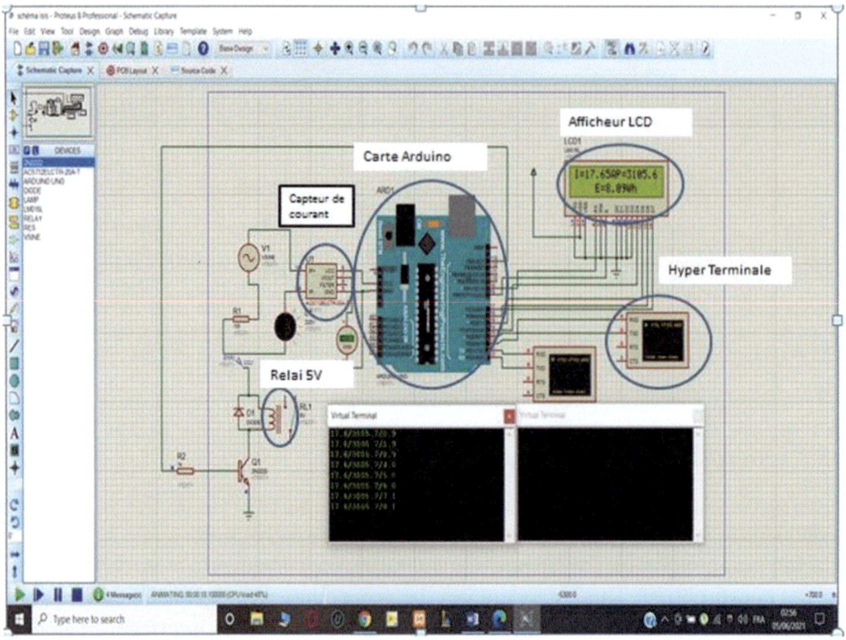

Fig. 3 The map produced under ISIS-PORTEUS

4 Conclusion

In this article, we have made a study on the practical realization of an electronic card to control a smart socket. These commands are managed from a PC (Development of a Labview application) and an Arduino UNO card.

References

1. Tanana M (2009) Evaluation formative du savoir-faire des apprenants à l'aide d'algorithmes de classification: application à l'électronique numérique. Diss. Rouen, INSA
2. Ghazal M et al (2016) "Smart plugs: perceived usefulness and satisfaction: evidence from United Arab Emirates. Renew Sustain Energy Rev 55:1248–1259
3. Ling Z et al (2017) Security vulnerabilities of internet of things: a case study of the smart plug system. IEEE Internet Things J 4(6):1899–1909
4. Suryadevara NK, Biswal GR (2019) Smart plugs: paradigms and applications in the smart city-and-smart grid. Energies 12(10):1957
5. Srinivasan A, Baskaran K, Yann G (2019) IoT based smart plug-load energy conservation and management system. In: 2019 IEEE 2nd international conference on power and energy applications (ICPEA). IEEE

6. Ramapatruni S et al (2019) Anomaly detection models for smart home security. In: 2019 IEEE 5th international conference on big data security on cloud (BigDataSecurity). IEEE International conference on high performance and smart computing (HPSC) and IEEE International conference on intelligent data and security (IDS). IEEE
7. Mtshali P, Khubisa F (2019) A smart home appliance control system for physically disabled people. In: 2019 conference on information communications technology and society (ICTAS). IEEE
8. Carry E, Friedt J-M. L'environnement Arduino est-il approprié pour enseigner l'électronique embarquée?
9. Wang L, Peng D, Zhang T (2015) Design of smart home system based on WiFi smart plug. Int J Smart Home 9(6):173–182

Waste-Free Energy Management in an Islanded Microgrid with Hydrogen Storage

Adrien Gauché, Raphaël Chenouard, David Morin, and Malek Ghanes

Abstract This paper addresses the challenges associated with modeling storage systems and characteristics of hydrogen in the planning of island microgrids powered solely by renewable generation. We propose a novel Mixed-Integer Linear Programming (MILP) formulation that accounts for the non-constant efficiency of hydrogen chain storage, aiming to minimize error between Energy Management System (EMS) control and grid performance. Furthermore, our formulation incorporates flexible constraints to effectively manage surplus energy and mitigate energy wastage in scenarios with limited available resources. Through this research, we contribute to the development of optimized planning strategies for island microgrids, facilitating the integration of renewable energy sources and enhancing their operational efficiency.

Keywords Energy management · Hydrogen storage · Microgrid · Renewable generation · Waste-free · Islanded constraints

This work is supported by the National Association of Research and Technology (Association Nationale Recherche et Technologie) of France through the program CIFRE under grant 2020/0995.

A. Gauché (✉) · R. Chenouard · M. Ghanes
Nantes Université, École Centrale Nantes, CNRS, LS2N, UMR 6004, 44000 Nantes, France
e-mail: adrien.gauche@ls2n.fr

R. Chenouard
e-mail: raphael.chenouard@ls2n.fr

M. Ghanes
e-mail: malek.ghanes@ls2n.fr

A. Gauché · D. Morin
S.A. PowiDian, 37700 La Ville-aux-Dames, France
e-mail: david.morin@powidian.com

© The Author(s), under exclusive license to Springer Nature Singapore Pte Ltd. 2024

A. Mellit et al. (eds.), *Proceedings of the 2nd International Conference on Green Energy Conversion System*, Green Energy and Technology,
https://doi.org/10.1007/978-981-97-6148-7_22

1 Introduction

1.1 Context and Motivation

Decarbonization and the adoption of renewable energy are crucial for addressing climate change. Smart Microgrids (MGs) provide a solution for powering isolated sites by integrating renewable generation, energy storage, and advanced control mechanisms [6]. Predictive control strategies optimize microgrid operations, improving energy generation, storage, and consumption efficiency [3]. Designing effective Energy Management Systems (EMS) is vital for remote locations with limited access to renewable energy and utility grids [7]. This paper focuses on energy management in isolated microgrids with renewable generation and hydrogen storage.

1.2 Literature Review

Raya-Armenta [7] reviews EMS optimization challenges and advancements for islanded microgrids. It explores frameworks, algorithms, uncertainty handling, and constraints. Optimization techniques like Dynamic Programming (DP), Reinforcement Learning (RL) and Rolling Horizon Control (RHC) are discussed, with RHC being resilient against failures. However, challenges remain in forecasting, modeling, algorithms, and proof of concept. It suggests exploring new metaheuristics such as swarm and evolutionary algorithms but caution is advised regarding their novelty, Camacho-Villalon [2] recommend mathematical formulation when possible to guarantee optimality [1] and to efficiently solve with Linear Programming (LP) techniques [4]. Erdinç [3] formulate a Mixed-Integer Linear Programming (MILP) RHC approach to enhance grid independence, reduce costs, emissions, and optimize renewable energy utilization. Logical constraints apply on the battery to prevent simultaneous charging and discharging. The RHC control effectively handles uncertainties. Tostado-Veliz [8] develops a MILP on an islanded microgrid with hydrogen storage and Demand Side Management (DSM). The EMS constrains the storage state to the same value for the first and last horizon step to avoid emptying the reservoir, this can lead to energy wastage. The authors show that flexible demand with robust optimization reduces total cost better than hydrogen storage and DSM are called more often using hydrogen, which can cause premature wear.

1.3 Contributions

Our proposed MILP formulation addresses gaps in the modeling of storage and the characteristics of hydrogen in planning island microgrids with only renewable generation. It considers the non-constant efficiency of the hydrogen chain storage to

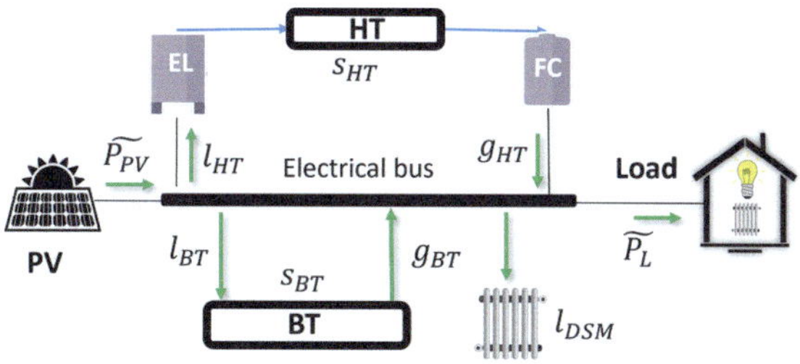

Fig. 1 Microgrid schematic [5], CC BY-SA 4.0 Adrien Gauché

minimize the error between EMS control and the real grid. Additionally, we introduce flexible constraints to manage surpluses effectively and prevent energy wastage in all boundary cases, considering limited energy and computing resources available.

2 Microgrid and Management

2.1 Microgrid Description

The microgrid case study is illustrated in Fig. 1. It consists of various components: PV generation $\widehat{P_{PV}}$ (kW), non-interruptible load consumption $\widehat{P_L}$ (kW) shiftable consumption l_{DSM} (kW), battery (BT) with power consumption (charge) l_{BT} and generation (discharge) g_{BT} (kW) with State Of Charge (SOC) s_{BT} (%), electrolyzer (EL) with power consumption l_{HT} (kW) for hydrogen (H2) production, hydrogen tank (HT) with H2 storage at mass s_{H2} (kg), and fuel cell (FC) for power generation g_{HT} (kW) using H2. These variables are defined for each time slot t, over the entire horizon $T = 48$ (number of slots), such that $t \in [[1; T]]$. For ease of reading, the variable t does not always appear, and we use lower case for optimization variables and upper case for fixed parameters. The controllable equipment in the MG consists of the FC, the EL and the shiftable load, represented by the variables g_{HT}, l_{HT}, l_{DSM}, respectively.

2.2 Optimization Model

The MG EMS utilizes the linear unit commitment problem formulation introduced in [4]. The formulation incorporates constraints on power generation and consumption, ensuring that the minimum and maximum power limits of each component are bounded. Unit commitment u_i^g and consumption u_i^l govern the operation status of generation and consumption units. g_u representing the under generation (unmet power) and l_s surplus power, bounded by the maximum demand power and maximum power availability. Logical constraints prevent simultaneous charge/discharge of the battery and simultaneous production/consumption of hydrogen. Then, generation is limited by raw demand and consumption by raw production to make the MILP model tighter. Variables $v(t)$ and $w(t)$ track the starting and stopping of equipment, following the constraints outlined in [4]. Constraint (1a) tracks filling of each storage. If f_i is true, storage i is full, thus storage commissioning (BT charge, H2 production) is deactivated (1b). Conversely, BT discharge or FC are deactivated when empty (1c). HT constraint is the mass of H2 as a function of temperature.

$$\frac{s_i(t)}{S_i^{MAX}(t)} - 1 \leq f_i(t) \leq \frac{s_i(t)}{S_i^{MAX}(t)}; \forall i \in \{BT, HT\} \tag{1a}$$

$$u_i^l(t) \leq 1 - f_i(t); \forall i \in \{BT, HT\} \tag{1b}$$

$$u_i^g(t) \leq \frac{s_i(t)}{S_i^{MIN}(t)}; \forall i \in \{BT, HT\} \tag{1c}$$

The waste commissioning variable u_s is true if and only if all the storage facilities are full (2a) and (2b). (2c) constrains wasted power to be less than surplus power, leading back out-of-set variables to the correct operating range.

$$u_s(t) \leq f_i(t); \forall i \in \{BT, HT\} \tag{2a}$$

$$u_{BT}(t) + u_{HT}(t) - 1 \leq u_s(t) \tag{2b}$$

$$l_s(t) \leq u_s(t) * |\max(P_{PV}(t) - P_L(t), 0)| \tag{2c}$$

The power flow in the electrical bus is the balance between intermittent solar generation and consumption, added by controllable generators and subtracted by controllable consumers. The stored energy is obtained by integration, adding energy of the previous time step $s_i(t-1)$, subtracting/adding the conversion efficiency θ_i, the consumption power (storage) l_i, the generation power g_i and the time step $dt = 0.5h$. HT storage is obtained from (3a) with hydrogen flux in ϕ^l_{HT} and out ϕ^g_{HT}. $\theta_i^{c\_a}$ and $\theta_i^{c\_b}$ are piecewise efficiency coefficients in (3b) and (3c) determined by error minimization on the manufacturer's curves. The error decreases from −5.2 to 1.4% for EL and from 14.4 to −5.5% for FC compared to constant coefficient while maintaining a compact model.

$$s_{HT}(t) = s_{HT}(t-1) + \phi^l_{HT}(t)dt - \phi^g_{HT}(t)dt; \forall t \in [[2; T]] \tag{3a}$$

$$\phi^l_{HT}(t) = \left(\theta_{HT}^{a\_l}\frac{l_{HT}(t)}{L_{HT}^{MAX}} + \theta_{HT}^{b\_l} * u^l_{HT}(t)\right) L_{HT}^{MAX} \tag{3b}$$

$$\phi^g_{HT}(t) = \left(\theta_{HT}^{a\_g}\frac{g_{HT}(t)}{G_{HT}^{MAX}} + \theta_{HT}^{b\_g} * u^g_{HT}(t)\right) G_{HT}^{MAX} \tag{3c}$$

The MILP problem minimizes the cost of energy over the horizon by minimizing component operating time with cost of equipment over guaranteed lifetime C^{time}, C^u the unmet power penalty, C^s the wasted power penalty, C^{up} cost of equipment over number of guaranteed start-ups and losses, all under the above and [4] constraints.

$$\min_{v_i,u_i,g_i,l_i} \sum_{t=1}^{T} \left(\sum_{i\in\{g_{BT},l_{HT},g_{HT},g_u\}} C_i^l l_i(t) + C_i^g g_i(t) \right) * dt + \sum_{i\in\{v_{BT}^g,v_{HT}^l,v_{HT}^g\}} C^{up} v_i(t) \tag{4}$$
$$\text{s.t.} \quad (1), (2), (3)$$

3 Experiments

3.1 Cases Studied and Results Analysis

The microgrid studied corresponds to an isolated grid in mountain refuge (Col du Palet, Vanoise National Park) with a PV generation of 4.8 kW, BT of 14.8 kWh, H2 nominal capacity of 6 kg, EL and FC of 2.5 kW with low power limit at 20%. Power flow parameters is solar generation and consumption predicted over the horizon of 24 h, composed of $T = 48$ slots with time steps of $dt = 0.5h$, starting in the middle of the night. 2 scenarios are tested on the limits of the EMS, full storage (BT 95%, HT 39 bar) in Figs. 2a and b and empty (BT 25%, HT 1 bar) Figs. 2c and d. In the full storage scenario 2a and b, the BT SOC is above 88% during the first half of the horizon, slightly charged by solar and FC generation 2a. Generator FC activates at the end of the night and EL at the beginning of the morning 2b, then the generation is clipped with l_s. BT supplies the energy demand for the next night. In the empty storage scenario Figs. 2c and d, there is an energy deficit during the night, virtually feeds by g_u. The battery takes charge during the day, followed by H2 production through the operation of the EL. After a decrease in generation during the midday period, BT charge to 92% and then discharges during the night to meet the energy demand.

3.2 Discussion and Improvements

In the full storage scenario Fig. 2b, H2 storage is very active (FC during night, EL during day), with the BT adjusting. EL is launched before BT is fully charged. With hard constraints, optimization would look to waste energy and finds no solution when storage starts out of limit. This problem is solved by the flexible constraint (1). Then, for energy shortage Fig. 2, the battery is not fully charged at the end of the day. The most efficient storage should be prioritized to manage unexpected power deviation and improve the overall efficiency, this could be integrated in the cost function (4). Moreover, the EMS optimizes only on expected values with exact amount of energy,

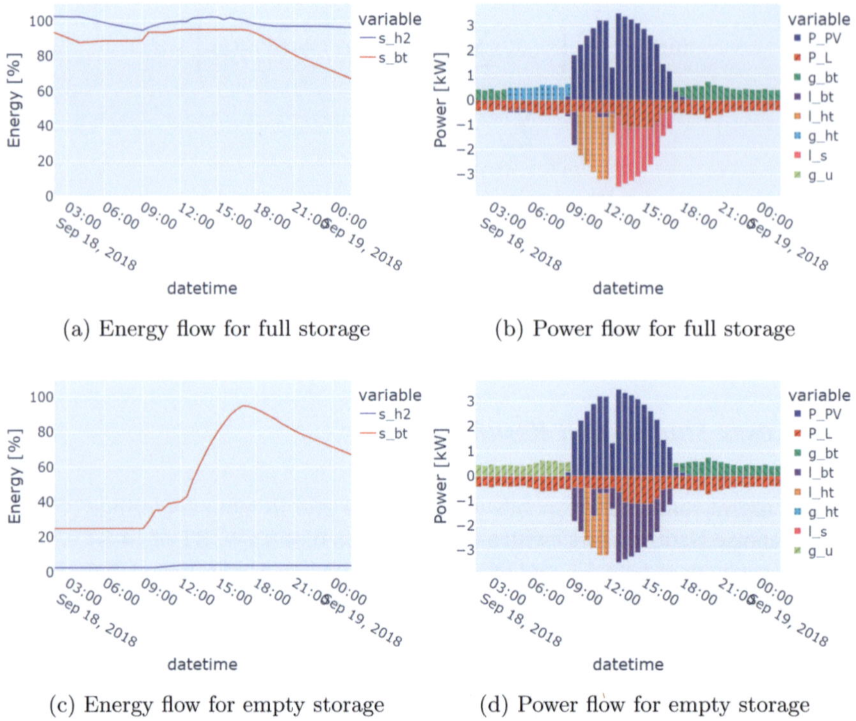

(a) Energy flow for full storage (b) Power flow for full storage

(c) Energy flow for empty storage (d) Power flow for empty storage

Fig. 2 Numerical experiments

without taking into account possible uncertainties. Integrating deviations would make it possible to provide more robust decisions. Then to test non-constant efficiency (3), a more detailed simulation model or real-world experimentation is needed.

4 Conclusion

In this paper, we introduce a MILP compact formulation to enhance the management of full and empty storage in isolated microgrids with non-constant H2 chain efficiency by reusing variables already in unit commitment. The proposed formulation ensures the proper functioning of the system under boundary conditions. Flexible constraints make it possible to plan even when storage is out of bounds while returning to the correct operating range and filling to maximum capacity. This generic model avoids wasting energy at isolated sites by taking into account physical and logical constraints. The next step of this work is to make intensive simulations and experiments on real systems over long periods to validate its performance. Additionally, integrating prediction errors into the planning process can enhance the robustness of the control strategy. These improvements will contribute to the development of a more reliable EMS for isolated microgrids.

References

1. Boukouvala F, Misener R, Floudas CA (2016) Global optimization advances in mixed-integer nonlinear programming, MINLP, and constrained derivative-free optimization. CDFO. Eur J Oper Res 252(3):701–727. https://doi.org/10.1016/j.ejor.2015.12.018
2. Camacho-Villalón CL, Dorigo M, Stützle T (2022) Exposing the grey wolf, moth-flame, whale, firefly, bat, and antlion algorithms: six misleading optimization techniques inspired by bestial metaphors. Int Trans Oper Res. https://doi.org/10.1111/itor.13176
3. Erdinç FG (2023) Rolling horizon optimization based real-time energy management of a residential neighborhood considering PV and ESS usage fairness. Appl Energy 344:121275. https://doi.org/10.1016/j.apenergy.2023.121275
4. Knueven B, Ostrowski J, Watson JP (2020) On mixed-integer programming formulations for the unit commitment problem. INFORMS J Comput. https://doi.org/10.1287/ijoc.2019.0944
5. Morin D, Stevenin Y, Brault P (2019) Energy management of isolated DC microgrids with hybrid batteries-hydrogen storage system using model predictive control and wavelet neural networks based forecasts. In: 2019 21st European conference on power electronics and applications (EPE '19 ECCE Europe). IEEE, Genova, Italy, pp P.1–P.10. https://doi.org/10.23919/EPE.2019.8915136
6. Olivares DE, Mehrizi-Sani A, Etemadi AH, Canizares CA, Iravani R, Kazerani M, Hajimiragha AH, Gomis-Bellmunt O, Saeedifard M, Palma-Behnke R, Jimenez-Estevez GA, Hatziargyriou ND (2014) Trends in microgrid control. IEEE Trans Smart Grid 5(4):1905–1919. https://doi.org/10.1109/TSG.2013.2295514
7. Raya-Armenta JM, Bazmohammadi N, Avina-Cervantes JG, Sáez D, Vasquez JC, Guerrero JM (2021) Energy management system optimization in islanded microgrids: an overview and future trends. Renew Sustain Energy Rev 149:111327. https://doi.org/10.1016/j.rser.2021.111327
8. Tostado-Véliz M, Rezaee Jordehi A, Fernández-Lobato L, Jurado F (2023) Robust energy management in isolated microgrids with hydrogen storage and demand response. Appl Energy 345:121319. https://doi.org/10.1016/j.apenergy.2023.121319

Robust PI Multiobserver for Discrete Nonlinear Singularly Perturbed System

Ltifi Marwa, Bahri Nesrine, and Ltaief Majda

Abstract This paper deals with the observability bound problem of a discrete-time nonlinear singularly perturbed systems (SPSs) subject to $\mathcal{L}_2$ disturbances. The considered system is represented by a coupled state multimodel (MM). A H_∞ proportional-integral (PIMO), features by its robustness, is designed, ensuring that the state estimation error is asymptotically stable and satisfies the H_∞ performance constraint for a proper bound of the singular perturbation parameter based on a quadratic Lyapunov function and the H_∞ norm. Before designing this observer, a sufficient conditions described in terms of linear matrix inequalities (LMIs) are developed to ensure the robust stability bound of the considered system. Then, using the results of these two tasks, the observability bound of the system is determined.

Keywords SPS · PIMO · Observability bound

1 Introduction

Singularly perturbed systems (SPSs) are systems with two time scales (slow and fast dynamics). SPSs are widely covered in the literature in term of modelling, identification, control and observation. Since the identification, control and diagnosis of systems often requires the knowledge of all state variables, which are not always measurable, state estimation of non linear SPSs subject to disturbances has long been an important research issue over the last few decades [1–5]. State estimation of standard non linear (SPSs) can be solved considerably using the two classical techniques of decomposition and reduction [1, 2, 5].

L. Marwa (✉) · L. Majda
National Engineering School of Gabes, University of Gabes, Street Omar Ibn Khattab, 6029 Gabes, Tunisia
e-mail: marwaltifidoctorante@gmail.com

B. Nesrine · L. Majda
Research Laboratory: Numerical Control of Industrial Processes (CONPRI), National Engineering School of Gabes, University of Gabes, Street Omar Ibn Khattab, 6029 Gabes, Tunisia

© The Author(s), under exclusive license to Springer Nature Singapore Pte Ltd. 2024

A. Mellit et al. (eds.), *Proceedings of the 2nd International Conference on Green Energy Conversion System*, Green Energy and Technology,
https://doi.org/10.1007/978-981-97-6148-7_23

In [6, 7], the authors have been developed a H_∞ and $\mathcal{H}_2$ filters to solve the state estimation problem of non linear SPSs. In these works, sufficient conditions are established for the resolution of the filtering problem in SPSs; Although, the approximation of the filter gains is not treated. In [5], a sliding mode observer was designed to solve a stabilization problem when the output of the system depends on the slow state which is unavailable.

However, this strategy suffers from the problem with local linearization.

All the limitations included in decomposition, reduction and linearization approaches make the synthesized estimator not always applicable. Moreover, even these techniques are applicable against standard non linear SPSs, they neither preserve the dynamics of the system nor provide a precise state estimate bound. An interesting method to solve the synthesis problems for nonlinear systems is to represent the considered system using multimodel (MM) approaches. However, no authors have considered designing a multiobserver system for nonlinear SPSs using these approaches.

Therefore, we consider to solve this problem in this paper.

In fact, using this approach to reduce the complexity of a nonlinear system into different linear submodels connected by activation functions called weighting functions [8–10]. In the literature, two basic state MM structures that are studied: the coupled state structure [9, 10], and the decoupled state structure [8]. The coupled structure is characterized by a global state x that couples all the submodels and a common perturbation parameter for all the submodels. The coupled state MM structure for nonlinear SPSs can be obtained using either the identification technique [11] or the Convex Polytopic Transformation (CPT) technique [12]. The analysis and synthesis of nonlinear SPSs can therefore be developed using the linear-SPSs state estimation theory [13].

Moreover, and thanks to their two degrees of freedom, H_∞ PI linear observers have been successfully developed for nonlinear systems represented by MM [14].

Then, the purpose of this paper is to design a H_∞ PI multiobserver (PIMO) for nonlinear SPSs with the consideration of improving the bound of the singular-perturbation parameter ε.

First, using the CPT approach, a coupled state MM representing the nonlinear SPS has been obtained. Then, the observability bound of the considered MM is determined by calculating the PIMO gains.

2 System Description, Preliminaries and Problem Statement

2.1 System Description

The general form of the discrete time non linear SPS subject to disturbances can be represented by two different models [11, 12, 15]:

$$\begin{array}{cc} model\ 1 & model\ 2 \\ \left\{\begin{array}{c} x_1(k+1) = f(x_1(k), x_2(k), u(k), w(k) \\ x_2(k+1) = \varepsilon g(x_1(k), x_2(k), u(k), w(k) \\ y(k) = h(x_1(k), x_2 \end{array}\right. & \left\{\begin{array}{c} x_1(k+1) = f(x_1(k), \varepsilon x_2(k), u(k), w(k) \\ x_2(k+1) = g(x_1(k), \varepsilon x_2(k), u(k), w(k) \\ y(k) = h(x_1(k), \varepsilon x_2(k) \end{array}\right. \end{array} \quad (1)$$

$x_1(k) \in R^{n_1}$, $x_2(k) \in R^{n_2}$ are respectively slow and fast states of the system, $n = n_1 + n_2$ is the order of the system, $u(k) \in R^r$ is the system control input, $y(k) \in R^p$ is the system output, $w(k) \in R^m$ is the disturbance signal whose elements belong to $\mathcal{L}_2$ $(w \in \mathcal{L}_2)$.

Coupled Multimodel representation

Using CPT approach, all the system non-linearities are considered bounded, then, if the SPS contains n_r non linearities, the MM is composed of 2^{n_r} linear submodels. The discrete time non linear SPS represented by (1) can be represented by a coupled state MM using CPT approach as follows:

$$\begin{aligned} x(k+1) &= \sum_{i=1}^{N} \alpha_i(\zeta(k))(A_i(\varepsilon)x(k) + B_i(\varepsilon)u(k) + E_i(\varepsilon)w(k) \\ y(k) &= \sum_{i=1}^{N} \alpha_i(\zeta(k))C_i(\varepsilon)x(k) \end{aligned} \quad (2)$$

where N is the number of the submodels.

$x(k)$ is the global state that couples all the submodels with $x(k) = \left[\, x_1(k)\ x_2(k) \,\right]^T$.

$A_i(\varepsilon)$, $B_i(\varepsilon)$, $\mathrm{C}_i(\varepsilon)$, E_i are constant matrices related to the ith submodel.

$\alpha_i(\zeta(k))$ are the weighting functions depending on the premise variable $\zeta(k)$. These functions must satisfy the following convex sum proprieties:

$$\left\{\begin{array}{l} \sum_{i=1}^{N} \alpha_i(\zeta(k)) = 1, \quad \forall k \\ 0 < \alpha_i(\zeta(k)) < 1, \quad \forall k, \quad i = 1, \ldots, N \end{array}\right. \quad (3)$$

2.2 Preliminaries

The following definitions, assumptions and lemmas, will be used in building our main results:

Lemma 1 [13] *For a scalar* $\varepsilon_0 > 0$, *if the following conditions are satisfied:*

$$\begin{array}{l} T_3 \geq 0 \\ \varepsilon_0^2 T_3 + \varepsilon_0 T_2 + T_1 < 0 \\ T_1 < 0 \end{array}$$

Then $\varepsilon^2 T_3 + \varepsilon T_2 + T_1 < 0, \ \forall \varepsilon \in [0\ \varepsilon_0]$.

Lemma 2 *Schur complement [13]. For the matrices L, H and S, if L and H are symmetric matrices, then the following LMIs are equivalents:*

$$\begin{pmatrix} L & H \\ H^T & S \end{pmatrix} < 0 \quad (4)$$

and

$$S < 0, L - HS^{-1}H^T < 0 \quad (5)$$

Assumption 1 For the considered system, the following signals are satisfied:

$$\|u\|_2{}^2 < \infty, \ \ \|w\|_2{}^2 < \infty.$$

2.3 Problem Statement

The PIMO is characterized by the use of an additional integral term $z(k)$ which provides a robustness degree of the state estimation with respect to the disturbances. In order to synthesize a PIMO, the considered state MM given by Eq. (2) is modified as follows:

$$\begin{cases} x(k+1) = \sum\limits_{i=1}^{N} \alpha_i(\zeta(k))(A_i(\varepsilon)x(k) + B_i(\varepsilon)u(k) + E_i w(k) \\ y(k) = \sum\limits_{i=1}^{N} \alpha_i(\zeta(k))(C_i(\varepsilon)x(k)) \\ z(k+1) = y(k) \end{cases} \quad (6)$$

Considering the augmented vector $x_a(k) = \begin{bmatrix} x(k) \ z(k) \end{bmatrix}^T$, Eq. (6) is written as:

$$\begin{cases} x_a(k+1) = \sum\limits_{i=1}^{N} \alpha_i(\zeta(k))(A_{ai}(\varepsilon)x_a(k) + D_1 B_i(\varepsilon)u(k) + E_{ai}(\varepsilon)w(k)) \\ y(k) = \sum\limits_{i=1}^{N} \alpha_i(\zeta(k))(C_i(\varepsilon)D_1{}^T x_a(k) \\ z(k) = D_2{}^T x_a(k) \end{cases} \quad (7)$$

where $A_{ai}(\varepsilon) = \begin{bmatrix} A_i(\varepsilon) \ 0_{n\times p} \\ C_i(\varepsilon) \ 0_{p\times p} \end{bmatrix}$, $E_{ai}(\varepsilon) = \begin{bmatrix} E_i \\ 0p \times m \end{bmatrix}$, $D_1 = \begin{bmatrix} I_n \\ 0_{p\times n} \end{bmatrix}$, $D_2 = \begin{bmatrix} 0_{n\times p} \\ I_p \end{bmatrix}$.

The state estimation of the coupled MM given by (6) is realized using the following PIMO:

$$\begin{cases} \hat{x}_a(k+1) = \sum_{i=1}^{N} \alpha_i(\zeta(k))(A_{ai}(\varepsilon)\hat{x}_a(k) + D_1 B_i(\varepsilon)u(k) + K_{P_i}(y(k) - \hat{y}(k)) \\ \qquad +K_{I_i}(z(k) - \hat{z}(k))) \\ \hat{y}(k) = \sum_{i=1}^{N} \alpha_i(\zeta(k))C_i(\varepsilon){D_1}^T \hat{x}_a(k) \\ \hat{z}(k) = {D_2}^T \hat{x}_a(k) \end{cases} \tag{8}$$

where $\hat{x}_a\ (k)$ is the estimator of the augmented vector $x_a\ (k)$, $\hat{y}(k)$ is the estimated output, $\hat{z}(k)$ is the estimated integral output, K_{Pi} is the local proportional gain observer calculated for the ith submodel, K_{Ii} is the local integral gain observer calculated for the ith submodel.

The state estimation error is defined by the following equation:

$$e_a(k) = x_a(k) - \hat{x}_a(k) \tag{9}$$

Our objective is to design a PIMO given by (8) for system given by Eq. (7), such that the error (9) converges to zero while the impact of the disturbance noise $w(k)$ on the state estimation error $e_a(k)$ is minimised for a bound of the singular perturbation parameter.

3 Observability Bound of the Discrete Time Nonlinear SPS

Before achieving the robust PIMO design task, the robust stability bound of the coupled MM given by (2) must be studied first.

3.1 Robust Stability Bound of the Discrete Time Nonlinear SPS

According to Assumption 1, the input $u(k)$ is bounded, so to study the robust stability bound of the coupled state MM given by (2), it is sufficient to study the robust stability bound of the following autonomous MM subject to noises:

$$\begin{cases} x(k+1) = A(\varepsilon)x(k) + Ew(k) \\ y(k) = C(\varepsilon)x(k) \end{cases} \tag{10}$$

where

$$\begin{cases} A(\varepsilon) = \sum_{i=1}^{N} \alpha_i(\zeta(k))A_i(\varepsilon),\ C(\varepsilon) = \sum_{i=1}^{N} \alpha_i(\zeta(k))C_i(\varepsilon), \\ E(\varepsilon) = \sum_{i=1}^{N} \alpha_i(\zeta(k))E_i \end{cases} \tag{11}$$

The following theorem provides adequate conditions for achieving the robust stability bound of the considered MM given by (2).

Theorem 1 *For a parameter* $\varepsilon_0 > 0$*, a disturbance* $w \in \mathcal{L}_2$*, the considered coupled MM given by (2)is robustly stable and satisfies the* H_∞ *performance with a minimal attenuation index* η $\forall\ \varepsilon \in [0\ \varepsilon_0]$ *if there exists a matrix* $Q = Q^T > 0$*, satisfying the following LMIs:*

Minimizing η^2 *subject to satisfy the following LMIs*

$$\begin{array}{l} \varphi_{ii}{}'(0) < 0, i = 1, 2, \ldots, N \\ \varphi_{ii}{}'(\varepsilon_0) < 0, i = 1, 2, \ldots, N \\ \text{and} \\ \varphi_{ij}{}'(0) + \varphi_{ji}{}'(0) < 0,\ j > i = 1, 2, \ldots, N \\ \varphi_{ij}{}'(\varepsilon_0) + \varphi_{ji}{}'(\varepsilon_0) < 0,\ j > i = 1, 2, \ldots, N \end{array} \tag{12}$$

where

$$\varphi_{ij}(\varepsilon) = \begin{bmatrix} -Q + C_i(\varepsilon)^T C_j(\varepsilon) & 0 & A_i(\varepsilon)^T Q \\ 0 & -\eta^2 I & E_i(\varepsilon)^T Q \\ QA_i(\varepsilon) & QE_i(\varepsilon) & -Q \end{bmatrix} \tag{13}$$

Proof Consider the following quadratic Lyapunov function:

$$V_1(k) = x^T(k)Qx(k) \quad with\ Q = Q^T > 0,$$

The system is robustly stable if the the following condition is satisfied:

$$\Delta V_1(k) + y^T(k)y(k) - \eta^2 w^T(k)w(k) < 0\ \forall\ \varepsilon \in [0\ \varepsilon_0] \tag{14}$$

After calculation done, the condition given in (14) becomes:

$$\begin{bmatrix} x(k) \\ w(k) \end{bmatrix}^T \varphi(\varepsilon) \begin{bmatrix} x(k) \\ w(k) \end{bmatrix} < 0\ \forall\ \varepsilon \in [0\ \varepsilon_0] \tag{15}$$

$$\text{with } \varphi(\varepsilon) = \begin{bmatrix} A^T(\varepsilon)QA(\varepsilon) + C(\varepsilon)^T C(\varepsilon) - Q & A(\varepsilon)^T QE(\varepsilon) \\ E(\varepsilon)^T QA(\varepsilon) & E(\varepsilon)^T QE(\varepsilon) - \eta^2 I \end{bmatrix} \tag{16}$$

$\Delta V_1(k) + y^T(k)y(k) - \eta^2 w(k)^T w(k)$ is a quadratic function in $\varphi(\varepsilon)$, so the condition (14) is guaranteed if and only if:

$$\varphi(\varepsilon) < 0 \quad \forall\ \varepsilon \in [0\ \varepsilon_0]. \tag{17}$$

Using the schur complement in (17), and substituting (11) into (17), the last inequality will be written as:

$$\varphi'(\varepsilon) = \sum_{i=1}^{N} \alpha_i(\zeta(k))^2 \varphi_{ii}'(\varepsilon) + \sum_{j>i=1}^{N} \alpha_i(\zeta(k))\alpha_j(\zeta(k))(\varphi_{ij}'(\varepsilon) + \varphi_{ji}'(\varepsilon)) < 0, \quad \forall\, \varepsilon \in [0 \;\; \varepsilon_0] \tag{18}$$

where

$$\varphi'_{ij}(\varepsilon) = \begin{bmatrix} -Q + C_i(\varepsilon)^T C_j(\varepsilon) & 0 & A_i(\varepsilon)^T Q \\ 0 & -\eta^2 I & E_i(\varepsilon)^T Q \\ QA_i(\varepsilon) & QE_i(\varepsilon) & -Q \end{bmatrix} \tag{19}$$

Using the second weighting function's property given by Eq. (3), and using Lemma 1, it is easy to demonstrate that the condition given in (14) is verified if:

$$\begin{cases} \varphi'_{ii}(0) < 0 \\ \varphi'_{ii}(\varepsilon_0) < 0 \end{cases} , i = 1, \ldots, N$$
$$\text{and} \tag{20}$$
$$\begin{cases} \varphi'_{ij}(0) + \varphi'_{ji}(0) < 0 \\ \varphi'_{ij}(\varepsilon_0) + \varphi'_{ji}(\varepsilon_0) < 0 \end{cases} , \quad 1 \le i < j \le N$$

Therefore, the considered nonlinear SPS described by the coupled state MM is robustly stable $\forall\, \varepsilon \in [0 \;\; \varepsilon_0]$ if the LMIs (12) are satisfied.

3.2 State Estimation Bound of the Discrete Time Nonlinear SPS

In this section, the proposed PIMO given by equation (8) has been synthesised for a proper bound of singular perturbation parameter.

Theorem 2 *For a parameter $\varepsilon_* > 0$, a scalar $\gamma > 0$, a disturbance $w \in \mathcal{L}_2$, the state estimation error between the considered coupled MM (7) and the proposed PIMO (8) is asymptotically stable and satisfies the H_∞ performance constraint with a minimal attenuation index γ $\forall\, \varepsilon \in [0 \; \varepsilon_*]$, if there exist $P = P^T > 0$, two matrices G_{Pi} G_{Ii}, $i = 1, \ldots, N$, and a scalar $\gamma > 0$ of appropriate dimensions satisfying the following LMIs:*

Minimizing γ^2 subject to satisfy the following LMIs:

$$\begin{aligned} &\phi_{ii}'(0) < 0, i = 1, 2, \ldots, N \\ &\phi_{ii}'(\varepsilon_0) < 0, i = 1, 2, \ldots, N \\ &\textit{and} \\ &\phi_{ij}'(0) + \phi_{ji}'(0) < 0, \; j > i = 1, 2, \ldots, N \\ &\phi_{ij}'(\varepsilon_0) + \phi_{ji}'(\varepsilon_0) < 0, \; j > i = 1, 2, \ldots, N \end{aligned} \tag{21}$$

where

$$\phi_{ij'}(\varepsilon) = \begin{bmatrix} -P + I^T I & 0 & {\Gamma'_{ij}}^T \\ 0 & -\gamma^2 I & {\Upsilon'_{ij}}^T \\ \Gamma'_{ij} & \Upsilon'_{ij} & -P \end{bmatrix} \tag{22}$$

$\Gamma'_{ij} = PA_{ai}(\varepsilon) - G_{Pi}C_j(\varepsilon){D_1}^T - G_{Ii}{D_2}^T,\ \Upsilon'_{ij} = E_{ai}(\varepsilon)^T P.$

The observer gains are: $K_{Pi} = P^{-1}G_{Pi},\ K_{Ii} = P^{-1}G_{Ii},\ i = 1, \ldots, N$

$$k_p = \sum_{i=1}^{N} \alpha_i(\zeta(k))K_{pi},\ \ K_I = \sum_{i=1}^{N} \alpha_i(\zeta(k))K_{Ii}$$

Proof The state estimation error $e_a(k)$ tends to zero if

$$\Delta V(k) + e_a(k)^T(k)e_a(k)(k) - \gamma^2 w^T(k)w(k) < 0 \qquad \forall\ \varepsilon \in [0\ \varepsilon_*] \tag{23}$$

Referring to Eqs. (7) and (8), the dynamic of the state estimation error is described:

$$\begin{aligned} e_a(k+1) &= x_a(k+1) - \hat{x}_a(k+1) \\ &= A_{obs}(\varepsilon)e_a(k) + E_{obs}(\varepsilon)w(k) \end{aligned} \tag{24}$$

where

$$\begin{cases} A_{obs}(\varepsilon) = \sum\limits_{i,j=1}^{N} \alpha_i(\zeta(k))\alpha_j(\zeta(k))A_{obsij}(\varepsilon),\ E_{obs}(\varepsilon) = \sum\limits_{i,j=1}^{N} \alpha_i(\zeta(k))\alpha_j(\zeta(k))E_{obsij}(\varepsilon) \\ A_{obsij}(\varepsilon) = A_{ai}(\varepsilon) - K_{Pi}C_j(\varepsilon){D_1}^T - K_{Ii}{D_2}^T,\ E_{obsij}(\varepsilon) = E_{ai}(\varepsilon) \end{cases} \tag{25}$$

After calculation done, the condition (23) becomes

$$\begin{bmatrix} e_a(k) \\ w(k) \end{bmatrix}^T \phi(\varepsilon) \begin{bmatrix} e_a(k) \\ w(k) \end{bmatrix} < 0\ \ \forall\ \varepsilon \in [0\ \varepsilon_*] \tag{26}$$

where

$$\phi(\varepsilon) = \begin{bmatrix} A_{obs}(\varepsilon)^T PA_{obs}(\varepsilon) - P + I^T I & {A_{obs}}^T(\varepsilon)PE_{obs}(\varepsilon) \\ E_{obs}(\varepsilon)^T PA_{obs}(\varepsilon) & E_{obs}(\varepsilon)^T PE_{obs}(\varepsilon) - \gamma^2 \end{bmatrix} \tag{27}$$

$\Delta V(k) + e_a^T(k)e_a(k) - \gamma^2 w^T(k)w(k) < 0$ is a quadratic function in $\phi(\varepsilon)$, so, the condition (23) is guaranteed if and only if:

$$\phi(\varepsilon) < 0\ \ \forall \varepsilon \in\ \ \left[0\ \varepsilon_*\right] \tag{28}$$

Applying the Schur Complement on (28), we obtain:

$$\phi'(\varepsilon) = \begin{bmatrix} -P + I^T I & 0 & A_{obs}(\varepsilon)^T P \\ 0 & -\gamma^2 I & E_{obs}(\varepsilon)^T P \\ P A_{obs}(\varepsilon) & P E_{obs}(\varepsilon) & -P \end{bmatrix} < 0 \forall \varepsilon \in \begin{bmatrix} 0 & \varepsilon_* \end{bmatrix} \quad (29)$$

Replacing (25) into (29), and we pose, $K_{Pi} = P^{-1}G_{Pi}$, $K_{Ii} = P^{-1}G_{Ii}$, Eq. (29) will be written as:

$$\phi'(\varepsilon) = \sum_{i=1}^{N} \alpha_i(\zeta(k))^2 \varphi_{ii}'(\varepsilon) + \sum_{j>i=1}^{N} \alpha_i(\zeta(k))\alpha_j(\zeta(k))(\varphi_{ij}'(\varepsilon) + \varphi_{ji}'(\varepsilon)) \; \forall \; \varepsilon \in [0 \quad \varepsilon_0]. \quad (30)$$

$$\text{with } \phi_{ij'}(\varepsilon) = \begin{bmatrix} -P + I^T I & 0 & {\Gamma'_{ij}}^T \\ 0 & -\gamma^2 I & {\Upsilon'_{ij}}^T \\ \Gamma'_{ij} & \Upsilon'_{ij} & -P \end{bmatrix} \quad (31)$$

$$\Gamma'_{ij} = P A_{ai}(\varepsilon) - G_{Pi} C_j(\varepsilon) {D_1}^T - G_{Ii} {D_2}^T, \; \Upsilon'_{ij} = P E_{ai}(\varepsilon)$$

Using the second weighting function's propriety given by Eq. (3), and Lemma 1, it is easy to show that the condition given in (23) is verified if:

$$\begin{cases} \phi'_{ii}(0) < 0 \\ \phi'_{ii}(\varepsilon_0) < 0 \end{cases} \quad i = 1, \ldots, N$$
$$\text{and} \quad (32)$$
$$\begin{cases} \phi'_{ij}(0) + \phi'_{ji}(0) < 0 \\ \phi'_{ij}(\varepsilon_0) + \phi'_{ji}(\varepsilon_0) < 0 \end{cases} \quad 1 \le i < j \le N$$

Then, the state estimation error is asymptotically stable and satisfies the H_∞ performance constraint with a minimal attenuation index $\gamma \; \forall \; \varepsilon \; \in \; [0 \; \varepsilon_*]$ if the LMIs (21) are satisfied.

Remark 1 From Theorem 1, we deduce that the considered system is robustly stable $\forall \varepsilon \; \in \; [0 \; \varepsilon_0]$. From Theorem 2, we deduce that the state estimation error of the considered system is asymptotically stable and satisfies the H_∞ performance constraint $\forall \varepsilon \; \in \; [0 \; \varepsilon_*]$. Then, as a conclusion, we conclude that the SPS given by (1) is completely observable $\forall \; \varepsilon \; \in \; \begin{bmatrix} 0 \; \varepsilon_1 \end{bmatrix}$ with $\varepsilon_1 = \min(\varepsilon_0, \varepsilon_*)$.

ε_1: is the observability upper bound of the system.

4 Example

Consider an example of discrete non linear SPS represented by the following model:

$$\begin{cases} \begin{bmatrix} x_1(k+1) \\ x_2(k+1) \end{bmatrix} = \begin{bmatrix} 0.7 & -0.9 \\ \varepsilon \sin(x_1(k)) & 1.8\varepsilon \end{bmatrix} \begin{bmatrix} x_1(k) \\ x_2(k) \end{bmatrix} + \begin{bmatrix} -0.5 \\ -\varepsilon \end{bmatrix} u(k) + \begin{bmatrix} 0.2 \\ 0.1\varepsilon \end{bmatrix} w_1(k) \\ y(k) = x_1(k) \end{cases} \tag{33}$$

The considered $\mathcal{L}_2$ disturbance is: $w_1(k) = \begin{cases} 0 & \text{pour } 0 < \mathrm{k} < 100 \\ 0.1 & \text{pour } 100 < \mathrm{k} < 170 \\ 0 & \text{pour } 170 < \mathrm{k} < 250 \end{cases}$. We have one non linearity; $\xi(k) = \sin(x_1(k))$, or $y(k) = x_1(k)$, then, $\xi(k) = \sin(y(k))$, then, the obtained coupled MM is contained 2 submodels:

Sub model 1:

$$\begin{cases} A_1(\varepsilon) = \begin{bmatrix} 0.7 & -0.9 \\ \varepsilon & 1.8\varepsilon \end{bmatrix}, B_1(\varepsilon) = \begin{bmatrix} -0.5 \\ -\varepsilon \end{bmatrix} \\ C_1(\varepsilon) = [1\ 0],\ E_1(\varepsilon) = \begin{bmatrix} 0.2 \\ 0.1\varepsilon \end{bmatrix} \end{cases}$$

Sub model 2:

$$\begin{cases} A_2(\varepsilon) = \begin{bmatrix} 0.7 & -0.9 \\ -\varepsilon & 1.8\varepsilon \end{bmatrix}, B_2(\varepsilon) = \begin{bmatrix} -0.5 \\ -\varepsilon \end{bmatrix} \\ C_2(\varepsilon) = [1\ 0],\ E_2(\varepsilon) = \begin{bmatrix} 0.2 \\ 0.1\varepsilon \end{bmatrix} \end{cases} \tag{34}$$

The weighting functions are:

$$\begin{aligned} \alpha_1(k) &= 0.5(1 + \sin(y(k))) \\ \alpha_2(k) &= 0.5(1 - \sin(y(k))) \end{aligned} \tag{35}$$

Applying Theorem 1 with $\varepsilon_0 = 0.2082$, we obtain the positive definite matrix:

$$Q = \begin{bmatrix} 23.7284 & -14.2995 \\ -14.2995 & 75.6154 \end{bmatrix}$$

We note that, the non linear SPS given by (33) is robustly stable $\forall\ \varepsilon \in \left[0\ 0.2082\right]$.

Then, resolving Theorem 2, with $\varepsilon_* = 0.3658$, we obtain: $P = \begin{bmatrix} 16.8906 & 25.7067 & 0.1907 \\ 25.7067 & 80.7499 & 0.4428 \\ 0.1907 & 0.4428 & 2.3592 \end{bmatrix}$ and, $\gamma = 0.6241$, we deduce that the state estimation error tends to zero $\forall\ \varepsilon \in [0\ \varepsilon_*]$.

Finally, we conclude from theorem 1 and 2, that the considered system (33) est observable $\forall \varepsilon \in [0\ \varepsilon_1]\, with\ \varepsilon_1 = \min(0.2082, 0.3658) = 0.2082$.

The proportional and integral gains for $\varepsilon_1 = 0.2082$ are:

$$\left\{ \mathrm{K}_I = \begin{bmatrix} 0.0814 \\ -0.0088 \\ -0.0905 \end{bmatrix} K_P = \begin{bmatrix} 1.0259 \\ -0.0257 \\ 1.0008 \end{bmatrix} \right.$$

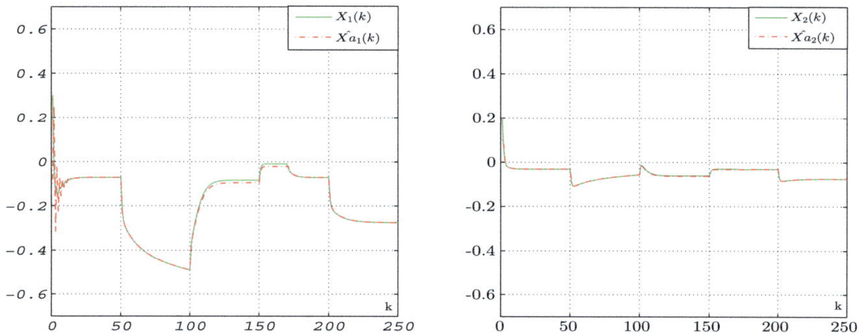

Fig. 1 Evolution of the state $x_1(k)$ $x_2(k)$ and their estimates $\hat{x}a_1(k)$ $\hat{x}a_2(k)$

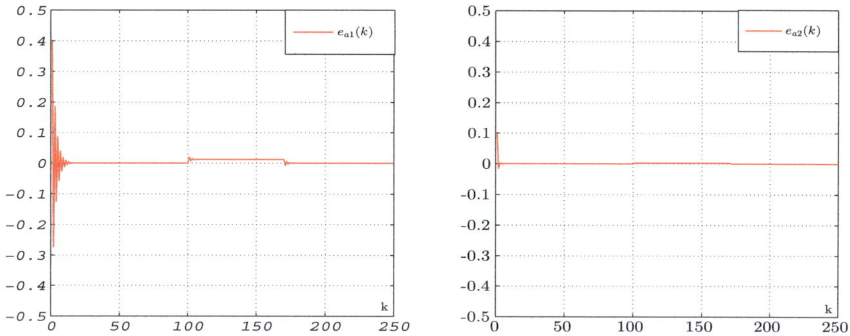

Fig. 2 Evolution of the state estimation errors $e_{a1}(k)$, and $e_{a2}(k)$

Figure 1 shows the evolution of the real states and their estimates and Fig. 2 shows the evolution of the state estimation errors.

5 Conclusion

This paper presents the observability bound of discrete-time nonlinear SPS subject to $\mathcal{L}_2$ disturbances which was described by a coupled MM. A PIMO which can estimate the states and attenuate the effect of the disturbances is synthesized based on an appropriate Lyapunov function. The considered numerical example demonstrates the effectiveness of the proposed strategy.

References

1. Radisavljevic V (2015) Two-stage feedback control design for a class of linear discrete-time systems with slow and fast modes. J Dyn Syst Meas Control 137. https://doi.org/10.1115/1.4030088
2. Cuevas l, Neic D, Manzie C (2020) Robustness analysis of nonlinear observers for the slow variables of singularly perturbed systems. Int J Robust Nonlinear Control 30:5628–5656. https://doi.org/10.1002/rnc.5100
3. Liu W, Wang Y, Wang Z (2021) Robust stability of discrete-time singularly perturbed systems with nonlinear perturbation. https://doi.org/10.21203/rs.3.rs-1045973/v1
4. Cuevas L, Nesic D, Manzie C (2020) Robustness analysis of nonlinear observers for the slow variables of singularly perturbed systems. Int J Robust Nonlinear Control 30:5628–5656. https://doi.org/10.1002/rnc.5100
5. Castro-Linar R, Alvarez-Galle J, Vasquez-Lopez V (2001) Sliding mode control and state estimation for a class of nonlinear singularly perturbed systems. Dyn Control 11:25–46
6. Aliyu M, Boukas E (2011) H_2 filtering for discrete-time nonlinear singularly perturbed systems. IEEE Trans Circuits Syst I: Reg Pap 58:1854–1864
7. Aliyu M, Perrier M, Baron L (2012) H_∞ filtering for discrete-time nonlinear singularly perturbed systems. Nonlinear Dyn Syst Theory 12:19–36
8. Tahri F, Ben Atia S, Messaoud A, Ben Abdennour R (2022) An interval multiobserver for nonlinear systems subject to internal and external disturbances. Real time experimental validation. J Process Control 116:199–208
9. Ichalal D, Marx B, Mammar S, Maquin D, Ragot J (2018) How to cope with unmeasurable premise variables in Takagi-Sugeno observer design: dynamic extension approach. Eng Appl Artif Intell 67:430–435
10. Nagy-Kiss, Anca M, Ichalal D, Schutz G, Ragot J (2015) Fault tolerant control for uncertain descriptor multi-models with application to wastewater treatment plant. In: 2015 American control conference (ACC). IEEE, pp 5718–5725. https://doi.org/10.1109/ACC.2015.7172235
11. Ben Rajab A, Bahri N, Ltaief M (2021) Discrete-time nonlinear singularly perturbed system identification using coupled multimodel representation. J Dyn Syst Meas Control 143:02100–92021. https://doi.org/10.1115/1.4048933
12. Abdeljawad R, Bahri N, Ltaief M (2019) Stabilization of discrete nonlinear singularly perturbed system with time-delay represented by a coupled multimodel. Energy Procedia 162:191–200. https://doi.org/10.1016/j.egypro.2019.04.021
13. Ltifi M, Bahri N, Ltaief M (2020) Observability bound of discrete linear singularly perturbed system. In: 2020 17th international multi-conference on systems, signals & devices (SSD). IEEE, pp 109–114. https://doi.org/10.1109/SSD49366.2020.9364161
14. Li X, Zhu F (2015) Simultaneous time-varying actuator and sensor fault reconstruction based on PI observer for LPV systems. Int J Adapt Control Signal Process 29:1086–1098
15. Ben Rajab A, Bahri N, Ltaief M (2017) Multimodel identification of discrete-time nonlinear singularly perturbed system. In: 2017 25th Mediterranean conference on control and automation (MED). IEEE, pp 1–6. https://doi.org/10.1109/MED.2017.7984086

Two Phase Interleaved DC-DC Converter-Based V2G/G2V Bi-directional On-Board EV Charger

Gaied Hajer and Flah Aymen

Abstract This paper presents a control and implementation of a bidirectional on-board charger for V2G applications; An interleaved buck/boost converter is used in this system to reduce the size of the passive components of its filters and to improve the current ripple when the battery is being discharged. The DC-DC converter and the DC-AC inverter make up the core topology of the bidirectional on-board charger, which is connected in a cascading form. The border controller is the controller utilized in this suggested system. For V2G functionality, the converter will be operated in boost mode to step up voltage from a battery pack's low voltage (placed in a vehicle) to an appropriate high voltage at the DC-link capacitor before converting to AC using a single-phase inverter. PLL is used as a grid synchronization method to connect to the electric grid. The validation of this proposed system is validated through simulation and experimental results. Ultimately, the simulation and experimental results in MATLAB/Simulink show that the proposed technique is effective and fast response.

Keywords Bidirectional interleaved converter · Vehicle to grid (V2G) · Electric vehicles

1 Introduction

An electric vehicle (EV) is a mode of transportation that uses one or more electric motors to propel the vehicle instead of a traditional internal combustion engine. These vehicles generally use rechargeable batteries to store electrical energy. Currently, there is a significant increase in the demand for electric vehicles, as many countries have set targets to ban internal combustion engine (ICE) vehicles [1].

G. Hajer (✉) · F. Aymen
Processes, Energy, Environment and Electrical Systems (Code: LR18ES34), National Engineering School of Gabes, University of Gabes, Gabes, Tunisia
e-mail: hajergaied.enig@gmail.com

© The Author(s), under exclusive license to Springer Nature Singapore Pte Ltd. 2024
A. Mellit et al. (eds.), *Proceedings of the 2nd International Conference on Green Energy Conversion System*, Green Energy and Technology,
https://doi.org/10.1007/978-981-97-6148-7_24

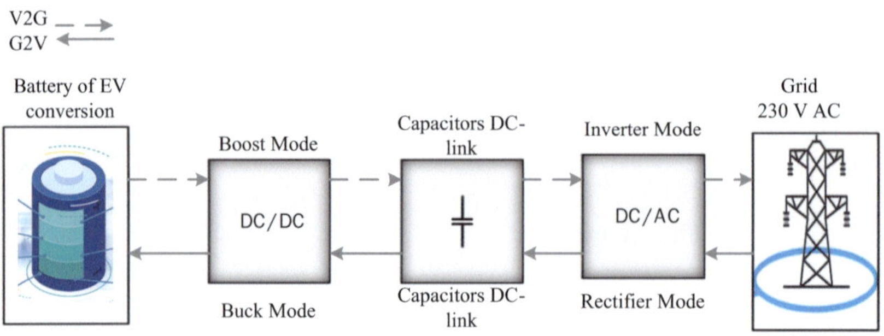

Fig. 1 General G2V and V2G construction block diagram

However, Vehicle to Grid (V2G) offers a number of benefits, such as improved grid flexibility through two mechanisms: (1) providing ancillary services to the power grid and (2) offering side-by-side control. Load balancing request. Leveraging these benefits for a smart grid using mobile energy storage systems in electric vehicles is a very intriguing prospect. The concept of using EV energy storage to power the grid during peak demand and replenish EV storage during periods of low demand holds great promise [2], particularly because most vehicles are parked in homes or office buildings for longer periods than they are driven on the roads.

A generic G2V (Grid-to-Vehicle) and V2G (Vehicle-to-Grid) block diagram is shown in Fig. 1. The two stages of the bi-directional on-board charger typically consist of AC-DC and DC-DC converters... For G2V functionality, the AC-DC inverter operates in rectifier mode, converting grid power to AC and storing it in DC link capacitors. The voltage is then regulated by a DC-DC converter to match the voltage of the EV battery. Depending on the desired battery voltage and DC link voltage, this converter can operate in either buck or boost mode.

2 Bidirectional Converter for V2G Charger

The suggested charger's circuit topology is depicted in Fig. 2, and it has two stages: an interleaved buck boost stage and an inverter stage.

An inverter stage and a boost stage are both present in this. An interleave boost or buck converter handles the boost stage, while a generic inverter with four switches handles the single-phase inverter connected to the single-phase grid [3].

C_1, L_1 and L_2 are filter parameters at the DC side, where C_{dc} is a capacitor at dc-link, and L_c, L_{c1} and C_{AC} are filter parameters at AC side.

In V2G mode of operation, the switches Sb and Sd are controlled, and the interleaved boost converter will be operated in boost mode to step up 160 V of battery voltage to be at least 400 V at DC-link voltage before converting to AC [1].

The interleaved boost converter's duty ratio equation is identical to that of a traditional boost converter. Therefore, Eq. (1), which is illustrated below [4], gives

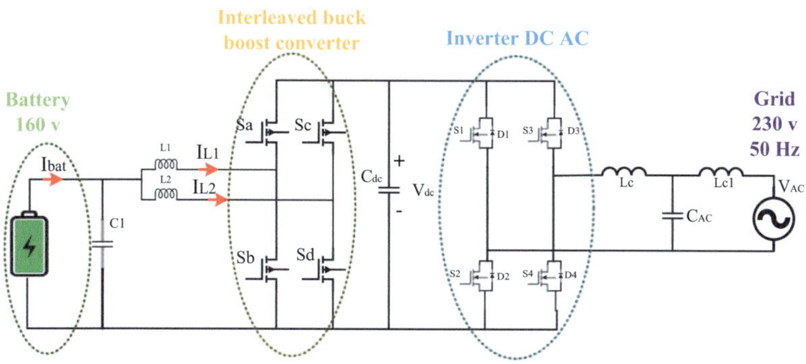

Fig. 2 Topology for a bidirectional on-board charger for V2G

the duty ratio.

$$D = 1 - \frac{V_{bat}}{V_{dc}} \tag{1}$$

where V_{bat} is the battery voltage and V_{dc} is the voltage of the dc-link capacitor. The peak-to-peak current ripple of inductors can be determined by Eq. (2).

$$\Delta I_{L1,L2} = \frac{V_{bat}.D}{f_{sw}.L} \tag{2}$$

Thereby, inductor value can be calculated by Eq. (3).

$$L = \frac{V_{dc} - 2V_{bat}}{f_{sw}.\Delta I_{bat}}(1 - D) \tag{3}$$

3 The Control System

The control concept is presented in Fig. 3. Where there are two voltage-feedback signals consisting of V_{dc} and grid voltage V_{bat} used in this control [5].

To generate control signals for the interleave converter, the control diagram to create PWM with two-phases shifted 180°. This modulation concept provides the control waveforms [6].

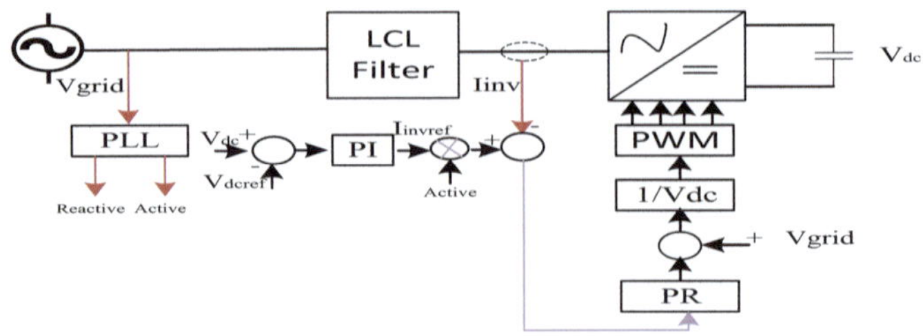

Fig. 3 Block diagram control concept

4 Simulation Results

The proposed system is a bi-directional on-board charger for electric vehicles (EVs) that uses a two-phase interlaced DC-DC converter. This system allows an EV to charge its battery from the power grid (Grid-to-Vehicle, G2V) and also to return the energy stored in the EV battery to the power grid (Vehicle-to-Grid, V2G).

Verify the proposed control concept two-way on-board charger for V2G application, the simulation was performed in the MATLAB/Simulink software and the results of the network voltage, voltage and battery current are shown in Figs. 4a, b and 5a, b, where the parameters and the specifications are detailed in Table 1.

The system provides dual functionality by allowing the EV to be charged from the power grid and by providing the ability to return excess energy from the EV battery to the power grid. The interlaced bending joint allows to optimize the use of the energy and to contribute to the management of the demand and the balancing of the electricity network.

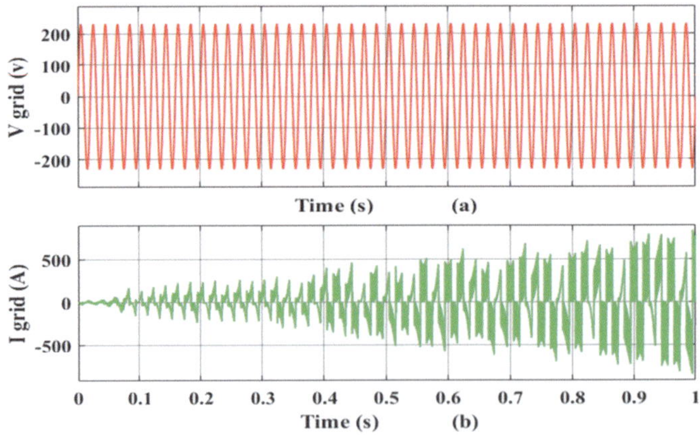

Fig. 4 Voltage and current waveform of grid

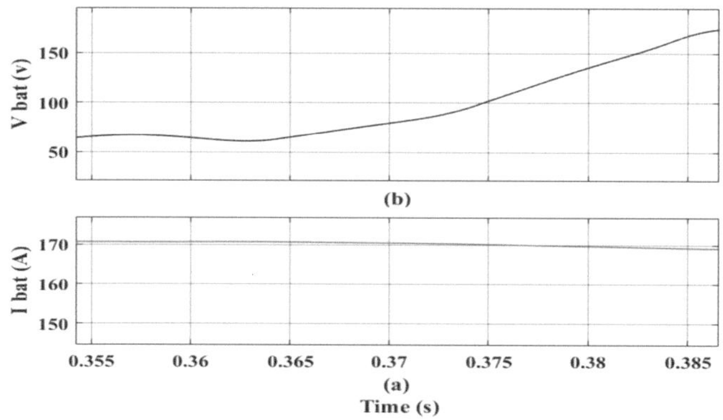

Fig. 5 Zoom of voltage and current waveform battery

Table 1 Parameters of the experimental system

Parameters	Symbol	Value
Battery voltage	V_{bat}	160 V
Filter capacitance	C_1	6.23 μF
DC-link capacitance	C_{dc}	5600 μF
Filter inductance	L_1, L_2	20 mH
Grid voltage	V_{grid}	230 V
Grid frequency	f	50 Hz
Switching frequency	f_{sw}	10 kHz
DC-link voltage	V_{dc}	400 V
Coupling inductance	L_c, L_{c1}	4.36 mH
Coupling capacitance	C_{AC}	6.23 μF
Rated power	P	2 kW

5 Conclusion

The interlaced two-phase DC-DC converter is connected to an external power source, usually the power grid. Alternating current (AC) from the power source is converted to direct current (DC) through a rectifier. The DC-DC converter then takes the input DC current and converts it into an output DC current at the appropriate voltage to charge the VE battery. The DC output is then sent to the EV battery for charging.

When the EV is connected to the power grid and its battery is sufficiently charged, the system allows the energy stored in the battery to be returned to the power grid. The two-phase interleaved DC-DC converter changes the voltage and current of the battery to match the requirements of the power grid. The DC current of the battery is converted into AC through an inverter. The alternating current is then synchronized with the frequency and voltage of the power grid, and then injected into the power grid for use by other electricity consumers.

References

1. Adaikkappan M, Sathiyamoorthy N (2022) Modeling, state of charge estimation, and charging of lithium-ion battery in electric vehicle: a review. Int J Energy Res 46(3):2141–2165
2. Martins CF, Sousa TJ, Pedrosa D (2021, Nov) Electric vehicle battery charger with vehicle-to-vehicle (V2V) operation mode. In: International conference on sustainable energy for smart cities. Springer International Publishing, Cham, pp 173–185
3. Gupta AK, Jha AK, Singh S, Saha SN, Singh SK (2022, Oct) A comparative review on electric vehicles and hybrid vehicles. In: 2022 1st IEEE international conference on industrial electronics: developments & applications (ICIDeA). IEEE, pp 83–88
4. Saeseiw C, Pachanapan P, Somkun S, Srita S, Kaewchum T (2022, Nov) Interleaved bidirectional buck-boost DC/DC converter for high voltage battery application. In: 2022 25th international conference on electrical machines and systems (ICEMS). IEEE, pp 1–6
5. Chaitanya MK, Vinayak AS, Pavan BJG, Lakshmi NN, Sandeep K (2023, Feb) A review of bidirectional DC-DC, DC-AC converters for V2G and G2V applications. In: 2023 International conference on power, instrumentation, energy and control (PIECON). IEEE, pp 1–6
6. De Melo RR, Tofoli FL, Daher S, Antunes FLM (2020) Interleaved bidirectional DC–DC converter for electric vehicle applications based on multiple energy storage devices. Electr Eng 102:2011–2023

Modeling and Simulation of a Proton Exchange Membrane Fuel Cell Stacks in MATLAB/SIMULINK Under Conditions Variations

Salah Beni Hamed and Mouna Ben Hamed

Abstract The Proton Exchange Membrane Fuel Cell (PEMFC) technology is the fastest growing technology. Nowadays, the latter is widely chosen as renewable energy source. Fuel cells are expensive to manufacture. Hence, modelling and simulation of its behavior are necessary in the aim to optimize its performance. In this work, we are focused on the PEMFC stack electric characteristic study. The concerned PEMFC stack produces a power of 100KW. The PEMFC stack are investigated by connecting a specific number of fuel cells in series. At the aim of accomplishing this step, a detailed model is investigated. It is developed based on the equivalent electric circuit of the PFMFC. A most significant one is adopted. Based on the PEMFC power, the number of PEMFC cells and the whole area of the PEMFC stack are suitably computed. The developed equivalent circuit model integrates the fundamental parameters of each sub-cell on the one hand. On the other hand, it involves the impact of temperature, oxygen partial pressure, hydrogen partial pressure and membrane water content. To demonstrate the performance of the used model, a simulation work using Matlab/Simulink tools is carried out. In fact, the PEMFC stack conceived equations are implemented in the form of masked blocs by using the subsystem concept in Matlab/Simulink environment. The electrical characteristics of the PEMFC stack behavior against temperature, oxygen partial pressure, hydrogen partial pressure and membrane water content are investigated. Most significant results are given and discussed.

Keywords Proton exchange membrane fuel cell · Modeling · Simulation · Matlab\Simulink

S. Beni Hamed (✉)
Tunis University, Tunis, Tunisia
e-mail: salahbenihamed@yahoo.fr

M. Ben Hamed
Gabes University, Gabes, Tunisia

© The Author(s), under exclusive license to Springer Nature Singapore Pte Ltd. 2024
A. Mellit et al. (eds.), *Proceedings of the 2nd International Conference on Green Energy Conversion System*, Green Energy and Technology,
https://doi.org/10.1007/978-981-97-6148-7_25

1 Introduction

Nowadays, climate change over time is something we cannot overlook. This requires a great deal of attention and planning. Furthermore, finite energy resources have given rise to many problems and worries. Technological progress has led to a tenfold increase in the use of different types of fuels. Thus, people rely on fossil fuels, especially oil, gas, and coal, to meet their energy needs.

Burning fossil fuels generates carbon dioxide (CO_2) emissions, causing the greenhouse effect. Indeed, CO_2 emissions rose by 1.5% in 2018 [1]. The greenhouse effect increases the temperature of the Earth, resulting in a warming of the planet as it is called global warming.

With the aim to save the earth and human health, the need for clean and substantial energy is needed. Renewable energy resources like solar energy, wind energy …etc. are the most cleanly and green energy resources of energy [2–5]. The generated power is influenced by numerous external elements, such as solar radiation, wind energy, etc. [6]. These external factors can't be controlled. They're totally dependent on the environment.

An alternative source of energy for producing clean energy is the use of fuel cells [7]. The fuel cell is a top-notch green energy source that is becoming more popular due to its heavy weight, small size, low maintenance requirements, high efficiency, and dependability [8–10]. It is sufficiently used to generate electricity for both grid-connected and stand-alone applications.

Alkaline fuel cells, direct methane fuel cells, proton exchange membrane fuel cells (PEMFC), phosphoric acid fuel cells, molten carbonate fuel cell, and solid oxide fuel cells are the most popular forms of fuel cells.

The PEMFC outperforms all other fuel cell types in terms of efficiency because of its performant features like low temperature, high power density, quick starting, compact weight, and no emissions it is when pure hydrogen. It is employed in numerous applications, particularly those related to transportation and housing.

A mathematical model describing the PEMFC's operating principle is required to develop it for such an application. There has been a lot of research done on the steady state, dynamic, and thermodynamic modeling of fuel cells based on their equivalent electric circuits. Various PEMFC parameters, including power output, stack voltage, etc., were examined. Only the ambient temperature and gas pressure at the inlet have an impact on a PEMFC's performance, according to the findings of his research.

The characteristic of the PEMFC varies especially with temperature, membrane water content and gas pressure.

In the most of established model, the membrane water content effect is omitted. In this study, a mathematical simulation mathematical model of the PEMFC is put out that takes into account the impacts of various dynamic situations in pressure reactant gazes for all operating temperatures as well as the water content of the membrane.

The Matlab/Simulink platform is used to theoretically validate the investigated model. Simulation results under various operating settings are displayed and discussed. The paper is organized as follows. In Sect. 2, a generality on the PEMFC stack is given. Simulation results are reported in Sect. 3. Finally, conclusions and some suggested prospects are given in Sect. 4.

2 PEMFC Stack Mathematical Model

The PEMFC is an electrochemical converter. The chemical energy is directly transformed to electrical energy and heat via a chemical reactions. In fact, hydrogen and oxygen gases act separately to form the water. The PEMFC structure integrates two electrodes: one for the anode and the other for the cathode. The two electrodes are separated by a solid membrane called the electrolyte. It is made up of a porous material mixed with a platinum substance. The porous electrode makes easily the flow of gases through it. A chemical reaction happens when the hydrogen injected by the fuel inlet port reaches the anode. The following Eq. (1) states that the molecular hydrogen is split into two protons (H^+) and two electrons

$$H_2 \rightarrow 2H^+ + 2e^- \tag{1}$$

When electrons migrate over the external circuit, a current flows. Through the inlet port at the cathode, oxygen is injected. The protons and molecule O_2 are mixed with the electrons from the outside circuit and the water is formed (2).

$$H_2 + \frac{1}{2}O_2 \rightarrow H_2O \tag{2}$$

When the outer circuit of the fuel cell is opened, an ideal voltage is measured. It is deduced based on the Nernst Eq. (3).

$$V_{oc} = \frac{\Delta Ge}{2F_c} + \frac{\Delta E_r}{2F_c}\left(T_{oc} - T_{refc}\right) + \frac{U_{gs}T_{oc}}{2F_c}\left[Ln(P_{GP\,H_2}) + 0.5Ln(P_{GP\,O_2})\right] \tag{3}$$

In which ΔGe represents the deviation in the free Gibbs energy of the reaction (J/mol), ΔE_r designs the entropy of the reaction (J/mol), $P_{GP\,H_2}$ and $P_{GP\,O_2}$ are respectively the hydrogen and the oxygen partial pressure (atm), T_{oc} and T_{refc} are respectively the cell operating temperature and the cell reference temperature (K), U_{gs} is the universal gas constant (J mol^{-1} K^{-1}), and F_c is the Faraday's constant.

The PEMFC generates a voltage potential differential and a current flow when an external load is attached to it. The following Eq. (4) provides the single cell's output voltage.

$$V_{FC} = V_{oc} - \Delta_{act} - \Delta_{ohm} - \Delta_{con} \tag{4}$$

In which Δ_{act} is the activation polarization losses, Δ_{ohm} is the ohmic polarization losses and Δ_{con} is the concentration polarization losses.

The voltage drop brought on by the anode and cathode being activated is known as the activation overvoltage. Both the Butter-Volmer equation and the Tafel equations are typically used to represent them. The following Eq. (5) can be created by simplifying and extending both of these equations.

$$\Delta_{act} = \sigma_1 + \sigma_2 T_o + \sigma_3 T_o Ln(C_{O_2}) + \sigma_4 T_o Ln(I_{fc}) \tag{5}$$

where I_{fc} is the fuel cell current (A) and the C_{O_2} is the concentration of the oxygen dissolved in a water film interface in the catalytic surface of the cathode (mol/m^3) estimated on the basis of the oxygen partial pressure and cell temperature by the law of Henry.

$$C_{O_2} = \frac{P_{GP\,O_2}}{5.0810^6 \exp\left(\frac{-498}{T_o}\right)} \tag{6}$$

The ohmic losses are caused by the resistance that the protons and electrons encounter while moving through the proton exchange membrane and the outside circuit, respectively. Therefore, the sum of the electron and proton ohmic losses can be used to simulate the overall ohmic resistance. Ohmic's law can be used to design this voltage drop.

$$\Delta_{ohm} = Ifc\left(R_{ef} + R_p\right) \tag{7}$$

With R_{ef} is the equivalent resistance of the electrons flow and R_p is the proton resistance and it is considered to be constant.

The equivalent resistance is given with (8).

$$R_{ef} = \frac{S_r Th_{fc}}{S_{fc}} \tag{8}$$

where S_r is the specific resistance of the membrane (Ωcm), S_{fc} is the fuel cell membrane active area (cm^2) and is the thickness of the membrane (cm).

The following expression for the specific resistance is used.

$$S_r = \frac{181.6\left[1 + 0.03\left(\frac{I_{fc}}{S_{fc}}\right) + 0.062\left(\frac{T_o}{303}\right)^2\left(\frac{I_{fc}}{S_{fc}}\right)^{2.5}\right]}{\left[W_{mc} - 0.634 - 3\left(\frac{I_{fc}}{S_{fc}}\right)\right]\exp\left(4.18\left(\frac{T_o - 303}{T_{refc}}\right)\right)} \tag{9}$$

where W_{mc} is the water membrane content.

The reduction in the concentration of the reactant, oxygen, and hydrogen, which changes during the reaction, is what causes the concentration polarization losses. The concentrations of hydrogen and oxygen at the anode and cathode, respectively, are connected to the concentration polarization losses, also known as mass transport losses. Thus, it can be represented by chemical reaction losses.

$$\Delta_{con} = C_{mt}Ln\left(1 - \frac{I_{fc}}{I_{max}}\right) \quad (10)$$

With I_{max} is the limiting current in the fuel cell, C_{mt} is a parameter that depends the cell.

An individual fuel cell is grouped together into a fuel cell stack. Equation (11) is used to calculate its output voltage.

$$V_{fcstack} = N_{sfc}V_{FC} \quad (11)$$

where N_{sfc} designs the number of fuel cells in stack.

3 Simulation Results

In Matlab/Simulink, the PEMFC stack mathematical model was implemented in the form of masked blocks. Equations (3) through (11) were used to build the models.

Figure 1 depicts the ideal voltage. The graph shows a straight line. The output voltage remains constant despite of the current. If the fuel cell is operated under open circuit conditions where no losses occur, this graph will be achieved.

Figures 2, 3 and 4 shows the three losses: activation polarization losses, ohmic losses and concentration polarization losses. It can be noticed that at lower current, the activation losses (Fig. 2) dominate and a steep drop can be noticed in the beginning. This is because of the slow start of the reaction. Additionally, the initial voltage output is used to overcome the activation energy of the reaction. Furthermore, the reaction at the cathode is a slow process and this is why the activation losses is influenced by the concentration of oxygen.

The graph of ohmic losses against current is shown in Fig. 3. The ohmic losses curve is linear. As the reaction continues and the current increases, the resistance of

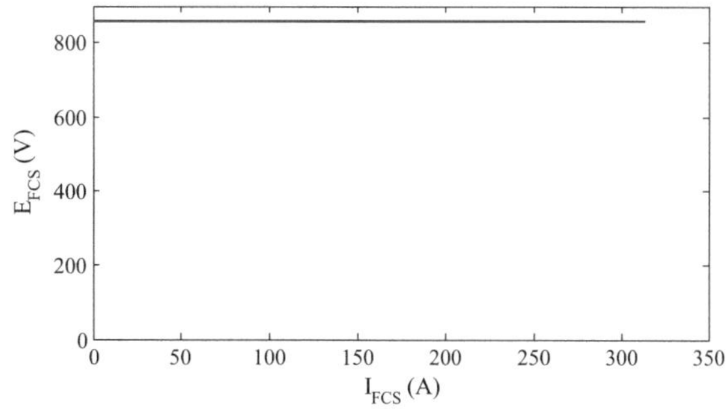

Fig. 1 PEMFC stack ideal voltage

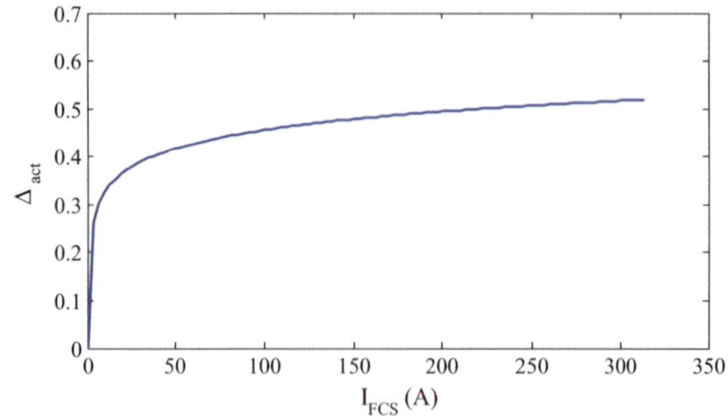

Fig. 2 Activation losses of the PEMFC stack

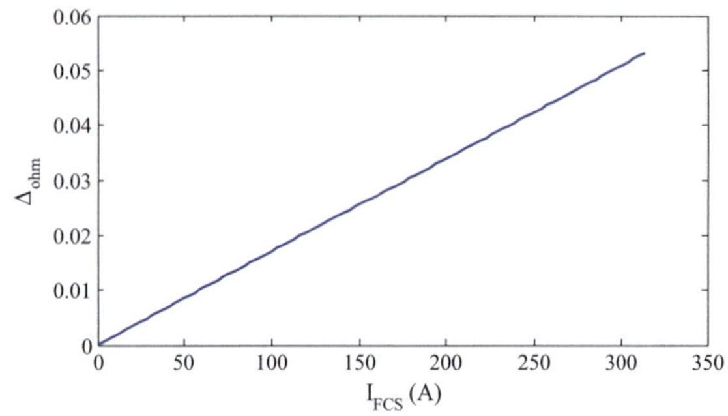

Fig. 3 Ohmic losses of the PEMFC stack

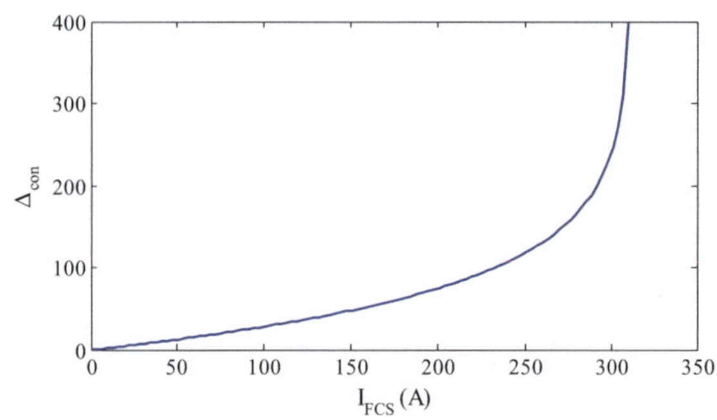

Fig. 4 Concentration polarization losses of the PEMFC stack

the fuel cell components increases. The resistance of the wire, the bipolar plate, and the proton exchange membrane is the main reason for these phenomena.

Figure 4 displays the concentration losses toward the polarization curve. It is evident that these losses increase as the current increases. At the limit current, they suddenly decrease. At this point, the reactant concentration at the electrode equals zero. The reactants being consumed exceed the reactants being supplied when the current from the fuel cell rises above the limiting current. When this happens, the fuel cell stops operating, which causes a step-down in voltage.

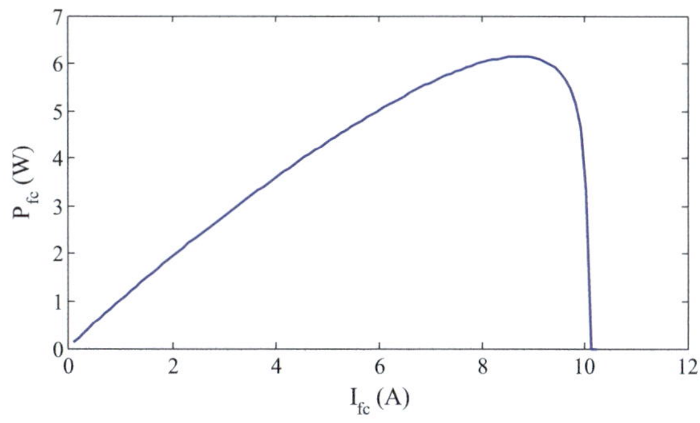

Fig. 5 Power of a single PEMFC cell

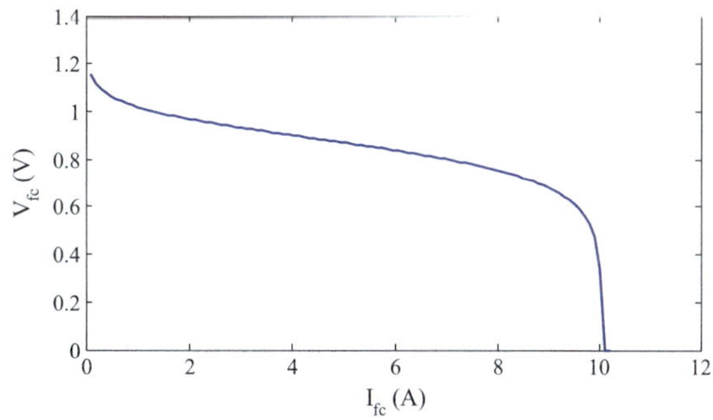

Fig. 6 Voltage polarization of a single PEMFC cell

Performance of a single PEMFC cell

The PEMFC's polarization power curve is shown in Fig. 5 for the specified values of the inputs variables. On the maximum value, the curve begins. The voltage is reduced as a result of the three potential losses. The overall voltage of the fuel cell drops with increasing current, as can be seen from the voltage polarization curve (Fig. 6).

Performance of PEMFC stack

Under operating conditions

Figure 7 displays the PEMFC stack's steady power output. Up to a certain amount of current, the stack power output indicates a rising pattern. The highest power production occurs at this level of rated current. Beyond this point, increasing the current causes the power to decrease. This is due to the fact that the current is becoming constrained and starting to diminish as concentration losses take over. A maximum power of 100 kW is achieved. After that, the power starts to decline.

Under temperature variation

A step change in the cell temperature is used to test the performance of the constructed model under temperature fluctuation with constant water membrane content and

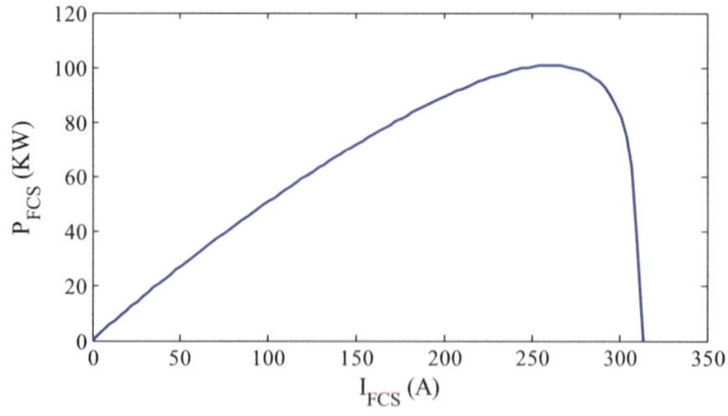

Fig. 7 Power of the PEMFC stack

constant partial gases. The acquired findings are presented in Figs. 8 and 9, which demonstrate how temperature affects the PEMFC stack's efficiency.

Under water membrane content variation

In this case, the membrane water content at constant cell temperature, and constant hydrogen and oxygen partial pressure values are taken into account.

Figures 10 and 11, respectively, show the power-current and voltage-current characteristics of the fuel cell stack depending on the amount of water in the membrane. These charts demonstrate that the output power of the fuel cell stack is a nonlinear

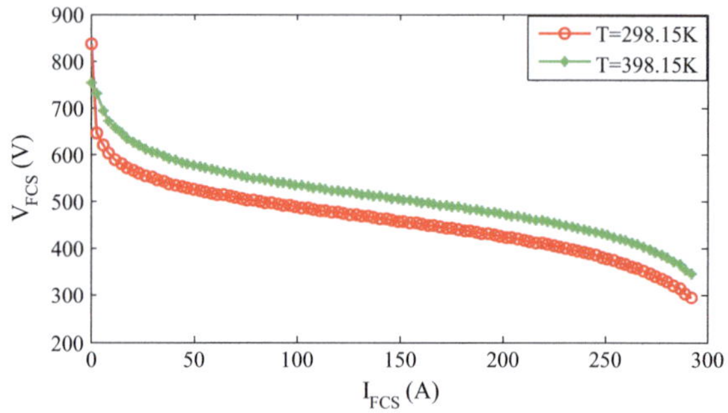

Fig. 8 Polarization of the PEMFC stack under temperature variation

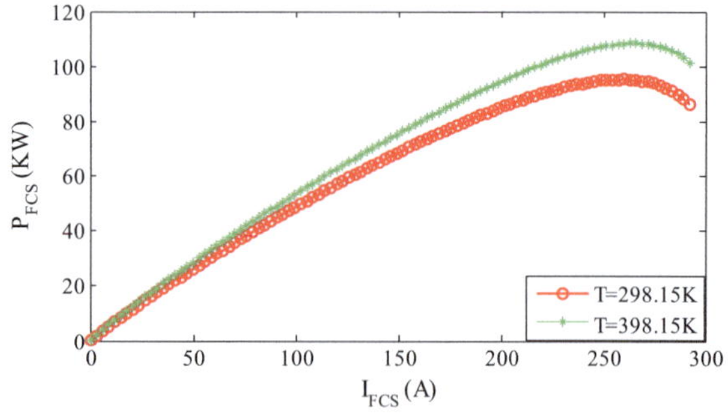

Fig. 9 Power of the PEMFC stack under temperature variation

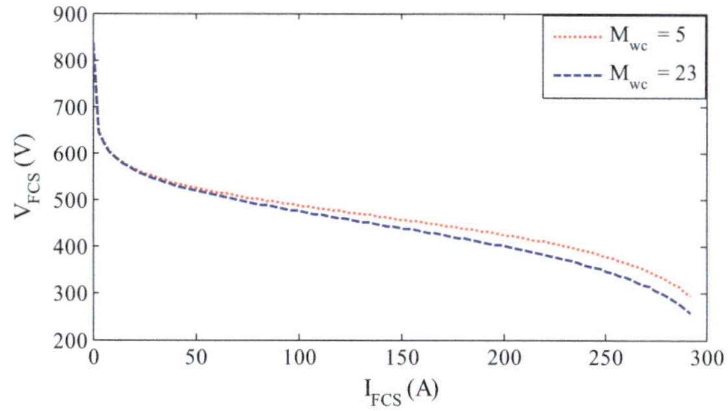

Fig. 10 Voltage of the PEMFC stack under water membrane content variation

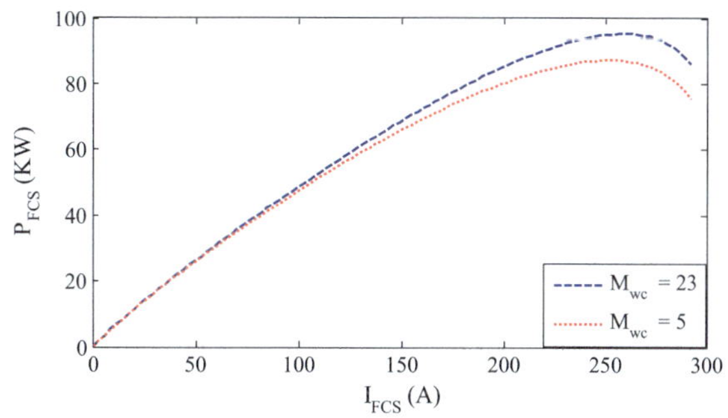

Fig. 11 Power of the PEMFC stack under water membrane content variation

function of current and that membrane water content has a significant impact. The PEMFC stack's power production efficiency increases as the water content of the membrane rises. Obtained results illustrates how water content affects the membrane polarization curves. Ohmic losses in the membrane are reduced because the proton conductivity of the membrane rises with increasing water content. As a result, increasing water content leads to better performance.

Under anode and cathode partial pressure variation

To investigate the impact of the PEFMC stack partial pressure with constant water membrane water and constant temperature, increasing values for hydrogen partial pressure with constant oxygen partial pressure are successively applied in the first step. The second stage involves applying increasing oxygen pressure values while maintaining a constant hydrogen partial pressure. The power-current and voltage-current curves are shown in Figs. 12, 13, 14 and 15.

It is noticed that as the partial pressure of hydrogen and oxygen increases, the voltage losses in the PEMFC decrease on the one hand. On the other hand, the open circuit voltage increase. Thus, better PEMFC efficiency is achieved with values of anode and cathode partial pressure.

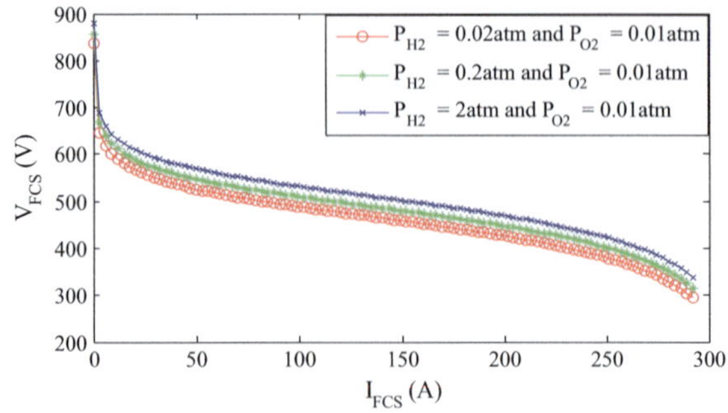

Fig. 12 Voltage of the PEMFC stack under oxygen pressure variation

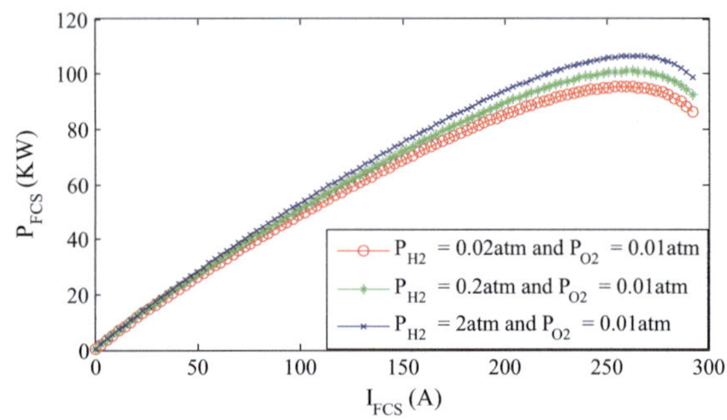

Fig. 13 Power of the PEMFC stack under oxygen pressure variation

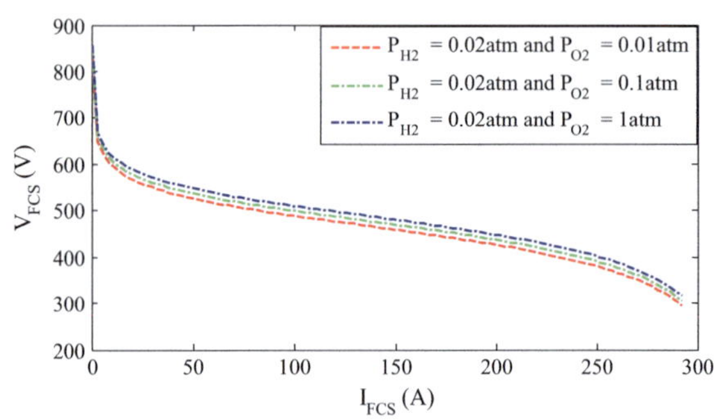

Fig. 14 Voltage of the PEMFC stack under hydrogen pressure variation

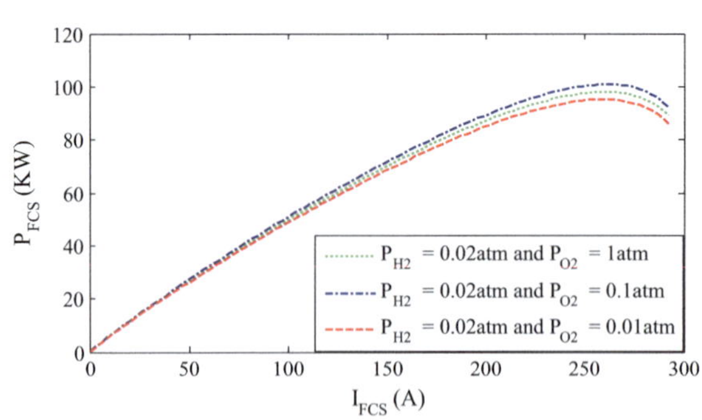

Fig. 15 Power of the PEMFC stack under hydrogen pressure variation

4 Conclusion

In this work, a PEMFC stack mathematical model is established in such a way as to simulate the working principle of the PEMFC stack. The influence of temperature variation on the PEMFC stack's efficiency is included in the model under investigation. The PEMFC stack is implemented using the single PEMFC model as masked blocks in the Matlab/Simulink software. The generated model's simulation results in the Matlab/Simulink environment are presented and discussed. According to the PEMFC stack's obtained curves for various input variable values, voltage losses are decreased and PEMFC efficiency is increased when the PEMFC stack is operated at higher input variable levels. Additionally, we see that there is one operating point when production is at its highest level for each value of the input variables. Additionally, the values of the input variables may change over time. Thus, the maximum power point will also change. As a result; a maximum power point tracking algorithm seems to be the most suitable solution to extract the maximum power point on the one hand. On the other hand, to better manage the energy flow and protect the system, an energy management strategy has to be integrated. In this work, only the steady state model is studied. This may be ameliorate by using the dynamic model.

The high efficient models can be effectively seemed from an experimental work. All these mentioned points are a good prospects to be followed by the author in the future work.

References

1. Minpeng C, Yanrong C, Shan J, Nicklas F (2023) Toward carbon neutrality before 2060: trajectory and technical mitigation potential of non-CO_2 greenhouse gas emissions from Chinese agriculture. J Clean Prod 368(417):1–12
2. Jinlan B, Chunyu W (2023) Renewable energy resources and sustainable development goals: evidence based on green finance, clean energy and environmentally friendly investment. Resour Policy 80:1–20
3. Muhammad I, Khalid Z, Abdelmohsen A, Nassani B, Gheorghiţa D, Haroon K, Mohamed H (2023) Does nuclear energy reduce carbon emissions despite using fuels and chemicals? Transition to clean energy and finance for green solutions. Geosci Front 11:1–13
4. Author F (2016) Article title. Journal 2(5):99–110
5. Rizwana Y, Xuhui Z, Rui T, Wasi Ul Hassan S (2023) The impact of green technology, environmental tax and natural resources on energy efficiency and productivity: perspective of OECD rule of law. Energy Rep 9(10):1308–1319
6. Bade Venkata S, Yegireddi S, Teegala Srinivasa K, Gaurav D, Ali Torabi H, Epari Ritesh P (2023) Natural energy materials and storage systems for solar dryers: state of the art. Solar Energy Mater Solar Cells 255(259):1–24
7. Muhammad I, Khalid Z, Abdelmohsen A, Gheorghiţa D, Haroon ur Rashid K, Mohamed H (2023) Does nuclear energy reduce carbon emissions despite using fuels and chemicals? Transition to clean energy and finance for green solutions. Geosci Front 14(6):1–13
8. Fereshteh N, Bengt S (2023) Review of exergy and energy analysis of fuel cells. Int J Hydrogen Energy 48(65):1–68

9. Li N, Zofia L, John S (2023) An approach for sizing a PV–battery–electrolyzer–fuel cell energy system: a case study at a field lab. Renew Sustain Energy Rev 181:1–13
10. Yaqoob M, Muhammad Usman K, Muhammad W, Umair Z, Faisal M, Faizan M, Muhammad S, Ali R (2023) Renewable energy as an alternative source for energy management in agriculture. Energy Rep 10:344–359

Modeling and Controlling of Wind Turbine Based on Permanent Magnetic Synchronous Generator

Rania Harizi and Mouna Benhamed

Abstract In contemporary times, there exists a significant rise in the demand for electricity. This surge necessitates a robust demand for natural resources and fossils. Consequently, it is incumbent upon us to integrate novel renewable and clean energy sources. The focus of this paper centers on wind turbines, which provide electrical energy through a dual conversion process, involving the conversion of kinetic energy of the blade to mechanical energy and the mechanical to electrical energy. The permanent magnetic synchronous generator (PMSG) is the device responsible for the second energy conversion. Firstly, we present a model of the wind turbine to enhance our comprehension of the energy conversion process. Additionally, we provide an illustration of another model for the PMSG. We then proceed to present an optimization of the generator speed aimed at maximizing the system's performance. This is achieved through the application of the strategy of maximum power point tracking (MPPT) with a PI regulator. Finally, we translate the mathematical model of the wind turbine and the PMSG into block diagrams using Simulink Matlab.

Keywords Wind turbine · Conversion energy · PMSG · MPPT · PI regulator

1 Introduction

The extensive usage of natural resources continues to harm our world by increasing greenhouse gas emissions. Therefore, to solve these issues, it is essential to replace these environmentally damaging fuels with an efficient and environment-friendly energy source. One of these renewable and cleanest energy sources is wind power [1].

The wind turbine, otherwise known as a wind generator, is a device that converts the kinetic energy of wind, which is harnessed by the blades, into mechanical energy

R. Harizi (✉) · M. Benhamed
Processes, Energy, Environment and Electrical Systems (Code: LR18ES34), National Engineering School of Gabes, University of Gabes, Gabes, Tunisia
e-mail: raniaharizi1@gmail.com

© The Author(s), under exclusive license to Springer Nature Singapore Pte Ltd. 2024
A. Mellit et al. (eds.), *Proceedings of the 2nd International Conference on Green Energy Conversion System*, Green Energy and Technology,
https://doi.org/10.1007/978-981-97-6148-7_26

that is made available through a transmission shaft. This mechanical energy is then converted into electrical energy through the use of either a synchronous or asynchronous electromagnetic generator especially the asynchronous generator with dual power supply and synchronous generator with permanent magnet [1, 2].

In this paper, we focus on variable speed wind turbine based on the permanent magnetic synchronous generator PMSG that received a lot attention due to his advantages such as the high density, low failure rate, the simple control, the absence of excitation circuit and the low maintenance [1, 3].

The paper is organized as follow: Sect. 2 represents the wind turbine model, the Sect. 3 describes the model of PMSG, the Sect. 4 refers to the strategy control of the turbine speed and finally a conclusion is given in Sect. 5.

2 Model of Wind Turbine

2.1 Wind Model

The wind speed is modeled as a deterministic sum of the frequency harmonics (Fourier series representation) [4]. It is given by

$$\boldsymbol{V_v} = \boldsymbol{A} + \sum_{k=1}^{i} \boldsymbol{a_k} * \sin(\boldsymbol{W_k t}) \tag{1}$$

where $\boldsymbol{V_v}$ is the wind speed, A is the average wind speed, $\boldsymbol{a_k}$ harmonic amplitude of the order k, $\boldsymbol{W_k}$ is the proper frequency of the proper oscillation k.

2.2 Conversion of Kinetic Energy to Mechanical Energy

The aerodynamic energy is given by [4]:

$$P_m = \frac{dE_c}{dt} = \frac{1}{2} \cdot \rho \cdot \mathrm{S} \cdot \mathrm{V}^3 \tag{2}$$

where E_c: the kinetic energy, ρ: the air density, S: the circular area, V is the wind speed.

2.3 Characteristics of the Turbine

The characteristic of the wind turbine requires defining some concepts; the Tip Speed Ratio and the power coefficient.

Tip Speed Ratio: The tip speed ratio λ is a ratio without dimension of tangential blade-end speed ($R\Omega_t$) to instantaneous speed V. This factor is a specific factor for wind generators. The expression of Tip speed ratio

$$\lambda = \frac{R\Omega_t}{V} \tag{3}$$

where R: the radius of the turbine and Ω_t is the turbine rotor speed.

The power coefficient C_p has a theoretical limit of 0.593 called the "Betz limit". This limit is never reached in practice. It depends mainly on two main factors: the pitch angle β and the tip speed ratio λ. The C_p of the turbine studied has as equation [3]:

$$C_p(\lambda, \beta) = (0.5 - 0.00167(\beta - 2)) \cdot \sin\left[\frac{\pi(\lambda + 0.1)}{18.5 - 0.3(\beta - 2)}\right] - 0.00184(\lambda - 3)(\beta - 2) \tag{4}$$

Figure 1 illustrates the variation of power coefficient C_p by varying the tip speed ratio λ. It can be seen that when β increases, the maximum efficiency of the turbine decreases.

Then for $\beta = 2$ the power coefficient C_p varies as a function of the specific speed λ. So, the top point of the curve represented in Fig. 2 indicates the optimal values of power coefficient ($C_{p\,max}$) and the optimal specific speed (λ_{opt}). The aerodynamic power of wind turbine is given by [5]:

$$P_{aer} = P_t = \rho \cdot \pi \cdot R^2 \cdot V_v^3 \cdot C_p(\lambda) \tag{5}$$

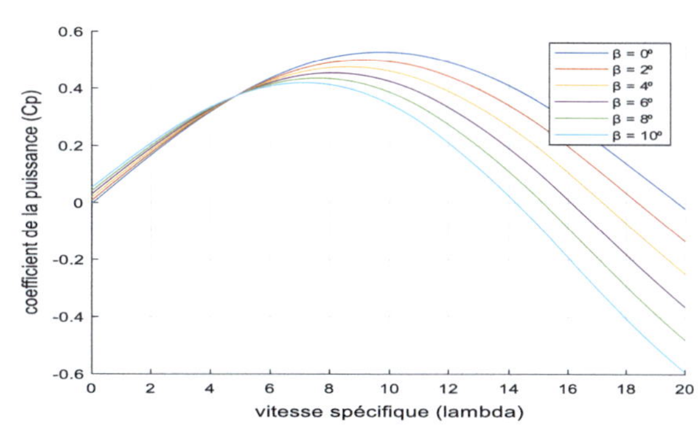

Fig. 1 Power coefficient as a function of specific speed λ

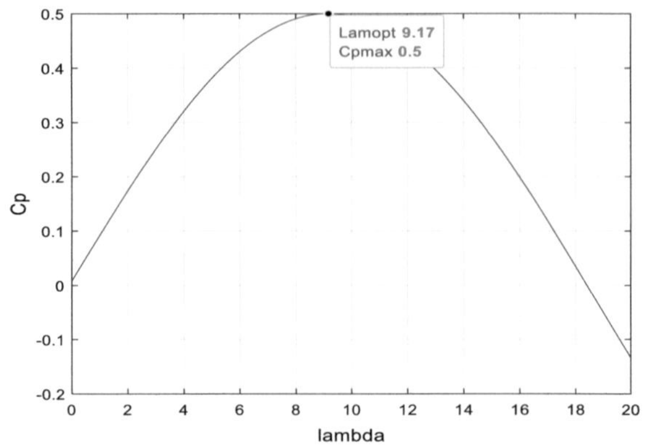

Fig. 2 Power factor coefficient C_p

2.4 *Aerodynamic Torque*

The wind turbine is designed to convert wind energy into mechanical torque that is written as follows:

$$C_{aer} = \frac{P_{aer}}{\Omega_t} = \frac{\rho \cdot \pi \cdot R^2 \cdot V_v^2 \cdot C_p(\lambda)}{2 \cdot \lambda} \tag{6}$$

3 PMSG Model

The voltage equations of PMSG respectively in d axis and q axis are presented in (7) [5]

$$\begin{cases} V_d = -R_s \cdot I_d - \frac{d\Psi_d}{dt} + \omega \cdot \Psi_q \\ V_q = -R_s \cdot I_q - \frac{d\Psi_q}{dt} - \omega \cdot \Psi_d \end{cases} \tag{7}$$

Where

$$\Psi_d = -L_d \cdot I_d + \Psi_f \tag{8}$$

$$\Psi_q = -L_q \cdot I_q \tag{9}$$

L_q = q axis inductance, L_d = d axis inductance, I_q = q axis current, I_d = d axis current, R_s = resistance of the stator winding, ω = angular velocity of the rotor, Ψ_d = flux of d axis, Ψ_q = flux of q axis, Ψ_f = flux of magnets.

The dynamic of the GSAP is given by the following mechanical equation:

$$C_m - C_{em} - f\Omega = J\frac{d\Omega}{dt} \tag{10}$$

where $f\Omega$ is the friction torque, J is the total moment of inertia of the machine.

The electromagnetic torque is express [5]

$$C_{em} = \frac{3}{2}P\left[\left(\mathrm{L}_d - \mathrm{L}_q\right) \cdot I_q \cdot I_d - \Psi_f \cdot I_q\right] \tag{11}$$

P is the number of pole pairs.

4 Control of the Conversion System

We discuss the MPPT (Maximum Power Point Tracking) in the region where the speed of the wind exceeds the initial speed and lower than the nominal wind speed. In order to achieve maximum gain and generate the greatest electrical energy, the system must uphold a power coefficient $C_{pmax} = 0.5$ and $\lambda_{opt} = 9.17$ [6].

Thus, to optimize the mechanical power (Pm) production, we must regulate its associated speed while simultaneously monitoring its electromagnetic torque (C_{em}).

This will necessitate the utilization of a type PI speed controller to guarantee adequate speed control and consequently, a suitable reference electromagnetic torque [3, 6, 7]. Figure 3 shows the implementation of the control device.

Thus, the electromagnetic torque is written as follows:

$$C_{m-ref} = PI(\Omega_{ref} - \Omega_r) \tag{12}$$

The expression of the PI regulator is given below

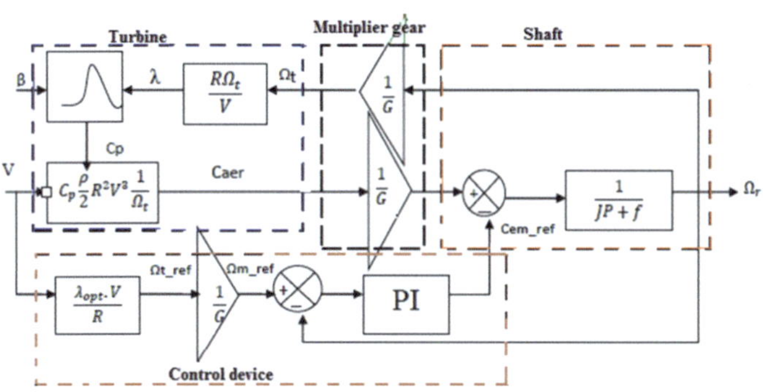

Fig. 3 Block diagram of the maximization of the extracted power MPPT with speed control

$$PI = \left(K_i + \frac{K_p}{s}\right)(\Omega_{ref} - \Omega_r) \tag{13}$$

$$K_i = w_n^2 J \tag{14}$$

$$K_p = 2\xi w_n J - f \tag{15}$$

where K_i is the integral gain, K_p is the proportional gain, ξ is the damping factor, w_n is the proper pulsation.

The part concerning the speed control loop has as a set Ω_{ref}, an output Ω_r (or Ω_m) and the mechanical torque C_m as a disturbance. The transfer equation for this segment can be expressed as follows

$$\Omega_r = F(s)\Omega_{ref}(s) - H(s)C_m(s) \tag{16}$$

where

$$F(s) = \frac{K_p s + K_i}{Js^2 + (f + K_p)s + K_i} \tag{17}$$

and

$$H(s) = \frac{s}{Js^2 + (f + K_p)s + K_i} \tag{18}$$

In the steady state the aerodynamic torque is zero due to the regulator's integral function,

$$\Omega_m = \Omega_r = F(s)\Omega_{ref}(s) \tag{19}$$

Figures 4, 5, 6 and 7 show, respectively, the variations in mechanical speed, electromagnetic torque, the speed of rotation of the turbine and the power extracted.

5 Conclusion

In this paper, both of the wind turbine and the PMSG are modeled by mathematical equations. Then, we proposed a strategy for controlling the speed of the generator through the optimal reference of its speed using the technique MPPT to reach a maximum power in the case of low winds and to adapt this power to the limit of the nominal power supported by the electric generator. It is then necessary to control and limit the converted mechanical power.

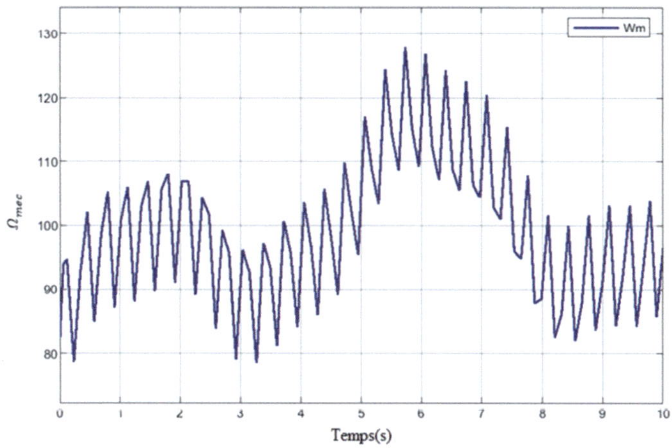

Fig. 4 Mechanical speed

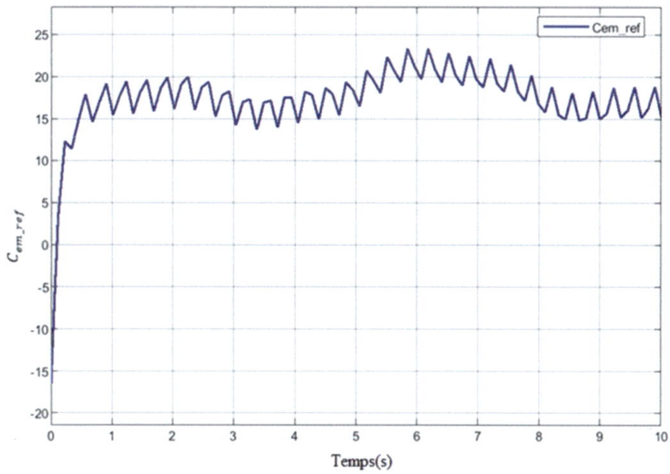

Fig. 5 Referential electromagnetic torque

For this, we used a speed controller of type PI to ensure a proper control of the speed and subsequently a suitable reference electromagnetic torque.

Finally, the simulation scheme and the simulation results have been demonstrated on the virtual platform: Matlab/Simulink.

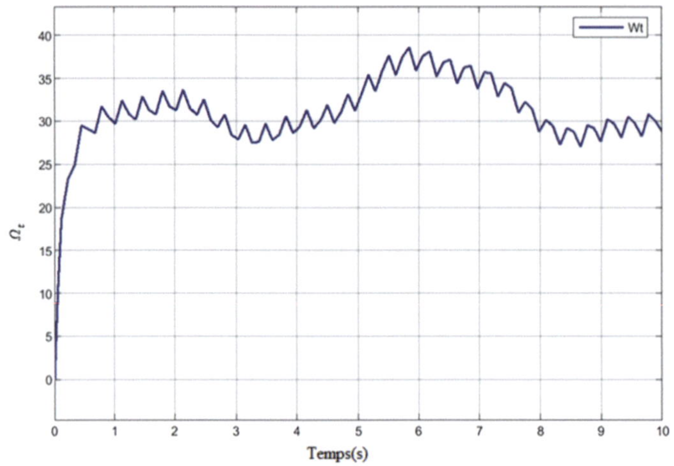

Fig. 6 Rotation speed of turbine

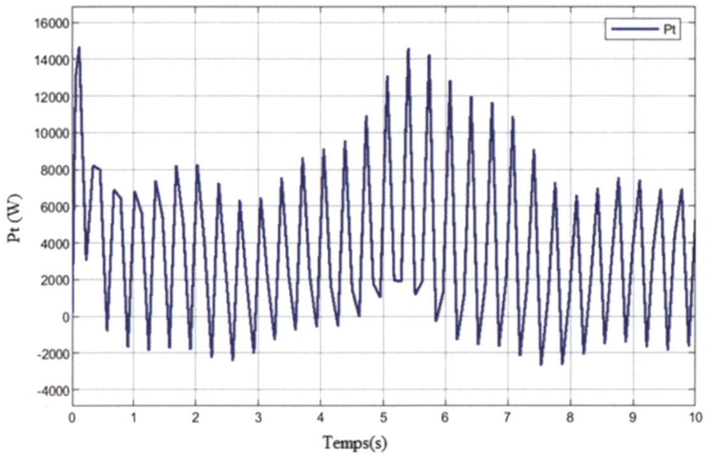

Fig. 7 The extracted power

References

1. Xing P, Fu L, Wang G, Wang Y, Zhang Y (2018) A compositive control method of low-voltage ride through for PMSG-based wind turbine generator system. IET Gener Transm Distrib 12(1):117–125
2. Heng TY, Ding TJ, Chang CW, Ping TJ, Yian HC, Dahari YM (2022) Permanent magnet synchronous generator design optimization for wind energy conversion system: a review. Energy Rep 8:277–282
3. Camara MS, Camara MB, Dakyo B, Gualous H (2014) Modélisation et commande d' une génératrice synchrone à aimant permanent pour la production et l' injection de l' énergie offshore dans un réseau. Ccm, pp 8–10

4. Elbeji O, Hamed MB, Sbita L (2014) PMSG wind energy conversion system: modeling and control. Int J Mod Nonlinear Theory Appl 03(03):88–97
5. Badoni P, Prakash SB (2014) Modeling and simulation of 2 MW PMSG wind energy conversion systems. IOSR J Electr Electron Eng 9(4):53–58
6. Hannachi M, Elbeji O, Benhamed M, Sbita L (2021) Optimal torque maximum power point technique for wind turbine: proportional–integral controller tuning based on particle swarm optimization. Wind Eng 45(2):337–350
7. Deng F, Liu D, Chen Z, Su P (2017) Control strategy of wind turbine based on permanent magnet synchronous generator and energy storage for stand-alone systems. Chin J Electr Eng 3(1):51–62

Polyisoprene Based Composite Layers: Electrical Properties for Stretchable High Storage Energy Materials

Douniazad Mezdour, Mohamed Tabellout, and Jean-François Pilard

Abstract In the present work, the dielectric relaxation spectroscopy technique (DRS) is used to investigate the electrical properties of synthetic photoreticulated Polyisoprene films containing conductive particles of core–shell structure. The conduction threshold of these materials is situated below 0.5 wt% of Polyaniline. Such a low percolation threshold is desired to maintain the matrix flexibility. For the 40 wt% loaded film, the conductivity reaches more than 10^{-5} S/cm. This value is suitable for semi-conductive applications. As expected, the conductivity was found to be thermally activated. The application of Efros-Shklovskii model gives a short localization length of about 4 A°. At low frequencies, the dielectric constant of these composites at room temperature is very high, a jump from nearly 10 to 10^7 was recorded for the 40 wt% loaded film. This is probably due to the interfacial polarization between the three constituents. Such high values of the dielectric constant allow using these materials as high k dielectrics for energy storage applications.

Keywords Polyisoprene · Polymer · Electrical properties · Energy storage

1 Introduction

Elastomeric polymers are commonly used for fabricating soft devices due to their well-suited mechanical behavior. Natural Polyisoprene (NR) and Synthetic Polyisoprene (IR) are rubber thermoset materials having interesting properties that include extremely high elongation, high values of tensile and tear resistance while also exhibiting excellent resilience. Natural rubber has unique mechanical properties while its synthetic counterpart is characterized by very high chemical uniformity, batch homogeneity and good processability [1]. Hence, composites based on IR

D. Mezdour (✉)
Jijel University, 18000 Jijel, Algeria
e-mail: d_mezdour@univ-jijel.dz

M. Tabellout · J.-F. Pilard
Le Mans Université, 72085 Le Mans, France

© The Author(s), under exclusive license to Springer Nature Singapore Pte Ltd. 2024

A. Mellit et al. (eds.), *Proceedings of the 2nd International Conference on Green Energy Conversion System*, Green Energy and Technology,
https://doi.org/10.1007/978-981-97-6148-7_27

could be the ideal alternative to NR in some fields [2]. The characteristics of the rubber recipe are adjusted using various polymers, fillers, plasticizers and other ingredients, to match the application criteria. Besides doping, intrinsically conductive polymers (ICPs) can be used to tailor their electrical properties. In polymer composite materials, the electrical conductivity is dependent on the formation of a conductive network in the insulating polymer matrix. The minimum amount of conductive additive that causes the abrupt increase of the electrical conductivity is known as the percolation threshold. This parameter is dependent on the properties of the conductive filler, such as the surface area, the aspect ratio and the properties of the insulating polymer matrix [3]. The percolation threshold is also affected by the interaction between the conductive filler and the polymer matrix, the manufacturing process and the processing conditions [4]. Polyaniline (PANI) is one of the most widely studied ICPs because of its remarkable and well-known properties, such as high electrical conductivity, chemical and environmental stability and ease of synthesis.

Few works were devoted to Polyaniline/polyispoprene composites. The focus of this study was to investigate the potential use of synthetic Polyisoprene/Polyamide@PANI composites in electronic applications as high energy storage materials.

2 Experimental Procedure

The Polyamide@PANI powder of core–shell type has been synthetized by oxidative polymerization of aniline around micrometer sized particles of polyamide 12 (PA12) in presence of a doping acid. As an organic acid, the Dodecyl benzene sulfonic dopant (DBSA) gives better dissolution properties to the resulting composite. The synthesis details are described in [5].

For the synthesis of Photoreticulated Oligoisoprene, the synthetic cis-1, 4-Polyisoprene is first purified by dissolving it in dichloromethane (CH_2Cl_2), followed by its precipitation in methanol. The epoxidation of the cis-1, 4-polyisoprene is performed following the procedure described in [6]. The produced carbonyl telechelic PI (5% epoxidized) is mixed with the PA12@PANI particles at different ratios and subjected to UV irradiation to obtain three phase layers.

3 Results and Discussion

Using the Novocontrol broadband dielectric spectrometer, the electrical data were recorded in the frequency range lying between 0.1 and 10^6 Hz. Figure 1 displays the frequency dependence of AC conductivity for photoreticulated layers of Polyisoprene containing various concentrations of PA12@PANI starting from 5 wt% to 40 wt%. As expected, increasing the concentration enhances the electrical conductivity. A

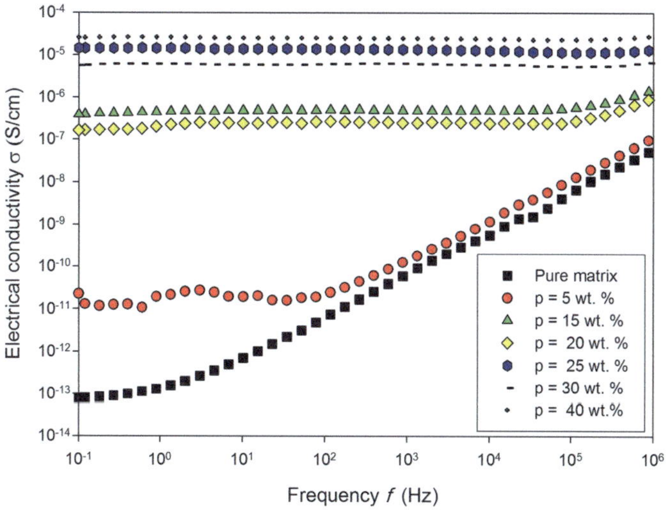

Fig. 1 Evolution of the electrical conductivity σ versus the frequency *f* for various concentrations p of PA12@PANI particles

relatively constant conductivity is obtained at high levels of PA12@PANI filler. The conductivity reaches nearly 10^{-5} S/cm in magnitude at 40 wt% of loading.

A low-frequency plateau appeared at 5 wt%. This is indicative of the transition from insulating to conductive behavior, due to the percolation of the PANI shells, creating an electrically conductive network inside the insulating matrix. The percolation threshold is then located below 5 wt% as this concentration is the first to mark the occurrence of a plateau corresponding to a DC conductivity.

Since the PA12@PANI powder contains 10 wt% of PANI, the 5 wt% of PA12@PANI content corresponds to 0.5 wt% of PANI in the composite layer.

The effect of the temperature on the electrical conductivity was investigated by the DRS technique in the temperature range of [−100 °C, + 30 °C].

The obtained results are plotted in Fig. 2 illustrating the evolution of the electrical conductivity of a Polyisoprene layer containing 40 wt% of PA12@PANI particles over all the DC plateau (up to 10^4 Hz). As expected, the Dc conductivity increases versus temperature. At −100 °C, the conductivity is four orders of magnitude lower than its value at room temperature, then it jumps abruptly when increasing the temperature to −80 °C. A monotonical increase is observed each 20 °C until reaching a value above 10^{-6} S/cm at 30 °C.

This value is one order of magnitude lower than its initial value even when exceeding the room temperature. The material doesn't recover its conductivity after this cooling/heating process.

In Fig. 3, the DC conductivity at 0.15 Hz is plotted as a function of $(1/T)^{1/2}$ for the 40 wt% loaded sample in the temperature range lying between—183 and 303 K, following the Efros-Shklovskii model [7]:

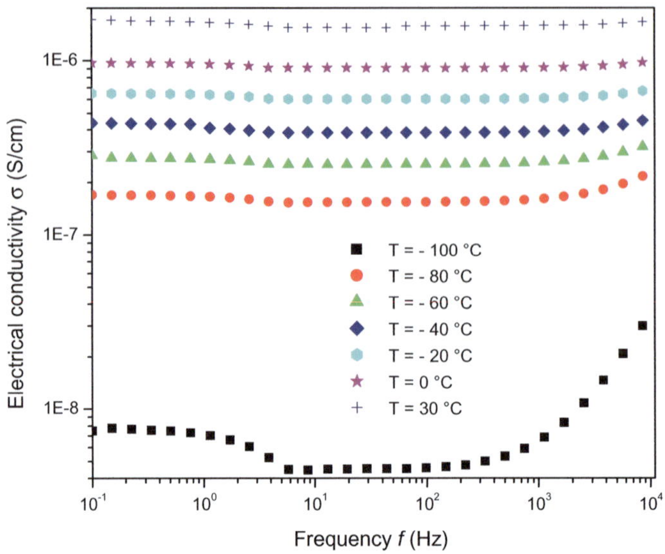

Fig. 2 Effect of the temperature on the electrical conductivity σ of a polyisoprene layer containing 40 wt% of PA12@PANI

$$\sigma = \sigma_0 \cdot exp\left[-\left(\frac{T_0}{T}\right)^{1/2}\right] \tag{1}$$

where σ is the electric conductivity, σ_0 a constant; T_0 is the characteristic temperature of Efros and Shklovskii linked to the relative permittivity of the material and the localization length ξ by the equation:

$$\xi = \frac{2.8e^2}{4\pi T_0 K_B \varepsilon_0 \varepsilon} \tag{2}$$

where e is the electron charge, K_B is the Boltzman constant, ε_0 is the vacuum permittivity and ε the relative permittivity of the material. T_0 describes the energy barrier of an electron to hop from one localized state to another one.

Extrapolation of the linear fit depicted in Fig. 3 gives a T_0 value of 2.46×10^4 K and σ_0 of 1.31×10^{-2} S/cm. The dielectric constant is extracted from the dielectric constant measurements versus frequency at a temperature of −60 °C. It was found to be of 458.38 at 0.15 Hz and considered to be a static value at such a low frequency.

At relatively high temperatures, the calculated localization length is about few Angströms and does not exceed 5 A° (Table 1).

A common feature of almost all organic semiconductors is the localization of charge carriers at band edge energies which arises due to a certain type of disorder in the material. It is assumed that the deeper the energy level of an electron state in the

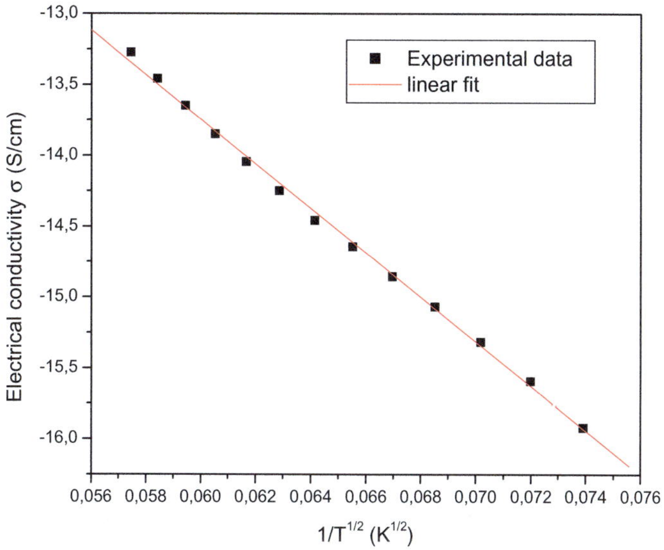

Fig. 3 Log plot of the electrical conductivity σ as a function of $(1/T)^{1/2}$ for the sample loaded with 40 wt% of PA12@PANI

Table 1 The transport parameters extracted from the experimental data using Efros-Schklovskii model

Load (wt%)	Electrical conductivity σ_0 (S/cm)	Temperature T_0 (K)	ε	Localization length ξ (A°)
40	1.31×10^{-2}	2.46×10^{4}	458.38	4.13

band gap, away from the valence- or conduction-band edges, the stronger the degree of spatial localization of the associated electron wave function.

The real part ε′ of the dielectric constant of IR/PA12@PANI composites at different filler's loadings is represented in Fig. 4. The Rubber film had a low dielectric constant, in the range of 4–6, ascribed to the non-polar nature of Rubber [8]. The IR composite films exhibited a higher dielectric constant at lower frequencies reaching a maximum value in the range of 10^6 for the 30 and 40 wt% loadings. At high frequencies, molecular movements arrest or decrease orientation polarization, leading to the decrease of the dielectric constant [9]. The curves show increasing dielectric constants versus increasing PA12@PANI concentrations in the synthetic rubber composite films over all the frequency range.

As Polyamides and PANI contain polar amide and NH groups respectively, increasing PA12@PANI particles in IR composites induces higher dipole polarization and thus enhanced dielectric constant [8]. In addition, the increase in PA12@PANI loading results in an increased interfacial polarization due to the conductivity contrast of the two media [9].

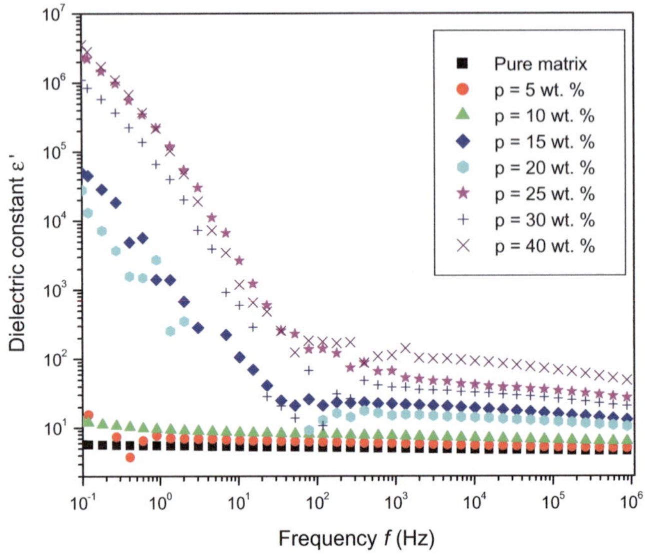

Fig. 4 Evolution of the dielectric constant ε′ versus the frequency *f* for various concentrations p of PA12@PANI particles

As the dielectric constant is the amount of energy from an external electrical field stored in materials, this composite type is a potential candidate as high energy storage material.

4 Conclusion

In this preliminary study, it has been shown that the DC conductivity of Polyisoprene based composite layers can be tuned by varying the embedded PA12@PANI particles concentration. The conductivity levels are strongly dependent on the resulting dispersion levels of core–shell particles. Percolation threshold is located below 0.5 wt% of Polyaniline. The conductivity reaches more than 10^{-5} S/cm for the 40 wt% loaded sample. This value is suitable for semi-conductive applications. Consequences of charge carrier localization on the electrical transport properties of these materials were also analyzed. Application of Efros and Shklovskii model gave a short localization length of about 4 A° for a 40 wt% of PA12@PANI content. This value is a measure of chain disorder between the ordered regions. Dielectric behavior of composite materials depends on the dielectric properties of the matrix and the filler, structure, shape and morphology of particles and filler dispersion in the composite. High values of the dielectric constant were obtained and reached $3 \cdot 10^6$ for the layer containing 40 wt% of PA12@PANI. This allows using these materials as high k dielectrics for energy storage applications.

References

1. Coi Rubber Home page. https://www.coirubber.com/natural-rubber-vs-synthetic-rubber/. Accessed 5 Aug 2021
2. Lockhorn D, Klüppel M (2019) Structure–property relationships of silica/silane formulations in natural rubber, isoprene rubber and styrene-butadiene rubber composites. J Appl Polym Sci 137:48435–48446
3. Li J, Kim J-K (2006) Percolation threshold of conducting polymer composites containing 3D randomly distributed graphite nanoplatelets. Compos Sci Technol 67(10):2114–2120
4. De Souza VL, Ribeiro dos Anjos EG, Almeida Verginio GE, Oyama IC, Ferreira Braga N, Ferreira da Silva T, Montagna LS, Passador FR (2021) A review concerning the main factors that interfere in the electrical percolation threshold content of polymeric antistatic packaging with carbon fillers as antistatic agent. Nano Select 3:248–260
5. Mezdour D, Tabellout M, Sahli S, Fatyeyeva K (2010) Electrical properties investigation in PA12/PANI composites. Macromol Symp 290:15–24
6. Kébir N, Campistron I, Laguerre A, Pilard J-F, Bunel C, Couvercelle JP, Gondard C (2005) Use of hydroxytelechelic cis-1,4-polyisoprene (HTPI) in the synthesis of polyurethanes (PUs). Part 1. Influence of molecular weight and chemical modification of HTPI on the mechanical and thermal properties of Pus. Polymer 46(18):6869–6877
7. Shklovskii BI, Efros AL (1975) Percolation theory and conductivity of strongly inhomogeneous media. UFN 117(3):401–435
8. Jacob M, Varughese KT, Thomas S (2006) Dielectric characteristics of sisal–oil palm hybrid biofibre reinforced natural rubber biocomposites. J Mater Sci 41:5538–5547
9. Louis N, Maria S, Banu A, Sudha S (2014) Effect of rice husk ash and hydrated silica on the dielectric properties of LDPE composites. Mater Sci Indian J 11:0974–7486

Diagnosis and Isolation Methodology for Faults Detection in the Pitch Angle Sensors of a Wind Power System

Bouaziz Laila, Dhaoui Mehdi, and Ben Hamed Mouna

Abstract Wind energy plays a major role in the production of electrical energy in the industrial world. Like any industrial system, the wind turbine can be threatened by anomalies likely to affect its performance or even cause a total shutdown of the installation. Therefore, fault diagnosis methods are necessary. In this work, in order to avoid the occurrence of a malfunction, either to avoid them or to limit their consequences, we used a model-based fault diagnosis technique to detect and isolate sensor faults and actuators in pitch angle systems. To do this, we used an observer applied to a "Benchmark" model for fault detection and isolation (FDI). Simulation results under the MATLAB/Simulink environment are presented to validate the effectiveness of this diagnostic method.

Keywords Wind turbine · Pitch angles · Fault detection · Fault isolation · Observer

1 Introduction

The global demand for energy is increasing day by day and the need for new sources of energy is necessary for the production of electricity. Due to the drying up of fossil energy in the world, researchers are increasingly interested in other so-called renewable energy sources to the detriment of traditional energy sources. Wind energy is one of the first forms of energy exploited by man. This energy ranks second after photovoltaic. During its operation, it causes no discharge (no greenhouse effect or acid rain) and no waste. However, the main disadvantage is the lack of flexibility and inconstancy. The wind system is increasingly complex and the diagnostic operation, which consists in detecting and locating the faults which affect the system has become essential to ensure the operational safety and availability of these systems.

B. Laila (✉) · D. Mehdi · B. H. Mouna
L/R: Processes, Energy, Environment and Electrical Systems (PEESE), Electrical Department of Engineering, National Engineering School, University of Gabes, 6029 Gabes, Tunisia
e-mail: bouazizlaila9@gmail.com

© The Author(s), under exclusive license to Springer Nature Singapore Pte Ltd. 2024

A. Mellit et al. (eds.), *Proceedings of the 2nd International Conference on Green Energy Conversion System*, Green Energy and Technology,
https://doi.org/10.1007/978-981-97-6148-7_28

In general, the wind turbine converts the kinetic energy of the wind into mechanical energy and then into electricity [1]. The kinetic energy of the wind is captured by the blades of the wind turbine, which transforms it into mechanical energy via the gearbox. The latter transforms the slow speed of the turbine shaft into a much higher speed compatible with the operating speed range of the electric generator allowing the generation of electric current, which will be delivered to the network. On the other hand, the penetration rate of wind farms in electrical networks is becoming increasingly important. Therefore, when connecting wind farms to the grids, problems of instability, adjustment of active and reactive power, frequency and voltage may arise. Faced with these problems, network managers must plan a cut-off procedure, which leads to the forced disconnection of wind turbines. For these reasons, many countries take the specific regulations seriously and define the rules necessary to connect farms to the networks. In addition, it appears necessary to regulate the power of wind farms in order to connect them to the network without affecting the quality of the energy produced. It is in this context that fits this work, which deals with the detection and localization of faults in the wind system.

2 Description of the Wind System

The purpose of this section is to present a mathematical model of the wind turbine. The model should be well detailed to use it as a simulation model. This model is derived and developed by Odgaard et al. [2] to be able to apply diagnostic methods to it. The considered model reflects a horizontal axis wind turbine with three blades, variable speed, with a complete converter and power of 4.8 MW. This "Benchmark" model contains four sub-models: the blades and the pitch system (Blade and Pitch System), the drive train (Drive Train), the converter and the generator (Converter and Generator) and finally the controller (Controller). These sub-models are shown in Fig. 1, where the variables involved in these subsystems are defined in [2–4].

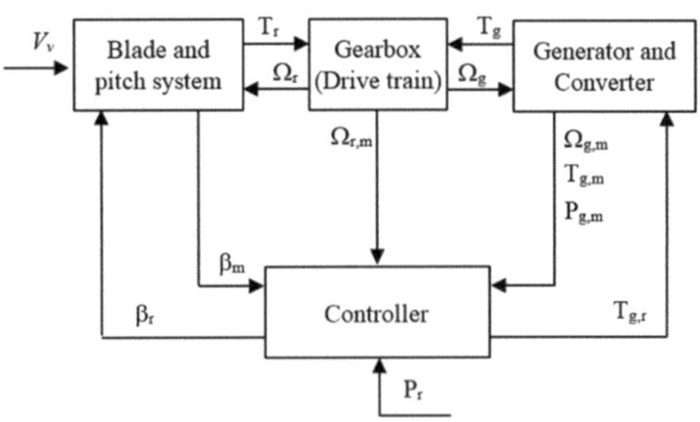

Fig. 1 Diagram of wind system sub-models [5]

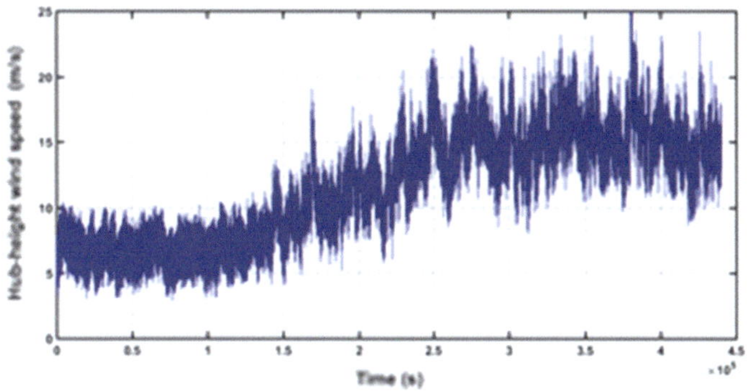

Fig. 2 Representation of the used wind speed sequence

2.1 Wind Model

To have a good coverage of normal operation of the wind system, we chose the predefined wind model shown in Fig. 2 [3, 4]. From this figure, it can be seen that the wind speed varies in the range of 5–20 m/s, with some peaks at 25 m/s.

2.2 Aerodynamic Model

The wind turbine captures the kinetic energy of the wind and converts it into torque that spins the rotor blades.

- The total power available in the wind likely to be collected by the blades can be written in the form:

$$P_v = \frac{1}{2} \cdot \rho \cdot \pi \cdot \mathrm{R}^2 \cdot \mathrm{V}_v^3 \tag{1}$$

where ρ: air density (kg m^{-3}), R: blade radius (m) and V_V: wind speed (m/s).

- The power captured by the wind turbine is expressed by Eq. (2) [1]:

$$P_T = C_p \cdot P_v = \frac{1}{2} \cdot C_p(\lambda, \beta) \cdot \rho \cdot \pi \cdot \mathrm{R}^2 \cdot \mathrm{V}_\mathrm{v}^3 \tag{2}$$

where C_p (λ, β): the power coefficient, β: the pitch angle, λ: turbine speed ratio and Ω_r being the angular speed of the rotor, with:

$$\lambda = \frac{\Omega_r \cdot R}{V_v} \tag{3}$$

- The aerodynamic torque relationship is:

$$T_r = \frac{1}{2} \cdot \frac{C_p(\lambda, \beta) \cdot \rho \cdot \pi \cdot \mathrm{R}^3}{\lambda} \cdot \mathrm{V}_\mathrm{v}^2 \tag{4}$$

- Knowing the rotational speed of the turbine, the aerodynamic reference torque is therefore expressed by Eqs. (3) and (4) [2].

$$T_r = k_{opt} \cdot \Omega_r^2 \tag{5}$$

$$k_{opt} = \frac{1}{2}\rho\pi R^5 \frac{C_{p\max}}{\lambda_{opt}^3} \tag{6}$$

2.3 Model of the Pitch Angle System

The system consists of three identical actuators, each having an internal controller [2]. This actuator is usually of the hydraulic type, adjusts the pitch angle of the blades by rotating them. It is modeled as a second-order transfer function:

$$\frac{\beta(s)}{\beta_r(s)} = \frac{\omega_n^2}{s^2 + 2 \cdot \zeta \cdot \omega_n \cdot s + \omega_n^2} \tag{7}$$

2.4 The Gearbox (Drive Train) Model

The aerodynamic torque is transferred to the generator via the Gearbox to increase the rotational speed of the rotor, to a higher speed required by the generator. The mechanical model is built of a slow shaft and a fast shaft linked by the gearbox [6].

$$\begin{cases} J_r \frac{d\Omega_r(t)}{dt} = T_r(t) - (B_{dt} - f_r) \cdot \Omega_r(t) + \frac{B_{dt}}{G} \cdot \Omega_g(t) - K_{dt} \cdot \theta_\Delta(t) \\ J_g \frac{d\Omega_g(t)}{dt} = \frac{\eta_{dt} \cdot B_{dt}}{G} \cdot \Omega_r(t) - \left(\frac{\eta_{dt} \cdot B_{dt}}{G^2} + f_g\right) \cdot \Omega_g(t) + \frac{\eta_{dt} \cdot K_{dt}}{G} \cdot \theta_\Delta(t) - T_g(t) \\ \frac{d\theta_\Delta(t)}{dt} = \Omega_r(t) - \frac{1}{G} \cdot \Omega_g(t) \end{cases} \tag{8}$$

where Ω_r and Ω_g are the rotor and generator speed, both measured by two sensors $\Omega_{r,i}$ and $\Omega_{g,mi}$ (with $i = 1; 2$). Moreover, θ_Δ denotes the torsion angle of the drive train. T_g and T_r are the converter and aerodynamic torques. J_r and J_g are the moment of inertia of the low- and high-speed shafts, respectively. f_r and f_g are the viscous frictions of the low- and high-speed shafts, respectively. K_{dt} is the torsion stiffness of the drive train, B_{dt} is the torsion-damping coefficient of the drive train, G is the gear ratio and η_{dt} is the efficiency of the drive train.

2.5 Model of the Generator and Converter

In this section, the model of the converter and generator is presented. The generator converts mechanical energy into electrical energy, while it is charged by a torque delivered by the converter. It is described as a first-order system [2]:

$$\frac{T_g(s)}{T_{g,r}(s)} = \frac{\alpha_{gc}}{s + \alpha_{gc}} \tag{9}$$

The power produced by the generator depends on the generator rotational speed and the applied load, as described in the following equation:

$$P_g(t) = \eta_g \cdot \Omega_g(t) \cdot T_g(t) \tag{10}$$

where: $P_g(t)$ is the power produced by the generator and η_g is the generator efficiency.

2.6 Controller Model

The wind turbine operates in four zones (Fig. 3), [1]: Zone I: The wind speed is too low. The turbine can rotate but the energy to be captured is insufficient. Zone II: The maximum power is captured in this zone for each wind speed. Different methods exist to optimize the extracted energy. This zone corresponds to operation at partial load (CP). Zone III: The available power becomes too great. The power extracted is therefore limited while remaining as close as possible to the nominal power. This zone corresponds to full load operation (PC). Zone IV: The wind speed becomes too strong. The turbine is stopped and the power extracted is zero. The controller is active in zones II and III. In zone II the optimum rotor speed is obtained by using the converter torque as the control signal. Zone III corresponds to constant power production. In this work, only zones II and III are considered [3].

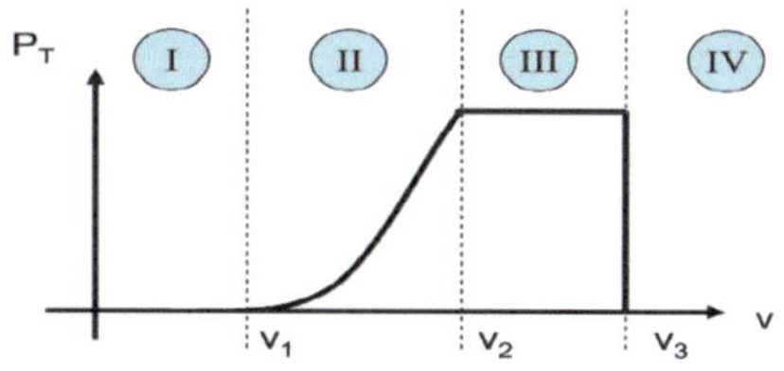

Fig. 3 Operating zones of a wind turbine [1]

3 Fault Scenarios

In order to identify the possible faults in a wind turbine and to determine their effect on the behavior of the system, faults that can arise in a wind turbine are chosen so as to cover the different types. Additionally redundancy in the system is identified to determine fault detection [6].

1. Fixed value type fault in the 2nd blade position sensor 1, $\beta_{1,m2} = 23°$ occurs between 3500 and 4100 s.
2. Additive bias fault in the 1st blade position sensor 2, $\beta_{2,m1} = 8$ occurs at times 3000 and 3400 s.
3. Gain type fault in the 1st blade position sensor 3, $\beta_{3,m1} = 0.5\ \beta_{3,m1}$ occurs at times 1200 and 1600 s.
4. Fixed value type fault in the 1st rotor speed sensor, $\Omega_{r,m1} = 2$ occurs at times 1000 and 2000s.
5. Gain type fault in the 2nd generator speed sensor, $\Omega_{g,m2} = 5.\ \Omega_{g,m2}$ occurs at times 1000 and 2000s.
6. Stall actuator dynamic slow change fault, the stall actuator parameters 2 (ξ, ω_n) change from [11.11, 0.6] to [5, 0.8] at times 2400 and 3400 s.
7. Abrupt change in the stalling actuator dynamics fault, the stalling actuator parameters 3 (ξ, ω_n) change from [11.11, 0.6] to [9, 0.32] at instants 3000 and 3200 s.
8. Offset type fault in the converter actuator, $T_g = 5000 + T_g$ at instants 3100 and 4100 s.
9. Multiplier dynamic change default, $\eta_{dt} = 0.92$ between 3000 and 4200 s.

4 Observer-Based Fault Detection

4.1 Observer Model

Figure 4 Principle of the pitch angle system control with Lumberger observer.

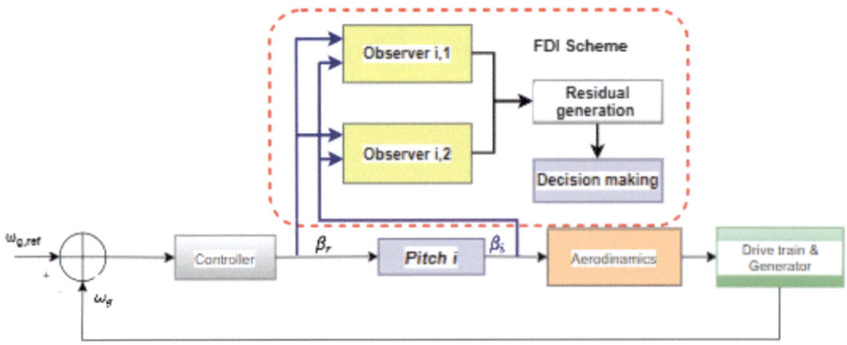

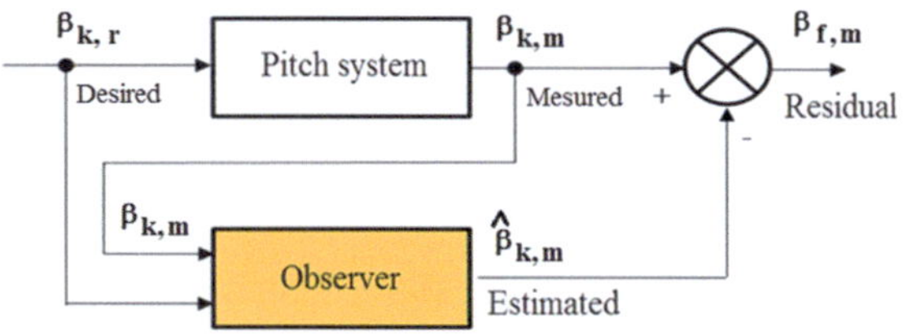

Fig. 4 (a) describe the diagram operating principle of the pitch angle and (b) show the structure of the Lumberger observer

4.2 *Methodology for Faults Detection in the Pitch Angle System*

The goal of the pitch angle system fault detection and isolation (FDI) algorithm is to detect and isolate the fault based on the two pitch angle sensor measurements (i.e. fault in $\beta_{i,m1}$ or $\beta_{i,m2}$). Fault detection can be accomplished by a simple check between the two sensor measurements [7]. After fault detection, and in order to determine which sensor is faulty, a fault isolation step (by observer) is necessary. Each observer is based on one of two scenarios:

- Scenario 1: Sensor 2 is faulty and Sensor 1 is healthy;
- Scenario 2: Sensor 1 is faulty and sensor 2 is healthy.

In its normal state, the pitch angle system is expressed by:

$$\begin{aligned} \dot{x}_{ob} &= A_{ob} \cdot x_{pb} + B_{ob} \cdot \left(\beta_r + \beta_f\right) \\ y_{ob} &= C_{ob} \cdot x_{ob} \end{aligned} \tag{11}$$

where: the state vector $x_{ob} = \left[\dot{\beta}_i \; \beta_i\right]^T$ contains the angular speed of setting $\dot{\beta}_i$ and the position β_i (i = 1, 2, 3). A_{ob}, B_{ob} and C_{ob} are the system matrices, y_{ob} is the pitch angle measurement, β_r the position reference, β_f is the position error caused by the fault defined as:

$$\beta_f = \beta_i - \frac{1}{2}\left(\beta_{i,m1} + \beta_{i,m2}\right) \tag{12}$$

So,

$$\dot{x}_{ob} = A_{ob} \cdot x_{ob} + B_{ob} \cdot \beta_r + B_{ob} \cdot \left[\beta_i - \frac{1}{2}\left(\beta_{i,m1} + \beta_{i,m2}\right)\right] \tag{13}$$

From scenario 1, we have: $\beta_{i,m1} = \beta_i = C_{ob} \cdot x_{pb}$.
The dynamical system of Eq. (13) under scenario 1 becomes:

$$\begin{aligned} \dot{x}_{ob} &= \left(A_{ob} + 0.5 \cdot B_{ob} \cdot C_{ob}\right) \cdot x_{ob} + B_{ob} \cdot \left(\beta_r - 0.5 \cdot \beta_{i,m2}\right) \\ \beta_{i,m1} &= C_{ob} \cdot x_{ob} \end{aligned} \tag{14}$$

Similarly, for scenario 2, the system becomes: $\beta_{i,m2} = \beta_i = C_{ob} \cdot x_{pb}$

$$\begin{aligned} \dot{x}_{ob} &= (A_{ob} + 0.5 \cdot B_{ob} \cdot C_{ob}) \cdot x_{ob} + B_{ob} \cdot \left(\beta_r - 0.5 \cdot \beta_{i,m1}\right) \\ \beta_{i,m2} &= C_{ob} \cdot x_{ob} \end{aligned} \tag{15}$$

According to the systems of Eqs. (14) and (15), the two observers OBS1 and OBS2 are described respectively by the following equations:

$$\begin{aligned} \dot{\hat{x}}^1_{ob} &= (A_{ob} + 0.5 \cdot B_{ob} \cdot C_{ob} - L_{ob} \cdot C_{ob}) \cdot \hat{x}^1_{ob} + B_{ob} \cdot \left(\beta_r - 0.5 \cdot \beta_{i,m2}\right) \\ &\quad + L_{pb} \cdot \beta_{i,m1} \\ \dot{\hat{\beta}}^1_{i,m1} &= C_{ob} \cdot \hat{x}^1_{ob} \end{aligned} \tag{16}$$

And

$$\begin{aligned} \dot{\hat{x}}^2_{ob} &= (A_{ob} + 0.5 \cdot B_{ob} \cdot C_{ob} - L_{ob} \cdot C_{ob}) \cdot \hat{x}^2_{ob} + B_{ob} \cdot \left(\beta_r - 0.5 \cdot \beta_{i,m1}\right) \\ &\quad + L_{pb} \cdot \beta_{i,m2} \\ \dot{\hat{\beta}}^2_{i,m2} &= C_{ob} \cdot \hat{x}^2_{ob} \end{aligned} \tag{17}$$

where $\hat{\beta}_{i,m1}$ and $\hat{\beta}_{i,m2}$ are the estimated outputs of $\beta_{i,m1}$ and $\beta_{i,m2}$ obtained from the two observers and L_{ob} is the observer gain matrix chosen to ensure the stability of the matrix $(A_{ob} + 0.5 \cdot B_{ob} \cdot C_{ob} - L_{ob}C_{ob})$.

We denote the estimation error by: $\tilde{\beta}_{i,m1} = \beta_{i,m1} - \tilde{\beta}_{i,m1}$ and $\tilde{\beta}_{i,m2} = \beta_{i,m2} - \tilde{\beta}_{i,m2}$.

The use of observers built from only the input/output measurements of the system makes it possible to detect faults. The residuals generated are compared to fault thresholds. We considered a structure of observers: The GOS structure (Generalized Observer Scheme): where the ith observer is controlled by all the outputs (inputs) except the ith and all the inputs (outputs).

- If $\left|\tilde{\beta}_{i,m1}\right| = 0$ and $\left|\tilde{\beta}_{i,m2}\right| \neq 0$, it is concluded that sensor 1 is faulty;
- If $\left|\tilde{\beta}_{i,m2}\right| = 0$ and $\left|\tilde{\beta}_{i,m1}\right| \neq 0$, we therefore conclude that sensor 2 is faulty;
- If $\left|\tilde{\beta}_{i,m2}\right| \neq 0$ and $\left|\tilde{\beta}_{i,m1}\right| \neq 0$, then it is concluded that there is an actuator fault.

5 Simulation Results

In this section, we will apply the observer-based flaw detection method to the two-pitch angle system β_1, β_3.

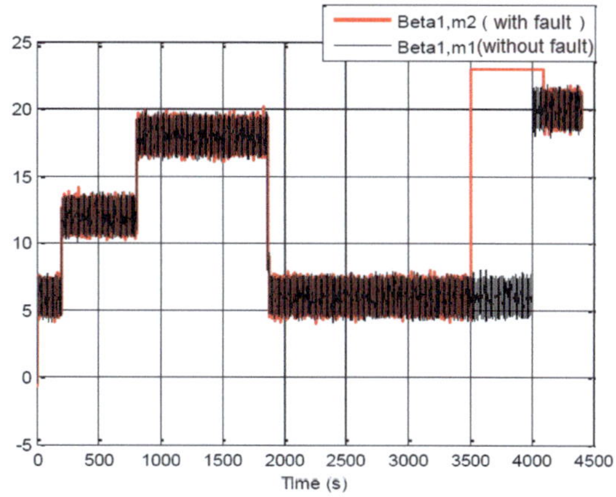

Fig. 5 Evolution of measurements $\beta_{1,m1}$ and $\beta_{1,m2}$ of blade 1

5.1 System 1 (Fixed Value Fault in $\beta_{1,m2}$)

For β_1, a fixed value type sensor fault (fault 1) has been introduced in the second sensor measurement $\beta_{1,m2}$ for the interval of 3500–4100 s.

Figure 5 represents the two measurements $\beta_{1,m1}$ (without fault) and $\beta_{1,m2}$ (with fault) obtained by sensor 1 and sensor 2. In the general case, in the absence of a fault, the two sensor outputs ($\beta_{1,m1}$ and $\beta_{1,m2}$) are almost identical, but after the appearance of a fault, it can be noticed that there is at least one of the components of these residues, which reacts to the presence of this anomaly. The fault is detectable within the previously mentioned time interval and the detection time is about 0.02 s. Figure 6 represents the two residues delivered by OBS1 and OBS2 in the presence of a defect. As we can see, due to the presence of the sensor noise, the residual is not correctly equal to zero in the absence of fault. The residual changes considerably after the occurrence of the fault. The residue delivered by OBS2 (bottom) converges towards zero, on the other hand the residue delivered by OBS1 (top) diverges from zero between 3500 and 4100 s (sensitivity to the fault), and according to the fault location structure indicated in the previous section we can conclude that there is a sensor fault in the second measurement $\beta_{1,m2}$ with an isolation time of 0.5 s.

5.2 System 3 (Fault 3 and 7 in $\beta_{3,m1}$)

For this system, the sensor fault has been associated with an actuator fault:

- Gain type sensor fault (fault 3) in the first sensor measurement ($\beta_{3,m1}$) and in the interval of 1200–1600 s;
- Actuator fault caused by low oil pressure (fault 7) which develops gradually in the interval between 3000 and 3200 s.

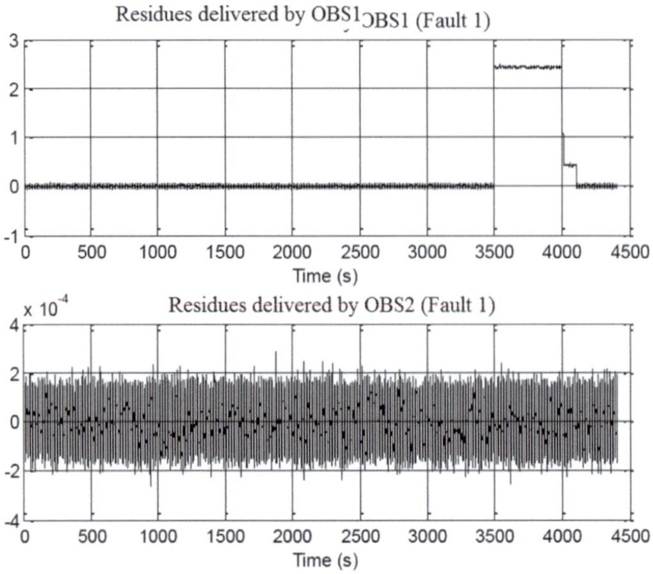

Fig. 6 Residues delivered by OBS1 and OBS2 in the presence of fault 1

Figure 7 represents the two measurements $\beta_{3,m1}$ and $\beta_{3,m2}$ with the evaluation of the defects. The results clearly show the changes from a normal regime to a faulty regime, which therefore allows the detection of faults 3 and 7, with a detection time of about 0.02 s and 0.6 s, respectively. Figure 8 represents the two residues delivered by OBS1 and OBS2. Both faults are detectable within the previously mentioned time intervals. The simulation time is divided into two parts:

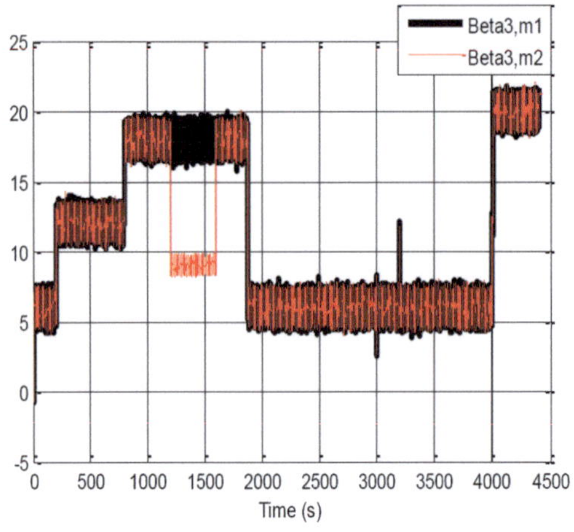

Fig. 7 Evolution of measurements $\beta_{3,m1}$ and $\beta_{3,m2}$ of blade 3

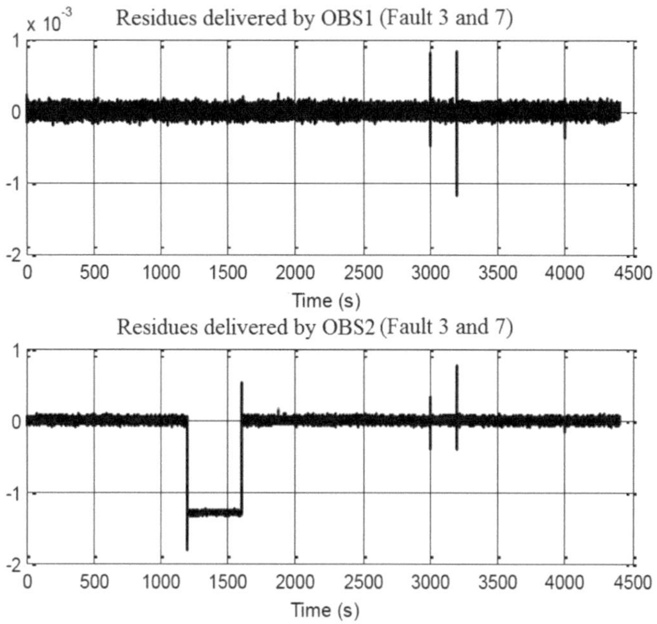

Fig. 8 Residues delivered by OBS1 and OBS2 in the presence of fault 3 and 7

- From 0 to 2000s: the residual delivered by OBS2 (bottom) deviates significantly from zero between 1200 and 1600 s, so it is sensitive to the fault, on the other hand the residual delivered by OBS1 (top) always remains average value substantially zero, and according to the fault location structure indicated in Sect. 4.1 it can be seen that there is a sensor fault in the first measurement ($\beta_{3,m1}$) with an isolation time of 0.6 s.
- From 2000 to 4500 s, the residuals delivered by OBS1 and OBS 2 are different from zero between 3000 and 3200 s, it can therefore be seen that there is an actuator fault with an insulation time equal to 1,6 s.

6 Conclusion

Today the wind turbine brings major developments to increase the power produced by adjusting the pitch angle of the blades in order to capture a lot of wind. Therefore, the blades of the wind turbine play a very important role in the energy conversion chain. In this work, an observer-based method for the detection and localization of sensor and actuator faults in the wind turbine blade pitch system is presented. The results obtained by simulation show that the diagnostic methods based on observers of the calibration system are very encouraging, mainly in the presence of noise. This observer allows good performance for the detection of defects thanks to a good robustness with respect to uncertainties, noises and disturbances in addition to a rapid detection of defects and the generation of several residues.

References

1. Laverdure N (2005) Sur l'intégration des générateurs éoliens dans les réseaux faibles ou insulaires, thèse de Doctorat de l'institut national polytechnique de Grenoble
2. Odgaard PF, Stoustrup J, Kinnaert M (2009) Fault tolerant control of wind turbines—a benchmark model. In: Proceedings of the 7th IFAC Symposium on Fault Detection, Supervision and Safety of Technical Processes, Barcelona, Spain, pp 155–160
3. Odgaard PF, Stoustrup J (2012) Fault tolerant control of wind turbines using unknown input observers. In: Proceedings of the 8th IFAC Symposium on Fault Detection, Supervision and Safety of Technical Processes, August 29–31. National Autonomous University of Mexico, Mexico City, Mexico
4. Odgaard PF, Stoustrup J (2012) Fault tolerant wind speed estimator used in wind turbine controllers. In: The 8th IFAC Symposium on Fault Detection, Supervision and Safety of Technical Processes, August 29–31. National Autonomous University of Mexico, Mexico City, Mexico
5. Odgaard PF, Stoustrup J (2012) Results of a wind turbine FDI competition. In: Proceedings of the 8th IFAC Symposium on Fault Detection, Supervision and Safety of Technical Processes, August 29–31. National Autonomous University of Mexico, Mexico City, Mexico
6. Esbensen T, Sloth C (2009) Fault diagnosis and fault-tolerant control of wind turbines. Master's Thesis, Aalborg University, Denmark
7. Zhang X, Zhang Q, Zhao S, Ferrari R, Polycarpou MM, Parisini T (2011) Fault detection and isolation of the wind turbine benchmark: an estimation-based approach. In: Proceedings of the 18th International Federation of Automatic Control World Congress (IFAC), August 28–September 2, Milan, Italy

Comparing Particle Swarm and Grey Wolf Optimization Algorithms to Increase the Photovoltaic System's Battery Autonomy

Yahya Lobna, Gaied Hajer, Habib Kraiem, and Flah Aymen

Abstract This research focuses on a photovoltaic system that powers an Electric Vehicle when moving in realistic scenarios with partial shading conditions. The main goal is to find an efficient control scheme to allow the solar generator producing the maximum amount of power achievable. The first contribution of this paper is the mathematical modelling of the photovoltaic system, its function and its features, considering the synthesis of the step-up converter and the maximum power point tracking analysis. This research looks at two intelligent control strategies to get the most power out, even with shading areas. Specifically, we show how to apply two evolutionary algorithms for this control. They are the "particle swarm optimization method" and the "grey wolf optimization method". These algorithms were tested and evaluated when a battery storage system in an Electric Vehicle is fed through a photovoltaic system. The Simulink/Matlab tool is used to execute the simulation phases and to quantify the performances of each of these control systems. Based on our simulation tests, the best method is identified.

Keywords Optimization algorithm · Control system · Renewable energy · PSO · GWO · Battery storage energy · Electric vehicle

1 Introduction

Due to its availability and the good conversion factor, solar energy technology has advanced at an exponential rate in the last few years [1]. As a renewable energy source, generation from solar energy eliminates pollution caused by traditional energy industries by lowering air novice emissions [2]. Moreover, electricity generation from this resource is quite viable for a variety of uses. In particular, the rapid development of solar energy instruments gives a complete kit of tools that can be directly applied

Y. Lobna (✉) · G. Hajer · H. Kraiem · F. Aymen
Processes, Energy, Environment and Electrical Systems (Code: LR18ES34), National Engineering School of Gabes, University of Gabes, Gabes, Tunisia
e-mail: lobnayahya2@gmail.com

© The Author(s), under exclusive license to Springer Nature Singapore Pte Ltd. 2024
A. Mellit et al. (eds.), *Proceedings of the 2nd International Conference on Green Energy Conversion System*, Green Energy and Technology,
https://doi.org/10.1007/978-981-97-6148-7_29

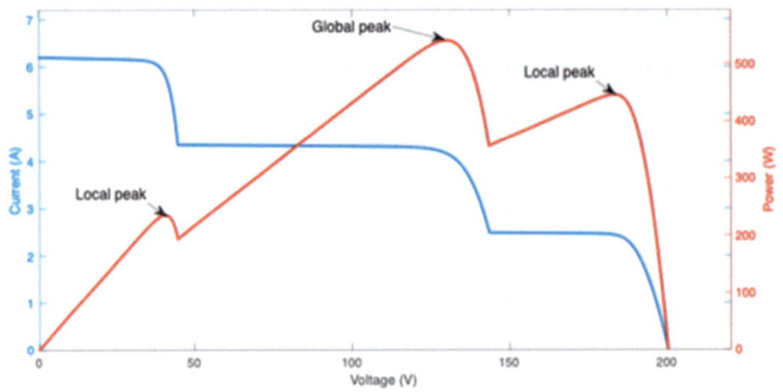

Fig. 1 Typical curve of a shaded PV panel

into the field of Electric Vehicles (EV) [2, 3]. Figure 1 depicts an N-phase interleaved bidirectional DC–DC converter with N converter modules connected in parallel.

This research paper is organized to initially present an introduction section, which describes the objective of the paper and explains the state-of-the-art in this application field. In Sect. 2, the PV system is modeled with the necessary equations that regroup all the parameters and constants that define this physical system. In Sect. 3, the two optimization algorithms are explained. Their flowcharts are exposed and the principle running of each one is described for the MPPT algorithm. In the next section, the simulation conditions and the obtained results are shown for each irradiance case. In the end, a conclusion section is formatted for resuming the paper and giving some perspectives of this work [4].

2 Model of the PV System

A PV module consists of several solar cells connected in series and in parallel to achieve the desired voltage and current levels. A solar panel cell is essentially a p-n semiconductor junction. When exposed to light, a direct current is generated. For simplicity, the single diode model of Fig. 2 is used in this document [5]. This model offers a good compromise between simplicity and precision with the basic structure.

In Fig. 2, we apply Kirchhoff's law. The current will be obtained by the following equation:

$$I = I_{ph} - I_D - I_p \tag{1}$$

where I_{ph} the current generated by light or photocurrent and I_p the current flowing in the parallel resistor, which can be computed as:

$$I_p = \frac{V + R_s I}{R_p} \tag{2}$$

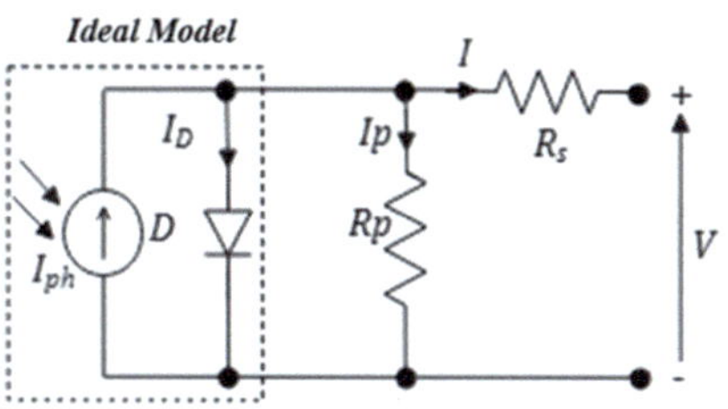

Fig. 2 Simplified equivalent circuit of solar cell

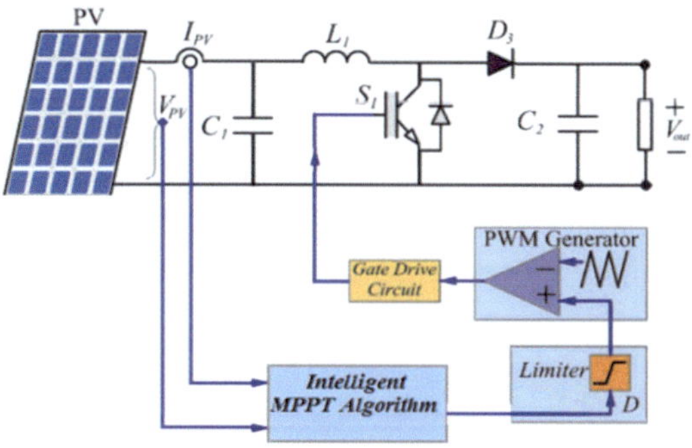

Fig. 3 Block diagram of a PV system using an intelligent MPPT algorithm

I_D is the current of the diode, which is proportional to the saturation current. The following equation expresses the value of this magnitude

$$\mathrm{I_D} = \mathrm{I_{SD}}\left(\mathrm{e}^{\left(\frac{\mathrm{q(V+R_SI)}}{\mathrm{nkT}}\right)} - 1\right) \tag{3}$$

Figure 3 shows the proposed block diagram of the PV system. In this model.

3 Optimization Algorithms

This paper focuses on two swarm-intelligence based algorithms. Specifically, they are the Particle Swarm Optimization and the Grey-Wolf Optimization algorithms. They have been selected for this PV system due to its demonstrated capability to adjust power converter [6, 6].

The flowchart of this algorithm is illustrated in Fig. 4.

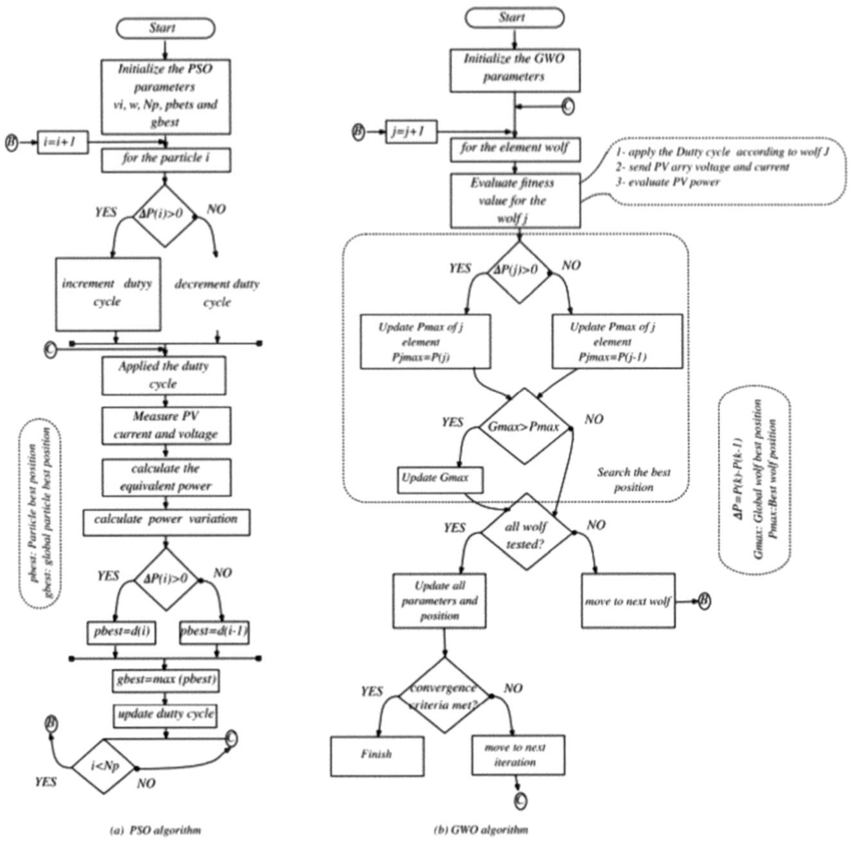

Fig. 4 Algorithm PV system using an intelligent MPPT algorithm

4 Implementation and Simulation Results

Indeed, to compare the efficiency of MPPT PSO and GWO we considered four simulation tests so that a wide variety of conditions are modelled. As can be observed, the simulation results from Figs. 5, 6, and 7 show the evolution of the power, current, voltage and duty cycle of the PV system for the four types of shading distribution we have tested on the panels. From these results, it can be observed that the two solutions ensure a good MPP tracking. The advantage of the PSO MPPT over the GWO MPPT is related to two issues: (i) the amplitude of the oscillations at the transient state and (ii) the accuracy to track the point of maximum power. A high oscillation exists for the case of GWO, which can be one the weaknesses of this algorithm. There is also a small oscillation when executing the PSO at the beginning, when the radiation form changes. However, this will not cause a problem as in the real situation, the modification of the radiation comes very slow. So, we look a batter performance in a real situation.

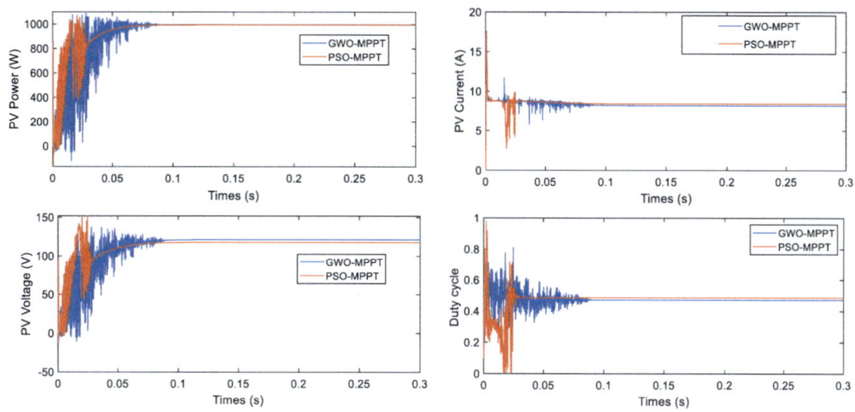

Fig. 5 PV characteristics under uniform irradiation: 0% partial

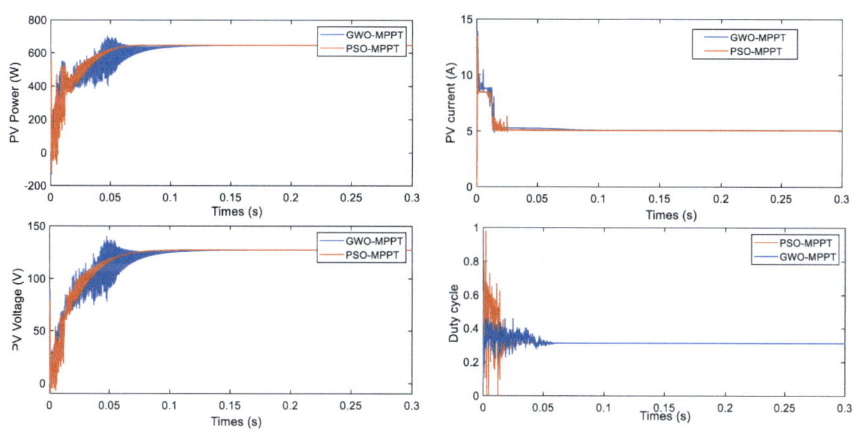

Fig. 6 PV characteristics under partial shading (600, 600, 1000, and 1000 W/m^2)

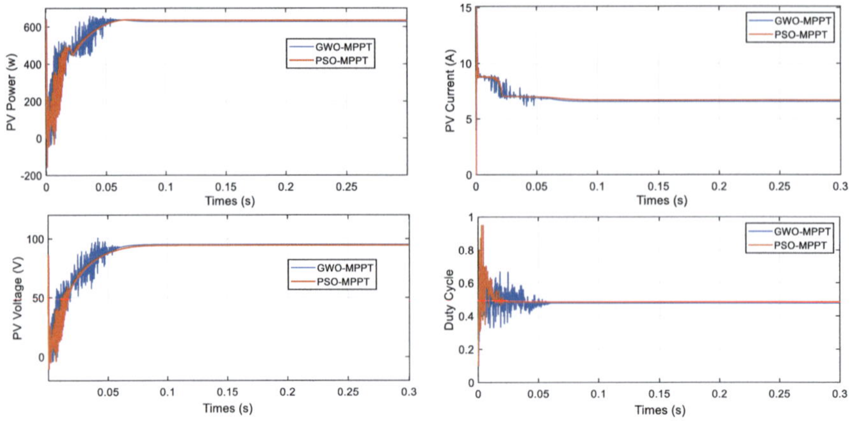

Fig. 7 PV characteristics under partial shading (500, 800, 1000, and 1000 W/m^2)

5 Conclusion

This work attempts to study and discuss two MPPT techniques based on two meta-heuristic optimization algorithms, i.e., PSO and GWO. These new techniques of MPPT overcome the problems of classic MPPT strategies (e.g., perturb and observe and incremental) when tracking the maximum power point, even in the presence of sudden changes of irradiation and shadows on the photovoltaic modules. The MPPT techniques studied show good behavior and better performance. A comparative study of simulation results for a different type of sharing shows PSO-MPPT effectiveness compared to GWOMPPT from the point of view of speed and oscillation during the transient state. In addition, a simulation test shows the efficiency of PSO MPPT versus GWO MPPT in terms of storage charge in the battery under uniform irradiation. As a future work, we would like to analyze how to adapt the algorithm's parameters (which are new constant to the dynamic lighting conditions.

References

1. Mokarram M, Mokarram MJ, Khosravi MR, Saber A, Rahideh A (2020) Determination of the optimal location for constructing solar photovoltaic farms based on multi-criteria decision system and Dempster-Shafer theory. Sci Rep 10(1):8200
2. Lv Z, Qiao L, Cai K, Wang Q (2020) Big data analysis technology for electric vehicle networks in smart cities. IEEE Trans Intell Transp Syst 22(3):1807–1816
3. Mobarak MH, Kleiman RN, Bauman J (2020) Solar-charged electric vehicles: a comprehensive analysis of grid, driver, and environmental benefits. IEEE Trans Transp Electrif 7(2):579–603
4. Mohamed N, Aymen F, Ali ZM, Zobaa AF, Abdel Aleem SHE (2021) Efficient power management strategy of electric vehicles based hybrid renewable energy. Sustainability 2021(13):7351

5. Habib AA, Hasan MK, Mahmud M, Motakabber SMA, Ibrahimya MI, Islam S (2021) A review: energy storage system and balancing circuits for electric vehicle application. IET Power Electron 14(1):1–13
6. Flah A, Majed A, Bajaj M, Naveen KS, Mishra S, Sharma SK (2021) Electric vehicle model based on multiple recharge system and a particular traction motor conception. IEEE Access 9:49308–49324
7. Kariem H, Touti E, Fetouh T (2020) The efficiency of PSO-based MPPT technique of an electric vehicle within the city. Meas Control 53(3–4):461–473

Impact of Pollution on PV Systems and Proposed Solutions

Dhaoui Mehdi, Hatira Bacem, and Miled Faouzi

Abstract The work presented in this paper aims to study the impact of pollution on the production of photovoltaic systems. The work proposes the establishment of a supervision and control system allowing real-time knowledge of the different sizes of the PV system as well as automatic intervention in the event of accumulation of dust or other. The communication with the operator is ensured by the GSM network. The proposed work is presented as a simple and easy solution that can be used to ensure good control of photovoltaic systems in polluted areas.

Keywords Photovoltaic system · Impact of pollution · Remote control · Supervision · Android application

1 Introduction

Progress and developments in the photovoltaic field always come up against a certain number of difficulties, among which the high cost of the kWh produced and the long period of return on investment. In terms of operation, photovoltaic modules need to be well exposed in order to operate under the most direct sunlight possible. Such equipment is therefore subject to an environment causing multiple sources of pollution and fouling. Among the most common are; industrial and automotive pollutants, acid rain, chimneys, pollen, dust, sand, tree leaves, etc. These pollutants directly reduce the efficiency of photovoltaic panels and can also lead to premature wear of the modules by causing intense heating phenomena via the "hot spot" effect. Dust is a term generally applied to solid particles with a diameter of less than 500 μm, which may be chemical, biological, or other in nature. In this context, many factors

D. Mehdi (✉) · H. Bacem
L/R: Processes, Energy, Environment and Electrical Systems (PEESE), Electrical Department of Engineering, National Engineering School, University of Gabes, 6029 Gabes, Tunisia
e-mail: dhaouim@yahoo.fr

M. Faouzi
National Superior Engineering School of Tunis (ENSIT), University of Tunis, Tunis, Tunisia

© The Author(s), under exclusive license to Springer Nature Singapore Pte Ltd. 2024

A. Mellit et al. (eds.), *Proceedings of the 2nd International Conference on Green Energy Conversion System*, Green Energy and Technology,
https://doi.org/10.1007/978-981-97-6148-7_30

influence dust deposition, such as grain size, physical properties of module surfaces, tilt and orientation of the PV module, type of glass, wind speed, and many other factors. Among the most polluted places in Tunisia, we cite the Gafsa basin and the areas adjoining the Gabes chemical group. This pollution affects the state of the solar panels, which affects the production capacity. Phosphate-based aerosols mainly influence this region. This type of aerosol has a major role in air pollution and in the physical processes of the diffusion of solar radiation. Many articles have attempted to study the phenomenon and the laws governing the transport and deposition of dust in the context of photovoltaic installations. In a study of dust accumulation on a solar PV village in Riyadh indicated a 32% reduction in yield after eight months. A reduction in the power produced by the PV panels of 17% due to the accumulation of sand on the panels in Kuwait City after 6 days. The organization Google, famous in information technology, studied the effects of dirt on the panels of a 1.6 MW solar installation in its headquarters. The company did a comparison on two different sets of solar panels in the Google campus, horizontal panels and inclined panels. Theoretically, dirt collects more on horizontal panels, while rain washes most dirt off angled panels and leaves some build-up in the corners. Google teams cleaned them as part of this study, 15 months after the panels were installed. For the flat panels, the cleaning resulted in the doubling of energy production. However, for angled panels, the difference turns out to be relatively small. In the same context, another work in Kuwait, stated that about 2.5 g/m^2/day of dust was collected between April and June. The new investigation into the effect of dust accumulation on tilted glass panels revealed a reduction in transmittance ranging from 64 to 17%, for tilt angles of 0° to 60° respectively after 38 days of exposure. A 30% reduction in the energy produced by a horizontal PV system was observed after three days of dust accumulation. In another study on the influence of physical properties and deposition density on the performance of a solar generator. Artificial dusts which included limestone, cement and carbon particles were used. It was revealed in the study that the deposition of the cement particles (at 73 g/m^2) would lead to the most significant reduction in short-circuit current of around 80%. Interestingly, it was found that the smaller the particle size for a fixed deposition density, the greater would be the reduction in solar intensity received by the solar panels. This was probably due to the greater ability of small particles to minimize inter-particle gaps [1].

2 Experimental Tests

To study the impact of dust on the performance of solar panels; 2 mono-crystalline silicon PV modules were used (Table 1) with a power of 400 Wp with a fixed inclination of 30° and a full South orientation. The characteristics and technical data of the PV panels used are described in Table 1.

The two modules are put in series and the measurements are taken during the day 06/19/2022 for a temperature $T = 53$ °C and an irradiation $I_r = 970$ W/m^2.

Table 1 PV module specifications

MODULE: PEAK SOLARPS72M400 at 1000 W/m^2 and 25 °C	
P_{mpp} (W)	400
V_{mpp} (V)	41.12
I_{mpp} (A)	9.73
V_{oc} (V)	49.17
I_{sc} (A)	10.31

Fig. 1 Photo of cleaned PV modules

2.1 Ist Experiment (Healthy Mode)

In this experiment, the two modules were cleaned and measurements of the quantities (current, voltage) were taken. The photo shows the two panels after cleaning (Fig. 1).

Table 2 shows the measurements taken on the PV generator with cleaned panels for $I_r = 970$ W/m^2 and T = 53 °C.

The experimental characteristics I (V) and P (V) are given by Fig. 2.

2.2 2nd Experiment (Sand Dust Impact)

In the second experiment, we cleaned one module and put sand dust on the second module. The PV generator is tested under the same metrological conditions. We took the measurements of the different sizes. Figure 3 shows the two photovoltaic modules with sand dust deposit.

Table 3 gives the measurements made on the PV generator with sand deposit.

Table 2 Quantities measured during experiment no. 1

V_{pv} (V)	I_{pv} (A)	P_{pv} (W)
0	10	0
10	9.99	99.9
15	9.98	190
25	9.95	265
33	9.92	350
42.8	9.83	430
45	9.7	520
50.5	9.57	580
68	9.35	635.8
72.6	9.1	653.37
75.1	8.7	660.66
76.6	8.25	631.95
78.2	7.7	602.14
78.4	7	550
78.6	6	475.8
78.8	4.5	360
78.9	2.6	200
79	0	0

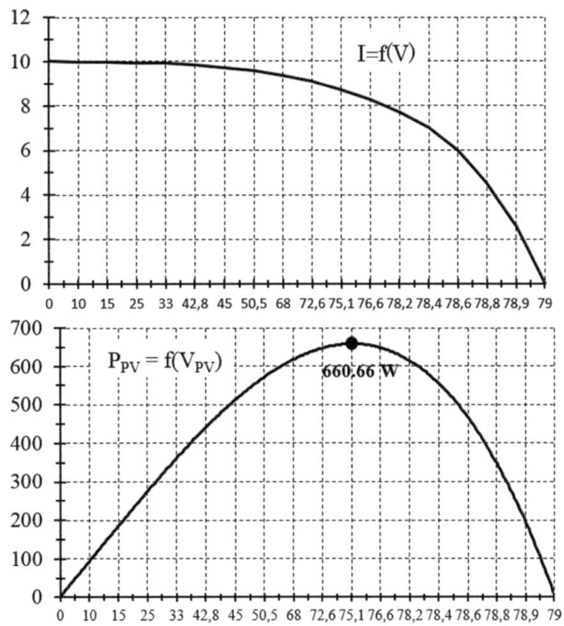

Fig. 2 Experimental I (V) and P (V) characteristics of the cleaned PV system

Fig. 3 2nd experiment with sand dust deposition

Table 3 Electrical quantities measured during the second experiment

V_{pv} (V)	I_{pv} (A)	P_{pv} (W)
0	9.8	0
21	9.7	203.7
31.7	9.5	301.15
33.8	9	304.2
35.2	8.5	299.2
36.4	8	291.2
37.5	7.5	281.25
48	7	336
61.5	6.5	399.75
68.3	6	409.8
71	5.8	411.8
76	0	0

The experimental characteristics I (V) and P (V) recorded during experiment 2 are given by Fig. 4.

According to Fig. 4, we notice a reduction in the voltage and the power supplied by the PV generator. This reduction is about 38% compared to the cleaned panels. This fact is due to the effect of sand dust deposition which is explained by the phenomenon of shading.

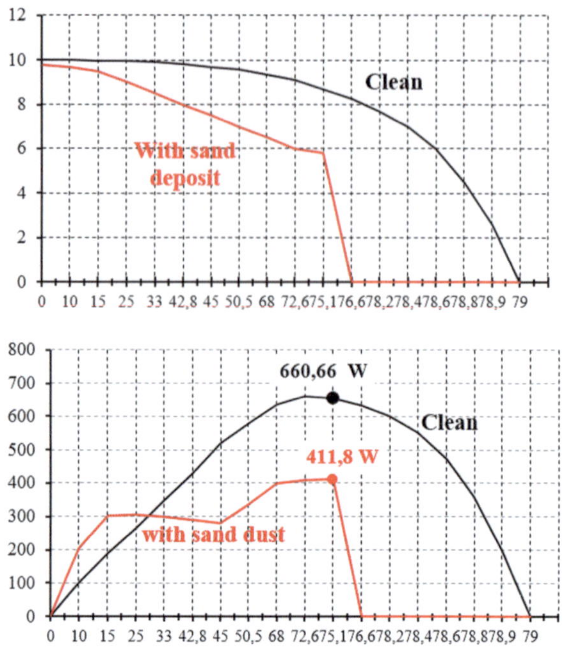

Fig. 4 I (V) and P (V) characteristics in clean mode and with sand dust deposition

2.3 3rd Experiment (Impact of Phosphate Dust)

During experiment 3, the first PV module was cleaned and phosphate dust was put on the second module. Figure 5 presents the PV generator tested in the experiment.

Fig. 5 Experiment no. 3 with phosphate dust deposition

Table 4 Quantities measured during experiment no. 3

V_{pv} (V)	I_{pv} (V)	P_{pv} (V)
0	9.75	0
31.5	9.5	299.25
34	9	306
35.5	8.5	301.75
33.5	8	268
37.5	7.5	281.25
37.7	7	263.9
39	6.3	245.7
47.5	5.8	275.5
66.1	5.5	363.55
66	5.42	357.72
66.2	0	0

Table 4 shows the measurements taken during this experiment.

The experimental characteristics I (V) and P (V) of experiment 3 are shown in Fig. 6.

The characteristics in Fig. 6 show a strong reduction in the voltage and the power supplied by the PV generator. This decrease of around 45% is the cause of the impact of phosphate dust deposition, which is again explained by a fairly strong shading phenomenon.

3 Supervision and Control of the PV System

3.1 Structure of Proposed Control System

The proposed system allows the remote supervision and control of the photovoltaic generator by GSM network. The system is composed of an Arduino card, floors of measurements, a display unit and one module for communication via the GSM network. The structure of the system is shown in Fig. 7 [2, 3].

3.2 Description

To measure the voltage and current from the PV generator, we used sensors, electrically isolated, outputting analog signals corresponding to the inputs of the Arduino. For data acquisition, we used an Arduino card. This device (Fig. 7), built around

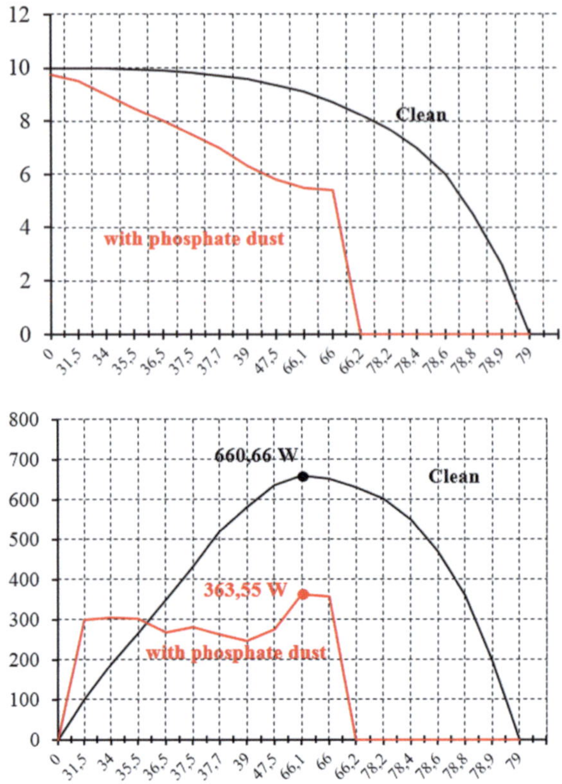

Fig. 6 I (V) and P (V) characteristics in clean mode and with phosphate dust deposition

a microcontroller Atmel ATMEGA 328P and is based on a simple input/output interface.

In this work, communication has established with a GSM module IComsat used for remote control of the station. With this system, the user can easily using his mobile phone, for controlling the system through by sending a single SMS that contains a control command, or the receipt of a report message informing him about the recent state of the monitored system [4–9].

3.3 Android Application

To facilitate the supervision and control of the state of the photovoltaic system, we have created an Android application that allows the user to receive all the measurements as well as the automatic remote control of the PV panel cleaning system by a single SMS. Figure 8 shows three menus of the proposed Android application.

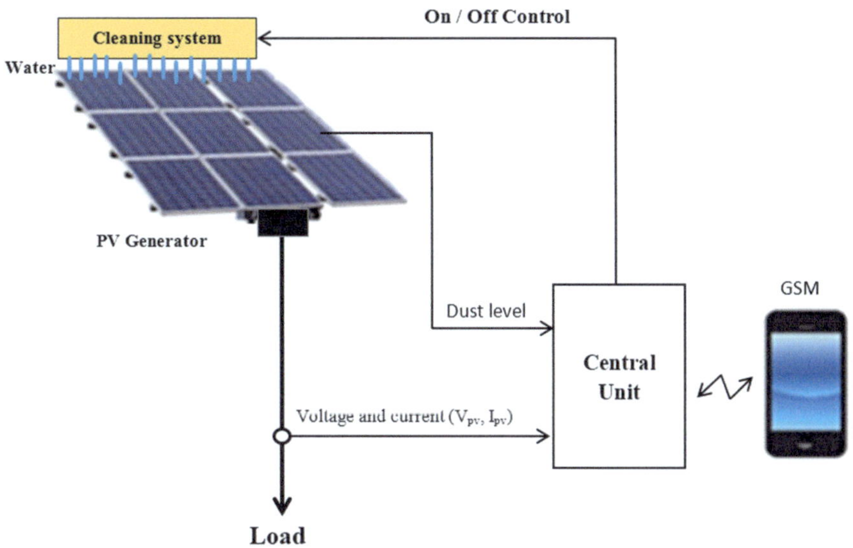

Fig. 7 Structure of the proposed system

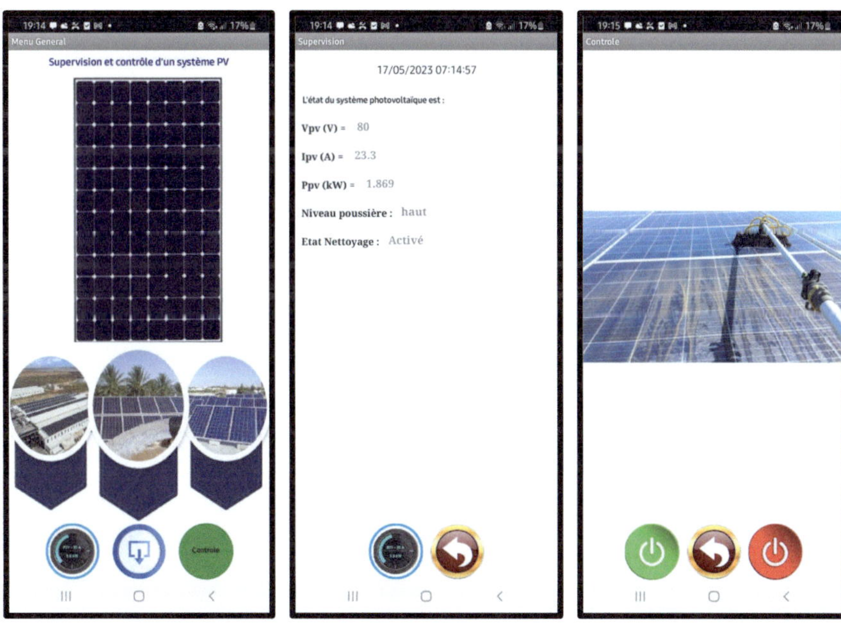

(a) Main menu (b) Supervision menu (c) Control menu

Fig. 8 Android application for acquisition and control

4 Conclusion

In this work, we carried out the modeling of the photovoltaic generator in healthy operation then failing. To represent the behavior of the cell in normal and/or faulty operation, the single diode model is used. Then, we presented the most frequent faults in a photovoltaic installation, such as the shading fault, the sunshine effect, the temperature effect and the dust effect. Finally, we presented the experimental results carried out to study the influence of the deposition of sand dust and that of phosphate on the production of photovoltaic panels. The results obtained clearly show the impact of dust (sand, phosphate, or other) on the production of the photovoltaic installation.

To limit the impact of pollution, a remote control and cleaning system via GSM is presented. The usefulness of this system is that it allows the user to receive all the station data at any time by SMS. He is well informed by SMS in the event of an anomaly or pollution of the PV panels. The designed system also allows remote control of the photovoltaic generator by sending a simple SMS.

References

1. Stéphane P (2009) Nouvelles architectures distribuées de gestion et de conversion de l'énergie pour les applications photovoltaïques. Doctorat de l'université de toulouse, France
2. Najari H, Houita M (2016) Contrôle et Supervision à distance d'une station de pompage Photovoltaïque. Mémoire de PFE, Institut Supérieur des Etudes Technologiques, Médenine Tunisie
3. Ait-Cheikh SM (2007) Etude, Investigation et conception d'algorithmes de commande appliqués aux systèmes photovoltaïques. Thèse de Doctorat d'état, Ecole Nationale Polytechnique, Alger, Algérie
4. Koutroulis E, Kalaitzakis K, Voulgaris NC (2001) Development of a microcontroller-based, photovoltaic maximum power point tracking control system. IEEE Trans Power Electron 16(1):46–54
5. Nesrine Fatima B (2013) Commande d'un Système Photovoltaïque via Satellite. Master automatique en contrôle de processus, Tlemcen, Algérie
6. Safari A, Mekhilef S (2011) Simulation and hardware implementation of incremental conductance MPPT with direct control method using CUK converter. IEEE Trans Indus Electron 58(4):1154–1161
7. Meekhun D (2010) Réalisation d'un système de conversion et de gestion de l'énergie d'un système photovoltaïque pour l'alimentation des réseaux de capteurs sans fil autonomes pour l'application aéronautique. Thèse de doctorat de l'université de Toulouse, France
8. Notice d'utilisation (2012) Système de commande/ de mesure/ d'alarme à distance avec GSM GX 106, version 2012
9. Bacem H, Mehdi D (2021) 'Détection et localisation de défauts pour les systèmes photovoltaïques isolés. Mémoire de master—énergies renouvelables, Ecole Nationale d'ingénieurs de Gabès

Comparison of PI/Fuzzy Control of Pitch Angle for Wind System Using GSAP Generator

Dhaoui Mehdi and Bouaziz Laila

Abstract The fast development of intelligent techniques brings new method for wind turbine control. For adjusting the aerodynamic torque of the wind turbine when winds speed inconsistent, we use the pitch angle control. As conventional pitch control customarily use classic controller, the mathematical model of the system should be well known. In this works, a fuzzy logic control pitch angle controller is developed. The obtained results show that the fuzzy logic controller give better control performances than conventional control strategies.

Keywords PI controller · Fuzzy logic controller · Pitch angle · Variable speed · Wind turbine

1 Introduction

Nowadays, among the all renewable energy sources, wind systems are more economic in compare with other. Variable wind speed systems deliver 20–30% more energy in compare with constant speed systems [1]. Many developed generation systems are used to extract maximum wind energy. Optimum wind energy extraction system is achieved by running wind turbine generator in a variable speed mode because of the higher energy gain and reduced stresses [4].

Pitch-adjusting variable-speed wind turbines have become the dominating type of yearly installed wind turbines in recent years. Pitch angle regulation is required in conditions above the rated wind speed when the rotational speed is kept constant.

D. Mehdi (✉) · B. Laila
L/R: Processes, Energy, Environment and Electrical Systems (PEESE), Electrical Department of Engineering, National Engineering School, University of Gabes, 6029 Gabes, Tunisia
e-mail: dhaouim@yahoo.fr

© The Author(s), under exclusive license to Springer Nature Singapore Pte Ltd. 2024
A. Mellit et al. (eds.), *Proceedings of the 2nd International Conference on Green Energy Conversion System*, Green Energy and Technology,
https://doi.org/10.1007/978-981-97-6148-7_31

2 Wind Turbine Model

Wind turbine converts wind energy into mechanical energy. For a variable-speed, mechanical power available at the output of the wind turbine could be expressed by:

$$P_T = \frac{1}{2} \cdot \rho \cdot \pi \cdot R^2 \cdot C_p(\beta, \lambda) \cdot v^3 \tag{1}$$

where ρ is the air mass density, V is the wind speed, R is the rayon of pale, and C_P is a power coefficient.

The mechanical torque captured by the wind turbine is obtained as:

$$C_T = \frac{1}{2} \cdot \rho \cdot \pi \cdot R^3 \cdot \frac{C_P(\beta, \lambda)}{\lambda} \cdot v^2 \tag{2}$$

The power coefficient C_P (β, λ) (Fig. 1), is a nonlinear function of the tip speed ration λ and the pitch angle β given by:

$$C_p(\beta, \lambda) = 0.5176\left(\frac{116}{\lambda_i} - 0.4 \cdot \beta - 5\right)e^{-\left(\frac{21}{\lambda_i}\right)} + 0.0068 \cdot \lambda \tag{3}$$

where: $\frac{1}{\lambda_i} = \frac{1}{\lambda + 0.008\beta} - \frac{0.035}{\beta^3 + 1}$.

The tip speed ratio (TSR) is defined as:

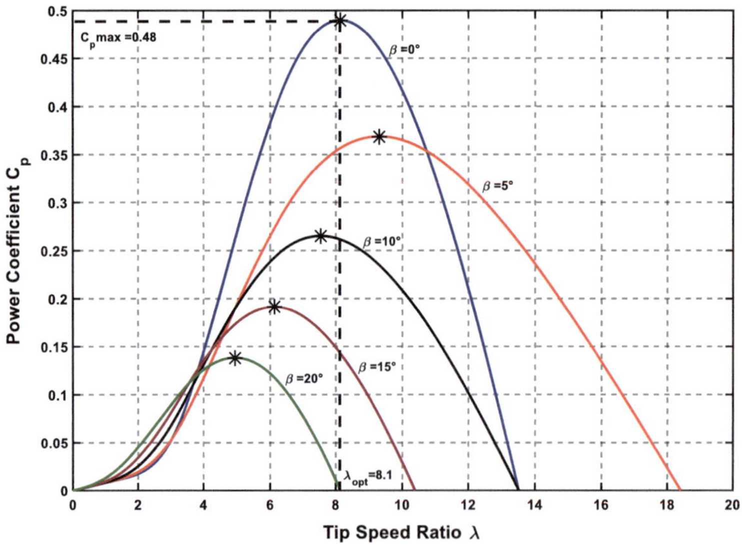

Fig. 1 Typical CP versus (β, λ) curve

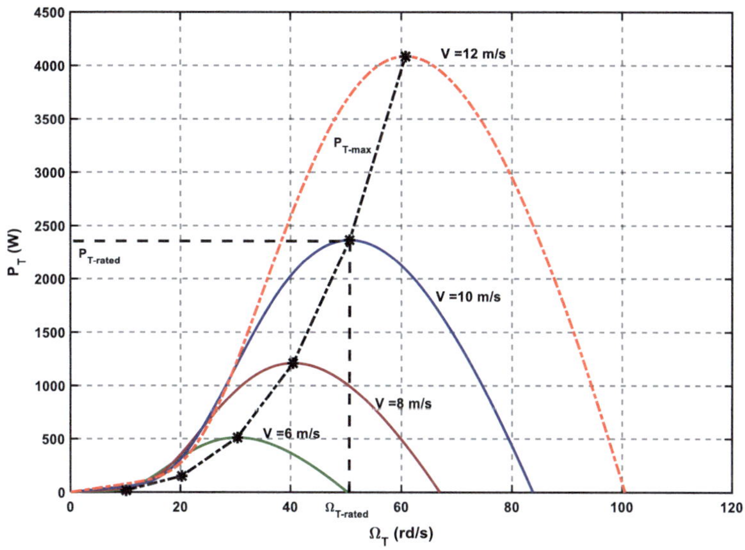

Fig. 2 Turbine power versus mechanical angular velocity characteristic

$$\lambda = \frac{R \cdot \Omega_T}{v} \tag{4}$$

From the relations (1), (2) and (4), the mechanical power of the turbine can be expressed by:

$$P_T = \frac{1}{2 \cdot \lambda^3} \cdot \rho \cdot \pi \cdot C_p \cdot R^5 \cdot \Omega_T^3 \tag{5}$$

The relationship between the turbine power and speed at maximum power output (Fig. 2) are given by:

$$P_{T\max} = \frac{1}{2 \cdot \lambda_{opt}^3} \cdot \rho \cdot \pi \cdot C_{popt} \cdot R^5 \cdot \Omega_T^3 = k_{opt} \cdot \Omega_T^3 \tag{6}$$

3 Pitch Angle Control

Wind turbines are generally designed to deliver maximum output at wind speeds around 15 m/s [2]. According to Eq. (1), for a given wind speed, the only parameter that allows to control extracted power is the C_p parameter. This parameter depends on the type of turbine and the type of design from the manufacturer. It is also a function of the pitch angle control. Pitch angle control is mainly used to limit the mechanical

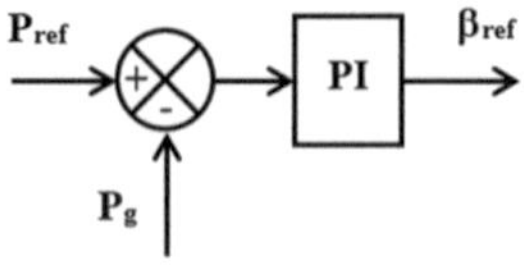

Fig. 3 PI controller of the pitch angle

power during strong wind speed or to adapt power production. The aerodynamic behavior of the wind turbine can be controlled actively by varying the turbine blade angle around their longitudinal axis [3].

In low and medium wind speed, the pitch angle control is rarely actuated and only set to the optimum value. In high wind speeds, the pitch angle control effects to regulate the extracted wind power so that the design limits of the turbine will not be exceeded. Rotor speed, torque or power control can be taken to regulate the pitch angle [5].

In this paper, the power error criteria are used. The error between the reference and the measured power at the output of the PMSG is sent to a PI controller to generate the reference value β_{ref} of the pitch angle [6].

3.1 PI Control

The control system structure used to generate the pitch angle reference is depicted in Fig. 2.

The feed forward term assumes that all the components are ideal and does not account for the losses in the system. The feedback path compensates for the losses by decreasing the pitch angle, if the output power is less than the desired power, to increase the power captured (Fig. 3).

3.2 Fuzzy Logic Control

Fuzzy logic control is one of the most powerful control methods where based on fuzzy set theory and associated techniques. The fuzzy logic algorithm is suitable for wind turbine control with complex nonlinear models and parameters variation. The fuzzy controller uses the perturbation and observation to track the optimal pitch angle without knowledge of wind turbine characteristic. Basic structure of the fuzzy logic controller consists of three important stages: Fuzzification, Decision Making Unit and Defuzzification Unit. The inputs of the Fuzzification are selected as error and rate of change of error and the output is the pitch angle reference.

The two input variables are calculated at every sampling instant say k [6].

$$e(k) = P_{ref}(k) - P_g(k)$$

$$\Delta e = \frac{e(k) - e(k-1)}{T_s} \tag{7}$$

where, $P_{ref}(k)$ and $P_g(k)$ are respectively the reference and generated power at instant n. In Fuzzification stage the crisp variables e(k) and Δe(k) are converted into fuzzy variables which can be identified by membership function (Fig. 4).

The proposed controller uses following linguistic labels: NB (Negative Big), NS (Negative Small), Z (Zero), PM (Positive Small), PB (Positive Big).

Fuzzification table.

The input and output variables map into the following fuzzy set (Table 1; Fig. 5).

The triangular membership functions are used to define both inputs (error, rate of change of error) and output (pitch angle).

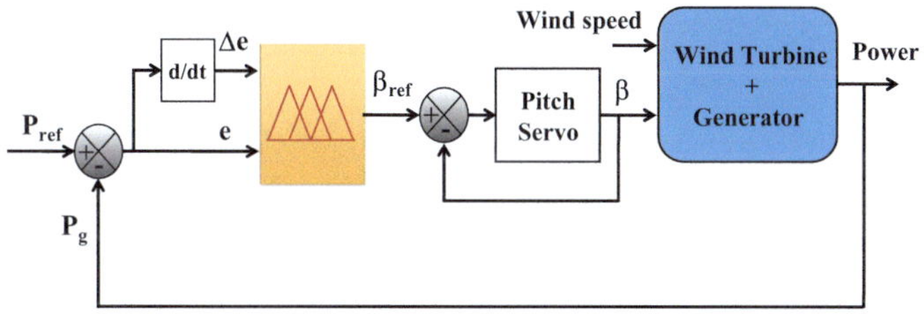

Fig. 4 Fuzzy logic control strategy

Table 1 Rules table

	e				
Δc	NB	NS	Z	PS	PB
NB	NB	NB	NS	NS	Z
NS	NB	NS	NS	Z	PS
Z	NB	NS	Z	PS	PS
PS	NS	Z	PS	PS	PB
PB	Z	PS	PS	PB	PB

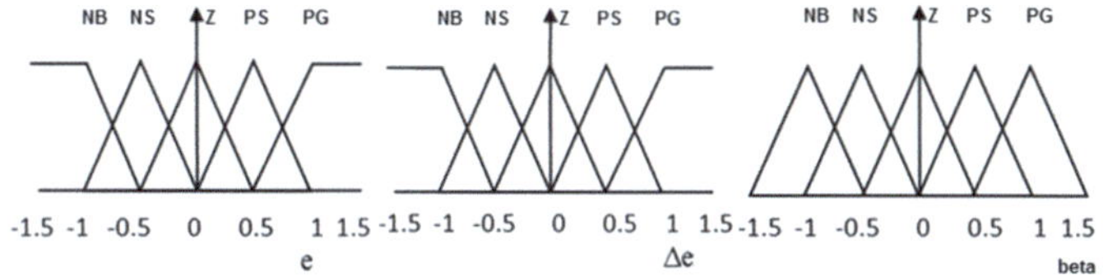

Fig. 5 Fuzzy inference system

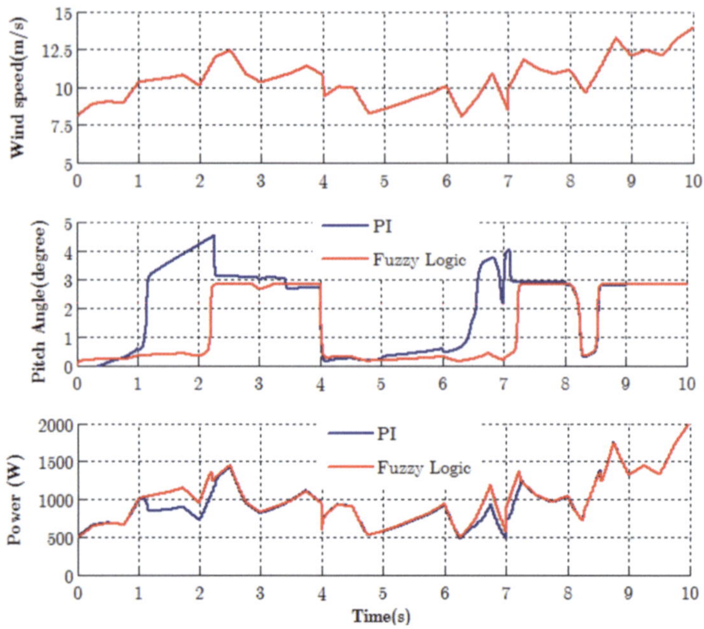

Fig. 6 Comparisons of pitch angle control strategies

4 Simulation Results

For evaluating the robustness of the fuzzy logic controller compared with the conventional PI controller at variable wind speed, the simulations have been carried out using MATLAB/Simulink software on a wind turbine whose characteristics are given in the Appendix (Table 2).

Figure 6 shows the results of the conventional pitch angle control strategy where power is used as the controlling variable. The comparisons of different strategies show that proposed fuzzy logic control strategies is more robust in pitch angle control.

5 Conclusion

The proposed method is based on fuzzy logic and PI controller for tracking the optimal pitch angle of wind turbine with a variable speed. The capability of the fuzzy logic controller is verified and it is found that it has a fast response than the conventional PI controller.

Table 2 Wind turbine technical parameter

Model NE-2000L—Generator type: 3 ~ PMSG			
Rated power	2 kW	Cut wind speed	3 m/s
Maximum power	2.2 kW	Rated wind speed	10 m/s
Rated speed	500 rpm	Max wind speed	45 m/s
Rated voltage (AC)	220 V	Rayon of pale	1.6 m
Pairs of poles	$p = 6$	Nombre of pale	3
Stator resistance	$R_s = 3.8\ \Omega$		
Stator inductance	$L_s = 2.7$ mH		
Permanent magnetic flux	$\Phi_f = 341$ mWb		

Appendix

See Table 2.

References

1. Weisser D, Garcia RS (2005) Instantaneous wind energy penetration in isolated electricity grids: concept and review. Renew Energy 30(8):1299–1308
2. Lupu L, Boukhezzar B, Siguerdidjane H (2006) Pitch and torque control strategy for variable speed wind turbines. In: Energy Conference Proceedings, Athens, Greece
3. Saikuma S, Saravanan S, Sandip RV (2011) Modelling and control of a wind turbine using permanent magnet synchronous generator. IJEST
4. Yin M, Li G, Zhou M, Zhao C (2007) Modeling of the wind turbine with a permanent magnet synchronous generator for integration. In: Power Engineering Society General Meeting, 2007. IEEE
5. Yao X, Liu Y, Xing Z (2008) Active vibration control strategy based on expert PID pitch control of variable speed wind turbine. In: Electrical Machines and Systems International Conference
6. Veeramani C, Mohan G (2013) A fuzzy based pitch angle control for variable speed wind turbines. Int Eng Technol (IJET)

Application of Stochastic Methods to Estimate Fouling Heat Exchanger

Ridha Zitouni, Ons Ghriss, Ali Fguiri, and Mohamed-Razak Jeday

Abstract Among the problems often encountered in the industry is the fouling of heat exchangers which can be defined as the deposition of undesirable materials, such as scale, suspended solids, and even insoluble salts on the internal surfaces of the heat exchanger. The presence of these deposits represents a resistance to fluid flow and heat transfer and increases the pressure loss and reduces the efficiency of the exchanger. Fouling of heat exchangers is unavoidable but can be mitigated by. Effective implementation of appropriate cleaning techniques. Periodic cleaning of heat exchangers is one option among the techniques to minimize losses due to fouling. However, cleaning activities in heat exchangers are costly. Consequently, the use of the reverse method with a fouling model to describe the fouling behavior and the experimental data makes it possible to obtain a better estimate of the value of the thermal resistance of fouling and thus a precise date for the cleaning. Our work consists in estimating the fouling resistance of the exchangers of the phosphoric acid concentration unit of the Tunisian chemical group using stochastic methods such as the genetic algorithm (GA) and the special swarm optimization (PSO). This method demonstrates a convergence towards the global minimum, even in the presence of several local minima, with great precision.

Keywords Heat exchanger · Fouling · Optimization · Inverse problem

1 Introduction

Increasing energy use is a significant challenge and has implications for economic growth, environmental sustainability and energy security. According to the International Energy Agency (IEA), global energy demand is expected to increase by

R. Zitouni (✉) · O. Ghriss · A. Fguiri · M.-R. Jeday
Laboratory of Process Research, Energy, Environment and Electrical Systems (PEESE), National School of Engineering of Gabes, University of Gabes, Street of Omar Ibn El Khattab, 6029 Zrig Gabes, Tunisia
e-mail: ridhazit440@gmail.com

© The Author(s), under exclusive license to Springer Nature Singapore Pte Ltd. 2024

A. Mellit et al. (eds.), *Proceedings of the 2nd International Conference on Green Energy Conversion System*, Green Energy and Technology,
https://doi.org/10.1007/978-981-97-6148-7_32

about 2.8% per year until 2050 [1]. In this context, it is essential to emphasize the importance of efficient energy use to optimize the performance of the devices and minimize energy losses. A major concern is the fouling of heat exchangers, which compromises their efficiency and impairs their performance [2].

The fouling process can be broken down into five basic steps; initiation, transport, attachment and aging [3]. Fouling can take different forms such as particulate fouling, corrosion, scaling, chemical and biofouling [4]. To prevent the formation of fouling deposits in the heat exchanger, it is essential to carry out regular cleaning operations. These operations play a crucial role in increasing the life of the device, minimizing disruptions and avoiding frequent maintenance outages. By eliminating accumulated deposits, cleaning helps maintain the optimum performance of the heat exchanger, ensuring efficient operation and increased productivity [5]. Therefore, cleaning operations are an essential practice to optimize the operation of the exchanger and extend the intervals between maintenance periods. However, the costs associated with cleaning and production shutdowns lead us to recognize that it is critically important to detect fouling in the device. Since fouling on the surface of the heat exchanger is a time-consuming process, it is difficult to observe its formation in real time and propose effective measures to prevent it. Consequently, many researchers [6–8] have turned to the prediction of the resistance to fouling on the surface of the heat exchanger, in order to better understand the formation of fouling. We are interested in the work of Fguiri et al. [8] focuses on the accurate estimation of the resistance to fouling in a heat exchanger used for the treatment of phosphoric acid by steam. The researchers used an inverse method approach to estimate fouling resistances from experimental data. In this study, Fguir et al. used the Levenberg–Marquardt method in the inverse method to estimate model parameters. This approach has produced satisfactory results. However, a major challenge is the need to have an initial estimate very close to the final result. This has led to the exploration of new methods for combining the inverse method with other approaches. So stochastic techniques are distinguished by their ability to seek the global optimum without requiring an initial starting point or knowledge of the gradient of a function [9]. These methods provide an effective exploratory approach to solving optimization problems, randomly exploring the solution space to find the best possible outcomes. Among the different approaches available to solve optimization problems, the GA and the PSO stand out as the most commonly used [10]. Their effectiveness has been demonstrated in finding optimal solutions for a variety of problems, including predicting and studying soiling parameters. Much research has been done on the use of these algorithms to analyze and predict associated parameters [11, 12]. In this study, we should estimate the thermal resistance to fouling for a unit of phosphoric acid concentration installed in the Tunisian chemical group. Using an industrial database and a thermal resistance fouling model, we will try to estimate the fouling of this heat exchanger using optimization methods such as genetic algorithm and optimization by particle swarm.

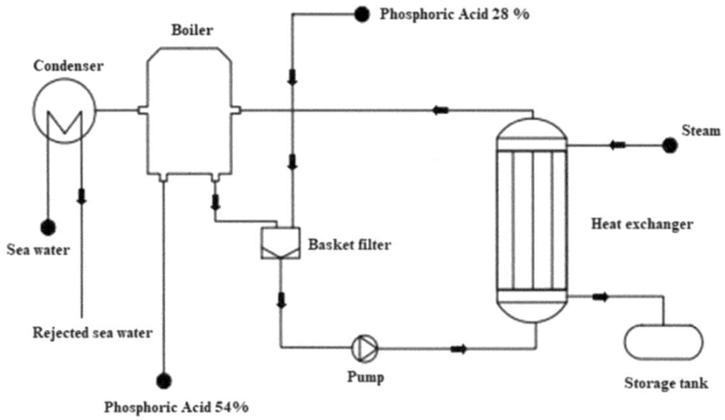

Fig. 1 Schematic of the phosphoric acid concentration unit

2 Materials and Methods

This study is based on a combination of experimental data, a mathematical model of fouling, and optimization methods.

2.1 Experimental Process

The experimental data was done on the level of the heat exchanger of the concentration phosphoric acid process. The phosphoric acid mixture (diluted and recycled) comes out of the basket filter; a centrifugal pump returns it to the heat exchanger. Phosphoric acid passes through a heat exchanger, 70 °C to approximately 80 °C. The mixture of superheated phosphoric acid passes through the boiler where some of the water evaporates and the concentrated acid is produced by overflow into the piping system inside the boiler. The remaining amount of phosphoric acid is recovered. The condenser also ensures the formation of unconquerable gases leaving the boiler by the action of deceiving the water created by the waterfall (Fig. 1) [13].

2.2 Fouling Model

The Kern and Seaton model best describes the progression of thermal resistance to fouling. The resulting model—led by Kern and Seaton—assumes that two processes work in a similar way [14].

$$R_f(t) = R_f^*\left(1 - e^{-\frac{t}{\tau}}\right) \quad (1)$$

The Kern and Seaton model is an expression of two parameters which are the R_f asymptotic fouling resistance and characteristic time τ. The values of these two parameters are important for the industrial to program the unit's shutdown and start the exchanger's cleaning.

2.3 Optimization Methods

Genetic algorithm method (GA). Works mainly on the valorization of each solution integrating exploitation and exploration. The farm is added by crossing two best-adapted parental solutions chosen by the selection approach. Exploration is added by the mutation approach [15]. Finally, the enhanced solution is added to the genetic algorithm population. The GA work strategy is presented in Fig. 2.

Particle swarm optimization method (PSO). Based on the best personal and global perception that has improved its positions by vector updating. The vector reserves its best global and personal positions. The best overall solution indicates the best value of the objective function that is exposed by collecting particles in the swarm. On the other hand, the best personal resolution indicates the best value of the objective function established by an individual particle [16]. The flowchart of the PSO algorithm is presented in Fig. 3.

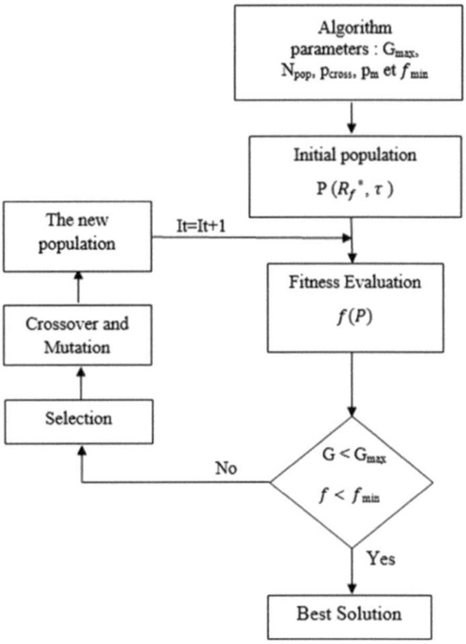

Fig. 2 Genetic algorithm flow chart

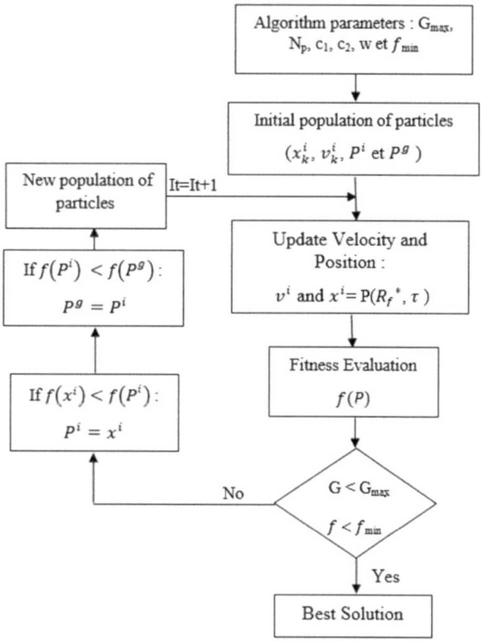

Fig. 3 Particle swarm algorithm flow chart

2.4 *Objectif Function*

Suppose that P = (R_f, τ) the vector of the estimated fouling parameters and f(P) is the sum of the quadratic difference between the experimental measurements and the mathematical modeling of the fouling thermal resistance so it is defined by:

$$f(P) = \sum_{i=1}^{m} \left[Y_i - Rf_i(P) \right]^2 \tag{2}$$

Y and R_f represent the measured and estimated fouling resistance.

We did the precision analysis with the mean square error (RMSE). This error gives a synthetic measure of the global error in a single value, which is written:

$$RMSE = \sqrt{\frac{\sum_{i=1}^{M} \left(Y_i - Rf_i^{method} \right)^2}{M}} \tag{3}$$

With Rf_i^{method}. is the Kern and Seaton model of the resistance to fouling determined by both algorithms.

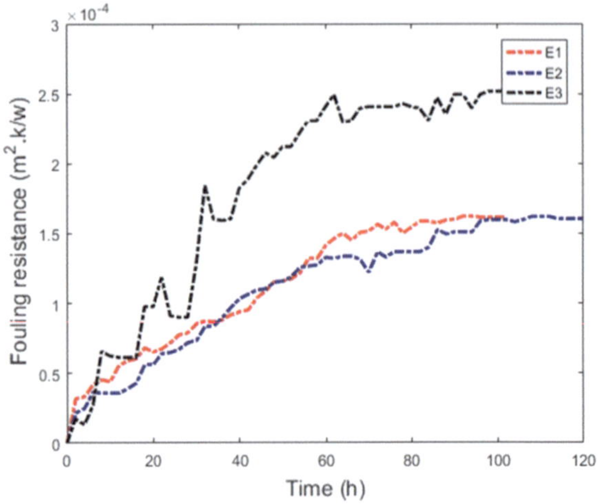

Fig. 4 Fouling resistance for the three heat exchangers E1, E2 and E3

3 Results and Discussion

3.1 Experimental Data

Figure 4 shows asymptotic evolutions of the thermal resistance of fouling, except that in some experimental points there are deviations that are attributed to the precision at the time of recording the operating data. For each heat exchanger, the fouling resistance reaches a constant value called the asymptotic fouling resistance. These asymptotic values vary from one heat exchanger to another, for our heat exchangers tubular (E1), graphite «Supplier A» (E2) and graphite «Supplier B» (E3) is equal to 1.61 10^{-4} m^2 K/W, 1.79 10^{-4} m^2 K/W and 2.71 10^{-4} m^2 K/W respectively for 102, 116, and 120 h of work.

3.2 Validation of the Estimate

The genetic algorithm and particle swarm are two stochastic methods although these types of methods do not give the same result from one run to another, they are very effective in overall optimization. So, we did a statistical analysis of those two methods. This analysis provides that run 20 times the two algorithms with the three heat exchangers to find the best, the means, the bad cases, and also the standard deviations between these executions. This analysis is done with RMSE values.

As shown in Table 1 we found that the standard deviation between the 20 runs for the three exchangers is acceptable, between 1.0618 10^{-6} and 3.4855 10^{-7} m^2 K/W for the genetic algorithm and between 4.4236 10^{-6} and 3.3671 10^{-6} m^2 K/W for the particle swarm algorithm. The two algorithms given near results at best case

Table 1 RMSE results of 20 runs of GA and PSO

Heat exchanger	Method	Best case	Mean case	Worst case	STD
E1	GA	8.60e−06	8.92e−06	9.81e−06	3.48e−07
	PSO	8.58e−06	1.31e−05	2.07e−05	4.28e−06
E2	GA	5.42e−06	6.17e−06	8.26e−06	7.48e−07
	PSO	5.41e−06	8.76e−06	1.59e−05	3.36e−06
E3	GA	1.54e−05	1.65e−05	1.88e−05	1.06e−06
	PSO	1.54e−05	1.99e−05	2.87e−05	4.42e−06

levels for the three exchangers such as 8.6036 10^{-6} and 8.5861 10^{-6} m^2 K/W for stainless steel tubular heat exchanger (E1), 5.4201 10^{-6} and 5.4188 10^{-6} m^2 K/W for the graphite poly block heat exchanger "Supplier A" (E2), 1.5417 10^{-5} and 1.5404 10^{-5} m^2 K/W for the graphite poly block heat exchanger "Supplier B" (E3).

3.3 Experimental Validation

Table 2 presents a comparison between the graphical method and the inverse estimation method based on the gene algorithm technique or partial swarm optimization, which aims to evaluate simultaneously the asymptotic value of the resistance to fouling and the characteristic time. The characteristic time can be understood as the time during which the industrialist decides to stop the process and start cleaning the exchanger. It can be seen that the simultaneous estimation of the two parameters is carried out successfully for the three heat exchangers.

We noted that the mean square error (RMSE) values are relatively low for the methods. In particular, for the graphic method, it varies between 7.63 10^{-6} and 2.06 10^{-5} m^2 K/W, for the GA method and the PSO method, comprised between 5.41 10^{-6} and 1.53 10^{-5} m^2 K/W. We found that the Kern and Seaton model is well-suited to the

Table 2 The precision of the GA and PSO algorithms

		Graphic method	GA method	PSO method
E1	R_f^* (m^2 K/W)	0.0001615	0.000191	0.00019
	τ (h)	44	49.2837	48.8795
	RMSE (m^2 K/W)	1.59e−05	8.58e−06	8.58e−06
E2	R_f^* (m^2 K /W)	0.00016	0.00018	0.000179
	τ (h)	40	50.2268	49.6773
	RMSE (m^2 K/W)	7.63e−06	5.41e−06	5.41e−06
E3	R_f^*(m^2 K/W)	0.000254	0.000279	0.000278
	τ (h)	38	40.5516	40.0898
	RMSE (m^2 K/W)	2.06e−05	1.53e−05	1.53e−05

measurement data of the three exchangers. So, the stochastic methods such as GA and PSO are fairly robust and stable estimation procedures whose main advantage is a good convergence rate and essentially it guarantees convergence towards the optimum minimum.

4 Conclusion

Based on different optimization methods, an asymptotic fouling resistance estimation problem, as well as the time constant τ was solved by relying on an industrial database and choosing the Kern and Seaton model as the mathematical model of fouling. The experimental study of the thermal resistance of fouling showed us the asymptotic evolution of this phenomenon as a function of time which justifies our choice of the mathematical model of Kern and Seaton. The estimation with the genetic algorithm and the particle swarm optimization was successfully carried out which led to the conclusion that stochastic methods provide accurate estimates for asymptotic fouling resistance and characteristic time for the three heat exchangers. These advantages offer the possibility of more accurate cleaning planning and optimization of heat exchanger performance. Thus, these methods prove to be more advantageous than the graphic approach traditionally used by industrialists for the exchangers of the phosphoric acid concentration unit of the Tunisian Chemical Group.

References

1. International Energy Agency. https://www.iea.org/reports/world-energy-outlook-2022/the-global-energy-crisis. Accessed 25 Jun 2023
2. Awais M (2019) Recent advancements in impedance of fouling resistance and particulate depositions in heat exchangers. Int J Heat Mass Transf 141:580–603
3. Jradi R (2019) Tubular heat exchanger fouling in phosphoric acid concentration process. In: Inverse Heat Conduction and Heat Exchangers. IntechOpen, London
4. Bott TR (1995) Fouling of heat exchangers. Theodore Reginald. Elsevier
5. Patil P (2022) Monitoring fouling in heat exchangers under temperature control based on excess thermal and hydraulic loads. Chem Eng Res Des 181:41–54
6. Wang J (2023) Fouling prediction of heat exchanger surface under alternating magnetic field based on IGWO–SVR. Int J Therm Sci 184:108018
7. Schlüter F (2022) Application of experimental data to model local fouling resistances. Heat Mass Transf 58(1):29–40
8. Fguiri A (2021) Estimation of fouling resistance in a phosphoric acid/steam heat exchanger using inverse method. Appl Therm Eng 192:116935
9. Arora JS (1995) Global optimization methods for engineering applications: a review. Struct Optim 9:137–159
10. Dokeroglu T (2019) A survey on new generation metaheuristic algorithms. Comput Ind Eng 137:106040
11. Costa ALH (2013) Parameter estimation of fouling models in crude preheat trains. Heat Transfer Eng 34(8–9):683–691

12. Sun L (2009) Research on the fouling prediction of heat exchanger based on Support Vector Machine optimized by particle swarm optimization algorithm. In: International Conference on Mechatronics and Automation, pp 2002–2007. IEEE (2009)
13. Fguiri A (2020) Heat exchangers fouling in phosphoric acid concentration. Heat Mass Transf 56:2313–2324
14. Müller-Steinhagen H (2011) Heat transfer fouling: 50 years after the Kern and Seaton model. Heat Transf Eng 32(1):1–13
15. Das P (2022) Evaluating the prospect of utilizing excess energy and creating employments from a hybrid energy system meeting electricity and freshwater demands using multi-objective evolutionary algorithms. Energy 238:121860
16. Biyanto TR (2015) Cleaning schedule optimization of heat exchanger networks using particle swarm optimization. arXiv preprint arXiv:1512.00883

Design of Efficient Off-Grid Solar Photovoltaic Water Pumping System

Lityem Wejden, Sbita Lassaad, and Mnif Farid

Abstract Today, most countries rely on renewable energy sources (RES), such as solar, wind, hydroelectric, biomass, geothermal, etc.... energies. Solar and wind energies in particular are widely used because they are naturally abundant and environmentally friendly. Electricity generation based on renewable energy is growing. Alternative energy sources and system development systems tend to develop traditional energy sources dependent on the greenhouse effect. Photovoltaic systems are used mainly because of their economic efficiency. The main application of off-grid solar photovoltaic (SPV) systems is water harvesting in rural areas that cannot be connected to the grid. In this application, photovoltaic (PV) and pump system control are key to improving overall efficiency. At this against this background, this work proposes a simple and efficient off-grid solar photovoltaic water pumping system (SPVWPS). The designed system is based on a DC-DC boost converter, a three-phase DC-AC inverter, and a three-phase asynchronous motor coupled to a centrifugal pump. P&O Method for Maximum Power Point Tracking (MPPT). Finally, the effectiveness of the proposed off-grid SPVWPS is verified using a 4-KVA prototype.

Keywords Maximum Power Point Tracking (MPPT) · Perturb and Observe (P&O) · Two level inverter · Centrifugal pump · Solar photovoltaic Water Pumping System

---

L. Wejden (✉) · S. Lassaad
Processes, Energies, Environment and Electrical Systems (ENIG, LR18ES34), National Engineering School of Gabes, University of Gabes, Gabes, Tunisia
e-mail: lityemwejden@gmail.com

M. Farid
Institute of Technological Studies of Gabes, University of Gabes, Gabes, Tunisia

© The Author(s), under exclusive license to Springer Nature Singapore Pte Ltd. 2024
A. Mellit et al. (eds.), *Proceedings of the 2nd International Conference on Green Energy Conversion System*, Green Energy and Technology,
https://doi.org/10.1007/978-981-97-6148-7_33

1 Introduction

The use of coal and oil as a source of energy harms the climate by increasing carbon emissions [1]. Due to the fluctuation in oil prices, developing nations' ability to install solar photovoltaic (PV) panels has grown dramatically during the past ten years. However, due to their numerous advantages, solar photovoltaic (SPV) generators can be used to produce energy in a variety of ways. For a simple, secure, and affordable decentralized energy supply option, consider SPV-isolated systems. They enable the installation of safe and decentralized electrical sources far from power plants. In telecommunications, agriculture, street lighting, signage, control, and rural development, isolated PV systems are used. The pumping of water systems is one of the most vital uses of PV systems in agriculture. Irradiation, load type, and atmospheric temperature are the three key variables that affect a PV system. Therefore, as meteorological conditions change, maximum power point tracking techniques are employed to derive the most power from PV.

2 System Design

In this fragment, the design of the PV pumping system consists of PV modules. It was built based on fourteen panels connected in series, and their main specifications are given in Table 1. A boost DC/DC converter, MPPT algorithms to control PV modules, a two-level inverter, and a centrifugal pump driven by ASM, as shown in Fig. 1. It also demonstrated the mathematical models of all functional blocks related to this SVPWPS that can be realized for experimental validation.

2.1 *PV Source Model and Simulation*

The PV cell's electrical equivalent circuit consists of a current source connected to a diode, a shunt resistor, and a string of resistance elements, as presented in Fig. 2 [2–5].

Table 1 PV module specifications

Characteristics	Value
Maximum power (P_{mpp})	260 W
Optimal voltage (V_{mpp})	31.7 V
Optimal current (I_{mpp})	8.20 A
Open circuit voltage (V_{oc})	37.8 V
Photocurrent (I_{sc})	9.07 A

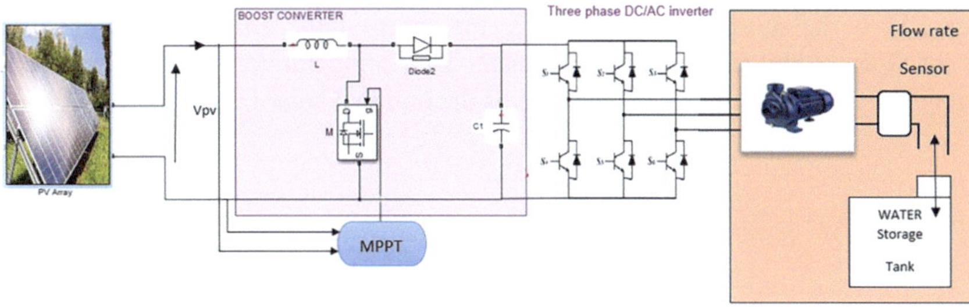

Fig. 1 The system architecture of the proposed off-grid water pumping system

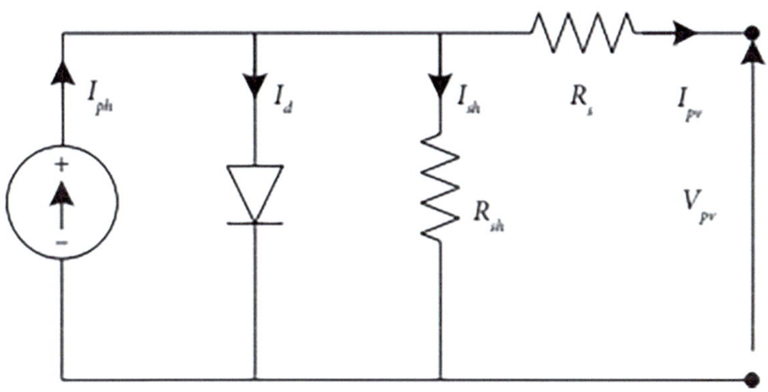

Fig. 2 Solar cell equivalent circuit

The equation of the current tension characteristic of the solar cell is given by the Eq. (1):

$$I_{pv} = I_{ph} - I_0\left(\exp\frac{q(V + R_s \cdot I)}{akTN_s} - 1\right) - \frac{(V + R_s \cdot I)}{R_{sh}} \tag{1}$$

where I_{ph} is the photo-current, I_0 the saturation current, R_s the series resistance of the cell, R_{sh} the parallel resistance of the cell and a is the diode ideality factor.

The expressions of photo current, and saturation current are given by the Eqs. (2) and (3).

$$I_{ph} = (I_{sc} + k_i(T - 298.15))\frac{G}{G_n} \tag{2}$$

$$I_0 = \frac{I_{sc} + k_i(T - 298.15)}{\exp\left(\frac{q(V_{oc} + K_V(T - 298.15))}{akTN_s}\right) - 1} \tag{3}$$

where I_{sc} is the short circuit current, K_i the short-circuit temperature coefficient, K_v the Open circuit temperature coefficient, T the temperature, V_{oc} the open circuit voltage, G is the solar irradiance and G_n is the nominal solar irradiance.

2.2 DC/DC Boost Chopper Model

A DC–DC boost converter is employed to maintain a consistent load voltage. Because the voltage produced from the PV system is insufficient to meet the load requirement, it is essential in PV applications. Figure 3 illustrates a DC/DC boost converter [6–8].

The capacitor is modeled by its capacitance C and series resistance R_C while the inductor is modeled by its inductance L and series resistance R_L, R_C and R_L are assumed to be insignificant in this design. Also, we should note that there are two types of waveforms in the inductor current. These modes are called discontinuous conduction mode (DCM) or continuous conduction mode (CCM). In the case of DCM The current iL(t) will be canceled prior to the cancellation commutation period. In contrast, in the case of CCM, the minimum value of iL(t) is more significant, or at the limit, it is equal to zero. In this work, we assume that our boost converter works in CCM mode. The output power VO of the converter is regulated as follows:

$$V_0 = \frac{1}{1-D} V_{in} \quad (4)$$

The value of the inductor L_{min} is chosen based on the estimated inductor ripple current ΔI_{I_L} at the maximum input voltage V_{in}, as shown in Eq. (5), where the quantity f_s represents the switching frequency.

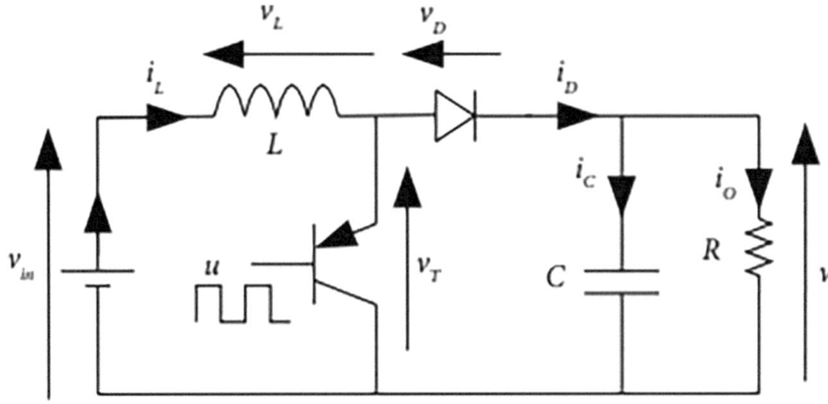

Fig. 3 Schematic diagram of DC/DC Boost converter

$$l_{min} = \frac{D}{f_s \cdot \Delta I_{I_L}} V_{in} \tag{5}$$

The resulting minimum value of capacitance C_{min}. The output voltage ripple ΔVO is given by

$$C_{min} = \frac{D}{f_s \cdot R \cdot V_0} V_0 \tag{6}$$

2.3 The Algorithm P&O-Perturb and Observe

The P&O approach implies periodic perturbations (increase or decrease in current or voltage produced by a PV source), with the evaluation of the power transfer of the current cycle to the power output of the previous disturbance cycle. If the working voltage of the module changes as the voltage rises power, the controller changes the operating point. Otherwise, it moves in the opposite direction. Flow chart of the P&O MPPT technique is shown (Fig. 4) [9–11].

2.4 Modeling of Centrifugal Pumping

The delivery capacity of the centrifugal pump is represented by the H-Q curve: flow Q (e.g., m^3/h) and head H (e.g., m). Varying the pump speed gives us a number of properties (H Q) as described below:

$$H = B_1 \cdot w_r^2 - B_2 \cdot w_r \cdot Q - B_3 \cdot Q^2 \tag{7}$$

Using the law of similarity, the characteristics (P, H, and Q) of a centrifugal pump set to speed N can be obtained as the new characteristics (P′, H′, and Q′) for any speed N′ according to the equation [12, 13].

$$Q' = Q \cdot \left(\frac{N'}{N}\right),\ H' = H \cdot \left(\frac{N'}{N}\right)^2,\ P' = P \cdot \left(\frac{N'}{N}\right)^3 \tag{8}$$

2.5 Modeling of Two-Level Inverter

The two-level voltage inverter is represented in the Fig. 5. It consists of three arms. Commute to transistors. Each arm consists of two cells, each containing a diode and

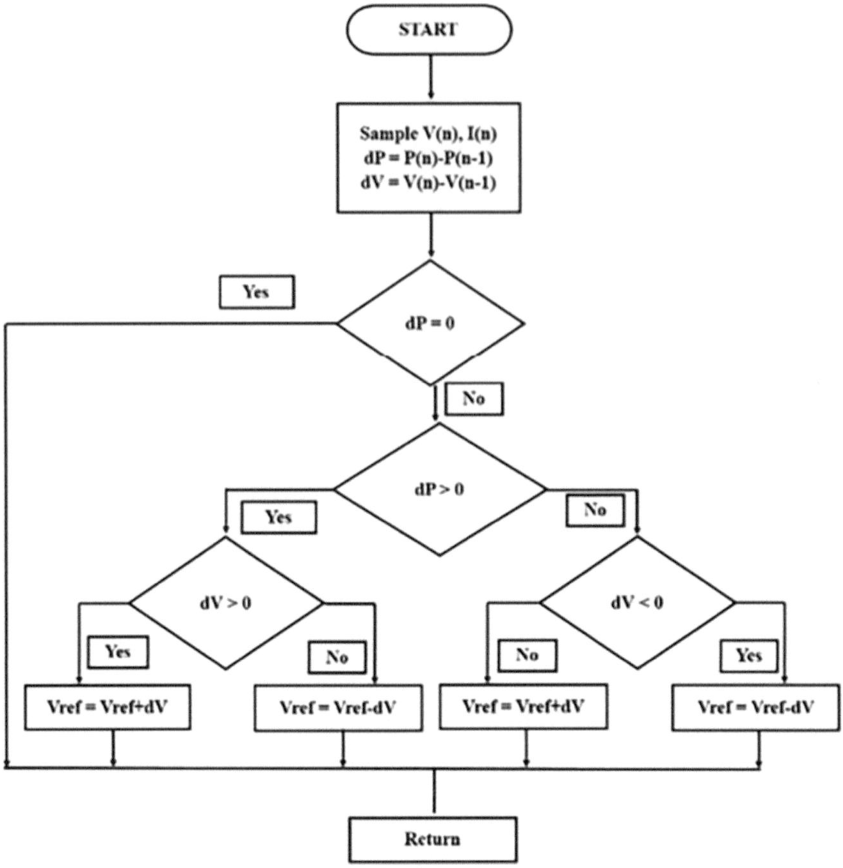

Fig. 4 Flowchart of P&O MPPT technique

a transistor. Work in forced commutation. All these elements are considered ideal switches. In fashion Commandable, the inverter arm is a two-position switch that allows you to obtain two levels of power at the output Tensions. The matrix equation is as follows [14]:

$$\begin{bmatrix} V_A \\ V_B \\ V_C \end{bmatrix} = E \begin{bmatrix} 2 & -1 & -1 \\ -1 & 2 & -1 \\ 1 & -1 & 2 \end{bmatrix} \begin{bmatrix} S_a \\ S_b \\ S_c \end{bmatrix} \tag{9}$$

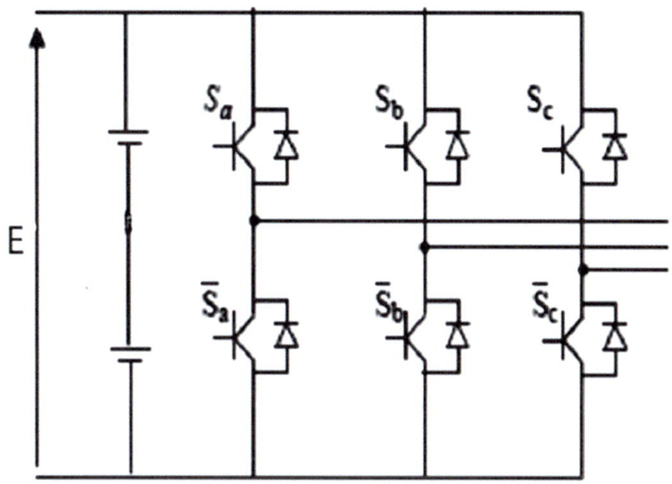

Fig. 5 Three-phase two-level voltage source inverter

3 Result

The following figures show the experimental and simulated results obtained for a temperature equal to 22 °C and an insolation variation equal to 955 W/m^2.

Figure 6 represents the comparison between the experimental and simulation measurements of the voltage produced by the PV array. The SPV offers 457 V and 3640 W.

Figure 7 shows the use of the MPPT and boost algorithm power is tracked and oscillations of solar energy are also reduced to the efficiency of tension of the

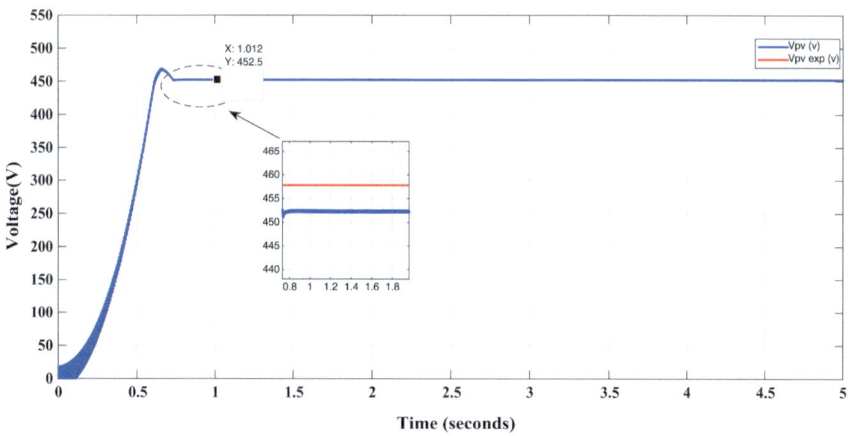

Fig. 6 Result of experimental and simulation of voltage PV

maximum power point. The P&O algorithm gives an excellent response in case of constant irradiation and constant temperature.

In Figs. 8 and 9, the graphs resulting from the output voltage and current for the inverter are presented based on simulated and experimental results.

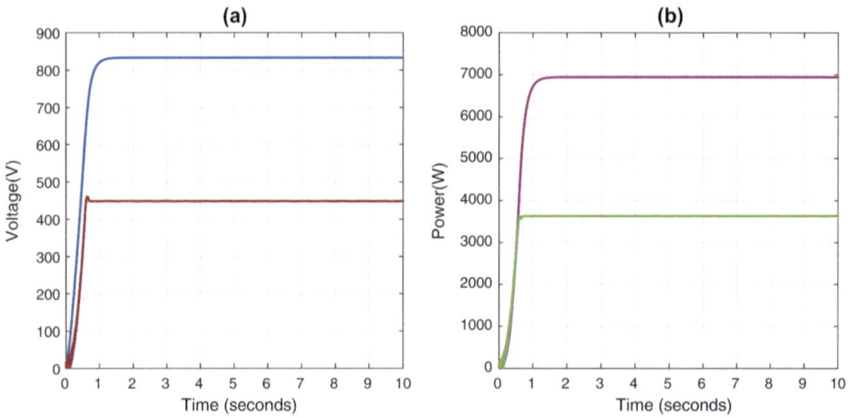

Fig. 7 The input and output of boost converter: **a** voltage, **b** power

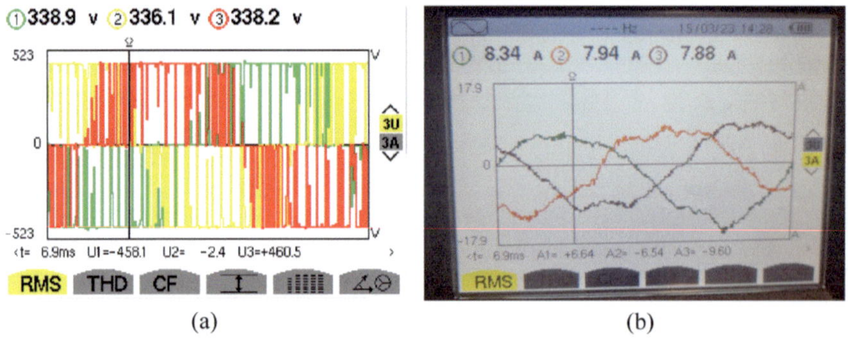

Fig. 8 The experimental output result of a 3-phase inverter: **a** voltage, **b** current

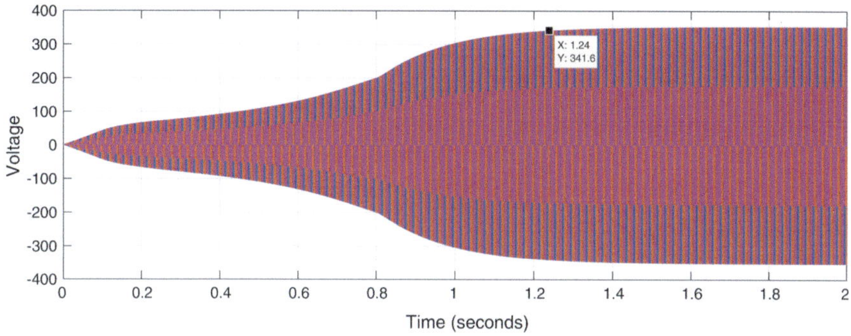

Fig. 9 3-Phase inverter output voltage waveforms for PV

4 Conclusion

The Matlab/Simulink model of a solar PVWPS has been implemented with an ASY motor operating the centrifugal pump. The model implements the P&O MPPT technique for regulating the DC-DC boost converter duty cycle. The Simulink model is tested under varying solar radiation and partial shading conditions. Under STC (1000 W/m^2 and 25 °C), the PV array produces 3640 W that generates a motor speed of 2800 rpm, delivering a water discharge of 4 m^3/h.

References

1. Hmidet A, Subramaniam U, Elavarasan RM, Raju K, Diaz M, Das N et al (2021) Design of efficient off-grid solar photovoltaic water pumping system based on improved fractional open circuit voltage MPPT technique. Int J Photoenergy 2021:1–18
2. Das A, Peu S (2022) Modeling of PV cell, PV module, PV array and PV IV characteristics analysis using MATLAB/SIMULINK
3. Pop T, Buzduga C, Pentiuc RD, Ifrim VC, Ungureanu C (2022) Mathematical modeling in MATLAB of a photovoltaic panel. IOP Conf Ser Mater Sci Eng 1256(1):012037
4. Selmy MS, Alsherif MZ, Noah MSE, Abdelqawee IM (2023) Standalone PV-based water pumping system optimization using jellyfish technique. Eng Res J Faculty Eng (Shoubra) 52(1):94–100
5. Nasirudeen SA, Haruna MS, Aminu MA, Osanaiye O (2021) Mathematical modelling, simulation and analysis of solar PV module in Simulink. In: 2021 1st International Conference on Multidisciplinary Engineering and Applied Science (ICMEAS), pp 1–5. IEEE
6. Gopal Y, Birla D, Lalwani M (2020) Reduced switches multilevel inverter integration with boost converters in photovoltaic system. SN Appl Sci 2:1–15
7. Prabaharan N, Palanisamy K (2016) Analysis and integration of multilevel inverter configuration with boost converters in a photovoltaic system. Energy Convers Manage 128:327–342
8. Necaibia S, Kelaiaia MS, Labar H, Necaibia A (2017) Implementation of an improved incremental conductance MPPT control based boost converter in photovoltaic applications. Int J Emerg Electr Power Syst 18(4)
9. Kumar Bindal R (2022) A comparative analysis of effective MPPT technology for photovoltaic system with boost converter. Mater Today Proc 69:A1–A5

10. Saady I, Karim M, Bossoufi B, El Ouanjli N, Motahhir S, Majout B (2023) Optimization and control of photovoltaic water pumping system using Kalman filter based MPPT and multilevel inverter fed DTC-IM. Results Eng 17:100829
11. Patra B, Nema P, Khan MZ, Khan O (2023) Optimization of solar energy using MPPT techniques and industry 4.0 modelling. Sustain Oper Comput 4:22
12. Saoudi A, Krim S, Mimouni MF (2021) Enhanced intelligent closed loop direct torque and flux control of induction motor for standalone photovoltaic water pumping system. Energies 14(24):8245
13. Shankar VKA, Umashankar S, Paramasivam S, Hanigovszki N (2016) A comprehensive review on energy efficiency enhancement initiatives in centrifugal pumping system. Appl Energy 181:495–513
14. Fernandez S (2013) Comparaison des performances électriques d'un onduleur triphasé deux niveaux à commandes directes et indirectes connecté au réseau. Doctoral dissertation, Lille 1

Enhanced Parameter Extraction of Photovoltaic Models Through the Hybridization of Ant Lion Optimizer and Cuckoo Search (ALOCS)

Rehouma Youssef, Naoui Mohamed, Degla Aicha, Danoune Mohammed Bilal, Hamida Mohamed Assad, Sbita Lassaad, and Gougui Abdelmoumen

Abstract The identification of photovoltaic (PV) model unknown parameters using measured characteristic curves is of great importance for designers involved in modelling, simulation, and control of PV systems. This paper introduces an enhanced heuristic algorithm, namely ALO-CS (Ant Lion Optimizer-Cuckoo Search), which combines the strengths of both ALO and CS optimization algorithms. The primary objective is to accurately extract the parameters of various commercial PV modules. To the best of our knowledge, this novel approach has not been previously explored in the literature, making it a unique contribution. In order to evaluate its effectiveness, a comparative analysis is conducted against several state-of-the-art methods. The experimental and analytical results demonstrate that the proposed Hybrid ALO-CS algorithm outperforms improved JAYA and other methods in terms of accuracy and quality, establishing its competitive advantage.

R. Youssef (✉) · D. M. Bilal · H. M. Assad · G. Abdelmoumen
LAGE Laboratory, University of Ouargla, 30000 Ouargla, Algeria
e-mail: rehouma.youssef@univ-ouargla.dz

R. Youssef
LEVRES Laboratory, University of El Oued, 39000 EL Oued, Algeria

N. Mohamed · S. Lassaad
Research Unit of Energy Processes Environment and Electrical Systems, ENIG, University of Gabés, 6029 Gabés, Tunisia
e-mail: mednaouiing@yahoo.com

D. Aicha
Centre de Dévelopment des Energies Renouvelables—CDER, 62 Route de l'observatoire, Bouzareah, Algeria

D. M. Bilal
Sustainable Development and Computer Science Laboratory SDCS-L, Ahmed Draia University, Adrar, Algeria

© The Author(s), under exclusive license to Springer Nature Singapore Pte Ltd. 2024
A. Mellit et al. (eds.), *Proceedings of the 2nd International Conference on Green Energy Conversion System*, Green Energy and Technology,
https://doi.org/10.1007/978-981-97-6148-7_34

Keywords Extraction · Cuckoo search optimization · Ant lion optimizer · Photovoltaic cell · PV panel

1 Introduction

For the past ten years, considerable attention has been directed to photovoltaic (PV) systems, which directly produces electricity from solar insolation. Moreover, it is universally abundant, inexhaustible, affordable, and environmentally profitable.

Unfortunately, PV arrays generally operate under harsh outdoor conditions, since they are exposed to the environment. This may cause serious degradation of the harvested power from the system or even a probable system failure due to unfixed and low power values. Thus, researchers should assess the performance of their PV systems under these critical conditions using accurate modeling and simulation before any real implementation in order to guarantee optimal design.

The performance and behavior of nonlinear PV systems are described by a number of mathematical models that have been published in scientific literature. Both the available single-diode model (SDM) and double-diode model (DDM) are frequently used [1]. The model parameters effect on the PV model's accuracy. However, these parameters are usually not always accessible as explained in [2]. Consequently, the modeling, assessment, and control of PV systems require the accurate extraction of parameters. To this matter, various methods of parameter identification have been developed recently. There have been a number of efforts devoted to the use of deterministic algorithms for parameter optimization, which rely on the minimizing of a suitable function [3–5]. The parameters of the PV models have been widely identified using meta-heuristic methods inspired by diverse natural phenomena. They impose no limitations on the issue characteristic; accordingly, their implemented is easy for many real-world problems.

In this paper, an improved hybrid heuristic algorithm based on hybrid Ant Lion optimizer and Cuckoo search optimization (ALOCS) algorithm is discussed and used for the parameter extraction of the different mathematical formulas of PV models additionally to a comparative results study performed using real operating measurement of current and voltage. The outline of this paper is as follows: Sect. 2 describes the adapted method and the optimization technique, followed by Sect. 3, which discusses the problem connected to PV models. Then, Sect. 4 highlights the experimental findings of the various models, and finally a conclusion.

2 Meta-Heuristic Methods Based PV Parameters Estimation Models

A brief description of the applied meta-heuristic in this section (ALO, CS and Hybrid ALO-CS) is provided below.

2.1 *Ant Lion Optimization (ALO)*

ALO is a recent stochastic search algorithm that translates how ants search in nature. Ants and lions are suggested as search agents in this novel way to obtain solutions. In actuality, they are in charge of the prey hunting, which entails the ants' haphazard foraging, the building of traps, the ants' entrapment in the traps, the capture of prey, and the rebuilding of the traps [6]. The mathematical model of ALO is given as follows: In nature, ants walk in a random way in search of food. The following definitions apply to each step in the optimization process:

$$X_i = \left[0; r(1); r(1) + r(2); t - 1 \sum j = 1\ r(j); T \sum j = 1\ r(j)\right] \quad (1)$$

where = 1, …, D. The ant or Ant lion dimension is D, and the maximum number of iterations is t.

$X = [X_1; \ldots; X_D]$, X_i is a $(t + 1) \times 1$ matrix, and r(j) presents a stochastic function which can be done as:

$$r - (1\ i\ f\ rand > 0.5\ and\ r - -1\ i\ f\ rand \leq 0.5) \quad (2)$$

2.2 *Cuckoo Search Optimization (CS)*

The optimization technique known as "Cuckoo search" was created in 2009 by Yang and Deb [7, 8] and was motivated by the required parasitism of some cuckoo species that lay their eggs in the host birds' nests of different species. Invading Cuckoos may lead to conflict with some host birds [9].

As an illustration, if a host bird learns that the eggs are not its own, it may either discard the foreign eggs or simply depart its nest and start a new one elsewhere. Some cuckoo species, such as the New World brood parasite Tapera, have progressed to the point where female parasitic cuckoos are tremendously specialized in duplicating the hues and patterns of the eggs of a few particular host species.

Using the population of eggs as a search engine is a CS approach. A new solution X_i^{t+1} (the cuckoo's egg) is computed using the Lévy flight operator during the search process as follows [10]:

$$X_i^{t+1} = \alpha \otimes \text{levy}(s, \lambda) + X_i^t \quad (3)$$

The symbol $\otimes$ is the input-wise product, and Γ presents the gamma function in these equations. $\alpha > 0$ is the scaling factor of the step size s.

2.3 Hybrid ALO-CS

The detail the implantation of the proposed ALO-CS algorithm employing hybrid cuckoo search and Ant lion optimization [10]. The flowchart in Fig. 1 demonstrates the process of estimating solar PV parameters by means of the proposed ALO-CS algorithm. The cascading structure is used by the ALO-CS algorithm to carry out its two search stages. Additionally, a productive balance between exploration and exploitation can be established by working together with the cuckoo search approach and Ant lion.

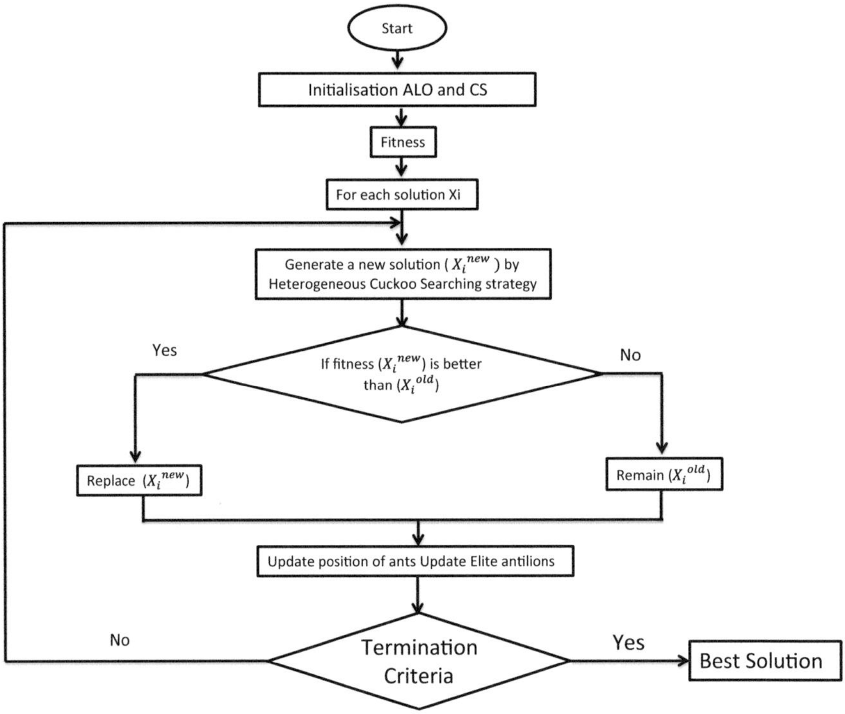

Fig. 1 Flowchart algorithm of hybrid Ant lion optimization based cuckoo search (ALO CS)

3 Photovoltaic Models and Problem Statement

The various cell/module PV model mathematical formulas are first presented in this section. The objective function for extracting SDM, DDM and PV model parameters is then detailed.

3.1 Solar Cell Model

Circuit models are typically used to identify the properties of the I–V curve. The simplest one is a parallel diode current source with resistance series (see Fig. 2) [11, 12]. The four-parameter model is easy to understand but less to use This model includes five parameters, which are the photo-generated current Iph, I_{s1} is the reverse saturation current of the diode, the ideality factor n1, the series resistance Rs, the parallel resistance Rsh.

There are five parameters in this model, including the photo-generated current Iph, a diode's reverse saturation current I_{s1}, n1 the ideality factor, R_S and R_p are the resistance series, parallel respectively.

The photovoltaic cell of the SDM's current-voltage relationship is provided by:

$$I = Iph - I_{d1} - I_p = Iph - I_{d1}\left(e^{\left(\frac{q(V+Rs.I)}{n1.K.T}\right)} - 1\right) - \frac{V + Rs * I}{Rp} \quad (4)$$

The double diode model (DDM), which is a more accurate, can be derived when considering this loss since the SDM excludes the influence of the depletion area recombination current loss. Figure 3 depicts the identical circuit of the DDM. The cell output current in this model can be done as follows [13, 14].

$$I = Iph - I_{d1}\left(e^{\left(\frac{V+Rs*I}{n1.Vth}\right)} - 1\right) - I_{d2}\left(e^{\left(\frac{V+Rs*I}{n2.Vth}\right)} - 1\right) - \frac{V + Rs * I}{Rp} \quad (5)$$

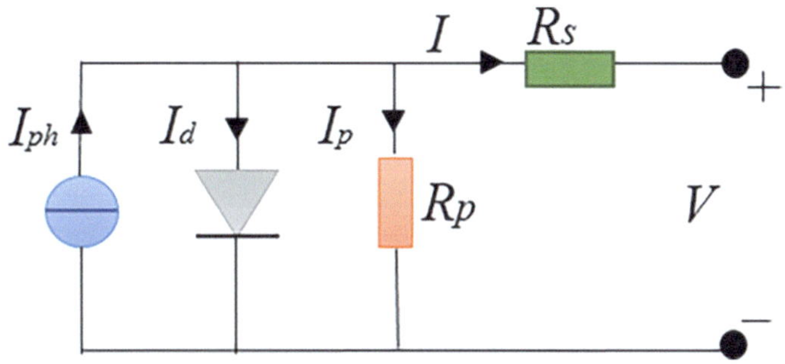

Fig. 2 Equivalent circuit of one diode model cell solar

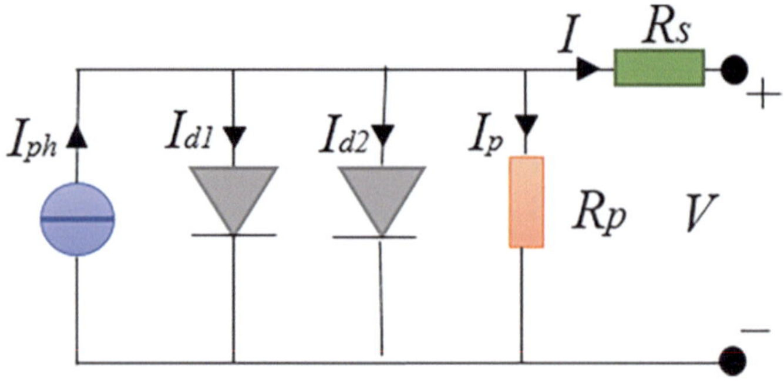

Fig. 3 Equivalent circuit of double diode model

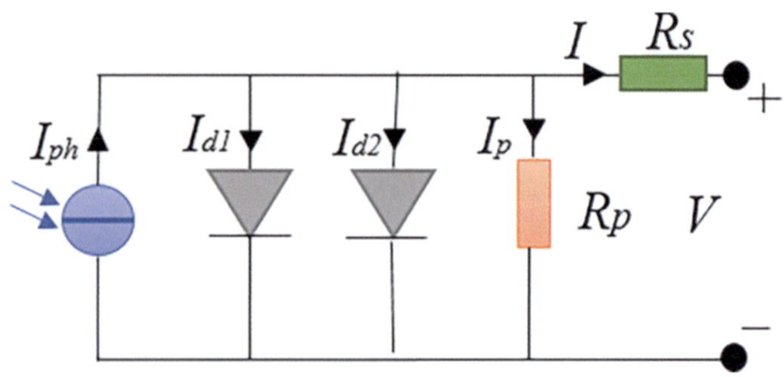

Fig. 4 Equivalent circuit of double diode module model

The first and second diode currents are I_{d1} and I_{d2} represent the first diffusion and second saturation currents diode current, the ideality factors for the diffusion and recombination diodes are n1 and n2, respectively. There are seven variables that must be extracted into the DDM from the I − V actual of the solar cell: I_{ph}, I_{d1}, I_{d2}, R_S, R_p, n_1 and n_2.

3.2 PV Module

M cells are series-connected to form a PV panel. The DDM is commonly used for all of element in the PV panel, and the PV panel I-V relationship can be written as:

$$I = Iph - Isd1\left(e^{\left(\frac{V+Rs.I.M}{n1.Vth}\right)} - 1\right) - Isd2\left(e^{\left(\frac{V+Rs.I.M}{n2.Vth}\right)} - 1\right) - \frac{V + Rs.I.M}{Rp} \quad (6)$$

where $\left[\text{Iph , Isd1, Isd2, n1, n2, Rs , Rp }\right]$ are the parameter module PV, M is 34 cell (Fig. 4).

3.3 Problem Formulation

We select the main objective of the modification of solar cells by decreasing the difference between the experimental data and those identified in environmental conditions by extracting the values of the parameters unknown from the models. So that optimum values can be obtained from certain unknown model parameters. The main target function, frequently used in this context, is the Root-Mean-Square-Error (RMSE). At each point of the I-V curve (1) or (2) was measured using the meta-heuristic method to reach I calculated at any voltage for a solar cell.

For Single-diode model

$$\text{Fiteness(5 paramter)} = \text{Iph} - \text{Isd}\left(e^{\left(\frac{\text{Vm}+\text{Rs}*\text{Im}}{\text{Vth}}\right)} - 1\right) - \frac{\text{Vm} + \text{Rs} * \text{Im}}{\text{Rsh}} - \text{Im} \tag{7}$$

For double diode model

$$\text{Fiteness(7 parameter)} = \text{Iph} - \text{Isd1}\left(e^{\left(\frac{\text{Vm}+\text{Rs}*\text{Im}}{\text{Vth}}\right)} - 1\right) - \text{Isd2}\left(e^{\left(\frac{\text{Vm}+\text{Rs}*\text{Im}}{\text{Vth}}\right)} - 1\right) - \frac{\text{Vm} + \text{Rs} * \text{Im}}{\text{Rsh}} - \text{Im} \tag{8}$$

For PV double diode model Module

$$\text{Fiteness(7 parameter)} = \text{Iph} - \text{Isd1}\left(e^{\left(\frac{\text{Vm}+\text{Rs}*\text{Im.M}}{\text{Vth}}\right)} - 1\right) - \text{Isd2}\left(e^{\left(\frac{\text{Vm}+\text{Rs}*\text{Im.M}}{\text{Vth}}\right)} - 1\right) - \frac{\text{Vm} + \text{Rs} * \text{Im}}{\text{Rsh}} - \text{Im} \tag{9}$$

where Im, Vm: represent measured current, voltage respectively.

The definition of an objective function is essential for estimating the parameters of PV cells. Therefore, the goal in this situation is to reduce the difference in inaccuracy between the simulated and actual currents. In order to accomplish this, the following function is used.

$$\text{RMSE} = \sum_{K=1}^{N} \frac{(\text{Fiteness})^2}{N} \tag{10}$$

where, N: is the actual data size.

4 Results and Analysis of the Experiments

In this section, in order to verify a solar cell in better different models by the ALO-CS algorithm, the models are applied to solving problems in estimating solar PV parameters. The I-V experimental data is obtained from a paper [15, 16]. The remaining second problem is there circumstances on solar PV panel models taken from [16, 17]. A PV module named XD12-12P-Poly-Crystalline silicon (under 431 W/m^2 at 24 °C).It is composed of 34 polycrystalline silicon cells that are linked in series to produce 63 pairs of current and voltage values [18]. Table 1 displays the lower and upper bounds for every parameter, which match those applied in other studies.

Table 2 reports the parameters for the designed algorithms in accordance with advice from relevant literature.

4.1 Results of Single Diode Model (SDM)

This subsection evaluates the ALO-CS algorithm's accuracy, dependability, and convergence performance using statistical data and convergence graphs. Table 3 submits the statistical findings for each model cell over 20 separate runs. The mean RMSE numbers show the average accuracy, whereas the worst RMSE values show the accuracy. The experimental findings for the SDM from ALO-CS and the comparative different algorithms are shown in Table 3.Take note that 20,000 iterations of function evaluations were used by IJAYA, MLBSA, BHCS, and ALO-CS. Clearly, Hybrid ALO-CS has better computational efficiency than IYAJA, MLBSA and BHCS.

Table 1 Parameter upper and lower constraints of solar cells

Parameter	Cell (SDM/DDM)		Double diode module PV	
	Lower	Upper	Lower	Upper
Iph (A)	0	1	0	1.5
Id1, Id2 (uA)	0	1	0	50
Rs (Ω)	0	0.5	0	2.5
Rp (Ω)	0	100	0	1500
n1, n2	1	2	1	2.5

Table 2 Setting parameters for the algorithms being compared

Algorithms	Parameter
ALO CS	NP = 20, pa = 0.25, $\alpha = 0.01$, $\beta = 1.5$
IJAYA	NP = 20
MLBSA	NP = 30
BHCS	NP = 20, pa = 0.3, $\alpha = 1.1$, $\beta = 1.7$, $\delta = 1.6$, E = 1

Table 3 Model parameters estimated by different algorithms for SDM

Algorithms	Iph (A)	Id (μA)	Rs (Ω)	Rsh (Ω)	a	RMS
IJAYA	0.75970043193	0.380932844	0.0356333061011	64.385538986706	1.49800462857331	1.186639e−03
MLBSA	0.76058342926	0.340050452	0.0362500040803	57.853457288735	1.48632376302086	1.001473e−03
BHCS	0.76077553173	0.323020756	0.0363770093316	53.718514958312	1.48118357295284	9.860219e−04
ALOCS	0.76077550908	0.323021058	0.0363770090039	53.718697435393	1.48118366851573	9.860219e−04

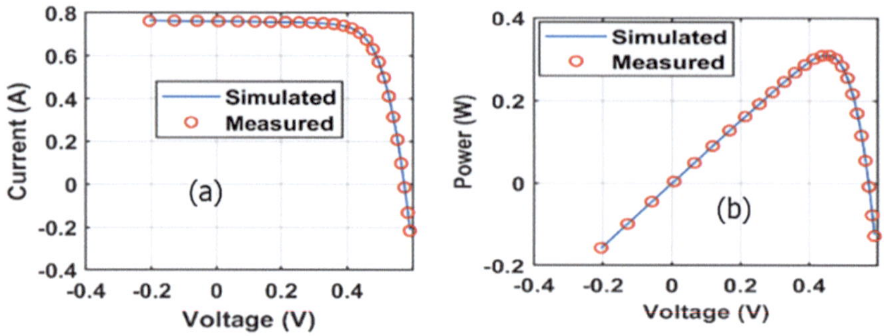

Fig. 5 Results of the ALOCS algorithm for extraction measured and estimated values are compared using characteristic **a** I-V curve, **b** P-V curve

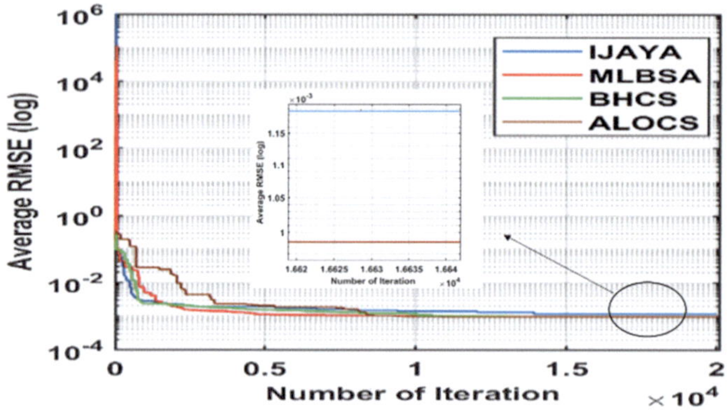

Fig. 6 Convergence of different algorithms for SDM

Resulting that the differential vectors in ALO-CS is very efficient for the extraction of parameters from the Single Diode Model (SDM) (Figs. 5 and 6).

This figure demonstrates that the ALOCS, BHCS have better convergence than the ALOCS algorithm and existing optimization algorithms listed in the literature (Fig. 7).

This figure displays how the different parameters of the ALOCS hybrid algorithm parameters (Rs and Rsh) and (Is and Iph) change with the number of iterations, as different kinds of objective functions were analysed.

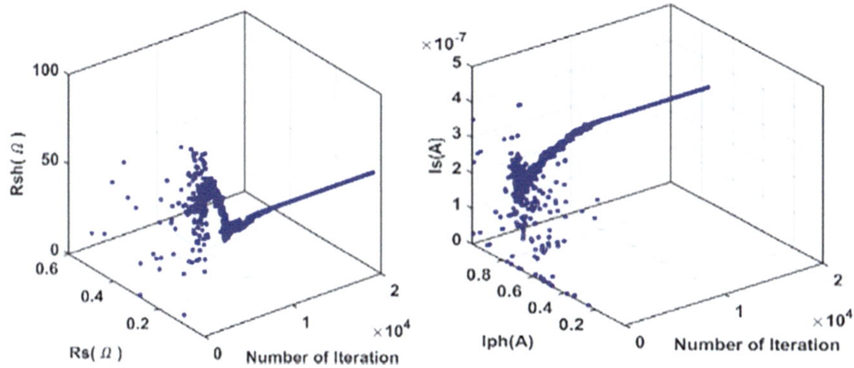

Fig. 7 Convergence graph of parameter identifier (Is-Iph) and (Rs-Rsh) for the single diode model

4.2 *Resultat of Double Diode Model*

The statistical results obtained by the proposed hybrid ALO-CS and the other algorithms are presented in Table 4 for the double diode model (Figs. 8, 9, and 10).

Table 4 Model parameters estimated using different algorithms for double diode cell

Algorithms	Iph (A)	I_{01} (μA)	I_{02} (μA)	n1	n2	Rs (Ω)	Rsh (Ω)	RMSE
IJAYA	0.7607	0.2101	0.2032	1.450	1.6769	0.0366	52.8834	9.953997e−04
MLBSA	0.7606	0.0753	0.2786	1.6160	1.4734	0.0363	55.0455	9.877638e−04
BHCS	0.7607	0.3231	0.002	1.4811	1.0870	0.03637	53.7224	9.860221e−04
ALOCS	0.7607	0.2394	0.6396	1.4558	1.9999	0.0366	55.2273	9.825830e−04

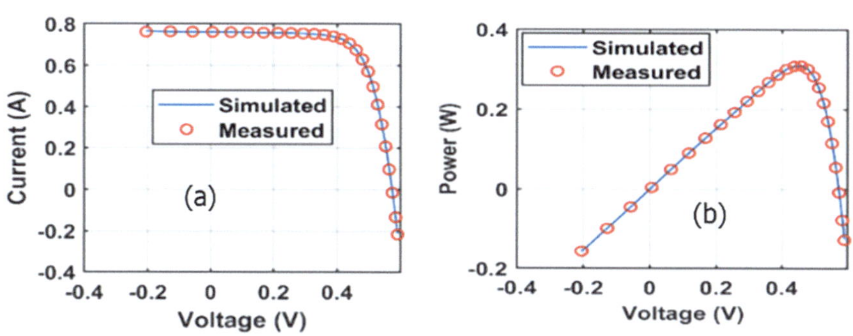

Fig. 8 Results of the ALOCS Algorithm for extraction measured and estimated values are compared using **a** the I V curve and **b** the P V curve

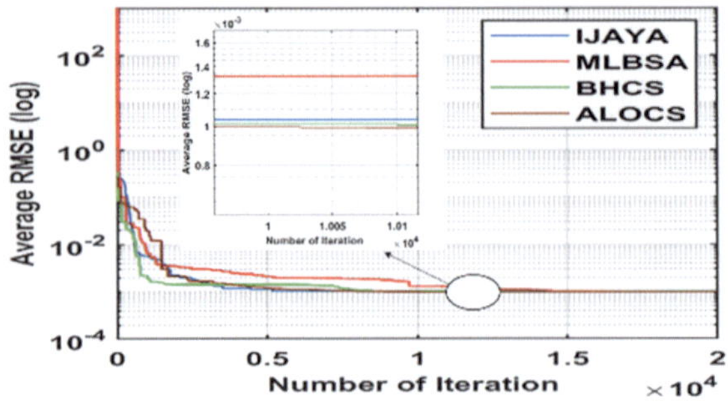

Fig. 9 Different algorithms convergence results for double diode model (DDM) cell

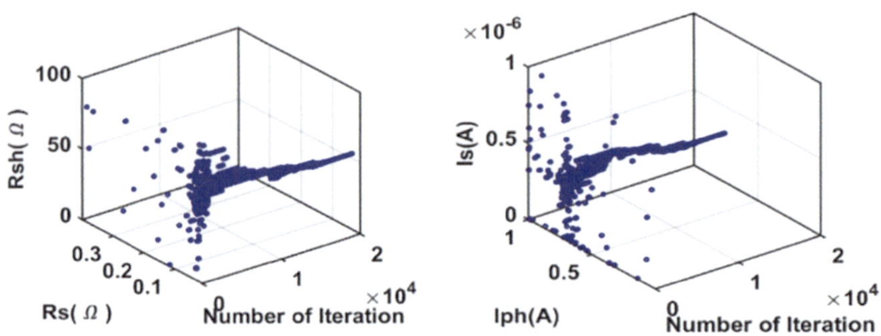

Fig. 10 Convergence graph of parameter identifier (Is-Iph) and (Rs-Rsh) for the double diode model

4.3 Resultats of PV Double Diode Model

The results of Table 5 clearly show that, when solving objective functions, the suggested algorithm came in first place on average (6.420972e−03) (Figs. 11, 12, and 13).

The hybrid ALOCS is the most effective in terms of solution convergence and quality, according to the results. while delivering quality to fix the parameter setting concerns for multiple PV models, at various Models SDM, DDM Cell, and PV of the study.

Table 5 The estimated model parameters by different algorithms for double diode model module: experimental resultats

Algorithms	Iph (A)	I01 (μA)	I02 (μA)	n1	n2	Rs (Ω)	Rsh (Ω)	RMSE
IJAYA	0.66192	5.90394	50 0	1.97574	2.49151	0.03816	65.00747	6.60e−03
MLBSA	0.66250	2.80897	40 10849	2.49998	2.24378	0.03410	55.39070	6.56e−03
BHCS	0.66298	1.804759	40 28614	2.26487	2.24878	0.03461	52.08694	6.55e−03
ALOCS	0.666916	12.03348	0 7.0145	2.071925	2.101320	0.042137	22.645927	6.420972e−03

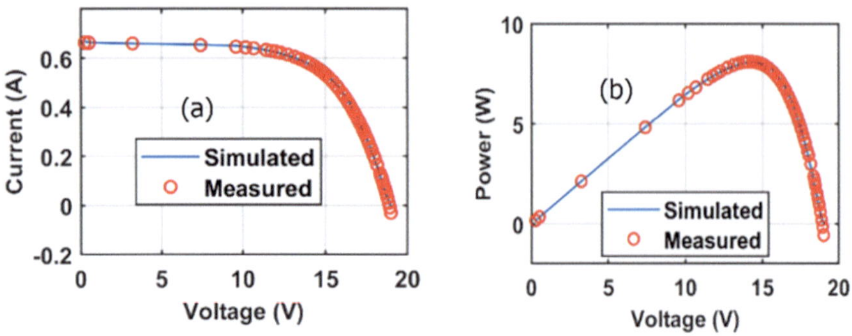

Fig. 11 Results of the ALOCS algorithm for extraction measured and estimated values are compared for module PV using **a** I-V curve and **b** P-V curve

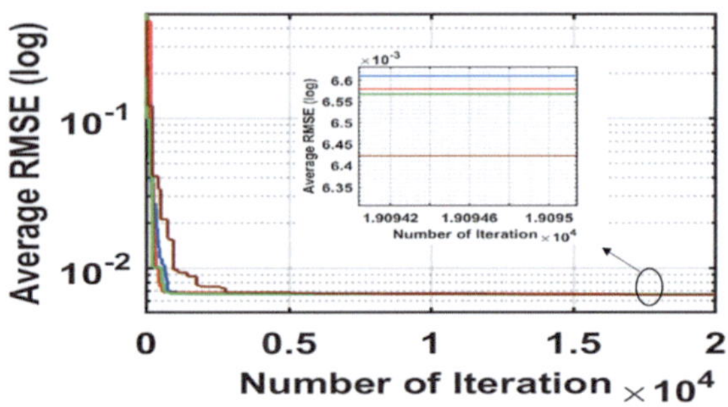

Fig. 12 Different algorithms convergence for double diode module panel

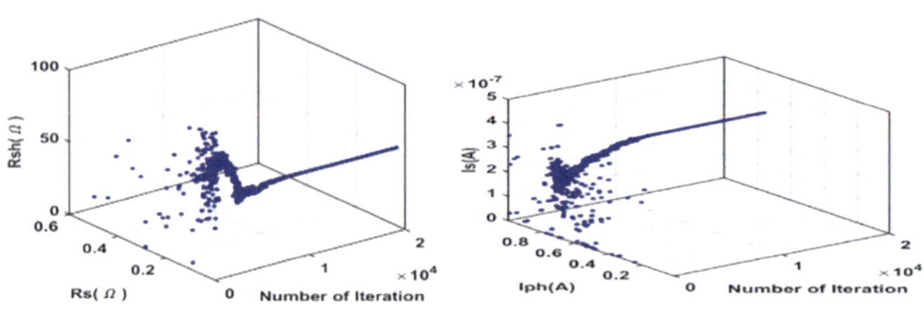

Fig. 13 Convergence graph of parameter identifier (Is-Iph) and (Rs-Rsh) for the double diode model module

5 Conclusion

In order to consistently and accurately estimate the parameters of various PV models, a new hybrid heuristic ALOCS Algorithms is developed in this study. The proposed algorithm will not really add any parameter settings that need to be adjusted. As a result, it is simple to apply and tune.

Results show that the discussed algorithm has superior performance in terms of quality and accuracy compared to other well-established algorithms. Accordingly, the hybrid ALOCS algorithm may be a promising method for resolving parameter identification with high accuracy.

A selection of the appropriate PV model is a major step to predict the PV system's behavior, and therefore a comparison has been made with experimental validation of the characterization of the PV panel using a SDM cell and DDM module.

The estimated execution time in a double diode is 1.248 times needed in a single diode. In the case of a large system, this difference in time could be a problem. The selection of the model, therefore, depends on whether the designer needs fast results, can either use the single diode model or use a dual diode model for greater accuracy. It will be interesting to see whether the results of this study are to be found as future studies, under partially shaded conditions.

References

1. Alam DF, Yousri DA, Eteiba MB (2015) Flower pollination algorithm based solar PV parameter estimation. Energy Convers Manag 101:410–422
2. Degla A, Saidj SH, Boumechcta S, Hazil O, Gherbi A, Arab AH (2021) Comparison and analysis of five MPPT techniques for boost converter based PV system under real input deviations. In: 2020 6th international symposium on new and renewable energy (SIENR), pp 1–6. https://doi.org/10.1109/SIENR50924.2021.9631910
3. Romero B, Del Pozo G, Arredondo B (2012) Exact analytical solution of a two diode circuit model for organic solar cells showing S-shape using Lambert W-functions. Sol Energy 86:3026–3029
4. Ortiz-Conde A, Sánchez FJG, Muci J (2006) New method to extract the model parameters of solar cells from the explicit analytic solutions of their illuminated I-V characteristics. Sol Energy Mater Sol Cells 90:352–361
5. Chen Y, Wang X, Li D, Hong R, Shen H (2011) Parameters extraction from commercial solar cells I-V characteristics and shunt analysis. Appl Energy 88:2239–2244
6. Mirjalili S (2015) The ant lion optimizer. Adv Eng Softw 83:80–98
7. Yang X-S, Deb S (2009) Cuckoo search via Lévy flights. In: World congress on nature & biologically inspired computing (NaBIC 2009). IEEE Publications, pp 210–214
8. Inderscience (2010) Cuckoo designs spring. Alphagalileo.org. Accessed 27 May 2010
9. Payne RB, Sorenson MD, Klitz K (2005) The cuckoos. Oxford University Press
10. Merah H et al (2022) Sizing and sitting of static VAR compensator (SVC) using hybrid optimization of combined cuckoo search (CS) and ant lion optimization (ALO) algorithms. Energies 15(13):4852
11. Lia S, Gonga W, Yana X, Hua C, Baib D, Wangc L, Gaod L (2019) Parameter extraction of photovoltaic models using an improved teaching learning-based optimization. Energy Convers Manage 186:293–305

12. Beigia AM, Maroosib A (2018) Parameter identification for solar cells and module using a hybrid firefly and pattern search algorithms. Solar Energy 171, 435–446
13. AlRashidi MR, AlHajri MF, El-Naggar KM, Al-Othman AK (2011) A new estimation approach for determining the I-V characteristics of solar cells. Sol Energy 85(7):1543–1550
14. Gong W, Cai Z (2013) Parameter extraction of solar cell models using repaired adaptive differential evolution. Sol Energy 94:209–220
15. Yu K, Liang J, Qu B, Cheng Z, Wang HJA (2018) e: multiple learning backtracking search algorithm for estimating parameters of photovoltaic models 226:408–422
16. Chen X, Yu KJSE (2019) Hybridizing cuckoo search algorithm with biogeography-based optimization for estimating photovoltaic model parameters 180:192–206
17. Abbassi A, Abbassi R, Heidari AA, Oliva D, Chen H, Habib A, Jemli M, Wang MJE (2020) Parameters identification of photovoltaic cell models using enhanced exploratory salp chains-based approach 198:117333
18. Rehouma Y, Tir Z, Gacem A, Rehouma F, Hamida MA, Gougui A (2021) Comparison study between improved JAYA and particle swarm optimization PSO algorithms for parameter extraction of photovoltaic module based on experimental test. In: Hatti M (ed) Artificial intelligence and renewables towards an energy transition. ICAIRES 2020. Lecture notes in networks and systems, vol 174. Springer, Cham. https://doi.org/10.1007/978-3-030-63846-7_21

Control of a Photovoltaic Pumping System with BLDC Motor

Hatira Bacem, Dhaoui Mehdi, and Sbita Lassaad

Abstract The increasing demand for energy as well as the growing requirements for ecological conservation have led to the study and expansion of the use cleaner, unconventional and less environmentally friendly alternative energy sources. Photovoltaic energy is now used as a more valuable natural energy source, for it is free, plentiful, fresh and dispersed on Earth. The work aims to study an autonomous photovoltaic pumping system using a BLDC motor, a three-phase voltage inverter and a DC-DC zeta converter. The intelligent control of the zeta converter with the incremental conductance function (INC-MPPT) will track the maximum power point, while the algorithm will ensure a smooth start of the BLDC motor used to drive the centrifugal pump. The obtained simulation results confirm the efficiency and good performance of the BLDC motor.

Keywords PV pumping system · BLDC motor · Zeta converter · Three-phase voltage inverter · Sizing · Modelling · Control

1 Introduction

Recourse to renewable energies is now becoming a vital necessity due to the increase in energy consumption and the depletion of conventional resources. Photovoltaic energy is one of the promising resources that has many electrical and environmental advantages [1]. This kind of energy is widely used in various fields, which can be used as grid-connected or isolated system. In the field of water pumping, photovoltaic energy is regaining its place given its ease of use and reduced maintenance cost. In this article, we chose to work with a brushless DC motor (BLDC), used for water pumping [2]. The BLDC motor has several advantages given its simple construction, low maintenance, synchronous speed operation and high torque [3]. Since the conversion

H. Bacem (✉) · D. Mehdi · S. Lassaad
L/R: Processes, Energy, Environment and Electrical Systems (PEESE), Electrical Department of Engineering, National Engineering School, University of Gabes, 6029 Gabes, Tunisia
e-mail: hatira.bacem@gmail.com

© The Author(s), under exclusive license to Springer Nature Singapore Pte Ltd. 2024
A. Mellit et al. (eds.), *Proceedings of the 2nd International Conference on Green Energy Conversion System*, Green Energy and Technology,
https://doi.org/10.1007/978-981-97-6148-7_35

efficiency of PV panels is low, a power maximization algorithm (MPPT) is generally inserted into the control chain. In addition, a DC/DC converter is essential to adapt the DC voltage level [4]. In the proposed work, the topology containing a Zeta type DC/DC converter followed by a three-phase inverter which are suitable for BLDC motor in water pumping application are considered. In addition, a control system is required to control the output of the converter operated with MPPT systems. Therefore, a control system using the PI controller is applied to achieve steady state operation in MPPT. The downside is such that the peak overshoot problem leads to the discovery of optimization-based methods.

2 Modeling of the Pumping System

The block diagram of the proposed system is shown in Fig. 1. It includes a PV generator, a Zeta-type DC/DC converter equipped with its MPPT algorithm, and a three-phase inverter controlled by the PWM technique, a BLDC motor driving a pump, sensors and a control system.

2.1 The BLDC Motor

The BLDC motor can operate in sensored or sensorless mode.

The equivalent circuit model of the motor shown in Fig. 2 can be represented by the following two equations:

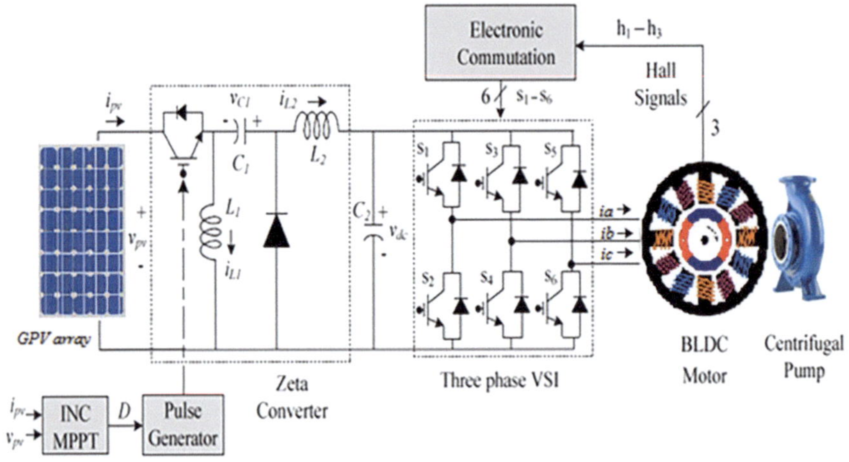

Fig. 1 Block diagram of PV pumping system with BLDC motor

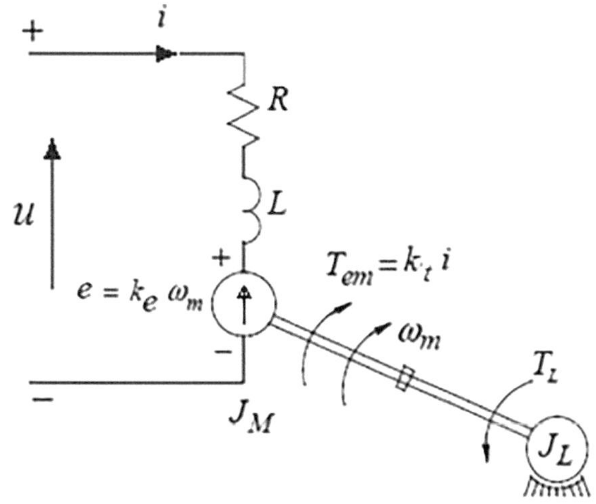

Fig. 2 BLDC motor equivalent circuit model

$$v(t) = R.i(t) + L.\frac{di(t)}{dt} + e(\theta_r) \tag{1}$$

$$e(\theta_r) = k_e.\Omega_m(t) \tag{2}$$

where, $e(\theta_r)$ is the induced EMF in armature (V), k_e is motor back-EMF constant (V.s/rad) and k_t is motor back-torque constant (N.m/A).

The characteristics of the BLDC motor used in this work are given in the appendix (Table 3).

(a) *Electrical equations*

The electrical equations that govern the operation of the BLDC motor are given by the following equations [5]:

$$v_a(t) = R.i_a(t) + L.\frac{di_a(t)}{dt} + e_a(\theta_r) \tag{3}$$

$$v_b(t) = R.i_b(t) + L.\frac{di_b(t)}{dt} + e_b(\theta_r) \tag{4}$$

$$v_c(t) = R.i_c(t) + L.\frac{di_c(t)}{dt} + e_c(\theta_r) \tag{5}$$

The expressions of the emf are defined as follows:

$$e_a(\theta_r) = \frac{k_e}{2}\Omega_m(t).Tra(\theta_e) \tag{6}$$

$$e_b(\theta_r) = \frac{k_e}{2}\Omega_m(t).Tra\left(\theta_e - \frac{2\pi}{3}\right) \quad (7)$$

$$e_c(\theta_r) = \frac{k_e}{2}\Omega_m(t).Tra\left(\theta_e + \frac{2\pi}{3}\right) \quad (8)$$

Where k_e is the back-EMF constant and θ_e the electrical angle which is calculated as follows $\theta_e = p.\theta_m$ with p the number of pair poles and $\theta_{r,}$ the rotor angular position and $Tra(\theta_e)$ is a trapezoidal wave function defined as follows:

$$Tra(\theta_e) = \begin{cases} 1, 0 \le \theta_e \le \frac{2\pi}{3} \\ 1 - \frac{6}{\pi}\left(\theta_e - \frac{2\pi}{3}\right), \quad \frac{2\pi}{3} \le \theta_e \le \pi \\ -1, \pi \le \theta_e \le \frac{5\pi}{3} \\ -1 - \frac{6}{\pi}\left(\theta_e - \frac{5\pi}{3}\right), \quad \frac{5\pi}{3} \le \theta_e \le 2\pi \end{cases} \quad (9)$$

The trapezoidal curves of the EMF are given by the following Fig. 3.

(b) *Mechanical equations*

The mechanical equation that governs the operation of a BLDC motor is the following [5]:

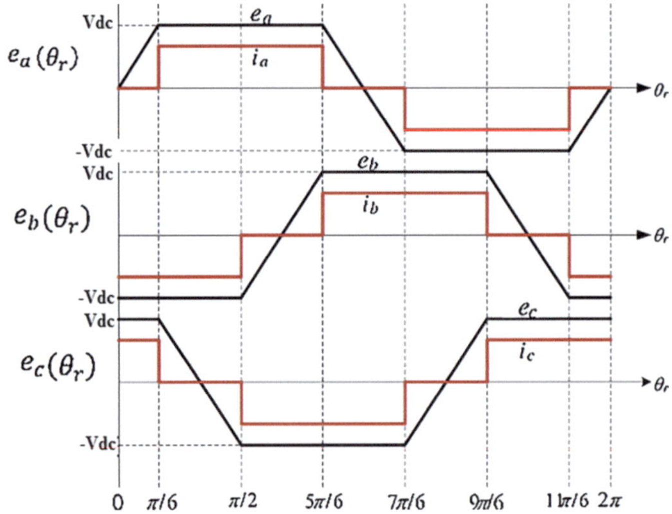

Fig. 3 Trapezoidal curves of the emf

$$J\frac{d\Omega_m}{dt} + B.\Omega_m = T_e - T_L \tag{10}$$

$$\frac{d\theta_r}{dt} = \frac{p}{2}.\Omega_m \tag{11}$$

The electromagnetic torque of a BLDC motor is calculated as follows [5]:

$$T_e = \frac{e_a.i_a + e_b.i_b + e_c.i_c}{\Omega_m} \tag{12}$$

Using the relations (6), (7) and (8), the electromagnetic torque is expressed by:

$$T_e = \left(Tra(\theta_e).i_a + Tra\left(\theta_e - \frac{2\pi}{3}\right).i_b + Tra\left(\theta_e + \frac{2\pi}{3}\right).i_c\right) \tag{13}$$

2.2 Inverter and Hysteresis Current Blocks

In this works, Inverter implementation is combined with current controller subsystem (Fig. 4) [6–10].

Hysteresis current control technique is considered as the main current control strategy. The Inverter phase voltages are as given in Eqs. (14–16) and switching of inverter devices based on rotor position is shown in Table 1.

$$h_a = \left(\frac{\pi}{6}<\theta_r<\frac{5\pi}{6}\right).S_1 + \left(\frac{5\pi}{6}<\theta_r<\frac{11\pi}{6}\right).S_2 \tag{14}$$

$$h_b = \left(0<\theta_r<\frac{\pi}{2}\right).S_3 + \left(\frac{5\pi}{6}<\theta_r<\frac{9\pi}{6}\right).S_3 + \left(\frac{11\pi}{6}<\theta_r<2\pi\right).S_4 \tag{15}$$

$$h_c = \left(0<\theta_r<\frac{\pi}{6}\right).S_5 + \left(\frac{\pi}{2}<\theta_r<\frac{7\pi}{6}\right).S_6 + \left(\frac{9\pi}{6}<\theta_r<2\pi\right).S_5 \tag{16}$$

$$\begin{cases} v_a(t) = \frac{1}{2}.V_{dc}.h_a \\ v_b(t) = \frac{1}{2}.V_{dc}.h_b \\ v_c(t) = \frac{1}{2}.V_{dc}.h_c \end{cases} \tag{17}$$

where S_1, S_2 belongs to phase-a of stator winding. Similarly (S_3, S_4) and (S_5, S_6) are for switching Phase-b and Phase-c respectively.

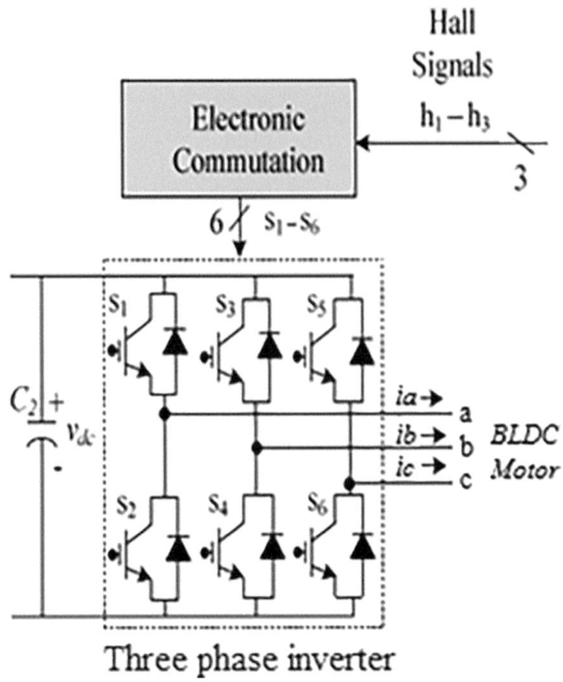

Fig. 4 Inverter and hysteresis current blocks

Table 1 Switching of inverter devices based on rotor position

Interval	Phase—A	Phase—B	Phase—C
$0 < \theta < \pi/6$	$s1 = 0; s2 = 0$	$s3 = 0; s4 = 1$	$s5 = 1; s6 = 0$
$\pi/6 < \theta < \pi/2$	$s1 = 1; s2 = 0$	$s3 = 0; s4 = 1$	$s5 = 0; s6 = 0$
$\pi/2 < \theta < 5\pi/6$	$s1 = 1; s2 = 0$	$s3 = 0; s4 = 0$	$s5 = 0; s6 = 1$
$5\pi/6 < \theta < 7\pi/6$	$s1 = 0; s2 = 0$	$s3 = 1; s4 = 0$	$s5 = 0; s6 = 1$
$7\pi/6 < \theta < 9\pi/6$	$s1 = 0; s2 = 1$	$s3 = 1; s4 = 0$	$s5 = 0; s6 = 0$
$9\pi/6 < \theta < 11\pi/6$	$s1 = 0; s2 = 1$	$s3 = 0; s4 = 0$	$s5 = 1; s6 = 0$
$11\pi/6 < \theta < 2\pi$	$s1 = 0; s2 = 0$	$s3 = 0; s4 = 1$	$s5 = 1; s6 = 0$

2.3 *Zeta DC/DC Converter*

The DC/DC converter used in this work is a non-isolated Zeta-type step-down converter [10–14].

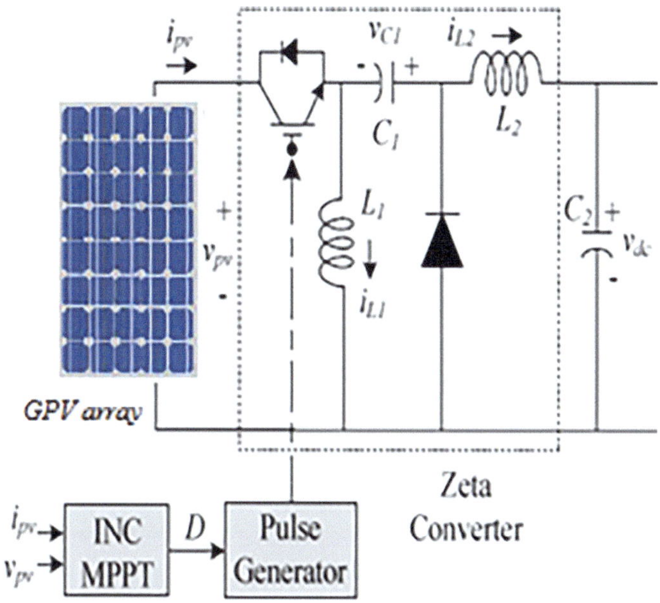

Fig. 5 Diagram of Zeta DC-DC converter

Figure 5 shows the structure of the proposed converter.

The Zeta converter receives the V_{PV} voltage from the photovoltaic generator and supplies a compatible DC voltage V_{dc} to the input of the inverter. The operation of zeta converter is designed in Continuous Conduction Mode (CCM) and the circuit operation can be defined by two modes of operation.

(a) *Mode 1*

In this mode, the switch $Q1$ is ON and the diode $D1$ is reverse biased. Inductors $L1$ and $L2$ is charged from the source and the inductor current $IL1$ and $IL2$ increases linearly. Also, discharging of $C1$ and charging of $C2$ take place. By Kirchhoff's Voltage law,

$$L_1 \frac{di_{L1}}{dt} = V_{PV} \tag{18}$$

$$L_2 \frac{di_{L2}}{dt} = V_{PV} + V_{C1} - V_{C2} \tag{19}$$

By Kirchhoff's current law,

$$i_{L1} = C_2 \frac{dv_{C2}}{dt} \tag{20}$$

(b) *Mode 2*

In this mode, the switch $Q1$ is OFF and the diode $D1$ is forward biased. During this interval, previously charged inductor $L1$ starts to discharge. So stored energy in $L1$ and $L2$ are discharged through capacitors $C1$ and $C2$. Therefore, the inductor currents i_{L1} and i_{L2} decrease gradually. By Kirchhoff's voltage law, voltage across inductor ($L1$) is given by,

$$L_1 \frac{di_{L1}}{dt} = -V_{L1} \tag{21}$$

Voltage across inductor ($L2$) is given by,

$$L_2 \frac{di_{L2}}{dt} = -V_{L2} \tag{22}$$

By applying Kirchhoff's current law, current through the capacitor $C1$ is,

$$i_{L1} = C_1 \frac{dv_{C1}}{dt} \tag{23}$$

The relation between input/output voltage and the duty cycle (D) of zeta converter in CCM is given by,

$$D = \frac{V_{dc}}{V_{PV} + V_{dc}} \tag{24}$$

Therefore,

$$\frac{V_{dc}}{V_{PV}} = \frac{I_{PV}}{I_{dc}} = \frac{D}{1-D} \tag{25}$$

By volt second balance,

$$V_{PV}.t_{ON} + (V_{PV} - V_{C1}).t_{OFF} = 0 \tag{26}$$

By applying volt-second balance, the relation between output voltage and input voltage is given by,

$$V_{dc} = \frac{D}{1-D} V_{PV} \tag{27}$$

The estimate of the duty cycle, D, initiates the design of the Zeta converter, which is estimated to be [15].

$$D = \frac{V_{dc}}{V_{dc} + V_{mpp}} = \frac{48}{48 + 189} = 0.2 \tag{28}$$

where V_{dc} is the average value of the output voltage of the zeta converter (DC bus voltage of the inverter) equal to the rated voltage of the BLDC motor.

The average direct current I_{dc} of the DC bus is estimated as follows [15]:

$$I_{dc} = \frac{P_{mpp}}{V_{dc}} = \frac{4860}{48} = 101.25\ A \tag{29}$$

The parameters of the converters L_1, L_2 and C_1 are estimated as being:

$$L_1 = \frac{D.V_{mpp}}{f_{sw}.\Delta I_{L1}} = \frac{0.2 \times 189}{20.10^3 \times 0.06 \times 25.71} = 1.2\ mH \tag{30}$$

$$L_2 = \frac{(1-D).V_{dc}}{f_{sw}.\Delta I_{L2}} = \frac{0.8 \times 48}{20.10^3 \times 0.06 \times 101.25} = 0.3\ mH \tag{31}$$

$$C_1 = \frac{D.I_{dc}}{f_{sw}.\Delta V_{C1}} = \frac{0.2 \times 101.25}{20.10^3 \times 0.1 \times 48} = 211\ \mu F \tag{32}$$

where f_{sw} is the switching frequency of the Zeta converter, ΔI_{L1} is the variation amplitude allowed in the I_{L1} current, identical to I_{mpp}; ΔI_{L2} is the variation amplitude allowed in the I_{L2} current, identical to I_{dc}; ΔV_{C1} is the amplitude of the authorized variation in the voltage across C_1, identical to V_{dc} [15].

2.4 PV Generator Modeling

In our case, the PV generator consists of 18 modules divided into 3 parallel branches, and 6 modules in series (Table 2). The equation of the voltage-current characteristic of the PV field is given by the equation:

$$I = N_p.I_{ph} - N_p.I_0\left[\text{Exp}\left(\frac{V + (N_S.R_S/N_p).I}{a.V_T}\right) - 1\right] - \frac{V + (N_S.R_S/N_p).I}{N_S.R_p/N_p} \tag{33}$$

With: I_{ph} the photo-generation current, I_o: the reverse saturation current of the diode, V and I are the voltage and current of the PV generator, R_s and R_p are respectively the series and parallel resistance of a PV module, a the ideality factor of the diode and V_t the thermal voltage.

The photocurrent and the saturation current are given by the attached equations:

$$I_{ph} = \frac{G}{G_n}(I_{SC} + k_i.\Delta T) \qquad I_0 = \frac{I_{SC} + k_i.\Delta T}{\text{Exp}\left(\frac{V_{OC}+k_V.\Delta T}{a.V_T}\right) - 1}$$

I(V) and P(V) characteristics of the PV generator tested under Standard Test Conditions (STC), (1000W/m^2 and 25 °C) are given in Fig. 6. At its optimum operating point (PPT), PV generator can provide a peak power $P_{mpp} = 4.860$kWp at voltage $V_{mpp} = 189$ V and a current $I_{mpp} = 25.71$ A.

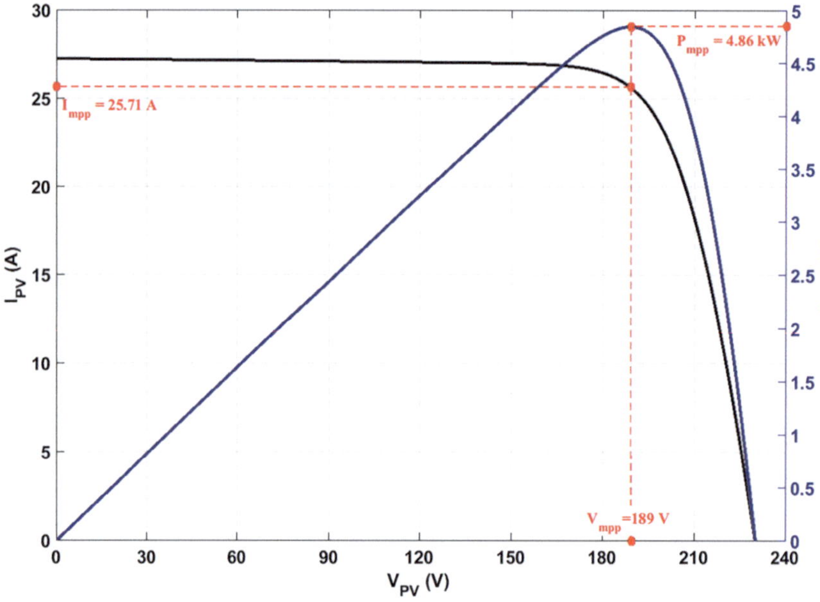

Fig. 6 I(V) and P(V) characteristics of the PV generator

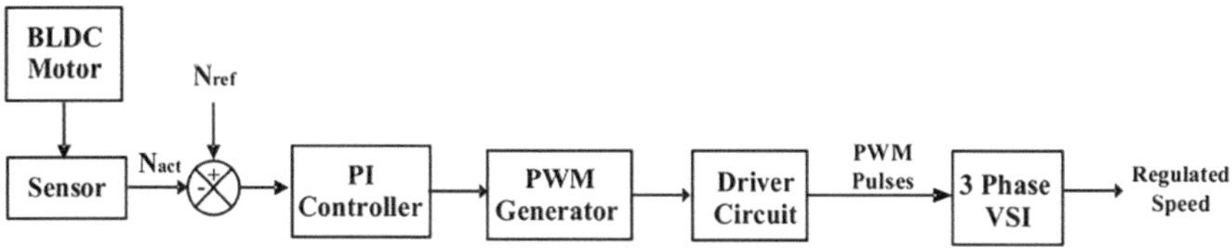

Fig. 7 Speed controller diagram

3 Control of the Pumping System

3.1 BLDC Motor Speed Control

For efficient operation of the centrifugal water pumping system, speed control of the BLDC motor is essential. This speed control can be done with or without a speed sensor. Although the sensorless method has some advantages such as reduced size and less expensive, it requires electronic circuits and complex algorithms. Therefore, in this method, a Hall Effect speed sensor is used to measure the actual speed of the BLDC motor. This speed is compared with the reference speed and the error value is transmitted to the PI controller. This provides the reference stator current, which compared to the measured currents. The error currents pass through a cascaded hysteresis controller that provides the PWM pulses. The driver circuit will transmit the emitted PWM pulses to the three-phase voltage inverter. The speed control scheme is shown in Fig. 7.

The k_p and k_i parameters of the speed controller are calculated according to the characteristics of the BLDC motor as follows:

$$k_p = \frac{L}{T_i} \; and \; k_i = \frac{R}{T_i} \tag{34}$$

where: $T_i = 3.\frac{L}{R}$.

3.2 Back-Emf Observer for Sensorless Speed Estimation

Speed sensorless control performs well for motor drive applications that do not require precise control, below 10% of rated speed. The pump drive that operates in a speed range of 20 to 100% is the good example for sensorless control.

The basic methodology of sensorless motor control revolves around counter-emf. According to the basic laws of electromagnetic theory, the back-emf generated inside the winding is proportional to the speed of the rotor.

Therefore, sensorless speed estimation is divided into two basic categories:

- Detection or measurement of back-emf of armature terminals
- Back-emf observer based on mathematical equations describing motor behavior.

The algorithm based on back-emf detection is 120 degree switching or trapezoidal control of a BLDC motor in which the back-emf of the unpowered phase is measured for the rotor position. This helps to switch the motor efficiently.

Closed-loop back-emf observers such as the Luenberger observer rely on computing the back-emf.

If the back-emf e_{ab}, e_{bc} and e_{ca} are known, the velocity can be calculated because k_e is already a known parameter. The state model relating to Eq. 1 is written in the following form:

$$\begin{cases} \dot{x} = A.x + B.u \\ y = C.x \end{cases} \tag{35}$$

With:

$x = \begin{bmatrix} i \\ e \end{bmatrix}$: Represents the vector of variables.

u: the command and y: The output to enslave.

$A = \begin{bmatrix} -\frac{R}{L} & -\frac{1}{L} \\ 0 & 0 \end{bmatrix}$: The state matrix and $B = \begin{bmatrix} \frac{1}{L} \\ 0 \end{bmatrix}$: The command vector.

$C = [1\ 0]$: The output vector.

The equation of state of the observer can therefore be described by the following system:

$$\begin{cases} \dfrac{d\hat{i}}{dt} = -\dfrac{R}{L}\hat{i} - \dfrac{1}{L}\hat{e} + \dfrac{1}{L}u + K_1\left(i - \hat{i}\right) \\ \hat{e} = K_2 \displaystyle\int \left(i - \hat{i}\right) \end{cases} \tag{36}$$

The convergence of the observer (Fig. 8) is ensured by taking into account the difference between the current measured and that estimated. The coefficients K_1 and K_2 ensure the dynamics of the observer. The coefficient K_1 ensures the acceleration or the slowing down of the estimated magnitudes towards the real states. The integral coefficient K_2 ensures the reduction of the established static error of the observer.

The f.e.m e_{bc} and e_{ca} are estimated in the same way.

After estimating the f.e.m, we can then determine the rotational speed of the BLDC motor.

Figure 9 shows the speed estimation diagram.

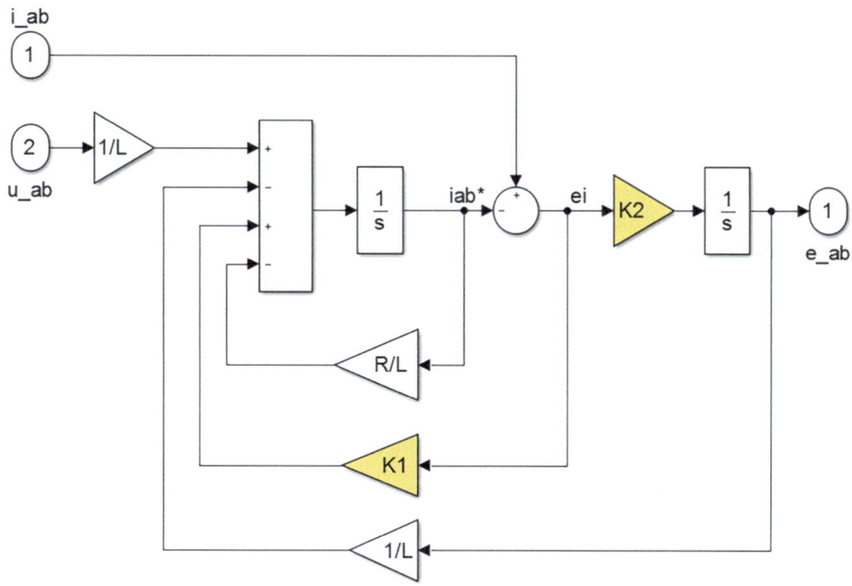

Fig. 8 Back-emf observer diagram (i.e. e_{ab})

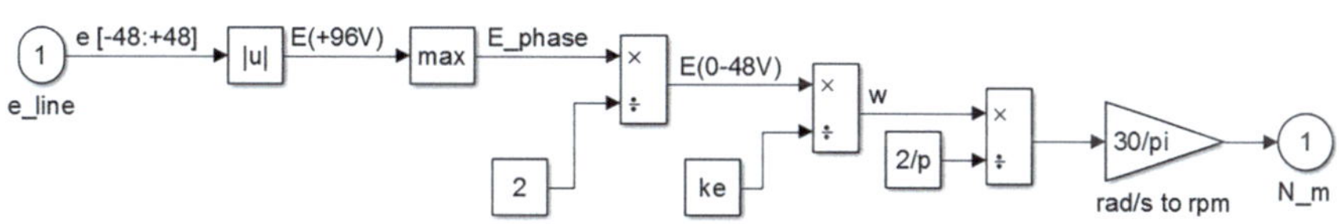

Fig. 9 Speed estimation diagram

4 Results and Discussion

Performance evaluation of PV pumping system with BLDC motor using zeta converter and three-phase voltage inverter is performed using simulated results in MATLAB/Simulink. The proposed system is designed, modeled and simulated taking into account the random and the instantaneous variation in the level of solar irradiance and its adequacy is demonstrated by testing startup, steady state, and dynamic behavior. The relevance of the system proposed under dynamic condition, the level of solar irradiance is varied over time.

The behavior of the different stages such as the GPV and the BLDC motor pump are shown in Figs. 10, 11 and 12.

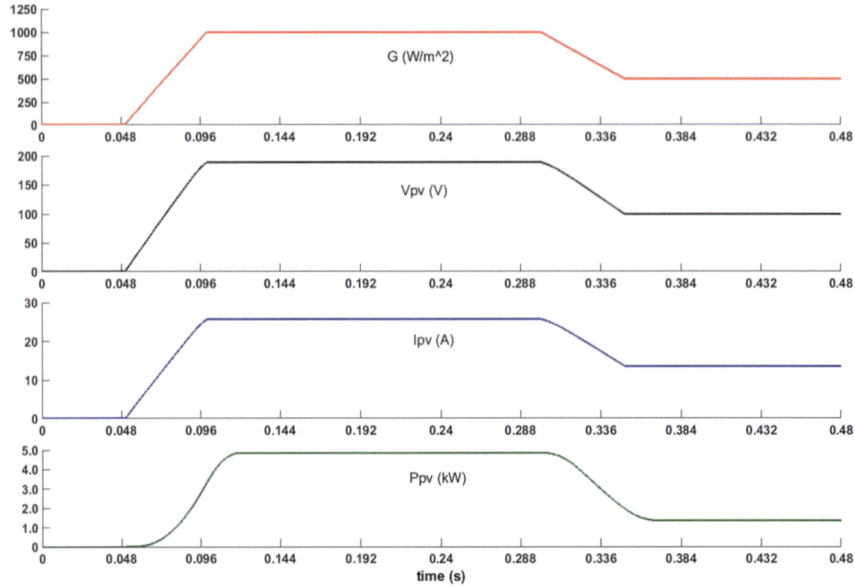

Fig. 10 Evolution of the variables (irradiation, voltage, current and power) of the PV generator

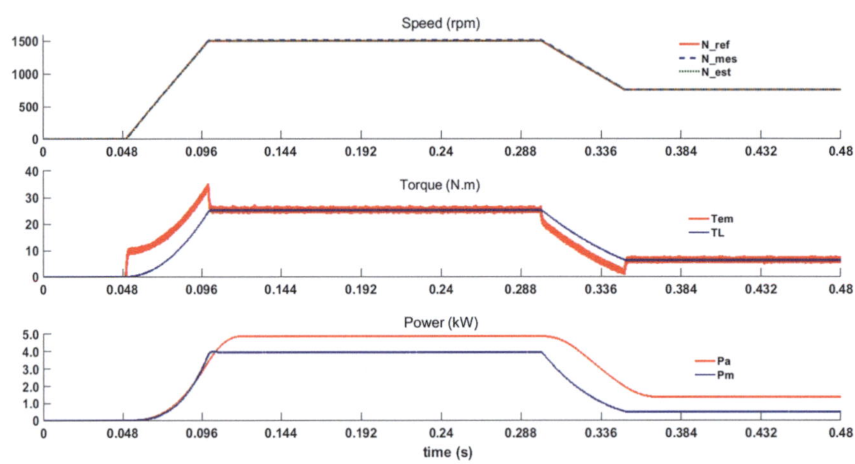

Fig. 11 Performances of the BLDC motor-pump variables (speed, torque and power)

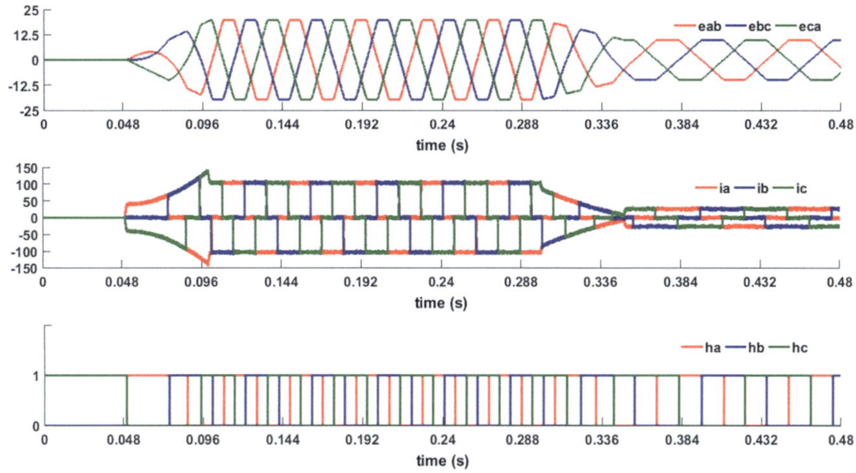

Fig. 12 Evolution of the BLDC motor-pump variables (emf, current and hall signals)

5 Conclusion

The GPV array-zeta converter fed Inverter-BLDC motor-pump for water pumping has been proposed and its suitability has been demonstrated by simulated results using MATLAB/Simulink. First, the proposed system has been designed logically to fulfil the various desired objectives and then modelled and simulated to examine the various performances under starting, dynamic and steady state conditions. The performance evaluation has justified the combination of zeta converter and BLDC motor drive for GPV array based water pumping. The system under study availed the various desired functions such as MPP extraction of the GPV array, soft starting of the BLDC motor, fundamental frequency switching of the inverter resulting in a reduced switching losses, reduced stress on IGBT switch and the components of zeta converter by operating it in continuous conduction mode and stable operation. Moreover, the proposed system has operated successfully even under the minimum solar irradiance.

Annexes

See Tables 2 and 3.

Table 2 Electrical specifications of Aleo Solar S79L270 module

P_{mpp} [W]	270
V_{mpp} [V]	31.49
I_{mpp} [A]	8.57
V_{oc} [V]	38.35
I_{sc} [A]	9.05
Efficiency	17%

Table 3 BLDC motor characteristics

Model	210BL0404-15
Rated voltage	48 VDC
Rated power	4 kW
Rated speed	1500 RPM
Rated torque	25.5 N.m
Maximum torque	75 N.m
Rated current	104 A
Resistance	0.029 Ω
Inductance	0.26 mH
Torque constant	0.241 N.m/A
Inertia	1.17×10^{-2} kg.m^2
f.e.m constant	25.3 V/krpm

References

1. Dubey M, Sharma S (2014) Solar PV stand-alone water pumping system employing PMSM drive. In: Member IEEE and Rakesh Saxena electrical engineering department 2014 IEEE students' conference on electrical, electronics and computer science
2. Bouzeriaetal H (2014) Speed control of photovoltaic pumping system. Int J Renew Energy Res 4(3)
3. Teodorescu R, Liserre M, Rodriguez P (2011) Grid converters for photovoltaic and wind power systems, 1st edn. John Wiley, UK
4. Mayssa F, Aymen F, Lassaad S (2012) Influence of photovoltaic DC bus voltage on the high speed PMSM drive. In: IECON 2012-38th annual conference on IEEE industrial electronics society, pp 4489–4494
5. Xuesong Z, Daichun S, Youjie M, Deshu C (2010) The simulation and design for MPPT of PV system based on incremental conductance method. WASE Int Conf Inf Eng (ICIE) 2:314–317
6. Moussa H, Fadel M, Kanaan H (2012) A single stage DC-AC boost topology and control for solar PV systems supplying a PMSM. In: international conference on renewable energies for developing countries (REDEC), pp 1–7
7. Shen JM, Jou HL, Wu JC (2012) Novel transformer less grid connected power converter with negative grounding for photovoltaic generation system. IEEE Trans Power Electron 27(4):1818–1829
8. Verma AK, Singh B, Shahani T (2012) Grid interfaced solar photovoltaic power generating system with power quality improvement at ac mains. In: IEEE third international conference on sustainable energy technologies (ICSET), pp 177–182

9. Moussa H, Fadel M, Kanaan H (2012) A single-stage DCAC boost topology and control for solar PV systems supplying a PMSM. In international conference on renewable energies for developing countries (REDEC), pp 1–7
10. Ahmed KH, Hamad MS, Finney SJ, Williams BW (2010) DC-side shunt active power filter for line commutated rectifiers to mitigate the output voltage harmonics. In: IEEE energy conversion congress and exposition (ECCE), pp 151–157
11. Reisi AR, Moradi MH, Jamasb S (2013) Classification and comparison of maximum powerpoint tracking techniques for photovoltaic system: a review. Renew Sustain Energy Rev 19:433–443
12. Shahin A, Payman A, Martin J-P, Pier Federici S, Meibody-Tabar F (2010) Approximate novel loss formulae estimation for optimization of power controller of DC/DC converter. In: 36th annual conference on IEEE industrial electronics society, pp 373–378
13. Singh B, Bist V (2012) A single sensor based PFC zeta converter fed BLDC motor drive for fan applications. In: Fifth IEEE power India conference, pp 1–6
14. Uno M, Kukita A (2014) Single-switch voltage equalizer using multi-stacked buck-boost converters for partially-shaded photovoltaic modules. In: IEEE transactions on power electronics, No 99
15. Jones WV (2013) Motor selection made easy: choosing the right motor for centrifugal pump applications. IEEE Ind Appl Mag 19(6):36–45
16. Satapathy S, Dash KM, Babu BC (2013) Variable step size MPPT algorithm for photo voltaic array using zeta converter—a comparative analysis. In: Students conference on engineering and systems (SCES), pp 1–6

Carbon Footprint Assessment of Di-Calcium Phosphate Production from the Tunisian Chemical Group

Douha Alaya, Ons Ghriss, Ali Fguiri, Elhem Zrigui, and Mohamed-Razak Jeday

Abstract Increased industrial activities are highly linked to increased carbon emissions, which have been identified as the main cause of the increase of global warming and contribute of a substantial amount of greenhouse gases into our environment. As such, an average carbon footprint of di-calcium phosphate (DCP) production from the Tunisian Chemical Group has been calculated and analyzed. The model established using the SimaPro software considers for carbon footprint evaluation using the "IPCC GWP 2013 100 a V1.03" for the impact assessment methodology. The specificity of this method is to assess the "climate change" impact. This study showed that direct emissions contributing for 1350.1 kg CO_2 eq/t of DCP which 64.96% is attributed to the synthesis reaction of (DCP), with a significant proportion of 46.14% of phosphoric acid used as a raw material. Natural gas contributes with a percentage of 32.6%of total GHG emissions. While, greenhouse gas emissions from transportation and electricity represents 1.2% of total emission.

Keywords Greenhouse gases · Di-calcium phosphate · Carbon footprint · SimaPro

1 Introduction

For thousands of years, our planet has experienced major climate changes such as melting ice and rising sea and ocean levels. These anthropogenic changes affect human and natural systems and can have significant impacts on resource availability, economic activity and well-being human, making it one of the greatest short- and

D. Alaya · O. Ghriss (✉) · A. Fguiri · M.-R. Jeday
Research Laboratory "Processes, Energetics Environment and Electrical Systems", National Engineering School of Gabes (ENIG), Gabes University, Omar Ibn Kattab ZRIG, Gabes 6029, Tunisia
e-mail: onsgh@hotmail.fr

E. Zrigui
Tunisian Chemical Group (GCT), B.P.72, Gabes 6000, Tunisia

© The Author(s), under exclusive license to Springer Nature Singapore Pte Ltd. 2024
A. Mellit et al. (eds.), *Proceedings of the 2nd International Conference on Green Energy Conversion System*, Green Energy and Technology,
https://doi.org/10.1007/978-981-97-6148-7_36

long-term challenges facing the world. Certainly, the greenhouse effect plays an essential role in the development of life on Earth. This natural process increases the temperature on the Earth's surface and helps maintain ecological balance. The average temperature of the Earth's surface between 2011 and 2020 was 1.1 °C (2 °F) warmer than the average temperature in the late nineteenth century (before the industrial revolution) and warmer than at any time in the last 100,000 years [1]. Global warming threatens not only human health but also animal and plant species, leading to water and food shortages and geological changes to coasts (partial or total disappearance). For this reason, WHO and other national organizations have set limit values for gaseous emissions to the atmosphere. In this context, the IPCC aimed at reducing GHG emissions in the various sectors with the aim of developing an environmental management scheme, in order to keep the global temperature rise below 2 °C and even to reduce it to 1.5 by 2030 [2], in partnership with the Kyoto Protocol, COP, the Paris Agreement... These actions require an inventory of emissions, which is somewhat like rewinding the clock on emissions or going back in time to prevent these catastrophic effects on earth human activities affect the chemical composition of the atmosphere and contribute to increasing levels of greenhouse gases (carbon dioxide (CO_2), methane, nitrous oxide, etc.), thus causing also the climate change. The combustion of oil, natural gas and coal (electricity, heat and industry) accounts for two-thirds of total global CO_2 emissions to the atmosphere. According to the IPCC's 6th report, methane emissions have increased by 6% over the last decade and by 156% since 1750 [3]. This gas stays in the atmosphere less than carbon dioxide, but has a much higher warming potential.

In this regard, in 1997, the Kyoto Protocol for the Reduction of Greenhouse Gases (GHGs) was signed by 184 countries. It complements the United Nations Framework Convention on Climate Change, which aims to reduce greenhouse gas emissions among signatories. Tunisia signed the United Nations Framework Convention on Climate Change (UNFCCC) in Rio in 1992 and ratified it in July 1993.

Tunisia's gross emissions increased from 29 Mt CO_2 eq in 1994 to 46.6 Mt CO_2 eq in 2012, an average increase of 2.7% per year [4]. The carbon accounting is one of the quantification methods that allows companies to estimate their "carbon footprint" to properly control their emissions; it is a diagnostic tool that analyzes direct and indirect greenhouse gas emissions activities within a given range. Several companies and organizations around the world are currently taking this approach. The list of greenhouse gases considered in the GHG emissions inventory includes carbon dioxide (CO_2), methane (CH_4), nitrous oxide (N_2O), hydrofluorocarbons (HFCs), perfluorocarbons (PFCs), sulfur hexafluoride (SF_6), nitrogen trifluoride (NF_3). The latter will be taken into account for the greenhouse gas emissions balance sheets from 1 July 2016. The CO_2 tax came into effect on January 1, 2008, and is intended to exempt companies that voluntarily take measures to reduce CO_2 emissions from the combustion of fossil fuels (fuels and fuels) [5].

The objective of this part of the work is to apply the carbon footprint method according to ISO 14067:2018 for a di-calcium phosphate production system to account for the GHG emissions generated by the production of one ton of this product. However, the tool chosen to calculate the carbon footprint is the software "SimaPro".

2 Materials and Methods

2.1 Production Process of Di-Calcium Phosphate

The di-calcium phosphate produced through a reaction between defluorinated and decadmiated phosphoric acid (AP) and calcium carbonate. The main reaction that leads to the formation of DCP is as follows:

$$CaCO_3 + 2H_3PO_4 \rightarrow 2CaHPO_4 + H_2O + CO_2.$$

The production process of di-calcium phosphate goes through the following steps (see Fig. 1).

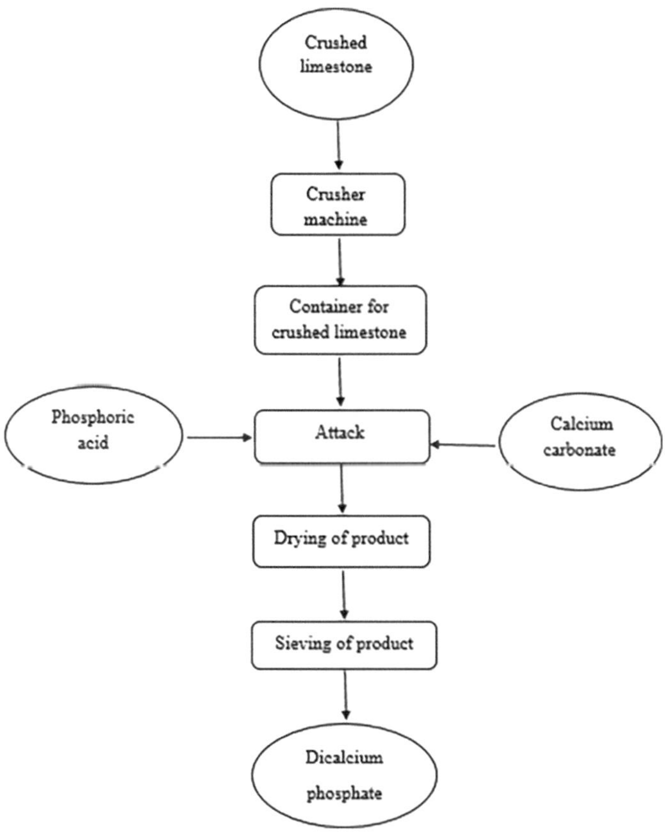

Fig. 1 Diagram of the production steps of di-calcium phosphate

2.2 *Carbon Footprint Calculation Approach with SimaPro Software*

The objective of this part of the work is to apply the carbon footprint method according to ISO 14067:2018 for the DCP production system to account for the GHG emissions generated by the production of one ton of this product with SimaPro software. In this context, four steps are included and defined:

Definition of objectives and scope of study. This first step includes two essential points.

Definition of objectives. The objective is to conduct a carbon footprint study of dicalcium phosphate produced at the phosphoric acid factory in Gabes, in order to present its impacts on climate change.

Definition of the scope of study

- Functional Unit (UF)

The functional unit (UF), designated in this study, is one ton of dicalcium phosphate produced. All data required for this study reported to this UF.

- System boundaries and assumptions

Defining the boundaries of the system helps determine the steps taken into account in calculating the carbon footprint. The physical boundaries of the system must be defined to include all processes that significantly affect the analytical results (see Fig. 2).

The DCP production system is divided into 6 modules in the following table:

Module 1: Raw Materials

- Limestone: After grinding, thin crystals of small masses are obtained which are then used in the process.

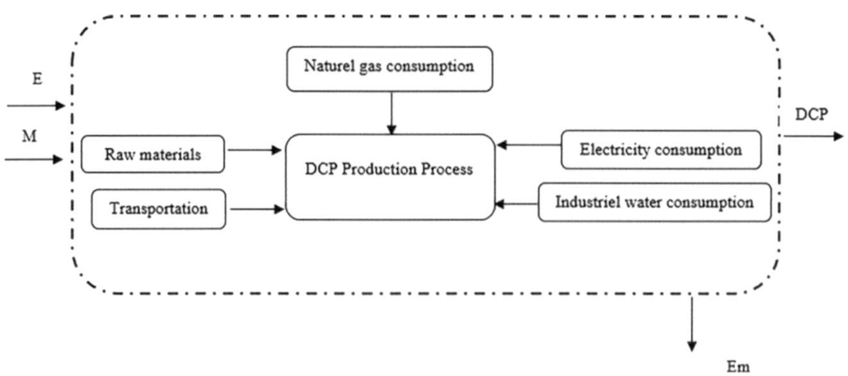

Fig. 2 DCP production system boundaries, E: energy, M: materials, Em: emission

- Phosphoric acid: The phosphoric acid produced in the plant.

Module 2: DCP Production Process

- This module includes data about the DCP production process. Data is retrieved from the factory.

Module 3: Electricity Consumption

- To ensure operation of equipment and lighting.

Module 4: Natural Gas Consumption

- The drying operation is carried out by a dryer connected to a natural gas burner.

Module 5: Industrial Water Use

- Water is used in daily plant activities such as washing equipment.

Module 6: Transportation

- Limestone is transported by bulk truck.

Analysis of the inventory. This second step involves the collection of data and calculations to define the inputs and outputs associated with the production of the functional unit, namely, one ton of DCP. The various data have been collected from industry personnel.

Assessment of environmental impacts. The third step in calculating the carbon footprint is the assessment of the environmental impact, which was conducted using the software 'SimaPro®9.4.0' and applying the impact assessment method 'IPCC GWP 2013 100a V1.03',the specificity of this method lies in evaluating the 'climate change' impact.

Results and interpretations. In this step, we will interpret the results obtained by the software.

3 Results and Discussion

The results show that the entire DCP system emits 1350.1 kg CO_2 eq/t of DCP. The contribution of each module to the impact of climate change illustrated in the following figure (see Fig. 3).

As clearly observed, the synthesis reaction (AP and $CaCO_3$) represents the highest environmental burden of the system, accounting for 64.96 of the total, with a significant proportion for AP at 46.14% since it is used as a raw material. Therefore, it is essential to seek optimization in phosphoric acid consumption. Implementing a CO_2 capture system could help reduce the overall environmental impact of the system. Natural gas contributes to 32.6% of the emissions, mainly due to the combustion

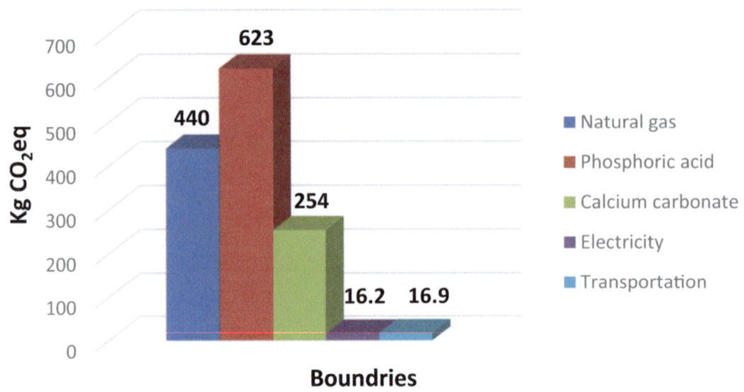

Fig. 3 The results of the environmental characterization of carbon footprint for one ton of DCP

emissions during the drying process. The transportation and electricity consumption modules account for 1.2% of the total emissions, primarily due to greenhouse gas emissions from transportation distances and the operation of process equipment, respectively.

4 Action Plan to Reduce GHG Emissions

Tunisian Chemical Group (GCT) has planned to implement an environmental improvement program to reduce greenhouse gas emissions. There are many simple measures that enable us to reduce our greenhouse gas emissions, which is achievable thanks to everyone's commitments. Aware of our carbon footprint on the environment, each employee strives to minimize its carbon footprint. It is not always easy to know how to reduce our CO_2 emissions to fight against the imminent climate change. Many solutions are proposed, including:

- Catalytic reduction of CO_2 emissions is an approach to fight against global warming because it recovers CO_2 instead of releasing it into the atmosphere
- The catalytic destruction of CO_2 that aims to break down the CO_2 molecule into other less harmful compounds this approach aims to completely remove CO_2 from the atmosphere but this method is energy intensive and can present technical and economic challenges on a large scale.
- Energetically: as a radical solution and based on the weather conditions of the Gabes region the only renewable energy that can be exploited, on the site, is solar energy. The use of renewable energy reduces not only GHG emissions (lack or low emissions) but also the cost of purchasing fuel (natural gas).
- And among the solutions often mentioned, there is reforestation, the idea is to plant green curtains around the plant (trees that store carbon) to offset our CO_2 emissions. We note for example: Paulownia Tomentosa, Bamboo, Iroko…

5 Conclusion

In an increasingly difficult socio-economic environment at the national and international level, particularly as regards the quality of products/services, price and environmental impact, Product manufacturers and service providers are constantly looking for new approaches and methods for analysis and optimization. This allows us to study the impact of different parameters on the performance of our activities (raw materials, energy, environmental emissions, social responsibility, etc.) in order to maximize our performance and meet the needs of our customers. Meet and anticipate new market needs. The main objective of this study was to quantify the carbon footprint, which required in-depth investigations on temporal aspects, In order to optimize the collection of data on emissions of different gases from the di-calcium phosphate production unit. For production, 1 ton of DCP has a climate change value of 1350.1 kg CO_2 eq. The main contributors to this climate change value are DCP synthesis reaction and Combustion reaction during DCP drying. We proposed an action plan to minimize greenhouse gas emissions from DCP:

- Catalytic reduction of CO_2 emissions in Gabes.
- Catalytic destruction of CO_2 emissions in Gabes
- Green curtain around plant.
- Use of renewable energy.

References

1. Connors S et al (2022) IPCC, climate change: summary for all 2021. V3. www.ipcc.ch/report/ar6/wg1/downloads/outreach/IPCC_AR6_WGI_SummaryForAll.pdf
2. Masson D et al (2019) IPCC, 2018: summary for policymakers, global warming of 1, 5 °C
3. J-P V Y, PM, PL, ER (2021) 6th IPCC assessment report climate change: the science. National Meteorological Institute
4. Ministry of Local Affairs and Environment and United Nations Development Programme (UNDP). In: Tunisia's third national communication under the United Nations framework convention on climate change
5. Fink M et al (2016) Réseau action climate (RAC-F). In: Corporate climate responsibility broadening it to indirect emissions

Improvements of an IG Wind Turbine System with Constant Switching Frequency Power Control for Microgrid Applications

Radia Abdelli, Djamila Rekioua, Toufik Rekioua, Zahra Mokrani, and Pierre-Olivier Logerais

Abstract This paper introduces a novel approach for the fixed-switching frequency of a wind turbine-based induction generator in microgrid. The proposed strategy combines power control and modulated hysteresis control current. One of the advantages of this method is the constant switching frequency of the converter switches, which eliminates a major drawback of the classical hysteresis control. This approach also allows to obtain an AC current with sinusoidal forms and a constant frequency while ensuring minimal total harmonic distortion (THD) of the current. Simulations were conducted using Matlab/Simulink, and the results demonstrated considerable improvements with regard to the classical method, including reduced current THD supplied to the AC grid, and, improved dynamic response performance as well.

Keywords Wind turbine · Constant switching frequency · Flywheel energy storage system

1 Introduction

Wind turbines with Self-Excited Induction Generators (SEIG) have been studied in several papers [1–4]. When supplying the grid, the currents and voltages need to have constant magnitudes and frequencies with a weak harmonic distortion and a unit power factor. Therefore, these limitations must be considered in the grid-side control of the converter. Several studies have examined the Hysteresis Current Control (HCC) and Sine Pulse Width Modulation (SPWM) inverter control techniques [5–8]. In [5],

R. Abdelli · D. Rekioua (✉) · T. Rekioua · Z. Mokrani
Laboratoire de Technologie Industrielle et de l'Information, Faculté de Technologie, Université de Bejaia, Bejaia, Algeria
e-mail: djamila.ziani@univ-bejaia.dz

P.-O. Logerais
Univ Paris Est Créteil, CERTES, IUT de Sénart-Fontainebleau, 36, rue Georges Charpak, 77567 Lieusaint, France

© The Author(s), under exclusive license to Springer Nature Singapore Pte Ltd. 2024
A. Mellit et al. (eds.), *Proceedings of the 2nd International Conference on Green Energy Conversion System*, Green Energy and Technology,
https://doi.org/10.1007/978-981-97-6148-7_37

the authors give a comparison between the two strategies. They found that if a null wire connection was accessible, the HCC, which could be easily implemented, would be a good choice for the grid-connected inverters because of its transient response, although it could cause an undesirable harmonic spectrum.

This undesirable harmonic spectrum is essentially due to the variable switching frequency. Since the amplitude of the current ripples is relative to the inverse of the switching frequency, the increase of this parameter is necessary, which could never be achieved with the classical HCC strategy implying a constant bandwidth, as the switching frequency is variable and changes in a random way. A simple strategy, that can give either the possibility to adjust the switching frequency or maintain it constant with a very easy implementation, is the Modulated Hysteresis Current Control (MHCC). This method relies on superposing a triangle signal with the same frequency as the current reference. A hysteresis controller then compares the modulated reference current with the measured one, just like in the conventional technique.

In this work, the MHCC strategy is applied to control the grid-side converter in a WEC system. It gives the possibility to adjust the switching frequency to high values in order to decrease the THD and satisfy the IEC harmonic standards.

2 Studied System Modeling

Figure 1 depicts the analyzed wind conversion system in association with the flywheel energy storage system (FESS).

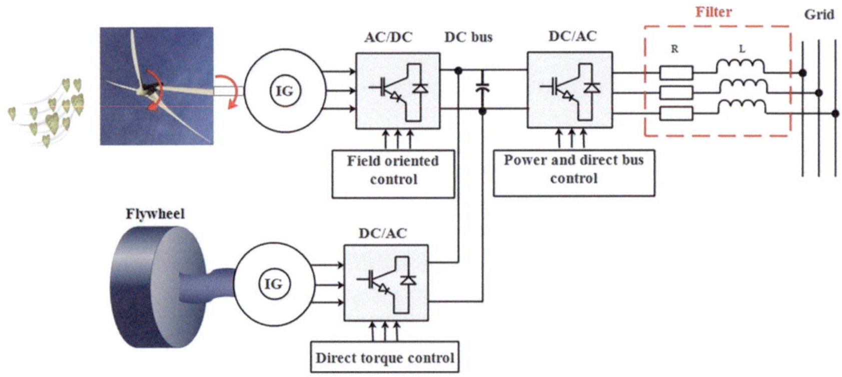

Fig. 1 Studied system

2.1 Wind, Generator and Rectifier Mode and Control

The following Eq. (1), gives the mechanical torque produced by the wind turbine generator [4–10]:

$$T_g = \frac{1}{2} C_p(\lambda) \rho S \frac{v^3}{G\omega_t} \tag{1}$$

where v is the averagewind velocity, G is a multiplier gain, S stands for the turbine section and ρ air density, ω_t (rad/s) and R are the turbine angular speed andblade radius. $C_p(\lambda)$ is the performance coefficient and λ is the tip-speed ratio which is outlined as follows:

$$\lambda = \frac{\omega_t.R}{v} \tag{2}$$

In this study, $C_p(\lambda)$ is provided via the Maximum Power Point Tracking (MPPT) algorithm, which is utilized to regulate the generator speed since each given wind speed v corresponds to an appropriate optimal value of the speed ratio λ_{opt} and then to $C_{p-\max}$.

The electromagnetic torque reference T_{e-ref} is expressed in terms of $C_{p-\max}$ by:

$$T_{e-ref} = \frac{C_{P-\max}}{\lambda_{opt}^3} \frac{\rho}{2} \pi \frac{R^5}{G^3} . \omega_m^2 \tag{3}$$

In the following equation, the function of the multiplier is to match the speed of the generator to the speed of the turbine [10–13].

$$\begin{cases} T_t = \frac{T_g}{G} \\ \omega_t = \frac{\omega_g}{G} \end{cases} \tag{4}$$

with T_t the turbine torque.

The mechanical equation is specified by the following classical expression [4]:

$$T_{e-ref} - T_g = J \frac{d\omega_m}{dt} + f\omega_m \tag{5}$$

where f is the friction coefficient and J is the shaft moment of inertia.

The IG model in the d–q frame is expressed with the matrix equation below [12]:

$$\begin{bmatrix} V_{sd} \\ V_{sq} \\ 0 \\ 0 \end{bmatrix} = \begin{bmatrix} R_s & -\omega_s.l_s & 0 & -\omega_s.m \\ \omega_s.l_s & R_s & \omega_s.m & 0 \\ 0 & -\omega_r.m & R_r & -\omega_r l_r \\ \omega_r.m & 0 & \omega_r.l_r & R_r \end{bmatrix} . \begin{bmatrix} I_{sd} \\ I_{sq} \\ I_{rd} \\ I_{rq} \end{bmatrix}$$

$$+\begin{bmatrix} l_s & 0 & m & 0 \\ 0 & l_s & 0 & m \\ m & 0 & l_r & 0 \\ 0 & m & 0 & l_r \end{bmatrix} \cdot \frac{d}{dt} \begin{bmatrix} I_{sd} \\ I_{sq} \\ I_{rd} \\ I_{rq} \end{bmatrix} \tag{6}$$

The stator or rotor quantities are indicated by the subscripts s and r, where R_s, R_r and ω_s, ω_r respectively, represent the phase resistances and electrical pulsations. The currents and voltages along the d and q axes are represented by (i_{ds}, i_{qs}) (i_{dr}, i_{qr}) (V_{ds}, V_{qs}) as the Park components. The phase leakage inductances for the rotor and stator are l_r and l_s, respectively, and m is the mutual inductance between the two.

The equations underneath define the AC/DC rectifier model [13].

$$\begin{pmatrix} V_a \\ V_b \\ V_c \end{pmatrix} = \frac{V_{dc}}{3} \begin{pmatrix} -2 & 1 & 1 \\ 1 & -2 & 1 \\ 1 & 1 & -2 \end{pmatrix} \begin{pmatrix} S_a \\ S_b \\ S_c \end{pmatrix} \tag{7}$$

$$i_{dc} = S_a i_a + S_b i_b + S_c i_c \tag{8}$$

The application of the PWM method, as shown in Fig. 2, where the field-oriented strategy is used to regulate the induction generator, yields the logic functions (S_a, S_b, S_c) giving the switch states of the converter.

The direct bus voltage is given as:

$$V_{dc} = \int \frac{1}{C} i_c dt \tag{9}$$

$$i_c = i_{dc} - i_n \tag{10}$$

$$i_n = F_1 i_{n1} + F_2 i_{n2} + F_3 i_{n3} \tag{11}$$

where i_{n1}, i_{n2}, i_{n3} are the supplied currents and F_i (i = 1, 2, 3), the switch states achieved with the MHCC strategy as displayed in Fig. 2 in which the grid link control is summarized.

2.2 Flywheel Energy Storage System (FESS)

The storage system under investigation in this work is made up of an inertial flywheel that is mechanically coupled to an induction motor and wired to V_{dc} through a three-phase converter. The direct torque control approach was used to maintain V_{dc} at a constant level. The flywheel reference speed is determined by [1, 4, 10]:

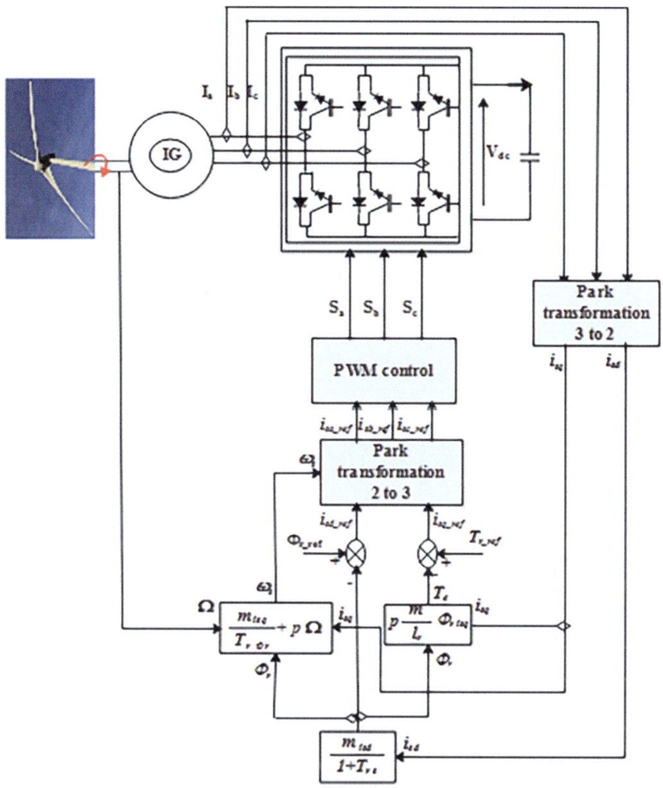

Fig. 2 The control of the generator-side converter

$$\Omega_{ref} = \sqrt{\frac{2.F_{c\ ref}}{J_F}}, E_{c-ref} = E_{ct_1} + \int_{t_1}^{t_2} P_{ref}\, dt \tag{12}$$

$$J_F = J_{machine} + J_{Flywheel} \tag{13}$$

The reference energy can be used to calculate the reference speed. The power assessment of the total system is given by [9]:

$$P_{ref} = Pl_{oad} - P - \Delta P \tag{14}$$

where ΔP is the power needed to maintain the DC voltage at a constant value, P is the wind power, and P_{ref} is the reference power.

2.3 Modeling and Control of the Inverter and the Grid Link

The inverter model is defined by the equation [4, 10–12]:

$$\begin{pmatrix} V_{in1} \\ V_{in2} \\ V_{in3} \end{pmatrix} = \frac{V_{dc}}{3} \begin{pmatrix} -2 & 1 & 1 \\ 1 & -2 & 1 \\ 1 & 1 & -2 \end{pmatrix} \begin{pmatrix} F_1 \\ F_2 \\ F_3 \end{pmatrix} \tag{15}$$

where V_{in1}, V_{in2}, V_{in3} are the phase voltages of the inverter.

The filter, which is composed of a resistance R_n and an inductance L_n, links the phase voltages of the grid V_{in1}, V_{in2}, V_{in3} to E_1, E_2, E_3, through the equation beneath:

$$\begin{pmatrix} V_{in1} \\ V_{in2} \\ V_{in3} \end{pmatrix} = R_n \begin{pmatrix} i_{n1} \\ i_{n2} \\ i_{n3} \end{pmatrix} + L_n \frac{d}{dt} \begin{pmatrix} i_{n1} \\ i_{n2} \\ i_{n3} \end{pmatrix} + \begin{pmatrix} E_1 \\ E_2 \\ E_3 \end{pmatrix} \tag{16}$$

The reactive power is adjusted to zero $\left(Q_{ref} = 0\right)$ and the supplied active one to its reference P_{ref} is given by [14, 15]:

$$P_{ref} = V_{dc}.\left(i_{dc} - i_{cref}\right) \tag{17}$$

$$P_{ref} = P_{dc} - P_{cref} \tag{18}$$

$$i_{cref} = PI.\left(V_{dc-ref} - V_{dc}\right) \tag{19}$$

P_{ref} and Q_{ref} are expressed in terms of Park components of the reference currents i_{nd-ref}, i_{nq-ref} by the equations below [16]:

$$P_{ref} = E_d i_{nd-ref} + E_q i_{nq-ref} \tag{20}$$

$$Q_{ref} = E_q i_{nd-ref} - E_d i_{nq-ref} \tag{21}$$

where E_d, E_q are the Park components of E_1, E_2, E_3.

By multiplying Eq. (20) by E_d and Eq. (21) by E_q, the two resulting equations are added and subtracted so as to find, the reference current expressions underneath [13, 16–19]:

$$\begin{aligned} i_{nd-ref} &= \frac{P_{ref} E_d + Q_{ref} E_q}{E_d^2 + E_q^2} \\ i_{nq-ref} &= \frac{P_{ref} E_q - Q_{ref} E_d}{V_d^2 + V_q^2} \end{aligned} \tag{22}$$

3 Simulation and Results

Simulations were run using Matlab/Simulink environment. The different results under a random variable wind velocity profile (Fig. 3) are presented in Figs. 4, 5 and 6 in the case of the PWM, HCC and MHCC strategies.

It is noticed that, under the MHCC strategy, the DC voltage bus is kept around its reference (465 V) with fewer oscillations than the other strategies, demonstrating the robustness of the suggested strategy. The power factor for the three strategies is around the unit for the PWM and the MHCC. It can be seen that under MHCC, the current is more sinusoidal than under HCC and looks like the one obtained under PWM.

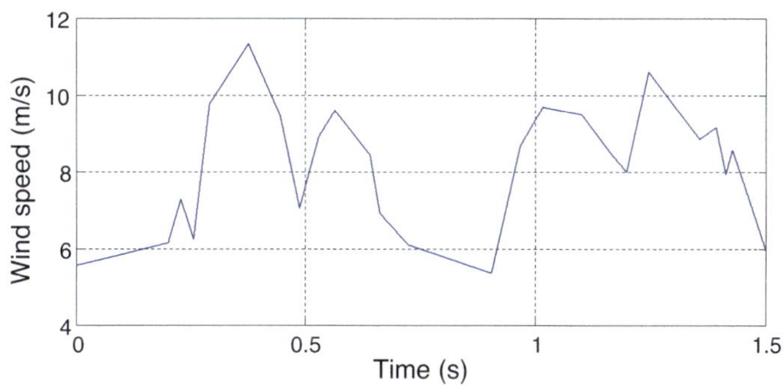

Fig. 3 Variable wind speed profile

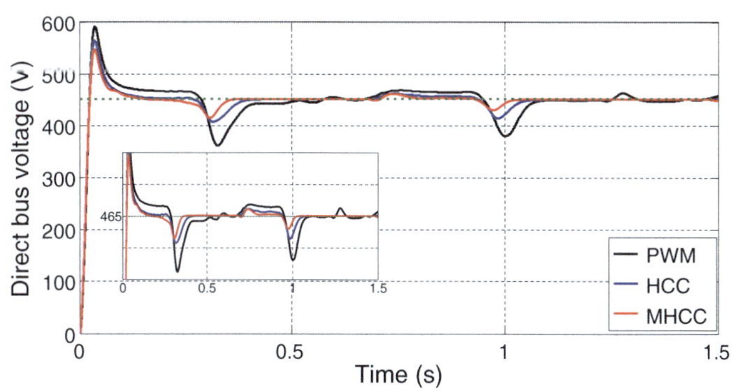

Fig. 4 Direct bus voltage

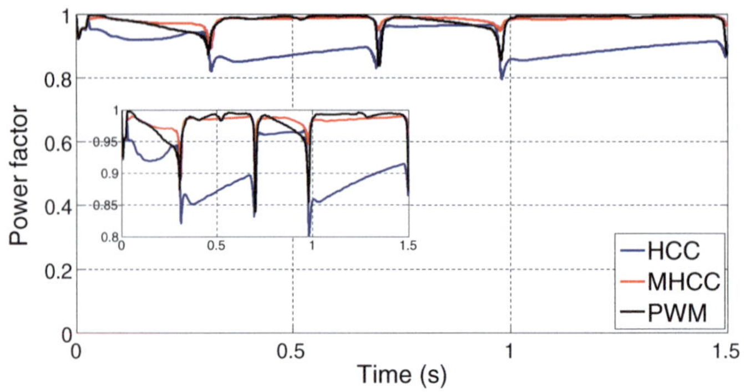

Fig. 5 Power factor

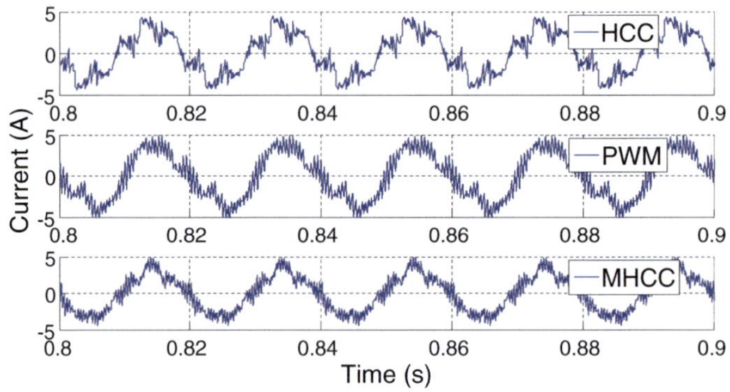

Fig. 6 Current sent to the grid

4 Conclusion

This work presents a study of an MHCC technique used for the grid-side converter in a wind conversion system. With the help of this technique, the conversion system is able to provide the grid with sinusoidal currents while maintaining a constant switching frequency, a unit power factor, and a phase voltage constant RMS. Simulated results have demonstrated the success of the MHCC approach in the investigated conversion system, and its simplicity in being applied.

References

1. Idjdarene K, Rekioua D, Rekioua T (2011) Tounzi A: wind energy conversion system associated to a flywheel energy storage system. Analog Integr Circ Sig Process 69:67–73
2. Kasprowicz AB, Husev O, Strzelecki R (2023) Induction generator with direct control and a limited number of measurements on the side of the converter connected to the power grid. Energies 16:63
3. Mebarki N, Rekioua T, Mokrani Z, Rekioua D (2015) Supervisor control for stand-alone photovoltaic/hydrogen/battery bank system to supply energy to an electric vehicle. Int J Hydrogen Energy 40(39):13777–13788
4. Rekioua D (2014) Wind power electric systems. Green Energy and Technology, London
5. Patel NA, Baria JC (2016) A hysteresis current control technique for electronics convertor. Int J Innov Res Sci Eng Technol 5(2):2203–2210
6. Chacko G, Syamala L, James N, Jos BM, Kallarackal M (2023) Switching frequency limited hysteresis based voltage mode control of single-phase voltage source inverters. Energies, 16
7. Idjdarene K, Rekioua D, Rekioua T, Tounzi A (2009) Direct torque control strategy for a variable speed wind energy conversion system associated to a flywheel energy storage system. In: Proceedings international conference on developments in eSystems engineering, DeSE, art. no. 5395083, pp 17–22
8. Rekioua D, Rekioua T, Idjdarene K (2005) Tounzi A: an approach for the modeling of an autonomous induction generator taking into account the saturation effect. Int J Emerg Electr Power Syst 4(1):1–18
9. Idjdarene K, Rekioua D, Rekioua T, Tounzi A (2007) Control strategies for an autonomous induction generator taking the saturation effect into account In: 2007 European conference on power electronics and applications EPE2007
10. Belaid S, Rekioua D, Oubelaid A, Ziane D, RekiouaT (2022) A power management control and optimization of a wind turbine with battery storage system. J Energy Storage 45:103613
11. Aissou R, Rekioua T, Rekioua D, Tounzi A (2014) Application of nonlinear predictive control for charging the battery using wind energy with permanent magnet synchronous generator. Int J Hydrogen Energy 41(45):20964–20973
12. Rekioua D, RekiouaT (2009) DSP-controlled direct torque control of induction machines based on modulated hysteresis control. In: 2009 international conference on microelectronics-ICM, pp 378–381
13. Rekioua D (2014) Control of wind turbine systems. In: Wind power electric systems. Green energy and technology. Springer, London
14. Ouari K, Rekioua T, Ouhrouche M (2014) Real time simulation of nonlinear generalized predictive control for wind energy conversion system with nonlinear observer. ISA Trans 53(1):76–84
15. Ziane H, Retif JM, Rekioua T (2008) Fixed-switching-frequency DTC control for PM synchronous machine with minimum torque ripples. Can J Electr Comput Eng 33(3/4):183–189
16. Hassani H, Zaouche F, Rekioua D, Belaid S, Rekioua T, Bacha S (2020) Feasibility of a standalone photovoltaic/battery system with hydrogen production. J Energy Storage 31:101644
17. Boutoubat M, Marmouh S, Hadjaissa A (2020) Grid connected WECS control for power production and power factor correction. In: 1st international conference on communications, control systems and signal processing (CCSSP), p 406
18. Mousa MG, Allam SM, Rashad EM (2017) A comparative study of vector-control strategies for maximum wind-power extraction of a grid-connected wind-driven brushless doubly-fed reluctance generator. Aust J Electr Electron Eng 14(1–2):1
19. Tamalouzt S, Benyahia N, Rekioua T, Rekioua D, Abdessemed R (2015) Wind turbine-DFIG/photovoltaic/fuel cell hybrid power sources system associated with hydrogen storage energy for micro-grid applications. In: 2015 3rd international renewable and sustainable energy conference (IRSEC), pp 1–6

Sliding Mode Control and Energy Management for Electrified Vehicles with Hybrid Energy Storage

Zahra Mokrani, Adel Oubelaid, Khoudir Kakouche, Djamila Rekioua, Nasser Eddine Mebarki, Toufik Rekioua, and Radia Abdelli

Abstract The growing interest in environmental sustainability and the permanent need for environmentally friendly transportation solutions have amplified the interest attributed to hybrid electric vehicles (HEVs) and accelerated their large commercialization. In this paper, the control and energy management of an electric vehicle powered by a fuel cell (FC) and supercapacitor (SC) is studied. The HEV is controlled using direct torque control and the vehicle speed is regulated using sliding mode controller (SMC). Simulation results yielded good vehicle performances and confirm the efficiency of the designed SMC controller. Moreover, robustness analysis was performed to evaluate HEV performance against possible variation in the parameters of its traction machine and the obtained results were satisfactory.

Keywords Hybrid electric vehicles · Fuel cell · Supercapacitor · Direct torque control · Sliding mode

1 Introduction

Hybrid electric vehicles (HEVs) have arisen as a significant technological innovation in the automobile industry, playing a novel and ecologically beneficial kind of transportation that is gaining popularity around the world [1, 2]. They are powered by electrical generators and provide a clean and sustainable alternative to traditional vehicles that use fossil fuels. Furthermore, HEVs, as symbols of the shift to more sustainable mobility, play an important role in mitigating climate change and safeguarding our planet for future generations [3–5].

Direct Torque Control (DTC) holds significant importance in the realm of electric vehicles. The DTC enables precise and responsive control of the motors by directly

Z. Mokrani (✉) · A. Oubelaid · K. Kakouche · D. Rekioua · N. E. Mebarki · T. Rekioua · R. Abdelli
Laboratoire de Technologie Industrielle et de l'Information, Faculté de Technologie, Université de Bejaia, 06000 Bejaia, Algeria
e-mail: zahra.mokrani@univ-bejaia.dz

© The Author(s), under exclusive license to Springer Nature Singapore Pte Ltd. 2024

A. Mellit et al. (eds.), *Proceedings of the 2nd International Conference on Green Energy Conversion System*, Green Energy and Technology,
https://doi.org/10.1007/978-981-97-6148-7_38

adjusting the torque produced, without the need for complex transmission systems. The incorporation of control mechanisms in HEV cars is critical for improving overall performance, efficiency, and safety [6–8]. Sliding mode control (SMC) holds significant importance when applied to direct torque control (DTC) in electric vehicles [8–12]. By implementing the SMC technique, the DTC system can enhance its ability to manage torque production of the electric motor, leading to improved performance and increased robustness [13, 14].

The control and energy management of a fuel cell-supercapacitor electric vehicle are the subject of this research. The energy management system employed in this study aims to harness the rapid charging and discharging capabilities of supercapacitors in order to mitigate the strain placed on fuel cells due to sudden power requirements. The HEV is controlled by a sliding mode controller with direct torque control. The simulation findings validate the efficacy of the provided energy management and sliding mode control solutions based on direct control.

2 HEV Modeling

2.1 Modeling of the Induction Motor

The induction motor is modeled by the following electrical and mechanical equations [1, 7, 15, 16]:

$$[\mathrm{V}_{\mathrm{abcs}}] = \mathrm{R}_{\mathrm{s}}[i_{abcs}] + \frac{d}{dt}[\varphi_{abcs}] \tag{1}$$

$$\mathrm{T}_{\mathrm{em}} = \frac{3\mathrm{PM}}{2}\left(\Phi_{s\beta} I_{s\alpha} - \Phi_{s\alpha} I_{s\beta}\right) \tag{2}$$

2.2 Electric Vehicle Model

The following equation governs the motion of the vehicle depicted in Fig. 1 [17–19]:

$$\sum \overrightarrow{F_{ext}} = \overrightarrow{F_{trac}} - \overrightarrow{F_r} \tag{3}$$

where $\overrightarrow{F_{ext}}$ represents the net external force acting on the vehicle body and $\overrightarrow{F_r}$ is the total resistive force.

The total resistive force is calculated as follow:

$$\overrightarrow{\mathrm{F_r}} = \overrightarrow{\mathrm{F_{aero}}} + \overrightarrow{\mathrm{F_{slope}}} + \overrightarrow{\mathrm{F_{rol}}} \tag{4}$$

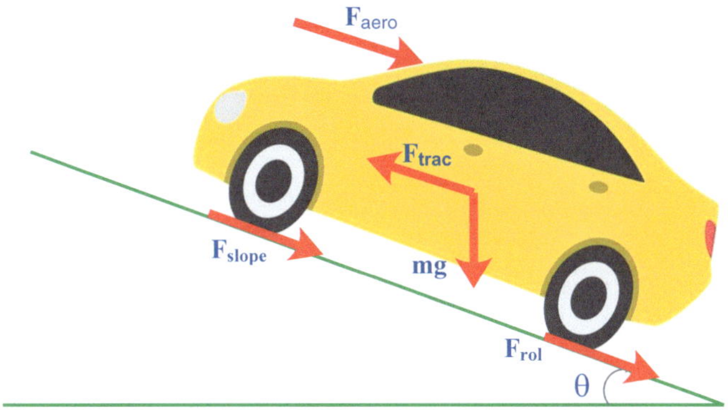

Fig. 1 The different forces applied to the vehicle

2.3 Fuel Cells

The PEM fuel cell's polarization curve is given by the Eq. (5), where $E_{Nernest}$ is the theoretical open-circuit voltage, V_{act} is the activation overvoltage, V_{ohm} is the ohmic overvoltage and V_{con} is the concentration surge voltage [1–17, 20–22]:

$$V_{PEMFC} = E_{Nernest} - V_{act} - V_{ohm} - V_{con} \tag{5}$$

2.4 Supercapacitors

The supercapacitor model is based on the two RC branches and is given by the equation below [1, 20, 23, 24]:

$$U_{sc} = N_{s\_sc} V_{sc} = N_{S\_sc}(V_1 + R_1 i_{sc}) = N_{s\_sc}\left(V_1 + R_1 \frac{I_{sc}}{N_{p\_sc}}\right) \tag{6}$$

3 SMC Design

Taking the speed derivative with respect to time using de conventional dot representation yields Eq. (7) shown below:

$$\dot{\Omega} = \left[(\mathrm{T_{em}} - \mathrm{T_r}) - f\Omega\right]/J \tag{7}$$

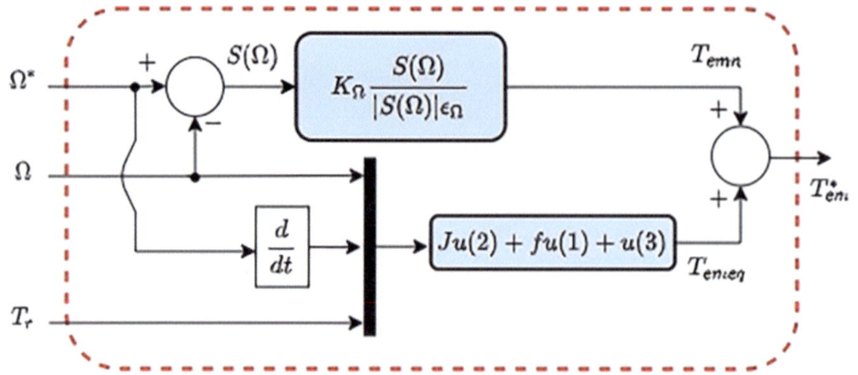

Fig. 2 Sliding mode controller scheme

The sliding surface and its derivative are given by Eq. (8)

$$\begin{cases} S(\Omega) = \Omega^* - \Omega \\ S(\dot{\Omega}) = \dot{\Omega}^* - \dot{\Omega} \end{cases} \tag{8}$$

We set the torque command expression as follow:

$$T^*_{em} = T_{emn} + T_{emeq} \tag{9}$$

The equivalent torque reference is highlighted in Eq. (10)

$$T_{emeq} = J\dot{\Omega}^* + T_{emn} + T_{emeq} + f\Omega \tag{10}$$

The simplified control law is given by Eq. (11) illustrated below:

$$T_{emn} = K_{wr} \frac{S(\Omega)}{|S(\Omega)| + \varepsilon_\Omega} \tag{11}$$

The overall scheme of the sliding mode controller is depicted in Fig. 2.

4 Simulation and Results

4.1 Driving Cycles

The simulation findings are derived from the global system represented in Fig. 3.

Figures 4, 5, and 6 depict the IM's properties. Motor currents are sinusoidal, and stator flux is kept constant.

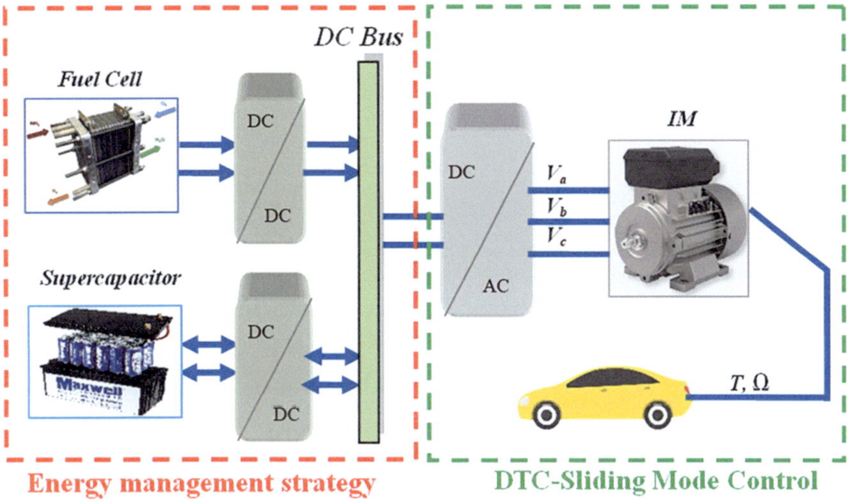

Fig. 3 HEV with DTC-sliding mode controller scheme

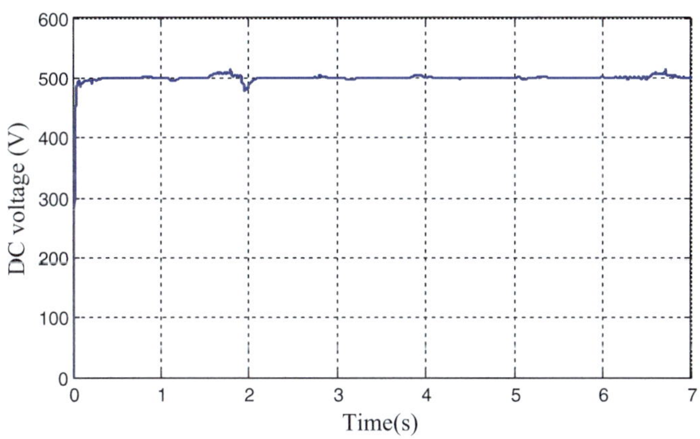

Fig. 4 DC bus voltage

The rotor speed is regulated to follow the reference velocity profile as show in Figs. 7 and 8 depicts that the electromagnetic torque achieves the load torque during the operation.

The proposed power management algorithm ensures continuous HEV power delivery, while the use of supercapacitors alleviates fuel cell stress during abrupt power demand variations as shown in Fig. 9 and the state of charge of the supercapacitors is kept between 90 and 30% (Fig. 10).

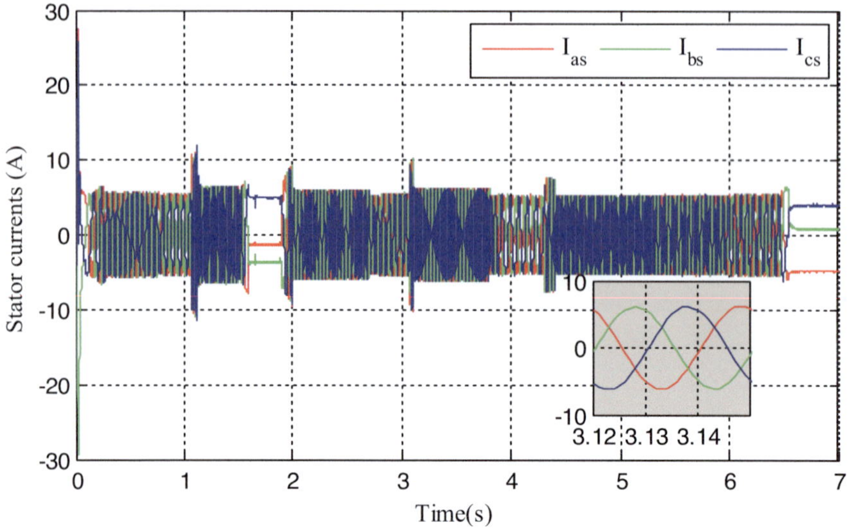

Fig. 5 Stator current waveforms

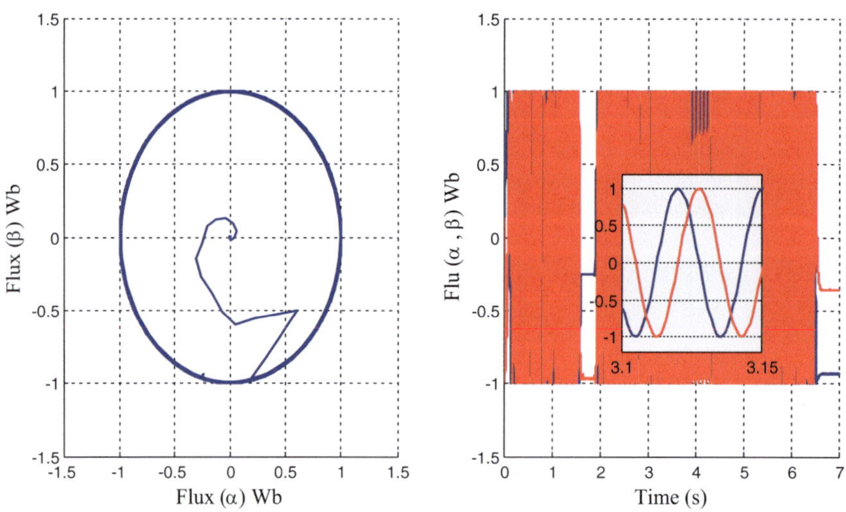

Fig. 6 Magnetic characteristics of the induction motor

4.2 Robustness Analysis

The effect of IM stator resistance fluctuation on HEV response is demonstrated in Fig. 11, where the HEV response is always stable with zero steady-state error, indicating that the HEV response is robust.

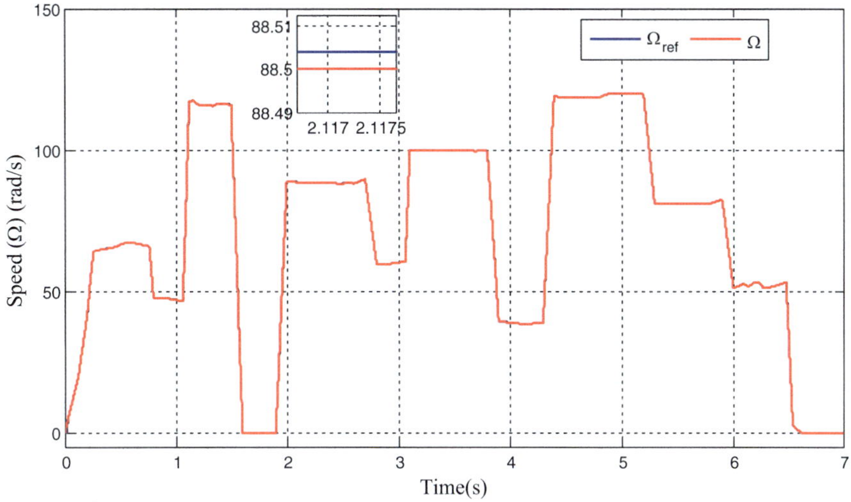

Fig. 7 Rotor speed waveform

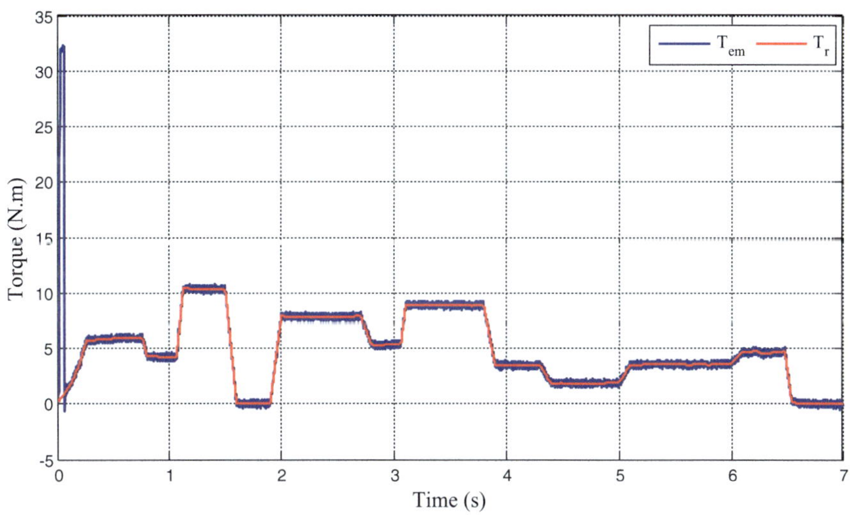

Fig. 8 Electromagnetic torque waveform

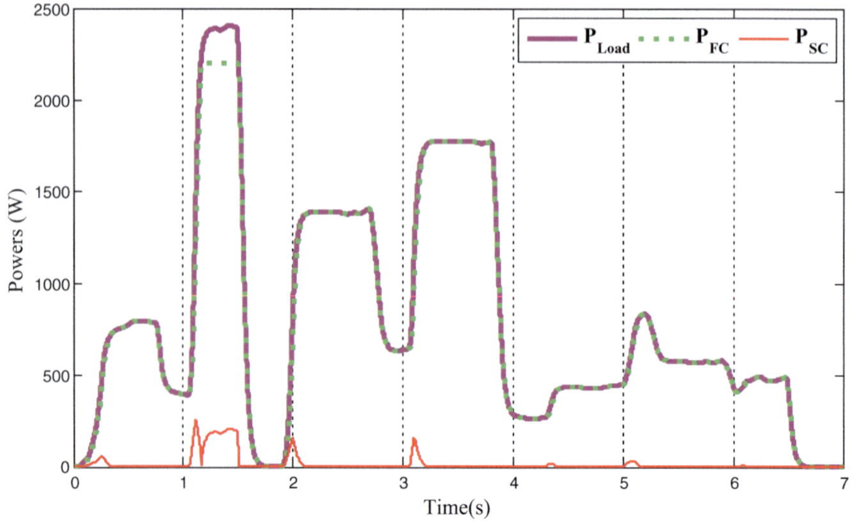

Fig. 9 Different powers

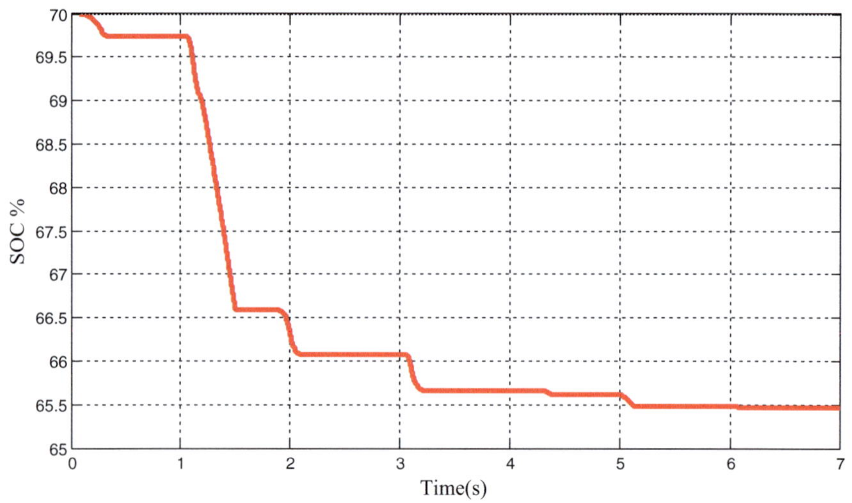

Fig. 10 Supercapacitor state of charge waveform

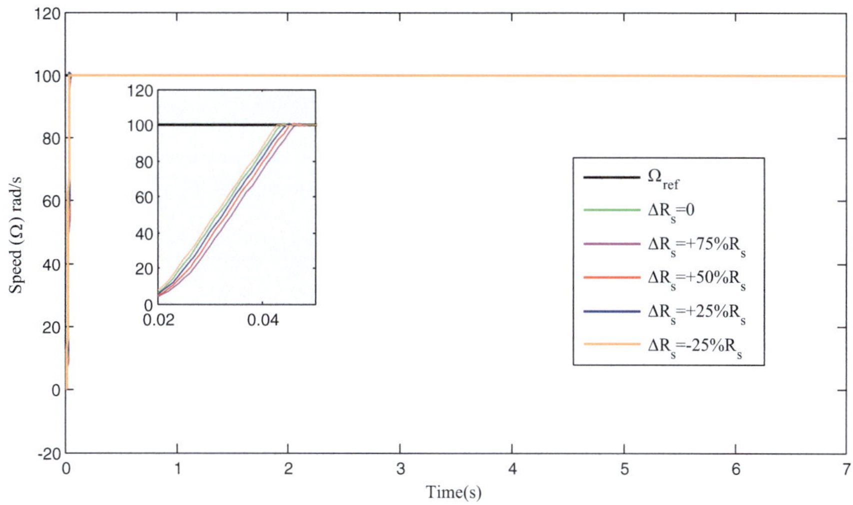

Fig. 11 Effect of variation of the stator resistance of the IM on HEV speed response

5 Conclusion

In this paper work, the control and energy management of a hybrid electric vehicle powered by a fuel cell and supercapacitor have been investigated. Through the application of direct torque control and a sliding mode controller, the HEV performance has been extensively enhanced. The simulation results confirmed the effectiveness of the designed SMC controller. Furthermore, the developed energy management has balanced the used of vehicle power sources. The conducted robustness analysis showed that the traction machine parameter variation doesn't affect the overall vehicle performance

References

1. Mokrani Z, Mebarki N, Oubelaid A, Kakouche K, Mohammedi A, Rekioua D, Rekioua T (2023) Direct torque control and power management strategy dedicated for fuel cell/supercapacitor electric vehicles. In: The 3rd conference on electrical engineering, Algiers
2. Oubelaid A, Taib N, Rekioua T (2022) Novel coordinated power sources switching strategy for transient performance enhancement of hybrid electric vehicles. COMPEL Int J Comput Math Electr Electron Eng 41(5):1880–1919
3. Gharibeh HF, Yazdankhah AS, Azizian MR (2020) Energy management of fuel cell electric vehicles based on working condition identification of energy storage systems, vehicle driving performance, and dynamic power factor. J Energy Storage, 31
4. da Silva SF, Eckert JJ, Silva FL, Corrêa FC, Silva LCA, Bueno AV, Dedini FG (2023) Aging-aware optimal power management control and component sizing of a fuel cell hybrid electric vehicle powertrain. Energy Convers Manag, 292

5. Lu Z, Zhang X (2022) Composite non-linear control of hybrid energy-storage system in electric vehicle. Energies 15:1567. https://doi.org/10.3390/en15041567
6. Rekioua D, Rekioua T (2009) DSP-controlled direct torque control of induction machines based on modulated hysteresis control. In: Proceedings of the international conference on microelectronics, ICM, art. no. 5418603, pp 378–381
7. Mokrani Z, Rekioua D, Rekioua T (2014) Modeling, control and power management of hybrid photovoltaic fuel cells with battery bank supplying electric vehicle. Int J Hydrogen Energy 39(27):15178–15187
8. Kakouche K, Rekioua T, Mezani S, Oubelaid A, Rekioua D, Blazek V, Prokop L, Misak S, Bajaj M, Ghoneim SSM (2022) Model predictive direct torque control and fuzzy logic energy management for multi power source electric vehicles. Sensors 22:5669
9. Ur Rahman A, Zehra SS, Ahmad I, Armghan H (2021) Fuzzy supertwisting sliding mode-based energy management and control of hybrid energy storage system in electric vehicle considering fuel economy. J Energy Storage 37:102468
10. Shtessel Y, Taleb M, Plestan F (2012) A novel adaptive-gain supertwisting sliding mode controller: methodology and application. Automatica 48(5):759–769
11. Shtessel YB, Tournes CH, Fridman L (2012) Advances in guidance and control of aerospace vehicles using sliding mode control and observation techniques. J Franklin Inst 349(2):391–740
12. Hassani H, Zaouche F, Rekioua D, Belaid S, Bacha B (2020) Feasibility of a standalone photovoltaic/battery system with hydrogen production. J Energy Storage 31:101–644
13. Tran M-T, Lee D-H, Chakir S, Kim Y-B (2021) A novel adaptive super-twisting sliding mode control scheme with time-delay estimation for a single ducted-fan unmanned aerial vehicle. Actuators 10:54
14. Ghanes M, Zheng G (2009) On sensorless induction motor drives: sliding-mode observer and output feedback controller. IEEE Trans Industr Electron 56(9):3404–3413
15. Rekioua D, Bensmail S, Bettar N (2014) Development of hybrid photovoltaic-fuel cell system for stand-alone application. Int J Hydrogen Energy 39(3):1604–1611
16. Djerioui A, Houari A, Machmoum M, Ghanes M (2020) Grey wolf optimizer-based predictive torque control for electric buses applications. Energies 13(19):5013
17. Oubelaid A, Taib N, Rekioua T, Bajaj M, Yadav A, Shouran M, Kamel S (2022) Secure power management strategy for direct torque controlled fuel cell/supercapacitor electric vehicles. Front Energy Res 10:971357
18. Dahmane Y, Chenouard R, Ghanes M, Alvarado Ruiz M (2023) Optimal electric vehicle fleet charging management with a frequency regulation service. World Electr Veh J 14(6):152
19. Al Attar H, Hamida MA, Ghanes M, Taleb M (2023) Review on modeling and control strategies of DC–DC LLC converters for bidirectional electric vehicle charger applications. Energies 16(9):3946
20. Oubelaid A, Taib N, Rekioua T, Bajaj M, Blazek V, Prokop L, Misak S, Ghoneim SSM (2022) Multi source electric vehicles: smooth transition algorithm for transient ripple minimization. Sensors 22(18):6772
21. Mebarki N, Rekioua T, Mokrani Z, Rekioua D (2015) Supervisor control for stand-alone photovoltaic/hydrogen/battery bank system to supply energy to an electric vehicle. Int J Hydrog Energy 40(39):13777–13788
22. Rekioua D (2018) Advances in renewable energies and power technologies, vol 1, pp 349–369
23. Zubieta L, Bonert R (2000) Characterization of double-layer capacitors (DLCs) for power electronics applications. IEEE Trans Ind Appl 36(1):199–205
24. Oubelaid A, Alharbi H, Humayd ASB, Taib N, Rekioua T, Ghoneim SSM (2022) Fuzzy-energy-management-based intelligent direct torque control for a battery—supercapacitor electric vehicle. Sustainability 14(14):8407

Generalization of Optimal Modulation Control for 3φ VSI Through Solving of Transcendent Equations

Henda Bouyahi and Adel Khedher

Abstract Pulse width modulation (PWM) techniques are primordial to high performance and great efficiency of inverters. The Selective Harmonic Elimination (SHE) technique determines in advance the switching angles, by taking into account the desired output voltage amplitude and harmonics to be eliminated. Thus, resolution of non-linear equation systems to set the switching angles poses crucial problem for the initial values choice. So, to remedy this problem, we propose in this work to formulate initial conditions by linear mathematical approximations, in order to approximate them to exact values that define pulse-width modulation. This formulation allows gain on the one hand the harmonics eliminated number (N) which will increase and on the other hand the computation time, which is often important. In this paper, we present transcendental equations resolution for switching angles (even or odd) and for two types of control patterns, namely: $\theta_N < 60°$ and $\theta_N < 90°$. In order to validate our study, we demonstrated switching angles variation law as a function of modulation index using MATLAB simulation.

Keywords Voltage source inverter · SHEPWM · Switching angles · Initial values · Harmonics eliminated

1 Introduction

In recent years, use of three-phase Voltage Source Inverters VSI in distributed power systems DGS and renewable energy applications has increased considerably. The flexibility of these power conversion devices makes them potential solution for many industrial applications such as variable speed drives. Indeed, VSI are applied to drive

H. Bouyahi · A. Khedher (✉)
LATIS-Laboratory of Advanced Technology and Intelligent Systems, Université de Sousse, Ecole Nationale d'Ingénieurs de Sousse, 4023 Sousse, Tunisia
e-mail: adel_kheder@yahoo.fr

H. Bouyahi
e-mail: bouyahihenda@gmail.com

© The Author(s), under exclusive license to Springer Nature Singapore Pte Ltd. 2024
A. Mellit et al. (eds.), *Proceedings of the 2nd International Conference on Green Energy Conversion System*, Green Energy and Technology,
https://doi.org/10.1007/978-981-97-6148-7_39

AC motor applications requiring high performance and efficiency. The absence of harmonics multiples three in lines quantities justifies the inverter advantage. Presence of superior order harmonics produces torque pulsations in some electric drive applications [1–3].

Nowadays, pulse-width modulation control technology is the most dominant approach to inverter control. Therefore, several techniques have been proposed to attenuate the harmonics presented in PWM inverter output waveform. Optimal Selective Harmonic Elimination (SHE) has become very efficient for inverter control to improve their output voltages quality their output voltages for low switching loss [4–6]. In addition, overvoltages, common-mode voltage and electromagnetic interference (EMI) problems caused by PWM switching can also be reduced [7–10]. Optimal modulation, otherwise known as pre-calculated or SHEPWM, consists to calculate the switching angles (switching sequences) of inverter switches in advance, in order to eliminate certain undesirable harmonics at specified order. The harmonic spectrum is closely controlled over a specified (desired) band frequency, and lower order harmonics are either minimized or eliminated. Subsequently, the switching angles are memorized and restored cyclically to ensure control. In fact, the PWM control can be implemented on a specific integrated circuit [11–14].

The angles real time optimisation requires resolution of non-linear transcendental equation system including trigonometric terms [15]. This set of non-linear equations can be solved by iterative numerical techniques such as Newton-Raphson method. Until now, the method difficulty resides in the judicious choice of initial conditions, the complexity which increases with harmonics number eliminated (N). When N gets important, the angles shrink and the initial conditions choice becomes difficult. To surmount this problem, which constitutes major difficulty associated to this modulation technique, a generalized formula of these initial conditions by linear mathematical approximations for approximate real (exact) values of angles has been proposed and demonstrated in this work. This formulation offers gain of the harmonics number eliminated, which will increase, and in computation time, which is often significant. Consequently, the proposed formulation offers higher bandwidth of converter and therefore better harmonic performances. This study is developed to optimise the switching angles for two-level three-phase inverter for any number of angles (even or odd) and for two types control patterns namely: $\theta_N < 60°$ and $\theta_N < 90°$. Simulation MATLAB is realized to determine the switching angles variation law as a function of modulation index in Sect. 2. In addition, simulation results for inverter and induction machine are shown and discussed in Sect. 3. Finally, some conclusions are given in Sect. 4.

2 Optimisation of Initial Conditions

2.1 SHEPWM Technique

Figure 1 depicts the output voltage waveform of SHEPWM technique. It is characterized by the harmonics number N per quarter period and switching angles θ_N [13, 15]. Generally, there are N switching angles, noted as $\theta_1, \theta_2, \ldots, \theta_N$ such that:

$$0 < \theta_1 < \theta_2 < \ldots < \theta_N < \frac{\pi}{2} \tag{1}$$

In order to calculate the switching angles, it's absolutely necessary to solve the following transcendent equation system (2):

$$\begin{cases} 1/2 - \cos(\theta_1) + \cos(\theta_2) - \cos(\theta_3) + \cos(\theta_4) + \cdots + \cos(\theta_N) - \frac{\pi}{8}\mathrm{Im} = 0 \\ 1/2 - \cos(5\theta_1) + \cos(5\theta_2) - \cos(5\theta_3) + \cos(5\theta_4) + \cdots + \cos(5\theta_N) = 0 \\ 1/2 - \cos(7\theta_1) + \cos(7\theta_2) - \cos(7\theta_3) + \cos(7\theta_4) + \cdots + \cos(7\theta_N) = 0 \\ \vdots \\ 1/2 - \cos((3i-1)\theta_1) + \cos((3i-1)\theta_2) - \cdots + \cos((3i-1)\theta_N) = 0 \text{ if i is even} \\ 1/2 - \cos((3i-2)\theta_1) + \cos((3i-2)\theta_2) - \cdots + \cos((3i-2)\theta_N) = 0 \text{ if i is odd} \end{cases} \tag{2}$$

To solve the non-linear equation system $f_i(\theta_1, \theta_2, \ldots, \theta_N) = 0$, when (i = 1, 2,…, N), we apply the Newton–Raphson algorithm shown in Fig. 2. The convergence condition will be reached when the value of all functions that appear in equation system (2) is smaller than desired precision value ε. A solution ($f(\theta) = 0$) can be achieved if the method converges. Otherwise, in the case of divergence versus the initial values $[\theta_1^0, \theta_2^0, \ldots, \theta_N^0]$, it is necessary to proceed using new initial value set. The correct solution must satisfy the condition represented by expression (1).

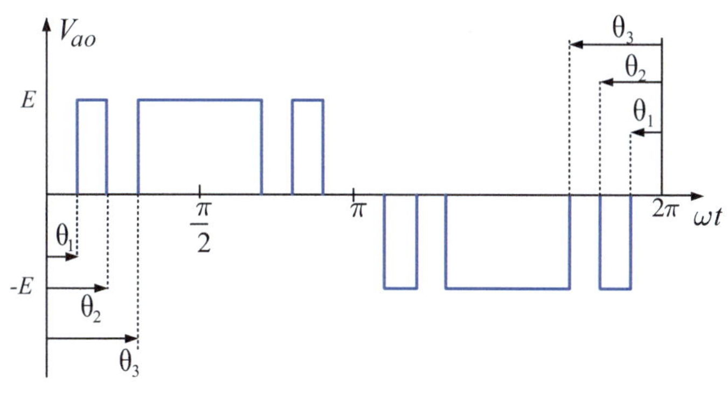

Fig. 1 Output voltage waveform for SHEPWM

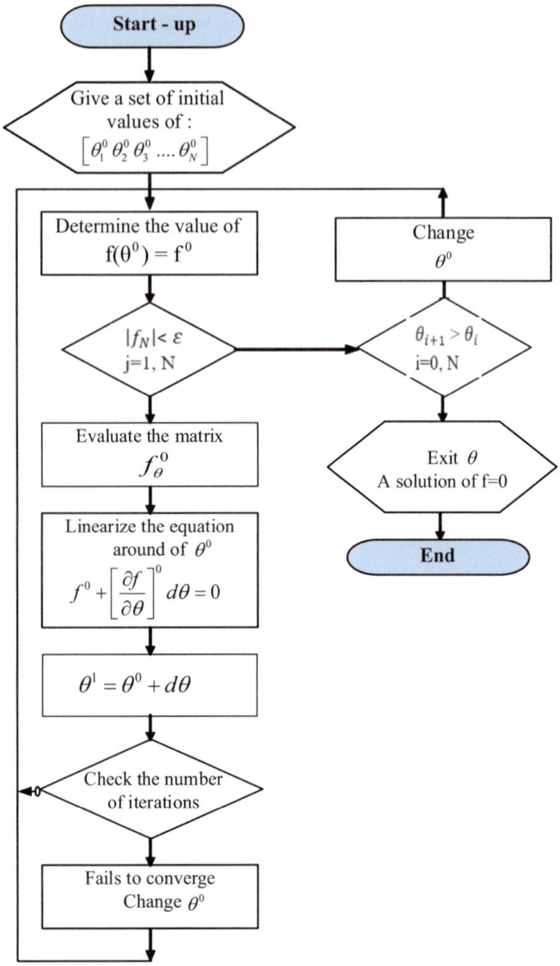

Fig. 2 Flowchart for calculating the switching angles

Where, $f = [f_1 f_2 \ldots f_N]$, $\theta = [\theta_1\, \theta_2\, \ldots\, \theta_N]$, $\theta^0 = [\theta_1^0\, \theta_2^0\, \ldots \theta_N^0]\left[\frac{\partial f}{\partial \theta}\right]^0 =$

$$\begin{bmatrix} \frac{\partial f_1}{\partial \theta_1} & \cdots & \frac{\partial f_1}{\partial \theta_N} \\ \vdots & \ddots & \vdots \\ \frac{\partial f_N}{\partial \theta_1} & \cdots & \frac{\partial f_N}{\partial \theta_N} \end{bmatrix} \left[\frac{\partial f}{\partial \theta}\right]^0 = \begin{bmatrix} \frac{\partial f_1}{\partial \theta_1} & \cdots & \frac{\partial f_1}{\partial \theta_N} \\ \vdots & \ddots & \vdots \\ \frac{\partial f_N}{\partial \theta_1} & \cdots & \frac{\partial f_N}{\partial \theta_N} \end{bmatrix}.$$

2.2 Elimination of an Odd Harmonics Number

In order to find the switching angles variation law θ_i as a function of modulation index Im, we present cases relative to N = 5, 7, 13, 21 for $\theta_N < 60°$ (Fig. 3) and

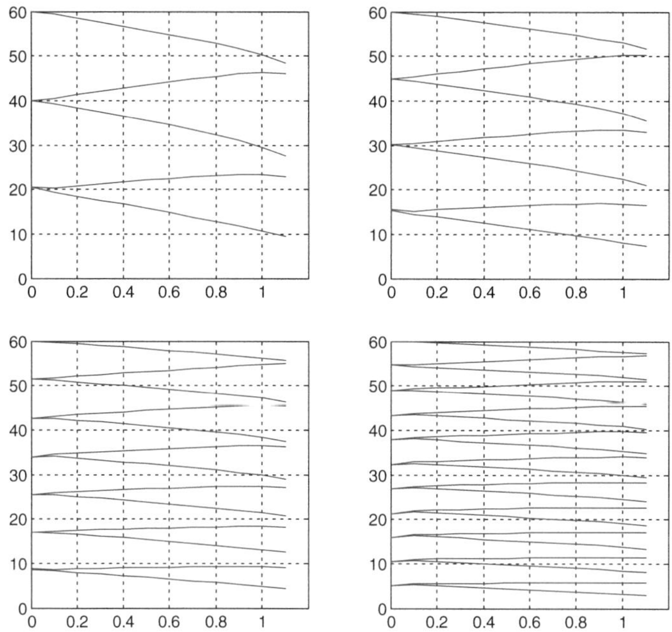

Fig. 3 Switching angles variation as a function of modulation index. For $\theta_N < 60°$, N = 5, 7, 13, 21

$\theta_N < 90°$ (Fig. 4). Referring to these results, it can be seen that switching angles varies almost linearly with Im. For Im = 0, the switching angles values θ_i and θ_{i+1} are equidistant and parallel. The slope value depends only on pulse number, not on the control type or the angle order. For $\theta_N < 60°$, the order angles (i and i + 1) varies with opposite sign slopes, as shown in Fig. 3. However, for $\theta_N < 90°$, only the first two angles θ_1 and θ_2 vary with a positive slope, while the other order angles (i and i + 1) vary with opposite sign slopes.

- **For control** $\theta_N < 60°$

$$\theta_i = p_p I_m + \theta_{i0}, \quad i = 2, 4, 6 \ldots \tag{3}$$

$$\theta_i = p_n I_m + \theta_{i0}, \quad i = 1, 3, 5, 7 \ldots \tag{4}$$

- **For control** $\theta_N < 90°$

$$\theta_i = p_p I_m + \theta_{i0}, \quad i = 1, 2, 4, 6 \ldots \tag{5}$$

$$\theta_i = p_n I_m + \theta_{i0}, \quad i = 3, 5, 7 \ldots \tag{6}$$

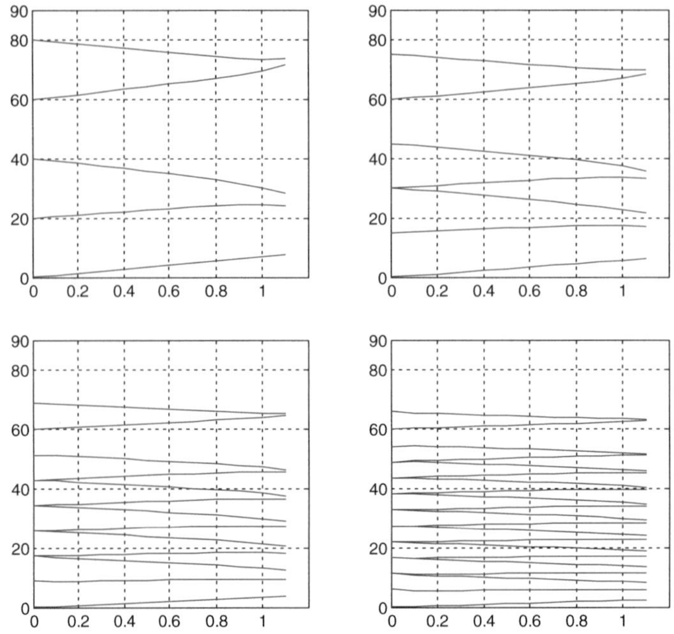

Fig. 4 Switching angles variation as a function of modulation index. For $\theta_N < 90°$, N = 5, 7, 13, 21

The variation of negative and positive slopes and angles at origin (Im = 0) as a function of N is shown in Fig. 5. The slope is estimated numerically using an exponential function, thus the negative slope is plotted as an absolute value. For the $\theta_N < 60°$, the significant angle is θ_{10}, however, for the control $\theta_N < 90°$ is θ_{20}, because the angle θ_{10} is zero.

The initial values of switching angles for the two control are estimated as follows:

- **For control** $\theta_N < 60°$

$$\begin{aligned} &\theta_{N0} = 60° \\ &\theta_{10} = \frac{120}{(N+1)} \\ &\theta_{i0} = \theta_{(i+1)0} = \frac{60(i+1)}{(N+1)} \quad \text{For } i = 1, 3, \ldots N-2 \end{aligned} \tag{7}$$

- **For control** $\theta_N < 90°$

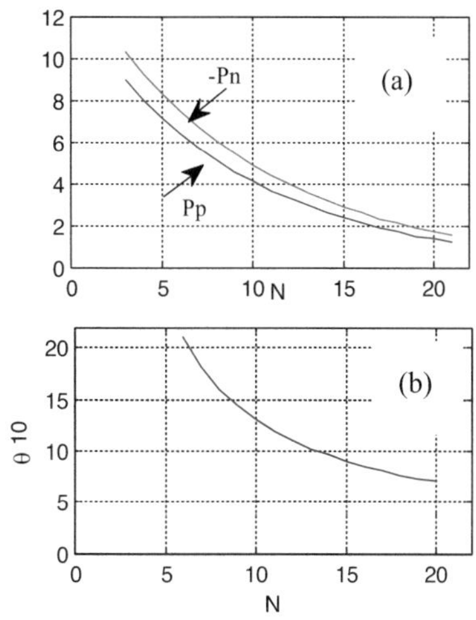

Fig. 5 Variation of N as a function: **a** Positive and negative slopes and **b** θ_{10}

$$\begin{aligned}
&\theta_{10} = 0^\circ \\
&\theta_{20} = \frac{120}{(N+1)} \\
&\theta_{(N-2)0} = \frac{60(N+1)}{(N+1)} \\
&\theta_{(N-1)0} = 60^\circ \\
&\theta_{N0} = \frac{60(N+3)}{(N+1)} \\
&\theta_i 0 = \theta_{(i+1)0} = \frac{60(i+1)}{(N+1)} \quad \text{For } i = 1, 3, 5 \ldots, N-4
\end{aligned} \tag{8}$$

2.3 *Elimination of an Even Harmonics Number*

Figures 6 and 7 present the results for N = 4, 8, 12, 20 respectively for $\theta_N < 60^\circ$ and $\theta_N < 90^\circ$. It can be seen that for $\theta_N < 60^\circ$, the slopes of two adjacent angles θ_i and θ_{i+1} are opposite in sign. Furthermore, for Im = 0, the angles 0° and 60° are respectively θ_{10} and θ_{N0} regardless of switching angles number. Thus, the angles θ_i and θ_{i+2} are parallel and equidistant at the origin. In addition, for $\theta_N < 90^\circ$, the slopes of angles θ_{10} and θ_{20} are the same sign, then the slopes of angles $\theta_{i0} = \theta_{(i+1)0}$ are the opposite sign. Moreover, the variation slopes do not assume constant values for well determined number of angles N; they vary according to the order of these angles.

Following these remarks, we can write the following equations:

$$\theta_i = p_p I_m + \theta_{i0}, \quad i = 1, 3, \ldots, N-1 \tag{9}$$

$$\theta_i = -p_p I_m + \theta_{i0}, \quad i = 2, 4, \ldots, N \tag{10}$$

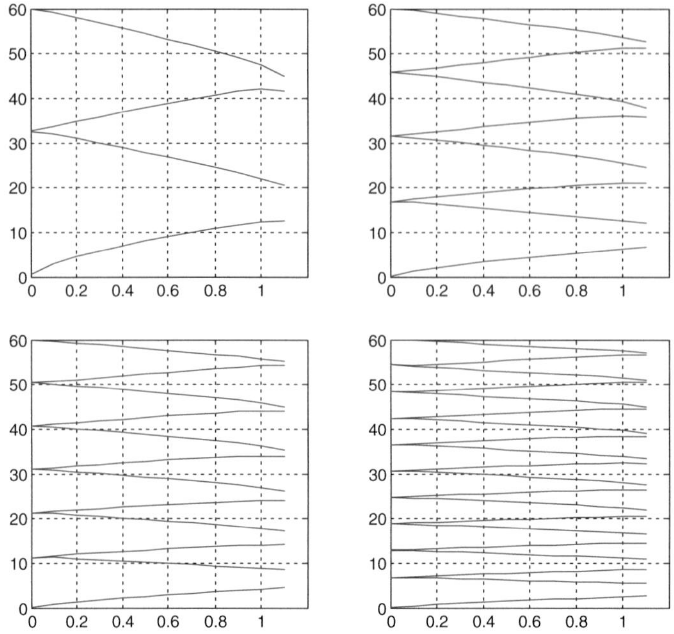

Fig. 6 Switching angles variation as a function of modulation index. For $\theta_N < 60°$, N = 4, 8, 12, 20

The curves of the positive slope p_p and the significant angles θ_{10} and θ_{20} are given by the following Fig. 8.

The initial values of the switching angles for the two controls are given by the following relationships:

- **For control** $\theta_N < 60°$

$$\begin{aligned} &\theta_{10} = 0° \\ &\theta_{i0} = \theta_{(i+1)0} = \frac{60(i-2)}{(N-2)} \quad \text{For } i = 2, 4, \ldots N-2 \\ &\theta_{N0} = 60° \end{aligned} \tag{11}$$

- **For control** $\theta_N < 90°$

$$\begin{aligned} &\theta_{10} = \frac{120}{(N+2)} \\ &\theta_{i0} = \theta_{(i+1)0} = \frac{60(i+2)}{(N+2)} \quad \text{For } i = 2, 4, 6\ldots, N-4 \\ &\theta_{(N-2)0} = 60\frac{N}{(N+2)} \\ &\theta_{(N-1)0} = 60° \\ &\theta_{N0} = \frac{60(N+4)}{(N+2)} \end{aligned} \tag{12}$$

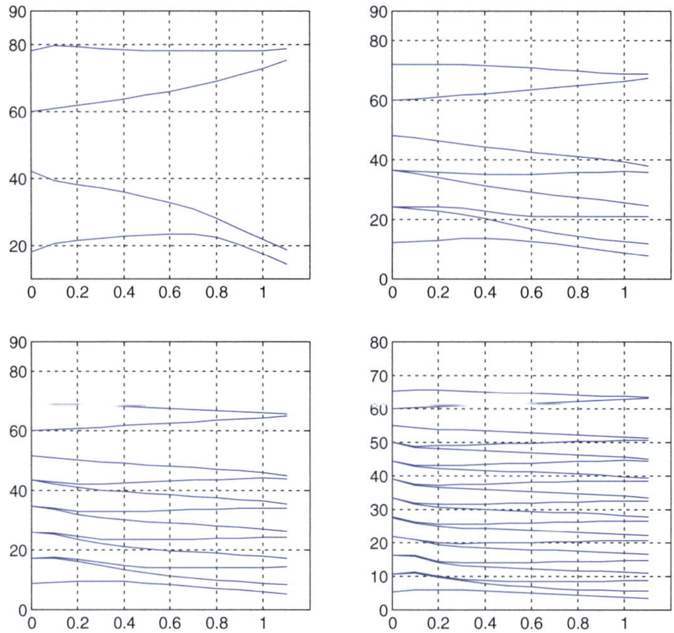

Fig. 7 Switching angles variation as a function of modulation index. For $\theta_N < 90°$, N = 4, 8, 12, 20

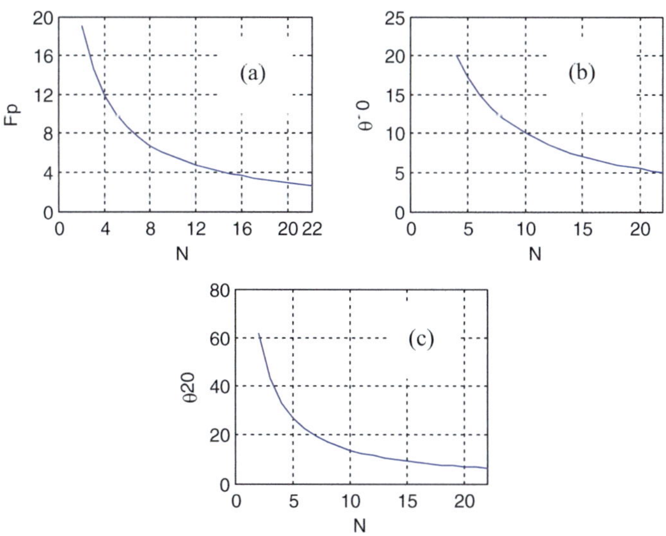

Fig. 8 Variation of N as a function: **a** positive slope p_p, **b** θ_{10} and **c** θ_{20} θ_{20}

3 Simulation Results

A numerical simulation model is developed and implemented on MATLAB/Simulink. The simulation results allow to study the characteristics of IM fed by a VSI two-level controlled by SHEPWM technique for $\theta_N < 90°$, N = 21 and Im = 1. The parameters of machine are shown in Table 1.

The stator current, speed and electromagnetic torque waveforms are depicted in Fig. 9. Referring to these results, we can see that the stator currents exhibit a peak of 6.7A at start-up. The machine speed oscillates for the first few moments, and then stabilizes around the nominal value.

In Fig. 10, we plot the variation of the first four significant harmonics as a function of modulation index Im relative to N odd. Then, for N = 21, the harmonics eliminated are of orders (5, 7, 11, 13, 17, 19, 23, 25, 29, 31, 35, 37, 41, 43, 47, 49, 53, 55, 59

Table 1 Parameters machine

Characteristic	Value
Power	375 W
Rotation speed	2850 tr/min
Stator resistance Rs	24 Ω
Stator inductance Ls	0.66 H
Rotor resistance Rr	10.88 Ω
Rotor inductance Lr	0.66 H
Mutual inductance Lm	0.63 H

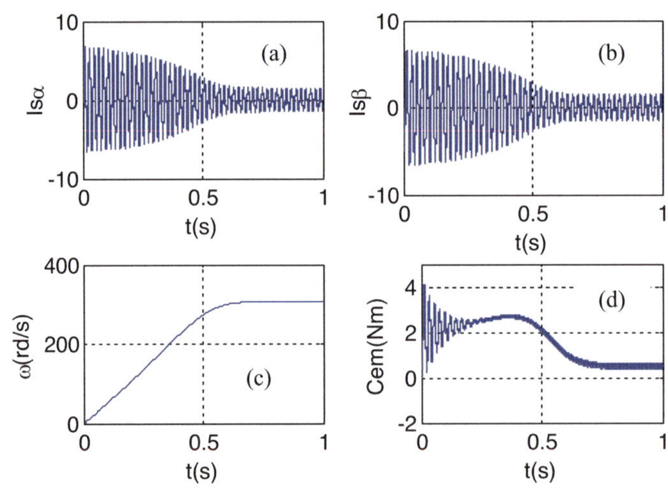

Fig. 9 Waveforms of: **a** $I_{s\alpha}$ stator current, **b** $I_{s\alpha}$ stator current **c** ω speed and **d** Cem, torque

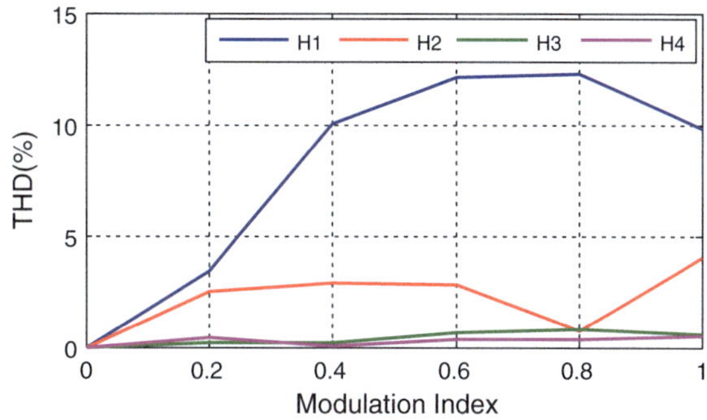

Fig. 10 Variation of the first four significant harmonics as a function of modulation index

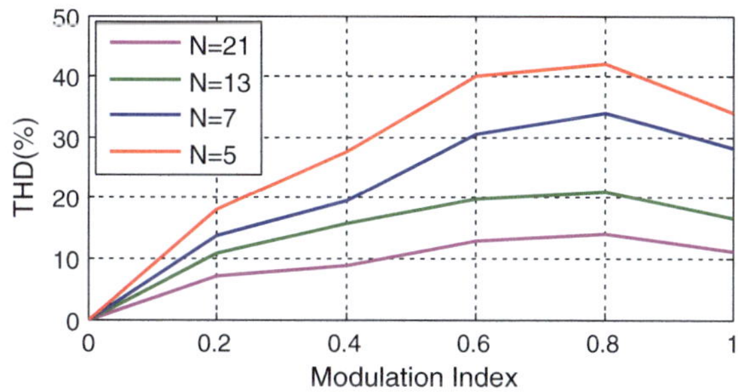

Fig. 11 THD variation as a function of modulation index for $\theta_N < 90°$ and N = 5, 7, 13, 21

and 61), the first four significant harmonics are of orders (H1 = 65, H2 = 67, H3 = 71 and H4 = 73).

The performance of the optimal control is also compared in terms of current harmonic distortion THD as a function of modulation index. For control $\theta_N < 90°$, and N = 5, 7, 13, 21, Fig. 11 shows the THD level of different values N as a function of the modulation index. It is clearly observed that the THD level decreases as the number of switching angles increases. Thus, when Im = 0.8, the THD value is high for all values of N. However, for Im = 1, the THD is low.

4 Conclusion

The main objective of this paper is to elaborate a general formulation of SHEPWM control of three-phase inverters. The proposed formulation ensures high bandwidth of VSI and therefore better harmonic performance. This study is carried out for any number of eliminated harmonics. In addition, the relations for obtaining initial values, which is the major difficulty associated with this modulation technique for solving the Newton–Raphson algorithm, are demonstrated. Furthermore, the variation laws of switching angles as a function of modulation index are validated and discussed.

A simulation model incorpored an IM has been designed and investigated. Series of simulation results are shown and discussed to validate our study.

References

1. Holtz J (1992) Pulse width modulation—a survey. IEEE Trans Industr Electron 39(5):410–420
2. Leon JI, Kouro S, Franquelo LG, Rodriguez J, Wu B (2016) The essential role and the continuous evolution of modulation techniques for voltage-source inverters in the past, present, and future power electronics. IEEE Trans Industr Electron 63(5):2688–2701
3. Haw LK, Dahidah MS, Almurib HA (2014) SHE–PWM cascaded multilevel inverter with adjustable DC voltage levels control for STATCOM applications. IEEE Trans Power Electron 29(12):6433–6444
4. Lee SS, Chu B, Idris NRN, Goh HH, Heng YE (2016) Switched-battery boost-multilevel inverter with GA optimized SHEPWM for standalone application. IEEE Trans Industr Electron 63(4):2133–2142
5. Zhang T, Du C, Qin C, Xing X, Chen A, Zhang C (2016) Neutral-point voltage balancing control for three-level T-type inverter using SHEPWM. In: 8th international power electronics and motion control conference (IPEMC), pp 1116–1122
6. Mohammadi HR, Falahati S, Zeraati M (2012) A new method for selective harmonic elimination in voltage source inverter using imperialist competitive algorithm. In Processdings of PEDSTC, pp 175–18
7. Balasubramonian M, Rajamani V (2014) Design and real-time implementation of SHEPWM in single-phase inverter using generalized hopfield neural network. IEEE Trans Industr Electron 61(11):6327–6336
8. Steczek M, Jefimowski W, Szeląg A (2020) Application of grasshopper optimization algorithm for selective harmonics elimination in low-frequency voltage source inverter. Energies 13(23):6426
9. Sharifzadeh M, Vahedi H, Portillo R, Franquelo LG, Al-Haddad K (2018) Selective harmonic mitigation based self-elimination of triplen harmonics for single-phase five-level inverters. IEEE Trans Power Electron 34(1):86–96
10. Najjar M, Moeini A, Bakhshizadeh MK, Blaabjerg F, Farhangi S (2016) Optimal selective harmonic mitigation technique on variable DC link cascaded H-bridge converter to meet power quality standards. IEEE J Emerg Sel Top Power Electron 4(3):1107–1116
11. Kumar MK, Sainath AM, Kumar VP (2013) Selective harmonic elimination by programmable pulse width modulation in inverters. Int J Eng Trends Technol 4(4):1196–1201
12. Kulkarni A, Upasani DE (2016) Implementation of SHEPWM in single phase inverter. Int J Sci Res Publ 6(1):431–434
13. Khedher A, Onduleur MLI-Machine à Induction: Etude et Optimisation de la Distorsion Harmonique, Diplôme d'Etudes Approfondies en Electrotechnique à l'Ecole Supérieur des Sciences et Techniques de Tunis, 19 Mars 1994
14. Tibola J, Pinheiro H, Figueiredo R (2011) Closed loop selective harmonic elimination applied to a grid connected PWM converter with LCL filter. In: IEEE conference on power electronics (COBEP), pp 746–752
15. Patel HS, Hoft RG (1973) Generalized techniques of harmonic elimination and voltage control in thyristor inverters: part I-harmonic elimination. IEEE Trans Ind Appl 3:310–317

Impact of FOC for Induction Machine Drive Using SVPWM and RPWM Strategies on Acoustic Noise

Henda Bouyahi and Adel Khedher

Abstract Electric motor control represents an important interest for industry. Thus, induction machines provide major source of acoustic noise. Field Oriented Control FOC offers an attractive solution to achieve high performance in variable-speed applications. This control consists to make the asynchronous machine control analogous to that of the DC machine where there is a natural decoupling between the rotor flux and electromagnetic torque. SVPWM and RPWM modulation techniques are also applied to supply systems in closed loop. The aim of this paper is to analyze the interaction between the inverter and the machine in closed loop and to estimate the level of noise emitted by the FOC control as a function of the modulation technique applied. For that we demonstrated and discussed the performance of the FOC control on the electrical and mechanical quantities produced by the machine. In addition, we have shown and compared the stator current harmonic spectra for each modulation technique in open and closed loop.

Keywords Field oriented control · Induction machine · RPWM · SVPWM

1 Introduction

Induction machine IM are promising electrical motor for Industrial and Electric Vehicle drives with their performances like high efficiency, low cost and almost free maintenance. Thus, induction motors are widely used in variable-speed electric drives where the machine is supplied by a voltage inverter. Nevertheless, feeding this machine by an inverter has been the root certain troublesome problems, such as acoustic noise. The overall noise level in an electrical machine originate from four

H. Bouyahi · A. Khedher (✉)
LATIS-Laboratory of Advanced Technology and Intelligent Systems, Université de Sousse, Ecole Nationale d'Ingénieurs de Sousse, 4023 Sousse, Tunisia
e-mail: adel_kheder@yahoo.fr

H. Bouyahi
e-mail: bouyahihenda@gmail.com

© The Author(s), under exclusive license to Springer Nature Singapore Pte Ltd. 2024
A. Mellit et al. (eds.), *Proceedings of the 2nd International Conference on Green Energy Conversion System*, Green Energy and Technology,
https://doi.org/10.1007/978-981-97-6148-7_40

main sources: magnetic, mechanical, aerodynamic and electronic [1, 2]. Indeed, the electronic noise in electrical machines is generally due to harmonics at inverter output. These harmonics create excitation forces that have a detrimental effect on the machine's stator. Flux harmonics in the air gap of an induction machine provoke to radial forces in the stator. These radial forces, in turn, lead to stator vibration. These vibrations are transmitted as acoustic waves by the stator frame. The noise emitted by an induction machine, fed by a voltage inverter, is an environmental problem, which is irritating to the human ear [3–7].

For several years, the research of the best operating performances of asynchronous machines powered by static converters has been the subject of several research and development projects. Hence, various control techniques have been developed and discussed in the literature [8–11]. The vector control FOC is a linear strategy, which uses linear controllers and a Pulse Width Modulation PWM to generate the voltages applied to the motors. Due to its high performance at both steady and transient states, FOC has been widely applied in power electronics [12–15].

Different Pulse Width Modulation strategies are applied to generate a variable voltage, variable frequency, three-phase AC using Voltage Source Inverters (VSIs) for variable speed AC drives. Sine PWM SPWM is widely used and popular PWM method. However, Space Vector PWM SVPWM strategy provides higher DC bus utilization and reduced the Total Harmonic Distortion THD compared to SPWM strategy [7]. The spectral analysis of constant switching frequency PWM schemes shows concentration of harmonics components in and around the integral multiples of the switching frequency [4, 7]. This has severe, impacts on the performance of induction machine in terms of acoustic noise. To improve the spread in spectrum, carrier frequency randomization is adopted in [3, 4, 6]. This technique utilized random carrier frequency or different shapes for modulating signal. In this paper, two modulation strategies SVPWM and RPWM are incorporated with control FOC for induction machine to test its suitability and to analyze the performance on terms of acoustic noise emitted.

Section 2 describes the dynamic modeling of induction motor. Field oriented control and Space vector and Random modulation techniques are presented in Sect. 3. The simulation results in MATLAB/ Simulink in open and closed loop are presented and discussed in Sect. 4. Section 5 concludes the present work.

2 Optimisation of Initial Conditions

The IM is composed of stator and rotor, as shown in Fig. 1. The three stator windings are presented by the axes (Sa, Sb, Sc), while the other axes are supported by rotor windings (ra, rb, rc), Fig. 1 depicts the relationships between the various machine angles. The electrical angle θ_s indicates the rotor position relative to stator, and the angles θ_r and respectively indicate the rotor and stator position relative to the reference frame (d, q). The relationship between these angles is given by the following expression:

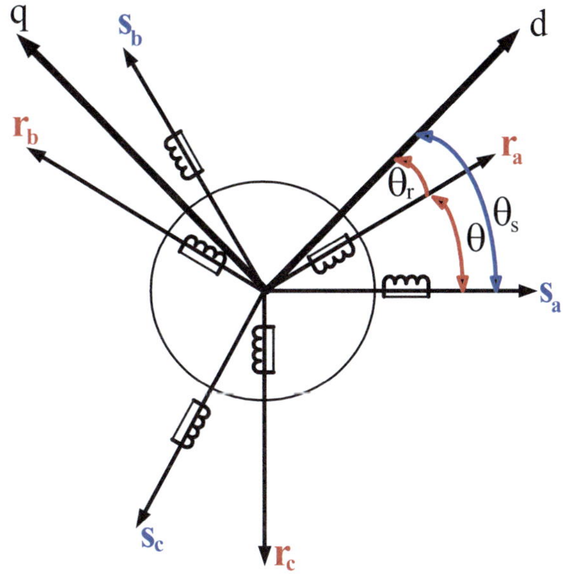

Fig. 1 IM representation in three-phase-diphase reference

$$\theta_s = \theta + \theta_r \tag{1}$$

The electrical equations are defined by:

$$\begin{cases} V_{sd} = R_s i_{sd} + \sigma L_s \frac{di_{sd}}{dt} + \frac{M}{L_r}\frac{d\varphi_r}{dt} - \omega_s \sigma L_s i_{sq} \\ V_{sq} = R_s i_{sq} + \sigma L_s \frac{di_{sq}}{dt} + \omega_s \frac{M}{L_r}\varphi_r + \omega_s \sigma L_s i_{sd} \end{cases} \tag{2}$$

Thus, the equations link the rotor electrical quantities are expressed by:

$$T_r \frac{d\varphi_r}{dt} + \varphi_r = Mi_{sd} \tag{3}$$

$$\omega_g = \frac{Mi_{sq}}{T_r \varphi_r} \tag{4}$$

The electromagnetic torque expression is given by the following form:

$$C_{em} = \frac{3}{2} n_p \frac{M}{l_r} i_{sq} \varphi_r \tag{5}$$

In the reference frame (d, q), the model (6) defines the machine state representation, input vector [U] is composed of stator voltages, state vector [X] is constituted by electrical quantities (stator currents) and magnetic quantities (rotor flux).

$$\frac{dX}{dt} = AX + BU \tag{6}$$

With: $X = \left[i_{sd}\ i_{sq}\ \varphi_{rd}\ \varphi_{rq}\right]^T$, $U = \left[v_{sd}\ v_{sq}\right]^T [A] = \begin{bmatrix} a_1 & a_2 & a_3 & a_4 \\ -a_2 & a_1 & -a_4 & a_3 \\ a_5 & 0 & a_6 & a_7 \\ 0 & a_5 & -a_7 & a_6 \end{bmatrix} [B] = \begin{bmatrix} b_1 & 0 \\ 0 & b_1 \\ 0 & 0 \\ 0 & 0 \end{bmatrix}$.

The state matrix coefficients are given by:

$$a_1 = -\frac{\left(R_s + \left(\frac{M}{l_r}\right)^2 R_r\right)}{\sigma l_s},\ a_2 = \omega_s,\ a_3 = \frac{M}{\sigma l_r l_s}\frac{1}{T_r}$$

$$a_4 = \frac{M}{\sigma l_r l_s}\omega,\ \ a_5 = \frac{M}{T_r},\ a_6 = -\frac{1}{T_r},\ a_7 = (\omega_s - \omega),\ b_1 = \frac{1}{\sigma l_s}.$$

where $\sigma = 1 - \frac{M^2}{l_r l_s}$ is the Blondel dispersion coefficient and $T_r = l_r/R_r$ is the rotor time constant.

3 Field Oriented Control of IM

3.1 FOC Control

Figure 2 presents the FOC control block diagram. The main elements are regulation loops for rotor flux and the speed, and two regulation loops for direct and quadrature stator current. Thus, the problem is addressed as a multi-variable system with two inputs (Stator voltages V_{sd} and V_{sq}) and two outputs (Stator currents i_{sd} and i_{sq}) [12–14].

The regulators used in this control system are proportional-integral (PI) type, of expression:

$$R(p) = k\left(1 + \frac{1}{\tau p}\right) \tag{7}$$

The constants k and τ allow act on the adjustment of system dynamics and static error. The pole compensation technique is used to determine PI regulators parameters.

- **Flux regulation:** The parameters k_f and τ_f of flux regulator are given by Eq. (8):

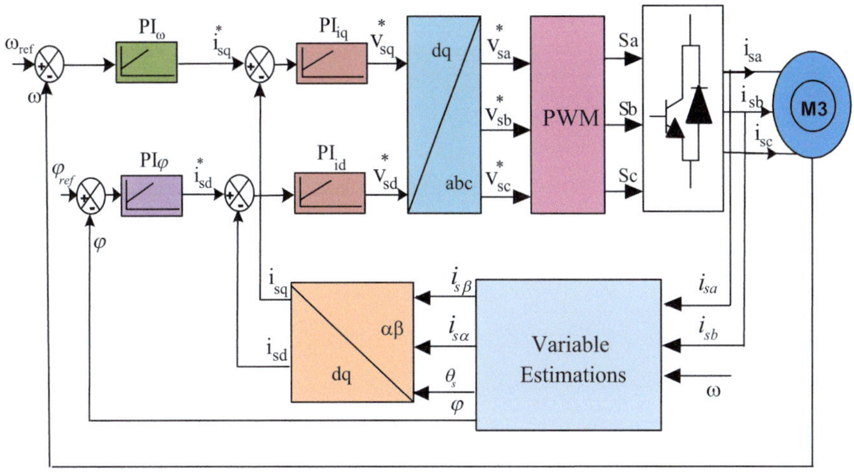

Fig. 2 FOC control block diagram

$$\begin{cases} T_f = T_r \\ k_f = \frac{4}{M} \end{cases} \tag{8}$$

- **Speed regulation:** The parameters k_v and T_v of speed regulator are determined by:

$$\begin{cases} T_v = \frac{J}{K_l} \\ k_v = 4K_l \end{cases} \tag{9}$$

- **Currents regulation:** The design of the current regulator is the same for both d-axis and q-axis current loops. The regulators parameters are expressed by:

$$\begin{cases} T_d = T_q = \frac{\sigma l_s}{R_s + \left(\frac{M}{l_r}\right)^2 R_r} \\ K_d = K_q = 4\left(R_s + \left(\frac{M}{l_r}\right)^2 R_r\right) \end{cases} \tag{10}$$

3.2 *SVPWM Strategy*

The SVPWM is a digital technique based on the representation of three reference voltages by space vector. Thus, the eight possible combinations of power switch states yield to eight vectors $\vec{V}_i(i = 0, 1, 2, 3, 4, 5, 6, 7)$ whose two vectors ($\vec{V}_0$ and $\vec{V}_7$) are inactive since they are identically zero. The six non-zero vectors in plane (V_α, V_β) are given in regular hexagon form who's the origin coincides with the zero-vectors, as shown in Fig. 3 [4].

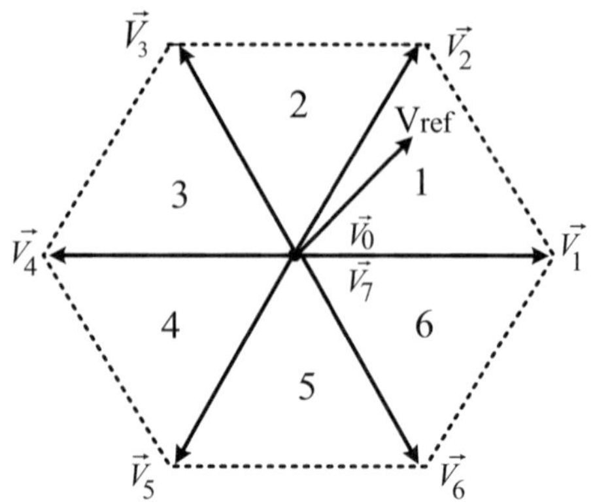

Fig. 3 Voltage space vectors of a two-level VSI

The eight vectors are defined in complex plane by the equations system (11).

$$\vec{V}_k = \begin{cases} \frac{2}{3}Ee^{j(k-1)\pi/3} & \text{if } k = 1, 2, 3, 4, 5, 6 \\ 0 & \text{if } k = 0, 7 \end{cases} \tag{11}$$

The SVPWM strategy uses projected signals in diphase plane deduced from Clarke transformation. The latter transforms a three-phase voltage system into a two-phase system. The reference vector, which characterizes the three-phase system, can be given by:

$$\vec{V}_{ref} = \begin{bmatrix} V_\alpha \\ V_\beta \end{bmatrix} \tag{12}$$

SVPWM is carried out by rotating the space vector. The generation of the control signals are based in the following three steps.

1. Sector number determination
2. Duty cycles determination
3. Switching sequences determination.

3.3 RPWM Strategy

Various RPWM strategies have been developed and proposed in literature for three-phase two-level inverters. The random frequency modulation can be considered as a special modulation technique by frequency displacement [4]. Instantaneous carrier period f_i is inversely proportional to switching period T_n.

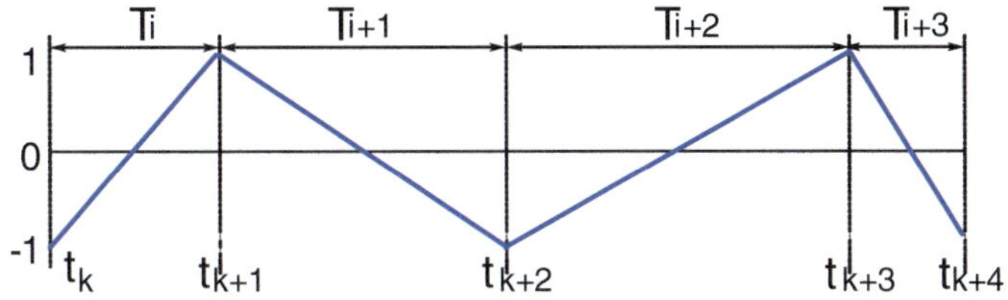

Fig. 4 Random-frequency carrier wave

$$\begin{cases} f_i = \dfrac{1}{T_n} \\ T_n = n(t_n) \end{cases} \quad (13)$$

where, n = 1,2….N, and Tn is the time period determined by random function $n(t_i)$.

An example of waveform RPWM triangular carrier is displayed in Fig. 4. The control signals are obtained by comparing the random-frequency carrier with sinusoidal reference signals.

4 Simulation Results

The simulation is carried out in MATLAB/Simulink to evaluate the performances of SVPWM and RPWM techniques for field-oriented control IM. The parameters induction machine are shown in Table 1. The load torque is assumed proportional to speed. The latter is described by slope straight line equal to 300 rad/s. The machine operates at switching frequency Fc = 5 kHz and voltage VDC = 560 V.

In fact, the stator current harmonics are the main reason for the acoustic noise increase. Thus, PWM modulation technique applied to modulate inverter is key factor that determines harmonics magnitude. Consequently, acoustic noise reduction consists to create spreading in the stator current harmonic spectrum. First, we present simulation results of PWM inverter-IM association in open-loop. Figure 5 presents stator currents and their harmonic spectra respectively for SVPWM and RPWM techniques. THD values are: %, 3.33% and 4.05%. Referring to simulation results

Table 1 Parameters machine

Characteristic	Value	Characteristic	Value
P	375 W	M	0.63 H
ω	2850 rpm	J	0.004 kg m^2
R_s	24 Ω	f	0.00159
l_s	0.66 H	n_p	1
R_r	10.88 Ω	T_n	1.2565
l_r	0.66 H	ψ_{sn}	0.7321

for SVPWM technique, we note that dominant harmonic components concentrate around switching frequency and its integer multiples. For RPWM technique, the harmonic spectrum is spread over large frequency range, and consequently reduce noise components magnitude.

Secondly, we present the vector control simulations results corresponding to temporal evolution of direct and quadrature stator current components, real and estimated rotor flux, real and reference speeds and electromagnetic torque, respectively using SVPWM and RPWM techniques. It can be seen that quadrature current presents peak during start-up phase and it's proportional to load torque variation, then the direct current remains constant. The real and estimated flux responses for SVPWM and RPWM presented are well established and of first-order type. Furthermore, real speed follows perfectly machine reference speed for both techniques. However, we observe torque oscillations for SVPWM technique compared with RPWM (Fig. 6 and 7).

Figure 8 depicts stator current waveform and harmonic spectra, pertaining to FOC control using SVPWM and RPWM techniques respectively. THD values are: %, 0.6% and 0.53%. In comparison with open-loop control, modulation techniques offer lower THD level, which proves the effectiveness of vector control in closed-loop. Referring to simulation results, we note that lower-order harmonic are concentrated around switching frequency and its integer multiples, for SVPWM technique. However, for RPWM, harmonic spectra spread over large frequency range, and therefore reduces the dominant noise component magnitude. Conclusively, RPWM possesses better performance both in open and closed loop.

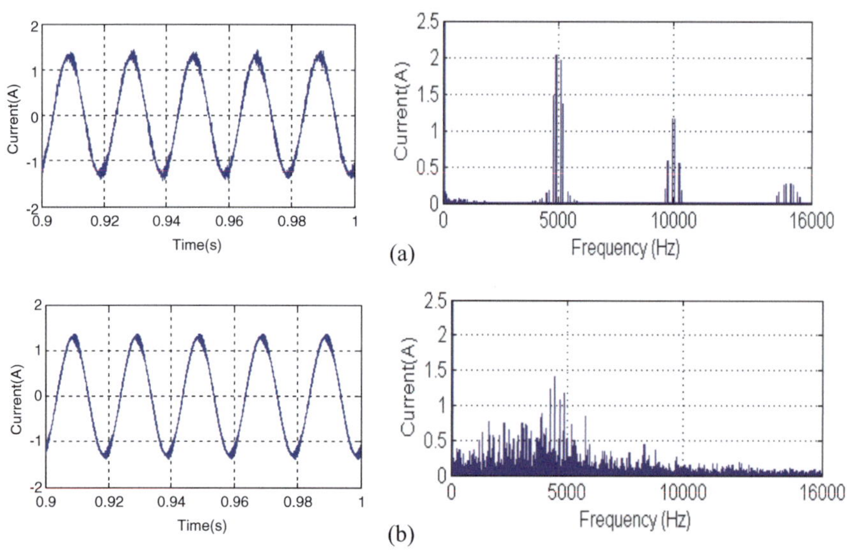

Fig. 5 Current waveform and its harmonic spectra in open loop for: **a** SVPWM and **b** RPWM

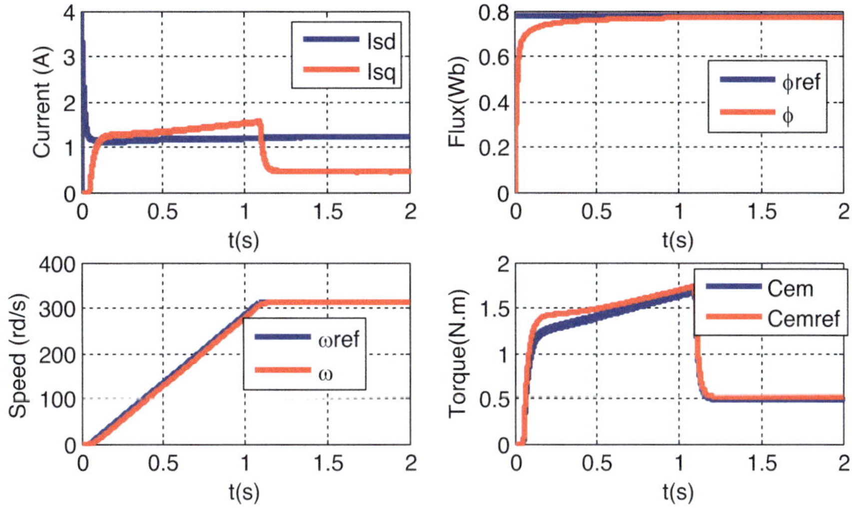

Fig. 6 Machine response for FOC control using SVPWM

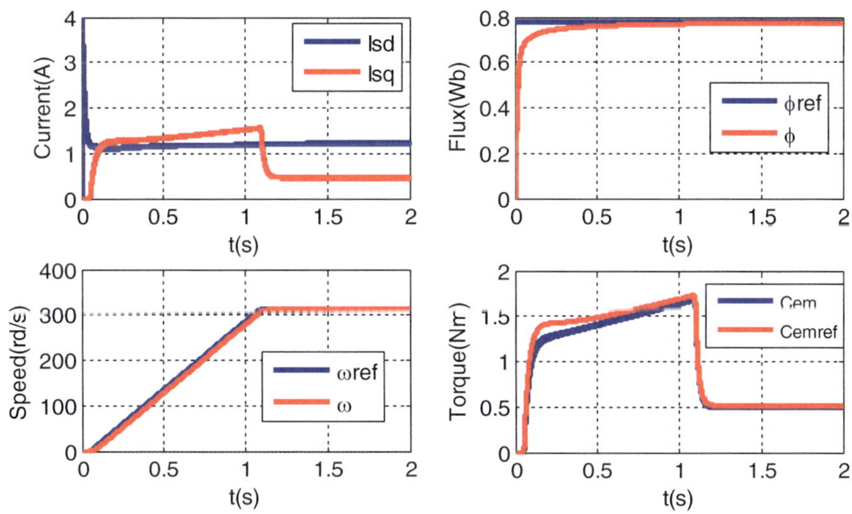

Fig. 7 Machine response for FOC control using RPWM

The noise intensity of electronic origin depends largely on stator current quality, which is closely related to modulation technique used. To take account of human exposure to the acoustic noise emitted by IM, stator current is converted into dB scale. Figures 9 and 10 illustrate stator current noise levels respectively in open and closed loop with SVPWM and RPWM techniques. From these results, we can clearly see that noise level is significantly reduced in closed loop compared to open,

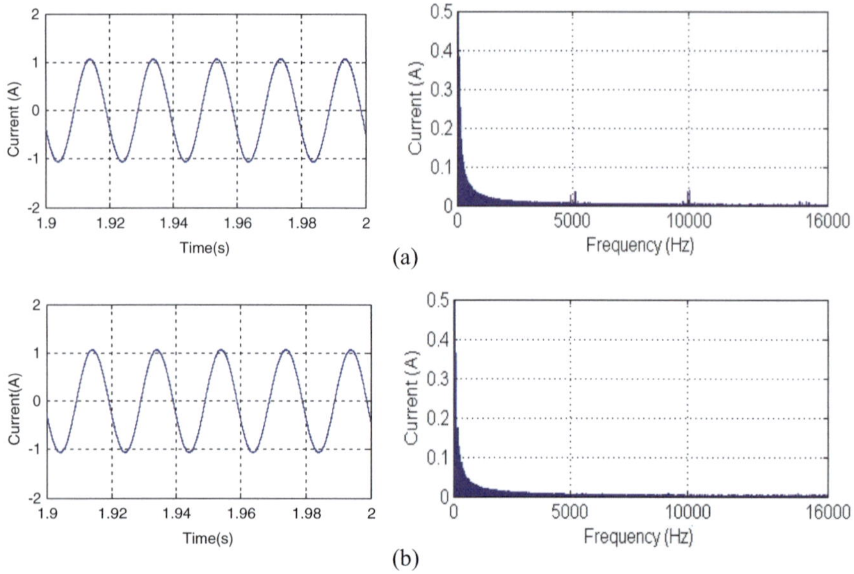

Fig. 8 Current waveform and its harmonic spectra for FOC control using: **a** SVPWM and **b** RPWM

for both modulation techniques. Thus, for SVPWM, lower-order noise components are concentrated around integer multiples of switching frequency, which is irritating for human ear. In addition, noise spectra is well dispersed over large frequency range for RPWM technique, thus low noise level in both open and closed loop.

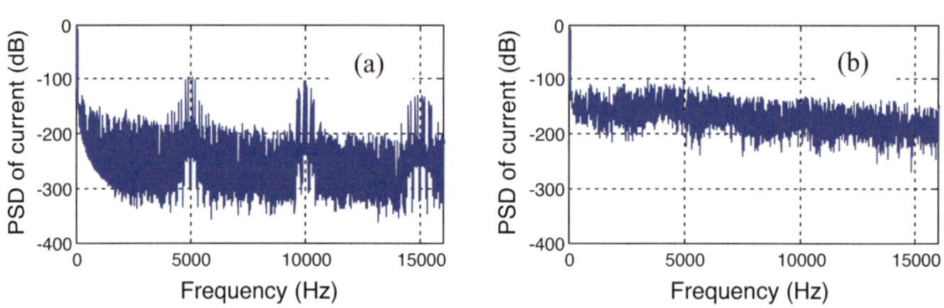

Fig. 9 Stator current spectra in dB in open loop using: **a** SVPWM and **b** RPWM

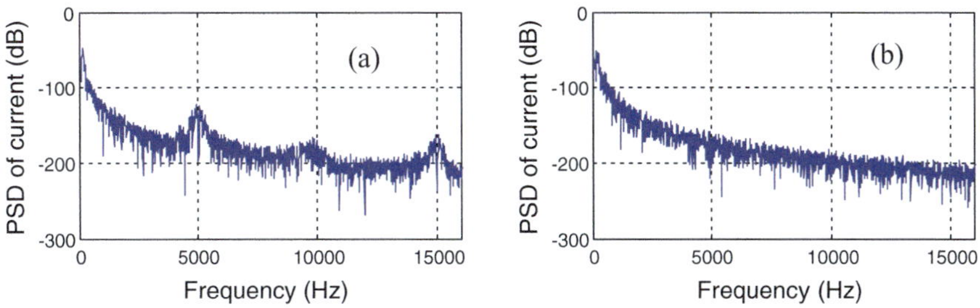

Fig. 10 Stator current spectra in dB for FOC control using: **a** SVPWM and **b** RPWM

5 Conclusion

The continuous exposure to acoustic noise emitted by IM causes illnesses and psycho-acoustic problems for human ears. Then, the research of efficient control techniques for variable speed drives has become promising research area. In this paper, the acoustic noise of induction machine is reduced by FOC control. Thus, FOC control with SVPWM and RPWM modulation techniques present improved performance in terms of THD in closed loop compared to open loop. Simulation results also show that for SVPWM technique, stator current dominant noise components are concentrated around switching frequency, which is irritate the human ear. However, for RPWM technique, stator current spectra is spread, and consequently human ear is more relieved through acoustic noise, both in open and closed loop.

References

1. Vijayraghavan P, Krishnan R (1999) Noise in electric machines: a review. IEEE Trans Ind Appl 35(5):1007–1013
2. Fang Y, Zhang T (2018) Sound quality investigation and improvement of an electric powertrain for electric vehicles. IEEE Trans Ind Electron 65(2):1149–1157
3. Khalid M, Mohan A, Pullukkara PP, Binojkumar AC (2022) A modified variable switching frequency spread-spectrum PWM technique with reduced torque ripple for a vector-controlled PMSM drive. IEEE Access 10:132800–132813
4. Bouyahi H, Ben Smida K, Khedher A (2020) Experimental study of PWM strategy effect on acoustic noise generated by inverter-fed induction machine. Int Trans Electr Energy Syst 30(3):e12249
5. Kroneisl M, Šmídl V, Peroutka Z, Janda M (2020) Predictive control of IM drive acoustic noise. IEEE Trans Ind Electron 67(7):5666–5676
6. Kumar A-B, Narayanan G (2016) Variable-switching frequency PWM technique for induction motor drive to spread acoustic noise spectrum with reduced current ripple. IEEE Trans Ind Appl 52(5):3927–3938
7. Binojkumar AC, Saritha B, Narayanan G (2015) Acoustic noise characterization of space-vector modulated induction motor drives—an experimental approach. IEEE Trans Ind Electron 62(6):3362–3371
8. Chan T, Shi K (2011) Applied intelligent control of induction motor drives. Wiley, Asia

9. Mohan A, Khalid M, Binojkumar AC (2021) Performance analysis of permanent magnet synchronous motor under DTC and space vector-based DTC schemes with MTPA control. In: International conference on communication, control and information sciences (ICCISc), pp 1–8
10. Abassi M, Khlaief A, Saadaoui O, Chaari A, Boussak M (2015) Performance analysis of FOC and DTC for PMSM drives using SVPWM technique. In: 16th international conference on sciences and techniques of automatic control and computer engineering (STA), pp 228–233
11. Wang F, Zhang Z, Mei X, Rodríguez J, Kennel R (2018) Advanced control strategies of induction machine: field oriented control, direct torque control and model predictive control. Energies 11(1):120
12. Khalid M, Mohan A (2020) Performance analysis of vector controlled PMSM drive modulated with sinusoidal PWM and space vector PWM. In: IEEE international power and renewable energy conference, pp 1–6
13. Divyasree P, Binojkumar AC (2017) Vector control of voltage source inverter fed induction motor drive using space vector PWM technique. In: International conference on energy, communication, data analytics and soft computing (ICECDS), pp 2946–2951
14. Sulaiman M, Patakor FA, Ibrahim Z (2013) DSP based implementation of field oriented control of three-phase induction motor drives. Int J Innov Res Comput Commun Eng 2(09)
15. Heydari F, Sheikholeslami A, Firouzjah KG, Lesan S (2010) Predictive field-oriented control of PMSM with space vector modulation technique. Front Electr Electron Eng China 5:91–99

A PMSG Wind Turbine Fault Diagnosis via Sparse Recovery Method

Wafa Torki, Syrine Derbel, Jean-Pierre Barbot, and Lassaad Sbita

Abstract This paper presents a diagnosis method to detect, isolate and estimate faults occurring in permanent magnet synchronous generator (PMSG). This method is based on a dynamical sparse recovery (SR) algorithm that is able to reconstruct online and with finite-time convergence a sparse vector of numerous faults from few system measurements. The faulty PMSG system is affected by a partial inter-turn short circuit and voltage dip faults. Using Matlab/Simulink environment, some scenarios are presented to present the efficiency of SR method for diagnostic of PMSG system.

Keywords Diagnosis · Left-invertibility · Permanent magnet synchronous generator · Short-circuit sparse recovery · Voltage dip

1 Introduction

Permanent Magnet Synchronous Generator (PMSG) is widely used in wind turbines (WT) and becomes one of the important applications nowadays owing to its high energy efficiency and high performances [1]. However, it can be affected by mechanical defects (see, for example [2, 3]) and/or electrical faults (see, for more details [4, 5], etc.). The major source of the electrical faults is the stator windings. The insulation degradation can cause an inter-turn short circuit producing an overheat in the stator winding [6].

In this paper, the considered conversion system is a PMSG affected by a partial inter-turn short-circuit in three windings and a voltage dip faults.

W. Torki (✉) · L. Sbita
Laboratory of Processes, Energy, Environment and Electric Systems, ENIG, Gabes 6072, Tunisia
e-mail: tourkiwafa91@yahoo.fr

S. Derbel · J.-P. Barbot
QUARTZ Laboratory, ENSEA, EA7393 Cergy-Pontoise, France

J.-P. Barbot
LS2N, CNRS, Ecole Centrale de Nantes, Nantes, France

© The Author(s), under exclusive license to Springer Nature Singapore Pte Ltd. 2024
A. Mellit et al. (eds.), *Proceedings of the 2nd International Conference on Green Energy Conversion System*, Green Energy and Technology,
https://doi.org/10.1007/978-981-97-6148-7_41

The diagnosis of the considered faults has attracted considerably researchers attention. Corresponding to these types of faults, different diagnosis methods have been developed in the literature (see, for example, [7, 8], etc.). However, when a system is affected by several faults comparing to its low measurements number, which is our study case, a method that can be applied for one is not necessarily applied for another one. Thus, each fault requires its own diagnosis method and this makes the system much complex.

The aim of this paper is to present and to apply a diagnosis method called Sparse Recovery Diagnosis (SRD) for dynamical systems. Its particularity is to provide a global diagnostic for numerous faults when few of them appear simultaneously. The major advantages of the SRD, comparing to other methods, that is an online diagnosis method with finite-time convergence.

This paper is organized as follows: A model of the proposed system in healthy and faulty cases is first presented in Sect. 2; followed by a description of the target diagnosis method in Sect. 3. Section 4 presents a system model transformation. Simulations results are illustrated and discussed in Sect. 5.

2 System Modeling

2.1 Healthy Model

The PMSG electrical model in natural (abc) reference frame is given by Eqs. (1) and (2):

$$[v_{abc}] = R\,[i_{abc}] + \frac{d\,[\phi_{abc}]}{dt} \tag{1}$$

$$[\phi_{abc}] = [L]\,[i_{abc}] + [\phi_r] \tag{2}$$

where, $[v_{abc}] = [v_a\ v_b\ v_c]^T$ is the stator voltages vector, $[i_{abc}] = [i_a\ i_b\ i_c]^T$ is the stator currents vector, $[\phi_r] = \varphi_r \begin{bmatrix} \cos(\omega t) \\ \cos(\omega t - 2\pi/3) \\ \cos(\omega t + 2\pi/3) \end{bmatrix}$ is the PM flux linkage vector with ω the rotor speed and φ_r is magnet flux amplitude, R is the winding stator resistance, $[L] = \begin{bmatrix} \mathrm{l} & \mathrm{M} & \mathrm{M} \\ \mathrm{M} & \mathrm{l} & \mathrm{M} \\ \mathrm{M} & \mathrm{M} & \mathrm{l} \end{bmatrix}$ is the inductance matrix, with $l = l_s + l_m$ and $M = -\frac{1}{2} l_m$ where M, l_s and l_m represent, respectively, the mutual inductance, the stator leakage inductance and the average value of the winding magnetizing inductance.

2.2 Faulty Model

In this paper, we present the most frequently faults that affect the PMSG: the partial inter-turn short circuit and voltages dip faults.

$\longrightarrow$ Multi phase Inter-turn short circuit

In this paper, it is assumed that a partial inter-turn short circuit fault occurs in the stator winding of phase a. The shorted portion of the affected phase creates a steady magnetic field that modifies the main field by forming an additional branch in the magnetic system. The extra circuit in the phase stator winding generates a current i_{fa} and resistance R_{fa}.

The fault severity can be indicated by the ratio $\eta_a = \frac{n_f}{n_t}$ between the short turns and the total turns number of the affected winding.

The non-salient poles PMSG with the inter-turn short circuit fault is modeled in the natural (a, b, c) reference frame. Using the matrix notation, the electrical model is given by (3) (see [10], for more modeling details):

$$\begin{bmatrix} v_a \\ v_b \\ v_c \end{bmatrix} = [R] \begin{bmatrix} i_a \\ i_b \\ i_c \end{bmatrix} + [L] \begin{bmatrix} \frac{di_a}{dt} \\ \frac{di_b}{dt} \\ \frac{di_c}{dt} \end{bmatrix} - \begin{bmatrix} \varphi_r \omega \sin(\omega t) \\ \varphi_r \omega \sin(\omega t - \frac{2\pi}{3}) \\ \varphi_r \omega \sin(\omega t + \frac{2\pi}{3}) \end{bmatrix}$$
$$- \eta_a \left(\begin{bmatrix} R \\ 0 \\ 0 \end{bmatrix} i_{fa} + \begin{bmatrix} l \\ M \\ M \end{bmatrix} \frac{di_{fa}}{dt} \right) \tag{3}$$

where $[R] = \begin{bmatrix} R_{a1} & 0 & 0 & 0 \\ 0 & R_{a2} & 0 & 0 \\ 0 & 0 & R_b & 0 \\ 0 & 0 & 0 & R_c \end{bmatrix}$ is the resistance matrix with R_{a1} and R_{a2} the new resistance quantities that are defined, using the proportionality property:

$$R_{a1} = (1 - \eta_a) R_a$$

and

$$R_{a2} = \eta_a R_a,$$

$$[L] = \begin{bmatrix} (1-\eta_a) l_s + (1-\eta_a)^2 l_m & (1-\eta_a)\eta_a l_m & (1-\eta_a) M & (1-\eta_a) M \\ (1-\eta_a)\eta_a l_m & \eta_a l_s + \eta_a^2 l_m & \eta_a M & \eta_a M \\ (1-\eta_a) M & \eta_a M & l & M \\ (1-\eta_a) M & \eta_a M & M & l \end{bmatrix}$$ is the inductance matrix.

In order to determine the fault current i_{fa}, the following equation is considered:

$$R_{fa} i_{fa} = v_{a2} \tag{4}$$

That leads to (5):

$$0 = Ri_a + l\frac{di_a}{dt} + M\frac{di_b}{dt} + M\frac{di_c}{dt} - (\frac{R_{fa}}{\eta_a} + R)i_{fa} - (l_s + \eta_a l_m)\frac{di_{fa}}{dt} \tag{5}$$

By a simple change of index, the same procedure will be considered for a short circuit in the phase b or c.

⟶ Voltage Dip Fault

Taking the case when a healthy PMSG is connected to the grid in case of a voltage dip fault in one phase. When the fault is occurring in phase a, the phase voltage is expressed by (6) (see, for example, [9] for modelisation)

$$\begin{cases} v_a = (V - d_a)\cos(\omega t) \\ v_b = V\cos(\omega t - \frac{2\pi}{3}) \\ v_c = V\cos(\omega t + \frac{2\pi}{3}) \end{cases} \tag{6}$$

where V and d_a present respectively the healthy voltage amplitude and the fault magnitude in phase a. The same procedure is applied if the fault occurs in the phase b or c. To generalize, the three phase voltages can be described by $(1 - \alpha_i)v_i$, $i = a, b, c$, where, $\alpha_i, i = a, b, c$ presents the ratio between the fault magnitude and the healthy phase voltage amplitude.

Notice that the first contribution of this paper is to make the dynamical modeling of the faulty system adapted to the acquisition process given by (7):

$$\Gamma = \zeta(W + \varepsilon) \tag{7}$$

where $\Gamma \in \mathbb{R}^p$ gathers the p observed linear measurements vector, $\zeta \in \mathbb{R}^{p\times q}$ is the fault distribution matrix, ε is a measurement noise and $W \in \mathbb{R}^q$ is the unknown signal faults vector to be recovered.

3 Left-Invertibility Under Sparse Assumption

3.1 Left-Invertibility and Parcimony

Fault detection and estimation are highly simplified using the left-invertibility process. This is based on recovering the information on the failures input from the system measurements (see, for example, [11]). However, the left invertible problem was processed just only in the case when the number of the unknown inputs is equal or fewer than the number of the measurements. In the contrary case, this problem has infinity of solutions. In particular, the SRD method is able to find a simple representation of a signal using some specified sub-matrix. In order to solve the acquisition problem of (7), some assumptions are required:

Assumption 1 (*s-sparsity condition*) It is assumed that W is s-sparse vector, i.e., no more than s components in W are non-zero. Moreover, s must be lower than p ($s \ll p$) and verifies the following condition:

$$2s + 1 \leq p \tag{8}$$

Assumption 2 (*Restricted Isometry Property RIP*) (see [12, 13] for more details) $\forall x \in \mathbb{R}^n$ and $\forall t > 0$, the matrix ζ that depends on the time and the states ($\zeta(x, t)$) satisfies the s-order following condition: $\exists \delta_s \in (0, 1)$ such that for all s-sparse signal W:

$$(1 - \delta_s) \|W\|_2^2 \leq \|\zeta(x, t)W\|_2^2 \leq (1 + \delta_s) \|W\|_2^2 \tag{9}$$

The goal of SR diagnosis is to solve the following optimization program:

$$W^* = \arg \min_{W \in \mathbb{R}^q} \frac{1}{2} \|\Gamma - \zeta W\|_2^2 + \lambda \|W\|_1 \text{, where } \|W\|_1 \triangleq \sum_i |W_i| . \tag{10}$$

3.2 SRD Algorithm Presentation

In order to solve the optimization problem (10) and to obtain a finite-time convergence, a dynamical algorithm based on sliding mode techniques was proposed in [14] and it is defined by (11).

$$\begin{cases} \tau \dot{u} = -\left\lceil u(t) + (\zeta^T \zeta - I_q)\hat{W} - \zeta^T \Gamma \right\rfloor^{\alpha} \\ \hat{W} = \Lambda_\lambda(u) \\ \lceil . \rfloor^{\alpha} \triangleq |.|^{\alpha} \text{sgn}(.) \end{cases} \tag{11}$$

where $u \in \mathbb{R}^q$ represents the state vector, $\hat{W}$ is the estimation of the sparse signal W, I_q is the identity matrix, τ is a time constant which depends on the physical properties of the system and $\Lambda_\lambda(u)$ is a continuous soft-thresholding function that is defined by the system (12):

$$\Lambda_\lambda(u) = \begin{cases} 0 & if\ |u| \leq \lambda \\ (u - \lambda)sgn(u) & if\ |u| > \lambda \end{cases} \tag{12}$$

with a balancing parameter $\lambda > 0$.

3.3 System Transformation

As it is presented in Sect. 2, in presence of an inter-turn short circuit a leakage current i_f appeared. This type of fault is initialized by controlling the resistance R_f.

In fact, the healthy machine is presented by a high value of the leakage resistance R_f. However, the short circuit results a low resistance value ($R_f = 0\Omega$ for an outright fault).

Assumption 3 For the proposed algorithm, the inter-turn short circuit is assumed to be an outright fault, that means $i_{fa} \simeq i_a$, $i_{fb} \simeq i_b$, and $i_{fc} \simeq i_c$.

Including all the faults presented in the Section 2 and respecting Assumption 3, the faulty PMSG model transformed to the sparse modelisation (7) is presented by (13):

$$\underbrace{\begin{bmatrix} v_a \\ v_b \\ v_c \end{bmatrix} - R \begin{bmatrix} i_a \\ i_b \\ i_c \end{bmatrix} - [L]\frac{d}{dt}\begin{bmatrix} i_a \\ i_b \\ i_c \end{bmatrix} + \varphi_r \omega \begin{bmatrix} \sin(\omega t) \\ \sin(\omega t - \frac{2\pi}{3}) \\ \sin(\omega t + \frac{2\pi}{3}) \end{bmatrix}}_{\Gamma}$$

$$= \underbrace{\begin{bmatrix} -Ri_a - l\frac{di_a}{dt} & -M\frac{di_b}{dt} & -M\frac{di_c}{dt} & v_a & 0 & 0 \\ -M\frac{di_a}{dt} & -Ri_b - l\frac{di_b}{dt} & -M\frac{di_c}{dt} & 0 & v_b & 0 \\ -M\frac{di_a}{dt} & -M\frac{di_b}{dt} & -Ri_c - l\frac{di_c}{dt} & 0 & 0 & v_c \end{bmatrix}}_{\zeta \quad *} \underbrace{\begin{bmatrix} \eta_a \\ \eta_b \\ \eta_c \\ \alpha_a \\ \alpha_b \\ \alpha_c \end{bmatrix}}_{W} \quad (13)$$

where η_a, η_b and η_c are, respectively, the fault severity in the phases a, b and c, $\Gamma \in \mathbb{R}^3$ presents the measurement vector that includes the available measurements, its derivatives and the healthy operating system dynamics, $\frac{di_a}{dt}$, $\frac{di_b}{dt}$ and $\frac{di_c}{dt}$ are the derivatives of the three phase currents extracted in finite-time using a Super Twisting Differentiators. We name $W = [W_1 \quad W_2 \quad W_3 \quad W_4 \quad W_5 \quad W_6]^T$.

4 Application of the Proposed Method and Simulation Results

In our simulation, we consider: d-axis stator inductance $L_d = 0.0044$ H, q-axis stator inductance $L_q = 0.0044$ H, Stator resistance $R_s = 0.5\Omega$, Permanent magnetic flux $\Psi_f = 0.39$ Wb, DC-link capacitor $C = 2.5$ mF, Number of pole-pairs $n_p = 19$, Load resistance $= 100\,\Omega$. For the designed system, a solver $Ode1$(Euler) is chosen since Eq. (11) is not C^1-smooth, the constant α is equal to 0.5 ensuring a convergence in a finite-time, $\lambda = 10^{-3}$, $\tau = 10^{-3}s$ and $\mu = 10^4$.

The six faults modeled in (13), are implemented. We consider that only one fault ($s = 1$) is occurring in order to satisfy the s-sparsity condition ($s \leq p < q$) explained in (7). Recall that in our case, the measurements dimension is equal to 3 ($p = 3$) and

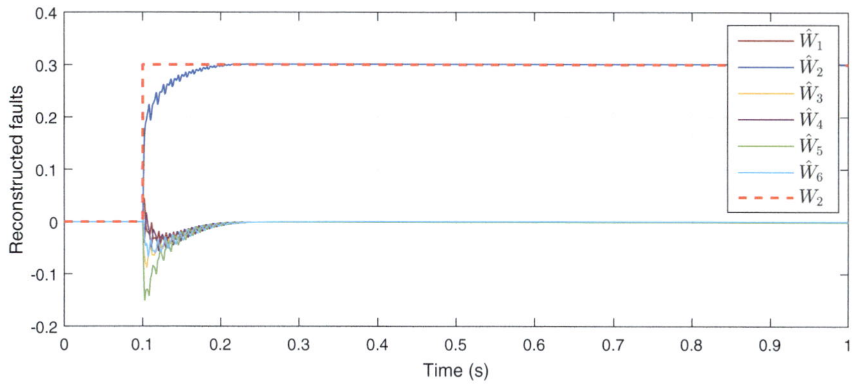

Fig. 1 Stator currents i_a, i_b and i_c ($R_{fb} = 0\Omega$) (**a**) detection of the inter-turn short circuit in the phase b (**b**)

the faults dimension is equal to 8 ($q = 6$) then the matrix ζ is (3×6) with a time varying elements.

In order to show the good performances and robustness of the SRD method, we present two scenarios of possible faults:

Scenario 1: This scenario corresponds to an inter turn short circuit that appears in the phase b at $t = 0.4$ s. It is an outright fault ($R_{f_b} = 0\,\Omega$). This means that only the second element of the faults vector W is activated ($W_2 \neq 0$). This fault is characterized by the presence of a constant value that represents the ratio between the short turns and the total turns number (Here, $\eta_b = 0.3$).

Looking at Fig. 1, we note that the SRD method estimates the considered faults where the estimated non zero fault $\hat{W}_2$, related to the short circuit in the phase b, converges to its real value η_b and the other estimated faults converge to zero.

Scenario 2: In this scenario, we consider a voltage dip fault appearing in the phase a. A step profile of the voltage dip ratio α_a is considered:

$$\alpha_a = \begin{cases} 0.4, & for \quad 0.2 \leq t \leq 0.5\,\text{s} \\ 0.2, & for \quad 0.5 < t \leq 0.7\,\text{s} \\ 0.3, & for \quad t > 0.7\,\text{s} \end{cases} \tag{14}$$

It is clearly shown in Fig. 2 that the fault $\hat{W}_4$ is accurately recovered in finite time. It convergences to its real value α_a that changes for each time interval. While the other non activated faults converge to zero.

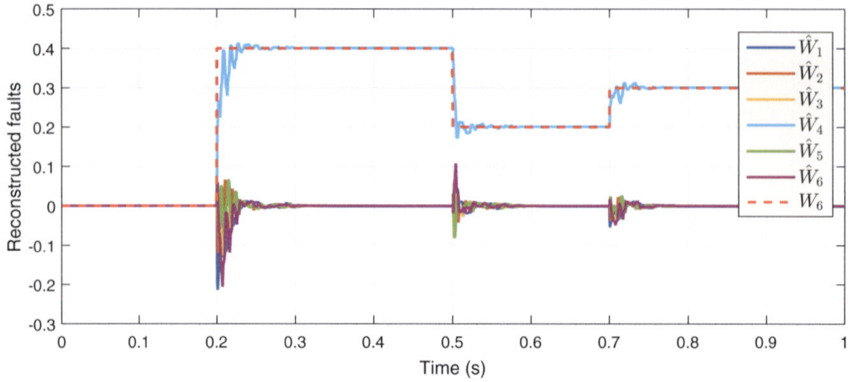

Fig. 2 Detection of the voltage dip fault in phase b

5 Conclusions

In this paper, a sparse recovery diagnosis method is applied to a PMSG wind turbine. It is used in order to detect, locate and estimate in finite-time convergence an inter-turn short circuit and voltage dip fault. In healthy and faulty cases, dynamical modelisations are presented. The simulation results, given in the last section, highlighted the well founded of the proposed algorithm and its effectiveness to recover a sparse fault vector even if the system is a dynamical one. A comparison with other diagnosis methods will be our future work and this in the stationary and non-stationary cases.

References

1. Gmez-Espinosa A, Hernndez-Guzmn VM, Jimnez-Hernndez BH, Rivas-Araiza EA, Rodrguez-Resndiz J, Herrera-Ruz G (2013) A new adaptive self-tuning Fourier coefficients algorithm for periodic torque ripple minimization in permanent magnet synchronous motors (PMSM). Sensors 13(3):3831–3847
2. Kang K, Song J, Kang C, Sung S, Jang G (2017) Real-time detection of the dynamic eccentricity in permanent-magnet synchronous motors by monitoring speed and back EMF induced in an additional winding. IEEE Trans Ind Electron 64(9):7191–7200
3. Ding P, Wang H, Bao W, Hong R (2019) HYGP-MSAM based model for slewing bearing residual useful life prediction. Measurement 162–175:141
4. Moon S, Lee J, Jeong H, Kim SW (2016) Demagnetization fault diagnosis of a PMSM based on structure analysis of motor inductance. IEEE Trans Ind Electron 63(6):3795–3803
5. Choi G, Jahns TM (2016) Investigation of key factors influencing the response of permanent magnet synchronous machines to three-phase symmetrical short-circuit faults. IEEE Trans Energy Convers 31(4):1488–1497
6. Sabir R, Rosato D, Hartmann S, Guhmann C (2019) Detection and localization of electrical faults in a three phase synchronous generator with rectifier. In: 2019 international conference on electrical drives and power electronics (EDPE). IEEE, pp 18–23

7. Grouz F, Sbita L, Boussak M (2012) Particle swarm optimization based fault diagnosis for non-salient PMSM with multi-phase inter-turn short circuit. In: International conference on communications, computing and control applications. IEEE, pp 1–6
8. Zhao KH, Chen TF, Zhang CF, He J, Huang G (2014) Online fault detection of permanent magnet demagnetization for IPMSMs by nonsingular fast terminal-sliding-mode observer. Sensors 14(12):23119–23136
9. El Malleh K, Chariag DE, Sbita L et al (2019) A control strategy for a three-phase grid connected PV system under grid faults. Electronics 8(8):906
10. Bechkaoui A (2016) Diagnostic des défauts dans les machines synchrones á vitesse variable. Doctorat, Université Badji Mokhtar-Annaba
11. Barbot JP, Boutat D, Floquet T (2009) An observation algorithm for nonlinear systems with unknown inputs. Automatica 45(8):1970–1974
12. Ren J, Yu L, Lyu C, Zheng G, Barbot JP, Sun H (2019) Dynamical sparse signal recovery with fixed-time convergence. Signal Process 162:65–74
13. Candes EJ (2008) The restricted isometry property and its implications for compressed sensing. CR Math 346(9–10):589–592
14. Yu L, Zheng G, Barbot JP (2017) Dynamical sparse recovery with finite-time convergence. IEEE Trans Signal Process 65(23):6146–6157

A Sparse Recovery Diagnosis Algorithm Applied on PMSG Wind Turbine

Wafa Torki, Syrine Derbel, Lassaad Sbita, and Jean-Pierre Barbot

Abstract Many researchers have paid great attention to fault diagnosis technologies in order to reduce worst operations and maintenance costs. Accordingly, this work presents a new method to detect, isolate and estimate faults occurring in permanent magnet synchronous generator (PMSG). This method is based on a dynamical sparse recovery algorithm that is able to reconstruct online and with finite time convergence a sparse vector of numerous faults from few system measurements. The faulty system is modeled in case of permanent magnet demagnetization and voltage dip faults and its dynamical modelisation is adapted to the proposed algorithm under some theoretical conditions. The system model with the proposed algorithm is implemented under Matlab/Simulink environment and its effectiveness is verified upon simulation results for such fault scenarios.

Keywords Demagnetization · Diagnosis · Left-invertibility · Permanent magnet synchronous generator · Restricted isometry property · Sparse recovery · Voltage dip

1 Introduction

Thanks to its high performances and its high energy performances, Permanent Magnet Synchronous Generator PMSG is, more and more attractive or wind turbines WT applications [1]. However, it can be affected by several defects which are, essentially, electrical faults that correspond to demagnetization fault [2], short or open circuit faults [3], voltage drop, etc. and mechanical faults that correspond to static

W. Torki (✉) · L. Sbita
Laboratory of Processes, Energy, Environment and Electric Systems, ENIG, Gabes 6072, Tunisia
e-mail: tourkiwafa91@yahoo.fr

S. Derbel · J.-P. Barbot
QUARTZ Laboratory EA7393, ENSEA, Cergy-Pontoise, France

J.-P. Barbot
LS2N, CNRS, Ecole Centrale de Nantes, Nantes, France

© The Author(s), under exclusive license to Springer Nature Singapore Pte Ltd. 2024
A. Mellit et al. (eds.), *Proceedings of the 2nd International Conference on Green Energy Conversion System*, Green Energy and Technology,
https://doi.org/10.1007/978-981-97-6148-7_42

or dynamic eccentricity [4] and bearing faults [5]. The stator windings are considered as the major source of those faults particularly electrical ones. The insulation degradation can cause an inter-turn short circuit producing an overheat in the stator winding [6]. Thus, the shorted turns can impair the magnetic coercivity and result a partial demagnetization of the permanent magnet (PM).

In this paper, the considered conversion system is a PMSG affected by A PM demagnetization and a voltage dip faults. The diagnosis of the considered faults has attracted considerably researchers attention. Corresponding to these types of faults, different diagnosis methods have been developed in the literature [7, 8], etc.. However, when a system is affected by several faults comparing to its low measurements number, which is our study case, a method that can be applied for one is not necessarily applied for another one. Thus, each fault requires its own diagnosis method and this makes the system much complex and fall down in big-data problem in case of various faults. Hence, the big-data becomes an issue that attracts researchers in various fields. The aim of this paper is to present and to apply a diagnosis method called Sparse Recovery Diagnosis (SRD) for dynamical systems. Its particularity is to provide a global diagnostic (Detection, isolation and estimation) for numerous faults when few of them appear simultaneously. The major advantages of the SRD, comparing to other methods, that is an online diagnosis method with finite-time convergence.

This paper is devided on 5 sections: Introduction is set in Sect. 1. Section 2 presents the modelling of PMSG in healthy and faulty cases. Section 3 describes the diagnosis method. Followed by a system model transformation in order to be adapted to the proposed algorithm. Simulations results are presented in Sect. 4. Conclusions are given in Sect. 5.

2 System Modeling

2.1 Healthy Model

The PMSG electrical model in natural (abc) reference frame is given by the Eqs. (1) and (2) [9]:

$$[v_{abc}] = R\,[i_{abc}] + \frac{d\,[\phi_{abc}]}{dt} \tag{1}$$

$$[\phi_{abc}] = [L]\,[i_{abc}] + [\phi_r] \tag{2}$$

where

- $[v_{abc}] = [v_a\ v_b\ v_c]^T$ is the stator voltages vector,
- $[i_{abc}] = [i_a\ i_b\ i_c]^T$ is the stator currents vector,

- $[\phi_r] = \varphi_r \begin{bmatrix} \cos(\omega t) \\ \cos(\omega t - 2\pi/3) \\ \cos(\omega t + 2\pi/3) \end{bmatrix}$ is the PM flux linkage vector with ω the rotor speed and φ_r is magnet flux amplitude,
- R is the winding stator resistance,
- $[L] = \begin{bmatrix} 1 & M & M \\ M & 1 & M \\ M & M & 1 \end{bmatrix}$ is the inductance matrix, with $l = l_s + l_m$ and $M = -\frac{1}{2} l_m$ where M, l_s and l_m represent, respectively, the mutual inductance, the stator leakage inductance and the average value of the winding magnetizing inductance.

2.2 *Faulty Model*

Numerous faults presented in the literature can affect the PMSG. Among them, we selected the most frequently one, *i.e.,* the PM flux demagnetization and voltages dip faults. These are described and modeled in the rest of this section.

$\longrightarrow$ Voltage Dip Fault

Taking the case when a healthy PMSG is connected to the grid in case of a voltage dip fault in one phase. When the fault is occuring in phase a, the pahse voltage is expressed by [8]:

$$\begin{cases} v_a = (V - d_a)\cos(\omega t) \\ v_b = V\cos(\omega t - \frac{2\pi}{3}) \\ v_c = V\cos(\omega t + \frac{2\pi}{3}) \end{cases} \tag{3}$$

where, V and d_a present respectively the healthy voltage amplitude and the fault magnitude in phase a. The same procedure is applied if the fault occurs in the phase b or c. To generalize , the three phase voltages can be described by $(1 - \alpha_i)v_i$, $i = a, b, c$, where, α_i, $i = a, b, c$ presents the ratio between the fault magnitude and the healthy phase voltage amplitude.

$\longrightarrow$ Demagnetization fault model

In the PMSG operation, the permanent flux magnitude and direction can vary in a vast range due to the error of the position feedback signal or various external effects, i.e., temperature rise and bad operation conditions that may lead to permanent magnet demagnetization. Thus, a deviation angle appears between the rotor flux directions and the d-axis of the d-q reference frame [7]. The PM flux linkage is illustrated in Fig. 1.

The rise of δ produces a rated flux drop value and hence a machine demagnetisation occurs.

In [7], a mathematical model of rotor PM flux-linkage demagnetization for PMSM in the d-q synchronous reference frame is introduced as follows:

Fig. 1 Permanent magnet flux-linkage deviation

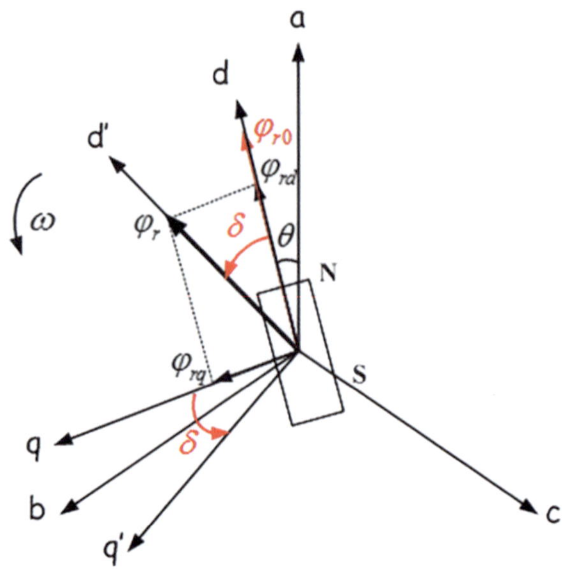

$$\begin{bmatrix} V_d \\ V_q \end{bmatrix} = R\begin{bmatrix} I_d \\ I_q \end{bmatrix} + \frac{d}{dt}\begin{bmatrix} \phi_d \\ \phi_q \end{bmatrix} + \begin{bmatrix} 0 & -\omega \\ \omega & 0 \end{bmatrix}\begin{bmatrix} \phi_d \\ \phi_q \end{bmatrix} \tag{4}$$

where, V_d and V_q are the d-q axis voltages, I_d and I_q are the d-q axis stator currents, ω is the rotor speed, R is the stator resistance and ϕ_d, ϕ_q are the stator flux-linkage that are defined by:

$$\begin{cases} \phi_d = L_d I_d + \varphi_{rd} \\ \phi_q = L_q I_q + \varphi_{rq} \end{cases}, \text{ with } \begin{cases} \varphi_{rd} = \varphi_r \cos\delta \\ \varphi_{rq} = \varphi_r \sin\delta \end{cases}$$

The d-q reference frame can not easily represent the machine model given simultaneous multi-faults consideration. Hence, an (abc) machine model is adopted to support demagnetization as indicated in (5):

$$\begin{aligned} \begin{bmatrix} v_a \\ v_b \\ v_c \end{bmatrix} = R\begin{bmatrix} i_a \\ i_b \\ i_c \end{bmatrix} + [L]\begin{bmatrix} \frac{di_a}{dt} \\ \frac{di_b}{dt} \\ \frac{di_c}{dt} \end{bmatrix} - \varphi_r\omega\left(\begin{bmatrix} \sin(\omega t) \\ \sin(\omega t - \frac{2\pi}{3}) \\ \sin(\omega t + \frac{2\pi}{3}) \end{bmatrix}\right. \\ \left. + \begin{bmatrix} \cos(\omega t) \\ \cos(\omega t - \frac{2\pi}{3}) \\ \cos(\omega t + \frac{2\pi}{3}) \end{bmatrix}\sin\delta - \begin{bmatrix} \sin(\omega t) \\ \sin(\omega t - \frac{2\pi}{3}) \\ \sin(\omega t + \frac{2\pi}{3}) \end{bmatrix}(1 - \cos\delta)\right) \end{aligned} \tag{5}$$

Notice that the first contribution of this paper is to make the dynamical modeling of the faulty system adapted to the acquisition process given by:

$$\Gamma = \zeta(W + \varepsilon) \tag{6}$$

where $\Gamma \in \mathbb{R}^p$ gathers the p observed linear measurements vector, $\zeta \in \mathbb{R}^{p\times q}$ is the fault distribution matrix, ε is a measurement noise and $W \in \mathbb{R}^q$ is the unknown signal faults vector to be recovered.

3 Left-Invertibility Under Sparse Assumption

3.1 Left-Invertibility and Parcimony

Fault detection and estimation are highly simplified using the left-invertibility process. This is based on recovering the information on the failures input from the system measurements [10]. However, the left invertible problem was processed just only in the case when the number of the unknown inputs is equal or fewer than the number of the measurements. In the contrary case, this problem has infinity of solutions. In particular, the SRD method is able to find a simple representation of a signal using some specified sub-matrix. In order to solve the acquisition problem of (6), some assumptions are required:

Assumption 1 (s-sparsity condition) It is assumed that W is s-sparse vector, *i.e.,* no more than s components in W are non-zero. Moreover, s must be lower than p ($s << p$) and verifies the following condition:

$$2s + 1 \leq p \tag{7}$$

Assumption 2 (Restricted Isometry Property RIP) [11, 12] $\forall x \in \mathbb{R}^n$ and $\forall t > 0$, the matrix ζ that depends on the time and the states ($\zeta(x,t)$) satisfies the s-order following condition: $\exists \delta_s \in (0, 1)$ such that for all s-sparse signal W:

$$(1 - \delta_s)\,\|W\|_2^2 \leq \|\zeta(x,t)W\|_2^2 \leq (1 + \delta_s)\,\|W\|_2^2 \tag{8}$$

Note that Assumption 2 is the classical RIP condition extended to matrix ζ depended to time and states.
The goal is to solve an optimization program of SR that minimizes a cost function composed of mean-squared error and the sparsity inducing terms. This sparse reconstruction problem is given by:

$$W^* = \arg\min_{W\in\mathbb{R}^q} \frac{1}{2}\,\|\Gamma - \zeta W\|_2^2 + \lambda\,\|W\|_1 \tag{9}$$

where $\|W\|_1 \triangleq \sum\limits_i |W_i|$.

3.2 SRD Algorithm Presentation

In order to solve the optimization problem (9) and converge to its equilibrium point in finite time, a dynamical algorithm based on sliding mode techniques was proposed in [13] and it is defined by 10.

$$\begin{cases} \tau\dot{u} = -\left\lceil u(t) + (\zeta^T\zeta - I_q)\hat{W} - \zeta^T\Gamma \right\rfloor^{\alpha} \\ \hat{W} = \Lambda_\lambda(u) \\ \lceil . \rfloor^{\alpha} \triangleq |.|^{\alpha}\mathrm{sgn}(.) \end{cases} \tag{10}$$

where $u \in \mathbb{R}^q$ represents the state vector, $\hat{W}$ is the estimation of the sparse signal W, I_q is the identity matrix, τ is a time constant which depends on the physical properties of the system and $\Lambda_\lambda(u)$ is a continuous soft-thresholding function that is defined by the system (11):

$$\Lambda_\lambda(u) = \begin{cases} 0 & if\ |u| \leq \lambda \\ (u-\lambda)\mathrm{sgn}(u) & if\ |u| > \lambda \end{cases} \tag{11}$$

with a balancing parameter $\lambda > 0$.
Under the assumption that M satisfies the s-sparsity and RIP condition, the output of the system (10) will converge to its equilibrium point, and $\hat{W}$ converges in finite time to W^* [13].

3.3 System Transformation

Including all the faults presented in Sect. 2, the global system model will be reduced to construct a simple model for the diagnosis algorithm. This model is detailed in (12) by Non Linear Functions (NLFs):

$$\begin{bmatrix} v_a \\ v_b \\ v_c \end{bmatrix} - R\begin{bmatrix} i_a \\ i_b \\ i_c \end{bmatrix} - [L]\frac{d}{dt}\begin{bmatrix} i_a \\ i_b \\ i_c \end{bmatrix} + \varphi_r\omega\begin{bmatrix} \sin(\omega t) \\ \sin(\omega t - \frac{2\pi}{3}) \\ \sin(\omega t + \frac{2\pi}{3}) \end{bmatrix} = -\varphi_r\omega\begin{bmatrix} \cos(\omega t) \\ \cos(\omega t - \frac{2\pi}{3}) \\ \cos(\omega t + \frac{2\pi}{3}) \end{bmatrix}\sin\delta$$
$$+\varphi_r\omega\begin{bmatrix} \sin(\omega t) \\ \sin(\omega t - \frac{2\pi}{3}) \\ \sin(\omega t + \frac{2\pi}{3}) \end{bmatrix}(1-\cos\delta) + \begin{bmatrix} \alpha_a & 0 & 0 \\ 0 & \alpha_b & 0 \\ 0 & 0 & \alpha_c \end{bmatrix}\begin{bmatrix} v_a \\ v_b \\ v_c \end{bmatrix} \tag{12}$$

The next step is to make a transformation of (12) to the form of the model (6). Thus, respecting the given conditions, a matrix ζ and a vector including faults informations must be well designed. However, regarding (12), we note that the demagnetization fault characterised by δ acts in a non linear way by the quantities $\sin\delta$ and $(1-\cos\delta)$. We are not able to gather these two different faults generated by the same defect which is the demagnetization. The considered fault vector dimension increases to eight.

In order to bring back our model to the form of the acquisition process given by (6), each term will be replaced by the adequate quantities as given bellow.

- $\Gamma = \begin{bmatrix} v_a \\ v_b \\ v_c \end{bmatrix} - R \begin{bmatrix} i_a \\ i_b \\ i_c \end{bmatrix} - [L] \frac{d}{dt} \begin{bmatrix} i_a \\ i_b \\ i_c \end{bmatrix} + \varphi_r \omega \begin{bmatrix} \sin(\omega t) \\ \sin(\omega t - \frac{2\pi}{3}) \\ \sin(\omega t + \frac{2\pi}{3}) \end{bmatrix}$
- $\zeta = \begin{bmatrix} -\varphi_r \omega \cos(\omega t) & \varphi_r \omega \sin(\omega t) & v_a & 0 & 0 \\ -\varphi_r \omega \cos(\omega t - \frac{2\pi}{3}) & \varphi_r \omega \sin(\omega t - \frac{2\pi}{3}) & 0 & v_b & 0 \\ -\varphi_r \omega \cos(\omega t + \frac{2\pi}{3}) & \varphi_r \omega \sin(\omega t + \frac{2\pi}{3}) & 0 & 0 & v_c \end{bmatrix} \in \mathbb{R}^{3\times 5}$

  and

$$\begin{aligned} W &= [W_1 \quad W_2 \quad W_3 \quad W_4 \quad W_5]^T \\ &= [\sin\delta \quad 1-\cos\delta \quad \alpha_a \quad \alpha_b \quad \alpha_c]^T \end{aligned} \tag{13}$$

The vector $\Gamma \in \mathbb{R}^3$ presents the measurement vector that includes the available informations about the system as the measurements, its derivatives and the healthy operating system dynamics. The derivatives of the three phase currents $\frac{di_a}{dt}$, $\frac{di_b}{dt}$ and $\frac{di_c}{dt}$ are extracted in finite-time using the Super Twisting Differentiators (STD).

In our case, three outputs system are available. Thus, satisfying the s-sparsity of parsimonious solution (see Assumption 1), we are able to diagnose only one of the five possible faults. By increasing the sensors number (measurements), we can obtain more system information and detect more simultaneous faults. The considered faults are assumed linearly independent, exceptionally, W_1 and W_2 can occur simultaneously since they are related to the same fault.

After verifying the ζ required conditions in a real-time, the dynamical algorithm will be applied and W can be estimated and exactly recovered from the vector measurements.

4 Application of the Proposed Method and Simulation Results

The structure of the proposed diagnosis method is summarized in Fig. 2. At first, a Super Twisting Differentiator is used to estimate the system outputs derivatives which are considered as virtual measurements. After that, a model of Non Linear Functions is constructed based on the system dynamics, outputs measurements and their derivatives [see (12)]. This model is then transformed to the adequate form required by the SRD algorithm [see model (6)]. Finally, the fault vector W is recovered by the proposed algorithm and its estimation is obtained.

In order to evaluate faulty system models and the proposed diagnosis algorithm, simulations are carried on via Matlab/Simulink. The system parameters operated in this simulation are given in Table 1. For the designed system, a solver $Ode1$ (Euler)

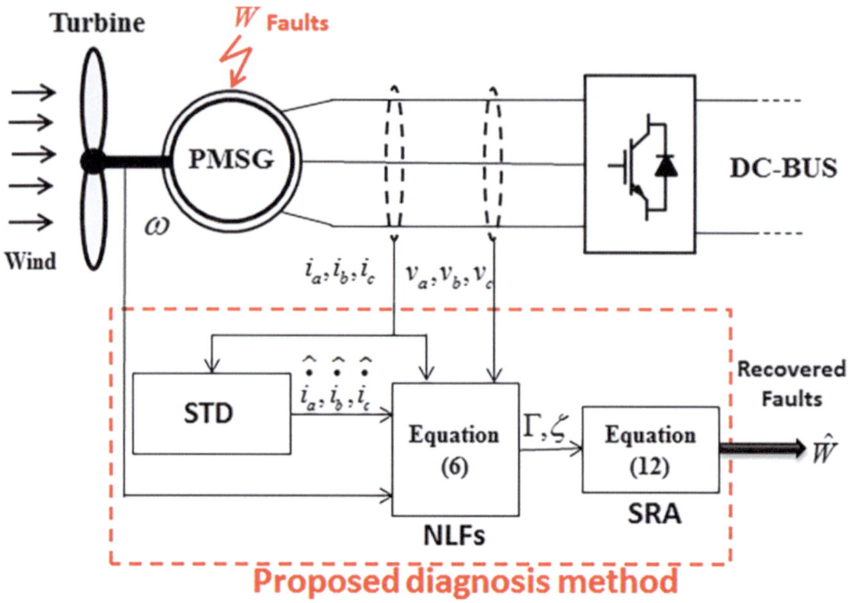

Fig. 2 Structure of the SRD method

Table 1 System parameters

Parameters	Values
d-axis stator inductance L_d	0.0044 H
q-axis stator inductance L_q	0.0044 H
Stator resistance R_s	0.5 Ω
Permanent magnetic flux Ψ_f	0.39 Wb
DC-link capacitor C	2.5 mF
Number of pole-pairs n_p	19
Load resistance	100 Ω

is chosen since the Eq. (10) is not C^1-smooth, the constant α is equal to 0.5 ensuring a convergence in a finite time, $\lambda = 10^{-3}$, $\tau = 10^{-3}$ s and $\mu = 10^4$.

The five faults modeled in (12), are implemented, although, we consider that only one fault ($s = 1$) is occurring in order to satisfy the s-sparsity condition ($s \leq p < q$) explained in (6). Recall that in our case, the measurements dimension is equal to 3 ($p = 3$) and the faults dimension is equal to 8 ($q = 5$) then the matrix ζ is 3×5 with a time varying elements.

Several scenarios are considered in order to show the good performances and robustness of the SRD method, two of them will be presented:

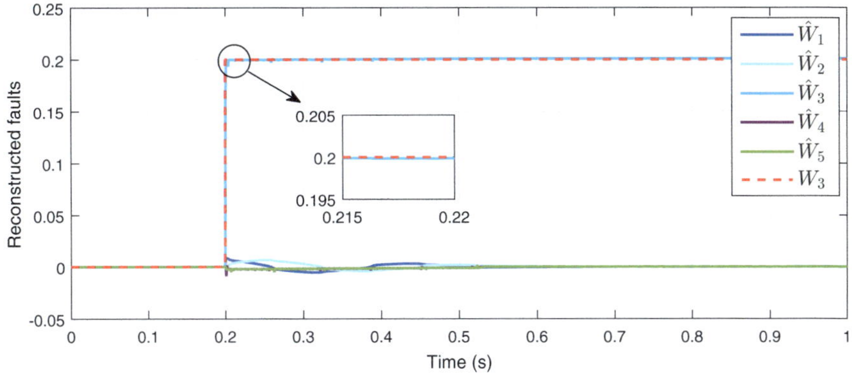

Fig. 3 Detection of the voltage dip fault in phase A

4.1 *Scenario* 1

In this scenario, we consider a voltage dip fault appearing in the phase a at $t = 0.2$ s. This fault is characterized by the voltage dip ratio α_a that is chosen equal to 0.2. The simulation results, obtained due to this scenario, are illustrated in Fig. 3. We note that the estimated fault $\hat{W}_6$ related to the activated fault converges to its real value α_a while the others non activated faults converge to zero.

4.2 *Scenario* 2

In this scenario,, the proposed method is applied to extract the demagnetization fault that affect the PMSG at $t = 0.2$ s. This fault is characterized by the deviation angle δ. In our simulation, it is set to 30°. As it is explained in Sect. 3, the two faults W_1 and W_2 are coordinated (*i.e.,* the same fault linked to the demagnetization generates the activation of W_1 and W_2). Thus, in this case, s will be equal to 2 ($s = 2$) that does not respect the s-sparsity condition [see Assumption 2]. Nevertheless, the two considered non-zero faults on W are accurately recovered in a finite time while the estimated values of other zero faults converge to zero. That's clearly shown in Fig. 4; only $\hat{W}_1$ and $\hat{W}_2$ related to the demagnetization fault converge to their values ($\sin\delta$ and $(1 - \cos\delta)$).

The detection of the demagnetization is ensured at a time by the activation of the two independent fault indicators W_1 and W_2. The accurate estimation of W_1 and W_2 yields to an accurate determination of the deviation angle δ. This later gives the field drop step. One can verify that by the Eq. (4).

In this scenario, we are achieved to recover simultaneously two elements of the vector faults W. This is due to the fact that the RIP is sufficient and not necessary condition.

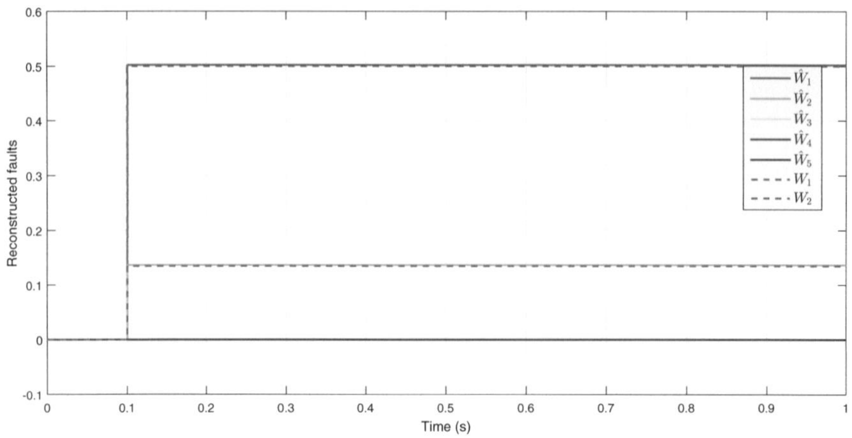

Fig. 4 Detection of the demagnetization fault

5 Conclusion

In this work, a sparse recovery diagnosis method is applied to a PMSG wind turbine. It is used in order to detect, locate and estimate in finite time convergence a demagnetization and voltage dip fault. In healthy and faulty cases, dynamical modelisations are presented and implemented on Matlab/Simulink. The simulation results, given in the last section, highlighted the well founded of the proposed algorithm and its effectiveness to recover a sparse fault vector. Our future work will be a comparison with other diagnosis methods in the stationary and non-stationary cases.

References

1. Gomez-Espinosa A, Hernandez-Guzman VM, Bandala-Sanchez M, Jimenez-Hernandez BH, Rivas-Araiza EA, Rodriguez-Resendiz J, Herrera-Ruiz G (2013) A new adaptive self-tuning Fourier coefficients algorithm for periodic torque ripple minimization in permanent magnet synchronous motors (pmsm). Sensors 13(3):3831–3847
2. Moon S, Lee J, Jeong H, Kim SW (2016) Demagnetization fault diagnosis of a PMSM based on structure analysis of motor inductance. IEEE Trans Industr Electron 63(6):3795–3803
3. Choi G, Jahns TM (2016) Investigation of key factors influencing the response of permanent magnet synchronous machines to three phase symmetrical short-circuit faults. IEEE Trans Energy Convers 31(4):1488–1497
4. Kang K, Song J, Kang C, Sung S, Jang G (2017) Real-time detection of the dynamic eccentricity in permanent-magnet synchronous motors by monitoring speed and back emf induced in an additional winding. IEEE Trans Industr Electron 64(9):7191–7200
5. Ding P, Wang H, Bao W, Hong R (2019) HYGP-MSAM based model for slewing bearing residual useful life prediction. Measurement 162–175:141

6. Sabir R, Rosato D, Hartmann S, Guhmann C (2019) Detection and localization of electrical faults in a three phase synchronous generator with rectifier. In: 2019 international conference on electrical drives and power electronics (EDPE). IEEE, pp 18–23
7. Zhao KH, Chen TF, Zhang CF, He J, Huang G (2014) Online fault detection of permanent magnet demagnetization for IPMSMS by nonsingular fast terminal-sliding-mode observer. Sensors 14(12):23119–23136
8. El Malleh K, Chariag DE, Sbita L et al (2019) A control strategy for a three-phase grid connected PV system under grid faults. Electronics 8(8):906
9. Grouz F, Sbita L and Boussak M (2012) Particle swarm optimization based fault diagnosis for non-salient PMSM with multi-phase inter-turn short circuit. In: International conference on communications, computing and control applications. IEEE, pp 1–6
10. Barbot JP, Boutat D, Floquet T (2009) An observation algorithm for nonlinear systems with unknown inputs. Automatica 45(8):1970–1974
11. Ren J, Yu L, Lyu C, Zheng G, Barbot JP, Sun H (2019) Dynamical sparse signal recovery with fixed-time convergence. Signal Processing 162:65–74
12. Candes EJ (2008) The restricted isometry property and its implications for compressed sensing. Comptes rendus, Mathematique 346(9–10):589–592
13. Yu L, Zheng G, Barbot JP (2017) Dynamical sparse recovery with finite-time convergence. IEEE Trans Signal Process 65(23):6146–6157

Design and Modeling of a PV Plant Connected to the Network for the Company Timab-Tunise

Zribi Hazem, Abid Aïcha, Dhaoui Mehdi, and Ben Hamed Mouna

Abstract The company "Timab Tunisie" wishes to set up a PV central with a power of 1 MW connected to the network. The objective is to compensate for its energy needs in terms of electricity. This central will make it possible to save primary energy about 465.43 toe/year and reduce interrupted CO_2 emissions by around 1108 tCO_2/year. The work presented in this article aims at the design and modeling of the PV generator. PV central is equipped with five independently controlled static converters. Digital MPPT controllers integrated in the inverters ensure the search for maximum power delivered. After the description of the structure of the system, we present the simulation results for different climatic conditions as well as an estimate of the annual energy production.

Keywords PV plant · Network · Modelling · Digital MPPT

1 Introduction

Timab Tunisie Company is one of the largest subsidiaries of the ROULLIER group specializing in the production and marketing of food phosphates with a production capacity of 250,000 tons/year with electrical energy consumption of around 4,405,770 kWh/year.

The company has approximately 100 employees and serves more than 300 customers in 82 countries.

The minimization of production downtime produced by power cuts and strict constraints on its customers makes the search for an alternative energy source imperative for the profitability of the company [1].

For all these reasons, the company aims to install a PV plant.

Z. Hazem (✉) · A. Aïcha · D. Mehdi · B. H. Mouna
L/R: Processes, Energy, Environment and Electrical Systems (PEESE), Electrical Department of Engineering, National Engineering School, University of Gabes, 6029 Gabes, Tunisia
e-mail: hazemzribi1978@gmail.com

© The Author(s), under exclusive license to Springer Nature Singapore Pte Ltd. 2024

A. Mellit et al. (eds.), *Proceedings of the 2nd International Conference on Green Energy Conversion System*, Green Energy and Technology,
https://doi.org/10.1007/978-981-97-6148-7_43

Having your own source of photovoltaic energy makes it possible to reduce your dependence on the STEG network and fluctuations in the price of energy. It also allows in some cases to be autonomous in terms of energy production. This is a laudable investment, since the cost of the proposed plant is made profitable by the savings made on the energy bill.

In the case of connected PV plants, the energy produced can be sold in full to network. A new financial income is thus realized by the resale of the electricity produced.

The study, design and modeling of a PV plant are essential steps to properly control and optimize the operation of the PV generator.

In this work, we have in a first step modeled the proposed PV plant. We also looked at the MPPT (Maximum Power Point Tracking) command.

Then, and to determine the static operating limits, we tested the system under different metrological constraints, whether in terms of irradiation level or temperature.

2 Design of the PV Plant

The structure of the studied PV plant (Fig. 1) consists of a photovoltaic generator, five converters (DC/AC) each equipped with 2 MPPTs and an electrical load. The assembly is connected to the medium voltage network.

3 Modeling of Photovoltaic Generator

The PV generator with a total peak power of 1.0 MW is made up of a set of 3300 photovoltaic modules divided into 5 PV fields each comprising 66 parallel branches and 10 modules in series.

The simplified model of the PV generator is described by the following equation:

$$i_{PV} = N_p \cdot I_{ph} \cdot \frac{G}{G_s} - N_p \cdot I_s \cdot \left(\exp\left(\frac{A_t \cdot N_p}{N_s} \right) - 1 \right) \tag{1}$$

where, I_{pv} is the current supplied by the PV generator, V_{pv} is the voltage at the generator terminals, I_{ph} is the photo current, directly proportional to the illumination, N_s is the number of modules in series, N_p is the number of parallel branches, G is the irradiation and G_r is the reference irradiation (1000 W/m^2).

PV module consists of 96 cells, with 12 cells in series and 8 in parallel and whose characteristics are given in Table 1. this module is tested for different metrological values.

Figure 2 show I(V) and P(V) characteristics for various irradiation and at 25 °C for one PV array.

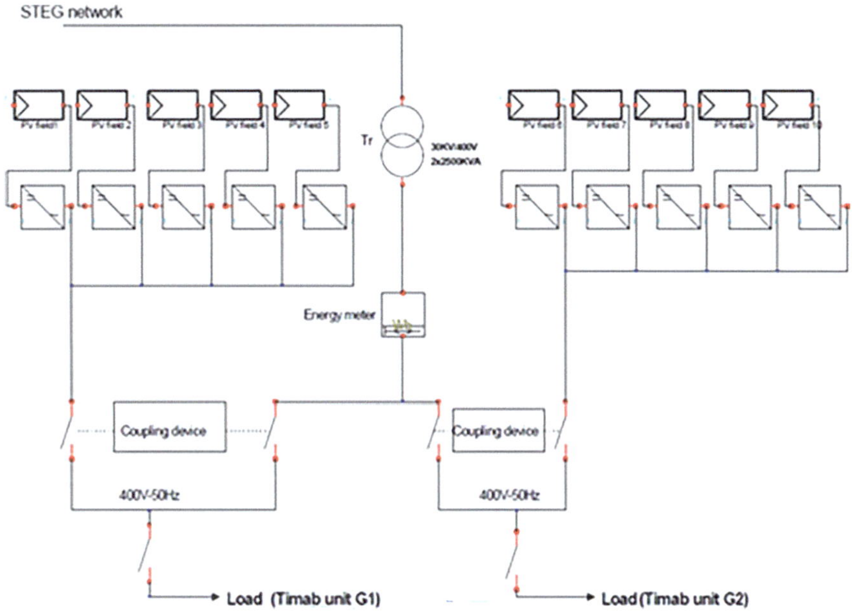

Fig. 1 Structure of the studied PV plant

Table 1 Characteristics of the PV generator for variable irradiation and temperature

Values	At T = 25 °C			At G = 1000 W/m^2		
	200 W/m^2	600 W/m^2	1000 W/m^2	15 °C	25 °C	45 °C
V_{mpp} (V)	526	543	547	565.5	547	510
I_{mpp} (A)	368	1102	1835	1826	1835	1851
P_{mpp} (kW)	193.5	599	1000	1033	1000	944

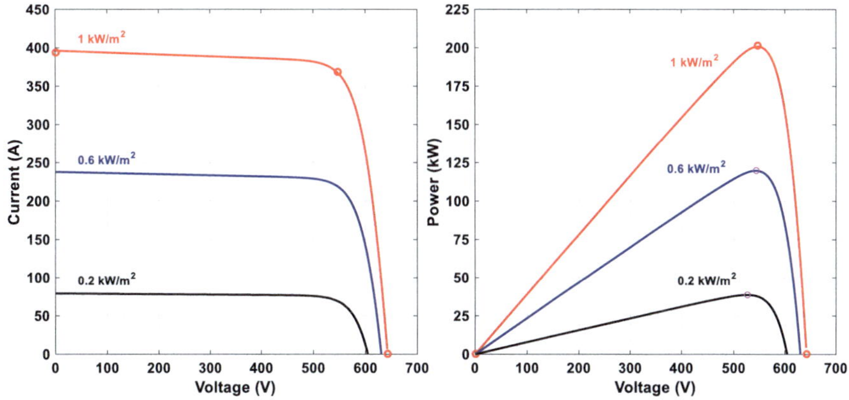

Fig. 2 Influence of irradiation on I(V) and P(V) characteristics of PV array

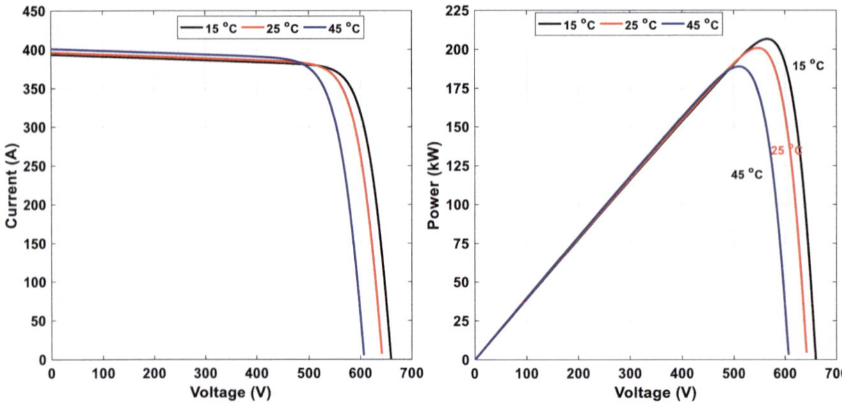

Fig. 3 Influence of temperature on I(V) and P(V) characteristics of array

From the obtained characteristics, it can be seen firstly, that the influence of irradiation is remarkable. Indeed, an increase in lift leads to the irradiation characteristic upwardly. Thus, this influence affects mainly the short-circuit current.

Typical characteristics (368 A, 547 V and 200 kW) of PV array is obtained for 1000 W/m^2 and a 25 °C, which shows good agreement with the experimental data of the manufacturer [2, 3].

Secondly, the effect of temperature is significant to the photovoltaic cell characteristics. Indeed, for the silicium, the current increases from about 0.025 mA/cm^2/°C, while the voltage decreases from 2.2 mV/°C/cell. This results a power reduction about 0.4%/°C [4, 5].

Figure 3 show the I(V) and P(V) characteristics for various values of temperature and for a standard irradiation (1000 W/m^2).

The results clearly show that a PV array output power decreases with temperature above 25 °C and increases with temperatures below 25 °C.

The PV generator has five PV array connected in parallel. This interconnect provides the same voltage with 5 times greater current and peak power.

Under Standard Test Conditions STC (1000 W/m^2 and 25 °C), the PV generator provides a current of around 1835 A with a power of 1.0 MW.

The operation of the PV generator under different levels of irradiation and temperature gave the results shown in Figs. 4 and 5.

From the obtained characteristics, it can be shown that in the conditions defined by a nominal irradition of 1000 W/m^2 and a temperature of 25 °C, the short-circuit current of the photovoltaic generator is about 1967 A and the open circuit voltage is 642 V. The optimal values is obtained for a voltage V_{mpp} = 827 V, a current I_{mpp} = 19.5 A, and a power P_{mpp} about 1.0 MW.

Table 1 summarizes the various measurements made on the PV generator for variable irradiation and temperature.

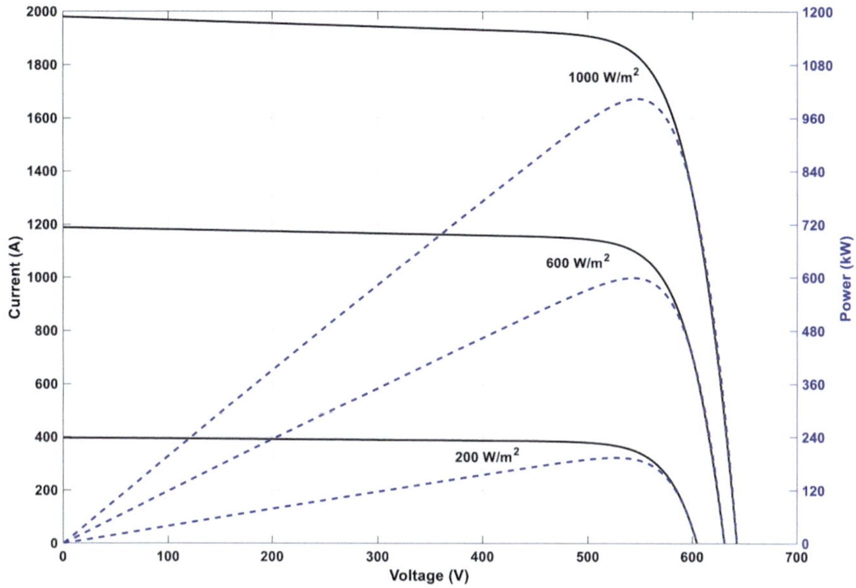

Fig. 4 Influence of irradiation on I(V) and P(V) characteristics of PV generator

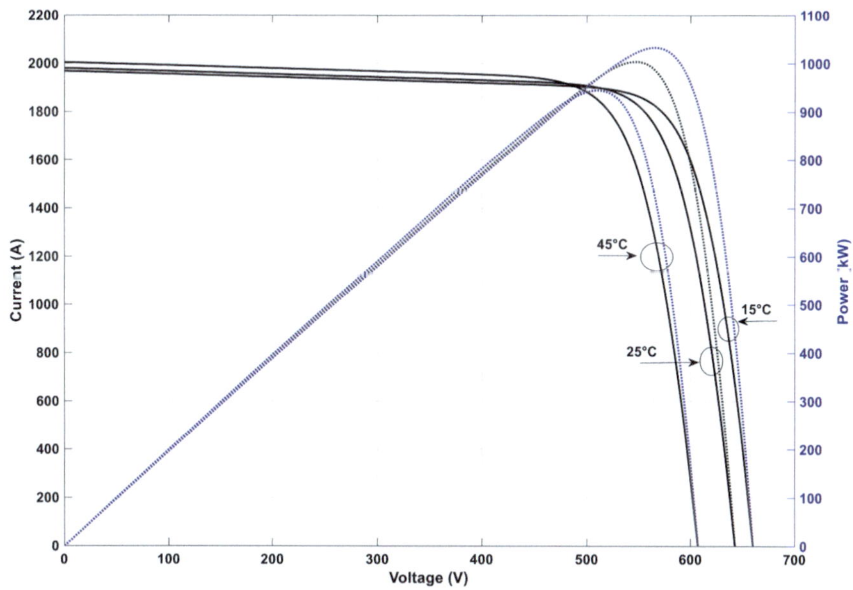

Fig. 5 Influence of temperature on I(V) and P(V) characteristics of PV generator

4 Optimization of the Provided Power

The chopper allows, from a continuous PV voltage to obtain a continuously variable voltage between V_{co} to V_{PV_opt} (Buck) and even beyond V_{PV_opt} (Boost) with the advantage of a quick control (Fig. 6). The mean value of the chopper output voltage (V_{sh}) is expressed in terms of the average value (V_{PV}) and the duty ratio α.

MPPT control is a functional body of the PV system according to meteorological variables, to seek the optimal operating point of the PV generator [6]. The principle of this regulation is based on the automatic variation of the duty cycle α to maximize the power P_{PV} at any time in the generator output.

$$P_{PV} = V_{PV} \cdot I_{PV} \tag{2}$$

The perturbation and observation method (P&O) is a widely used approach in the search for MPPT because it is simple and requires only measures voltage VPV and IPV current of the PV panel. It can track the maximum power point even when variations of illumination and temperature [7, 9, 10].

When the maximum power point is reached, VPV oscillates around the optimal value V_{PV_opt}. This causes a power loss which increases with no increment of the step. In the large increment, the MPPT algorithm responds quickly to sudden changes in operating conditions. On the other hand, if the step is small, the losses during conditions of slow or stable metrological changes will be lower, but the system cannot respond quickly to rapid changes in temperature or illumination [8, 11, 12].

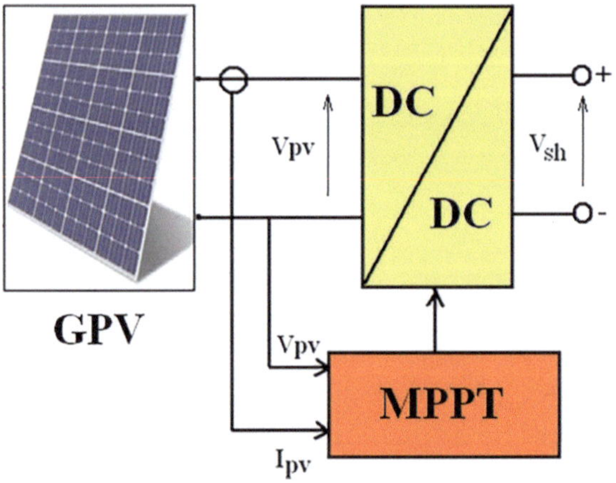

Fig. 6 Block diagram of a PV system with MPPT

5 AC Power Expected and Performance of the PV System

Determining the AC power expected out of a PV system requires additional considerations.

The output power of a PV module is specified in watts and with a manufacturing tolerance based on a cell temperature of 25 °C. Assuming the tolerance is ± 5%.

This power can be reduced because of a build-up of dirt on the surface of the module. Assuming power loss due to dirt is 5%.

A solar modules output power decreases with temperature above 25 °C and increases with temperatures below 25 °C. Assuming power loss due to temperature is 15%. Then, the "worst case" adjusted output of a 305 W PV module is therefore around.

$$P_{\text{module}} = P_C - \sum (DC_{loss}) = 0.234 \text{ kW} \tag{3}$$

The actual DC energy from the solar generator is equal to the derated output power of the module × number of modules × irradiation for the tilt and azimuth angle of the array. The average daily PSH is about 6.36 and that are 3300 modules in the PV generator. Therefore the DC energy output of the array is:

$$E_{DC} = P_{\text{module}} \times N_{\text{module}} \times PSH = 4911 \text{ kWh/day} \tag{4}$$

The DC energy output will be further reduced by the power loss (voltage drop) in the DC cable connecting the solar array to the grid connect inverter. Assume that the cable losses for the DC cables are 3%.

This energy, delivered to the input of the inverter will be further reduced by the power energy loss in the inverter. Assume that the inverter efficiency is 98%.

The AC energy output of the inverter will be further reduced by the power loss in the AC cable connecting the inverter to the grid, say switch board where it is connected. Assume that the cable losses are 1%.

Finally, the average daily AC energy that was delivered by the generator to the grid will be:

$$E_{AC} = E_{DC} - \sum (AC_{loss}) = 4621 \text{ kWh/day} \tag{5}$$

Our system is formed by 3300 modules, each with a STC rating of 305 Wp. Therefore, the rated output power of the array under STC (standard test conditions) is:

$$P_{C\text{-}STC} = 3300 \times 0.305 = 1006 \text{ kWp} \tag{6}$$

Therefore over a typical year of 365 days, then energy yield of the solar generator is:

$$E_{gen} = 365 \times E_{AC} = 1,686,665 \text{ kWh/year}$$

The specific energy Yield is expressed in kWh per kWp and it calculated as follows:

$$SY = \frac{E_{gen}}{P_{C\text{-}STC}} = 1677 \text{ KWh/KWp} \tag{7}$$

The performance ratio (PR) is used to access the installation quality. The PR provides a normalized basis so comparison of different types and sizes of PV systems can be under taken. The performance ratio is a reflection of the system losses:

$$PR(\%) = \frac{E_{gen}}{E_{ideal}} \cdot 100 = \frac{E_{sys}}{P_{C\text{-}STC} \cdot H_{tilt}} \cdot 100 = \frac{SY}{H_{tilt}} \cdot 100 \tag{8}$$

where E_{gen} is the actual yearly energy yield from the system; E_{ideal} is the ideal energy output of the generator and H_{tilt}: the yearly average daily irradiation, in kWh/m^2 for the specified tilt angle.

The average daily PSH at Gabes is about 5. Therefore, the yearly irradiation H_{tilt} would be 2322 KWh/m^2. Then, the performance ratio is:

$$PR(\%) = \frac{1318}{2322} \cdot 100 = 72.22\%. \tag{9}$$

6 Anticipated Production PV Generator

The PV system will be installed in Gabes.

The characteristics of the implantation site are given by the online software PVGIS (Table 2).

Table 2 Site implantation characteristics

Latitude	33.92° N
Longitude	10.09° E
Altitude	7 m
Inclination of the solar panel	35°

The average daily solar insolation in units of kWh/m^2 per day is sometimes referred to as "peak sun hours". The term "peak sun hours" refers to the solar insolation which a particular location would receive if the sun were shining at its maximum value for a certain number of hours. Since the peak solar radiation is 1000 W/m^2, the number of peak sun hours is numerically identical to the average daily solar insolation.

The site of Gabes receives 5 kWh/m^2/day. It can be said to have received 5 h of sun per day at 1000 W/m^2. Being able to calculate the peak sun hours is useful because PV modules are often rated at an input rating of 1000 W/m^2.

Figures 7 and 8 respectively shows the monthly irradiation and the energy production.

Table 3 give the various measurements made on the PV generator with the PVGis 2023.

It appears that the horizontal solar radiation is large relative to the tilted solar radiation. This behavior is observed during the sunniest months. For months 1, 2, 3, 10, 11 and 12, **the** trend reverses. This does not affect the operation of the photovoltaic station since the power supplied is low.

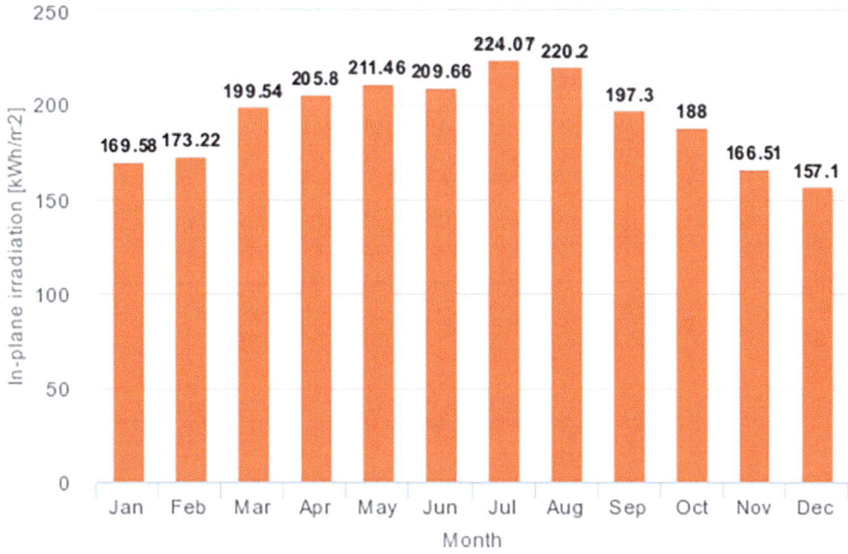

Fig. 7 Monthly in-plane irradiation for fixed angle (PVGIS 2023)

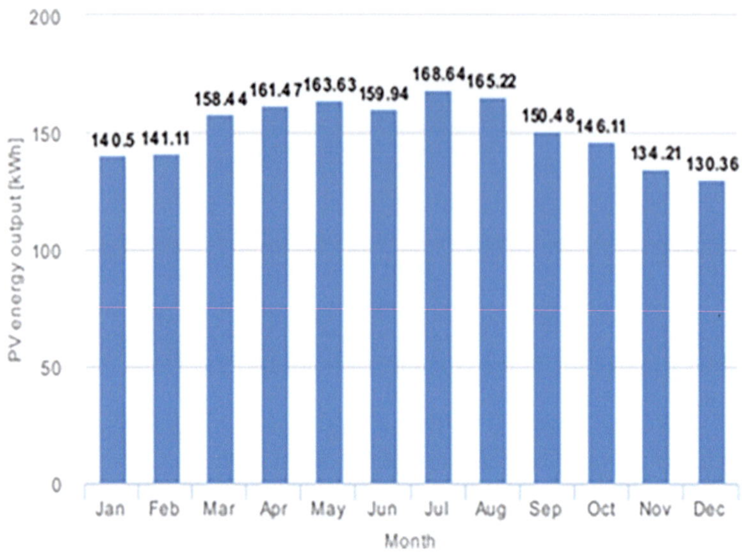

Fig. 8 Monthly energy output from fix-angle PV system (PVGIS 2023)

Table 3 Various measurements made on the PV generator (PVGis 2023)

Provided inputs	
Location (Lat/Lon)	33.920, 10.090
Horizon	Calculated
Database used	PVGIS-SARAH2
PV technology	Crystalline silicon
PV installed (kWp)	1000
System loss (%)	14
Simulation outputs	
Slope angle (°)	35
Azimuth angle (°)	0
Yearly PV energy production (kWh)	1,820,100.04
Yearly in-plane irradiation (kWh/m^2)	2322.45
Year-to-year variability (kWh)	21,043.49
Changes in output due to	
Angle of incidence (%)	– 2.53
Spectral effects (%)	0.63
Temperature and low irradiance (%)	– 7.09
Total loss (%)	– 21.63

When the horizontal solar radiation reaches the value 1000 W/m^2, total power output of the system is about 1.0 MW.

7 Conclusion

In this paper, the design and modeling of the installer photovoltaic generator for the company Timab-Tunisie were presented. The studied structure includes 3300 PV modules divided into 5 PV arrays and 5 PV inverters with MPPT. The system has been tested for different values of metrological quantities.

The results show that the performance of the PV generator degrades with the increase in temperature and increases with the evolution of the irradiation. The MPPT regulation optimizes the power delivered by the PV generator according to the irradiation. This allows the PV system to work at the optimum operating point.

In addition, the installation of the PV generator could improve the energy balance of the Timab Company, providing it with approximately 1.7 GWh, which represents a considerable part of the electrical energy needed.

All the results of this work show the efficiency of the photovoltaic plant.

Appendix

See Table 4.

Table 4 PV module specifications

Electrical typical data	Sun power SPR 305 E
Pmpp (W)	305
Vmpp (V)	54.7
Impp (A)	5.58
Voc (V)	64.2
Isc (A)	5.96
Practical module efficiency	17.3%
Specified Standard Temperature $\boldsymbol{T_s}$	25 (°C)
Specified Standard illumination G_s	1000 W/m^2
Voltage temperature coefficients	− 0.27269%/°C
Current temperature coefficients	0.061745%/°C
Dimensions	884 × 546 × 35 mm (L × W × H)
Temperature range	− 40 °C to + 85 °C

References

1. Document, 'Etude Stratégique sur le Développement des Energies Renouvelables en Tunisie', Agence Nationale des Energies Renouvelables, Novembre 2003
2. Zubi G, Bernal-Agustın JL, Fracastoro GV (2009) High concentration photovoltaic systems applying III–V cells. Renew Sustain Energy Rev 13(9):2645–2652
3. Takamoto T, Ikeda E, Kurita H, Ohmori M, Yamaguchi M (1997) Two-terminal Monolithic $In_{0.5}Ga_{0.5}P$/GaAs tandem solar cells with a high conversion efficiency of Over 30 %. Jpn J Appl Phys 36(10):6215–6220
4. Stan M, Aiken D, Cho B, Cornfeld A, Ley V, Patel P, Sharps P, Varghese T (2010) High efficiency quadruple junction solar cells using OMVPE with inverted metamorphic device structures. J Cryst Growth 312(8):1370–1374
5. King RR, Boca A, Hong W, Liu X-Q, Bhusari D, Larrabee D, Edmondson KM, Law DC, Fetzer CM, Mesropian S, Karam NH (2009) Band-gap-engineered architectures for high-efficiency multijunction concentrator solar cells. In: 24th European photovoltaic solar energy conference, 21–25 September 2009, Hamburg, Germany, 1AO.5.2, pp 55–61
6. Ait-Cheikh SM (2007) Etude, investigation et conception d'algorithmes de commande appliqués aux systèmes photovoltaïques. Thèse de Doctorat d'état, Ecole Nationale Polytechnique, Alger, Algérie
7. Jain S, Agarwal V (2007) Comparison of the performance of maximum power point tracking schemes applied to single-stage grid-connected photovoltaic systems. IEEE Trans Power Electron 1(5):753–762
8. Koutroulis E, Kalaitzakis K, Voulgaris NC (2001) Development of a microcontroller-based, photovoltaic maximum power point tracking control system. IEEE Trans Power Electron 16(1):46–54
9. Ishaque K, Salam et al (2012) An improved particle swarm optimization (PSO) based MPPT for PV with reduced steady state oscillation. IEEE Trans Power Electron 27(8):3627–3638
10. Safari A, Mekhilef S (2011) Simulation and hardware implementation of incremental conductance MPPT with direct control method using Cuk converter. IEEE Trans Indus Electron 58(4):1154–1161
11. Soon TK et al (2013) Simple and low cost incremental conductance maximum power point tracking using buck-boost converter. J Renew Sustain Energy 5:023106–023110
12. Rahmani R et al (2013) Hybrid technique of an colony and particle swarm optimization for short term wind energy forecasting. J Wind Eng Indus Aerodyn Elsevier Part A 123:163–170

Control Design of Two Boost Choppers in Parallel by Petri Net and Flatness Approaches

Souhir Messaoudi, Florentina Nicolau, Malek Ghanes, Lassaad Sbita, and Jean-Pierre Barbot

Abstract In this paper, we consider two boost choppers connected in parallel, with two different sources, and a load with low power demand. The two boost choppers can enter in a discontinuous mode of operation by providing a small current that approaches zero. To prevent this, we propose a new strategy for power distribution based on the value of the current at the limit between continuous and discontinuous modes. In the case of a very low power demand, one boost chopper can be disconnected, ensuring a continuous mode operation for the remaining boost chopper.

Keywords Petri nets · Power electronics · Average model · Flatness

1 Introduction

This article[1] is a continuation of [8], where a flatness-based approach (see [5], for a seminal reference for flatness) enabled efficient control of several boost choppers connected in parallel with different characteristics, the objective being to make them work on the same voltage bus. Under the assumption that all boost choppers are in continuous mode, in [8], we gave an in-depth study of flatness of the system and proposed a flat output for this type of boost choppers configuration, which ensures homogeneous power sharing among the choppers. However, in that paper the discontinuous mode operation of the boost choppers is not considered. Hence, the contribution of the present paper is to propose a new strategy for power distribution in the case of two boost choppers that takes into account the discontinuous mode

[1] This work is supported by Campus France via the PHC Maghreb 46003SM project.

S. Messaoudi (✉) · M. Ghanes · J.-P. Barbot
LS2N UMR 6004 CNRS, Ecole Centrale Nantes, Nantes, France
e-mail: Souhir.Messaoudi@ec-nantes.fr

F. Nicolau · J.-P. Barbot
QUARTZ EA 7393, ENSEA, Cergy-Pontoise, France

S. Messaoudi · L. Sbita
PEESE LR18ES34, ENIG, Université Gabes, Gabes, Tunisie

© The Author(s), under exclusive license to Springer Nature Singapore Pte Ltd. 2024
A. Mellit et al. (eds.), *Proceedings of the 2nd International Conference on Green Energy Conversion System*, Green Energy and Technology,
https://doi.org/10.1007/978-981-97-6148-7_44

operation, which was one of the main perspective of [8]. The strategy is based on the value of the current at the limit between continuous and discontinuous modes of the boost choppers. It is designed using Petri Net approach and allows the choice of the tracking objective for {the flatness-based controllers that were designed in [8]. The integration of Petri Nets in complex systems analysis and design has provided new opportunities for optimizing power flow, control strategies, and overall system performances. This approach has found applications in various areas, including renewable energy systems [6], power distribution networks [7, 9], and smart grid technologies [1, 3]. The paper is organized as follows. The averaged modeling of the two boost choppers system is presented in Sect. 2. Section 3 gives a flat output for the overall system, and presents the control laws. Section 4 describes the proposed control strategy based on a Petri Nets representation. Finally, simulation results and some future work directions are given in Sect. 5.

Notations: $i = 1, 2$ integer labeling the boost choppers, I_i current in the inductance of the ith boost chopper, I_{avi} averaged current in the ith inductance, $V_{C_{av}}$ averaged bus voltage, V_C bus voltage, C bus capacitance, T switching period, $F = \frac{1}{T}$ switching frequency, V_i source voltage of the ith boost chopper, L_i inductance of the ith boost chopper, P_{ri} rated power of the ith boost chopper, R load resistance, I_{di} desired averaged current, $\alpha_{i,c}$ duty cycle, $\alpha_{i,d}$ diode conduction ratio in discontinuous mode, V_{dC} desired averaged bus voltage, E_t total energy stored in booth choppers, E_{dt} desired total energy, P_{limc} power limit in continuous mode, P_{limh} power limit of homogeneous power distribution.

2 Two Boost Choppers Modeling

A boost chopper is a type of switching power supply commonly used to convert DC voltage into a higher regulated DC voltage [4, 10]. The scheme of a single boost chopper is depicted in Fig. 1a, and that of two boost choppers connected in parallel is given in Fig. 1b.

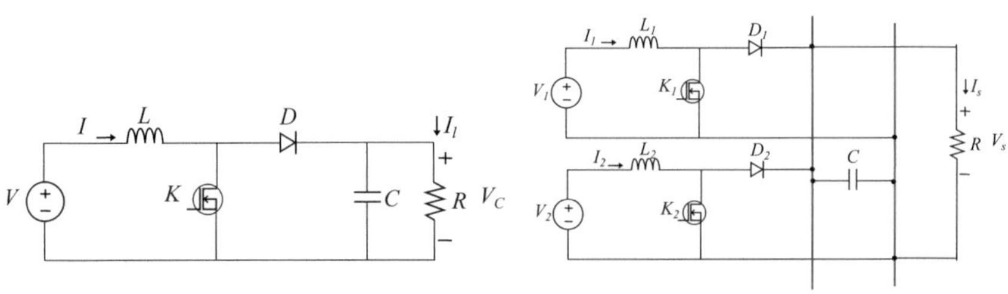

(a) One boost chopper. (b) Two boost choppers in parallel.

Fig. 1 Schemes of one and two boost choppers

To simplify the analysis and facilitate the transition to the average model, we assume that the system is operated using pulse width modulation (PWM) techniques. The boost chopper operates in two distinct modes that we explain below. In Sects. 2.1 and 2.2, only one boost is considered, see Fig. 1a.

2.1 Discontinuous Mode of One Boost Chopper

Discontinuous conduction mode occurs when the switching ripple of the inductance current or capacitor voltage is large enough to reverse the polarity of the applied switching current or voltage [4]. From time 0 to $\alpha_c T$, with $\alpha_c \in [0, 1[$, the switch K is closed then the diode D does not conduct, the energy provided by the DC voltage source $V > 0$ is being stored in the inductance L, then the capacitor C is being discharged through the load resistance R (see Fig. 2a). From $\alpha_c T$ to $(\alpha_c + \alpha_d)T$, the diode D conducts and the switch K is opened, the energy stored in the inductance L is transferred in the capacitor and the load (see Fig. 2b). From $(\alpha_c + \alpha_d)T$ to T, the diode does not conduct and the switch is opened (see Fig. 2c). Figure 2a gives the following equations for $K = 1$: $\dot{I} = \frac{V}{L}$, $\dot{V}_c = \frac{-V_c}{RC}$, while from Fig. 2b, we deduce the following equations for $K = 0$ and $I > 0$: $\dot{I} = \frac{V}{L} - \frac{I}{L}$, $\dot{V}_c = \frac{I}{C} - \frac{V_c}{RC}$. Figure 2c, yields the following equations for $K = 0$ and $I \leq 0$: $\dot{I} = 0$, $\dot{V}_c = \frac{-V_c}{RC}$.
The average model of the boost chopper of Fig. 1a in this mode is:

$$\begin{pmatrix} \dot{I}_{av} \\ \dot{V}_{C_{av}} \end{pmatrix} = \begin{pmatrix} 0 & \frac{-\alpha_d}{L} \\ \frac{\alpha_d}{C} & \frac{-1}{RC} \end{pmatrix} \begin{pmatrix} I_{av} \\ V_{C_{av}} \end{pmatrix} + \begin{pmatrix} \frac{(\alpha_c+\alpha_d)V}{L} \\ 0 \end{pmatrix} \tag{1}$$

2.2 Continuous Mode of One Boost Chopper

The continuous mode corresponds to $(\alpha_c + \alpha_d) = 1$, which gives:

$$\begin{pmatrix} \dot{I}_{av} \\ \dot{V}_{Cav} \end{pmatrix} = \begin{pmatrix} 0 & \frac{\alpha_c-1}{L} \\ \frac{1-\alpha_c}{C} & \frac{-1}{RC} \end{pmatrix} \begin{pmatrix} I_{av} \\ V_{Cav} \end{pmatrix} + \begin{pmatrix} \frac{V}{L} \\ 0 \end{pmatrix} \tag{2}$$

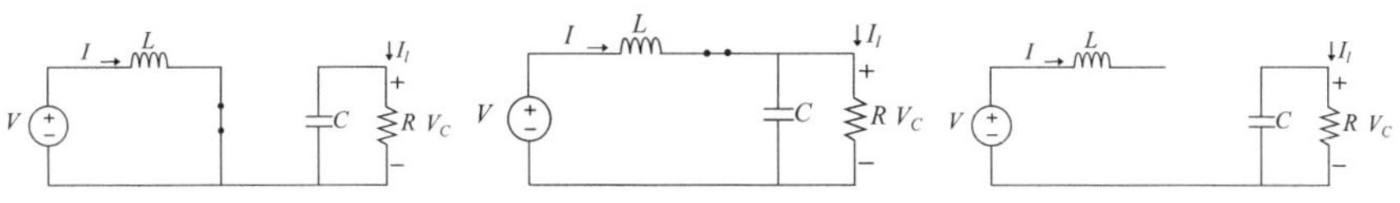

(a) On-mode: $0 < t \leq \alpha_c T$. (b) Off-mode: $\alpha_c T < t \leq (\alpha_c + \alpha_d)T$. (c) Off-mode: $(\alpha_c + \alpha_d)T < t \leq T$.

Fig. 2 Configuration-modes operation of the boost chopper

2.3 Model of the Two Boost Choppers System in Continuous Mode

In Fig. 1b, two different boost choppers with different sources are connected in parallel. Each boost and its characteristics are labeled using the integer i, with $1 \leq i \leq 2$. Note that the parallel connected capacitances C_i of each boost chopper have been regrouped into one total capacitance $C = C_1 + C_2$ leading to a state reduction. The continuous mode of the two boost choppers is given by:

$$\begin{pmatrix} \dot{I}_{av1} \\ \dot{I}_{av2} \\ \dot{V}_{Cav} \end{pmatrix} = \begin{pmatrix} 0 & 0 & \frac{\alpha_{1,c}-1}{L_1} \\ 0 & 0 & \frac{\alpha_{2,c}-1}{L_2} \\ \frac{1-\alpha_{1,c}}{C} & \frac{1-\alpha_{2,c}}{C} & \frac{-1}{RC} \end{pmatrix} \begin{pmatrix} I_{av1} \\ I_{av2} \\ V_{Cav} \end{pmatrix} + \begin{pmatrix} \frac{V_1}{L_1} \\ \frac{V_2}{L_2} \\ 0 \end{pmatrix} \tag{3}$$

3 Flatness Analysis

The proposed flat outputs are: $h_1 = I_{av1}$, and $h_2 = \sum_{i=1}^{2} \frac{L_i}{2} I_{avi}^2 + \frac{C}{2} V_{Cav}^2$, which correspond respectively to the current in the inductance of the first chopper boost and the total energy of the two boost choppers. From these flat outputs a control law is designed as follows (we send the reader to [8] for proofs and discussion on how these expressions are obtained):

$$\alpha_{1,c} = \frac{L_1}{V_{Cav}}\left(\frac{V_{Cav} - V_1}{L_1} - \lambda_1(h_1 - h_{d1}) + \dot{h}_{d1}\right), \tag{4}$$

$$\begin{aligned} \alpha_{2,c} = &\left(\frac{V_2 V_{Cav}}{L_2} + \frac{2V_{Cav} I_{av2}}{RC}\right)^{-1}\left(-\sum_{i=1}^{2}\frac{V_i(V_i - V_{Cav})}{L_i} + \frac{2V_{Cav}}{R}\left(\frac{h_1 + I_{av2}}{C} - \frac{V_{Cav}}{RC}\right)\right. \\ &\left. - \left(\frac{V_1 V_{Cav}}{L_1} + \frac{2V_{Cav} h_1}{RC}\right)\alpha_{1,c} - \lambda_{2,0}(h_2 - h_{d2}) - \lambda_{2,1}(\dot{h}_2 - \dot{h}_{d2}) + \ddot{h}_{d2}\right), \end{aligned}$$

where h_{di} denote desired trajectories of flat outputs h_i and are designed in the next subsections. The constant gains λ_1, and $\lambda_{2,j}$, $j = 0, 1$, are computed via a classical pole placement, and guarantee that the tracking error $h_i - h_{di}$, $1 \leq i \leq 2$, exponentially converges to 0. We also use a load power control loop (see [8] for the exact expression) that establishes a relationship between the bus voltage and the flat output reference trajectories.

3.1 Desired Flat Output Trajectories with Respect to a Homogeneous Power Distribution (HPD) Strategy

To ensure a homogeneous distribution of power in the two boost choppers system, the desired powers P_{di} in each boost are functions of the power demanded by the load P_l and are calculated as follows: $P_{di} = \frac{P_l P_{ni}}{P_{n1}+P_{n2}}$, $i = 1, 2$, where P_{ni} represents the nominal power of the ith source. From this we can determine the desired flat outputs as follows: $h_{d1} = I_{d1}$ and $h_{d2} = \sum_{i=1}^{2} \frac{L_i}{2} I_{di}^2 + \frac{C}{2} V_{dC}^2$, where $I_{di} = \frac{P_{di}}{V_i}$.

3.2 Desired Flat Output Trajectories with Respect to the Continuous Mode Limit (CML) Strategy

To calculate the average limit current value $I_{avi_{lim}}$ between continuous and discontinuous mode, ones needs to consider the operating continuous mode of each boost chopper, with $I_{avi_{lim}} = \frac{V_i}{2L_i F}(1 - \frac{V_i}{V_C})$. Then the desired limit power $P_{di_{lim}}$ assigned to the ith boost chopper is: $P_{di_{lim}} = \beta V_i I_{av_{lim}}$ leading to $I_{di_{lim}} = \frac{P_{di_{lim}}}{V_i}$, $1 \leq i \leq 2$. The first component of the desired flat output is $h_{d1} = I_{d1_{lim}}$ while the second one is the desired limit energy $h_{d2_{lim}}$ given by $h_{d2_{lim}} = \sum_{i=1}^{2} \frac{L_i}{2} I_{di_{lim}}^2 + \frac{C}{2} V_{dC}^2$.

4 Petri Nets Control Strategy

Petri Net control is a powerful graphical representation that utilizes a bipartite graph, composed of two distinct types of nodes: places and transitions [7]. These nodes are interconnected through direct arcs, forming a structured network. In the context of the two boost choppers system, we will use a Petri Net control design strategy (see Fig. 3 and Table 1) in order to carefully select the appropriate number of connected boost choppers and to choose the appropriate flat output reference trajectory (designed either using the HPD strategy or the CML strategy), such that the boost choppers are in continuous mode.

5 Simulation Results, Discussions and Future Work

Simulation tests were conducted in the Matlab/Simulink environment. The evaluation focused on the proposed strategy for connecting and disconnecting a boost chopper, in conjunction with the flatness-based control method. Table 2a displays the different parameters of the boost choppers, obtained from the reference [2], as well as the control loop parameters utilized in the simulations. Table 2b displays the

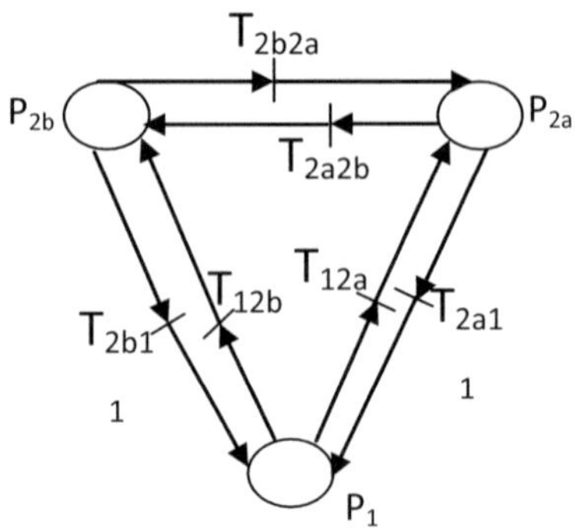

Fig. 3 Proposed control scheme based on petri nets representation

Table 1 Transitions and places

(a) *Transition validation table*

Transitions	Conditions
T_{2a2b}	$1.2 \times P_{limc} < P_l < 1.1 \times P_{limh}$
T_{2b2a}	$P_l > 1.2 \times P_{limh}$
T_{2a1}	$P_l < P_{limc}$
T_{2b1}	$P_l < P_{limc}$
T_{12a}	$P_l > 1.2 \times P_{limh}$
T_{12b}	$1.2 \times P_{limc} < P_l < 1.1 \times P_{lmih}$

(b) *Place active table*

Places	Designations
P_1	a single boost chopper is connected
P_{2a}	two boost choppers are connected with a homogeneous power distribution strategy
P_{2b}	two boost choppers are connected with respect to the continuous mode limit strategy

load variation. To obtain the average values, all measured variables were subjected to filtering using a second-order Butterworth low-pass filter with a cutoff frequency F_{cut}.

Figure 4d illustrates the measured bus voltage $V_{C_{av}}$, showing the accurate tracking of the desired reference constant trajectory V_{dC} through the implementation of the flatness-based control law. Additionally, it should be noted that the response shows an overshoot of less than 4.5% during load variations. Figure 4a, c show that the measured energy E_t and the measured current I_{av1} accurately track their respective reference trajectories E_{dt} and I_{d1} with respect to homogeneous power distribution for $t \in [0s, 0.5s[$ and $t \in [2s, 2.5s[$. For $t \in [0.5s, 1s[$ and $t \in [1.5s, 2s[$, E_t and I_{av1} accurately track their respective reference trajectories E_{dt} and I_{d1} with respect to the current limit between continuous and discontinuous mode. For $t \in [1s, 1.5s[$ the measured current I_{av1} equals to zero and the measured energy E_t is different

Table 2 Load variation, boost system and control parameters

(a) *System and control parameters*

Symbols	Values	Symbols	Values	Symbols	Values
L_1	10 mH	V_1	25 V	λ_1	1000
L_2	33 mH	V_2	30 V	$\lambda_{2,0}$	1,000,000
C	1 mF	P_{r1}	100 W	$\lambda_{2,1}$	2000
T	50 μs	P_{r2}	40 W	λ_3	2
P_l	100 W	F_{cut}	1.5 kHz	β	20

(b) *Resistance variation*

Resistance (Ω)	Interval
50	$0\,\mathrm{s} \leq t < 0.5\,\mathrm{s}$
100	$2\,\mathrm{s} \leq t < 2.5\,\mathrm{s}$
150	$0.5\,\mathrm{s} \leq t < 1\,\mathrm{s}$
200	$1.5\,\mathrm{s} \leq t < 2\,\mathrm{s}$
350	$1\,\mathrm{s} \leq t < 1.5\,\mathrm{s}$

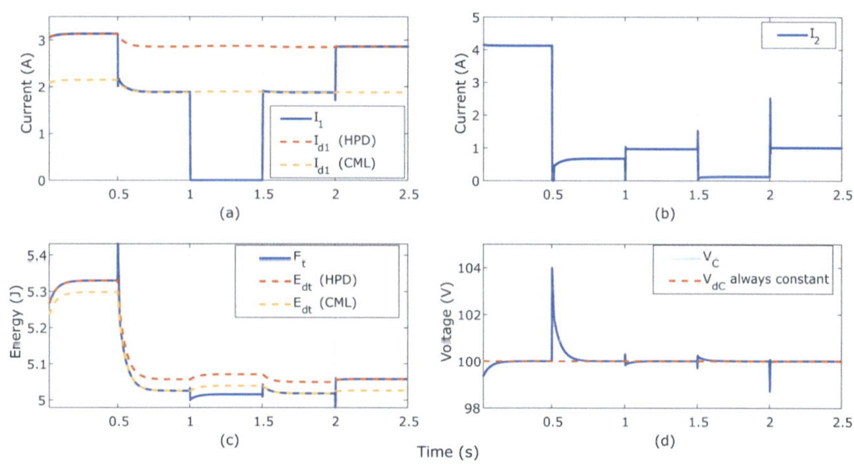

Fig. 4 Average current, voltage and global energy

from its desired reference computed using the two strategy because the first boost chopper is disconnected during this sub-interval.

Future work will consider experimental studies, boost choppers with different switching frequencies, and the inclusion of energy storage elements such as batteries or super-capacitors.

Acknowledgements We are most thankful for the "PARTENARIAT HUBERT CURIEN MAGHREB: PHC" which Tunisian higher ministry and scientific research DGRS code is PHC21MAG15. That has provided financial and technical supports for the larger project development from which this paper grew.

References

1. Castellanos Contreras JU, Rodríguez Urrego L (2023) Technological developments in control models using petri nets for smart grids: a review. Energies 16(8):3541
2. Chitra P, Seyezhai R (2014) Basic design and review of two phase and three phase interleaved boost converter for renewable energy systems. Int J Appl Sci 1(1):1–26
3. Dey A, Chaki N, Sanyal S (2011) Modeling smart grid using generalized stochastic petri net. arXiv:1108.4139
4. Erickson RW (2000) Fundamentals of power electronics, 2nd edn. Kluwer Academic Publishers, Secaucus, NJ, USA
5. Fliess M, Lévine J, Martin P, Rouchon P (1995) Flatness and defect of non-linear systems: introductory theory and examples. Int J Control 61(6):1327–1361
6. Lu D, Fakham H, Zhou T, Francois B (2010) Application of petri nets for the energy management of a photovoltaic based power station including storage units. Renew Energy 35(6):1117–1124
7. Matos LO, Sanchez, JWG (2016) Reconfiguration strategy for fault tolerance of power distribution systems using petri net. In: 2016 IEEE Ecuador technical chapters meeting (ETCM). IEEE, pp 1–6
8. Messaoudi S, Nicolau F, Ghanes M, Sbita L, Barbot J-P (2023) Flatness-based control strategy for n parallel connected boost choppers and n sources with differing characteristics. IEEE Control Syst Lett: 1
9. Mhadhbi Z, Zairi S, Gueguen C, Zouari B (2018) Validation of a distributed energy management approach for smart grid based on a generic colored petri nets model. J Clean Energy Technol 6(1)
10. Sira-Ramirez H, Silva-Ortigoza R (2006) Control design techniques in power electronics devices. Springer

Modeling and Control of a Tidal Turbine Based on a PMSG Generator

Zribi Hazem, Abid Aïcha, Dhaoui Mehdi, and Ben Hamed Mouna

Abstract In this work, we have studied a tidal energy conversion system using a synchronous generator with permanent magnets and connected to grid. To achieve the objectives, a Park modeling of the machine was carried out with an oriented flux vector control, which allowed the machine to operate at variable speed. For the control of the grid side converter, we used hysteresis current control loops. To guarantee a constant DC bus voltage level, a method based on the calculation of the TSR error equipped with a PID regulator was used. The results of the various simulations of the entire conversion chain, carried out under the MATLAB/Simulink environment, made it possible to evaluate the performance of the proposed control algorithms.

Keywords Tidal turbine · Permanent magnet synchronous generator · Park modeling · Vector control · Inverter

1 Introduction

A tidal turbine is a hydraulic turbine (underwater or floating) that uses the kinetic energy of sea or river currents. Compared to a wind turbine, tidal turbines take advantage of the density of water, 832 times greater than that of air. Despite a generally lower fluid velocity, the recoverable power per unit propeller area is much greater for a tidal turbine than for a wind turbine. The relatively recent awareness of the need to replace oil can explain the late development of tidal turbines as well as the fact that the marine environment is particularly demanding and aggressive (corrosion due to salt, "encrusting" organisms). The tidal turbine benefits from enormous technical efforts that have already been made in the development of wind power, but have been the subject of less research [1, 2]. The objective of this work consists in the

Z. Hazem (✉) · A. Aïcha · D. Mehdi · B. H. Mouna
L/R: Processes, Energy, Environment and Electrical Systems (PEESE), Electrical Department of Engineering, National Engineering School, University of Gabes, 6029 Gabes, Tunisia
e-mail: hazemzribi1978@gmail.com

© The Author(s), under exclusive license to Springer Nature Singapore Pte Ltd. 2024
A. Mellit et al. (eds.), *Proceedings of the 2nd International Conference on Green Energy Conversion System*, Green Energy and Technology,
https://doi.org/10.1007/978-981-97-6148-7_45

study, the modeling and the control of a tidal turbine conversion chain based on the synchronous machine with permanent magnets GSAP. The turbine and radial flow is offset on the periphery of the propeller. This so-called "Rim Driven" technology is envisaged for propulsion as well as for the recovery of the kinetic energy of currents.

The study is validated by the results of simulations carried out on the Matlab/Simulink environment.

2 Conversion of the Kinetic Energy of the Water Current

Consider a tidal turbine with a rotor of diameter D immersed in a current whose speed is V_w (Fig. 1).

The total kinetic power available on the tidal turbine [3] is given by the following expression (Eq. 1):

$$P_{tide} = \frac{1}{2}\rho\pi R^2 \cdot V_w^3 \tag{1}$$

where ρ (kg/m^3) is the water density (fresh water 1000 kg/m^3, sea water 1025 kg/m^3), R (m) is the turbine radius, V_w (m/s) is the tidal speed.

The modeling of the resource requires the knowledge of a certain number of data related to the site targeted for the installation of the tidal turbine (current, bathymetry, nature of the bottom). Hydrographic services such as SHOM in France have databases listing, for a given area, the values of the current in spring tides and neap tides. These values are indicated for each hour in relation to high tide, from H − 6 to H + 6. Knowing the tidal coefficient, it is possible to extrapolate the value of the current for a given hour from the following first-order model:

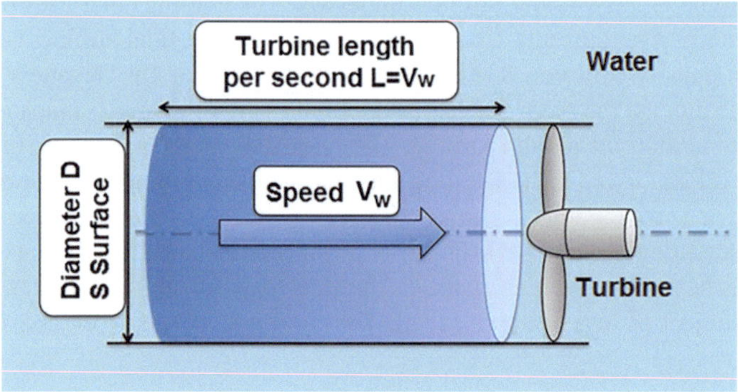

Fig. 1 Principle diagram of a tidal turbine immersed in a current of water

$$V_w = V_1 + \frac{(C - 45)(V_2 - V_1)}{95 - 45} \tag{2}$$

where 95 and 45 are respectively the mean tidal coefficients corresponding to spring tides and neap tides. However, the coefficients C, V_1 and V_2 are determined from the SHOM databases (1994). In order to simplify the modeling, it was necessary to adopt certain assumptions. Indeed, the flow is considered, initially, unmodified. In particular, the effects of the swell and limit sofas are stripped. It is also assumed that the speed of the water current is constant over one hour. In addition, the time slots in the SHOM currents book are tide times. Remember that one hour of tide equals 62 min. The approximation therefore remains correct. To illustrate the variation in energy production, we give the theoretical example of a 500 kW tidal turbine installed on a typical site in the English Channel [4, 5]. Figure 2 shows the tidal speed at this site during the month of October 2008 [6].

It should be noted that the variations of water are very slow, the gusts as in the case of the wind, do not have an equivalent in water. However, because of the computation time required, for numerical simulations, much faster water variations (approximately a minute and sometimes even a second) will be used. Figure 3 shows an example of a variable water profile used to drive the tidal turbine model.

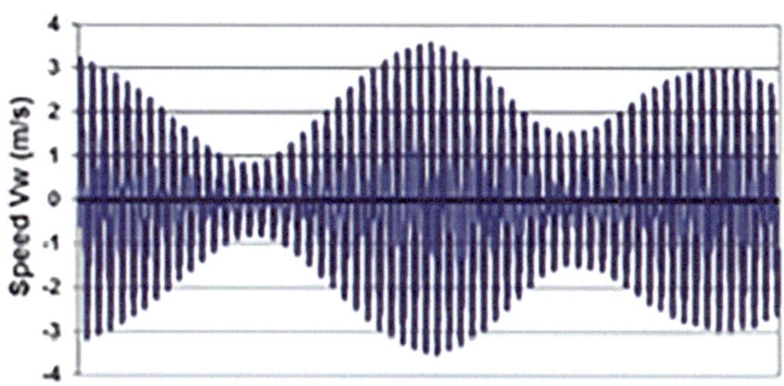

Fig. 2 Example of tidal current speed between 01/10/2008 and 31/10/2008

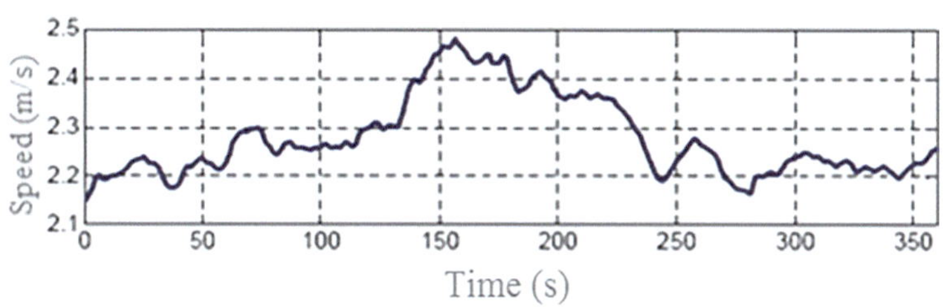

Fig. 3 Example of tidal current speed profile

3 Modeling of the Tidal Energy Conversion System

The energy conversion chain of the tidal system connected to the electrical network is represented by the Fig. 4.

The study of the global system requires the modeling of the various elements of the conversion chain.

3.1 Turbine Modeling

The tidal turbine can only capture part of the kinetic energy of the tidal speed.

$$P_t = P_{tid} C_p(\lambda) = \frac{1}{2}\rho\pi R^2 C_p(\lambda) \cdot V_w^3 \tag{3}$$

Expressions (2) and (3) are valid for any system that extracts kinetic energy from a moving fluid (wind turbines and tidal turbines). For tidal turbines, the power coefficient C_p is generally defined in the interval 0.35–0.5 [2]. It should be noted that to date, there is no mathematical model in the literature of the power extraction coefficient of these tidal turbines. For the purposes of the simulation, we extrapolated this coefficient from the curves of the experimental results of (4) in the form of a polynomial approximation which provided the best results. The power coefficient C_p is given by:

$$C_p(\lambda) = C_1 \cdot \lambda^4 + C_2 \cdot \lambda^3 + C_3 \cdot \lambda^2 + C_4 \cdot \lambda + C_5 \tag{4}$$

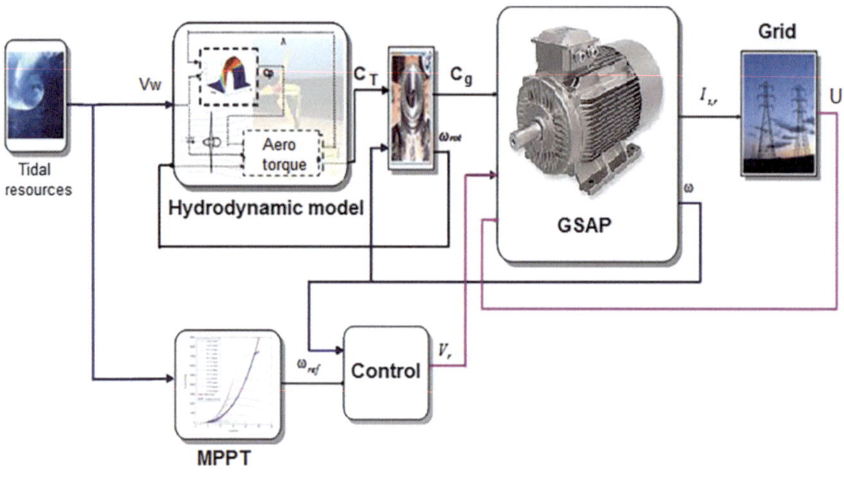

Fig. 4 Overall diagram of a hydrokinetic conversion chain using a GSAP tidal turbine

where: $C_1 = 1.92 \times 10^{-4}$, $C_2 = -5.2 \times 10^{-3}$, $C_3 = 2.55 \times 10^{-2}$,

$$C_4 = 5.79 \times 10^{-2} \text{ and } C_5 = 1.38 \times 10^{-4}$$

The tip-speed ratio is defined as:

$$\lambda = \frac{R \cdot \Omega_t}{V_w} \tag{5}$$

where Ω_t is the mechanical turbine speed (rad/s).

For different values of λ, the $C_p(\lambda)$ curves are shown in Fig. 5.

The fact that the power is proportional to the cube of the speed of the current has many consequences in the design of tidal turbines and the choice of sites conducive to the exploitation of this form of energy. For an electrical power of 200 kW thanks to a tidal turbine, the following table indicates the diameter of the rotor according to the maximum speed of the tidal current [6]. Table 1 also shows a comparison with a wind turbine of the same power.

Using the relation (2), we can obtain the characteristics (Fig. 6) of the mechanical power of the turbine for different water current speeds. These characteristics relate to a turbine with a rotor diameter of 4.4 m.

The hydrodynamic performance of the turbine is given by the block diagram illustrated in the Fig. 7.

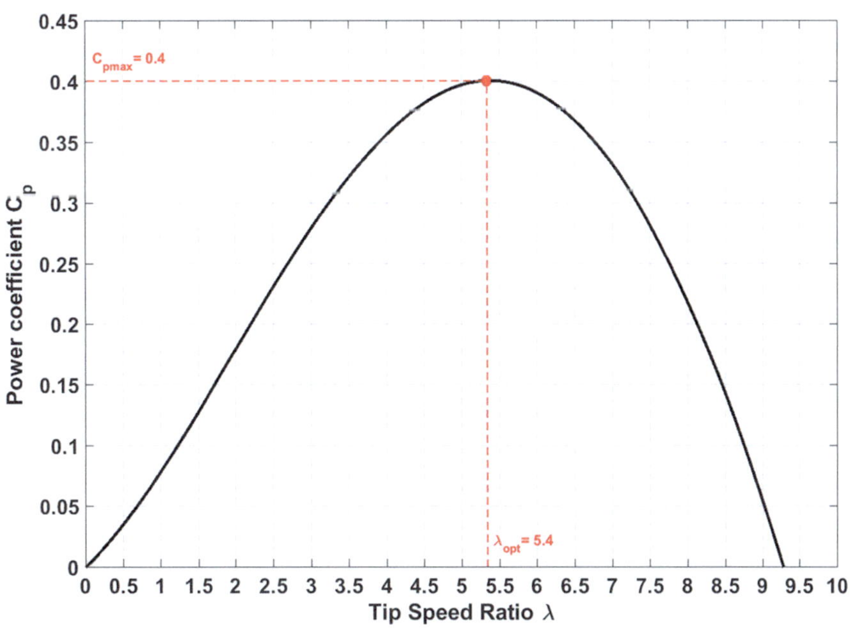

Fig. 5 Power coefficient characteristic versus speed ratio λ

Table 1 Dimensions of a 200 kW tidal turbine according to the current speed taken into account for the dimensioning ($C_p = 0.40$)

V_w (m/s)	S (m^2)	D (m)
4	15.3	4.4
3	36.2	6.8
2	122.1	12.5
1.4	355.9	21.3
1	976.6	35.3
Wind turbine (V = 13 m/s)	352.8	21.2

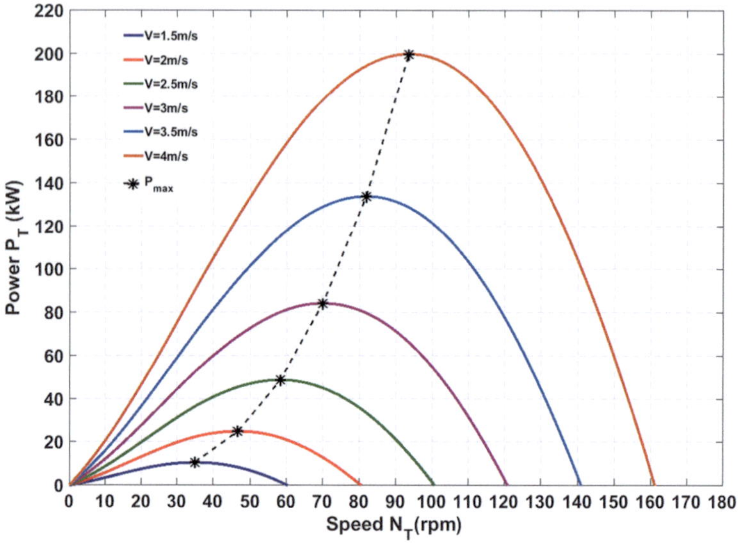

Fig. 6 Power characteristics of a tidal turbine as a function of current speed and rotation speed

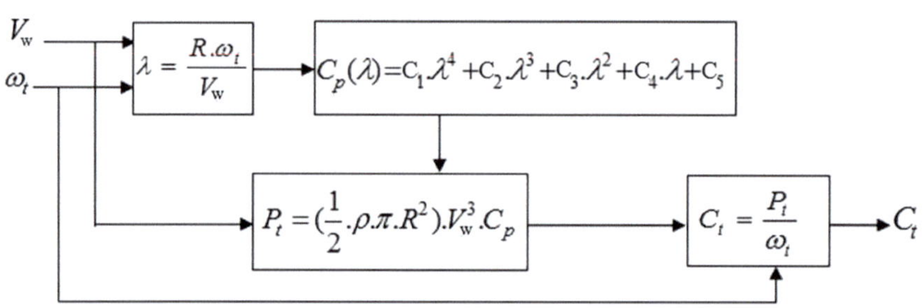

Fig. 7 Block diagram of the tidal turbine model

3.2 Model of the PMSG

The wind turbine drives a Permanent Magnet Synchronous Generator PMSG (Fig. 8).

The model of the machine in the Park frame is given by the following equations:

$$\begin{cases} V_{sd} = -R_s \cdot i_{sd} - L_s \cdot \dfrac{di_{sd}}{dt} + \omega_g \cdot L_s . i_{sq} \\ V_{sq} = -R_s \cdot i_{qs} - L_s \cdot \dfrac{di_{sq}}{dt} - \omega_g \cdot L_s \cdot i_{sd} + \omega_g \cdot \phi_f \end{cases} \tag{6}$$

where R_s is the phase resistance of the stator winding, L_s is the stator cyclic inductance, Φ_f is the flux of the permanent magnetic, v_{sd} and v_{sq} are the $d - q$ components of the stator voltages respectively, i_{sd} and i_{sq} are the d-q components of the stator currents respectively, Ω_g the speed generator and p the number of pairs poles.

Relation (7) express the mechanical equation of the system, where J_t and J_g present respectively the moments of inertia of the turbine and the generator; f is the coefficient of viscous friction of the generator; Ω_g is the mechanical speed of the generator.

$$\left(J_t + J_g\right) \cdot \frac{d\Omega_g}{dt} + f \cdot \Omega_g = T_g - T_{em} \tag{7}$$

The electromagnetic torque is defined by:

$$\begin{gathered} T_{em} = -\frac{3}{2} p \cdot \varphi_f \cdot i_{sq} \\ N = \frac{30\Omega}{\pi} \rightarrow \Omega = \frac{N \cdot \pi}{30} \end{gathered} \tag{8}$$

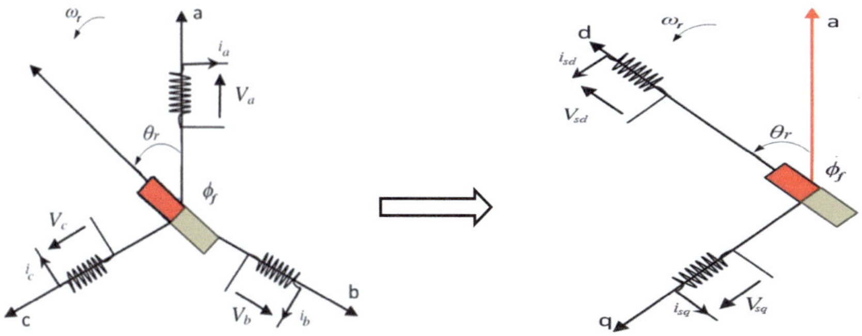

Fig. 8 Transformation of the real model (a, b, c) to Park's reference (d, q)

3.3 Rectifier Model

The generator is often coupled to a three-phase diode rectifier. The signals (voltages and current) supplied by the generator are linked to those output from the rectifier by the following matrix equation.

$$\begin{bmatrix} V_a \\ V_b \\ V_c \end{bmatrix} = \frac{V_d}{3} \cdot \begin{pmatrix} -2 & 1 & 1 \\ 1 & -2 & 1 \\ 1 & 1 & -2 \end{pmatrix} \cdot \begin{bmatrix} S_a \\ S_b \\ S_c \end{bmatrix} \tag{9}$$

$$I_d = S_a \cdot I_a + S_b \cdot I_b + S_c \cdot I_c \tag{10}$$

With: V_d and I_d are respectively, the voltage and the current modulated by the rectifier, S_a, S_b and S_c, are the logic functions corresponding to the state of the switch (equal to 1 if the switch is closed and zero where it is open).

3.4 Model of the "Buck-Boost" Converter

For proper operation of the tidal turbine at a wide band of tidal speed variation, the insertion of a static converter (DC/DC) in cascade with the rectifier will be the key element to have an acceptable energy yield (Fig. 9).

The Buck-Boost presents, by its characteristics, a rather interesting alternative to extract the maximum power such as:

- For low tidal speeds, the voltage V_d is also low, it is therefore necessary to impose a high voltage on the DC side by increasing the duty cycle, the Buck-Boost will operate in Boost (lift).

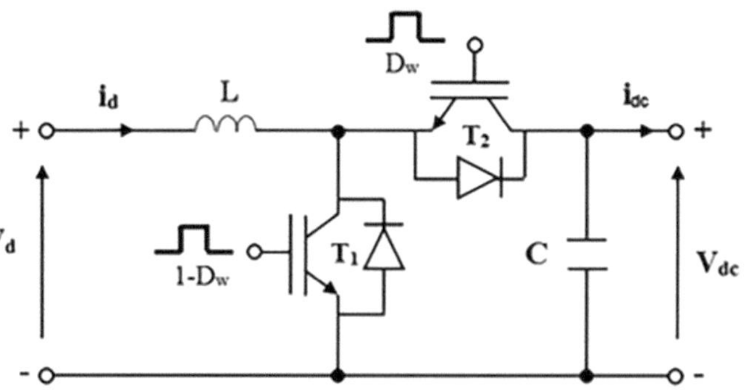

Fig. 9 Diagram of the buck-boost converter

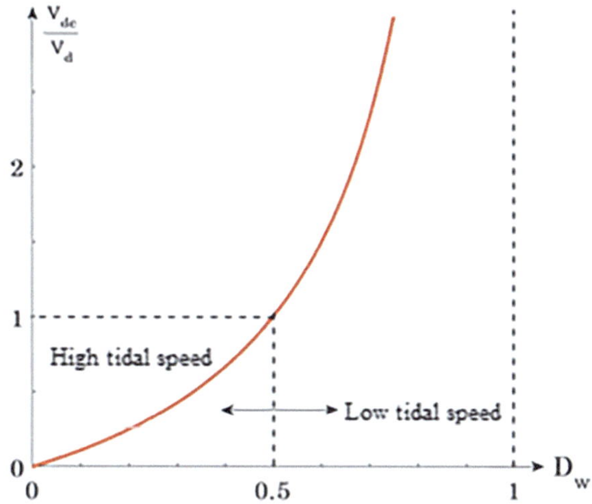

Fig. 10 Buck-Boost characteristic according to the duty cycle (D)

- For high speeds, the voltage V_d, is in this case high, it is necessary to impose a low voltage on the DC side by reducing the duty cycle, the Buck-Boost will operate in Buck (step down) (Fig. 10).

The output voltage V_{DC} of the buck-boost converter is linked to that of the input V_d by the following relationship:

$$V_{DC} = \frac{D_w}{1 - D_w} \cdot V_d \tag{11}$$

3.5 Inverter and Hysteresis Current Blocks

The scheme of the inverter with its control is shown in the Fig. 11.

Three inverter comprises 6 switches (IGBT) and controlled by PWM signals (S_i). The states S_i of the switches of the same arm in the power circuit of the three-phase inverter are complementary. The switch k_i is in the closed state if $S_i = 1$ and it is open if $S_i = 0$.

Using the S_i states, we can obtain the output string voltages of the inverter measured against the negative terminal of the DC voltage as follows:

$$\begin{cases} V_{an} = \dfrac{V_{DC}}{3} \cdot [2 \cdot S_1 - S_2 - S_3] \\ V_{an} = \dfrac{V_{DC}}{3} \cdot [2 \cdot S_2 - S_1 - S_3] \\ V_{cn} = \dfrac{V_{DC}}{3} \cdot [2 \cdot S_3 - S_1 - S_2] \end{cases} \tag{12}$$

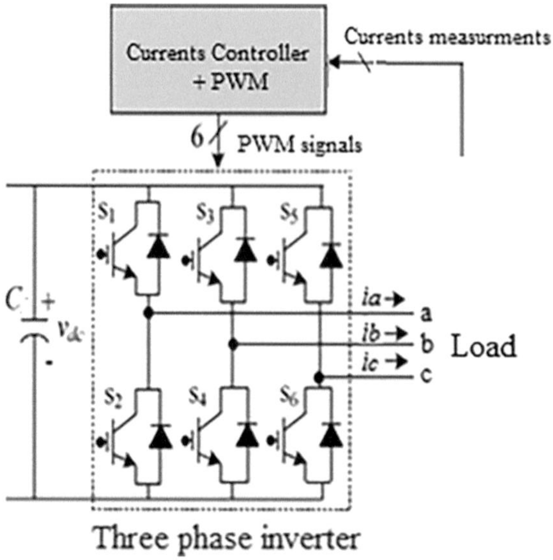

Fig. 11 Inverter and currents controller

The phase-to-phase voltages are:

$$\begin{cases} U_{ab} = V_{an} - V_{bn} = (S_1 - S_2) \cdot V_{DC} \\ U_{bc} = V_{bn} - V_{cn} = (S_2 - S_3) \cdot V_{DC} \\ U_{ca} = V_{cn} - V_{an} = (S_3 - S_1) \cdot V_{DC} \end{cases} \quad (13)$$

4 Tidal Turbine Control Strategy

The tidal turbine control diagram is given by the Fig. 12.

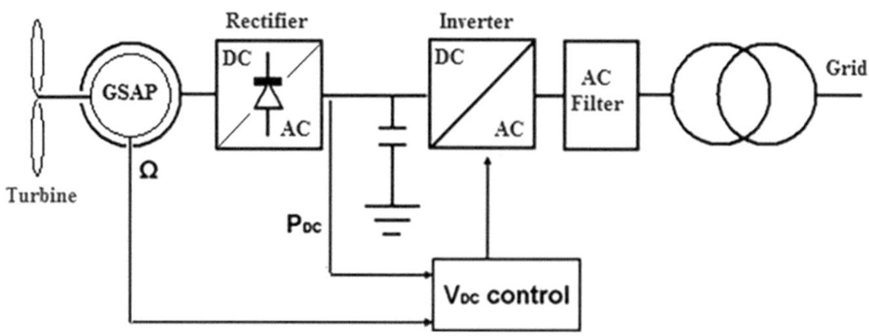

Fig. 12 Diagram of DC bus regulation

For a variable speed tidal turbine with a direct drive synchronous generator, the voltage varies with the speed of rotation of the turbine. The voltage in one of the phases, v_a can be described as follows:

$$v_a(t) = \omega_g \phi_m \cos(\theta_e) - R_s i_a(t) - L_s \frac{di_a(t)}{dt} \tag{14}$$

where ω_g is the angular electrical frequency, Φ_m is the maximum flux value.

The DC voltage reference can be calculated to operate the turbine optimally if the tidal speed as well as the relationship between the aerodynamic power coefficient, C_P, and the optimal speed ratio, λ_{opt}; the $C_P - \lambda_{opt}$ dependence of the turbine are known. However, accurate tidal measurements are difficult and expensive to achieve. A tidal velocity error has a great influence on control because the relationship between power and tidal velocity is cubic. To circumvent this problem, an MPPT (maximum power point tracking) algorithm can be used to find the optimal operating point by iteration. For this method, the tidal speed sensor in the control system is replaced by continuous power and rotational speed measurements. Another control method combining the MPPT with the control of the optimum speed ratio λ_{opt}, but using the measured tidal speed, is presented in [7]. The method of combined control of the MPPT and the optimum speed ratio is compared with a proposed method of optimal current given the MPPT control in [8, 9]. The method proposed here is not an MPPT, but focuses on maintaining the optimal speed ratio at a given level by comparing the speed of rotation to the speed of the predicted tide. Maintaining the optimal gear ratio at a fixed value should be an easier control strategy than torque control, which is based on a look-up table to match torque to speed.

For the 200 kW turbine, the dependence between C_P and tip speed ratio has been estimated with the aerodynamic models described in [10], see Fig. 4. This curve has not been verified and therefore the exact power-wind speed dependency, that is, the power curve, is not known. Hence, the tidal speed sensor independent method has been considered a good solution, as otherwise both an error in tidal speed and an error in C_P might affect the control and alter the results. Tip speed ratio control takes advantage of the fact that the $C_P - \lambda$ curve is quite flat close to its optimum value, so small errors in the estimated tip speed ratio should not influence the results too much.

The efficiency of the system from the power absorbed by the turbine to the DC bus is denoted η, that is,

$$P_{DC} = V_{DC} \cdot I_{DC} = \eta \cdot P_t \tag{15}$$

where P_{DC} is the power measured on the DC bus. The efficiency, η, takes into account the mechanical losses in couplings and bearings in the tidal turbine, generator losses, and rectification losses. The generator losses for this specific generator are discussed in [10]. The efficiency is dependent on rotational speed and current, but does not change much for the operational range considered here. Here, the efficiency is assumed to be fairly constant. However, the decreasing generator efficiency at low tidal speeds might influence results slightly.

The estimation of the tip speed ratio is based on Eq. 5 as,

$$\lambda_t = \sqrt[3]{\frac{0.5 \cdot \rho \cdot S \cdot R^3 \cdot C_p(\lambda_{t-1}) \cdot \Omega^3 \cdot \eta}{P_{DC}}} \tag{16}$$

The error, e(t), between the optimum top speed ratio, λ_{opt}, and the result of Eq. 17 is calculated as follows,

$$e(t) = \lambda_{opt} - \lambda_t \tag{17}$$

And is used in a PID (proportional-integral-derivative) regulator described as,

$$u(t) = k_p \cdot e(t) + k_i \cdot \int_0^t e(T)dT + k_d \cdot \frac{de(t)}{dt} \tag{18}$$

where k_p, k_i, and k_d are constants regulating the proportional, integral, and derivative parts of the regulator, respectively. A new set point for the DC voltage, V_{DC}, is calculated as,

$$V_{DC}(t) = V_{DC}(t-1)[1 + u(t)] \tag{19}$$

There are two reasons for using a peak speed ratio error-controlled regulator to adjust the DC voltage; a rapid voltage increase is interrupted under gusty tidal speed conditions and the influence of the current-dependent voltage drop across the generator terminals ($-R_s i_s - L_s dI_s/dt$ in Eq. 14) can be interrupted. The inverter is used to control the DC voltage based on the output of the regulator. Thus, when the speed of the tide fluctuates, the power drawn is changed, through the generator current, giving a new set point of the DC voltage regulator (for large enough fluctuations in wind speed).

5 Simulation Results and Discussion

To simulate the tidal system, we chose a GSAP permanent magnet synchronous generator, direct coupled to a tidal turbine. The characteristics of the system are given in Tables 2 and 3. Thus, in order to identify the magnitudes of the system, we tested the model for a variable tidal speed profile. In order to evaluate the performance of the control system for the variable speed DC bus reference concept, we have set the VDC reference to 490 v in order to have a phase-to-phase voltage of 400 V at the output inverter. Figures 13, 14 and 15 represent the simulation results of a tidal

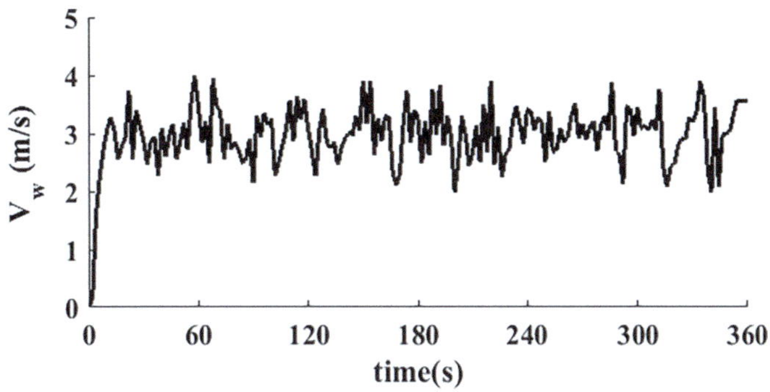

Fig. 13 The tidal current speed profile

turbine driving a 200 kW permanent magnets synchronous generator (PMSG) and connected to the grid.

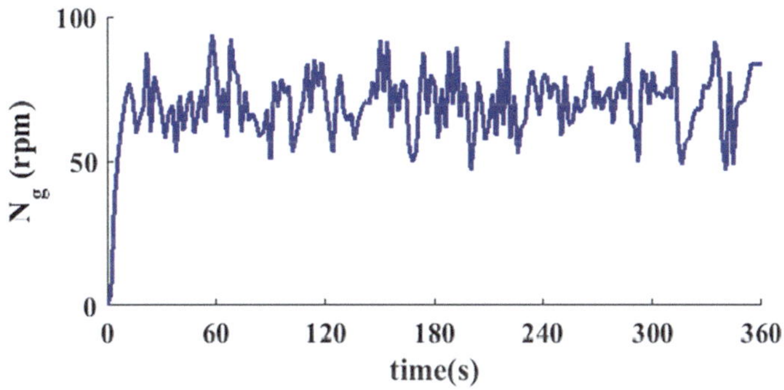

Fig. 14 PMSG generator rotation speed response

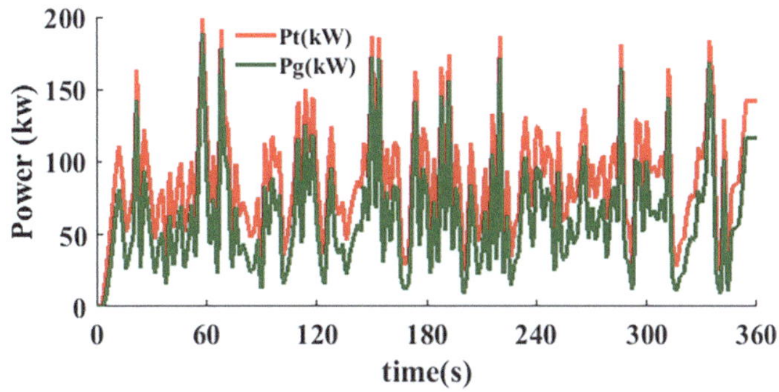

Fig. 15 Evolution of the mechanical and electrical powers of the generator

According to the results of the simulations in Fig. 15, it can be seen that the electric power of the generator follows the same profile as the mechanical power applied by the shaft of the tidal turbine. The difference between the two powers provides information on the efficiency of the synchronous generator. The voltages and currents supplied by the generator (Fig. 16) have sinusoidal shapes and variable amplitudes depending on the tidal speed. This there shows the good performance of the synchronous generator.

The simulation result obtained for the DC bus regulation (Fig. 17) shows a good maintenance of the DC bus voltage and power at a constant level, which confirms the good regulation and the best performances of the applied technique.

Proper control of the DC bus voltage leads to good performance in terms of the voltages and currents supplied by the inverter (Fig. 18). The objective set by the

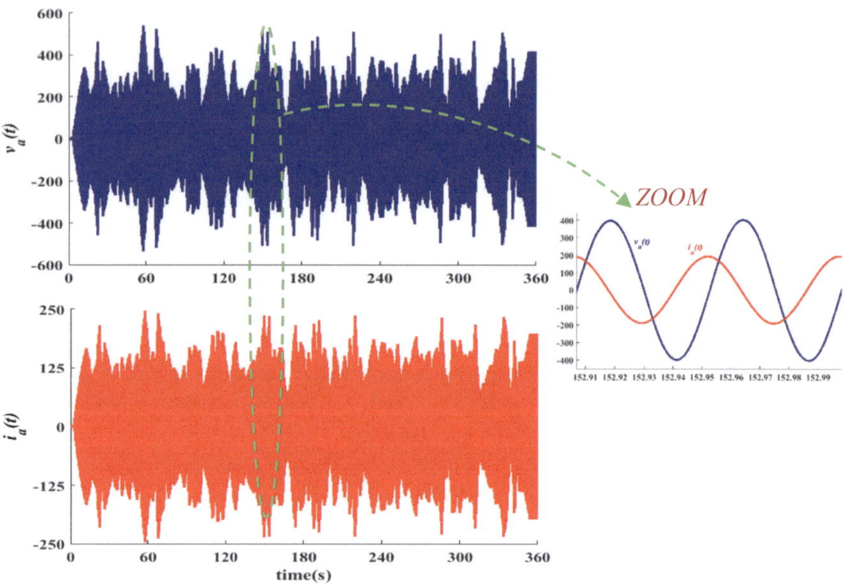

Fig. 16 Response of voltage and current supplied by the generator (phase a)

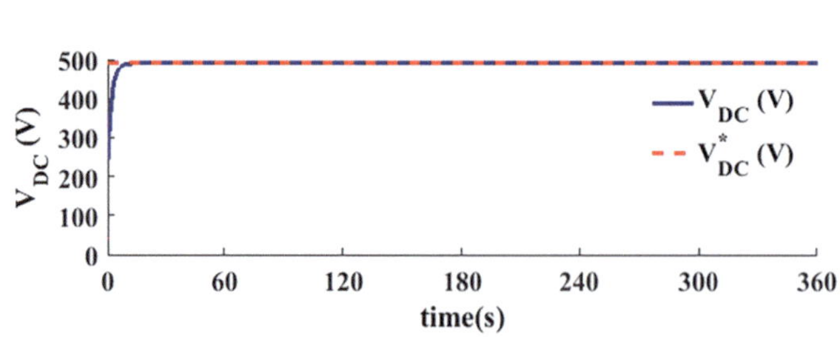

Fig. 17 Evolution of the DC bus voltage

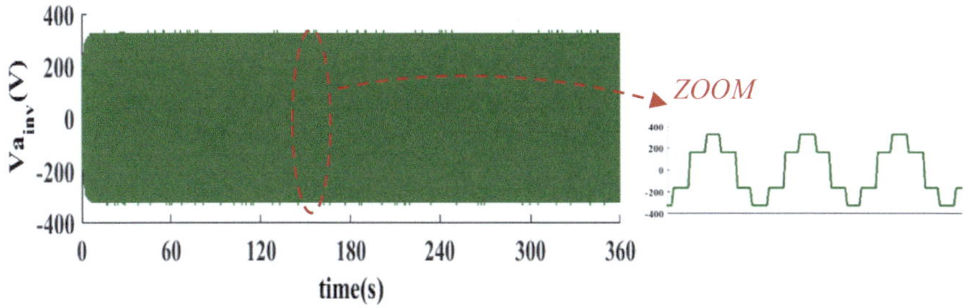

Fig. 18 Evolution of the voltage at the output of the inverter (phase a)

choice of the constant V_{DC} level is achieved with an rms value of the phase-to-phase voltage equal to 400V. The signals provided by the inverter have many harmonics which can affect the operation of the electrical network. This problem is solved with the insertion of an integrated filter in the inverter. The results (Fig. 19) clearly shows the improvements made by the built-in filter. Indeed, the voltages and currents supplied to the network are almost sinusoidal.

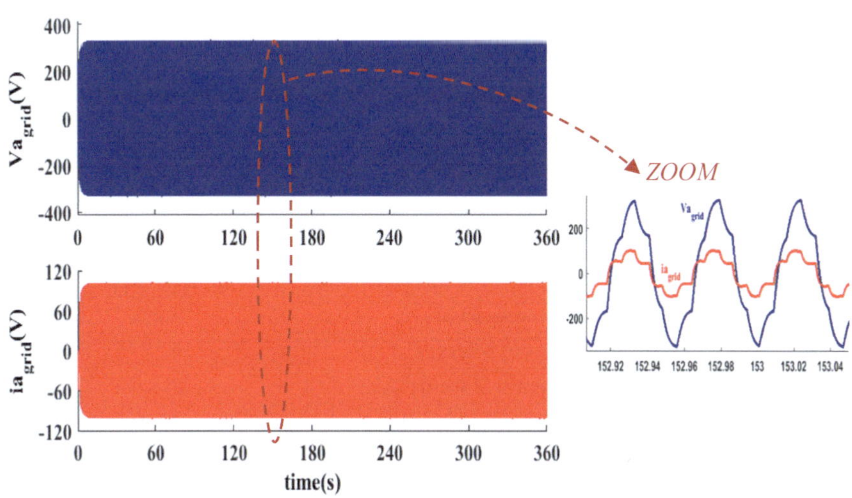

Fig. 19 Evolution of the voltage and current supplied to the grid (phase a)

6 Conclusion

In this work, a grid-connected power conversion system using a tidal turbine and a permanent magnet synchronous generator was presented and studied. From Park's model of the synchronous machine, we have defined an oriented flux vector control to have a variable speed operation of the machine. The use of current regulation loops (direct and quadrature) gave good performance at the PWM control level of the inverter. Thus, to adjust the DC bus voltage to a constant level in the face of tidal speed fluctuations, a method based on the calculation of the speed ratio error TSR and a PID type regulator was used. The used technique was eliminate, on the one hand, the rapid increases in generator voltage under gusty tidal speed conditions. On the other hand, the mastery and control of the power at the level of the DC bus. The capability of the proposed controllers is checked using Matlab/Simulink models.

Appendix

See Tables 2 and 3.

Table 2 Parameters of the tidal turbine

Power (P_t)	200 kW
Diameter of rotor (D)	4.4 m
Number of blades	3
Rated speed (N_t)	94 rpm
Rated tidal speed (V_w)	4 m/s
moment of inertia (J_t)	16 kg m^2

Table 3 Parameters of the PMSG generator

Rated power (P_g)	200 kW
Rated voltage (U_g)	660 V
Rated current (I_g)	175 A
Stator resistance (R_s)	1.0 Ω
Stator inductance (L_s)	0.72 mH
permanent magnetic flux (Φ_f)	0,81 Wb
Rated speed (N_g)	94 rpm
pairs of poles (p)	16
moment of inertia (J_g)	2.6 kg m^2

References

1. Ben Elghali SE (2008) Modélisation et Commande d'une Hydrolienne Equipée d'une Génératrice Asynchrone Double Alimentation. JCGE'08 LYON, 16 et 17 décembre 2008
2. Ben Elghali SE, Benbouzid MEH, Charpentier JF (2010) Generator systems for comparison of PMSG and DFIG for marine current turbine applications. In: XIX international conference on electrical machines. ICEM 2010, Rome
3. Batten WMJ, Blunden LS, Bahaj AS (2006) CFD simulation of a small farm of horizontal axis marine current turbines. In: Proceedings world renewable energy congress (WREC) VIII, Florence, pp 19–25
4. Hydro BC (2002) Tidal current energy. Green Energy Study for British Columbia Phase 2: Mainland, Triton Consultants Ltd
5. Bahaj AS, Molland AF, Chaplin JR, Batten WMJ (2007) Power and thrust measurements of marine current turbines under various hydrodynamic flow conditions in a cavitation tunnel and a towing tank. Renew Energy 32:407–426 (ELSEVIER)
6. L'énergie des courants de marée (Les hydroliennes), https://argonautes.club/l-energie-des-courants-de-maree.html (Consulted on 07/10/2023)
7. Jeong HG, Seung RH, Lee KB (2012) An improved maximum power point tracking method for wind power systems. Energies 5:1339–1354
8. Zhu Y, Cheng M, Hua W, Wang W (2012) A novel maximum power point tracking control for permanent magnet direct drive wind energy conversion systems. Energies 5:1398–1412
9. Andreica M, Bacha S, Roye D, Guiraud J (2008) Intégration d'une hydrolienne au réseau, MPPT et qualité de l'énergie. Revue des énergies renouvelables 11(4):493–502
10. Deglaire P (2010) Analytical aerodynamic simulation tools for vertical axis wind turbines. Doctoral dissertation, Uppsala University [Acta Universitatis Upsaliensis], Uppsala, Sweden. ISSN 1651-6214, p 774

A Novel MPPT Technique for Solar PV System to Maximize IoT Network Lifetime

Sirine Rabah, Aida Zaier, Jaime Lloret, and Hassen Dahman

Abstract The advent of the Internet of Things (IoT) has revolutionized various industries by connecting physical devices and enabling them to exchange data. However, One of the key challenges faced by IoT nodes is the limited availability of energy resources. Therefore, optimizing energy efficiency in IoT nodes has emerged as a crucial research area. This paper presents a solution aimed at addressing this challenge, an optimized solar energy harvesting system for IoT nodes, using solar photovoltaic energy. Our work introduces a novel approach called MPPT HHO-HBA. This approach combines two optimization techniques, Harris Hawks Optimization and Honey Badger algorithm. The aim of our approach is to optimize the Maximum Power Point Tracking (MPPT) of solar panels for IoT nodes to maximize their power output and increase the overall network's lifetime. To ensure that the harvested solar power is effectively used, a SEPIC converter is employed to boost the electrical energy generated by the solar panel. The proposed approach is implemented in MATLAB/Simulink, and the effectiveness of the methodology is assessed using appropriate metrics. The results demonstrate that the proposed method surpasses previous strategies in terms of efficiency. Additionally, our system has the capability to power numerous sensor nodes, contributing to enhanced sustainability and reduced carbon footprint compared to traditional energy sources.

S. Rabah (✉)
MACS Laboratory LR 16ES22, National Engineering School of Gabes, University of Gabes, Omar ibn Elkhattab Street, Zrig Eddakhlania, 6029 Gabes, Tunisia
e-mail: sirinerabah7@gmail.com

A. Zaier
bLR-11/TIC-03 Innov'COM Laboratory, Higher School of Communication of Tunis, University of Carthage, 2083 Ariana, Tunisia

J. Lloret
Instituto de Investigacion para la Gestion Integrada de Zonas Costeras, Universitat Politecnica de Valencia, 46730 Gandia, Spain

H. Dahman
Department of Electrical Engineering, National Engineering School of Gabes, University of Gabes, 6029 Gabes, Tunisia

LaPhyMNE Laboratory (LR05ES14), FS Gabes, University of Gabes, 6072 Gabes, Tunisia

© The Author(s), under exclusive license to Springer Nature Singapore Pte Ltd. 2024
A. Mellit et al. (eds.), *Proceedings of the 2nd International Conference on Green Energy Conversion System*, Green Energy and Technology,
https://doi.org/10.1007/978-981-97-6148-7_46

Keywords IoT · Energy harvesting · Solar · MPPT · HHO · HBA · Network lifetime

1 Introduction

An Internet of Things (IoT) network consists of sensor nodes strategically deployed in various fields to serve specific applications [1]. These spatially distributed sensor nodes are used for numerous purposes, including traffic management, medical monitoring, disaster management, structural health monitoring, environmental monitoring, weather forecasting, military surveillance, and more [2, 3]. To power IoT nodes, power-limited batteries and chemical batteries are commonly used due to their remote deployment, which makes battery replacement impractical [4]. The life span of the IoT network is determined by the battery life of each sensor node; however, replacing batteries in a large-scale, distributed IoT network can be challenging during deployment [5–7]. To address this issue and enhance the development of IoT, extensive research has been dedicated to energy harvesting techniques and energy-efficient batteries for IoT nodes [8]. Energy harvesting techniques are essential as IoT nodes often operate with limited and non-replaceable batteries [9]. By optimizing energy utilization, sensor nodes can be powered continuously, even without batteries [10]. One of the most promising approaches is the integration of an energy harvesting system with an energy storage device, particularly focusing on renewable energy sources such as solar energy, wind energy, and piezoelectric energy [11, 12].

The implementation of an efficient solar energy harvesting system is particularly crucial for ensuring a long network lifetime [13]. Solar energy harvesting is highly advantageous in the design of IoT nodes, as it significantly reduces the need for manual battery replacements in remote and challenging locations like battlefields, forests, glaciers, and volcanoes [14]. By utilizing solar PV energy, IoT nodes powered by solar energy harvesting systems can measure pressure, humidity, light, and temperature simultaneously, contributing to an extended overall sensor network operation lifetime [15, 16]. This technology not only optimizes the performance of IoT networks but also reduces the efforts and resources required for maintenance and battery replacement. Various factors, including radiation levels and temperature, significantly influence the performance of Photovoltaic (PV) panels. Consequently, there is a pressing need for additional Maximum Power Point Tracking (MPPT) approaches to achieve higher efficiency, optimize power extraction from PV panels, and ensure optimal performance [17]. Several techniques have been developed for this purpose.

The constant voltage method is commonly used in solar energy systems to maintain a consistent voltage during battery charging [18]. However, this method often results in significant oscillations and requires a lengthy convergence time. In contrast, the Perturb and Observe (P&O) algorithm serves as an active power control system and exhibits strong dynamic performance in response to solar irradiation [19]. Another method used in solar system is the incremental conductance

(INC) method [20]. However, traditional MPPT approaches have certain drawbacks, including concerns about dynamic response and steady-state oscillations, which limit their effectiveness in adapting to changing environmental conditions [21].

To address these challenges, researchers have developed sophisticated processing systems that employ advanced techniques such as genetic algorithms (GA), neural networks (NN), and fuzzy logic controllers (FLC) for intelligent MPPT controllers in solar system. Nevertheless, the use of FLC, GA, and NN techniques may be limited due to factors such as a limited number of fuzzy deduction rules, small population sizes, and limited data availability [22]. While these intelligent techniques demonstrate effective performance under specific conditions, they may struggle to precisely track the Maximum Power Point (MPP) when confronted with fluctuations in the power curve [8].

In the context of PV systems, the efficient extraction of the maximum power point is of utmost importance. Therefore, the primary objective of this work is to develop an efficient and effective method to accomplish this goal.

The main contributions of this work are outlined below:

- The solar energy harvesting system generates power to supply the IoT node during communication, and the power is stored in a 3.6 V battery.
- A DC-DC SEPIC converter is employed to convert the voltage amplitude from one level to another, optimizing the power generated by the system.
- A novel hybrid optimization technique that combines the Harris Hawks Optimization (HHO) and Honey Badger Algorithm (HBA) is used to optimize the Maximum Power Point Tracking (MPPT) process, significantly improving the efficiency of battery charging.

The paper is organized as follows:

Section 2 provides a detailed explanation of the proposed solar energy harvesting system, encompassing the fundamental model and its constituent blocks. Subsequently, Sect. 3 delves into an analysis of the selected control algorithm. Following this, Sect. 4 showcases the simulation outcomes. Lastly, the paper concludes with Sect. 5, summarizing the findings and insights.

2 Proposed Methodology

The solar energy harvesting system comprises various components, including solar cells, DC-DC SEPIC converter, maximum power point tracking (MPPT), energy-efficient algorithms, and battery/supercapacitor storage. In our work, we have proposed an optimized solar energy harvesting-based wireless sensor network to maximize the lifetime of the IoT network. To achieve maximum power extraction from the solar panel, we employ a novel hybrid optimization technique that combines the Harris Hawks Optimization (HHO) and Honey Badger Algorithm (HBA) to optimize the MPPT process. The Single End Primary Inductor Current (SEPIC) converter is used to convert the energy from the PV panel. The solar energy harvesting system

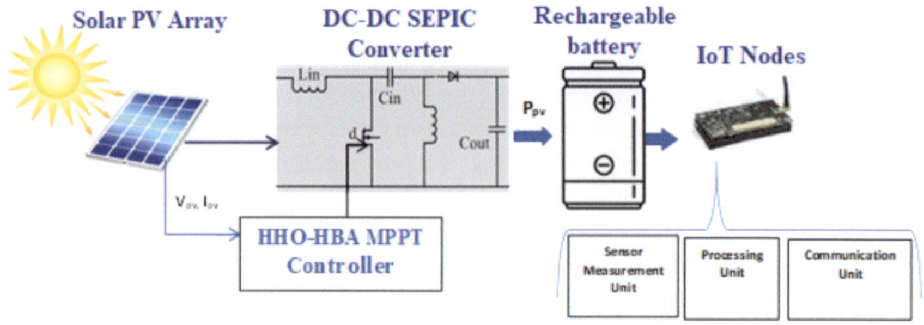

Fig. 1 Block diagram of the proposed system

provides a DC power supply to the IoT nodes, harnessing ambient sunlight through solar panels, which directly convert light energy into DC electrical energy. The SEPIC converter adjusts the DC voltage to charge the battery, which powers the IoT nodes. Figure 1 illustrates the workflow of our proposed methodology. The IoT network is composed of key components, including a sensing board measurement unit, a computational/processing unit, and a communication unit. The IoT node measures the desired physical quantities (e.g., pressure, humidity, light, and temperature) using the sensor measurement unit. The microcontroller processes the sensed data, and the transmitter unit sends the data wirelessly to nearby network nodes in the form of data packets. The data is further forwarded to the USB gateway node through cluster head nodes. This setup enables users to control and monitor various applications, such as forest monitoring, temperature monitoring, glacier monitoring, volcano monitoring, and battlefield monitoring.

2.1 *Modeling Solar Photovoltaic System*

The section introduces the mathematical models of the Solar PV system. PV modules are connected in series and parallel within the PV field to achieve the desired power output. Each module consists of individual cells [23]. The modeling begins with the basic element, the cell, which can be represented as an ideal current source providing a current, I_{ph}, proportional to the incident light power, in parallel with a diode (P-N junction). The output current equation based on Kirchhoff's current law is given by Eq. (1).

$$I = I_{ph} - I_D \tag{1}$$

The photo-current (I_{ph}) is directly proportional to the solar irradiation and temperature, as shown by Eq. (2):

$$I_{ph} = [I_{sc} - K_i(T - T_{STC})]\frac{G}{G_{STC}} \tag{2}$$

where (I_{sc}) is the short-circuit current, K_i represents the short-circuit current temperature coefficient, G_{STC} and T_{STC} denote the irradiation and temperature values under standard test conditions, while G and T represent the actual values of irradiation and temperature.

The module diode current is determined using Eq. (3), which depends on the saturation current I_0:

$$I_D = I_0 \exp\frac{V_D}{V_{th} - 1} \tag{3}$$

Here, V_D represents the diode voltage, and $V_{th} = \frac{(A\times K\times T)}{q}$ represents the thermal voltage.

The saturation current is given by Eq. (4):

$$I_0 = I_{rs}\left(\frac{T}{T_{STC}}\right)^3 \exp\left[\frac{q \times E_g}{A \times K}\left(\frac{1}{T_{STC}} - \frac{1}{T}\right)\right] \tag{4}$$

where $q = 1.6 \times 10^{-19} C$ is the electron charge, $K = 1.38 \times 10^{-23} J/K$ is the Boltzmann's constant, (A = 2) is the diode ideality factor, $E_g \cong 1.1\ eV$ at $T = 25$ °C is the semiconductor band-gap energy, and I_{rs} is the reverse saturation current.

Additionally, when selecting a solar panel for simulation several factors need to be taken into account. These factors encompass parameters such as Open Circuit Voltage (V_{oc}), short circuit current (I_{sc}), maximum output power (P_{max}) and surface area. Furthermore, the panel's durability is of utmost importance to withstand outdoor conditions and reduce potential replacement expenses. In light of these considerations, we have selected a solar panel to simulate a low-power device, adhering to the specifications outlined in Table 1.

Table 1 Characteristics of the solar cell used [13]

Parameter	Value
Open circuit voltage (V_{oc})	5.6 V
Short circuit current (I_{sc})	660 mA
Maximum output power (P_{max})	2.4 W
Voltage at maximum power (V_{mp})	4 V
Current at maximum power (I_{mp})	600 mA
Surface area	36 cm^2

2.2 SEPIC Converter Modeling

The output voltage of a SEPIC converter can deviate from the input voltage, potentially being either higher or lower [24]. Illustrated in Fig. 2, the SEPIC converter employs two inductors, L1 and L2, wound on the same core and exposed to identical voltages during the switching cycle. The MOSFET duty cycle is governed by the SEPIC converter's output voltage. Upon switch activation, the output voltage drops to 0 V. This converter proves particularly valuable in battery charging scenarios where the voltage may fluctuate below or exceed the regulator's output.

The output voltage of a SEPIC converter is commonly formulated as follows:

$$V_0 = V_{in} \times \left(\frac{D}{D-1}\right) \tag{5}$$

where V_0 represents the output voltage and V_{in} stands for the input voltage, and D denotes the duty cycle. The duty cycle (D) of the SEPIC converter operating in continuous conduction mode (CCM) is determined by the subsequent equation:

$$D = \frac{V_0 + V_D}{V_{in} + V_0 + V_D} \tag{6}$$

Here, V_D signifies the voltage across the diode.

The maximum attainable duty cycle D_{max}, where $V_{in(Min)}$ denotes the minimum input, can be articulated as:

$$D_{\max} = \frac{V_0 + V_D}{V_{in(\text{Min})} + V_0 + V_D} \tag{7}$$

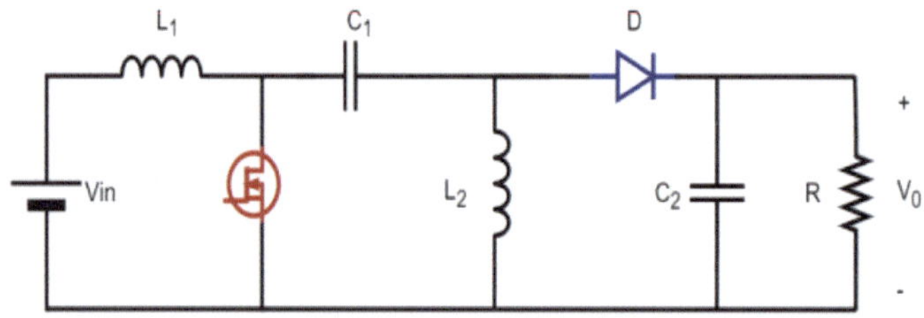

Fig. 2 SEPIC converter basic circuit

2.3 Energy Storage

Energy storage devices within IoT networks primarily consist of two categories: batteries and supercapacitors. Rechargeable batteries find extensive application in solar energy systems due to their superior energy densities compared to supercapacitors. In the realm of IoT networks, several rechargeable battery types are commonly employed, encompassing nickel-cadmium (NiCd), nickel-zinc (NiZn), nickel metal hydride (NiMH), and lithium-ion (Li-ion). In our proposed study, we utilize a lithium-ion battery featuring a capacity of 740 mAh and a nominal voltage of 3.7 V. Notably, this contemporary battery type offers numerous advantages over conventional rechargeable batteries, including heightened energy density, swifter discharge rates, elevated cell voltage, prolonged life cycle, and immunity to memory effects [25].

3 Energy Efficient MPPT Technique

3.1 MPPT Technique

The optimization of the efficiency of solar power systems often involves the utilization of the maximum power point tracking (MPPT) technique, which is specifically designed for tracking the maximum power peak of nonlinear electrical generators such as those found in solar photovoltaic (PV) systems. The power peak in such systems is influenced by various factors such as solar radiation and temperature variations. In general, the inability to effectively track power peaks in such systems results in energy loss and potential harm to the interconnected system. Recent research has concentrated on devising metaheuristic optimization methods for the creation of MPPT's in solar PV systems. However, these methods are accompanied by certain drawbacks, including complexity, restricted convergence, and additional constraints. To surmount these limitations, we introduce a groundbreaking hybrid approach. This methodology integrates the Honey Badger (HBA) [26] and Harris Hawks Optimization (HHO) [27] algorithms expedite the convergence rate, elevate the optimization efficacy, and augment the overall system functionality.

3.2 Implementation of HHO-HBA for MPPT in Solar System

The proposed method introduces a novel and innovative approach by synergizing the Honey Badger Algorithm (HBA) and the Harris Hawks Optimization (HHO) algorithm. These two algorithms, both operating on a population-based paradigm, bring their individual strengths to the table, resulting in a powerful fusion capable of addressing complex optimization challenges in energy harvesting systems. The

Honey Badger Algorithm (HBA) employs a hybrid strategy of both personal and global best positions during its updating process. It incorporates a random search mechanism alongside local searches to locate optimal solutions. HBA, inspired by the solitary foraging behavior of honey badgers, autonomously seeks out food sources, signifying a self-reliant and exploratory approach. On the other hand, the Harris Hawks Optimization (HHO) algorithm emulates the cooperative hunting strategy of hawks. In this approach, the search space is divided into sub-spaces, with each hawk assigned to explore one of these segments. Hawks communicate with each other to share information and collectively converge on the global optimum. The HHO algorithm leverages the collective intelligence of the swarm, akin to the collaborative hunting tactics of hawks. One key distinction between the HHO and HBA algorithms lies in their selection mechanisms. HBA employs random selection to determine which individuals will carry out the search operation. In contrast, the HHO algorithm utilizes a probability distribution function to assign individuals for the search operation. This variation in selection methods highlights the diverse strategies employed by these algorithms. The compatibility of HHO and HBA stems from their distinct qualities. The cooperative and collective nature of HHO complements the solitary and exploratory approach of HBA. It's worth noting that each algorithm's strengths can mitigate the weaknesses of the other. By combining their strategies, the hybrid approach capitalizes on their respective advantages, potentially leading to a more efficient and robust optimization process. This innovative fusion presents a promising avenue for enhancing the optimization performance in energy harvesting systems.

The HHO algorithm is regarded the main algorithm in this regard, and the HBA algorithm is added to it. Incorporating the HBA algorithm into the HHO algorithm can generate a diverse set of search space solutions that can enhance the performance of the HHO algorithm. The implementation of the HHO-HBA algorithm involves several essential steps is illustrated in Fig. 3. Firstly, the HBA algorithm is utilized to initialize the positions of the members of the HHO group. The initialization phase is crucial because it provides a varied set of initial solutions to the HHO algorithm, thereby increasing the chances of finding the optimal solution. Secondly, the HBA algorithm's search strategy is employed to update the positions of the HHO group members in each iteration, thus ensuring that the algorithm explores various regions of the solution space. Thirdly, the HHO algorithm's update rule is applied to update the positions of the HHO group members based on their fitness values. Finally, the HBA algorithm's search strategy is employed to perform local search operations on the best solution obtained by the HHO algorithm. This final step fine-tunes the solution and enhances the algorithm's overall performance. The integration of the HBA algorithm into the HHO algorithm has demonstrated that the resulting algorithm can provide improved solutions for different optimization problems.

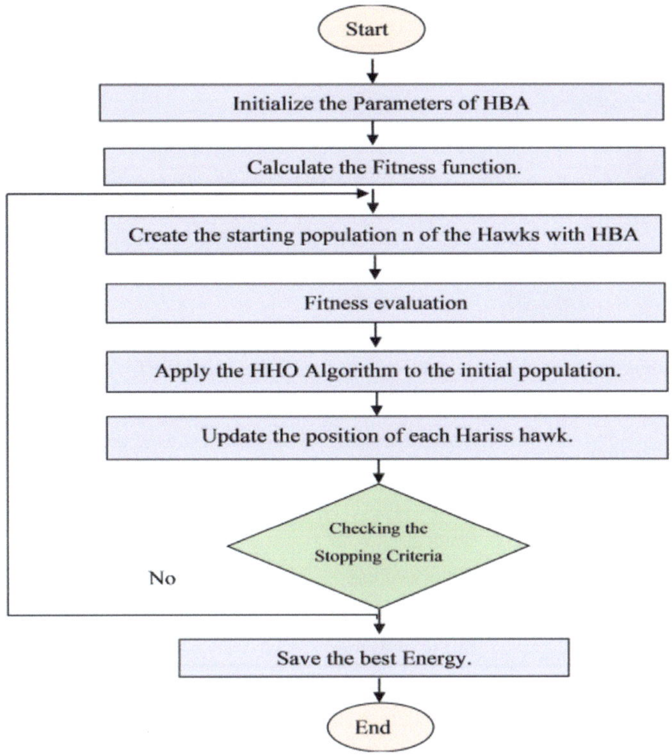

Fig. 3 Flowchart of the proposed methodology

4 Results and Discussion

To simulate a continuous power source for IoT nodes, a system using solar PV was tested using a MATLAB/Simulink R2021a platform. The generated energy is stored in a rechargeable battery, which acts as a storage medium. A SEPIC converter with an HHO-HBA MPPT controller was used to regulates the voltage supplied to the battery prior to being discharged, as depicted in Fig. 4.

The findings have been scrutinized with respect to the battery's current, voltage, state of charge, and efficiency. The efficiency of the proposed methodology has been juxtaposed with that of other prevalent controller algorithms, such as P&O-MPPT and HHO-MPPT controller. Table 2 displays the parameters used in the proposed work, as explained in the simulation. In Fig. 5, the battery's performance using the P&O control technique is depicted, showing SoC, current, and voltage over a 100-s simulation period. According to the findings, the State of Charge (SoC) of the battery varied from 0 to 87.02% over the given period. As illustrated in Fig. 6, the HHO-MPPT controller achieved an SoC range of 0 to 93% in the same simulation period.

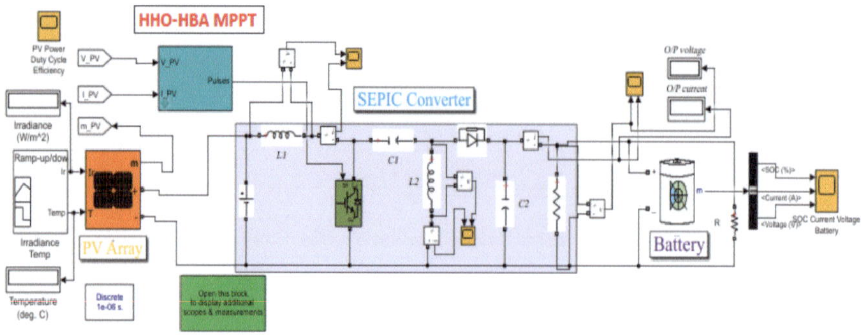

Fig. 4 Simulink model of the solar system using the HHO-HBA MPPT

Table 2 Simulation parameters

Parameter	Value
Temperature (T)	25 °C
Irradiance (W/m^2)	1000
Battery voltage	3.7 V
Battery capacity	740 mAh
SEPIC converter	Boost converter
Rechargeable battery type	Lithium-ion
Initial duty cycle	0.5
IoT load model	10-Ω resistor

Conversely, Fig. 7 showcases the outcomes of the suggested approach, demonstrating that the battery's SoC reached as high as 98%.

When the battery is being charged, the current flowing through it is negative, which indicates that the electrochemical cells are undergoing oxidation. On the other hand, when the battery is discharging, the current flowing through it is positive, indicating that the electrochemical cells are undergoing reduction.

In Fig. 8 a comparison was made between the HHO-HBA MPPT controller proposed in this study and a existing controller algorithm. Results showed that the HHO-HBA MPPT controller achieved an efficiency of 98.94%, while the efficiency of the existing P&O-MPPT and HHO-MPPT controllers were 89.72% and 97.6%, respectively. The findings imply that the introduced control algorithm could be a superior option for maximizing power output in photovoltaic systems. Simulation outcomes reveal the superior performance of the hybrid energy harvesting system for IoT nodes, which integrates the HHO-HBA MPPT controller, compared to conventional systems utilizing P&O-MPPT and HHO-MPPT methods.

The HHO-HBA MPPT algorithm attains enhanced solar power extraction from solar cells, thereby leading to improved battery efficiency and decreased charging duration. These results underscore the potential effectiveness of the proposed HHO-HBA MPPT controller in optimizing energy harvesting within IoT systems.

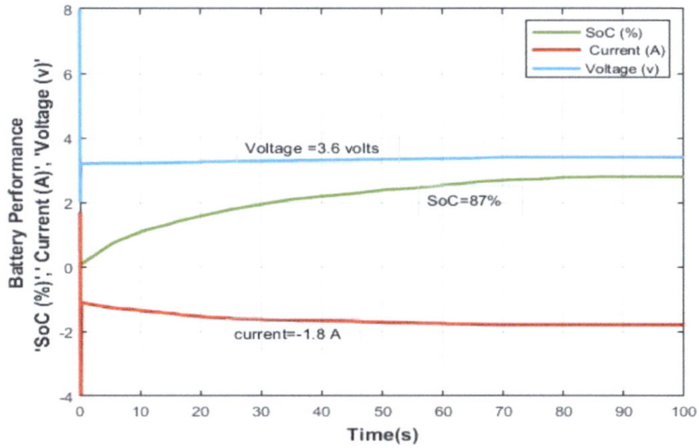

Fig. 5 The performance of battery using P&O-MPPT approach for 100 s

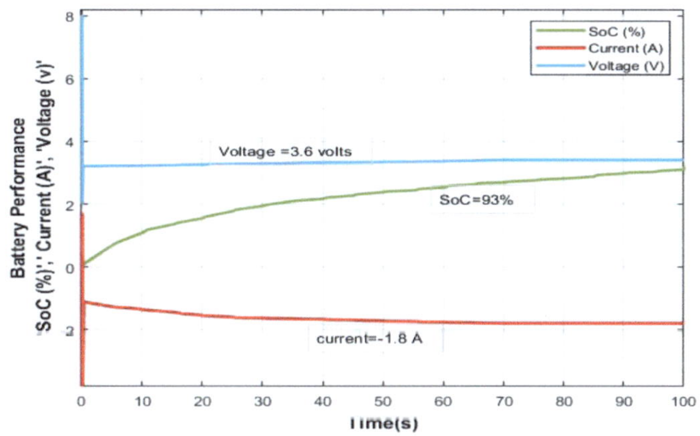

Fig. 6 The performance of battery using HHO-MPPT approach for 100 s

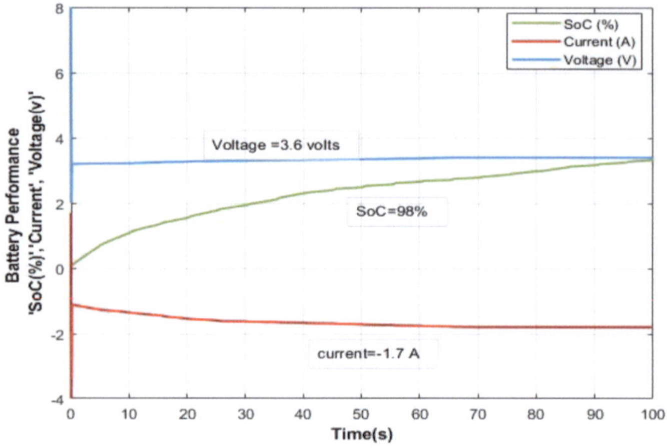

Fig. 7 The performance of battery using HHO-HBA MPPT approach for 100 s

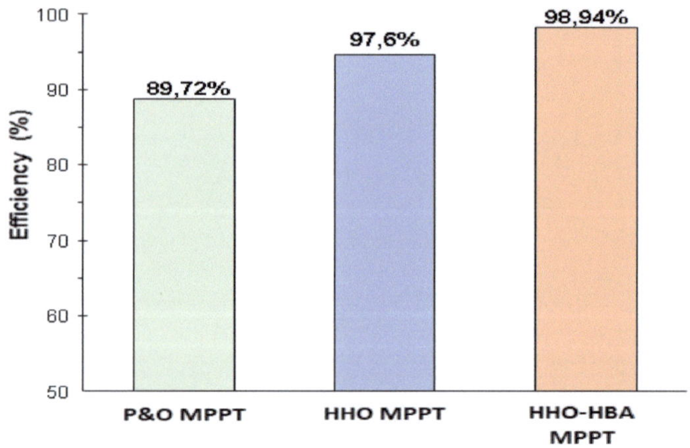

Fig. 8 Efficiency comparison

5 Conclusion

This paper presents a comprehensive study on the design, optimization, and simulation of a solar energy harvesting system with a SEPIC converter to power the IoT nodes. The system incorporates the novel HHO-HBA MPPT control technique, with the primary objective of maximizing power generation from the solar system.

To evaluate the performance of the proposed methodology, we used the MATLAB/Simulink platform and conducted a comparative analysis with conventional approaches, including P&O-MPPT and HHO-MPPT. Our investigation centered on critical parameters such as battery current, voltage, state of charge (SoC), and efficiency. The simulation outcomes demonstrate that through the optimization of the MPPT algorithm, enabling more effective utilization of solar energy harvesting, the rechargeable battery's capacity experiences a substantial increase. Consequently, this augmented energy reservoir serves to adequately power the IoT network nodes. Furthermore, the overall efficiency was computed, accounting for the efficiencies of both the SEPIC converter and the HHO-HBA MPPT in the solar system. Notably, the proposed methodology achieved an impressive overall efficiency rating of 98.94%. This study also points to various potential avenues for future research, including the integration of novel energy storage systems, exploration of advanced MPPT algorithms, and the implementation of real-time monitoring.

References

1. Bahadur DJ, Lakshmanan L (2023) A novel method for optimizing energy consumption in wireless sensor network using genetic algorithm. Microprocess Microsyst 96:104749
2. Fahmy HMA (2023) WSNs applications. In: Concepts, applications, experimentation and analysis of wireless sensor networks. Springer Nature Switzerland, Cham, pp 67–242
3. Abdelhafdh M, Fourati M, Fourati LC, Chouaya A (2019) Wireless sensor network monitoring system: architecture, applications and future directions. Int J Commun Netw Distrib Syst 23(4):413–451
4. Dash L, Khuntia M (2020) Energy efcient techniques for 5G mobile networks in WSN: a survey. In: 2020 international conference on computer science, engineering and applications (ICCSEA). IEEE, pp 1–5
5. Yasin S, Ali T, Draz U, Shaf A, Ayaz M (2019) A parametric performance evaluation of batteries in wireless sensor networks. In: Recent trends and advances in wireless and IoT-enabled networks. Springer, Cham, pp 187–196
6. Miao L, Huo ZM, Rong Y, Mu HW, Sun ZX (2023) IoT adaptive threshold energy management algorithm based on energy harvesting. Ad Hoc Netw 149:103241
7. Nguyen M, Nguyen C, Tran H, Nguyen H (2023) Energy harvesting for devices in wireless sensor networks: a Review. EAI Endorsed Trans IoT 9(2):e2
8. Ahmad FF, Ghenai C, Bettayeb M (2021) Maximum power point tracking and photovoltaic energy harvesting for Internet of Things: a comprehensive review. Sustainable Energy Technol Assess 47:101430
9. Vasanthi G, Prabakaran N (2022) An improved approach for energy consumption minimizing in WSN using Harris Hawks optimization. J Intell Fuzzy Syst 43(4):4445–4456
10. Hidalgo-Leon R, Urquizo J, Silva CE, Silva-Leon J, Wu J, Singh P, Soriano G (2022) Powering nodes of wireless sensor networks with energy harvesters for intelligent buildings: a review. Energy Rep 8:3809–3826
11. Zheng X, He L, Wang S, Liu X, Liu R, Cheng G (2023) A review of piezoelectric energy harvesters for harvesting wind energy. Sens Actuators A Phys, 114190
12. Singh J, Kaur R, Singh D (2021) Energy harvesting in wireless sensor networks: a taxonomic survey. Int J Energy Res 45(1):118–140
13. Rabah S, Zaier A, Lloret J, Dahman H (2023) Efficiency enhancement of a hybrid sustainable energy harvesting system using HHHOPSO-MPPT for IoT devices. Sustainability 15(13):10252
14. Ahmed R, Buchli B, Draskovic S, Sigrist L, Kumar P, Thiele L (2019) Optimal power management with guaranteed minimum energy utilization for solar energy harvesting systems. ACM Trans Embedded Comput Syst (TECS) 18(4):1–26

15. Chinipardaz M, Amraee S (2022) Study on IoT networks with the combined use of wireless power transmission and solar energy harvesting. Sādhanā 47(2):86
16. Kaur P, Singh P, Sohi BS (2020) Adaptive MAC protocol for solar energy harvesting based wireless sensor networks in agriculture. Wireless Pers Commun 111(4):2263–2285
17. Kaur H, Buttar AS (2021) Lifetime enhancement of wireless sensor network using solar energy harvesting technique. IET Wirel Sens Syst 11(1):54–65
18. Podder AK, Roy NK, Pota HR (2019) MPPT methods for solar PV systems: a critical review based on tracking nature. IET Renew Power Gener 13(10):1615–1632
19. Sharma H, Haque A, Jaffery ZA (2018) Modeling and optimisation of a solar energy harvesting system for wireless sensor network nodes. J Sens Actuator Netw 7(3):40
20. Mohamed SA, Abd El Sattar M (2019) A comparative study of P&O and INC maximum power point tracking techniques for grid-connected PV systems. SN Appl Sci 1(2):174
21. Pant S, Saini RP (2019) Comparative study of MPPT techniques for solar photovoltaic system. In: 2019 international conference on electrical, electronics and computer engineering (UPCON). IEEE, pp 1–6
22. Qais MH, Hasanien HM, Alghuwainem S (2020) Enhanced whale optimization algorithm for maximum power point tracking of variable-speed wind generators. Appl Soft Comput 86:105937
23. Martin AD, Cano JM, Silva JFA, Vazquez JR (2015) Backstepping control of smart grid-connected distributed photovoltaic power supplies for telecom equipment. IEEE Trans Energy Convers 30(4):1496–1504
24. Kircioğlu O, Ünlü M, Camur S (2016). Modeling and analysis of DC-DC SEPIC converter with coupled inductors. In: 2016 international symposium on industrial electronics (INDEL). IEEE, pp 1–5
25. Tuna G, Gungor VC (2016) Energy harvesting and battery technologies for powering wireless sensor networks. In: Industrial wireless sensor networks. Woodhead Publishing, pp 25–38
26. Hashim FA, Houssein EH, Hussain K, Mabrouk MS, Al-Atabany W (2022) Honey Badger algorithm: new metaheuristic algorithm for solving optimization problems. Math Comput Simul 192:84–110
27. Mansoor M, Mirza AF, Ling Q (2020) Harris Hawk optimization-based MPPT control for PV systems under partial shading conditions. J Clean Prod 274:122857

Uniform and Stable TEM_{00} Mode Multibeam Solar Laser Approach

Miguel Catela, Dawei Liang, Cláudia R. Vistas, Hugo Costa, Dário Garcia, Bruno D. Tibúrcio, and Joana Almeida

Abstract The utilization of TEM_{00} mode laser beams is critical in precise material processing, optical communication, and high-resolution imaging. While TEM_{00} mode solar laser beams have been achieved, the need for a highly accurate and continuously operating solar tracking system presents significant challenges. A solar laser pumping approach is here proposed to enable the production of four uniform and stable TEM_{00} mode solar laser beams even in the presence of solar tracking errors. The solar radiation was collected and concentrated using a heliostat—parabolic mirror system. An aspheric lens further concentrated the solar rays into a homogenizer that evenly distributed the solar power flux onto four Nd:YAG rods positioned within a single pump cavity. Through Zemax® and LASCAD™ software analysis, successful generation of four uniform and stable 1064 nm continuous-wave solar laser emissions was numerically demonstrated, exhibiting similar TEM_{00} power levels and beam quality factors. The system showed resilience against solar tracking errors of up to $\pm$ 0.2°.

Keywords Solar laser · Nd:YAG · Multibeam · Solar tracking error · Single mode

1 Introduction

Lasers are ubiquitous in modern society, playing a critical role in a wide range of applications such as manufacturing, material processing, telecommunications, defense, and healthcare. The TEM_{00} mode, known as the fundamental mode or single mode, is typically the most stable mode in a laser resonator, as it experiences minimal mode competition and exhibits a well-defined Gaussian shaped intensity profile over time [1]. These laser systems present also superior beam quality (M_x^2, M_y^2), being

M. Catela (✉) · D. Liang · C. R. Vistas · H. Costa · D. Garcia · B. D. Tibúrcio · J. Almeida
CEFITEC, School of Science and Technology, NOVA University of Lisbon, 2829-516 Caparica, Portugal
e-mail: m.catela@campus.fct.unl.pt

© The Author(s), under exclusive license to Springer Nature Singapore Pte Ltd. 2024

A. Mellit et al. (eds.), *Proceedings of the 2nd International Conference on Green Energy Conversion System*, Green Energy and Technology,
https://doi.org/10.1007/978-981-97-6148-7_47

particularly well-suited for applications that require high precision, such as micromachining, engraving, and lithography. Additionally, the industry has shown significant interest in TEM_{00} mode multibeam laser systems due to their potential for reducing processing time. By employing techniques such as beam overlapping or tailoring each beam for specific tasks within the overall procedure, these systems can rapidly perform tasks that would normally take days or even weeks using a single beam [2]. Still, intensive electricity consumption is a constraint for the above-mentioned laser large-scale applications considering the cost and environmental impacts [3]. As such, there has been a growing interest in the development of solar-pumped lasers [4]. In 2022, Liang et al. achieved the state-of-the-art multimode solar laser collection and solar-to-laser conversion efficiencies of 41.3 W/m^2 and 4.6%, respectively, by end-side-pumping three 2.5 mm diameter, 25 mm length Ce:Nd:YAG rods through an effective collection area of 0.40 m^2 [5]. While the state-of-the-art TEM_{00}-mode solar laser collection and solar-to-laser conversion efficiencies of 7.9 W/m^2 collection and 0.79% was attained in 2017 by end-side-pumping a single 4.0 mm diameter, 35 mm length Nd:YAG rod, through an effective collection area of 1.18 m^2 [6]. In order to achieve the previous results, is of paramount importance to incorporate an accurate solar tracking system (STS) [7]. Here, we proposed a solar laser pumping approach to achieve four uniform and stable TEM_{00} mode solar laser emissions even under solar tracking error.

2 NOVA Parabolic-Heliostat Mirror System

NOVA heliostat-parabolic mirror system is presented in Fig. 1. The system uses a dual-axis heliostat equipped with a set of large plane mirrors (93.5% reflectivity) to track the Sun's apparent path in both azimuthal and altitude directions and redirect the incoming solar radiation towards a stationary back-surface silver coated parabolic mirror (80% reflectivity) with 1.5 m diameter, 60° rim angle and 660 mm focal length.

In numerical analysis, the solar laser head was initially aligned with the Sun at solar noon and the STS was turned off. As the Sun apparently moves in the sky, the variations of both altitude (Δh) and azimuth ($\Delta\alpha$) angles can be described as altitude and azimuthal solar tracking errors. Still, the variation of the altitude angle can be considered approximately zero around solar noon at the latitude of Lisbon [7]. The standard direct solar spectrum for one-and-a-half air mass (AM1.5D) was used as the reference data for consulting the solar spectrum. By select the key solar wavelengths associated with the 22 peak absorption wavelengths responsible for 1.0% Nd:YAG lasing, computing time was reduced. Considering a typical solar irradiance of 950 W/m^2 in Lisbon, the 1.767 m^2 full collection area of the parabolic mirror and the 16% overlap between the Nd:YAG absorption spectrum and the solar emission spectrum, about 269 W (950 W/m^2 $\times$ 1.767 m^2 $\times$ 16%) effective input solar pump power was introduced in Zemax®. The combined reflectance of the NOVA heliostat-parabolic mirror system was also defined (93.5% $\times$ 80% = 74.8%).

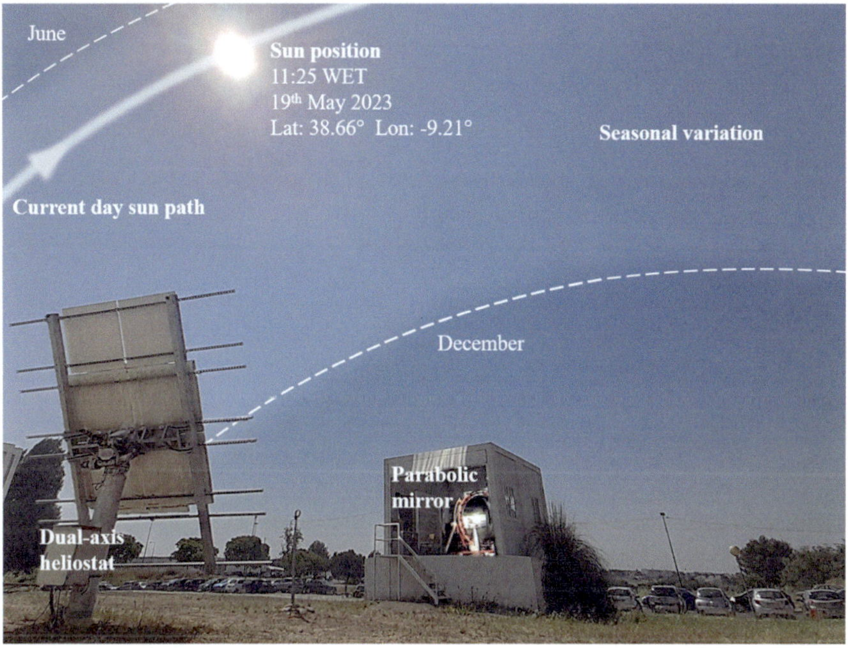

Fig. 1 NOVA heliostat-parabolic mirror system operating on 19th May 2023 at 11.25 WET

3 Optimization of a Four-Rod Solar Laser Head for Uniform Energy Distribution

A 28.5 mm thickness, 66.0 mm diameter fused silica aspheric lens further concentrated the solar radiation towards the input aperture of a flux homogenizer. The square-shaped homogenizer (95% reflectivity) with input aperture width $W_H = 13.0$ mm transmitted the concentrated solar radiation through its length $L_H = 0.0$ mm into a square-shaped pump cavity with input aperture width $W_{in} = 11.5$ mm, output aperture width $W_{out} = 6.5$ mm, and length $L_C = 11.5$ mm. Four Nd:YAG rods with diameter $D_R = 2.5$ mm and length $L_R = 15$ mm were fixed into a single holder and positioned within the pump cavity parallel to each other with rotational symmetry. Each Nd:YAG rod was high reflection (HR 1064 nm) coated on its front end and anti-reflection (AR 1064 nm) coated on its back end (Fig. 2).

Spectral variations, including transmission through the aspheric lens, reflectance of the homogenizer and pump cavity, the absorption spectrum and wavelength-dependent refractive indices of fused silica material for the aspheric concentrator and cooling water, along with the 22 peak absorption wavelengths and their respective coefficients for the 1.0% Nd:YAG laser medium were incorporated into the glass catalog data in Zemax®. For the ray-tracing analysis of the effective absorbed pump power and respective pump distribution within the laser media, a detector volume, divided into 75,000 voxels, was used for each rod. Figure 3 presents the effective

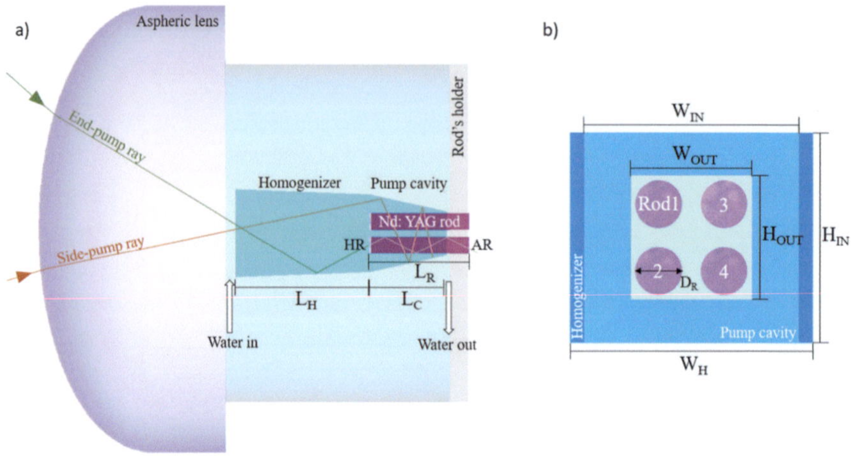

Fig. 2 Design of the solar laser head. **a** Top view; **b** Front view

pump power absorbed by each one of the four rods (P_A) using a homogenizer with length L_H for: a) the perfect alignment with $\Delta\alpha = 0.00°$ (Fig. 3a, b) single-axis misalignment with $\Delta\alpha = 0.15°$ (Fig. 3b), The transversal absorbed pump flux distribution on the top of the four Nd:YAG rods were also presented for $L_H = 0$ and 20 mm, where $L_H = 0$ means no homogenizer. The homogenizer was essential to evenly distribute the pump power among the four Nd:YAG rods, as demonstrated in Fig. 3. By increasing the L_H to 20 mm, it was possible to observe similar transversal absorbed pump flux distribution on the top of the four laser rods even under solar tracking errors.

4 Optimization of the Laser Cavity for TEM_{00} Mode Solar Laser Stable Operation

The effective absorbed power data from the Zemax® analysis was then imported into the LASCAD™ software. The material data of 1.0 at. % Nd:YAG crystal was programmed [7]. As represented in Fig. 4, each of the four optical resonators was comprised of two opposing mirrors: the HR 1064 nm 99.98% coating on the upper end face of each Nd:YAG rod and the PR 1064 nm output mirror, with reflectivity varying between 90 and 99%. L represents the separation length between the output end face of each Nd:YAG rod with AR coatings at 1064 nm and the respective PR 1064 nm output mirror and plays a crucial role in the extraction of the TEM_{00} mode laser beam. Long resonators were chosen so that only the TEM_{00} mode could oscillate, improving the beam quality factors (M_x^2, M_y^2).

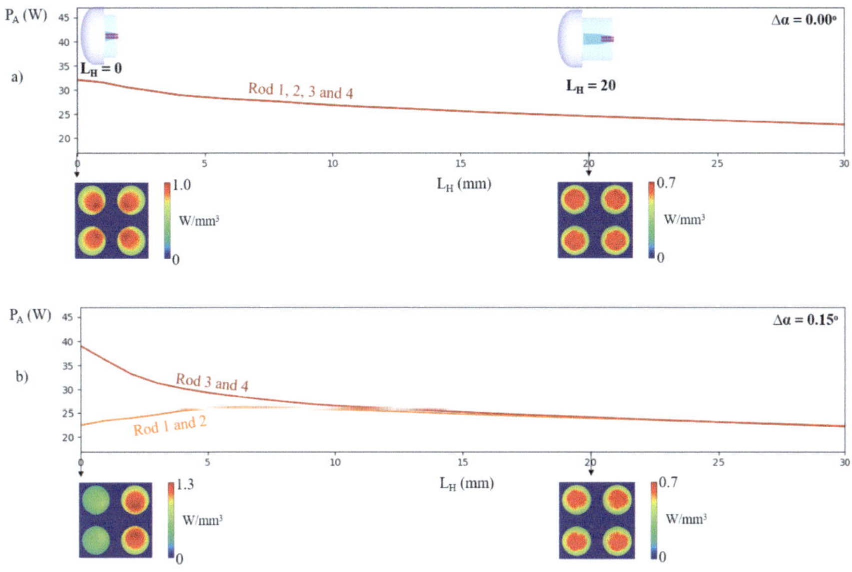

Fig. 3 Effective absorbed pump power (P_A) in function of the homogenizer length (L) for each laser rod (1–4). **a** $\Delta\alpha = 0.0°$; **b** $\Delta\alpha = 0.15°$

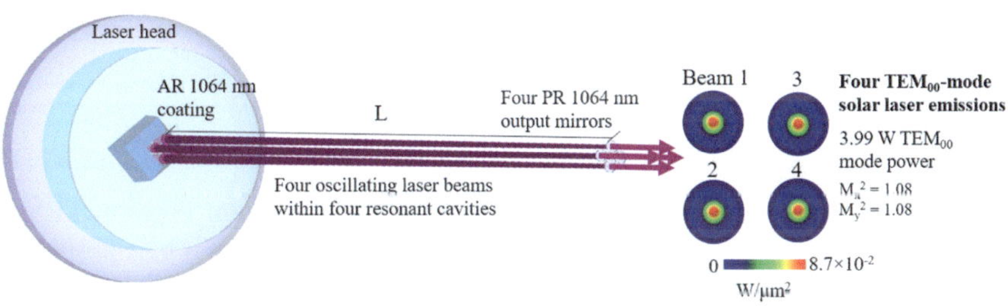

Fig. 4 Laser cavities for TEM$_{00}$ mode extraction. Four PR 1064 nm laser output couplers were positioned at a distance L from the AR end face of their respective Nd:YAG rod for the efficient extraction of four TEM$_{00}$-mode laser beams (1–4), here presented for L = 430 mm and RoC = − 2000 mm at the optimum alignment

The performance of the solar laser head in producing TEM$_{00}$ mode solar laser power was evaluated with LASCAD™ numerical analysis software for typical azimuthal solar tracking errors up to ± 0.3°, as presented in Fig. 5.

The LASCAD™ software computed the maximum TEM$_{nm}$ mode that can oscillate considering the apertures that limit the spatial extent of the beam, such as the rod and the PR mirror diameters. The maximum TEM$_{00}$ mode solar laser power of 3.99 W was found for RoC = − 2000 mm, maintaining the single mode oscillation at $\Delta\alpha = \pm 0.1°$. By flattened the PR mirror to RoC = − 6000 mm, it was possible to maintain single mode oscillation even with $\Delta\alpha = \pm 0.2°$.

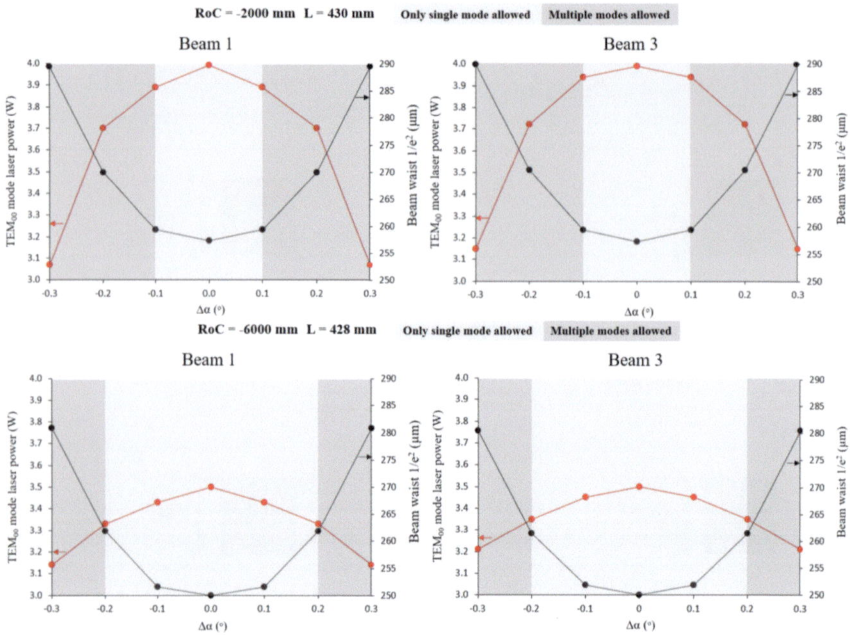

Fig. 5 TEM_{00} mode solar laser power versus azimuthal solar tracking error for beam 1 and 3. The $1/e^2$ beam waist is also presented for each case

5 Conclusion

A solar laser pumping approach to enable the production of four uniform TEM_{00} mode solar laser beams even in the presence of solar tracking errors is here proposed. The flux homogenizer was essential to uniformly distribute the pump power among the four Nd:YAG rods, as numerically demonstrated through Zemax® analysis. The optimization of the position and radius of curvature of the PR mirror in LASCAD™ software played a crucial role in ensuring stability of the laser resonator. Successful generation of four uniform and stable 1064 nm continuous-wave solar laser emissions was numerically demonstrated, exhibiting similar TEM_{00} mode power levels and beam quality factors even under solar tracking errors of up to $\pm$ 0.2°. The ability to control the power density of solar laser beams can progressively diminish laser based industry's carbon footprint, yielding long-term environmental benefits.

References

1. Koechner W (2006) Solid-state laser engineering; Springer Series in optical sciences, 6th edn. Springer, New York, NY
2. Wei J, Jin P, Cao X et al (2022) Direct generation of a stable multi-beam pulsed 355 nm UV laser based on a micro-lens array. Chin Opt Lett 20:041405. https://doi.org/10.3788/col202220.041405
3. Lv J, Peng T, Zhang Y, Wang Y (2021) A novel method to forecast energy consumption of selective laser melting processes. Int J Prod Res 59:2375–2391. https://doi.org/10.1080/00207543.2020.1733126
4. Laser Focus World (2022) Solar pumping converts broadband sunlight into efficient laser light. https://www.laserfocusworld.com/lasers-sources/article/14283698/solar-pumping-converts-broadband-sunlight-into-efficient-laser-light. Accessed 21 Dec 2022
5. Liang D, Vistas CR, Garcia D et al (2022) Most efficient simultaneous solar laser emissions from three Ce:Nd:YAG rods within a single pump cavity. Sol Energy Mater Sol Cells 246:111921. https://doi.org/10.1016/j.solmat.2022.111921
6. Liang D, Almeida J, Vistas CR, Guillot E (2017) Solar-pumped Nd:YAG laser with 31.5 W/m^2 multimode and 7.9 W/m^2 TEM_{00}-mode collection efficiencies. Sol Energy Mater Sol Cells 159:435–439. https://doi.org/10.1016/j.solmat.2016.09.048
7. Catela M, Liang D, Vistas C et al (2023) Stable emission of solar laser power under non-continuous solar tracking conditions. Appl Opt 62:2697–2706. https://doi.org/10.1364/AO.485158

Maximizing Power-Point Tracking with Machine Learning

Jouili Mbarka, Farhat Maissa, and Sbita lassaad

Abstract This paper discusses the use of machine learning to track the maximum power point. In order to train neural networks, the error back propagation approach is employed. The advantage of neural networks is their quick and accurate tracking of maximum power points. This method makes use of a neural network to provide the reference voltage for the maximum power point under various atmospheric circumstances. The maximum power point can be tracked by properly controlling the dc-dc boost converter. Simulation results are obtained using MATLAB/SIMULINK to verify theory analysis.

Keywords Deep learning · Solar system · ANN

1 Introduction

Recently, many nations throughout the world have focused heavily on the development of renewable energy to counteract the anticipated eventual depletion of fossil resources. Renewable energy has an energy density that is at least as high as that of fossil fuels, and clean energy doesn't produce any polluting materials like nitrogenous compounds, sulfate compounds, or dust. In terms of the environment and energy security, hydrogen is seen as a potential replacement for fossil fuels because it is inherently clean, sustainable, and emission-free. As a result, there are numerous active

J. Mbarka (✉) · S. lassaad
Processes, Energy, Environment and Electrical Systems (Code: LR18ES34), National Engineering School of Gabes, University of Gabes, Gabes, Tunisia
e-mail: mbarka.jouili@gmail.com

S. lassaad
e-mail: Lassaad.sbita@enig.rnu.tn

F. Maissa
American University of Ras Al Khaimah, 10021 Ras Al Khaimah, UAE
e-mail: maissa.farhat@aurak.ac.ae

© The Author(s), under exclusive license to Springer Nature Singapore Pte Ltd. 2024
A. Mellit et al. (eds.), *Proceedings of the 2nd International Conference on Green Energy Conversion System*, Green Energy and Technology,
https://doi.org/10.1007/978-981-97-6148-7_48

studies that are currently being conducted on the generation and use of hydrogen in our society [1].

The usage of photovoltaic systems is the primary technique for harvesting solar energy. Photovoltaic PV generates electricity using photons from the sun [1]. PV technologies are based on the photoelectric effect, which was initially introduced in 1839 by a French physicist named Edmund Becquerel. With no fuel requirements, no emissions, no moving components, and a potentially endless energy source—the sun—solar cells and modules that use this PV effect are the ultimate energy producers. The photoelectric effect happens when a portion of a pair of negatively charged metal plates is struck by an ultraviolet light beam made up of photons.

PV photovoltaic arrays typically convert sunlight into electricity. The amount of sun illumination and the fluctuating temperature of the environment affect how much DC power is produced. Additionally, it varies based on the weight of the load. A PV array displays a current-voltage characteristic with a distinct point, known as the maximum power point, where the PV array generates its maximum output power under uniform irradiance and temperature. For a PV array, the maximum-power-point-tracking MPPT algorithm is required to supply the greatest power to the load [1]. The instantaneous maximum power of the PV array [2] is continuously detected by converters under the control of an MPPT algorithm.

2 Photovoltaic Modelling

2.1 Ideal PV Cell Model

The equivalent circuit of the ideal PV cell is shown in (Fig. 1). The fundamental equation for the I-V property of the ideal PV cell, which is derived from the theory of semiconductors [3], is as follows:

$$I = I_{PV,cell} - I_{0,cell}\left[\exp^{\left(\frac{qV}{aK_T}\right)} - 1\right], \quad (1)$$

$$I_d = I_{0,cell}\left[\exp^{\left(\frac{qV}{aK_T}\right)} - 1\right], \quad (2)$$

where $I_{0,cell}$ is the reverse saturation or leakage current of the diode, q is the electron charge ($1.60217646 \times 10^{(-19)}C$), Id is the Shockley diode equation, and $I_{PV,cell}$ is the current created by incident light; it is directly proportional to the sun's irradiation, "a" stands for the diode ideality constant [3], "k" stands for the Boltzmann constant ($1.3806503 \times 10^{(-23)}J/K$), and "T" stands for the p-n junction's temperature in Kelvin.

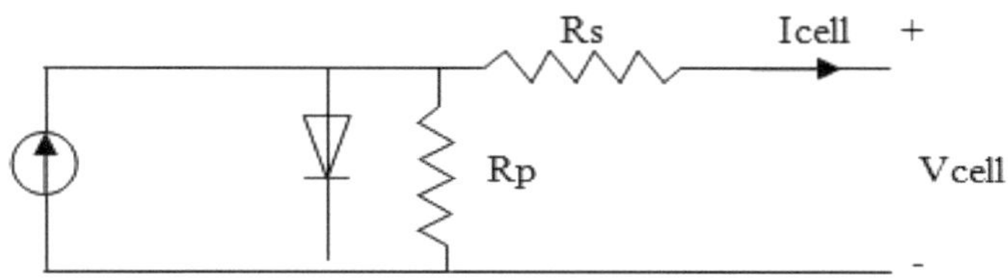

Fig. 1 Equivalent circuit of PV model

2.2 Modeling the PV Array

Equations 1 and 2 of the PV cell do not depict the V-I property of a realistic PV array. The observation of the features at the terminal of the PV array needs the introduction of extra parameters to the fundamental Eqs. 3 and 4, since practical arrays are made up of numerous connected PV cells.

$$I = I_{PV} - I_0\left[\exp\exp\left(\frac{(V + R_S I)}{V_t a}\right) - 1\right] - \left(\frac{(V + R_S I)}{R_P}\right), \tag{3}$$

In general, current or voltage sources include electric generators. The practical PV device exhibits hybrid behavior, which, depending on the operating point, may be of a current or voltage source. The actual PV device contains two resistances: R_s a series resistance whose effect is stronger when the device is operating in the voltage source region and R_p a parallel resistance whose influence is stronger when the device is operating in the current source region. The device's overall structural resistance is represented by the R_s resistance. R_s Primarily depends on the resistances of the p and n bodies, the contact resistance of the n layer with the top metal grid, the resistance of the grid, and the contact resistance of the metal base with the p semiconductor layer.

The leakage current of the p-n junction is primarily responsible for the R_p resistance, which relies on how the PV cell was made. In order to simplify the model, some authors ignore the resistance, which is typically large in value. R_s has a very low value and is occasionally overlooked as well. The internal features of the device R_s, R_p, as well as outside factors like temperature and radiation level affect the I-V characteristic of the PV array, which is depicted in (Fig. 2).

The creations of charge carriers and, as a result, the devices current are directly influenced by the amount of incident light.

It is challenging to determine the elemental cells' light-generated current I_{PV} without the interference of the series and parallel resistances. The maximum current accessible at the terminals of the practical device, the nominal short-circuit current $I_{SC,n}$ is the only information provided in datasheets.

Due of the low series resistance and large parallel resistance of actual devices, the assumption ($I_{SC} \simeq I_{PV}$) is typically utilized in PV device modeling. Equation 4 [2, 3], shows that the light-generated current of the PV cell relies linearly on solar radiation and is also influenced by temperature.

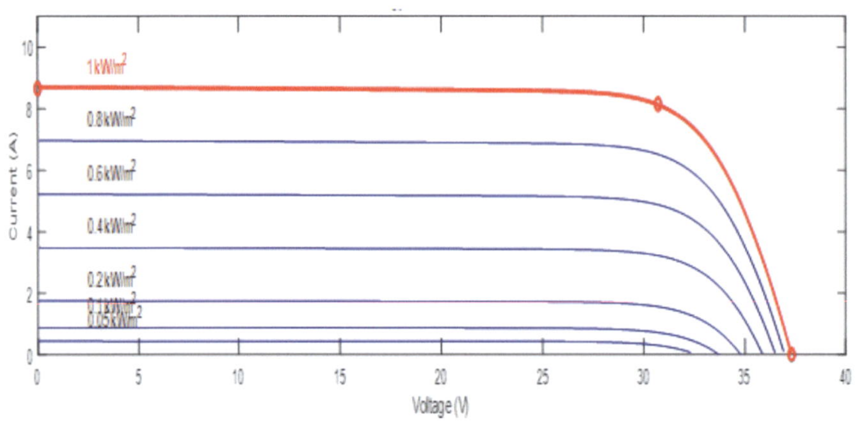

Fig. 2 I-V characteristic of PV

$$I_{PV} = \left(I_{PV,n} + K_I \cdot \Delta T\right)\frac{G}{G_n}, \tag{4}$$

3 Maximum Power-Point Tracking

3.1 Maximum Power-Point Tracking Methods

The maximum power is vary depending on the environment's temperature and irradiance (Fig. 3). A real-time maximum power-point tracker is an essential component of the PV system since the maximum accessible energy of solar arrays is always changing due to atmospheric conditions. The three categories of proposed maximum power point tracking MPPT schemes in the technical literature [4] are as follows [5]:

(1) **Direct methods.**
(2) **Artificial intelligence methods.**
(3) **Indirect methods.**

The MPP is searched using the direct methods, which are often referred to as true seeking methods, by repeatedly perturbing the PV array's operational point. Perturb and Observe P&O [6], Hill Climbing HC [7], and Incremental Conductance INC [8] methods are commonly used in PV systems under this category. To obtain the MPP, the P&O technique disturbs the PV array's operating voltage. Hill climbing approach disturbs the duty cycle of the dc-dc interface converter similarly to P&O scheme. The major characteristic of these systems is simplicity; nevertheless, inherent steady state oscillation restricts these methods to low-power applications. The incremental conductance approach, which relies on the fact that the slope of the power versus

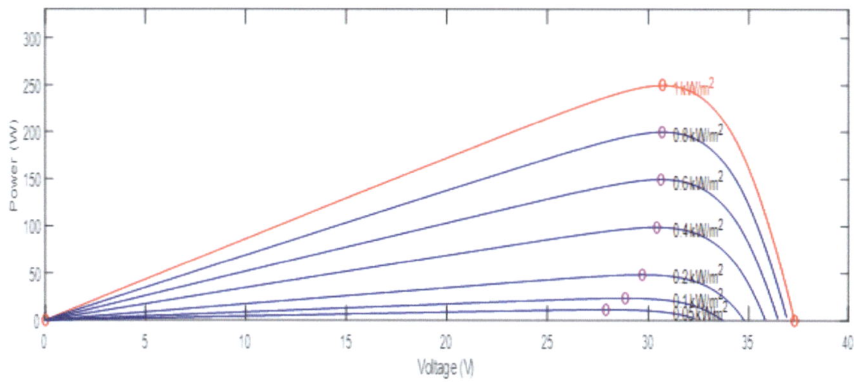

Fig. 3 P-V characteristic of PV

voltage is zero at the MPP, can reduce steady state oscillation. The dynamic performance of MPP tracking has been suggested to be enhanced by artificial intelligence and indirect techniques.

The artificial intelligence techniques offer a quick but computationally intensive solution for the MPPT problem by focusing on the nonlinear properties of PV arrays.

The indirect methods are based on extracting the MPP of the array from its output characteristics [9]. Fractional open-circuit voltage OCV and short-circuit current SCC schemes provide a simple and effective way to obtain the MPP.

3.2 *MPPT by Using Neural Network*

In this paper, an artificial neural network is employed to track the maximum power point. To achieve MPP, a three-layer neural network is required, as shown in (Fig. 4). Temperature and radiation are the two input variables, and the output variable of ANN is voltage of MPP V_{mpp}. To train the neural network, some data must be obtained as input and output variables. As a result, weights of neurons in various layers are obtained. To acquire data, PV model programming in MATLAB is employed. There are numerous ways to train an ANN [8]. In this study, the ANN is trained using the error back propagation method.

After training the ANN and specifying the weights of the neurons, the output of the ANN is the V_{mpp} for any T and G as inputs. Now, using the V-I characteristic of the modeled PV, it is possible to determine the current at the highest power point I_{mpp}. Consequently, by adding V_{mpp} and I_{mpp}, the maximum power $P_{\max}$ is obtained.

The PV and maximum power point tracker system, seen in (Fig. 5), consists of a dc-dc boost converter and a control unit powered by a neural network. To exert control at all times.

Chopper with specified V_{mpp} and I_{mpp} duty cycle of chopper is obtained by the following equation:

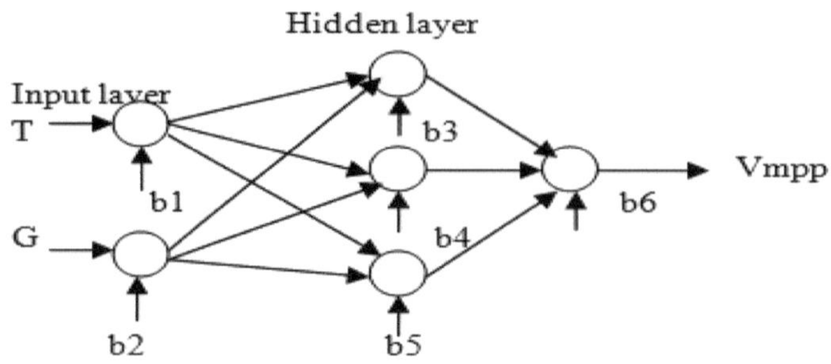

Fig. 4 Neural network structure

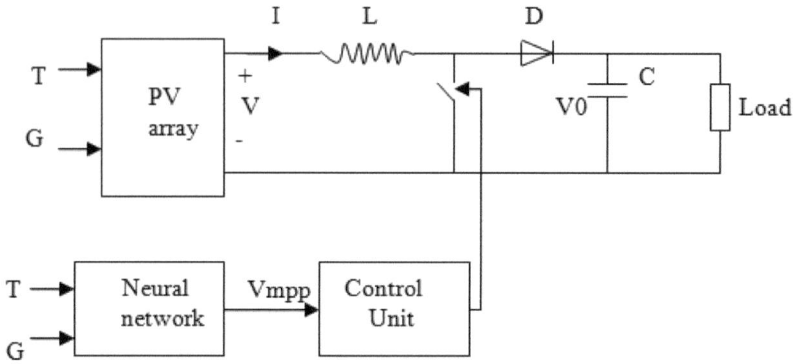

Fig. 5 DC-DC chopper includes PV as a input and its control units

$$D = 1 - \sqrt{\left(\frac{V_{mpp}}{I_{mpp}} \frac{I_{out}}{V_{out}}\right)}, \tag{5}$$

4 Simulation Results

Using MATLAB/SIMULINK, a standalone PV system coupled to a boost dc-dc converter is simulated in order to validate the theoretical analysis discussed in earlier parts. First, a set of data encompassing temperature, irradiance, and maximum power point voltage are used to train the neural network model. To train ANN, the error back propagation approach is employed. The neural network model is simulated using MATLAB/SIMULINK and connected to the control unit after receiving the weights of the neurons. The output of the ANN is now automatically connected to $V_{ref\,,\,mpp}$ for any T and G inputs.

Three states are used for simulation, each with a different temperature and amount of light. (Fig. 6a) depicts the output of the neural network for these three pairs of

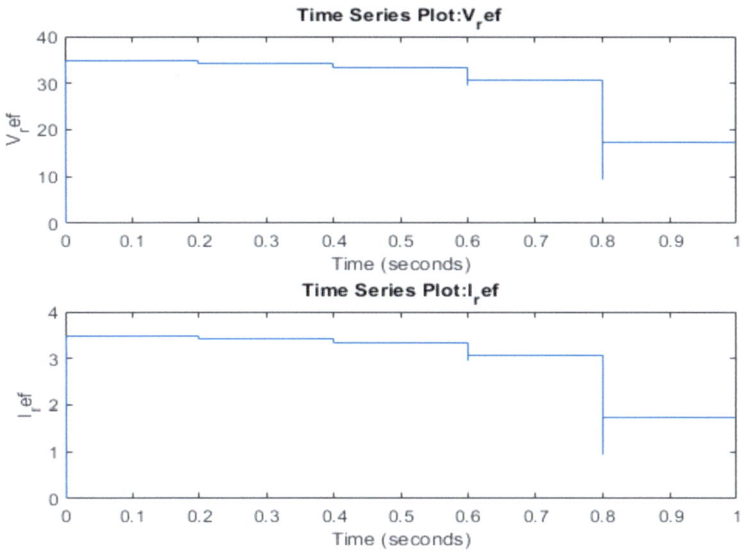

Fig. 6 **a** Reference voltage of maximum power point generated by ANN, **b** reference current of maximum power point

T and G, which is the reference voltage of the maximum power point ($V_{ref,mpp}$). The reference current of the power point $I_{ref,mpp}$ is received from the control unit in (Fig. 6b) by specifying $V_{ref,mpp}$.

Applying $V_{ref,mpp}$ and $I_{ref,mpp}$ to the chopper control unit results in switching pulses that enable PV's output voltage and current to follow $V_{ref,mpp}$ and $I_{ref,mpp}$ (Fig. 7) displays the PV's output voltage and current. This chart makes it clear that the PV system's output voltage and current track $V_{ref,mpp}$ and $I_{ref,mpp}$. It is possible to determine that the maximum power reference is tracked by comparing the maximum power reference to the power pulled from PV (Fig. 8).

The figures demonstrate that the control system has good dynamic performance in terms of voltage, current, and therefore maximum power-point tracking.

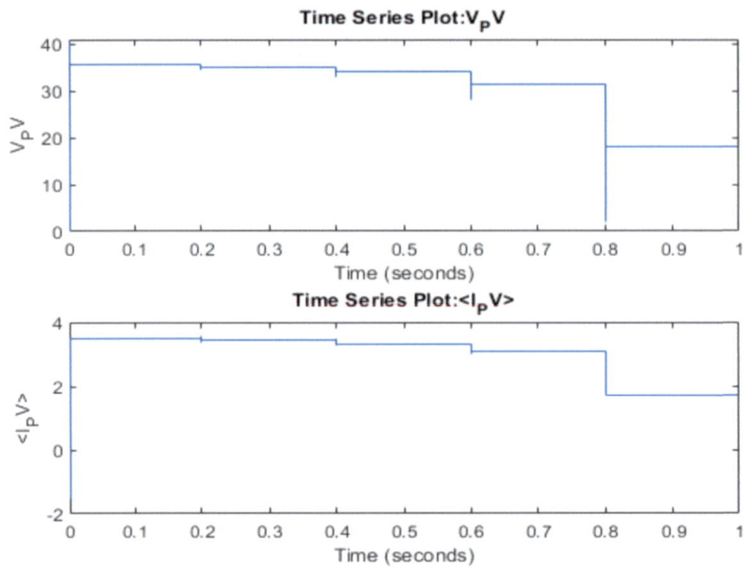

Fig. 7 **a** Output voltage of PV, **b** output current of PV

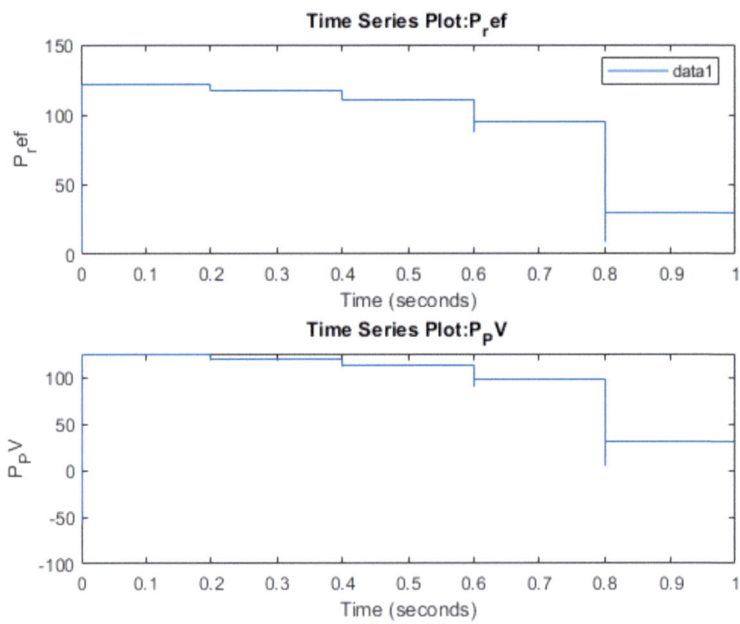

Fig. 8 **a** Reference maximum power, **b** drawn power from PV

5 Conclusion

This paper discusses MPPT that uses neural networks. The point of maximum power is rapidly and precisely specified using a neural network under any fluctuation in meteorological situations. The neural network has an additional benefit over other techniques in PV maximum power-point tracking due to its superior dynamic performance. The DC-DC boost chopper also keeps track of the peak power point. Consequently, the highest power and optimum efficiency of solar energy are attained.

References

1. Dehghani M, Taghipour M, Gharehpetian GB, Abedi M (2020) Optimized fuzzy controller for MPPT of grid-connected PV systems in rapidly changing atmospheric conditions. J Mod Power Syst Clean Energy 9(2):376–383
2. Fatima K, Alam MA, Minai AF (2019) Optimization of solar energy using ANN techniques. In: 2019 2nd international conference on power energy, environment and intelligent control (PEEIC). IEEE, pp 174–179
3. Bouraiou A, Hamouda M, Chaker A, Sadok M, Mostefaoui M, Lachtar S (2015) Modeling and simulation of photovoltaic module and array based on one and two diode model using Matlab/ Simulink. Energy Procedia 74:864–877
4. Singh P, Palwalia DK, Gupta A, Kumar P (2015) Comparison of photovoltaic array maximum power point tracking techniques. Int Adv Res J Sci Eng Technol 2(1):401–404
5. Salas V, Olias E, Barrado A, Lazaro A (2006) Review of the maximum power point tracking algorithms for stand-alone photovoltaic systems. Sol Energy Mater Sol Cells 90(11), 1555–1578
6. Zhao D, Qian M, Ma J, Yamashita K (2020) Photovoltaic generator model for power system dynamic studies. Sol Energy 210:101–114
7. Khosrojerdi F, Golkhandan NH (2018) Microcontroller-based maximum power point tracking methods in photovoltaic systems. In: 2018 9th annual power electronics, drives systems and technologies conference (PEDSTC). IEEE, pp 330–334
8. Harrag A, Messalti S (2019) PSO-based SMC variable step size P&O MPPT controller for PV systems under fast changing atmospheric conditions. Int J Numer Model Electron Networks Devices Fields 32(5):e2603
9. Zečević Ž, Rolevski M (2020) Neural network approach to MPPT control and irradiance estimation. Appl Sci 10(15):5051

Modeling and Simulation of a Triple Junction Photovoltaic Generator Feeding an Electric Vehicle Drive in MATLAB/SIMULINK

Salah Beni Hamed and Mouna Ben Hamed

Abstract In this research work, the control of an electric vehicle powered by a photovoltaic system is examined. A triple-junction solar cell made of InGaP/InGaAs/Ge with a high efficiency served as the foundation for the used photovoltaic generator. The fourth case studies that are examined involve standard test settings, a step change in solar radiation with a constant temperature, a step change in temperature with a constant sun radiation, simultaneous step changes in temperature, and solar radiation conditions. An electric vehicle's three-phase permanent magnet motor (PMSM) serves as the foundation for its traction system. The control approach relies on the field oriented control strategy. All of the control algorithms and the designed system are implemented in Matlab/Simulink. Significant results are provided and discussed for a wide range of speeds, including the two operating modes under step changes in environmental circumstances.

Keywords Electric vehicle · Field oriented control · Triple-junction solar cell · Photovoltaic system · Field weakening region

1 Introduction

Nowadays, the own vehicles' number on the road increases. This aspect contributes to the transportation sector's excessive oil consumption [1, 2]. Furthermore, too much carbon dioxide in the air traps heat, warming the globe [3, 4]. Electric vehicles in the transportation sector appear to be the best option for preserving the environment and human life when the electrical energy they consume is generated from renewable energy sources as solar photovoltaic energy, biomass, wind energy, geothermal

S. Beni Hamed (✉)
Tunis University, Tunis, Tunisia
e-mail: salahbenihamed@yahoo.fr

M. Ben Hamed
Gabes University, Gabes, Tunisia

© The Author(s), under exclusive license to Springer Nature Singapore Pte Ltd. 2024
A. Mellit et al. (eds.), *Proceedings of the 2nd International Conference on Green Energy Conversion System*, Green Energy and Technology,
https://doi.org/10.1007/978-981-97-6148-7_49

energy, etc. [5, 6]. Each of them possesses distinctive qualities and prominent merits. Solar energy, in particular photovoltaic (PV) systems, have recently acquired popularity as the most extensively utilized renewable energy source due to its environmental friendliness, safety, cleanliness, lack of noise, and low maintenance requirements [7, 8]. First utilized is the mono-junction solar cell [9]. The conversion efficiency remains low as less sunlight is transformed into electric energy. The multi-junction solar cell is then developed as a solution to this issue [10]. A significant proportion of sunlight is subsequently converted to electrical energy because the converted solar cell is made up of many semiconductors with varied wavelengths [10]. As a result, it performs better than mono-junction solar cells. The PV module is a photovoltaic system's basic power conversion component. Photovoltaic modules should be deployed in parallel or serial configurations to meet the demand for electric energy. Next, a PV generator is acquired.

The primary contribution of this research project is the control of the photovoltaic electric vehicle outfitted with a triple junction solar cell under various weather scenarios and operating modes.

The structure of the work is as follows. The modeling of the studied system is shown in the second section. The third section presents and discusses the simulation's findings. The conclusion and some potential futuristic perspectives are presented in section four.

2 Modelling of the Studied System

2.1 Triple-Junction Photovoltaic Generator Based on InGap/InGaAs/Ge Mathematical Model

The triple-junction InGap/InGaAs/Ge solar cell is made up of three sub-cells connected in series, as shown in Fig. 1. In the literature, many electrical equivalent circuits for photovoltaic solar cells have been used [11]. Figure 1 shows the one that was used in this research work. A light source, a diode, a parallel-connected resistor, and a series-connected resistor serve as the representational elements for each sub-cell. The electricity flow is denoted by (1) using Kirchhoff's current law.

$$I_{pv} = I_{phj} - I_{Dj} - I_{Rshj} \tag{1}$$

In which I_{phj} is the light generated current, I_{Dj} is the currents flows across the diode and I_{Rshj} is the current intensity moving into the shunt resistor.

For the top sub-cell j = 1, j = 2 for the medium sub-cell and j = 3 for the bottom sub-cell.

Incident light is absorbed, causing a stream of freed electrons to flow and creating an electricity flow. This is known as light-generated current. It is a function of the absorbed amount of sunshine as it is indicated in Eq. (2).

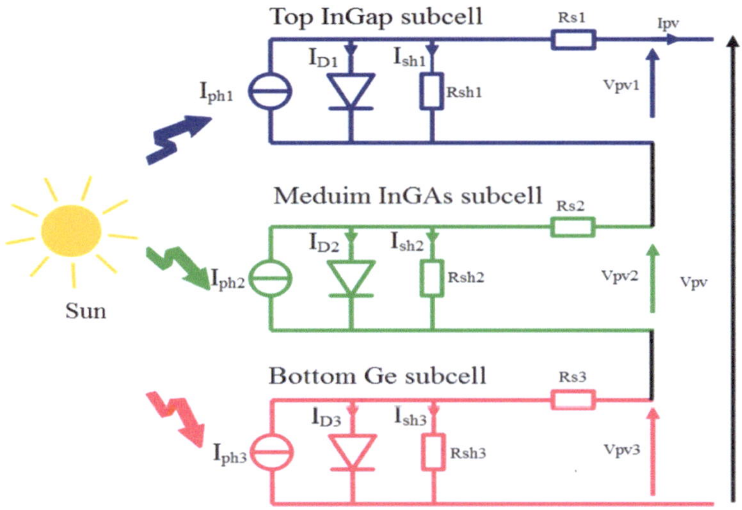

Fig. 1 Electrical equivalent circuit of multi-junction InGap/InGaA/Ge solar cell

$$I_{phj} = \frac{R_s}{R_{s\,STC}}\left[I_{ccj} + \gamma\left(T_{cj} - T_{STC}\right)\right] \tag{2}$$

where T_{STC} is the temperature solar cell at standard test conditions in °C, γ is the temperature coefficient of the actual short circuit current in A/°C, R_s and $R_{s\,STC}$ are the solar radiation and the solar radiation at standard test conditions in W/m2 respectively.

When a diode is forward biased, an electricity flows through it. The current intensity expression is given as (3).

$$I_{Dj} = I_{satj}\left[\exp\left(\frac{qV_{DJ}}{n_j BT}\right) - 1\right] \tag{3}$$

Its forward electric potential is given with Eq. (4).

$$V_{Dj} = V_{pvj} + R_{sj} I_{pv} \tag{4}$$

The mathematical behavior of the electrical flow caused by the thermal creation of electron-hole pairs inside the diode's depletion zone and the diffusion current caused by minority carriers flowing across the depletion region is described by Eq. (5).

$$I_{satj} = K_j\, T\left(\frac{3 + \delta_j}{2}\right)\exp\left(\frac{-E_{gj}}{n_j BT}\right) \tag{5}$$

where q is the electron charge, n_j is the diode ideality factor, B is the Boltzmann's constant, E_{gj} is the band-gap energy, K and δ are constant and T is the absolute temperature.

As seen in (6), the temperature of each layer affects the energy difference between the electrons' valence band and the bottom of the conduction band.

$$I_{satj} = K_j\, T\left(\frac{3+\delta_j}{2}\right)\exp\left(\frac{-E_{gj}}{n_j BT}\right) \tag{6}$$

Equation (7) illustrates the entire electric potential of the triple-junction solar cell.

$$\begin{aligned} V_{pv} = {} & \frac{n_1 BT}{q} Ln\left(\frac{I_{ph1} - I_{pv} - I_{Rsh1}}{I_{sat1}} + 1\right) + \frac{n_2 BT}{q} Ln\left(\frac{I_{ph2} - I_{pv} - I_{Rsh2}}{I_{sat2}} + 1\right) \\ & + \frac{n_3 BT}{q} Ln\left(\frac{I_{ph3} - I_{pv} - I_{Rsh3}}{I_{sat3}} + 1\right) - R_s I_{pv} \end{aligned} \tag{7}$$

In which R_s is the equivalent serial resistance.

2.2 *Modelling of Electric Vehicle*

The traction part of the electric vehicle is composed around the three phase permanent synchronous motor (PMSM). A nonlinear differential equation can be used to represent the (d-q) reference frame dynamic mathematical model that describes the electrical behavior of the PMSM, as shown by the system Eq. (8) [12].

$$\begin{cases} V_{ds} = R \times i_{ds} + L_{ds} \times \dfrac{di_{ds}}{dt} - \omega_s \times L_{qs} \times i_{qs} \\ V_{qs} = R \times i_{qs} + L_{qs} \times \dfrac{di_{qs}}{dt} + \omega_s \times L_{ds} \times i_{ds} + \omega_s \times \phi_a \end{cases} \tag{8}$$

In which, the direct and quadratic stator voltage and current components are represented respectively with V_{ds}, V_{qs} and i_{ds}, i_{qs} represent respectively. The stator resistance is designed with R. The d and q axis self-stator inductances are represented with L_{ds} and L_{qs}, synchronous speed is given with ω_s and ϕ_a designs the permanent magnet flux linkage.

The motor's mechanical response to applied forces is indicated with Eq. (9).

$$J_t \frac{d\omega_s}{dt} = n_p (T_m - T_r) - f\omega_s \tag{9}$$

In which T_m, J_t, f and n_p are the electromagnetic torque, the total moment of inertia, the viscous friction coefficient and the number of pair poles respectively.

The load torque is applied on the electric vehicle wheels. It is defined with (10).

$$T_L = \left[M_{ev} \times g \times f_r + \frac{1}{2} \times \rho_a \times A_{fr} \times C_{dr} \times V_{ev}^2 + M_{ev} \times g \times A \sin(\alpha_{ro}) \right] \times r_w \tag{10}$$

Which M_{ev} is the total mass of the electric vehicle, g is the gravity acceleration and f_r is the constant of the rolling resistance force, ρ_a is the density of the air, A_{fr} is the area of the electric vehicle's frontal part, C_{dr} is the coefficient of the aerodynamic drag and V_{ev} is the vehicle speed, α_{ro} represents the road slope angle and r_w is the radius of the vehicle's wheel.

The electromagnetic torque is defined with (11).

$$T_m = \frac{3}{2} \times n_p \times \left(\phi_a \times i_{qs} + \left(L_{ds} - L_{qs}\right) \times i_{qs} \times i_{ds}\right) \tag{11}$$

Using the field-oriented control strategy, Eq. (11) begins

$$T_m = \frac{3}{2} \times n_p \times \phi_a \times i_{qs} \tag{12}$$

For fixed flux value, the PMSM drive behavior is equivalent to that of the separately excited DC motor [13]. The field-oriented control approach integrates three control loops: one outer loop for speed control and two inner loops for current components control. The target direct component of the current is a value derived from the field weakening algorithm as it is detailed in [14] and the target quadratic component value is the output of the speed controller. In the three loops, the proportional and integral controller is used.

3 Simulation Results

The overall drive scheme, as shown with Eqs. (1) through (12), is implemented in Matlab/Simulink software to verify the efficiency and dependability of solar electric vehicles. Numerous simulations are considered, including ones that simulate high-speed operation and changes in the meteorological weather conditions.

Figures 2, 3, 4, 5, 6, 7, and 8 display the obtained simulation results under simultaneous abrupt temperature and radiation fluctuations.

The temperature and radiation fluctuations adopted in the simulation are shown in Figs. 2 and 3. Standard test settings, abrupt radiation and temperature variations, as well as simultaneous abrupt radiation and temperature fluctuation, are all variation scenarios that are taken into consideration.

For the electric vehicle application, an appropriate conduction drive cycle is selected, encompassing the two operating modes, as shown in Fig. 8. Mechanical and photovoltaic powers are given in Figs. 4 and 5 respectively.

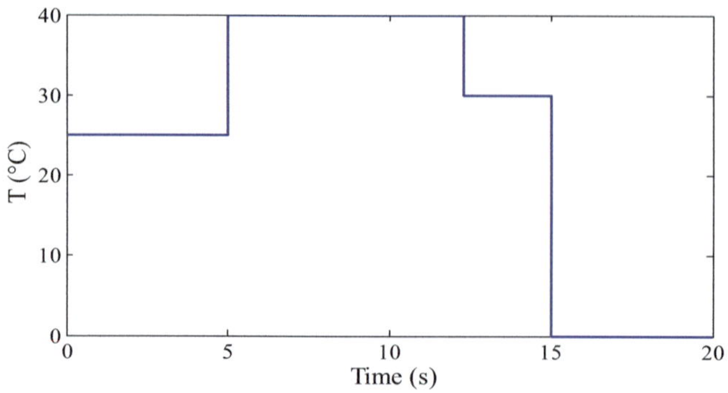

Fig. 2 Temperature trajectory

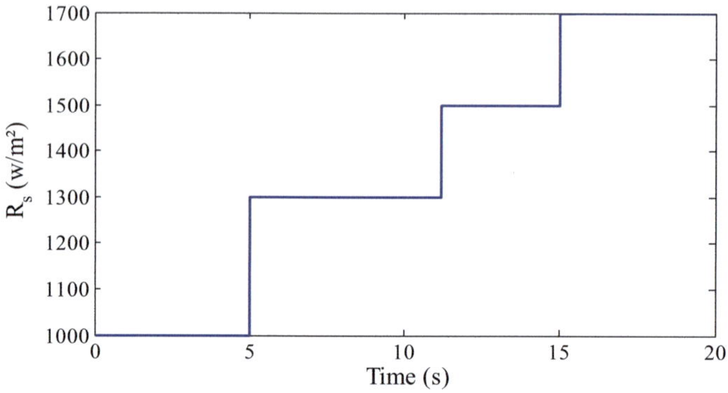

Fig. 3 Radiation trajectory

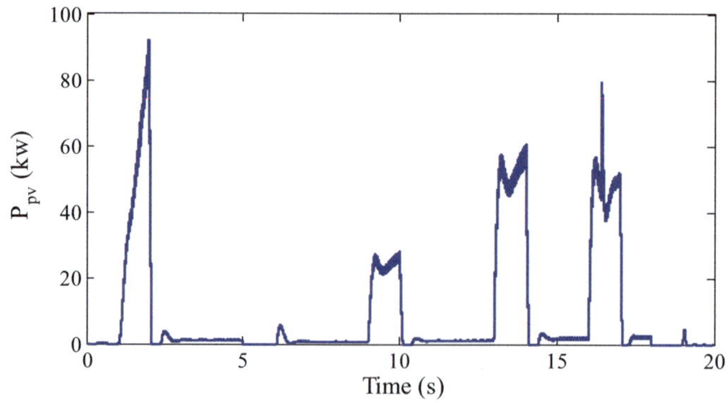

Fig. 4 Power of the triple junction photovoltaic generator

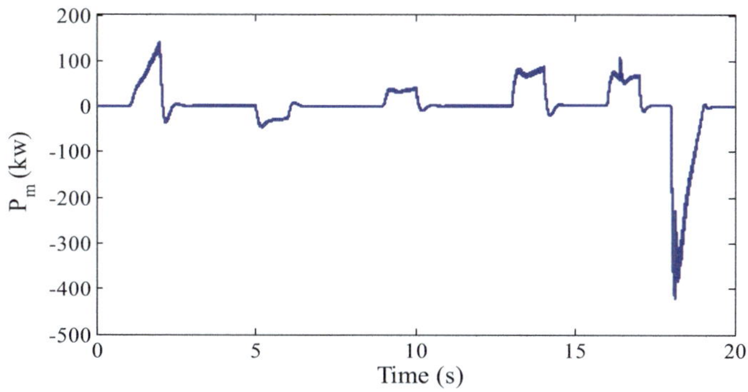

Fig. 5 Mechanical power of the electric vehicle

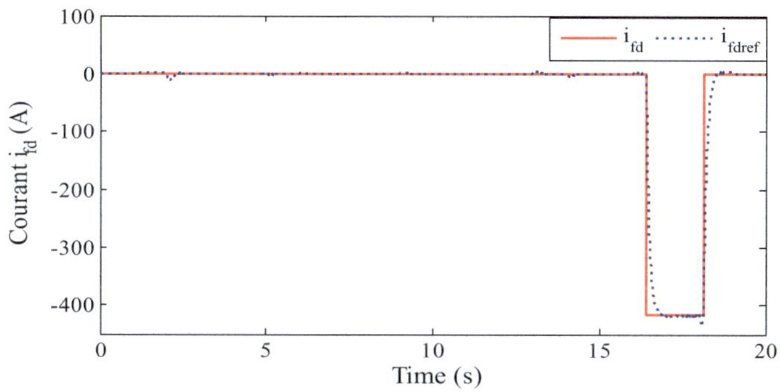

Fig. 6 Actual and target direct components of the PMSM stator current

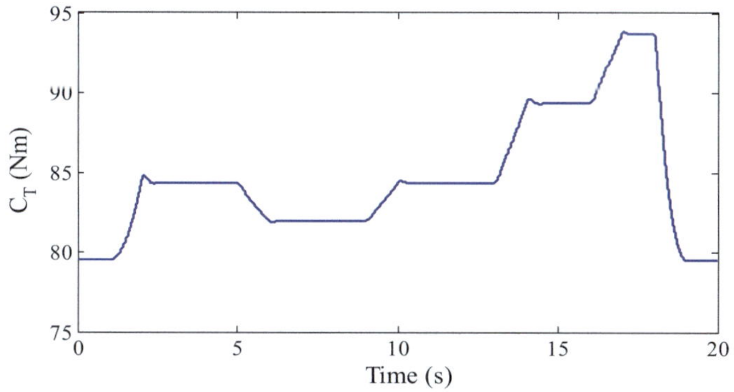

Fig. 7 Traction torque

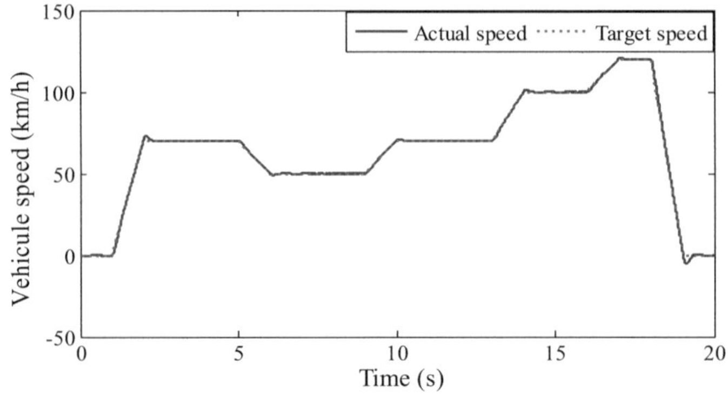

Fig. 8 Electric vehicle speeds

The electromagnetic torque and flux are each controlled separately by the motor's current components. As it is noticed from the evolution of the stator current components. Indeed, the torque is created by the reverse component, and the field weakening algorithm, in accordance with the speed operating mode, provides the direct component (Fig. 6).

4 Conclusion

In this paper, a high efficient photovoltaic generator is developed. It is used for feeding the electric vehicle. The traction part is conceived around a three phase permanent synchronous motor. The indirect field oriented control strategy is used in the flux and torque decoupling associated with a field weakening algorithm. Triple-junction solar cells with great efficiency form the basis of the PV generator. Different operating conditions are considered, including abrupt meteorological conditions variations and different working modes.

The performance of the investigated system is shown by the simulation results produced using the Matlab and Simulink software packages.

Despite the excellent performance achieved, the system as designed for direct connection faces significant issues. The most common operational point is a suboptimal one. The maximum power point tracking algorithm can be used to solve this issue. Future research will also focus on testing the entire PV electric vehicle in real-world scenarios. The aforementioned trend toward futuristic applications will be the subject of the author's future work.

References

1. Inglesi-Lotz R, Eyup D, Nel J, Panayiotis T (2023) Connectedness and spillovers in the innovation network of green transportation. Energy Policy 180:1–10
2. Zhihan L, Wenlong S (2023) Impacts of intelligent transportation systems on energy conservation and emission reduction of transport systems: a comprehensive review. Green Technol Sustain 1(1):1–16
3. Ji M, Gongxing Y, Azher M, Samia E, Mohamed Amine K, Amin J, Elhosiny A (2022) The effect of carbon dioxide emissions on the building energy efficiency. Fuel 326:1–10
4. Harsh B, Manan D, Tanmay J, Manan S, Ashish U (2023) Forecasting and mitigation of global environmental carbon dioxide emission using machine learning techniques. Clean Chem Eng 5:1–9
5. Alessandro M, Niccolò A, Claudio DP, Fabrizio L (2023) Photovoltaic-thermal solar-assisted heat pump systems for building applications: integration and design methods. Energy Built Environ 4(1):39–56
6. Ieva P, Marika K, Armands G, Ritvars F, Andra B (2022) Spatial analyses of smart energy system implementation through system dynamics and GIS modelling. Wind power case study in Latvia. Smart Energy 7:1–11
7. Yang B, Linen Z, Xiaoshun Z, Hongchun S, Tao Y, Haofei L, Lin J, Liming S (2019) Novel bio-inspired memetic salp swarm algorithm and application to MPPT for PV systems considering partial shading condition. J Clean Prod 215:1203–1222
8. Dawit DG (2018) Determinants of household adoption of solar energy technology in rural Ethiopia. J Clean Prod 2014:193–204
9. Miguel AR (2021) Quo Vadis solar energy research? Appl Sci 11(30):1–11
10. Aleksandra Z, Monika P, Ireneus Z (2021) Application of LCA to determine environmental impact of concentrated photovoltaic solar panels—state-of-the-art. Energies 14(31):1–20
11. Boussaada Z, Ben Hamed M, Sbita L (2014) Photovoltaic cell mathematical models: a comparative study. In: IREC international renewable energy congress (IREC). Hammamet, Tinisia
12. Salah L, Franck P, Alain G (2007) Higher order sliding mode control based on integral sliding mode. Automatica 43(3):531–537
13. Hamed SB, Aicha A, Hamed M, Sbita L, Mohit B, Sherif SMG, Hossam MZ, Salah K (2023) A robust MPPT approach based on first-order sliding mode for triple-junction photovoltaic power system supplying electric vehicle. Energy Reports 9:4275-4297
14. Aymen F, Sbita L (2013) A novel IMC controller based on bacterial foraging optimization algorithm applied to a high speed range PMSM drive. Appl Intell 38:114–129

An Intelligent Estimation Based on Multi-model and Neural Approach for an Isolated Wind System

Abid Aicha, Dhaoui Mehdi, Ben Hamed Mouna, and Sbita Lassaad

Abstract This paper presents an intelligent estimator for a wind system. This estimator is based on the combination of two powerful tools that are neural and multi model approach. This estimation approach based on artificial intelligence is exploited to estimate currents of an isolated wind system based on PMSG. The studied system contains a turbine which, under the wind, drives in rotation a PMSG that generates electricity towards an isolated site via multilevel controlled rectifier.

Keywords Intelligent estimator · Wind system · Multi-model approach · Neural approach · Multilevel controlled rectifier

1 Introduction

As a popular source of renewable and clean energy, wind system has attracted the most attention of the energy sources.

Wind systems face stochastic climate and difficulties of information measurement that can be analyzed to attain an optimum performance. Thus, to have exact and precise data information, the used sensors must be maintained and monitored regularly that is very difficult since wind turbines are usually installed in mountains and distant regions. Thus, in order to make wind system more efficient and more reliable and to improve their power production, the design of an estimator or an observer is necessary [1–4]. To solve this problem, an intelligent estimator has been proposed in this paper.

The intelligent estimator is based on the combination of two powerful tools that are multimodel and neural approach [5–8].

It is often difficult to synthesize a sufficient adequate estimator able to take into account not only the system's nonlinearity and complexity but also the different

A. Aicha (✉) · D. Mehdi · B. H. Mouna · S. Lassaad
L/R: Processes, Energy, Environment and Electrical Systems (PEESE), Electrical Department of Engineering, National Engineering School, University of Gabes, 6029 Gabes, Tunisia
e-mail: aicha.abid@gmail.com

© The Author(s), under exclusive license to Springer Nature Singapore Pte Ltd. 2024
A. Mellit et al. (eds.), *Proceedings of the 2nd International Conference on Green Energy Conversion System*, Green Energy and Technology,
https://doi.org/10.1007/978-981-97-6148-7_50

perturbation and climatic variation. Thus, many different approaches have been developed to deal with the estimation problem. As a solution, intelligent approaches such as multi-model and neural approaches are exploited in this work. Indeed, the multi-models facilitate the extension of some analysis tools developed in linear context to nonlinear context. The multimodel approach consists in representing the whole behavior of a nonlinear system by a set of simple local models. Generated sub models are then combined using validity function to contribute to the construction of the whole model as studied in [9, 10]. The major contributions of this paper consist, firstly, in using an improved Kmeans clustering algorithm and secondly, in replacing the validities function by a neural network.

The organization of this paper considers the different steps for the estimation of the wind system currents. The aerogenerator consists of a wind turbine connected to a PMSG and a multilevel rectifier that generated isolated sites.

Section 2 is dedicated to aerogenerator modeling. Section 3 deals with the creation of a neural-multi-model estimator respecting the different steps that lead to the realization of this task.

Simulation results for the wind system estimator using the two approaches are performed and implemented using the environment MATLAB/SIMULINK.

2 Aerogenerator Modeling

The wind system (see Fig. 1) considered in this paper holds a wind turbine, drive train, a per-manent magnet synchronous generator PMSG and MLI rectifier that feeds a RL charge via a filter. A vector control is used to pilot the wind system aiming not only to maximize the system power, but also to stabilize the DC link that generates a constant DC voltage (Vdc).

2.1 Wind Turbine Modeling

The turbine thanks to its blades convert the wind energy into mechanic energy [5].

The wind turbine model consists in modeling the aerodynamic, mechanical and electrotechnical characteristics.

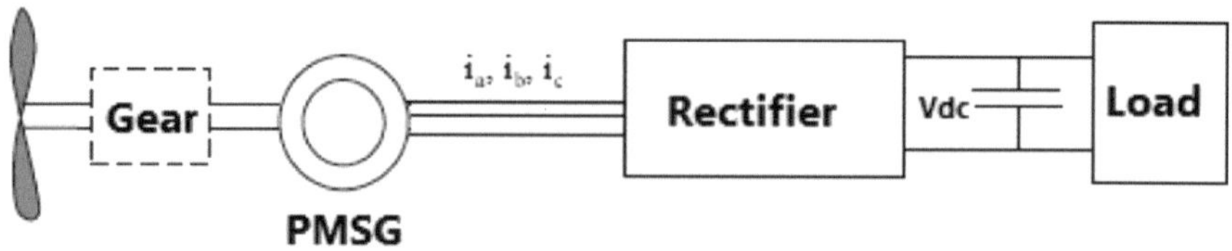

Fig. 1 Wind system

Equation (1) describes mathematically the wind turbine power P_{WT}.

$$P_{WT} = C_p(\beta, \lambda)P_{wind} = \frac{1}{2}\rho\pi R^2 C_p V_v^3 \tag{1}$$

The value of the power coefficient Cp depends on both the pitch angle β and the tip speed ratio λ that is expressed in Eq. 2.

$$\lambda = \frac{\omega R_t}{V_v} \tag{2}$$

In this work, the turbine studied is E40 that is constructed by the allemand ENERCON.

Equation (5) describes the mathematical model of the drive train.

The dynamic equation that characterizes the transmission system is given by the differential Eq. (3).

$$C_t = Cem + J\frac{d\omega}{dt} + f\omega \tag{3}$$

where ω is the turbine rotor speed.

2.2 Permanent Magnet Synchronous Generator Model

The mathematical model of the PMSG in dq-frame is described by Eq. (4).

$$\begin{cases} V_d = R_s i_d + L_d \frac{di_d}{dt} - \omega L_q i_q \\ V_q = R_s i_q + L_q \frac{di_q}{dt} + \omega L_d i_d + \omega \emptyset_f \end{cases} \tag{4}$$

The electromagnetic torque is written as (5).

$$C_{em} = \frac{3}{2}p(\varphi_d i_q - \varphi_q i_d) \tag{5}$$

with

Rs: The resistance of each stator winding.
Ld, Lq: The self-inductances of each stator winding.
p: The number of pole pairs.

φ_d and φ_q that are the flux vectors generated by the permanent magnet are given by Eq. (6).

$$\begin{cases} \varphi_d = L_d i_d + \varphi_f \\ \varphi_q = L_q i_q \end{cases} \tag{6}$$

3 The Wind Drive System Modeling via the Newneural-Multi-model Method

The objective of the proposed estimator is to reconstruct the state of an observable system from the measurement of inputs and outputs. It is used when all or part of the state vector cannot be measured.

Representing a complex system by a multimodel approach consist in replacing the whole nonlinear model by a set of local linear models. The multi-model approach requires a strategy of four steps that are database acquisition, clustering, identification and combination of the local models or fusion.

The acquired inputs/outputs database from the real system is clustered into N clusters using an improved Kmeans algorithm that is GBK-means. Then obtained clusters are identified parametrically to generate a set of linear local models. These models will be combined through validity function using fusion approach to create the multimodel. The validity functions must take its values in $\{0, 1\}$.

In this section we propose a novel method of modeling based on both the multi-model and neural approach. It consists in the combination of the two approaches. This modeling method exploits the first three steps of the multimodel approach while the fourth step that is the fusion approach is replaced by a neural network as depicted in Fig. 2.

The obtained neural multimodel requires fours steps.

3.1 Data Acquisition

An excitement process is performed through the application of a Pseudo Random Binary Signal (PRBS) added to the input signal that is the continuous bus reference voltage Vdc. The collection of the system's input/output signals help to construct a rich database.

The outputs are the three abc currents generated by the PMSG.

3.2 Database Clustering

The clustering consists in decomposing the collected database into L clusters using various clustering algorithms.

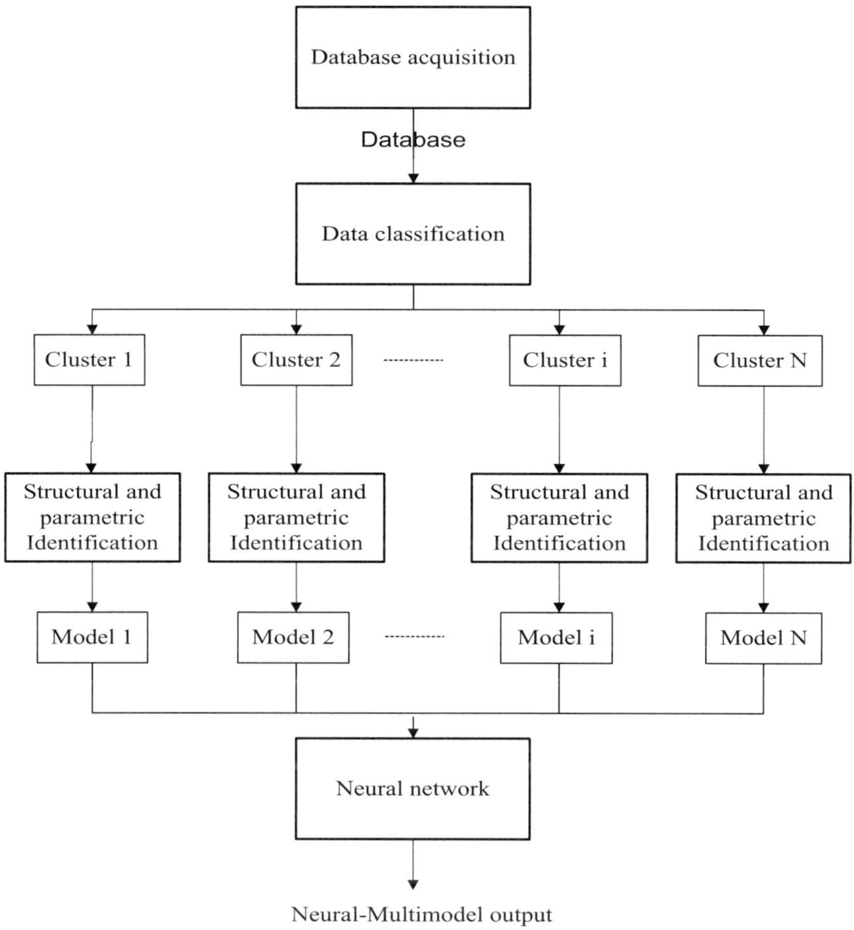

Fig. 2 The modeling proposed schema

We propose to use an improved recently used clustering algorithm that is GBKmeans algorithm.

The proposed clustering algorithm is presented.

GBK-means algorithm

Initialization phase

Fixing the number of clusters (L), computing the initial distance matrix of objects and breakdown points ($d_1,\ldots, d_L$) as well as the associated parameters

Repeat:

Assigning:

Attributing objects to clusters

Evaluating:

Assessing assigned objects to clusters pursuant to the bargaining game trade off Function (F)

Termination of the desired value check:

If the desired value is gained, exit the loop

Determining the best solution (according to PSO):

Decide the best-generated solution

If $F_{current} < F_{best}$:

Update velocity and position of particles in PSO

Updating:

Update $D_1,..,D_L$ and $d_1,..,d_L$

Calculate new distance matrix

Bargaining:

Change cancroids to new positions according to $D_1,..,D_L$ and $d_1,..,d_L$, as well as trade-off function

Until: termination condition is met

3.3 Local Model Identification

The structure identification is performed on each cluster using instrumental determinants ration (RDI) method or the general procedure for order estimation while the parameters identification of each sub model are identified using recursive least square (RLS) method.

3.4 Local Model Fusion

The obtained sub models are combined using the validity concept. The validities are calculated using two approach that are residue approach and neural approach.

Validities calculation via Residue approach

The residue r_i is calculated as the geometric distance between the system output $y(k)$ and the ith local model output $y_i(k)$.

$$r_i(k) = \|y(k) - y_i(k)\|, i \in [1, \ldots, N] \quad (7)$$

N is the number of local models.

Next, each residue has to be normalized using the expression (8).

$$r_{ni}(k) = \frac{r_i(k)}{\sum_{i=1}^{N} r_i(k)} \quad (8)$$

Then it has to be exploited in the validity calculation that is given in (9).

$$\vartheta_i(k) = \frac{1 - r_{ni}(k)}{N - 1} \quad (9)$$

These validities will be reinforced to reduce the perturbation phenomenon, as follow:

$$\vartheta_i^{renf}(k) = \vartheta_i(k) \prod_{j=1}^{N} \left(1 - e^{-\left(\frac{r_j}{\sigma}\right)^2}\right) \quad (10)$$

where σ is a positive constant.

The normalized reinforced validities are defined as follow.

$$\vartheta_i^{renfn}(k) = \frac{\vartheta_i^{renf}(k)}{\sum_{i=1}^{N} \vartheta_i^{renf}(k)} \quad (11)$$

Finally, the multimodel output is a combination of the local model outputs $y_{i(k)}$ weighted by the obtained calculated validities.

Validities Neural computation

In this work, the main objective of the neural network, known as a universal approximator, is to estimate the validities functions that are the network weights. Therefore the use of neural network consists in reaching the targets that are the network outputs from a set of inputs.

The Adeline-type neural network (see Fig. 3) is used to replace traditional validity functions. In our case, the network inputs are the local models outputs (y_i, i = 1,..,L). The network output is the global output i.e. the multi-model output (y_s).

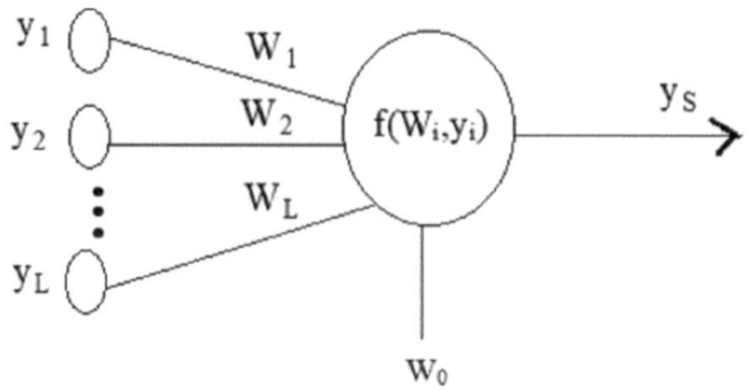

Fig. 3 The Adeline neural network

4 Simulation Results

Under MATLAB/ SIMULINK environment, the fourth steps of the creation of the intelligent observer are developed and implemented. A rich dataset is picked out of the wind system. The system's input is the desired DC bus voltage and the system's outputs are the three-phase currents (ia, ib and ic).

A multi-model regrouping five local models is obtained (Eq. 12),

$$\begin{cases} x_i(k+1) = A_i x_i(k) + B_i u(k) \\ y_i(k) = C_i x_i(k) \\ y(k) = \sum_{i=1}^{N} V_i(k) y_i(k) \end{cases} \quad (12)$$

where

u, x_i and y_i are respectively the local system's input, state and output.

N is the number of the local system, N = 5.

V_i is the i-th validity.

The different obtained system's matrix are expressed as,

$$A_1 = \begin{bmatrix} 0.0235 & 0 & 0 \\ 0 & 0.0125 & 0 \\ 0 & 0 & 0.034 \end{bmatrix}$$

$$A_2 = \begin{bmatrix} 0.06992 & 0 & 0 \\ 0 & 0.072 & 0 \\ 0 & 0 & 0.041 \end{bmatrix}$$

$$A_3 = \begin{bmatrix} 0.0072 & 0 & 0 \\ 0 & 0.0032 & 0 \\ 0 & 0 & 0.0085 \end{bmatrix}$$

$$A4 = \begin{bmatrix} 0.5626 & 0 & 0 \\ 0 & 0.455 & 0 \\ 0 & 0 & 0.009 \end{bmatrix}$$

$$A_5 = \begin{bmatrix} 0.0087 & 0 & 0 \\ 0 & 0.0072 & 0 \\ 0 & 0 & 0.0062 \end{bmatrix}$$

$$\boldsymbol{B}_1 = \begin{bmatrix} 0.0326\ 0.0351\ 0.0416 \end{bmatrix}$$

$$\boldsymbol{B}_2 = \begin{bmatrix} 0.0968\ 0.0829\ 0.0062 \end{bmatrix}$$

$$\boldsymbol{B}_3 = \begin{bmatrix} 0.0299\ 0.0542\ 0.0314 \end{bmatrix}$$

$$\boldsymbol{B}_4 = \begin{bmatrix} 0.0054\ 0.0064\ 0.0036 \end{bmatrix}$$

$$\boldsymbol{B}_5 = \begin{bmatrix} 0.0361\ 0.0487\ 0.0720 \end{bmatrix}$$

$$C_1 = \begin{bmatrix} 1\ 0\ 0 \\ 0\ 1\ 0 \\ 0\ 0\ 1 \end{bmatrix}$$

$C_1 = C_2 = C_3 = C_4 = C_5$.

To validate the estimation procedure, the simulation results are studied.

Figures 4, 5, and 6 approve the efficiency of the proposed method. Indeed the estimated currents follow the real ones in the three cases with feeble values of errors that are in the order of 10^{-4}.

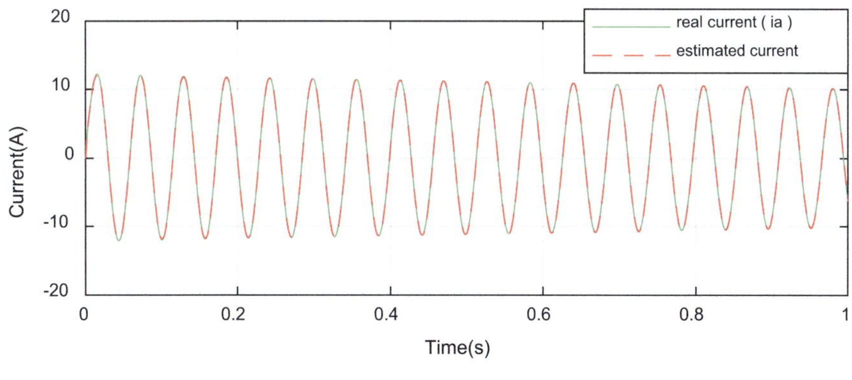

Fig. 4 Real and estimated current (ia)

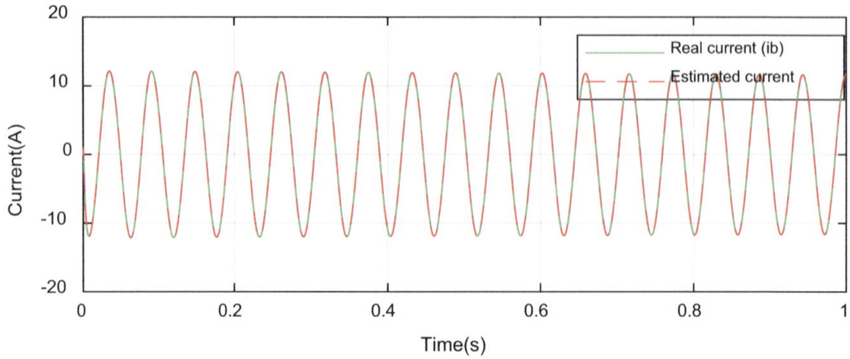

Fig. 5 Real and estimated current (ib)

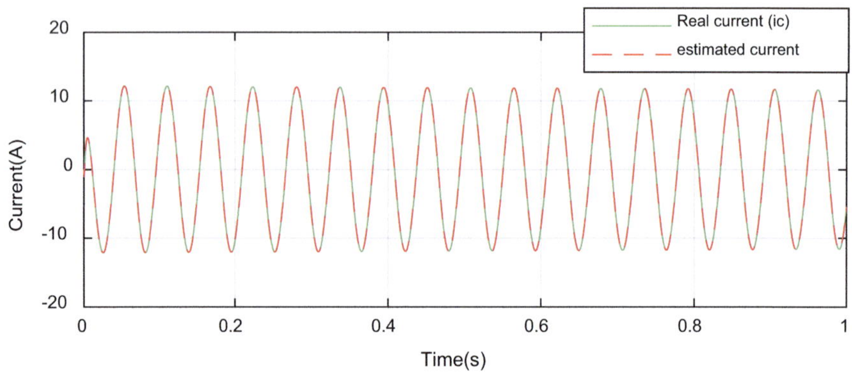

Fig. 6 Real and estimated current (ic)

5 Conclusion

In this paper, a multi-model neural estimator is developed. The main contribution of this work is the improvement of the multimodel approach by not onlt using an improved clustering algorithm but also using a neural network to calculate the validities functions. The proposed estimation method will be used in diagnosis context in future work.

References

1. Bian F, Meng K, Jia Y, Ma J, Hai R (2022) Application of effective wind speed estimation and new sliding mode observer in wind energy conversion system. Math Probl Eng 8863163:11
2. Lio AWH, Meng F (2020) Effective wind speed estimation for wind turbines in down-regulation. J Phys Conf Ser 1452:012008

3. Dora Arul Selvi B, Kannan S (2023) A performance comparison of machine learning methods for short-range wind power estimation. E3S Web Conf 387:05013
4. Zhang Y, Li P, Li H, Zu W, Zhang H (2023) Short-term power prediction of wind power generation system based on logistic chaos atom search optimization BP neural network. Int Trans Electr Energy Syst 6328119:11
5. Sassano M, Astolfi A (2019) A local separation principle via dynamic approximate feedback and observer linearization for a class of nonlinear systems. IEEE Trans Autom Control 64(1):111–126
6. Hua X, Huang D, Guo S (2020) Extended state observer based on ADRC of linear system with incipient fault. Int J Control Autom Syst 18:1–10
7. Maalej I, Ben Halima Abid D, Rekik C, Derbel N (2018) Single and multiple faults detection in nonlinear system using fuzzy Kalman observer, De Gruyter Oldenbourg. ASSD—Adv Syst Signals Devices 5:241–256
8. Rodrıguez del Nozal A, Orihuela L, Millan P (2019) Distributed estimation design for LTI systems: a linear quadratic approach. Int J Syst Sci
9. Ben Mabrouk Z, Ben Hamed M, Abid A, Sbita L (2020) Fault detection of nonlinear networked control system based on multimodal approach subject to induced delay. In : Series: systems engineering methods, developments and technology BISAC: TEC000000A. Closer look at fault-tolerant control, 348 p. 2020, Engineering & technology, nova, systems engineering, systems engineering methods, developments and technology. Tags: 9781536175288, 9781536175295, systems engineering
10. Ben Mabrouk Z, Abid A, Ben Hamed M, Sbita L (2020) Diagnosis of sensores failure in induction motor. In: Series: systems engineering methods, developments and technology BISAC: TEC000000A. Closer look at fault-tolerant control, 348 p. 2020, Engineering & technology, nova, systems engineering, systems engineering methods, developments and technology. Tags: 9781536175288, 9781536175295, systems engineering

Improve of a Hybrid PV System Performances Based on MPPT Algorithm and Multi-level Converter

Abid Aicha, Hayder Wafa, Ben Hamed Mouna, and Sbita Lassaad

Abstract This work present an improvement of a Hybrid PV system performance based on the use of both MPPT algorithm and multilevel inverters. Inverters are an essential element in renewable energy production systems. Conventional inverters have several drawbacks such as overvoltage problems at the semiconductor terminal in high-power applications, and the obtaining of a non-sinusoidal voltage that is rich in harmonics. Multilevel inverter topologies can be arranged in many different combinations; among the popular multilevel inverter topologies are Diode Clamped Multilevel Inverters (DCMLI). Two types are studied and compared that are three and five Diode Clamped Multilevel Inverter (DCMLI). The modeling and the control of the PV system are developed and implemented in Matlab/Simulink environment. Simulation results are presented and analyzed up.

Keywords Hybrid PV system · Diode clamped multilevel inverter (DCMLI) · Modeling · Control · Comparative study · Matlab/Simulink

1 Introduction

Renewable energy presents a necessity that responds the consumer's needs of electric energy thanks to its good performance and the benefits of renewable natural sources. Improving its performance is a promising task. Such renewable systems are based in the use of converters that are essential elements.

Recently, multilevel inverters have attracted largely attention and have gained great interest thanks to their ability to generate an output voltage or current waveform composed of a higher number of levels [1–4].

A. Aicha (✉) · H. Wafa · B. H. Mouna · S. Lassaad
L/R: Processes, Energy, Environment and Electrical Systems (PEESE), Electrical Department of Engineering, National Engineering School, University of Gabes, 6029 Gabes, Tunisia
e-mail: aicha.abid@gmail.com

© The Author(s), under exclusive license to Springer Nature Singapore Pte Ltd. 2024
A. Mellit et al. (eds.), *Proceedings of the 2nd International Conference on Green Energy Conversion System*, Green Energy and Technology,
https://doi.org/10.1007/978-981-97-6148-7_51

This high number of levels makes it possible to reduce the difference between the levels, which reduces the total harmonic distortion (THD) and consequently reduces the number of harmonics and provides a better quality of output power.

The inverter control circuit is based on the switches. The latter are subject, in the majority of cases, to high thermal stresses since they are called to block the full voltage imposed by the continuous source, since the phenomenon is repetitive so that the performance and service life of the components are effectively affected as well as the appearance of redhibitory harmonics that cause losses in the load and in the converter. These harmonics inject noises introducing non-linearities that make stabilization of the adjustment system more difficult [5–7].

As a relevant solution to overcome this constraint, the structure of the multi-level inverters brings considerable improvements thanks to the serialization of the semiconductors of power that allows having additional levels [8–10].

The use of this type of architecture makes it possible to limit the voltage stresses experienced by the switches by dividing the continuous voltage at the input of the inverter [9–14].

In this paper we will focus on the comparative of three and five level Diode Clamped Multilevel Inverter (DCMLI).

The modeling and the control of the two proposed multilevel inverter is implemented and validated in Matlab/Simulink. The paper is organized as follows. In Sect. 2, the two multilevel inverters description is presented. Simulation results and the comparative study are given consecutively in the third and the fourth section. Finally, conclusions and some suggested prospects are presented in Sect. 4.

2 Description and Modeling of the Studied System

The system (see Fig. 1) consists of a hybrid PV system that contains a PV generator connected to an AC load via DC-DC then DC-AC converters. A set of batteries are used to supply the load in the case of lack of PV production.

To improve the PV system performance, firstly an MPPT algorithm is applied to maximize the generated power, next multilevel inverter is used to improve signals quality, and reduce harmonics.

The used MPPT algorithm is the known perturb and observer algorithm.

3 Description of the Diode-Clumped Multilevel Inverter (DCMLI)

The diode-clumped multilevel inverter (DCMLI) consists in generating an output voltage at N levels by using (N-1) capacitors on the DC bus aiming to divide up the total DCbus voltage [15, 16]. The middle point of the N capacitors is defined as the

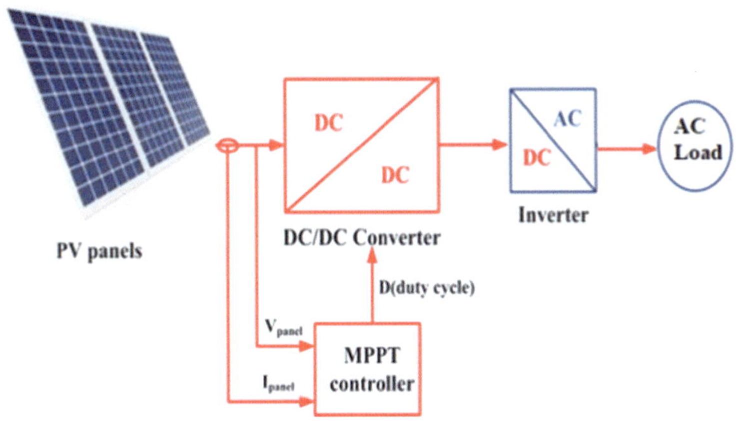

Fig. 1 The hybrid PV system

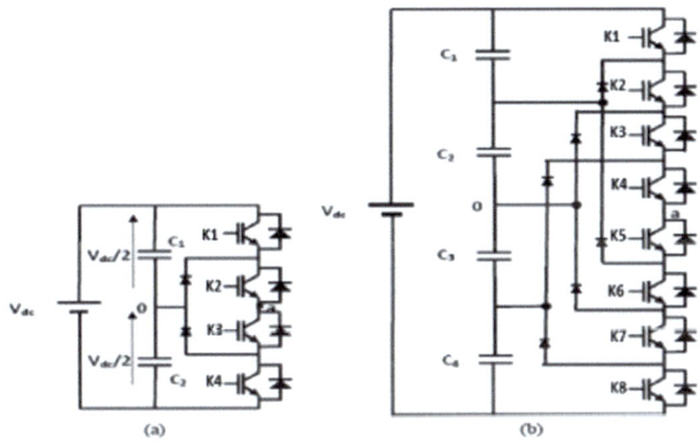

Fig. 2 Three-level and five-level diode-clumped multilevel inverter

neutral point that is connected to 2(N-2) clamped diodes. Each branch in the NPC inverter is composed of 2(N-1) switches as shown in Fig. 2.

3.1 *Three-Level Diode-Clumped Multilevel Inverter*

The three-level diode-clumped multilevel inverter as shown in Fig. 1a is composed of two identical capacitor C_1 and C_2, that help to divide the dc bus voltage Vdc, two clamped diodes and fours switched devices. Thus, to obtain the output of the multilevel inverter the three level voltage, different switches control are possible as shown in Table 1. The different combinations of commands (sequences) possible to have all the possible levels.

The three configurations are represented in Fig. 3. Configuration 1 is represented by Fig. 3a. Configuration 2, which is represented by Fig. 3b, has two possibilities to have the zero level. Configuration 3 is represented by Fig. 3c.

Table 1 Possible control combinations for three voltage levels

Configurations	K1	K2	K3	K4	Vao
1	1	1	0	0	$V_{DC}/2$
2	0	1	1	0	0
3	0	0	1	1	$-V_{DC}/2$

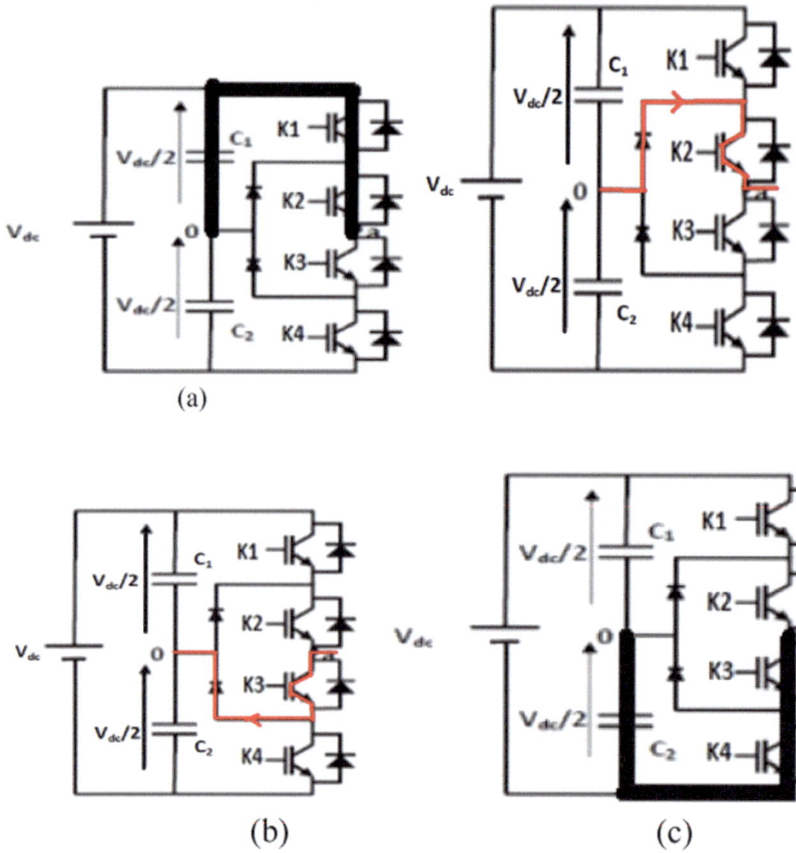

Fig. 3 The three possible configurations

Table 2 Possible control combination for five voltage levels

Configurations	K1	K2	K3	K4	K5	K6	K7	K8	Vao
1	1	1	1	1	0	0	0	0	$V_{DC}/2$
2	0	1	1	1	1	0	0	0	$V_{DC}/4$
3	0	0	1	1	1	1	0	0	0
4	0	0	0	1	1	1	1	0	$-V_{DC}/4$
5	0	0	0	0	1	1	1	1	$-V_{DC}/2$

4 Five-Level Diode-Clumped Multilevel Inverter

The five-level diode-clumped multilevel inverter as shown in Fig. 1b is composed of four identical capacitor C1, C2, C3 and C4 that help to devide the dc bus voltage Vdc, six clamped diodes and eight switched devices. Depending on the state of the switches K1, …, K8, the five levels voltage given by the multilevel inverter are illustrated by Table 2.

The five configurations are represented in Fig. 4. Configuration 1 is represented by Fig. 4a, configuration 2 is represented by Fig. 4b, configuration 3 is represented by Fig. 4c, and configuration 4 is represented by Fig. 4d.

5 The Control of the Proposed System and Simulation Results

The proposed system is a PV system with storage that supplies an isolated site represented by an alternating load. The modeling and the control of the different elements of the PV system are done under MATLAB/SIMULINK environment.

The used irradiation and temperature used for the simulation are represented by Fig. 5.

To optimize the PV system performance, a P&O algorithm is used to maximize the power by controlling the duty cycle of the dc-dc boost converter. Both the PV generator and the storage system supply the load via the inverter.

The DC bus voltage must be stable and well regulated and adjusted as shown in Fig. 6. The batteries are used both as an auxiliary source in the case of a shortage or absence of production and as a load in the event of overproduction. The load is always connected. Thus an algorithm is used for energy management.

Indeed, depending on the PV production, energy consumption and state of charge of the batteries, the operating scenario of the hybrid system is described by Fig. 7. In case of excess production (if (PV power) Ppv > PL (Load power)), the batteries are charged. On the other hand, in case of a power deficit (P_{PV} < PL), the batteries are partially discharged to meet the demand as approved by the evolution of the state of charge (see Fig. 8).

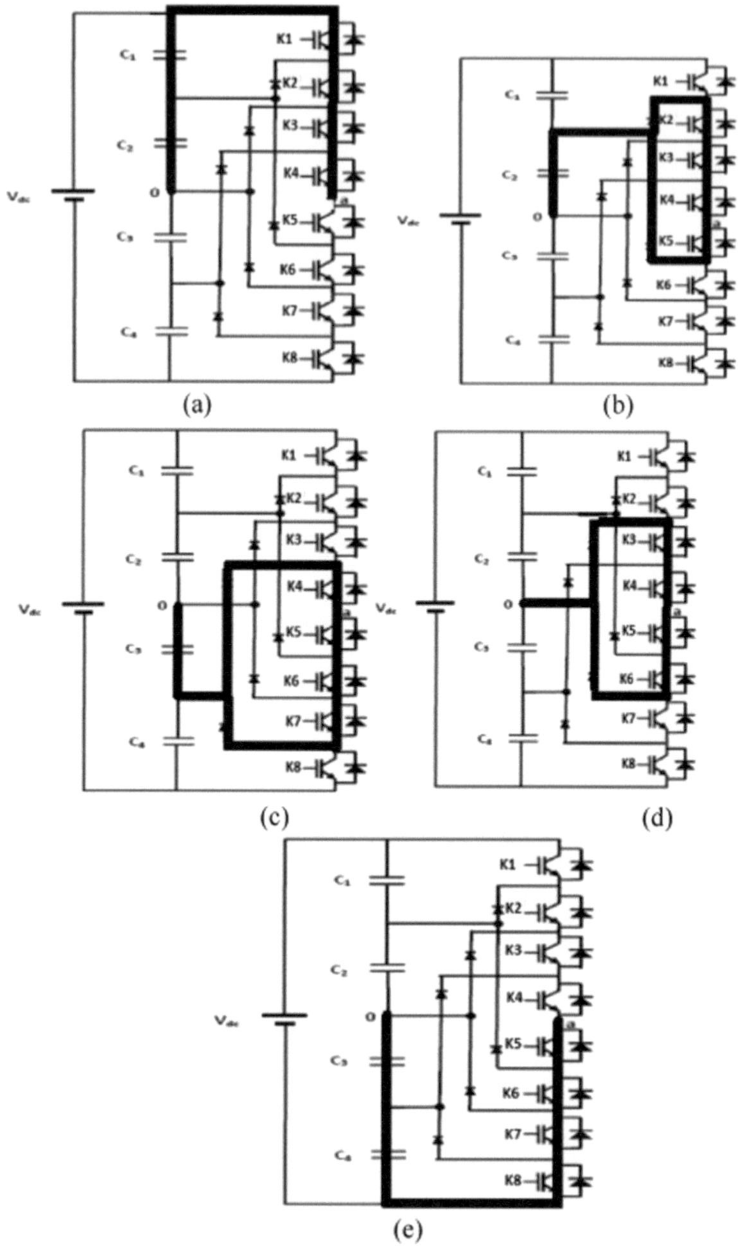

Fig. 4 The five possible configurations

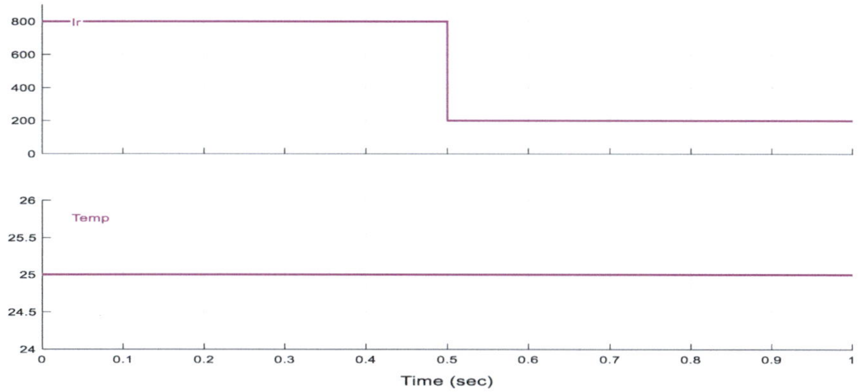

Fig. 5 Irradiation and temperature evolutions

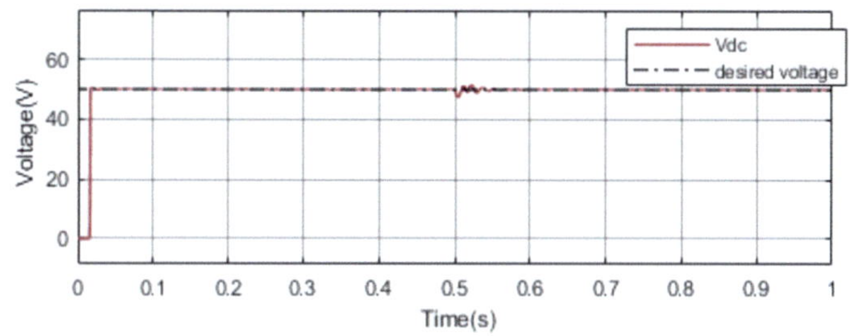

Fig. 6 DC bus voltage

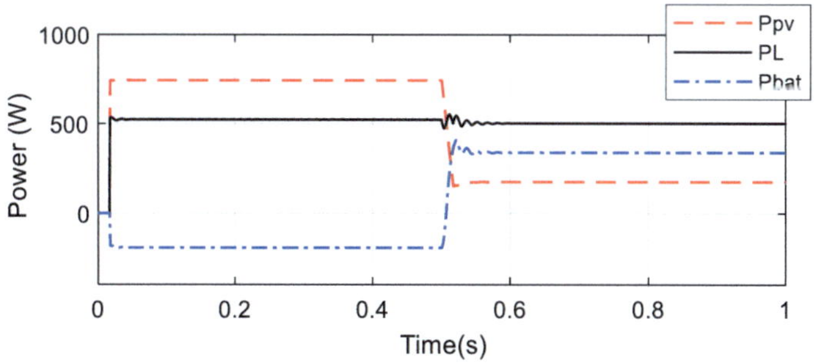

Fig. 7 The operating scenario of the hybrid system

The control of the two multilevel inverters is done using sinusoidal PWM techniques as an easy and efficient technique able to eliminate the harmonics. The modeling and control of the two proposed inverter are established using MATLAB/SIMULINK environment.

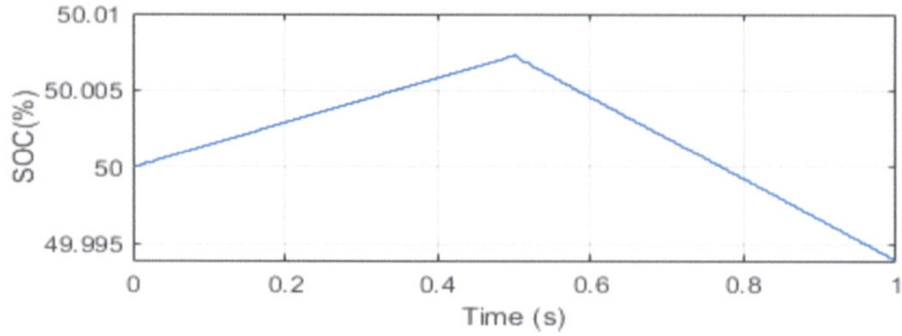

Fig. 8 State of charge of batteries

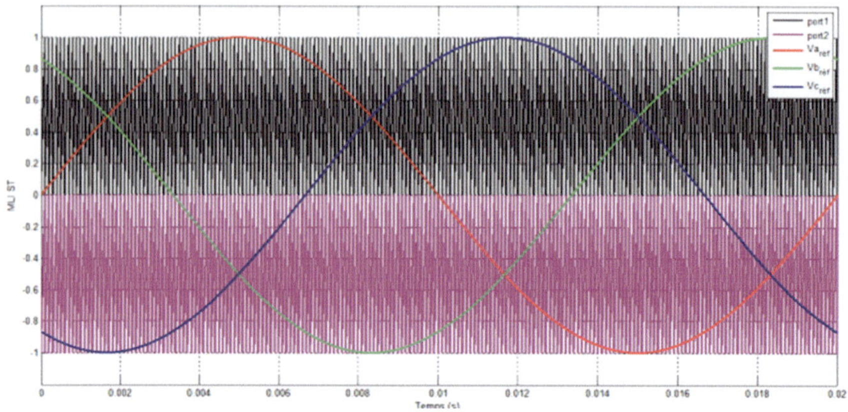

Fig. 9 The principle of the PWM based on two triangular carriers

To obtain the converter PWM control pulses at three voltage levels, it requires two triangular carriers that are compared to a reference sinusoidal signal.

5.1 *Simulation Results for Three Level Inverter*

To obtain the converter PWM control pulses at three voltage levels, it requires two triangular carriers that is compared to a reference sinusoidal signal as shown in Figs. 9, 10, and 11.

5.2 *Simulation Results for Five Level Inverter*

In order to generate PWM pulses with five voltage levels, four triangular carriers are required as described in Figs. 12, 13, and 14.

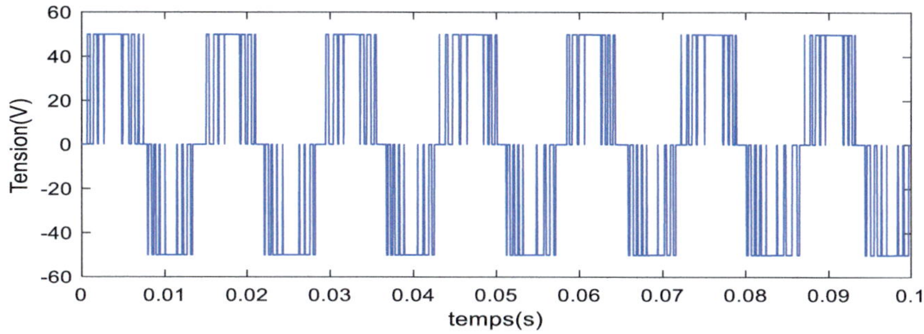

Fig. 10 Three-Level NPC inverter output signal

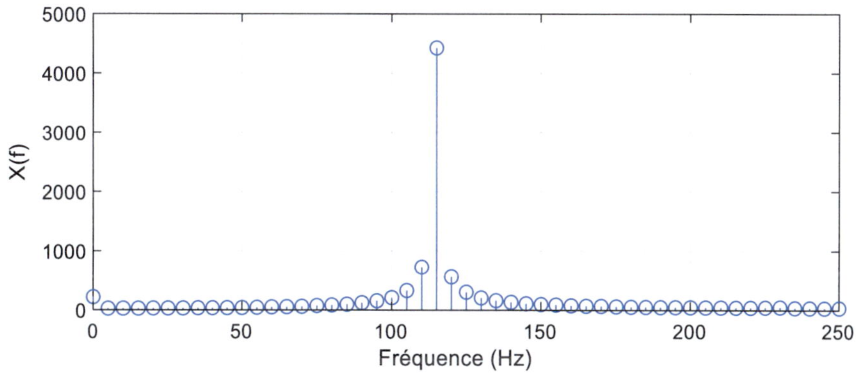

Fig. 11 Three three-level inverter harmonic spectrum

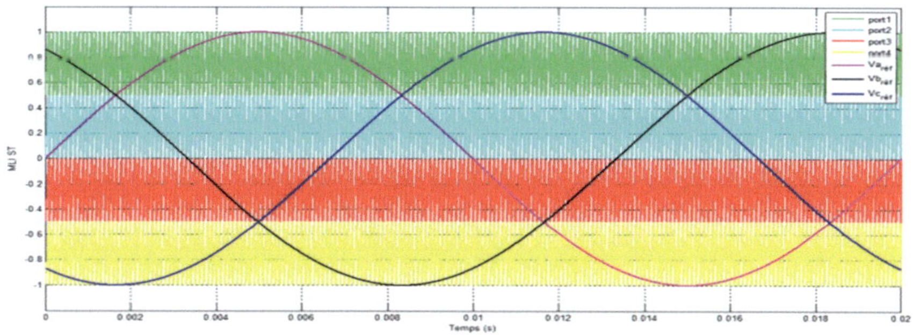

Fig. 12 The principle of the PWM based on four triangular carriers

5.3 Comparative Study

The different simulation results help to establish a comparative study that is summarized in Table 3.

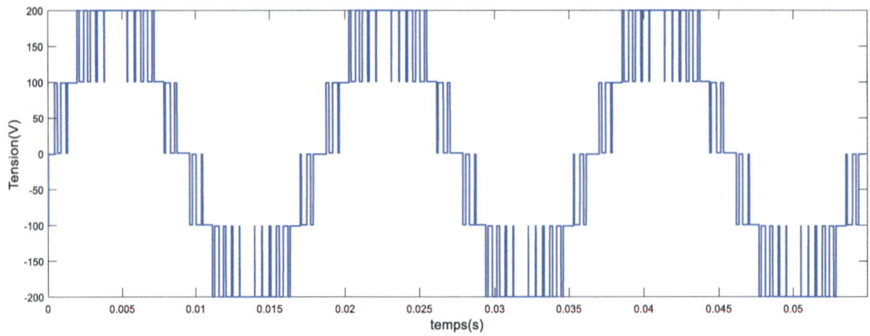

Fig. 13 Five-level NPC inverter output signal

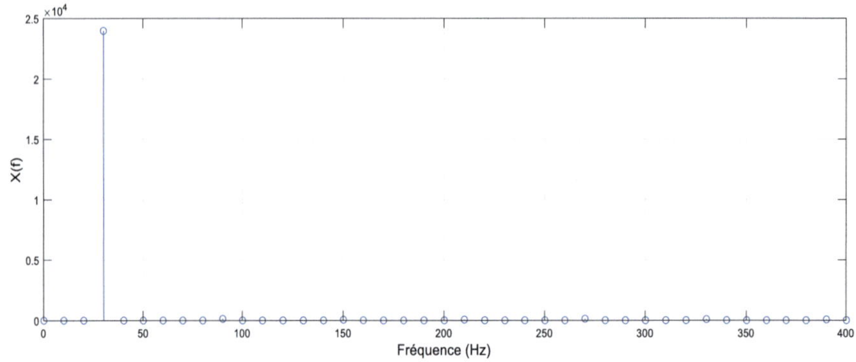

Fig. 14 Five-level inverter harmonic spectrum

Table 3 THD calculation for the two proposed inverters

Type of inverter	Three-level NPC inverter	Five-level NPC inverter
THD (%)	26.6117	10.505

As shown in Table 3 we can notice that the THD of the five levels inverter is better than the three levels, thus the harmonic are more eliminated using the five levels inverter.

6 Conclusion

In this paper, a hybrid PV system with storage is studied. The modeling and the control of the different elements are done using MATLAB/SIMULINK environment. The system performances are improved based not only on the use of an optimizing algorithm but also on the use of two multilevel inverters that can help to reduce the harmonics and improve the generated signal quality. The comparative study of two types of multilevel inverters was carried out. The comparison was made based on the THD calculation of the different voltage signals obtained by applying the PWM modulation technique. As an extended work the five multi-level inverter will be used in a grid connected renewable energy application in order to reduce the grid pollution.

References

1. Ibrahim ET, Gyang DD, Benson SA (2020) Review of multilevel inverters and their control techniques. EJERS—Eur J Eng Res Sci 5(6):659–664
2. Choudhury S, Bajaj M, Dash T, Kamel S, Jurado F (2021) Multilevel inverter: a survey on classical and advanced topologies, control schemes, applications to power system and future prospects. Energies 14:5773
3. Kumar NB (2021) Design and analysis of different multi-level inverter topologies for single phase im drive. Turk J Comput Math Educ (TURCOMAT) 10:4104–4117
4. Quach T-H, Do D-T, Nguyen M-K (2019) A PWM scheme for five-level H-bridge T-type inverter with switching loss reduction. Electronics 8(6):702
5. Dileep Krishna M, Vadhera S (2020) Comparative study of hybrid and cascaded H-bridge multilevel inverters. Int J Res Eng Technol
6. Kar PK, Priyadarshi A, Karanki SB (2020) Development of an enhanced multilevel converter using an efficient fundamental switching technique. Electr Power Energy Syst 119:105960
7. Zhang J, Xu S, Din Z, Hu X (2019) Hybrid multilevel converters: topologies, evolutions and verifications. Energies 12:615. https://doi.org/10.3390/en12040615
8. EL-Hosainy A, Hamed HA, Azazi HZ, El-Kholy EE (2017) A review of multilevel inverter topologies, control techniques, and applications. In: 2017 nineteenth international middle east power systems conference (MEPCON), Menoufia University, Egypt
9. Bhuvaneswari G, Nagaraju (2005) Multi-level inverters—a comparative study. IETE J Res 51(2):141–153
10. Ala G, Caruso M, Miceli R, Pellitteri F, Schettino G, Trapanese M, Viola F (2020) Fault diagnosis of cascaded H bridge multilevel inverter by DWPT multi resolution and ANN. In: First Int Conf Power Control Comput Technol (ICPC2T)
11. Anand A, Vinayak BA, Raj NGJ, George S (2020) A generalized switch fault diagnosis for cascaded H-bridge multi level inverters using mean voltage prediction. IEEE Trans Ind Appl 56(2):1563–1574
12. Jain S, Sonti V (2018) Fault tolerant operation of CHB multilevel inverters based on the SVM technique using an auxiliary unit. J Power Electron 18(1):56–69
13. Manjunath TG (2016) Ashok Kusagur, fault diagnosis and reconfiguration of multilevel inverter switch failure-A performance perspective. Int J Electr Comput Eng (IJECE) 6(6):2610–2620
14. Qasim MA, Velkin VI, Mohammed MF, Sammour AA, Du Y, Abdul-Adheem Salih S, Aljashaami SA, Gulmurodovich SP (2023) Design of a multi-level inverter for solar power systems with a variable number of levels technique. Int J Power Electron Drive Syst (IJPEDS) 14(2):1218–1229. ISSN: 2088-8694. https://doi.org/10.11591/ijpeds.v14.i2.pp1218-1229
15. Barnawi AB, Alfifi ARA, Elbarbary ZMS, Alqahtani SF, Shaik IM (2024) Review of multilevel inverter for high-power applications. Front Eng Built Environ 4(2): 77–89. https://doi.org/10.1108/FEBE-05-2023-0020
16. Tak N, Chattopadhyay SK (2024) Three-Phase NPC-LDN hybrid transformerless multilevel inverter. IEEE J Emerg Select Top Ind Electron 5(3): 917–929. https://doi.org/10.1109/JESTIE.2023.3346814

Flatness of the Three-Phase Grid Connected DC-AC Inverter with LCL Filter

Marouane Laaziz, Florentina Nicolau, Ghanes, Nadia Machkour, Jean-Pierre Barbot, and Robert Boisliveau

Abstract The three-phase inverter, connected to the grid by an LCL filter, is vulnerable to multiple faults, therefore an active fault-tolerant control is required to preserve the electrical power conversion between renewable resources and the grid. In this work, we provide a flatness analysis of the inverter and its LCL filter, the goal being to determine whether the flatness-based control structure stays accurate under both healthy and faulty circumstances.

Keywords DC-AC inverter · Grid · Flatness analysis · Fault control

1 Introduction

The paper investigates the flatness of the three-phase grid connected inverter under healthy and faulty modes operation, in the framework of distributed power generation systems (DGPS). Numerous defects, such as excessive or low DC voltage compared to the healthy constant value, Mosfet faults, and grid-side faults can cause the grid-connected inverter to fail completely and shut down the DGPS [10, 11, 13, 15, 20]. In this paper, we consider only the fault related to the incorrect closure of the Mosfet [1], the goal being to preserve the electrical power conversion between the renewable sources and the grid. To achieve this, a fault detection technique associated with a real-time active fault tolerant control is proposed. The problems of fault detection,

This work is partially supported by Campus France via the PHC Maghreb 46003SM project and the first author is also financed by the Moroccan National Center for Scientific and Technical Research (CNRST) via the 30UH2C2021 grant.

M. Laaziz (✉) · Ghanes · J.-P. Barbot · R. Boisliveau
LS2N UMR CNRS 6004, Ecole Centrale de Nantes, Nantes, France
e-mail: marouane.laaziz-etu@etu.univh2c.ma

M. Laaziz · N. Machkour
LCCPS, ENSAM, Hassan II University of Casablanca, Casablanca, Morocco

F. Nicolau · J.-P. Barbot
Quartz EA 7393, ENSEA, Cergy-Pontoise, France

© The Author(s), under exclusive license to Springer Nature Singapore Pte Ltd. 2024

A. Mellit et al. (eds.), *Proceedings of the 2nd International Conference on Green Energy Conversion System*, Green Energy and Technology,
https://doi.org/10.1007/978-981-97-6148-7_52

Table 1 Paper nomenclature

Acronyms	Meaning	Acronyms	Meaning
V_{dc}	DC bus voltage (constant)	R_{on}	Conduction resistance of the Mosfet
i_{dc}	DC current	Δv_{jl}, Δr_{jl}	Losses related to the low Mosfet
j	phase index, $j \in \{a, b, c\}$	Δv_{jh}, Δr_{jh}	Losses related to the high Mosfet
i_j	Inverter output current	L_1	Inverter output inductance
v_{fj}	Capacitor voltage	r_1	Internal resistor of L_1
i_{gj}	Grid current	L_2	Grid side inductance
v_{gj}	Grid voltage	r_2	Internal resistor of L_2
K_j	Switch state	C_f	Filter capacitor
f_{sw}	Switching frequency	R_d	Damping resistor
f_g	Grid frequency	f_g	Grid frequency
α_j	Average control ($\alpha_j \in (0, 1)$)	i_s, $i_{s'}$	Homopolar currents
P	Active power of the system	φ	Flat output

localization and estimation are not covered in this paper. In this contribution, we model the faulty system in the *abc* frame, that does not consider the system balanced compared to the *dq* frame. Even if the homopolar component of the unbalanced system is included in the $dq0$ frame, see [3, 18], this control does not localize the faults. The proposed active fault control is based on the differential flatness [7, 12] of the considered model, which allows all system states and controls to be parameterized as functions of the flat output and a finite number of its time-derivatives. There are various benefits of this control, such as a simple trajectory tracking [19] and the absence of zero dynamics [6]. Moreover, in the present study case, the structure of the flatness-based control is the same for both systems with or without faults compared to the other existing controls [2, 4, 5, 16, 17]. For example the PI controller has to adjust their gains during the occurred faults. From this last point of view, a general model that works for both healthy and faulty conditions of the three-phase grid connected inverter is proposed. The healthy and faulty models are both designed with respect to losses characterization and estimation. The difference between the models is only that the losses level is higher for the faulty model. Then it is demonstrated that a chosen flat output of the healthy model is also relevant for the faulty model.

The remaining paper is organised as follows: The flatness analysis of the considered system is given in Sect. 2. The simulation results are presented in Sect. 3. In Sect. 4, conclusions and perspectives are discussed.

The main notations used in this paper are given in Table 1.

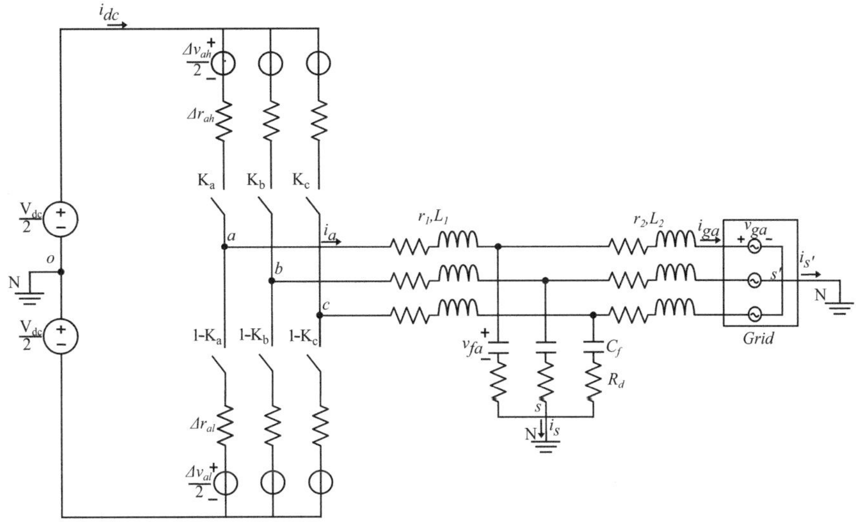

Fig. 1 Model of the three-phase grid connected inverter with LCL filter

2 Flatness of the Three-Phase Grid Connected Inverter with LCL Filter: Faulty Model

The considered model is a model that uses the losses to detect the fault, this model is presented in Fig. 1. The averaged equations of this model are presented in Sect. 2.1, which are suitable to analyze its flatness, see Sect. 2.2 (we work with the continuous-time averaged model instead of the switching model because the flatness theory for switching systems is more involved). The losses Δv_{jh}, Δr_{jh}, Δv_{jl}, Δr_{jl}, are supposed known. The healthy model of Fig. 1 is obtained with the condition that $\Delta v_{jh} = \Delta v_{jl}$ and $\Delta r_{jh} = \Delta r_{jl}$. The proposed model works for both healthy and faulty circumstances. As mentioned in the introduction, the fault considered in this paper is an inappropriately closed Mosfet. As an example, the ill component is supposed to be at the low Mosfet of the phase a. The fault is detected when the resistance Δr_{al} and the voltage drop Δv_{al} are high compared to the healthy levels.

2.1 The Average Model

The average model of phase a is described by:

$$\begin{bmatrix} \frac{d\bar{i}_a}{dt} \\ \frac{d\bar{i}_{ga}}{dt} \\ \frac{d\bar{v}_{fa}}{dt} \end{bmatrix} = \begin{bmatrix} -\frac{r_1+R_d+\Delta r_{al}}{L_1}\bar{i}_a + \frac{R_d}{L_1}\bar{i}_{ga} - \frac{1}{L_1}\bar{v}_{fa} \\ -\frac{V_{dc}-\Delta v_{al}}{2L_1} \\ \frac{R_d}{L_2}\bar{i}_a - \frac{r_2+R_d}{L_2}\bar{i}_{ga} + \frac{1}{L_2}\bar{v}_{fa} - \frac{1}{L_2}v_{ga} \\ \frac{1}{C_f}\bar{i}_a - \frac{1}{C_f}\bar{i}_{ga} \end{bmatrix} + \begin{bmatrix} \frac{1}{L_1}\Big(-(\Delta r_{ah} - \Delta r_{al})\bar{i}_a \\ -\frac{\Delta v_{ah}}{2} - \frac{\Delta v_{al}}{2} + V_{dc}\Big) \\ 0 \\ 0 \end{bmatrix} \alpha_a, \tag{1}$$

and is of the form

$$\dot{x}_a = f(x_a) + g(x_a)u_a, \tag{2}$$

with $x_a = [\bar{i}_a, \bar{i}_{ga}, \bar{v}_{fa}]^T \in \mathbb{R}^3$ and $u_a = \alpha_a \in (0, 1)$. Model (1) is a non-linear control-affine system.

The average model of the three-phase grid connected inverter consists of three identical and decoupled subsystems of the form (1) (one subsystem for each phase). The focus is made on phase a and the results can be expanded to other phases.

From now on, by faulty model we understand model (1) for which $\Delta v_{ah} \neq \Delta v_{al}$ and $\Delta r_{ah} \neq \Delta r_{al}$ (the losses level is higher for the faulty model), while for the healthy model, we will always have $\Delta v_{ah} = \Delta v_{al}$ and $\Delta r_{ah} = \Delta r_{al}$.

Remark 1 The approach that we employ in this work can also handle symmetrical faults (where both the high and low Mosfets are damaged). We only focus on asymmetrical faults because they are more probable to appear.

2.2 *Flatness Study of the Average Model*

Definition 1 (*Flatness* [7]) The system $\Xi : \dot{x} = F(x, u)$, where $x \in \mathbb{R}^n$, $u \in \mathbb{R}^m$ is flat, if there locally exist m smooth functions $\varphi_k(x, u, \dots, u^{(l)})$, for $1 \leq k \leq m$, such that locally

$$x = \gamma(\varphi, \dots, \varphi^{(\beta-1)}) \text{ and } u = \psi(\varphi, \dots, \varphi^{(\beta)}),$$

for a certain integer β, where $\varphi = (\varphi_1, \dots, \varphi_m)$ is called a flat output.

Proposition 1 *The following conditions hold:*

(i) The healthy system, given by (1)*, with* $\Delta v_{ah} = \Delta v_{al}$ *and* $\Delta r_{ah} = \Delta r_{al}$*, is flat with* $\varphi = \bar{i}_{ga} - \frac{R_d C_f}{L_2}\bar{v}_{fa}$ *being a flat output.*

(ii) The faulty system, given by (1) *and for which we suppose that*

$$-0.5\Delta v_{ah} - 0.5\Delta v_{al} + V_{dc} + (\Delta r_{al} - \Delta r_{ah})\bar{i}_a \neq 0, \tag{3}$$

is flat with $\varphi = \bar{i}_{ga} - \frac{R_d C_f}{L_2}\bar{v}_{fa}$ *being a flat output.*

Notice that since (1) is a single-input control system, the flat output φ has only one component.

Proof The proof of (i) follows the same line as that of condition (ii) that is given below.

Proof of (ii). Let us consider the faulty model given by (1). A necessary condition for φ to be a flat output of (1) is that the matrix $G = [g, ad_f(g), ad_f^2(g)]$ is invertible, see [19], where $ad_f g = [f, g]$ is the Lie Bracket [9] of the vector fields f and g describing the dynamics of (1) (recall that we rewrote system (1) into the compact form (2)), and $ad_f^2(g) = [f, ad_f(g)]$. By a straightforward computation it can be shown that the rank of the 3×3 - matrix G is three if condition (3) is satisfied.

The function $\varphi = \bar{i}_{ga} - \frac{R_d C_f}{L_2} \bar{v}_{fa}$ is a flat output if the following conditions are satisfied

$$\begin{cases} \pounds_g \varphi = 0 \\ \pounds_g \pounds_f \varphi = 0 \\ \pounds_g \pounds_f^2 \varphi \neq 0 \end{cases},$$

where $\pounds_g \varphi$ is the Lie derivative φ along the vector field g [9], which can be easily verified by a direct calculation. □

Assumption 1 From now on, we assume that the fault Δr_{al} is always such that (3) is satisfied (that is, we avoid the flatness singularity).

Assumption 1 guarantees that φ is a flat output for both the healthy and faulty models. Thus the same flatness-based control in both healthy and unhealthy circumstances is applied according to Proposition 1 and Assumption 1. The expression of u_a as function of $\varphi^{(3)}$ is: $u_a = \frac{1}{\pounds_g \pounds_f^2 \varphi}(\varphi^{(3)} - \pounds_f^3 \varphi)$.

The feedback $v_a = \pounds_f^3 \varphi + \pounds_g \pounds_f^2 \varphi u_a$, is applied to get $\varphi^{(3)} = v_a$ and only then the following feedback controller [9] is used for tracking the flat output reference φ^*:

$$v_a = -K_1(\varphi - \varphi^*) - K_2(\dot{\varphi} - \dot{\varphi}^*) - K_3(\ddot{\varphi} - \ddot{\varphi}^*) + \varphi^{*(3)}, \tag{4}$$

where K_1, K_2, K_3 are the gains of the feedback controller (they are chosen such that the feedback controller is stable).

Remark 2 Due to the fact that both the healthy and faulty models admit the same flat output φ (and will thus have the same reference trajectory φ^*), the feedback controller (4) is unaffected by healthy or unhealthy conditions. On the other hand, observe that the control u_a is dependent on the fault value through the terms $\pounds_f^3 \varphi$ and $\pounds_g \pounds_f^2 \varphi$.

3 Simulation Results

The Ode1 Euler solver is used in the Matlab/Simulink with a 10^{-7} s discretization step. The simulation parameters are described in Table 2. LCL parameters are determined in accordance with IEEE 1547 [8, 14].

The simulation results are presented in Fig. 2. Remark that, at 0.1 s a fault is applied by changing the losses values of the low Mosfet of phase a to the following values: $\Delta r_{al} = 5\Omega$ and $\Delta v_{al} = 4V$. As it can be seen in Fig. 2 (from 0.1 to 0.2 s), before applying the flatness based active control, the error $e_j = i^*_{gj} - i_{gj}$, where i^*_{gj} is the grid current reference (with a maximum value of 6 A), indicates that the fault occurs at the phase a. At 0.2 s, where the flatness based active control is applied, the error e_a is reduced, which compensates the influence of the fault.

4 Conclusion and Perspectives

The paper proposed a flatness-based approach for the three-phase grid connected inverter for which the faults are modeled as losses. The main contribution is to propose a flat output that is the same for both healthy and faulty models. This result allowed to work with the same active fault control structure at both regimes.

Table 2 Simulation parameters

Parameter	Value	Parameter	Value
V_{dc}	350 V	L_1	10 mH
P	1.1 kW	r_1, r_2	1 Ω, 0.5 Ω
f_g	50 Hz	L_2	1 mH
f_{sw}	10 kHz	C_f	4.5 µF
v_{gj}	68 V	R_d	1 Ω

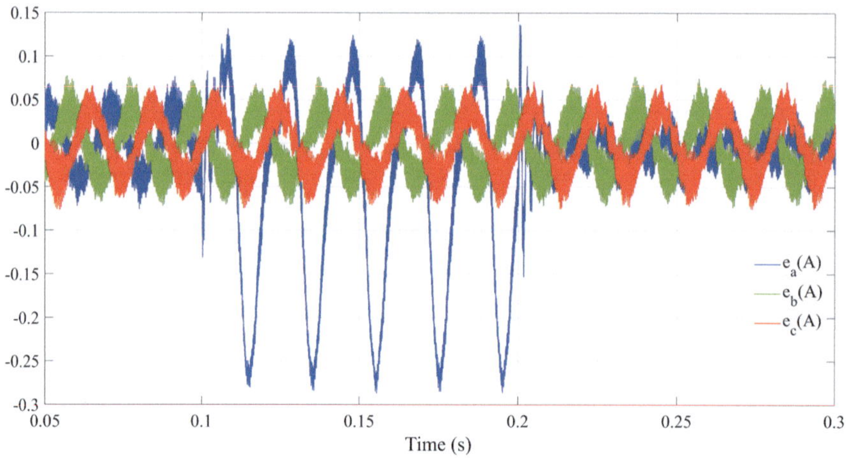

Fig. 2 The grid current errors e_j

The work perspectives involve coupling the losses and faults estimator with the flatness-based control including the stability analysis and making a comparison with other control methods.

References

1. Celaya JR, Saxena A, Wysocki P, Saha S, Goebel K (2010) Towards prognostics of power MOSFETs: accelerated aging and precursors of failure. In: Annual conference of the PHM society
2. Dang C, Tong X, Song W (2020) Sliding-mode control in DQ-frame for a three-phase grid-connected inverter with LCL-filter. J Franklin Inst 357:10159–10174
3. Djerioui A, Houari A, Saim A, Ait-Ahmed M, Pierfederici S, Benkhoris MF, Machmoum M, Ghanes M (2020) Flatness-based grey wolf control for load voltage unbalance mitigation in three-phase four-leg voltage source inverters. IEEE Trans Ind Appl 56:1869–1881
4. Elnozahy A, Yousef AM, Abo-Elyousr FK, Mohamed M, Abdelwahab SAM (2021) Performance improvement of hybrid renewable energy sources connected to the grid using artificial neural network and sliding mode control. J Power Electron 21:1166–1179
5. Etienne L, Acosta Lúa C, Di Gennaro S, Barbot JP (2020) A super-twisting controller for active control of ground vehicles with lateral tire-road friction estimation and CarSim validation. Int J Control Autom Syst 18:1177–1189
6. Fliess M, Sira-Ramírez H, Marquez R (1998) Regulation of non-minimum phase outputs: a flatness based approach. In: Perspectives in control: theory and applications. Springer, London, pp 143–163
7. Fliess M, Lévine J, Martin P, Rouchon P (1995) Flatness and defect of non-linear systems: introductory theory and examples. Int J Control 61:1327–1361
8. IEEE Standard for Interconnection and Interoperability of Distributed Energy Resources with Associated Electric Power Systems Interfaces. IEEE Std 1547-2018 pp. 1–138 (2018)
9. Isidori A (ed) (1985) Nonlinear control systems: an introduction, vol 72. Springer, Berlin, Heidelberg
10. Keller, J., Kroposki, B.: Understanding Fault Characteristics of Inverter-Based Distributed Energy Resources. Tech. rep., National Renewable Energy Lab. (NREL), Golden, CO (United States) (2010)
11. Khomfoi S, Tolbert LM (2007) Fault diagnostic system for a multilevel inverter using a neural network. IEEE Trans Power Electron 22:1062–1069
12. Levine J (2009) Analysis and control of nonlinear systems: a flatness-based approach. Springer, Berlin, Heidelberg
13. Lu B, Sharma SK (2009) A literature review of IGBT fault diagnostic and protection methods for power inverters. IEEE Trans Ind Appl 45:1770–1777
14. Mahamat C (2018) Analyse et commandes des convertisseurs multi-niveaux pour un générateur photovoltaïque connecté au réseau électrique. phdthesis, Université Paris Saclay (COmUE)
15. Plet CA, Graovac M, Green TC, Iravani R (2010) Fault response of grid-connected inverter dominated networks. In: IEEE PES general meeting, pp 1–8
16. Sanchis P, Ursaea A, Gubia E, Marroyo L (2005) Boost DC-AC inverter: a new control strategy. IEEE Trans Power Electron 20(2):343–353
17. Selvaraj J, Rahim NA (2009) Multilevel inverter for grid-connected PV system employing digital PI controller. IEEE Trans Ind Electron 56:149–158
18. Shahin A, Abulanwar S, Ghanem A, Rizk MEM, Deng F, Pierfederici S, Ismael I (2022) Sensorless robust flatness-based control with nonlinear observer for non-ideal parallel DC-AC inverters. IEEE Access 10:53940–53953
19. Sira-Ramírez H, Agrawal SK (2018) Differentially flat systems. CRC Press, Boca Raton
20. ZCS Azzurro, user manual 3PH 3.3KTL-V3/3PH 12KTL-V3 - Grid connected inverter

Fuzzy Logic for Improving DTC and DPC Controls in WECS Based on Doubly-Fed Induction Generator

Sofia Lalouni Belaid, Oualid Djoudi, and Salah Tamalouzt

Abstract This paper provides a performance enhancement for direct torque control (DTC) and direct power control (DPC) applied to two level back to back converter feeding the rotor of doubly fed induction generator (DFIG) powered by random speed wind turbine (WT). For this purpose, the sector number in the DTC and DPC is extended from 6 to 12 sectors. While, the switching table based on hysteresis controllers are replaced by fuzzy logic controller (FLC) in both DTC and DPC control. The designed fuzzy 12 sectors DTC (F12S-DTC) and fuzzy 12 sectors DPC (F12S-DPC) significantly reduce ripples in the system under consideration, thanks to error-level consideration and precise localization of the rotor flux vector. The simulation results obtained with the proposed control structure show satisfactory improvements in terms of torque/flux and GSC active/reactive power ripples reduction.

Keywords Doubly fed induction generator · Fuzzy logic control · Direct torque control · Direct power control · 12 sectors methodology

1 Introduction

Wind energy is the most popular kind of renewable energy, because of its wide availability and advanced exploitation technology [1, 2]. Wind power generation systems based on the doubly fed induction generator (DFIG) are an attractive option in the field of power generation because of its outstanding advantages. Compared to the

---

S. L. Belaid (✉) · O. Djoudi · S. Tamalouzt
Université de Bejaia, Faculté de Technologie, Laboratoire de Technologie Industrielle et de l'Information, 06000 Bejaia, Algeria
e-mail: sofia.lalouni@univ-bejaia.dz

O. Djoudi
e-mail: oualid.djoudi@univ-bejaia.dz

S. Tamalouzt
e-mail: salah.tamalouzt@univ-bejaia.dz

© The Author(s), under exclusive license to Springer Nature Singapore Pte Ltd. 2024
A. Mellit et al. (eds.), *Proceedings of the 2nd International Conference on Green Energy Conversion System*, Green Energy and Technology,
https://doi.org/10.1007/978-981-97-6148-7_53

other generators, DFIG has the opportunity to operate with small converters in a wide speed range [3]. Nevertheless, due to its non-linearity, dynamics and the variation of parameters during operation, its control is more difficult. To surpass these constraints, the torque and flux of this machine must be controlled by more sophisticated control algorithms [4, 5]. Among the algorithms proposed in the literature to control the wind turbine powered by the DFIG (WT-DFIG), direct torque control (DTC) introduced by Takahashi in 1985 is commonly used. This control outperforms the field oriented control (FOC) by its simplicity, better transient response and low dependence on machine parameters [6]. However, in this type of control, both electromagnetic torque and rotor flux are subject to nonlinear hysteresis controller, which imposes restrictions such as high and variable switching frequency [7]. In the last years, the fuzzy logic control (FLC) has emerged in the field of system control because of its advantages, notably the non-necessity of having the system mathematical model and its operating principle is inspired by human logic [8]. The fuzzy control decision is based on the degree of control variable error, which allows flexible and intelligent control, that justifies the adoption of this controller to replace the hysteresis controller in the classical DTC. A further approach to improving the DTC method's performance is to extend the number of sectors from 6 to 12, which allows the rotor flux vector to be well localized, so that electromagnetic torque and rotor flux ripples will be greatly reduced when the appropriate voltage vector is selected. In this article, an enhanced DTC control is designed by combining fuzzy DTC (F-DTC) and 12-sector DTC (12S-DTC). On the other hand, the power control at the point of connection to the grid side converter (GSC) is necessary [9]. For this purpose, an improved direct power control (DPC) is implemented to control the exchange of active and reactive power between the DFIG rotor and the grid. The proposed GSC control is based on DPC based on 12-sector fuzzy inference systems (F12S-DPC). MATLAB/SIMULINK software is used to evaluate the performance of the proposed control system (Fig. 1).

2 Proposed Fuzzy 12 Sectors DTC

The main drawback of conventional DTC (C-DTC) is the use of hysteresis controllers that generate larger ripples in the electromagnetic torque and rotor flux. To overcome this problem and at the same time to keep the advantages of C-DTC such as simplicity of implementation, fast response and low requirement of machine parameters, fuzzy inference system is the best alternative.

Instead of the switching look up table and hysteresis controllers, a fuzzy logic controller (FLC) is used. Besides, the sector number has been increased from six to twelve. Rotor flux error, electromagnetic error and rotor flux angle are the element of the fuzzification step. The torque error is converted into three five sets (PL: Positive large, PS: Positive Small, Z: Zero, NS: Negative Small and NL: Negative Large). While, the rotor flux error is composed of two fuzzy sets (P: Positive and N: Negative). The fuzzification process of the rotor flux angle can be obtained by using twelve fuzzy

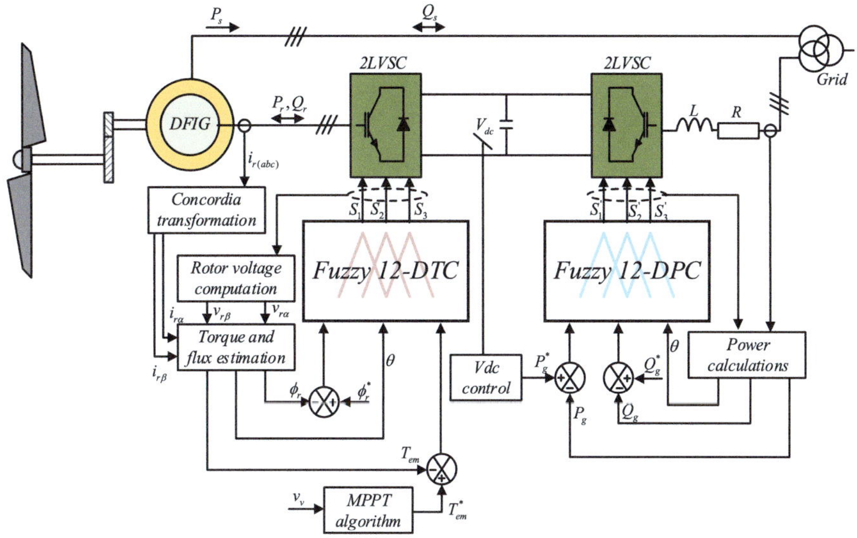

Fig. 1 Global diagram of the proposed system

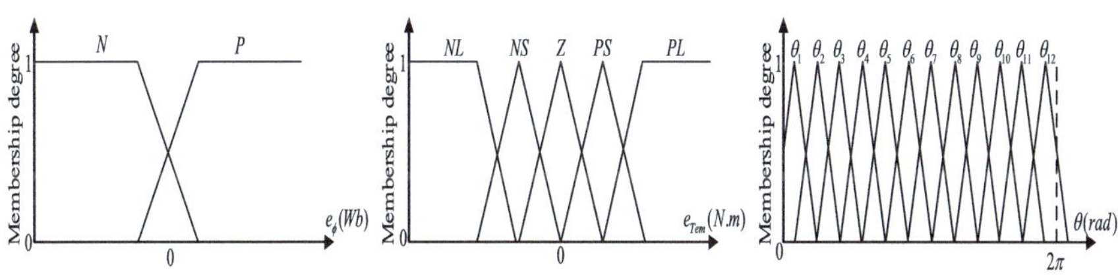

Fig. 2 Modified DTC input membership functions

membership functions. The membership functions used are represented in Fig. 2. The inference engine can determine the suitable voltage vector to be applied to the rotor side converter from a 120 rules fuzzy switching table [10].

3 Proposed Fuzzy 12 Sectors PDC for GSC

The GSC is the intermediary between the DFIG's rotor-side converter (RSC) and the grid. Therefore, its main task is to transport the active and reactive power involved between these two elements. The control of GSC has been treated several times in the literature [11–13]. Among the most commonly used controls in recent years is the DPC control. It doesn't require the Park transform and the PWM modulation block. To control the active and reactive power transmitted by GSC, it is enough to know the instantaneous values of the powers and the position of the voltage vector applied to the GSC. Based on these three parameters a switching table is constructed for direct selection of the appropriate voltage vector in the conventional DPC (C-DPC). In the proposed fuzzy 12 sectors DPC, the active and reactive powers are converted into two membership functions (P: Positive and N: Negative). Whereas, the GSC voltage vector angle is divided into twelve fuzzy membership functions.

The active/reactive power can be estimated from the GSC voltage and current according the following equation.

$$\begin{cases} P_g = V_{G\alpha} i_{\alpha'} + V_{G\beta} i_{\beta'} \\ Q_g = V_{G\alpha} i_{\beta'} - V_{G\beta} i_{\alpha'} \end{cases} \tag{1}$$

To better locate the position of GSC voltage vector, the stationary frame is divided into 12 sectors. The following equation yields the position of GSC voltage vector.

$$(n-2)\frac{\pi}{6} \le \gamma_n \le (n-1)\frac{\pi}{6}; n = 1, 2, \ldots, 12 \tag{2}$$

The fuzzy rules table [23] is used for selecting the appropriate voltage vector to compensate for the GSC active/reactive power error.

4 Simulation Results and Discussion

The results of the studied system performed by MATLAB/Simulink software are presented in this section. The WT-DFIG modeling and parameters can be found in [14]. The wind speed profile is illustrated in Fig. 3. An optimal power coefficient is achieved with the fuzzy MPPT method, as shows Fig. 4. Figure 5 depicts the electromagnetic torque waveform obtained with C-DTC, 12S-DTC and the proposed F12S-DTC.

The zoom area in the electromagnetic torque attests on the effectiveness of the proposed method. Torque ripples are greatly reduced compared to conventional DTC

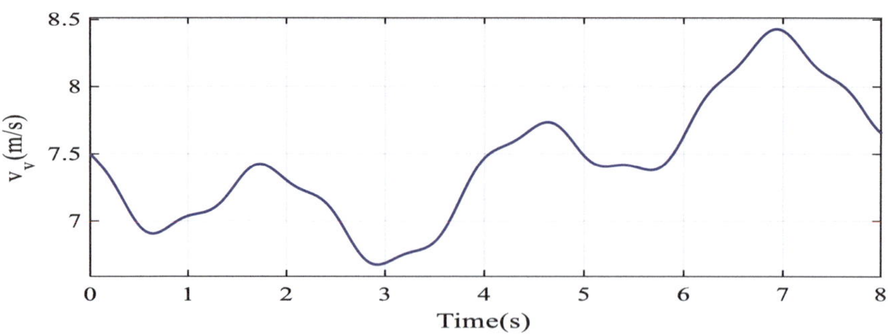

Fig. 3 Wind speed waveform

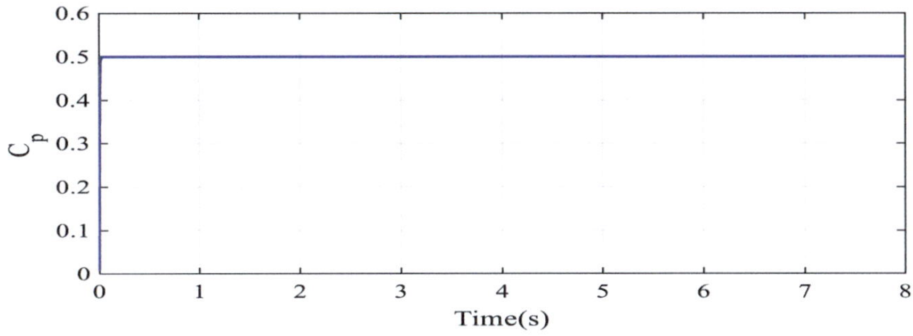

Fig. 4 Power coefficient evolution

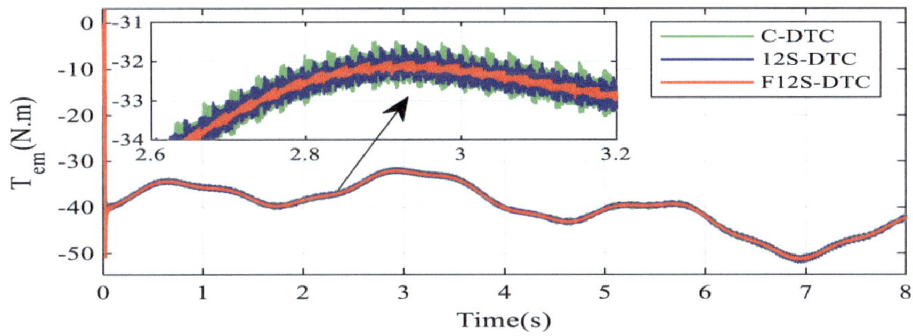

Fig. 5 Electromagnetic torque waveform with zoom

and 12S-DTC. The rotor flux trajectory is shown in Fig. 6. The ripples are significantly reduced with the propose F12S-DTC technique. Figure 7 shows the DC-link voltage waveform obtained with C-DPC, 12S-DPC and the suggested F12S-DPC method. The mentioned three methods allow tracking the DC-link voltage reference. Figures 8 and 9 show respectively the evolution of the GSC active and reactive power,. The proposed F12S-DPC method significantly improves power ripple compared to the traditional DPC method and the 12S-DPC method.

The rotor and the grid exchange active power in both directions. During the sub-synchronous mode, the rotor active power is positive; then the grid supplies the DFIG rotor. A reduction of torque/flux and GSC active/reactive power ripples by the proposed F12S-DTC appears in the above figures and the stator current quality returned to the grid. The stator current waveform and current spectral analysis, present an improvement of about 74% compared to the conventional DTC and 65% compared to 12S-DTC is obtained.

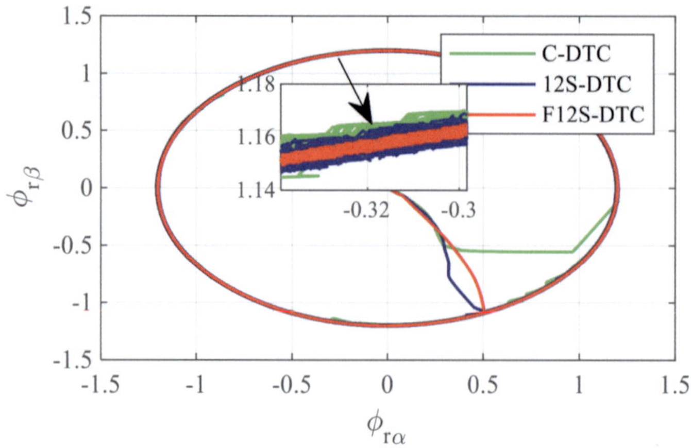

Fig. 6 Rotor flux trajectory

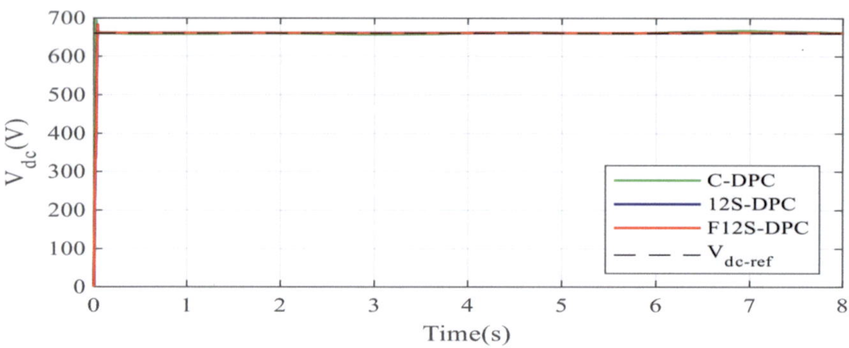

Fig. 7 DC-link voltage

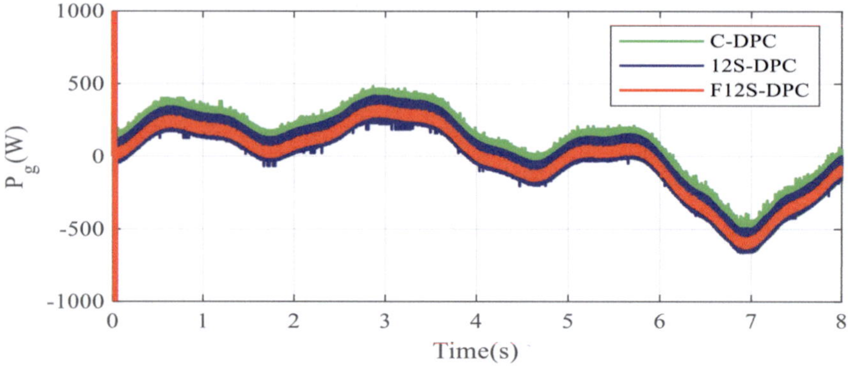

Fig. 8 GSC active power waveform

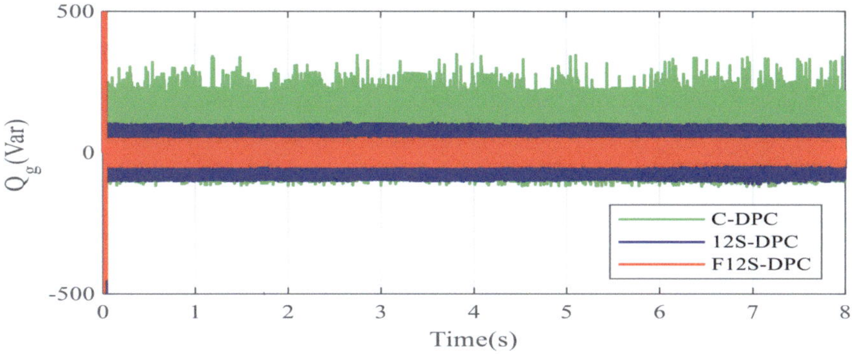

Fig. 9 GSC reactive power waveform

5 Conclusion

In this paper, an improvement of direct torque and power control is proposed for the DFIG in grid-connected wind energy conversion system. To this end, two fuzzy switching tables based on 12 sectors are proposed to replace switching tables based on hysteresis controllers in conventional DTC and DPC methods. The main objective is to reduce the ripples that occur in electromagnetic torque and rotor flux with C-DTC and active and reactive power with C-DPC. The simulation results show high performances, significant reduction of torque/flux ripples and GSC active/reactive power ripples by using the proposed control structure. The WT-DFIG supplies AC current to the grid with low THD.

References

1. Kelkoul B, Boumediene A (2021) Stability analysis and study between classical sliding mode control (SMC) and super twisting algorithm (STA) for doubly fed induction generator (DFIG) under wind turbine. Energy 214:118871
2. Gu H, Wang J (2013) Irregular-shape wind farm micro-siting optimization. Energy 57:535–544
3. Kadri A, Marzougui H, Aouiti A, Bacha F (2020) Energy management and control strategy for a DFIG wind turbine/fuel cell hybrid system with super capacitor storage system. Energy 92:116518
4. Zemmit A, Messalti S, Harrag A (2018) A new improved DTC of doubly fed induction machine using GA-based PI controller. Ain Shams Eng J 9(4):1877–1885
5. El Ouanjli N, Motahhir S, Derouich A, Ghzizal A, Chebabhi A, Taoussi M (2019) Improved DTC strategy of doubly fed induction motor using fuzzy logic controller. Energy Rep 5:271–279
6. Djoudi O, Lalouni Belaid S, Tamalouzt S (2022) A high degree of direct torque control applied to a grid-connected wind energy system based on a DFIG J Ren Energies A247:81–92
7. Benamor A, Benchouia MT, Srairi K, Benbouzid MEH (2019) A novel rooted tree optimization apply in the high order sliding mode control using super-twisting algorithm based on DTC scheme for DFIG. Int J Electr Power Energy Syst 108:293–302

8. Tamalouzt S, Belkhier Y, Sahri Y, Bajaj M, Ullah N, Shahariar Chowdhury MD, Titseesang T, Techato K (2021) Enhanced direct reactive power control-based multi-level inverter for DFIG wind system under variable speeds. Sustainability 13(16):9060
9. Mondal S, Kastha D (2015) Improved direct torque and reactive power control of a matrix-converter-fed grid-connected doubly fed induction generator. IEEE Trans Industr Electron 62(12):7590–7598
10. Lalouni Belaid S (2021) Improved IM DTC by using a fuzzy switching table in PV applications. J Electr Eng Comput Sci 88(1–2):26–32
11. Mohammadi J, Vaez-Zadeh S, Ebrahimzadeh E, Blaabjerg F (2018) Combined control method for grid-side converter of doubly fed induction generator-based wind energy conversion systems. IET Renew Power Gener 12(8):943–952
12. Zarei ME, Nicolás CV, Arribas JR, Ramírez D (2018) Four-switch three-phase operation of grid-side converter of doubly fed induction generator with three vectors predictive direct power control strategy. IEEE Trans Industr Electron 66(10):7741–7752
13. Gao S, Zhao H, Gui Y, Zhou D, Blaabjerg F (2020) An improved direct power control for doubly fed induction generator. IEEE Trans Power Electron 36(4):4672–4685
14. Djoudi O, Lalouni Belaid S, Tamalouzt S (2023) Multilevel converter and fuzzy logic solutions for improving direct control accuracy of DFIG-based wind energy system. Period Polytech Electr Eng Comput Sci 67(2):136–148

Thermal and Mechanical Properties of the Ti_2AlX (X = C, N) MAX Phases: A DFT Study

Lyacine Rabahi, Bellel Cheniti, Idir Hadji, Zoheir Boutaghou, and Riad Badji

Abstract The aim of the present work is to provide new insights on how the nitrogen concentration affects the stability, thermal, and mechanical properties of the Ti_2AlC MAX Phase. To this purpose, the DFT-based pseudopotential method combined with the Debye Quasi-harmonic model is applied. The results display a strong dependency of the studied properties on the nitrogen concentration; indeed, the stability and rigidity of the Ti_2AlC alloys are substantially improved by nitrogen insertion even at elevated temperatures. Ti_2AlC, Ti_2AlN, and $Ti_2AlC_{0.5}N_{0.5}$ are brittle and dynamically stable at 0 K with their hexagonal structures. The results are deeply discussed in light of the available theoretical and experimental results.

Keywords MAX phases · DFT calculations · Quasi-Harmonic model · Thermal properties · Mechanical properties

1 Introduction

The ternary carbides and nitrides commonly called MAX phases represent a new class of materials that combine the advantages of both Metals and Ceramics [1, 2]. Like metals, they are readily machinable and possess good thermal and electrical conductivities, owing to the substantial density of States at the Fermi Level [2–4] (resestivities ranging from 0.07 to 2 $\mu\Omega$.m [4, 5]). They are also relatively soft, and thermal shock resistant, with Vickers hardness values ranging from 1.4 to 8 GPa [5], and possess good plasticity and microscale ductility at room and finite temperatures [2, 3]. Like ceramics, they are good refractory materials with higher thermodynamic stability and relatively low thermal expansion coefficients. These properties make them potential candidates for automotive and aerospace applications.

The interest of the scientific community in the MAX phases is growing steadily. A few years ago, only 155 ternary MAX phases were synthesized and characterized,

L. Rabahi (✉) · B. Cheniti · I. Hadji · Z. Boutaghou · R. Badji
Research Center in Industrial Technologies (CRTI), Cheraga, P. O. Box 64, 16014 Algiers, Algeria
e-mail: rabahil@yahoo.fr

© The Author(s), under exclusive license to Springer Nature Singapore Pte Ltd. 2024
A. Mellit et al. (eds.), *Proceedings of the 2nd International Conference on Green Energy Conversion System*, Green Energy and Technology,
https://doi.org/10.1007/978-981-97-6148-7_54

involving 16 A elements (mostly groups 13 and 14) and 14 M (early transition metal) elements of the periodic table [4]. However, the recent discovery of both solid solutions and ordered MAX phases (in and out of plane), that contain either two M, two A or two X elements, has opened the door to the discovery of many new MAX phases. The diversity of the elements that can form MAX Phases provides this class of materials modulable properties depending on the intended application. Further, the formation of many MAX Phases was made possible by the substitution of one or more of its constituting elements, and this substitution may trigger new properties not observed in the ternary host system, thereby allowing the establishment of a close relationship between the composition and physical properties of this class of materials.

In this context, the main objective of the present work is to investigate the effects of nitrogen concentration on the the stability, thermal, and mechanical properties of the Ti_2AlC MAX Phase. To carry out the results, the pseudopotential method based on Density Functional Theory DFT was applied together with the Debye Quasi-Harmonic model. The paper is organized as follows: Sect. 2 is devoted to the technical details of DFT calculations. Results are presented and discussed in Sect. 3. In Sect. 4, concluding remarks summarize the main results.

2 Computational Details

The present study calculations were carried out using the Pseudopotential method based on Density Functional Theory (DFT) [6, 7] as implemented in the Quantum-Espresso package [8]. The generalized gradient approximation (GGA) of Perdew et al. (PBE) [9] was taken to include the exchange–correlation energy and ultra-soft pseudopotentials of Vanderbilt [10]. A kinetic energy cut-off of 40 Ry and a well-converged value of the k-point mesh ($8 \times 8 \times 2$) over the Brillouin zone were considered. All structures were allowed to fully relax using Broyden-Fletcher-Goldforb-Shanno (BFGS) scheme until the total energy has converged to less than 10^{-5} eV/atom. The maximum force has converged to lower than 0.004 eV/Å.

3 Results and Discussion

3.1 Structural Properties

In order to determine the structural properties of the Ti_2AlX MAX Phases at 0 K, a set of total energy versus total volume calculations was performed. The corresponding values were fitted with the universal Birch-Murnaghan equation of state. The obtained

Table 1 Cell parameters (a and c) bulk moduli, z internal parameter, and heats of formation of the Ti_2AlX (X = C and N) MAX Phases predicted at 0 K

		a (au)	c (au)	z	B (GPa)	ΔH_f (eV/f.u)
Ti_2AlC	This work	5.77	25.87	0.0836	138.5	– 0.741
	Theo	5.81[11]	26.03[11]	0.0849[12]	124.5[11]	– 0.76[11] – 0.83[13]
	Exp	5.75[14]	25.71[14]	0.085[15]	144.0[16]	–
$Ti_2AlC_{0.5}N_{0.5}$	This work	5.71	25.72	0.5839	146.0	– 0.964
	Theo	5.72[17]	25.97[17]			
	Exp	5.71[18]	25.73[18]			
Ti_2AlN	This work	5.64	25.67	0.0856	147.9	– 1.192
	Theo	5.66[19]	25.79[19]	0.0850[12]	161.9[19]	– 1.087[20]
	Exp	5.65[12]	25.73[12]	0.085[15]	169.0[21]	

lattice parameters, (c/a) ratios bulk moduli, heats of formation and internal parameter z obtained after relaxation are presented in Table 1. For comparison, some experimental and theoretical values are collected from the literature.

From the Table, the cell parameters (a and c) as well as bulk modulus and z internal parameter values of the Ti_2AlC and Ti_2AlN MAX Phases are very consistent with the theoretical [11–15] and experimental ones [12, 14–21]. Moreover, the studied phases possess considerable rigidity, as reflected by the higher bulk modulus values. Comparatively, Ti_2AlN is the most rigid material, followed by $Ti_2AlC_{0.5}N_{0.5}$ whereas Ti_2AlC is the least rigid. In terms of stability, the three studied MAX Phases possess negative values of the formation energies, meaning that their formation can take place spontaneously without any external energy supply. Interestingly, the partial or total substitution of Carbon by Nitrogen increases the stability of the Ti_2AlC Phase as already found by S. Aryal et al. [22].

3.2 *Mechanical and Thermal Properties*

The elastic constants are related to many physical properties of materials, such as melting point, Debye temperature, thermal expansion, and specific heat. Also, they provide information on materials structural and mechanical stabilities. The elastic constants C_{11}, C_{12}, C_{13}, C_{33}, and C_{44} of the Ti_2AlC, Ti_2AlN, and $Ti_2AlC_{0.5}N_{0.5}$ MAX Phases are calculated and shown in Table 2.

As can be seen, the predicted values are in fairly good agreement with the theoretical results [11, 14, 22]. However, according to our knowledge, there are no available experimental values of the elastic constants of the studied phases. Furthermore, the

Table 2 Calculated elastic constants of the Ti_2AlC, Ti_2AlN and $Ti_2AlC_{0.5}N_{0.5}$ phases

		C_{11}	C_{12}	C_{13}	C_{33}	C_{44}
Ti_2AlC	This work	302.5	56.9	58.4	267.6	115.6
	Theo	297.8[11]	55.5[11]	53.8[11]	258.3[11]	111.1[11]
	Exp					
$Ti_2AlC_{0.5}N_{0.5}$	This work	292.7	51.9	77.4	269.8	115.5
	Theo	297.8[22]	58.6[22]	86.2[22]	275.6[22]	122.3[22]
	Exp					
Ti_2AlN	This work	313.5	60.9	85.3	284.6	130.6
	Theo	309.1[19]	70.0[19]	84.8[19]	288.5[19]	126.8[19]
	Exp					

elastic constant values obtained for the Ti_2AlN MAX phase are clearly higher than those obtained for the Ti_2AlC phase, whereas the $Ti_2AlC_{0.5}N_{0.5}$ phase possesses intermediate values slightly close to those of the Ti_2AlN phase. Moreover, it is known that the C_{11} and C_{33} elastic constants measure the linear compression resistance along the < xx > and < zz > directions, respectively, whereas the C_{44} constant reflects the ability to resist shear strain in the (100) or (010) planes. From our results, it is clearly visible that C_{11} values are higher than those of C_{33} in the cases of Ti_2AlC, $Ti_2AlC_{0.5}N_{0.5}$, and Ti_2AlN alloys, meaning that the three materials are more compressible under uniaxial stress along the < zz > direction. This fact could be explained by the great difference between a and c cell parameter values, as well as by the strong Ti–C and Ti-N bonds with respect to the Al-C and Al-N ones.

Moreover, the relevant criteria of mechanical stability conditions known as Born stability criteria for a hexagonal phase [11] are:

$$C_{11} > 0;\ C_{11} - C_{12} > 0;\ C_{44} > 0;\ (C_{11} + C_{12})C_{33} - 2(C_{13})^2 > 0 \qquad (1)$$

From Table 2, the mechanical stability conditions are easily verified in the case of the three studied MAX Phases, consequently, it could be concluded that the Ti_2AlC, $Ti_2AlC_{0.5}N_{0.5}$, and Ti_2AlN MAX Phases are mechanically stable at 0 K with the hexagonal structure. From the elastic constant values of Table 2, the average Voigt-Reuss-Hill approximation [11] was used to deduce other mechanical properties such as Bulk modulus (B), Young modulus (E), shear modulus (G), and Poisson's ratios (n). The obtained values are reported in Table 3.

The bulk modulus values of the three MAX Phases are very close to those obtained from the Murnaghan equation of state fit, as well as those obtained theoretically by applying other methods or measured experimentally [11, 15, 22, 23]. A good agreement is also noticed between our results of the Young modulus (E), shear modulus (G), and Poisson's ratios (ν) and those collected from the literature [11, 12, 15, 16, 22, 23]. Comparatively, Ti_2AlN is more resistant to elastic deformation, as reflected by its relatively high Young modulus and shear modulus values, followed

Table 3 Averaged bulk moduli (B), Young moduli (E), shear moduli (G), Poisson's ratios (n), Anisotropic factor (A), as well as empirical Vickers hardness (Hv) for the three studied MAX Phases

		B	E	G	n	G/B	A	H_v
Ti_2AlC	This work	135.69	273.01	117.2	0.165	0.863	1.02	24.35
	Theo	141[11]	266.0[11]	114.5[11]	0.16[11]		1.04[25]	24.3[11]
	Exp	144[20]	278[20]	119[20]	0.17[20]			5.55[26]
$Ti_2AlC_{0.5}N_{0.5}$	This work	140.55	267.7	113.18	0.183	0.805	1.13	21.68
	Theo	148.12[22]	273.99[22]	114.95[22]	0.192[22]			
	Exp		290[12]	123[12]	0.18[12]			
Ti_2AlN	This work	152.05	289.40	122.33	0.183	0.804	1.22	22.81
	Theo	161.91[19]	283.48[14]	118.74[19]	0.193[14]		1.32[24]	21.1[27]
	Exp	169[23]	286[23]	124[23]	0.18[23]			

by Ti_2AlC. The ratio of shear modulus to bulk modulus introduced by Pugh [24] predicts the brittle (if $G/B > 0.57$) or ductile (if $G/B < 0.57$) behavior of materials. The obtained values of this ratio are higher than the critical value of 0.57, implying that Ti_2AlC, Ti_2AlN, and $Ti_2AlC_{0.5}N_{0.5}$ MAX Phases are brittle and nitrogen atoms slightly improve the ductility of the Ti_2AlC Phase.

The Poisson's ratio ν provides information about the nature of bonding forces: For covalent materials, The Poisson's ratio value is very small (generally around 0.1) whereas a typical value of 0.25 characterizes ionic materials [13]. In general, better plasticity is often characterized by higher values of the Poisson ratio (higher than the critical value of 0.26 [24]. From our results, the calculated values of ν range from 0.16 to 0.18, indicating a mixture of covalent and ionic bonding in the case of the three studied MAX Phases. Once again, Ti_2AlN and $Ti_2AlC_{0.5}N_{0.5}$ are slightly less brittle than Ti_2AlC, as already concluded from Bulk and Shear modulus analyses. The anisotropic factor A was obtained using the following formula [24]

$$A = \frac{4C_{44}}{(C_{11} + C_{33} - 2C_{13})} \tag{2}$$

For an isotropic material, the anisotropic factor value is equal to 1, whereas any smaller or higher value than 1 reflects a degree of anisotropy. From our results, it could be concluded that Ti_2AlC, Ti_2AlN and $Ti_2AlC_{0.5}N_{0.5}$ are anisotropic materials as already found by several studies. Moreover, the magnitude of the deviation from 1, is a measure of the degree of elastic anisotropy of the crystal, in our case, deviation from the unit is larger for Ti_2AlN compared to the $Ti_2AlC_{0.5}N_{0.5}$ and Ti_2AlC possesses the smallest deviation. Finally, the intrinsic Vickers hardness parameter provides information about mechanical applications of a given material. This parameter is calculated applying the empirical formula [13]:

$$H_V = 2\left[\left({}^{G}\!/_{B}\right)^2 G\right]^{0.585} - 3 \quad (3)$$

The H_v obtained values for Ti_2AlC, Ti_2AlN and $Ti_2AlC_{0.5}N_{0.5}$ are reported in the last column of Table 3. Our value for Ti_2AlC is much higher than the experiment one of 5.55 GPa [25], and the nitrogen insertion decreases the overall hardness of Ti_2AlC material. The discrepancy is already pointed out by several authors [13] and is attributed to the plastic deformation that ductile materials can locally accumulate prior to fracture. Consequently, this empirical model may not be accurate to predict the hardness of real materials.

To include the temperature effects on the volume expansion and bulk modulus of the Ti_2AlC, Ti_2AlN, and $Ti_2AlC_{0.5}N_{0.5}$ MAX Phases, the Quasi-Harmonic Debye model was considered [26]. It is known that this model is entirely valid, with weak or negligible an-harmonic effects appearing at high temperatures [27, 28]. Accordingly, the volume expansion and bulk modulus variations with temperature are computed from 0 to 1000 K and shown in Fig. 1a and b respectively.

As can be seen, the volume expansion of the three studied MAX phases increases with temperature. More precisely, after a sharp increase at low temperatures up to 300 K, which is due to the electronic contributions, the variation becomes less sensitive to the temperature and is roughly similar for all studied MAX phases. Thereby, indicating the absence of any significant effect of the nitrogen concentration on the volume expansion of the Ti_2AlC Phase. This fact could be attributed to the great similarity between carbon and nitrogen atoms. In terms of the bulk modulus, this parameter is continually decreasing across the entire temperature range. The nitrogen concentration effect is clearly visible in this case, and Phases containing nitrogen are relatively the most rigid at low and elevated temperatures. The rigidity of the three MAX Phases decreases with similar rates and drops by more than 12% while maintaining acceptable rigidity at high temperatures around 1000 K.

4 Concluding Remarks

The DFT-based pseudo-potential method as well as the Quasi-Harmonic Debye model are used to investigate nitrogen effects on the stability, mechanics and thermal properties of the Ti_2AC MAX phases. The obtained results can be summarized as follows:

- The structural parameters of the studied MAX phases, predicted at 0 K, display good agreement with the experimental and theoretical results.
- Nitrogen insertion improves the stability of the Ti_2AlC MAX Phase at 0 K and increases its rigidity.

Fig. 1 Variation with temperature of: **a** volume expansion and **b** bulk modulus of the Ti_2AlC, Ti_2AlN and $Ti_2AlC_{0.5}N_{0.5}$ MAX phases

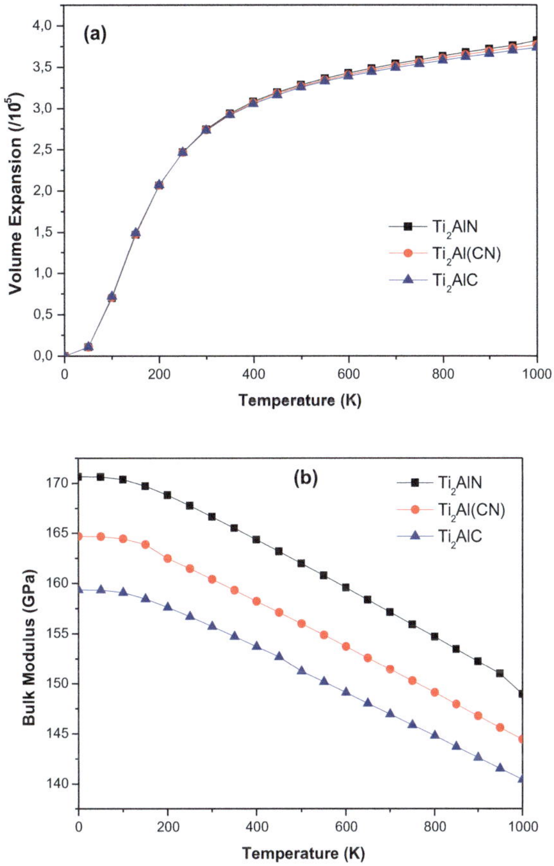

- Ti_2AlC, $Ti_2AlC_{0.5}N_{0.5}$, and Ti_2AlN MAX Phases are mechanically stable at 0 K with their hexagonal structure.
- Ti_2AlC, $Ti_2AlC_{0.5}N_{0.5}$, and Ti_2AlN are brittle materials.
- Nitrogen concentration has no significant effect on the volume expansion of the Ti_2AlC Phase. On the contrary, it increases its rigidity even at elevated temperatures.

References

1. Barsoum MW, Radovic M (2011) Elastic and mechanical properties of the MAX phases. Annu Rev Mater Res 41:195–227
2. Ding H, Glandut N, Fan X, Liu Q, Shi Y, Jie J (2016) First-principles study of hydrogen incorporation into the MAX phase Ti_3AlC_2. Int J Hydrog Energy 41:6387–6393

3. Radovic M, Barsoum MW (2013) MAX phases: bridging the gap between metals and ceramics. Am Ceram Soc Bull 92(3):20–27
4. Sokol M, Natu V, Kota S, Barsoum MW (2019) On the chemical diversity of the MAX phases. Trends Chem 1(2):210–223
5. Barsoum MW (2000) The Mn+1Axn phases: a new class of solids: thermodynamically stable nanolaminates. Prog Solid State Chem 28:201–281
6. Hohenberg P, Kohn W (1964) Inhomogeneous electron gas. Phys Rev 136:864–871
7. Kohn W, Sham L (1965) J: Self-consistent equation including exchange and correlation effects. Phys Rev 140:1133–1138
8. Baroni S, Dal Corso A, De Gironcoli S, Giannozzi P, Cavazzoni C, Ballabio G, Scandolo S, Chiarotti G, Focher P, Pasquarello A, Laasonen K, Trave A, Car R, Marzari N, Kokalj A. http://www.quntum-espresso.org/
9. Perdew J, Burke K, Ernzerhof M (1996) Generalized gradient approximation madesimple. Phys Rev Lett 77:0031–9007
10. Vanderbilt D (1990) Soft self-consistent pseudopotentials in a generalized eigenvalue formalism. Phys Rev 41:7892–7895
11. Meftah C, Iles N, Rabahi L, Gallouze M, Feraoun HI, Drir M (2022) A systematic DFT study of (Ti3/2RE1/2)AlC alloys: a new database for adjustable mechanical and electronic properties
12. Djedid A, Méçabih S, Abbes O, Abbar B (2009) Theoretical investigations of structural, electronic and thermal properties of Ti_2AlX(X=C, N). Physica B 404:3475–3482
13. Du YL, Sun ZM, Hashimoto H (2010) First-principles study on phase stability and compression behavior of Ti_2SC and Ti_2AlC. Physica B 405:720–723
14. Feng W, Cui S (2014) Mechanical and electronic properties of Ti_2AlN and Ti4AlN3: a first-principles study. Can J Phys 1652
15. Duong T, Gibbons S, Kinra R, Arroyave R (2011) Ab-initio approach to the electronic, structural, elastic, and finite-temperature thermodynamic properties of Ti2AX (A= Al or Ga and X = C or N). J Appl Phys 110:093504
16. Barsoum MW (2010) Ceramics science and technology, vol 2. Properties. ISBN: 978-3-527-31156-9
17. Hill R (1952) The elastic behaviour of a crystalline aggregate. Proc Phys Soc A 65:349
18. Hug G, Jaouan M, Barsoum MW (2005) X-ray absorption spectroscopy, EELS, and full-potential augmented plane wave study of the electronic structure of Ti_2AlC, Ti_2AlN, Nb_2AlC, and $(Ti_{0.5}Nb_{0.5})_2$ AlC. Phys Rev B 71:024105
19. Du YL, Sun ZM, Hashimoto H, Barsoum MW (2009) Theoretical investigations on the elastic and thermodynamic properties of $Ti_2AlC_{0,5}N_{0,5}$. Phys Lett A 374:78–82
20. Radovic M, Ganguly A, Barsoum MW (2008) Elastic properties and phonon conductivities of Ti_3Al $(C_{0.5}, N_{0.5})_2$ and Ti_2Al $(C_{0.5}, N_{0.5})$ solid solutions. J Mater Res 23:1517
21. Hettinger JH, Lofland SE, Finkel P, Meehan T, Palma J, Harrel K, Gupta S, Ganguly A, El-Raghy T, Barsoum MW (2005) Electrical transport, thermal transport, and elastic properties of M_2AlC (M = Ti, Cr, Nb, and V). Phys Rev B 72:115120
22. Aryal S, Sakiidja R, Ouyang L, Ching WY (2015) Elastic and electronic properties of Ti_2Al(CxN1-x) solid solutions. J Eur Ceram Soc 35:3219–3227
23. Lin Z, Zhuo M, Li M, Wang J, Zhou Y (2007) Synthesis and microstructure of layered-ternary Ti_2AlN ceramic. Scr Mater 56:1115
24. Wang S, Zhao Y, Du S, Zhao X, Zhang S, Wang R, Lu X, Guan C, Zhao Z (2023) Trends in structural, electronic and elastic properties of Ti_2AC (A= Al, Zn, Cu, Ni) based on late transition metals substitution
25. Cai L, Huang Z, Hu W, Hao S, Zhai H, Zhou Y (2017) Fabrication, mechanical properties and tribological behaviors of Ti_2AlC and $Ti_2AlSn_{0.2}C$ solid solutions. J Adv Ceram 6:90–99
26. Yu W, Jia W, Guo F, Ma Z, Zhang P, Tromas C, Brunet VG, Kent PRC, Sun W, Dubois S (2020) The correlation between N deficiency and the mechanical properties of the Ti_2AlNy MAX Phase. J Eur Ceram Soc 40:2279–2286
27. Blanco M, Francisco E, Luana V (2004) GIBBS: isothermal-isobaric thermodynamics of solids from energy curves using a quasi-harmonic Debye model. Comput Phys Commun 158(57):00104655

28. Rabahi L, Alili B, Bradai D, Grosdidier T, Kellou A (2017) DFT calculations of structural, magnetic and thermal properties of C15, C14 and C36 Laves phases in Fe-Nb-Zr. Intermetallics 83:92–100

Impact of MPPT Technique in Hybrid Photovoltaic-Wind Sources Connected to the Grid on the AC-DC Load Power Stability

Fethi Khlifi and Jamel Belhadj

Abstract This article discusses the proposal of hybrid systems utilizing various Renewable Energy Sources (RES), such as wind and solar energy conversion, to enhance system efficiency while reducing generating costs and emissions. These hybrid systems are connected through microgrids, providing a practical and efficient way to integrate renewable energy sources and ensure power supply in diverse scenarios. To achieve high reliability and efficiency, the article employs the P&O Maximum Power Point Tracking (MPPT) method for both Photovoltaic (PV) and Wind-Turbine systems. Simulation results using MATLAB / Simulink demonstrate the impact of climatic conditions on the dynamic behavior of the microgrid.

Keywords Renewable energy sources · Wind turbine · PV system · MPPT · Micro-grid

1 Introduction

Due to the inherent instability and unpredictability of Renewable Energy Sources (RES), innovative solutions are required to maintain grid stability and reliability. RES adopt various strategies to supply loads at any given moment or draw from energy storage systems, making microgrids an attractive option for enhancing grid dependability and stability in small areas [1]. By utilizing microgrids based on RES, communities can generate green power independently from the main grid, reducing costly grid connection expenses and cable installations while contributing to overall system stability [2].

F. Khlifi (✉) · J. Belhadj
Université de Tunis, ENSIT, BP 56, 1008 Tunis, Tunisie
e-mail: fethikhlifi@gmail.com

Université de Tunis El Manar, LSE-ENIT, 1002 Tunis, Tunisie

© The Author(s), under exclusive license to Springer Nature Singapore Pte Ltd. 2024

A. Mellit et al. (eds.), *Proceedings of the 2nd International Conference on Green Energy Conversion System*, Green Energy and Technology,
https://doi.org/10.1007/978-981-97-6148-7_55

Recent advancements in power electronic converters and high-performance controllers have allowed for the integration of different types of RES into microgrids [3]. Several converter designs and control methods have been developed specifically to incorporate wind and solar electricity [4]. In this paper, two different RES technologies are examined: PV solar systems and wind turbines.

PV solar systems offer benefits such as easy installation, minimal maintenance, and no moving parts. On the other hand, wind turbines are more cost-effective, especially when deployed extensively. However, they require skilled staff for operation and maintenance. Combining both technologies in a RES can provide continuous electricity in various scenarios [5].

Maximizing energy generation is crucial due to the varying output of RES throughout the day, affected by factors like irradiation and wind speed [6]. Effective control systems must adjust to these changing parameters to ensure the microgrid operates at Maximum Power Point Tracking (MPPT). Wind turbines' speed must be adjusted for different wind speeds, while the output DC voltage and current of PV array systems must also be modified to achieve MPPT.

For RES integrated with the grid, various converters are employed. PV systems use DC-DC boost converters to regulate the PV panels' volatile output voltage and DC-AC converters to generate the required voltage and frequency for grid connection. The wind turbine in this study is based on a Permanent Manganite Synchronous Generator (PMSG) and requires an AC-DC-AC converter due to the intermittent nature of wind energy [7].

2 Modeling of Hybrid System

The study explores the potential advantages of integrating photovoltaic and wind turbines in hybrid power generation systems compared to standalone PV or wind energy systems [8]. The research focuses on investigating the characteristics of wind and solar energy, as well as load considerations, within a microgrid context. A combined microgrid simulation is developed to assess the integration of wind and solar energy resources. The microgrid configuration consists of two buses: an AC bus and a DC bus, as illustrated in Fig. 1. The AC bus is connected to an alternating current source, while the DC bus accommodates sources providing direct current. This arrangement offers increased overall system efficiency by reducing the number of converters needed. This configuration is widely adopted due to its flexibility in combining various power sources and loads [9, 10].

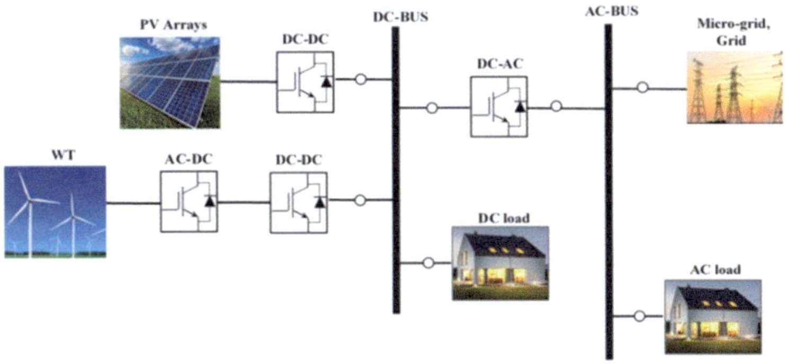

Fig. 1 PV/Wind system

2.1 Photovoltaic Source

The PV generator converts sunlight into direct current (DC). To achieve the required voltage and power levels, multiple solar cells are connected in parallel or series [11]. A DC-DC step-up converter is employed to obtain the necessary DC bus voltage. The maximum power generation of the PV solar system can be achieved by utilizing the MPPT algorithm approach. Various control techniques can be applied to achieve the maximum output power in PV systems. In this study, the P&O approach is utilized, where the operating point is on the right side of the MPPT when the P/V ratio is negative and vice versa. The tracking system adjusts its position based on whether the P/V ratio is negative or positive until power and voltage remain constant [12]. The power output of the solar cell is determined by the following equation:

$$P_{PV} = V \times I \tag{1}$$

2.2 Wind Turbine Modeling

The major contributors to wind energy production are three-phase Permanent Magnet Synchronous Generators (PMSGs) designed with adjustable pitch wind technology. These generators are renowned for their high efficiency and reliability. The generated AC power is then converted into DC electricity using a three-phase uncontrolled rectifier. The wind turbine system based on PMSG with Maximum Power Point Tracking (MPPT) techniques is depicted in Fig. 2, which also requires a step-up converter to elevate the voltage and obtain the required DC bus [13].

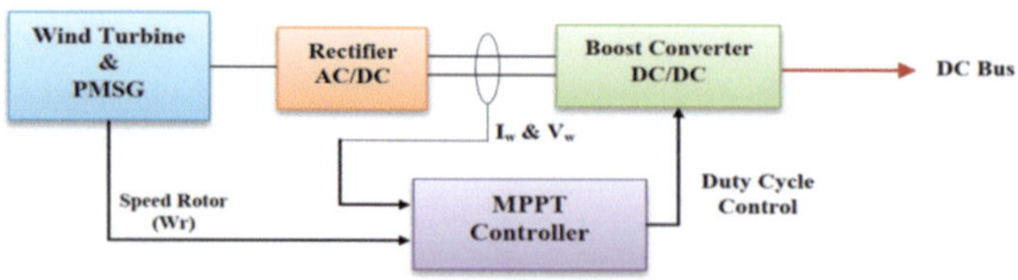

Fig. 2 PMSG wind turbine configuration with MPPT method

The mechanical energy is produced from the wind turbines after converted from wind energy, and their kinetic energy may be estimated using the Eq. (2) below [14]:

$$P_{wind} = \frac{1}{2} \times \rho \times C_P(\lambda, \beta) \times A \times V^3 \tag{2}$$

where "ρ" represents air density in (kg/m), "A" represents the total area produced by the blades or wings of the turbine in (m^2) at rotating, and "V" represents the wind speed (m/sec). The energy conversion efficiency of a wind turbine has been demonstrated to be an execution of the tip speed ratio, which can be defined also as "Cp" which represents the wind-turbine function of the tip speed ratio λ and it can be given in the next expression [15]:

$$\lambda = \frac{w \times r}{V} \tag{3}$$

where "w" represents the speed of the rotational turbine blades in (rad/sec) and "r" represent the radius of the area that is generated due to the rotating of the turbine blades.

C_P represents the power coefficient and it can be defined as the percentage of mechanical energy transferred to electrical energy received by a wind turbine and is computed using the formula below:

$$Cp\ (\lambda, \beta) = 0.5 \times \left[\frac{116}{\lambda_i} - (0.4 \times \beta) - 5\right] \times e^{\frac{21}{\lambda_i}} + 0.0068\lambda \tag{4}$$

$$\frac{1}{\lambda_i} = \frac{1}{(\lambda + 0.08 \times \beta)} - \frac{0.035}{1 + \beta^3} \tag{5}$$

where "β" is the pitch angle "λ" is the tip speed ratio, during the MPPT algorithm.

2.3 *MPPT for Wind Turbine and PV Systems*

Solar cells are responsible for converting solar energy into electrical power. The efficiency of solar panels is optimized when they are positioned perpendicular to the sun's rays. To further enhance power production, trackers are employed to rotate

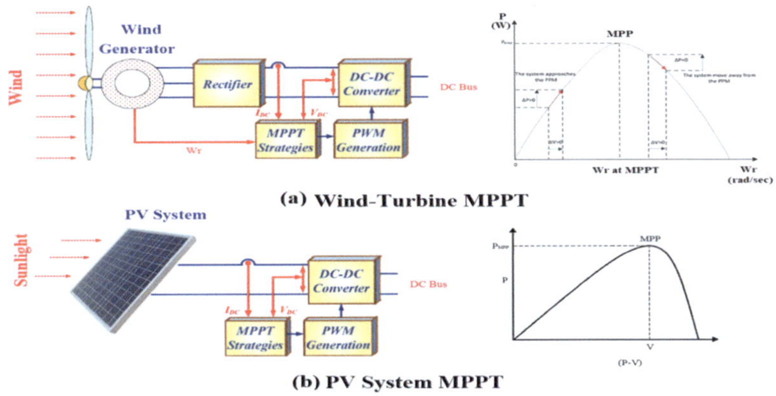

Fig. 3 **a** PMSG wind turbine, and **b** PV system MPPT configurations

the PV arrays and align them directly with the sun's rays. This approach has the advantage of elevating the PV arrays to maximize power generation. The tracking method improves the duration of peak power production, and a similar methodology is applied to the wind turbine system to boost its output power. In this research, the Maximum Power Point Tracking (MPPT) is achieved using the (P&O) algorithm, whose operational principle is depicted in Fig. 3 [16].

3 Simulation Results

The microgrid consists of a PV-generating system equipped with a farm of 15*5 "brightwatts BWI-60-M200" panels, producing approximately 15 kW of power at a solar irradiance of 1000 W/m2. Additionally, a PMSG wind turbine generates 18 kW of power when the wind speed reaches 10 m/s. The wind turbine is connected to a rectifier and DC-DC step-up converter, while the PV system is linked to a DC-DC step-up converter to increase the voltage to 800 at the DC bus voltage. To convert the DC power into AC power with a voltage of 380 AC volts and a frequency of 50 Hz, a three-phase voltage source inverter is employed. Detailed parameters of the proposed PV and wind turbine systems can be found in Tables 1 and 2.

At t = 7 s of the simulation, the planned input irradiation in Fig. 4a is adjusted from 1000 to 800 W/m^2, while the temperature is kept constant at 25 °C. The voltage and current of the PV system in the proposed setup, utilizing the P&O MPPT method, are displayed in Fig. 4b, while the corresponding output power is shown below.

Table 1 PMSG parameters

Parameters	Values
Rated power (kW)	18
Rated wind Speed (m/s)	10
Wind turbine radius (m)	4.75
Number of pair poles	8
Stator resistance (Ω)	0.5
Stator inductance (H)	0.0068
PMSG voltage (V)	318 V
Permanent magnet flux (Φ)	2.4

Table 2 PV parameters

Parameters	Values
Rated power of the PV system (kW)	15
Rated power of each PV panel (W)	200
No. of series panels	15
No. of parallel panels	5
Voltage of each PV panel (V)	28.9
Current of the each PV panel (A)	6.93
Total voltage of PV system (V)	144.5
Total current of the PV system (A)	103.9

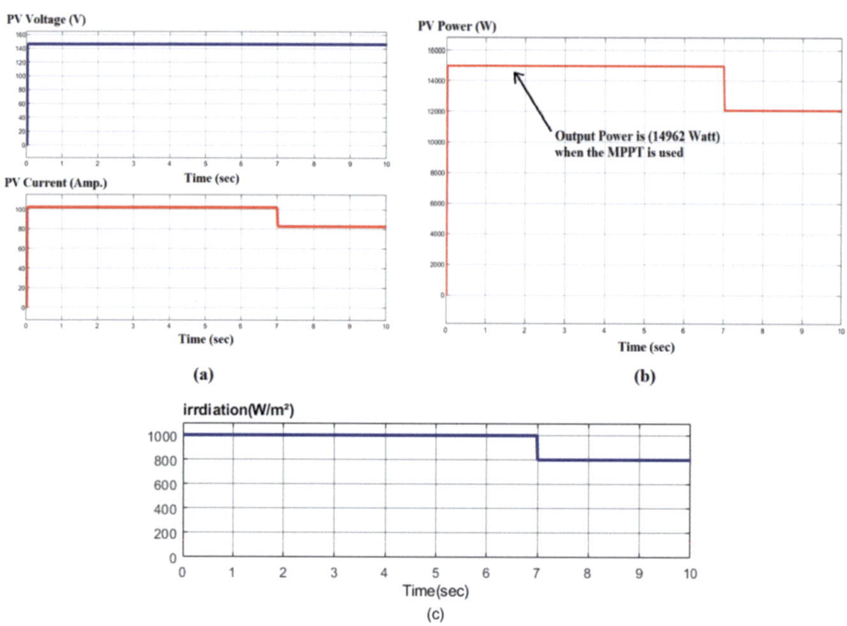

Fig. 4 **a** PV output voltage and current, **b** PV output power and **c** irradiation

The efficiency of the PV system when the MPPT algorithm is used can be given in the following equation:

$$\eta_{(P\&O)} = (\text{Actual Maximum output Power})/(\text{Maximum rated power}) * 100\% \quad (6)$$

$$\eta_{(P\&O)} = (14.962)/(15\text{ kW}) * 100\% = 99.74\%$$

The most significant and trustworthy wind energy generators are three-phase PMSGs (Permanent Magnet Synchronous Generators) operating on controlled pitch wind, ensuring high reliability and efficiency. To convert the generated AC power into DC power, a three-phase uncontrolled rectifier is employed. It is essential to position the operating point of the wind turbine system on the curve's peak between "Cp" and "λoptimal." This ensures that the Maximum Power Point Tracking (MPPT) is achieved, guaranteeing the wind turbine generator operates at its maximum power level, as illustrated in Fig. 5.

The proposed PMSG (Permanent Magnet Synchronous Generator) exhibits a tip speed ratio (λ) of approximately 8.1 and a power coefficient (Cp) of around 0.48, as shown in Fig. 5. The maximum power output is attained when the blade angle or pitch angle is set to zero ($\beta = 0$), as assumed in this study. The estimated tip speed ratio of the suggested PMSG wind turbine, obtained using MATLAB program, is presented in Fig. 6a, while the power coefficient is shown in Fig. 6b.

The maximum power point tracking (MPPT) for the wind turbine is reached at a wind speed of 10 m/s (beyond t = 3 s), as illustrated in Fig. 6. Figure 7 displays the input/output characteristics of the wind turbine, including wind speed, electrical torque, pitch angle, output power, turbine speed in rad/sec, and three-phase voltage.

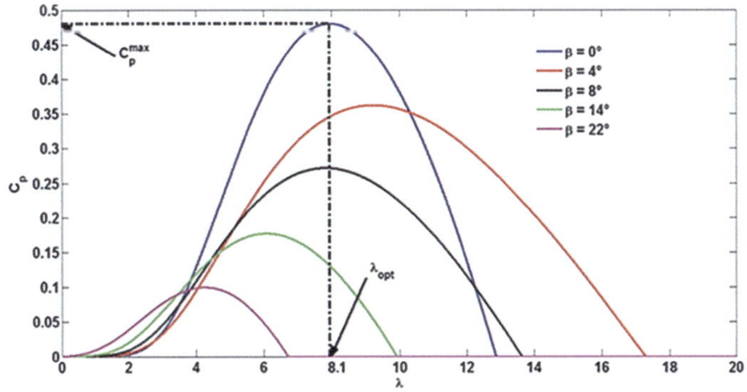

Fig. 5 λ versus Cp at different pitch angles

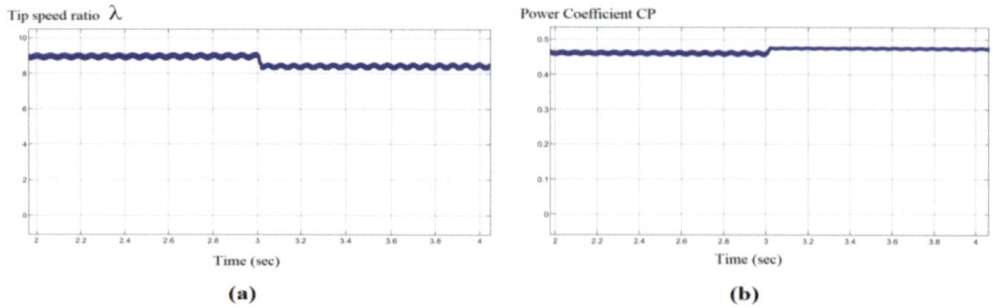

Fig. 6 **a** Tip speed ratio, and **b** power coefficient using MATLAB program

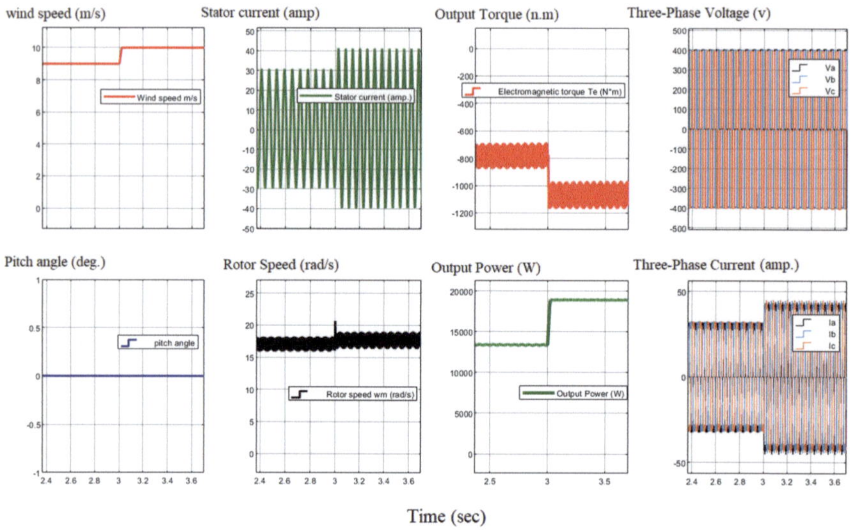

Fig. 7 The input/output characteristics of the wind turbine

The article presents two loads, DC and AC loads. The DC load remains constant at approximately 5 kW (Fig. 8a) and is connected to the DC bus as depicted in Fig. 1. On the other hand, the AC load is considered variable, ranging from 20 to 25 kW. The voltage and current of the AC load are illustrated in Fig. 8b throughout the simulation. To mitigate harmonics, the study utilizes an LC filter, resulting in a total harmonic distortion of less than 2.5% in both voltage and current waveforms.

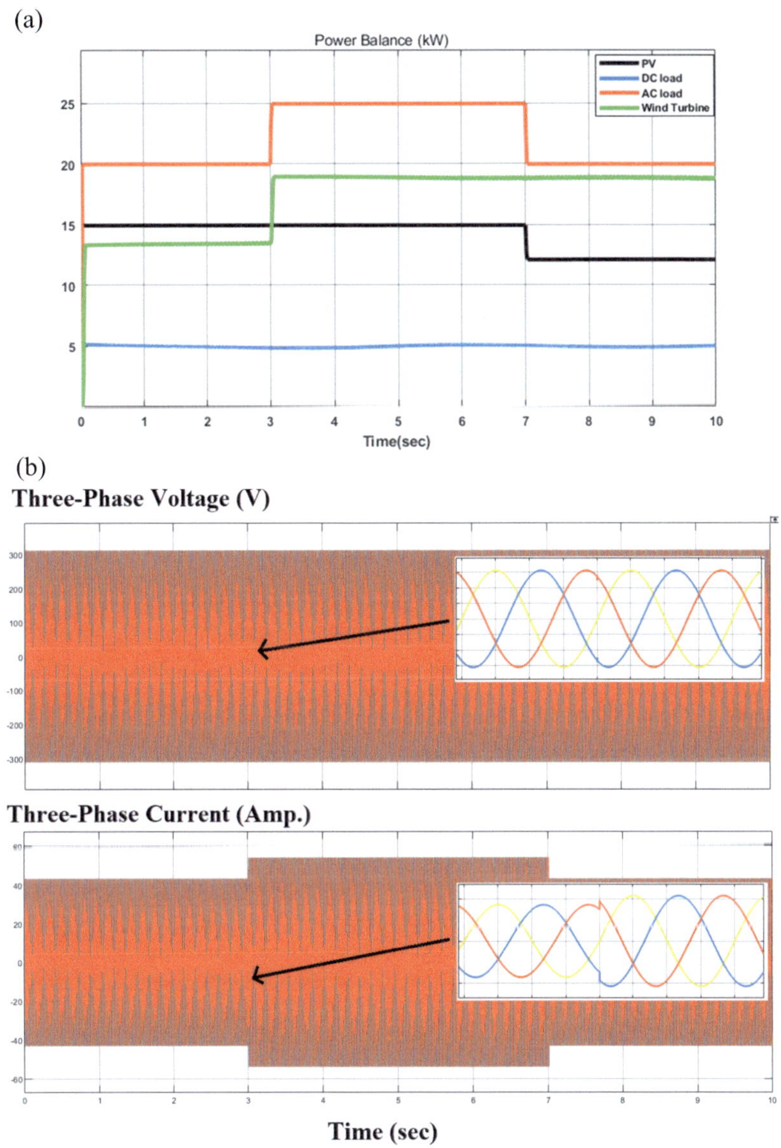

Fig. 8 **a** Different powers, **b** three-phase voltage and current at load side

4 Conclusion

This research proposes the implementation of two converters to optimize the output of PV and wind systems in a microgrid. These converters control the DC bus voltage system through PV-boost output control and PMSG-boost output regulation using the

P&O MPPT method. To ensure consistent simulation results, the study recommends varying irradiation, wind speed, and loads.

By applying the MPPT method, the photovoltaic and wind power systems achieve high efficiency. Any excess energy from the PV and wind sources will be fed into the public network when $P_{PV} + P_{WT} > P_{DC} + P_{AC}$. Conversely, if $P_{PV} + P_{WT} < P_{DC} + P_{AC}$, the network will compensate for the deficit to meet the load demand.

References

1. Barber S, Boller S, Nordborg H (2019) Feasibility study for 100% renewable energy microgrids in Switzerland. Wind Energ Sci. https://doi.org/10.5194/wes-2019-97
2. Bogdanov D, Ram M, Aghahosseini A, Gulagi A, Oyewo AS, Child M, Caldera U, Sadovskaia K, Farfan J, De Souza Noel Simas Barbosa L, Fasihi M, Khalili S, Traber T, Breyer C (2021) Low-cost renewable electricity as the key driver of the global energy transition towards sustainability. Energy. https://doi.org/10.1016/j.energy.2021.120467
3. Maurya R, Prakash S, Singh AK (2023) Challenges, configuration, control, and scope of DC microgrid systems: a review. J Electr Eng Technol 18:1655–1674. https://doi.org/10.1007/s42835-022-01304-y
4. Pathak G, Singh B, Panigrahi BK, Chandra A, Al-Haddad K (2016) Wind-PV based microgrid and its synchronization with utility grid. In: IEEE international conference on power electronics, drives and energy systems (PEDES), Trivandrum, India, pp 1–6. https://doi.org/10.1109/PEDES.2016.7914334
5. Hemeida AM, El-Ahmar MH, El-Sayed AM, Hasanien HM, Alkhalaf S, Esmail MFC, Senjyu T (2020) Optimum design of hybrid wind/PV energy system for remote area. Ain Shams Eng J 11(1):11–23. https://doi.org/10.1016/j.asej.2019.08.005
6. Qasim MA, Vladimir I V (2021) PWM effect on MPPT for hybrid PV solar and wind turbine generating systems at various loading conditions. Period Eng Nat Sci 9(2):581–592. https://doi.org/10.21533/pen.v9i2.1840
7. Zishan F, Tightiz L, Yoo J, Shafaghatian N (2023) Sustainability of the permanent magnet synchronous generator wind turbine control strategy in on-grid operating modes. Energies 16:4108. https://doi.org/10.3390/en16104108
8. Saleh AS, Antar RK, Ali AJ (2022) Design and simulation of hybrid renewable energy system for on-grid applications. In: IMDC-IST 2022, EAI. https://doi.org/10.4108/eai.7-9-2021.2314947
9. Roy P, He J, Zhao T, Singh YV (2022) Recent advances of wind-solar hybrid renewable energy systems for power generation: a review. IEEE Open J Ind Electron Soc 3:81–104. https://doi.org/10.1109/OJIES.2022.3144093
10. Chen C, Liu H, Xiao Y, Zhu F, Ding L, Yang F (2022) Power generation scheduling for a hydro-wind-solar hybrid system: a systematic survey and prospect. Energies 15:8747. https://doi.org/10.3390/en15228747
11. Alkassoum N, Seibou B, Mahamadou O, Soumaïla NT (2020) Comparative study of the models of the photovoltaic cell (1 diode & 2 diode) and efficiency optimization. J Sci Eng Res 7(11):110–124
12. Salman S, Ai X, Wu Z (2018) Design of a P-&-O algorithm based MPPT charge controller for a stand-alone 200W PV system. Prot Control Mod Power Syst 3:25. https://doi.org/10.1186/s41601-018-0099-8.
13. Zebraoui O, Bouzi M (2018) Comparative study of different MPPT methods for wind energy conversion system. In: International conference on renewable energies and energy efficiency, IOP conference series: earth and environmental science, vol 161, p 012023. https://doi.org/10.1088/1755-1315/161/1/012023

14. Jai Andaloussi Z, Raihani A, Elmagri A, Lajouad R (2022) Toward the optimization of a PMSG wind energy conversion system on-grid by a robust mixed controller. E3S Web Conf 336:00043. https://doi.org/10.1051/e3sconf/202233600043
15. Tran CH, Nollet F, Essounbouli N, Hamzaoui A (2018) Maximum power point tracking techniques for wind energy systems using three levels boost converter. In: 2018 7th international conference on clean and green energy-ICCGE 2018, IOP conference series: earth and environmental science, vol 154, p 012016. https://doi.org/10.1088/1755-1315/154/1/012016
16. Rekioua D, Rekioua T, Elsanabary A, Mekhilef S (2023) Power management control of an autonomous photovoltaic/wind turbine/battery system. Energies 16:2286. https://doi.org/10.3390/en16052286

New Control Hybrid Power

Achwak Alazrag and Lassaad Sbita

Abstract This paper proposed a method for simulating, modeling, and controlling a hybrid grid-connected power system—a system that is really rather complicated. We research effective ways to use two renewable energy sources without interfering with the main network. The photovoltaic source and the wind source are the two renewable energy sources that make up our hybrid energy system (HES). In order to recoup the system investment cost, it is preferable for these two systems to operate at full capacity. Where a fuzzy logic algorithm-controlled DC/DC boost converter comes after the solar generator (PVG). The induction motor utilized in the wind system serves as a variable speed generator that is directly linked to the turbine. Realizing that our study does not take pitch angle control into account. In spite of changes in wind speed, the MPPT control technique, which is based on the ideal peak speed ratio, is employed to achieve the highest level of energy efficiency. The Point of Common Collecting (PCC) is used to link the two systems. Finally, the results of the simulation demonstrate the viability of our method and the successful implementation of the suggested control technique for medium-sized HES in the production of electricity.

Keywords Hybrid system · Wind turbine · Photovoltaic · DC/DC · PI control · Induction motor

A. Alazrag (✉) · L. Sbita
Research Laboratory of Process, Energies, Environment and Electrical Systems, National Engineering School of Gabes, University of Gabes, Zrig Eddakhlania, Gabes 6029, Tunisia
e-mail: achwaklazrag@gmail.com

L. Sbita
e-mail: lassaad.sbita@enig.rnu.tn

© The Author(s), under exclusive license to Springer Nature Singapore Pte Ltd. 2024

A. Mellit et al. (eds.), *Proceedings of the 2nd International Conference on Green Energy Conversion System*, Green Energy and Technology,
https://doi.org/10.1007/978-981-97-6148-7_56

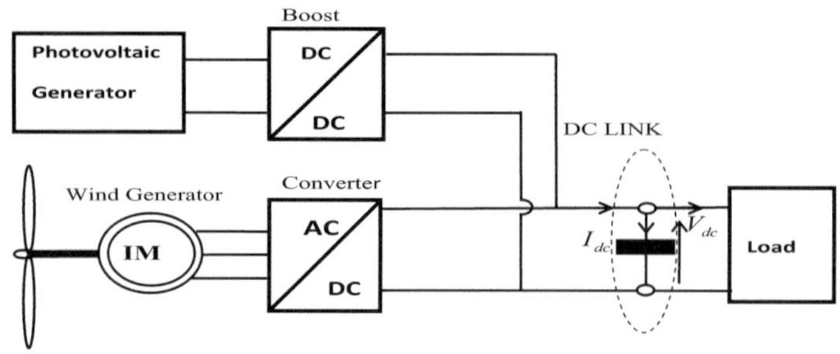

Fig. 1 Hybrid system PV wind

1 Introduction

The usage of renewable energy has increasingly become one of the fastest-growing and cleanest energy sources due to the crisis of the depletion of conventional energy sources and the environmental harm brought on by an increase in carbon dioxide emissions [1, 2]. The most promising methods for producing electricity are solar and wind power, respectively Solar radiation and wind, on the other hand, are sporadic and strongly reliant on climatic factors [3–5]. Renewable and traditional energy sources are combined in a hybrid energy system (HES). HES is useful since it responds to the demand of the AC loads and lowers the system's total cost. The suggested remedy is to use trackers to connect HES to the load through a DC bus.

In order to reduce the cost of the system, our analyzed HES system is defined by two energy sources without a storage system. The DC bus will connect these two power sources to the load. The fuzzy logic algorithm will be used to control the DC/DC converter, which follows the photovoltaic generator (PVG) in Fig. 1. MPPT (Maximum Power Point Tracking) is a methodology to control where many academics have focused on the implementation of this approach. The induction motor (IM), which is part of the wind turbine, is responsible for realizing the decoupled current control concept, which regulates both active and reactive power while maintaining a constant DC bus voltage [1, 3]. The Control Field-Oriented (FOC) is used to control the speed of the IM [6].

2 Hybrid System

2.1 PV Generator

Solar cells are arranged in series and parallel to create a PV array. Cells connected in series provide higher output voltages whereas those connected in parallel boost current [4]. The terminal voltage and current are denoted by V and I. The ideal PV cell's equivalent circuit, which includes series resistance (Rs) and shunt resistance (Rsh) [5].

Then deduce, the coordinates of photovoltaic generator:

$$\begin{cases} V_{pv\_pvG} = \frac{1}{2}.\frac{n_s.I_{ph}}{I_s}.V_T \\ I_{pv\_pvg} = \frac{n_p.I_{ph} - \frac{n_p.V_{pv\_PVG}}{n_s.V_T}.I_s}{1 + \frac{n_p.I_s.R_s}{n_s.V_T}} \\ P_{pv\_pvG} = \frac{1}{2}\frac{n_s.I_{ph}}{I_s}.V_T.\frac{n_p.I_{ph}.n_s.V_T - n_p.V_{pv\_PVG}.I_s}{n_s.V_T + n_p.I_s.R_s} \end{cases} \tag{1}$$

2.2 *DC-DC Converter with MPPT Tracking*

A DC-DC converter receives a DC input voltage and outputs a desired DC voltage. The voltage level of the output and the input are different. Additionally, DC-DC converters are utilized to regulate power buses and offer noise isolation [7].

Transformers are typically used to increase voltage, however they do have certain losses. To get around these issues and provide the necessary output, a DC-DC converter is employed. As a high frequency switch, it comprises of an inductor, a capacitor diode, and an IGBT. This sort of configuration allows for a higher voltage power supply to the load. The output voltage varies depending on the switch's duty cycle.

a/[0 to DT]

The inductor generates a magnetic field when the switch is closed, and current flows through it in a clockwise direction, storing some energy. The inductor's left side has positive polarity.

$$V_{pv} = L\frac{dI_{pv}}{dt} \tag{2}$$

b/[DT to T]

As the impedance increases when the switch is opened, current will decrease. To keep the current flowing toward the load, the energy of the magnetic field that was previously generated will be diminished. As a result, the polarity will be inverted, making the inductor's left side negative. Due to the series connection of the two sources, a larger voltage will be used to charge the capacitor through the diode.

$$V_{pv} = L\frac{dI_{pv}}{dt} + V_{dc} \tag{3}$$

With, D: The duty cycle.

Boost Control With Fuzzy logic control

Maximum Power Point Tracking (MPPT) is a technique used in photovoltaic (PV) systems to extract the maximum power available from a solar panel. MPPT algorithms are typically used in charge controllers or inverters to continuously adjust the operating parameters of the PV system to maintain the panel's output at its maximum power point (MPP) under changing environmental conditions.

MPPT, fuzzy logic can be applied to enhance the performance of the tracking algorithm by considering various input variables and their relationships. Here's a general overview of how fuzzy logic can be incorporated into an MPPT system [8]:

Fuzzification: The crisp input variables are converted into fuzzy values using membership functions associated with each linguistic variable. The membership functions determine the degree of membership of the input variables to each fuzzy set.

Inference Engine: The inference engine applies the fuzzy rules to the fuzzified input variables to generate fuzzy output values. The inference engine evaluates the antecedents of the rules and calculates the degree of fulfillment of each rule.

Defuzzification: The fuzzy output values obtained from the inference engine are then defuzzified to obtain a crisp output value. Defuzzification methods like centroid, mean of maximum, or weighted average can be used to determine the final output.

MPPT Adjustment: The crisp output value obtained from defuzzification represents the required adjustment to be made in the PV system's operating parameters, such as the duty cycle of a DC-DC converter. The MPPT controller uses this adjustment to maintain the PV system at its maximum power point (Fig. 2).

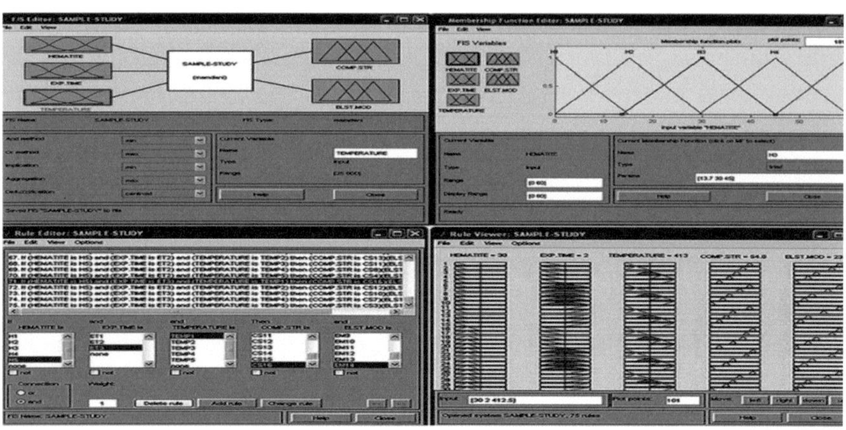

Fig. 2 Fuzzy logic steps

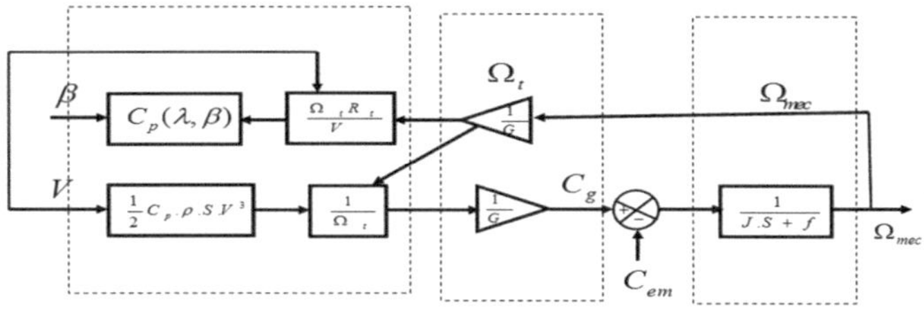

Fig. 3 Mechanical shaft modeling

2.3 *Wind Turbine*

The wind turbine is coupled to the shaft of asynchronous machine through a gear box that is inserted to adapt the low rotational speed of the wind turbine to the high speed of the generator. The IM is connected to the power grid through two back to-back converters [2, 3] (Fig. 3).

The power captured by a wind turbine is given by:

$$P_{wt} = \frac{1}{2}\rho S v^3 \tag{4}$$

where ρ is the air density, R is the radius of the blades, Cp is the wind turbine power coefficient, and u is the wind speed. For each blade pitch angle β, the value of the tip-speed ratio λ_{opt} is constant for all MPPs. the mechanical power obtained from wind turbine is expressed in Eq. (2):

$$P_m = P_{wt} C_P(\lambda, \beta) = \frac{1}{2} C_P(\lambda, \beta)\rho R^2 v^3 \tag{5}$$

The mechanical speed appearing on the turbine rotor can represented by:

$$\Omega_{em} = \frac{C_g - C_{em}}{J.s + f} \tag{6}$$

With:

$$C_P(\lambda, \beta) = 0.5176\left(\frac{116}{\lambda_v} - 0.4\beta - 5\right).e^{-\frac{21}{\lambda_v}} + 0.0068\lambda_v$$

$$\lambda_v = \frac{R_t.\Omega_t}{V_v}$$

The induction motor is characterized by the voltage and current equations [9]:

$$\begin{bmatrix} V_{as} \\ V_{bs} \\ V_{cs} \end{bmatrix} = \begin{bmatrix} R_s & 0 & 0 \\ 0 & \mathrm{R_s} & 0 \\ 0 & 0 & \mathrm{R_s} \end{bmatrix} \begin{bmatrix} I_{as} \\ I_{bs} \\ I_{cs} \end{bmatrix} + \frac{d}{dt} \begin{bmatrix} \phi_{as} \\ \phi_{bs} \\ \phi_{cs} \end{bmatrix}$$
$$\begin{bmatrix} V_{ar} \\ V_{br} \\ V_{cr} \end{bmatrix} = \begin{bmatrix} R_s & 0 & 0 \\ 0 & \mathrm{R_s} & 0 \\ 0 & 0 & \mathrm{R_s} \end{bmatrix} \begin{bmatrix} I_{ar} \\ I_{br} \\ I_{cr} \end{bmatrix} + \frac{d}{dt} \begin{bmatrix} \phi_{ar} \\ \phi_{br} \\ \phi_{cr} \end{bmatrix} \quad (7)$$

And:

$$\begin{bmatrix} [\phi_s] \\ [\phi_r] \end{bmatrix} = \begin{bmatrix} [L_s] & [M_{sr}] \\ [M_{sr}] & [L_r] \end{bmatrix} \begin{bmatrix} [I_s] \\ [I_r] \end{bmatrix} \quad (8)$$

Jgikholhjljljlk

$$\begin{bmatrix} i_{ds} \\ i_{dr} \\ i_{qs} \\ i_{qr} \end{bmatrix} = \frac{1/(L_s.L_r)}{\sigma_{sr}} \begin{bmatrix} L_r & -\mathrm{M_{sr}} & L_r & -\mathrm{M_{sr}} \\ -\mathrm{M_{sr}} & \mathrm{L_s} & -\mathrm{M_{sr}} & L_r \end{bmatrix} \begin{bmatrix} \phi_{\mathrm{ds}} \\ \phi_{\mathrm{dr}} \\ \phi_{\mathrm{qs}} \\ \phi_{\mathrm{qr}} \end{bmatrix} \quad (9)$$

The torque can be expressed as:

$$P = v_{ds} i_{ds} + v_{qs} i_{qs} \quad (10)$$

$$Q = v_{ds} i_{ds} - v_{qs} i_{qs} \quad (11)$$

3 Control Strategy

The control system employed in the examined setup incorporates two primary loops: an inner PI control loop and an outer control loop. The inner loop employs PI controllers to regulate the d- and q-axis current components. On the other hand, the outer control loop comprises two controllers: the Maximum Power Point Tracking (MPPT) controller and another PI controller responsible for power factor regulation.

The MPPT controller plays a crucial role in determining the optimal Iqs current to maximize the mechanical power transfer from the wind turbine to the electric generator. Its objective is to track the maximum power point, ensuring efficient utilization of the available wind energy.

Simultaneously, the second PI controller within the outer loop is responsible for regulating the power factor. It guarantees the provision of the required reactive power to the load, maintaining the desired power factor levels as per system requirements.

4 Simulation and Result

Vvmoy = 14 m/s.

The system has been exposed to a real wind speed profile, in order to observe the behavior of the SCIG and its control. These variations can be seen in the Fig. 4.

Figure 5 provides a graphical depiction of the progressive variations in irradiation, current, and voltage levels of the DC bus. As the irradiation levels increase, there is a simultaneous increase in both the current and voltage. Notably, it is observed that throughout the duration of the measurements, the DC bus maintains a stable voltage level of 500 V.

The simulation is started to monitor several signals related to the modified speed of the motor when a change in wind speed is introduced (Fig. 6).

Similarly, our focus was on tracking the temporal progression of the electromagnetic torque generated by an induction machine and its coupled aero turbine. This

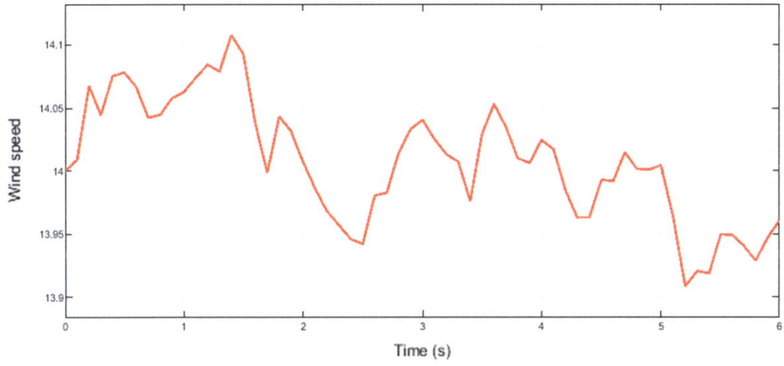

Fig. 4 Evolution of a wind speed

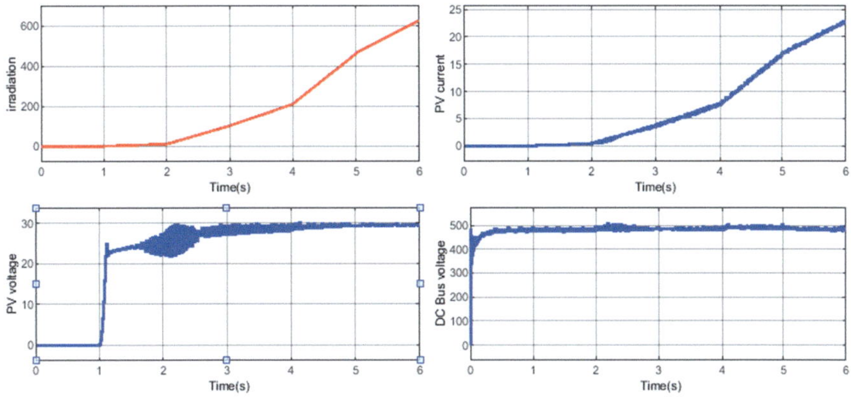

Fig. 5 PV characterized + DC bus voltage

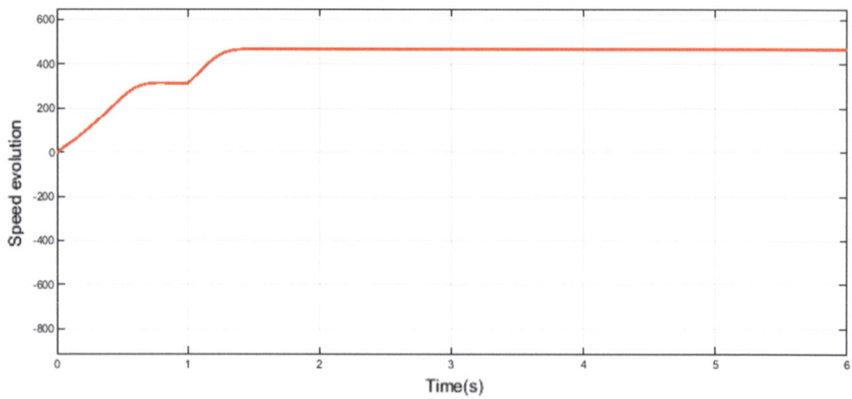

Fig. 6 Speed evolution of induction motor

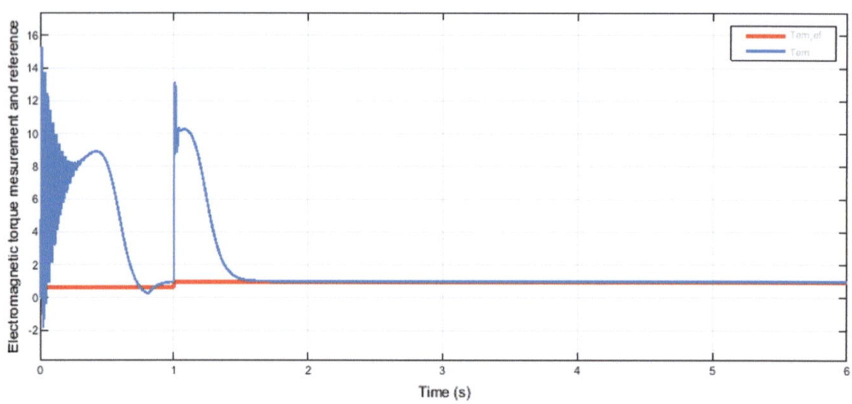

Fig. 7 Electromagnetic torque

investigation involved monitoring and analyzing the changes in torque over time. The corresponding results can be observed in Fig. 7.

The Fig. 8 is shown the evolution of the stator d axis current and their reference value. At the specific time point of 1.5 s, there is an occurrence of confusion or misalignment between the actual current values (Ids) and the reference current values (Ids ref). This discrepancy or mixing up of values may have implications for the control or regulation of the system being observed.

The active power and reactive are presented in Fig. 9.

The reactive is equal to zero and active power equal to1750W.

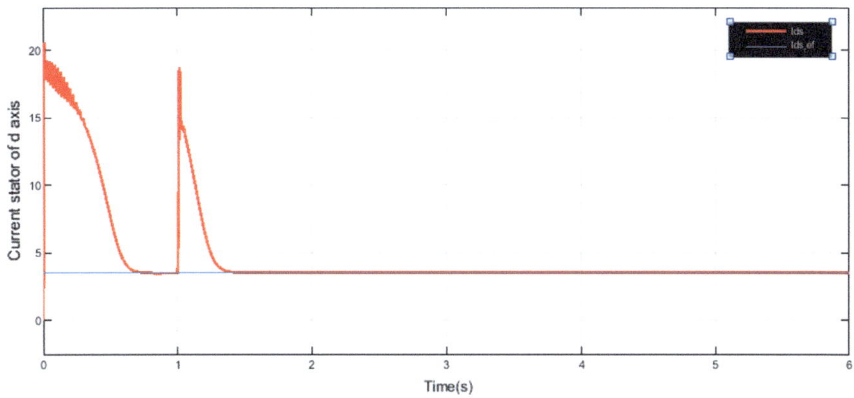

Fig. 8 Stator current of d axis

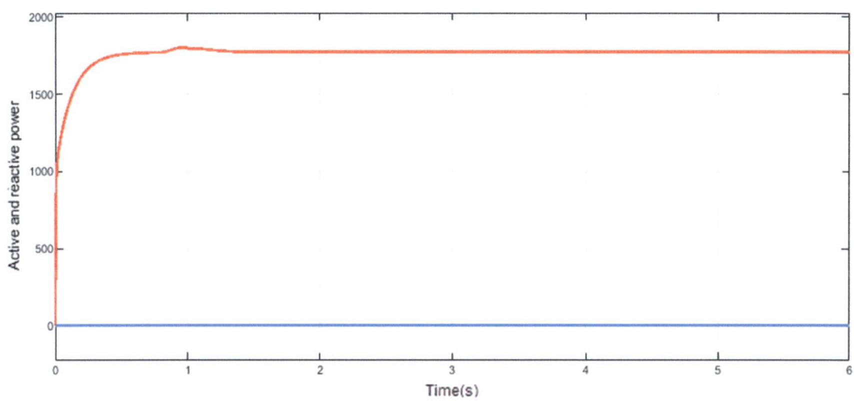

Fig. 9 Active and reactive power

5 Conclusion

The current effort focuses on the simulation and control of SEH coupled to the load, which is the most recent use of renewable energy. Two systems PV and Wind systems linked to the load through DC Bus were the only systems included in our analysis. Simulation and control methodologies are given.

The simulation findings demonstrate that the two PV and wind systems' active powers are efficiently augmented and injected into the load. Despite abrupt changes in solar radiation and wind speed, the combined reactive power of the two systems remains constant at zero. Despite the fact that this DC Bus design is more expensive than the DC Bus architecture, we chose it because of how easy it is to order.

References

1. Zigha, A (2013) étude et simulation d'un système hybride photovoltaïque-éolien» mémoire de master de l'université Constantine I.
2. Rouas R, Etude d'un système hybride éolien-photovoltaïque (Doctoral dissertation, Université de Tizi Ouzou-Mouloud Mammeri)
3. Gergaud, O (2002) Modélisation énergétique et optimisation économique d'un système de production éolien et photovoltaïque couplé au réseau et associé à un accumulateur (Doctoral dissertation, École normale supérieure de Cachan-ENS Cachan)
4. Djamila Rekioua TR (2015) Control of a grid connected photovoltaic system. In: 4th international conference on renewable energy research and applications. Palermo, Italy
5. Colin H, Duvauchelle C, Moine G, Tanguy Y, Gaiddon B, Tran-Quoc T (2010) Raccordement des installations photovoltaïques au réseau public de distribution électrique à basse tension: cadre réglementaire, impacts et préconisations. Projet ESPRIT, Avril 2010
6. Alazrag A, Sbita L (2022) Asynchronous machine with wind turbine IRFO control. WSEAS Trans Power System 17:141–154
7. Tigrine C, Ouali OA (2018/2019) Etude et simulation des techniques MPPT d'un système photovoltaïque. République Algérienne Démocratique et Populaire Ministère de l'Enseignement Supérieur et de la Recherche Scientifique UniversitéA. MIRA-BEJAIA
8. Ghabara C, Jouini H, Lahbib M, Mami A (2017) Stratégie d'optimisation floue du point d'extraction de la puissance maximale (MPPT) pour les petites éoliennes connectées à des sites isolés. In: International conference on green energy and environmental engineering, vol 5, pp 48–54
9. Sahraoui M (2013) Commandes non linéaires de la machine asynchrone (Doctoral dissertation, Université Mohamed Boudiaf M Sila)

Power Control Strategy for a SCIG Wind Turbine Generator

Achwak Alazrag and Lassaad Sbita

Abstract This research paper presents a new approach to examine the behavior of a grid connected Squirrel Cage Induction Generator (SCIG) system, both in steady state and dynamic condition. The SCIG which is driven by a wind turbine via a gearbox, is connected to the distribution system through Power Electronics Converters (PECS), filters, and step-up transformers. The main focus of this paper is on the vector strategy employed for the SCIG, which allows independent control of flux and electromagnetic torque on the PWM converter, while achieving maximum power point tracking (MPPT). The grid side converter plays a crucial role in regulating the DC link voltage, injecting active power through d-axis current, and controlling reactive power injection through q-axis current. The effectiveness of the proposed method is demonstrated through simulation results. The simulation of the entire system are conducted within the MATLAB/Simulink environment, and the obtained results are considered satisfactory.

Keywords Wind turbine · Electrical grid · Back to back converter · Fuzzy control · SCIG

1 Introduction

The global demand for electrical energy has significantly increased in recent decades due to population growth and industrialization, particularly in emerging countries [1].

Moreover, the rising cost of energy is attributed to the depletion of non-renewable resources used in power generation plants. To address these challenges, renewable energies, such as green wind energy, have emerged as viable solutions for sustainable

A. Alazrag (✉) · L. Sbita
Research Laboratory of Process, Energies, Environment and Electrical Systems, National Engineering School of Gabes, University of Gabes, 6029 Zrig Eddakhlania, Gabes, Tunisia
e-mail: achwaklazrag@gmail.com

L. Sbita
e-mail: lassaad.sbita@enig.rnu.tn

© The Author(s), under exclusive license to Springer Nature Singapore Pte Ltd. 2024

A. Mellit et al. (eds.), *Proceedings of the 2nd International Conference on Green Energy Conversion System*, Green Energy and Technology,
https://doi.org/10.1007/978-981-97-6148-7_57

electricity production [2]. However, wind speed variations, especially during gusts, pose a major issue for wind energy resources. These fluctuations impose substantial mechanical stresses on the system, which can be mitigated by utilizing asynchronous machines, like induction generators, as opposed to synchronous generators that operate at fixed speeds. Consequently, the use of variable speed wind turbines, employing induction generators (SCIG), has gained prominence over fixed-speed wind turbines.

This paper focuses on the study and description of a new model for SCIG wind turbines, aiming to analyze their steady-state and dynamic behavior.

The proposed system comprises a SCIG driven by a wind turbine through a gearbox. The SCIG's output terminals are connected to a distribution system via a step-up transformer, connection cables, and power electronic converters (PECs) with appropriate controls [2, 3]. To simulate the steady-state and dynamics of the SCIG wind turbine system, a comprehensive mathematical model is developed, encompassing variables such as wind speed, wind turbine, gearbox, SCIG, DC-link voltage, PECs, step-up transformer, and the grid system. The model is implemented using SIMULINK software. The structure of the paper is as follows: Sect. 2 introduces the mathematical models employed in the proposed SCIG wind farm system, which connects the wind turbine set to the distribution system through PECs and connection lines. Section 3 presents the control strategies employed, specifically utilizing fuzzy controllers. Section 4 provides dynamic simulation results of the analyzed wind turbine. Finally, Sect. 5 presents the key conclusions drawn from this study.

2 Squirrel Cage Induction Generator SCIG

The wind turbine is coupled to the shaft of a SCIG through a gear box that is inserted to adapt the low rotational speed of the wind turbine to the high speed of the generator. The SCIG is connected to the power grid through back to-back converter.

The power captured by a wind turbine is given by:

$$P_{wt} = \frac{1}{2}\rho S v^3 \tag{1}$$

where ρ is the air density, R is the radius of the blades, Cp is the wind turbine power coefficient, and u is the wind speed. For each blade pitch angle β, the value of the tip-speed ratio λ_{opt} is constant for all MPPs. the mechanical power obtained from wind turbine is expressed in Eq. (2):

$$P_m = P_{wt} C_P(\lambda, \beta) = \frac{1}{2} C_P(\lambda, \beta)\rho R^2 v^3 \tag{2}$$

The mechanical torque appearing on the turbine rotor can therefore be subsequently represented by:

$$T_m = \frac{1}{2\Omega_t} C_P(\lambda, \beta)\rho\Pi R^2 v^3 \tag{3}$$

With:

$$\begin{aligned} C_P(\lambda, \beta) &= 0.5176\left(\frac{116}{\lambda_v} - 0.4\beta - 5\right) \cdot e^{-\frac{21}{\lambda_v}} + 0.0068\lambda_v \\ \lambda_v &= \frac{R_t.\Omega_t}{V_v} \end{aligned} \tag{4}$$

The SCIG is characterized by the voltage and current equations:

$$\begin{cases} V_{ds} = R_s i_{ds} + \frac{d\phi_{ds}}{dt} - \omega_s \phi_{qs} \\ V_{qs} = R_s i_{qs} + \frac{d\phi_{qs}}{dt} + \omega_s \phi_{ds} \\ 0 = R_r i_{dr} + \frac{d\phi_{dr}}{dt} - \omega_r \phi_{qr} \\ 0 = R_r i_{qr} + \frac{d\phi_{qr}}{dt} + \omega_r \phi_{dr} \end{cases}$$
$$With: \tag{5}$$
$$\begin{cases} \phi_{ds} = L_s i_{ds} + M_{sr} i_{dr} \\ \phi_{qs} = L_s i_{qs} + M_{sr} i_{qr} \\ \phi_{dr} = L_r i_{dr} + M_{sr} i_{ds} \\ \phi_{qr} = L_r i_{qr} + M_{sr} i_{qs} \end{cases}$$

And:

$$\begin{bmatrix} i_{ds} \\ i_{dr} \\ i_{qs} \\ i_{qr} \end{bmatrix} = \frac{1/(L_s \cdot L_r)}{\sigma_{sr}} \begin{bmatrix} L_r & \mathrm{M_{sr}}\, L_r & -\mathrm{M_{sr}} \\ -\mathrm{M_{sr}}\, \mathrm{L_s} & -\mathrm{M_{sr}}\, L_r & \end{bmatrix} \begin{bmatrix} \phi_{\mathrm{ds}} \\ \phi_{\mathrm{dr}} \\ \phi_{\mathrm{qs}} \\ \phi_{\mathrm{qr}} \end{bmatrix} \tag{6}$$

The torque can be expressed as:

$$P = v_{ds} i_{ds} + v_{qs} i_{qs} \tag{7}$$

$$Q = v_{ds} i_{ds} - v_{qs} i_{qs} \tag{8}$$

3 Control Strategy

For normal operation of the AC/DC and the DC/AC of the studied wind power generation system, the DC link voltage Vdc must be properly maintained at a constant value under all operation conditions. A constant DC voltage for Vdc indicates a balanced condition of the active power flow between the AC/DC and the DC/AC. To achieve this balanced condition, the DC/AC is designed to control the DC-link voltage to ensure that the energy collected by the AC/DC is transmitted to the distribution system. The AC/DC is assigned to control the active power generated by the SCIG.

The control system of the studied setup consists of two main loops: an inner fuzzy control loop and an outer control loop. The inner loop utilizes fuzzy controllers to regulate the d- and q-axis current components. This regulation is essential for generating the corresponding d- and q-axis voltages through space vector modulation. The outer control loop comprises two controllers: the Maximum Electric Loss (MEL) controller and the Maximum Power Point Tracking (MPPT) controller. The MEL controller calculates the optimal Ids current to minimize the electric loss of the generator. On the other hand, the MPPT controller determines the optimal Iqs current to maximize the mechanical power transferred from the wind turbine to the electric generator.

For the line-side inverter, a voltage-oriented control strategy is employed using a double loop control system. Similar to the inner loop of the generator-side control, an inner fuzzy control loop regulates the d and q-axis current components to generate the corresponding d- and q-axis voltages through space vector modulation. The outer fuzzy control loop ensures decoupled control of the active and reactive power delivered to the grid. Specifically, one fuzzy controller in the outer loop maintains a constant DC bus voltage, there by controlling the active power. The other fuzzy controller regulates the power factor on the line side, ensuring the provision of the required reactive power to the grid [4, 5] (Fig. 1).

3.1 *Space Vector Modulation (SVM)*

The Space Vector Modulation component utilizes a two level converter shown. With this converter, there are eight possible states. The states are shown in Table 1 along with the Park's transformation of these states.

$$\begin{cases} V_{AN} = [2L_1 - L_2 - L_3]\frac{V_{dc}}{3} \\ V_{BN} = [-L_1 + 2L_2 - L_3]\frac{V_{dc}}{3} \\ V_{CN} = [-L_1 - L_2 + 2L_3]\frac{V_{dc}}{3} \end{cases} \quad (9)$$

Also:

$$\begin{cases} V_\alpha = \sqrt{\frac{2}{3}}\left[V_{AN} - \frac{1}{2}V_{BN} - \frac{1}{2}V_{CN}\right] \\ V_\beta = \sqrt{\frac{2}{3}}\left[\frac{\sqrt{3}}{2}V_{BN} - \frac{\sqrt{3}}{2}V_{CN}\right] \end{cases} \quad (10)$$

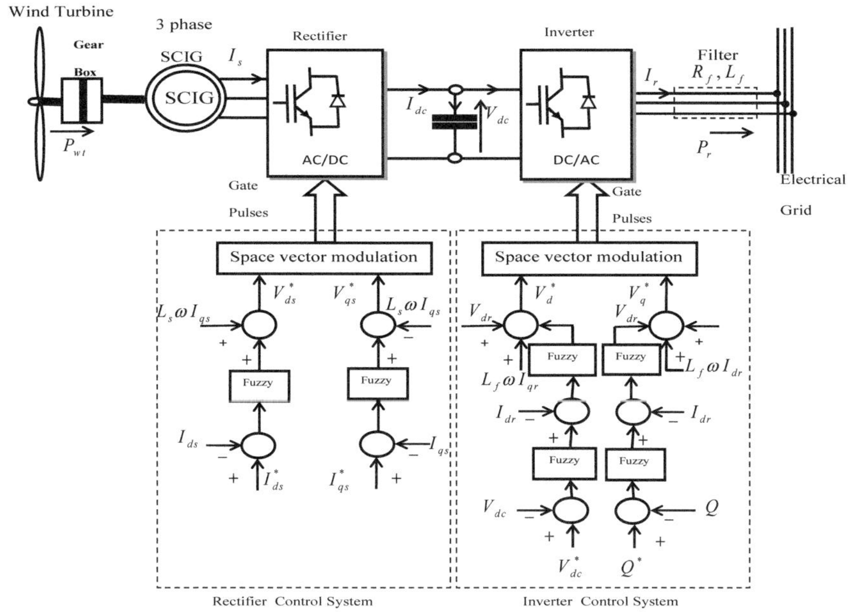

Fig. 1 Fuzzy control block diagram

Table 1 State of space vector modulation

L1	L2	L3	V_{AN}	V_{BN}	V_{CN}	V_α	V_β
1	1	1	0	0	0	0	0
1	0	0	$\frac{2V_{dc}}{3}$	$\frac{-V_{dc}}{3}$	$\frac{-V_{dc}}{3}$	$\sqrt{\frac{2}{3}}V_{dc}$	0
1	1	0	$\frac{V_{dc}}{3}$	$\frac{V_{dc}}{3}$	$\frac{-2V_{dc}}{3}$	$\sqrt{\frac{1}{6}}V_{dc}$	$\sqrt{\frac{1}{2}}V_{dc}$
0	1	0	$\frac{-V_{dc}}{3}$	$\frac{2V_{dc}}{3}$	$\frac{-V_{dc}}{3}$	$-\sqrt{\frac{1}{6}}V_{dc}$	\\$\sqrt{\frac{1}{2}}V_{dc}$
0	1	0	$\frac{-2V_{dc}}{3}$	$\frac{V_{dc}}{3}$	$\frac{V_{dc}}{3}$	$-\sqrt{\frac{2}{3}}V_{dc}$	0
0	0	1	$\frac{-V_{dc}}{3}$	$\frac{-V_{dc}}{3}$	$\frac{2V_{dc}}{3}$	$-\sqrt{\frac{1}{6}}V_{dc}$	$-\sqrt{\frac{1}{2}}V_{dc}$
1	0	1	$\frac{V_{dc}}{3}$	$\frac{-2V_{dc}}{3}$	$\frac{V_{dc}}{3}$	$\sqrt{\frac{1}{6}}V_{dc}$	$-\sqrt{\frac{1}{2}}V_{dc}$
0	0	0	0	0	0	0	0

3.2 DC Bus Control

The DC power is:

$$P_{dc} = I_{dc} \cdot V_{dc} \tag{11}$$

Then:

$$P_{dc} = \frac{1}{2} C \frac{dV_{dc}^2}{dt} \tag{12}$$

So:

$$V_{dc}(n) = \left[V_{dc}^2(n-1) + \frac{2 \cdot Te}{C} \left(P(n) - P_g(n) \right) \right]^{1/2} \tag{13}$$

C: The DC bus capacitor.

Te: sampling period.

The control strategy is divided into two blocks the first is for the calculation of the DC B us voltage and the second is reserved for the control (Fig. 2).

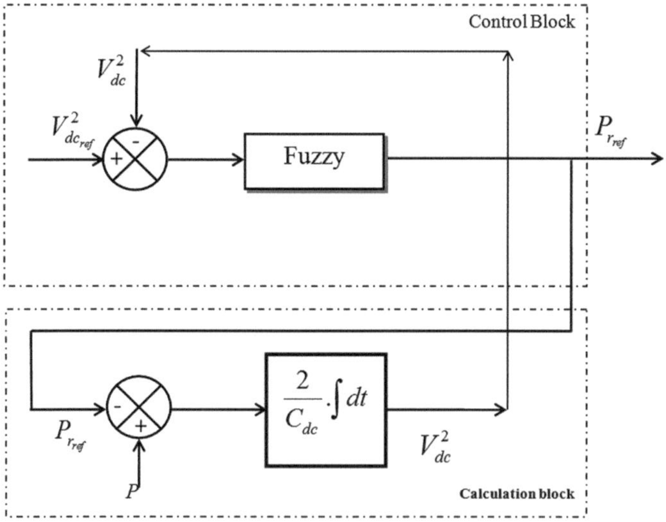

Fig. 2 Quadratic control of DC bus

4 Simulation and Result

$$\text{Vvmoy} = 10\ \text{m/s}$$

Is the mean component of the considered wind speed.

g(t)Is a gust superimposed on this average component.

The system has been exposed to a real wind speed profile, in order to observe the behavior of the SCIG and its control. These variations can be seen in the Fig. 3.

When a modification in the wind speed is introduced, the simulation is initiated to observe the signals related to the "Wind turbines" field. The control of active and reactive power, alternator speed, and wind turbine pitch angle is monitored. As the simulation progresses, the generated active power starts to gradually increase until it reaches its nominal value. At the beginning of the simulation, the wind turbine blade angle is set to zero degrees. It is noticeable that the absorbed reactive power increases in tandem with the increase in generated active power. The time evolution of the mechanical speed of an induction machine in the wind farm is depicted in Fig. 4.

In the same way, we were interested in the follow-up of the temporal evolution of the electromagnetic Torque developed successively by an induction machine and the aero turbine which is coupled to it, Fig. 5.

Figure 6 successively represent the evolution of the DC bus voltage and their reference level as well as the DC power the Dc bus voltage remains constant over the reference value. It is important to remark that the error is lower than 0.001%. From

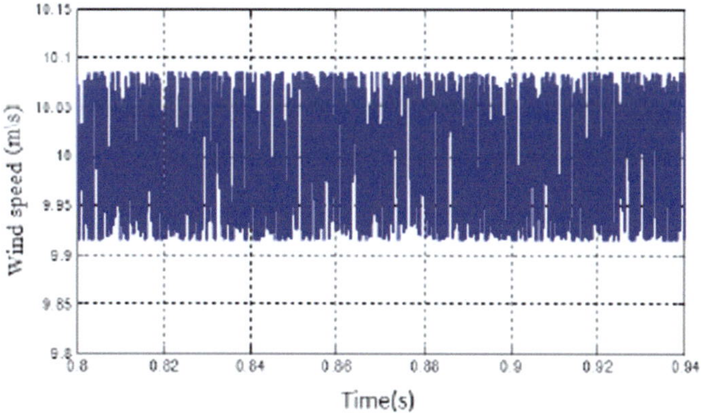

Fig. 3 Evolution of a wind speed

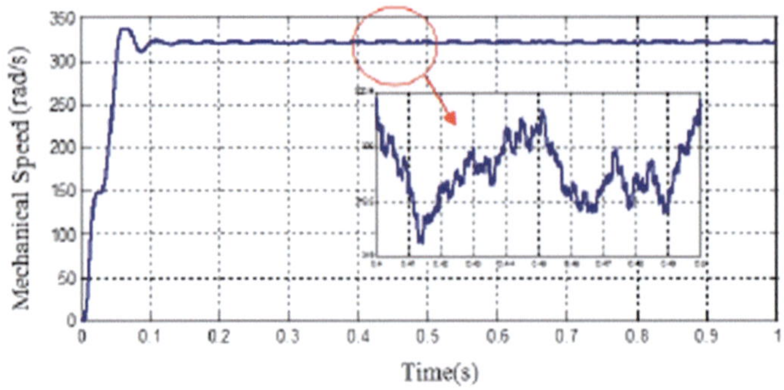

Fig. 4 Reference and real Mechanical speed evolution

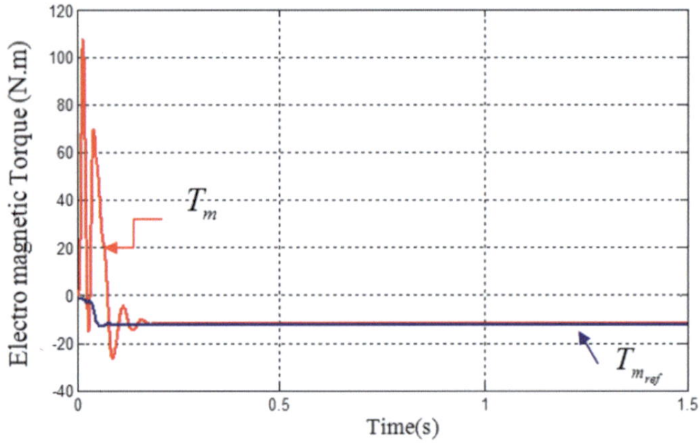

Fig. 5 Electromagnetic Torque

Fig. 6 it can be concluded that the Dc bus is able to transfer all power delivered by the stator-side converter (Figs. 7 and 8).

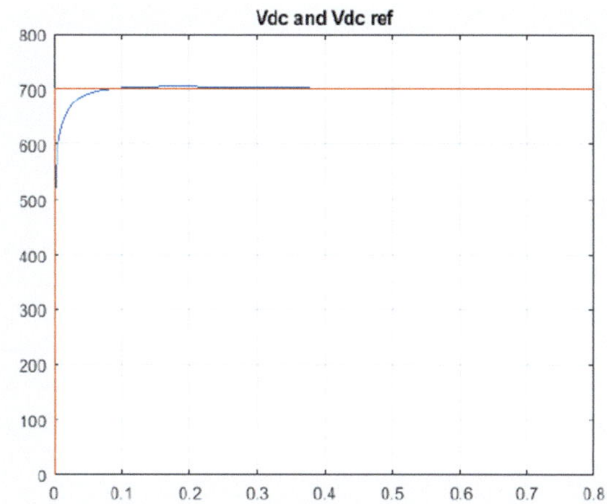

Fig. 6 The DC voltage measurement and reference

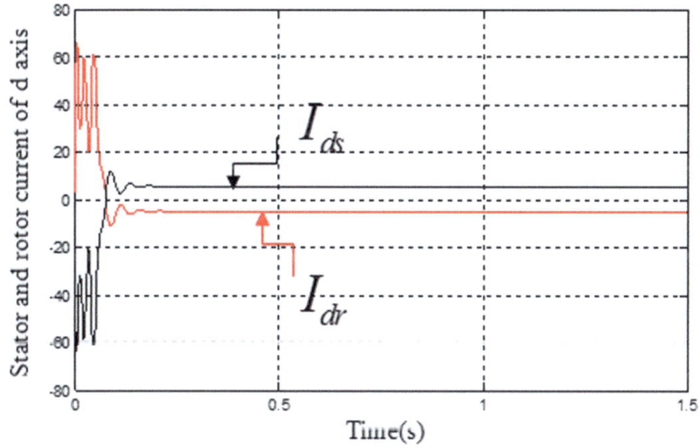

Fig. 7 Stator current (d, q) axis

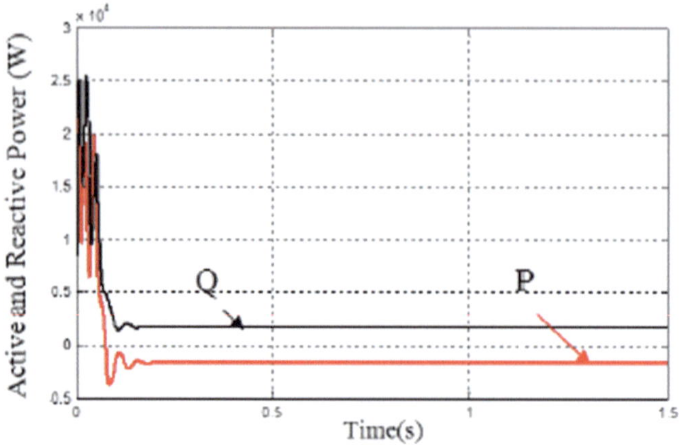

Fig. 8 Active and reactive power

5 Conclusion

This paper presents a control technique in order to deal with a SCIG connected to the grid through a full power converter. Control strategy for variable speed wind turbine with a SCIG is completely studied in this paper. A vector control strategy for the generator side converter and grid side converter to extract and exchange maximum power are discussed and implemented using MATLAB/Simulink. From the simulation results, the proposed model and controls can be used to study and simulate wind turbine based on SCIG.

References

1. Pahlevaninezhad, Safaee A, Eren S, Bakhshai A, Jain P (2009) Adaptive nonlinear maximum power point tracker for a WECS based on permanent magnet synchronous generator fed by a matrix converter. In: Energy conversion congress and exposition, ECCE 2009. IEEE, pp 2578–2583
2. Abo-Khalil G, Kim HG, Lee DC, Seok JK (2004) Maximum output power control of wind generation. Proc IEEE Int Conf IECON, 1676–1681
3. Abo-Khalil G (2011) Model-based optimal efficiency control of induction generator for wind power systems. Proc IEEE ICIT, pp 191–197
4. Mesemanolis A, Mademlis C, Kioskeridis I (2012) High-efficiency control for a wind energy conversion system with induction generator. IEEE Trans Energy Convers 27(4):958–967
5. Frédéric POITIERS Etude et commande de génératrices asynchrones pour l'utilisation de l'énergie éolienne. Machine asynchrone à cage autonome. Machine asynchrone à double alimentation reliée au réseau thèse doctorale, école polytechnique de l'Université de Nantes, soutenue le 19 décembre 2003

Modeling of BEM Type Shell and Tube Heat Exchanger Using the BELL-DELAWARE Method

Marwa Mami, Ali Fguiri, and Mohamed Razak Jeday

Abstract The shell-and-tube heat exchangers are widely used in various energy conversion systems. In this work, the modelling of the exchanger type of a Liquefied Petroleum Gas (LPG) production unit was carried out using the Bell Belaware method and the Taborek version by playing on the main geometric parameters such as the transverse flow area and leakage zones. Thus, the results obtained are validated by real exchanger study data. Due to the application of this method to shell and tube heat exchangers equipped with baffles, high precision is attained. The heat transfer coefficients obtained show good agreement with the study data.

Keywords Heat exchanger type E modelling · Bell-Delaware method · Taborek version

1 Introduction

Energy is the foundation of all industrial politics and economics. As a result, process optimization to achieve high yields and better management of available energy has always been the subject of scientific research. Considering all of its advantagess (heat recovery, machine cooling, etc.), the heat exchanger plays a key role in the energy management strategy. In particular, those of the shell and tube type are considered the most frequently used class of heat exchanger design in processes thanks to its excellent pressure resistance, ease of manufacturing, dependability, and inexpensive price. According to the literature, the countercurrent arrangement for some applications is the most efficient, producing the highest temperature variation for each fluid [1].

The improvement of heat transfer is the research focus of the whole world.

M. Mami (✉) · A. Fguiri · M. R. Jeday
Processes, Energetic, Environmental and Electrical Systems Research Laboratory (PEEES), National School of Engineers of Gabes, University of Gabes, Street Omar Ibn El Khattab, 6029 Gabes, Tunisia
e-mail: mami_marwa@live.com

© The Author(s), under exclusive license to Springer Nature Singapore Pte Ltd. 2024
A. Mellit et al. (eds.), *Proceedings of the 2nd International Conference on Green Energy Conversion System*, Green Energy and Technology,
https://doi.org/10.1007/978-981-97-6148-7_58

The Bell-Delaware method created between the years 1960 and 1963 [2] is based on the Tinker method which suggests a subdivision of the shell side flow pattern into a number of individual flows.

In short, according to Palen and Taborek [3] and Bell [4], the Delaware methods gave the best results compared to the other methods available in the literature.

In this article, we develop to calculate the tube side and shell side transfer coefficient for a Type E exchanger of the LPG process based mainly on the Delaware method with Taborek's version.

2 Materials and Method

The procedure presented in this article for calculating the shell and tube side transfer coefficient is based primarily on the Bell-Delaware method. Correction factors are thus used, as in the Delaware method to take into account the deviation of the flow inside the shell. However, the correlations presented in Taborek's version of the Bell-Delaware method are used. The shell-side transfer coefficient obtained was compared with the study data.

The modeling steps are as follows:

1st step: Insertion of the thermo-physical properties of the fluids inlet.
2nd step: Calculation of transfer coefficient on the tube side h_t by application of the Gnielinski correlation.
3rd step: Calculation of shell side transfer coefficient h_s by the Bell-Delaware method and its Taborek version.
4th step: Calculation of global transfer coefficient U.
5th step: Validation of the results.

$$\frac{1}{U} = \frac{1}{h_t} + \frac{1}{h_s} + \frac{e}{\lambda} + R_t + R_s \tag{1}$$

Through its simplicity, the BEM exchanger (Fig. 1) [5] occupies an increasingly important area in the heat exchanger market, this type is generally found in refrigeration cycles as well as petrochemical installations.

The thermo-physical properties of two fluids are shown in Table 1 and the geometric characteristics of this exchanger in Table 2 below.

2.1 Bell-Delaware Method Description

The Bell-Delaware approach is a conventional method known in shell-side transfer coefficient calculations. It is based on many parameters such as the number of tubes, the number and spacing of baffles, and the Reynolds number. In addition, many

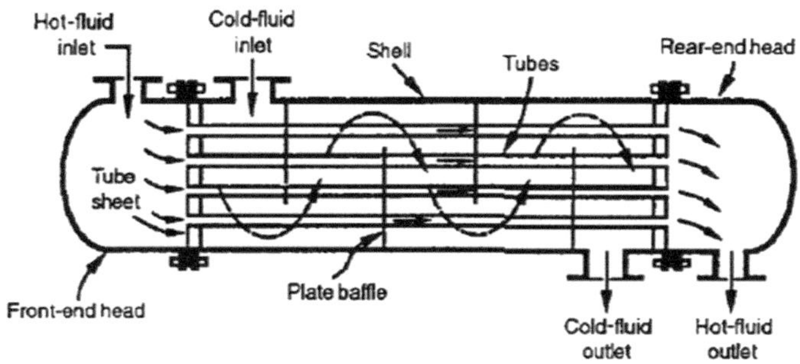

Fig. 1 BEM type heat exchanger with a shell side pass and tube side pass [5]

Table 1 Thermo-physical properties of fluids

Properties	Unit	Cold fluid (tube)	Hot fluid (shell)
Input temperature	°K	305	401
Output temperature	°K	319	328
Flow	kg/s	5.422	1.030
Pressure	bar	32.8	8.6
Specific heat Cp	kJ/kg.K	$26{,}776.10^5$	

Table 2 Geometric properties of the heat exchanger

Description	Unit	Value
Number of shell passes	–	1
Number of tube-side passes	–	1
Outside shell diameter	m	0.219
Internal shell diameter	m	0.203
Outer diameter of tubes	m	0.019
Number of tubes	–	44
Step tubes	m	0.024
Length	m	3.6
Chicane cutting	%	20
Exchange surface	m^2	9.1

factors affecting the complex flow distribution in the shell are given consideration to obtain more accurate results in the Bell-Delaware method [6]. In an ideal heat exchanger where the openings between the shell elements are neglected, there is a single flow between the baffles which are perpendicular to the tubes and follow a path partially parallel to the tubes in the window zones (Fig. 2).

Consequently, the fluid flowing on the tube side is divided into several streams, contrary to the ideal.

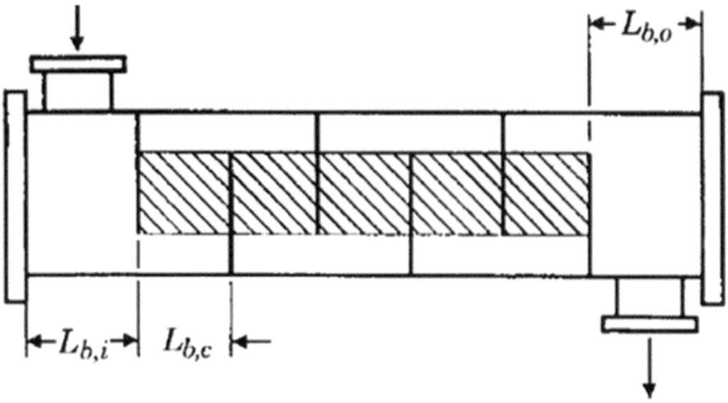

Fig. 2 The input and output sections

In addition, a single type E tube and shell exchanger maybe divided into three sections:

- The input and output sections

The spaces between two baffles at the entrance, center and exit, are defined by L_{bi}, L_{bc}, L_{bo}, respectively, and are shown in the Fig. 3. L_{bi}, L_{bo} are equal with L_{bc}.

- The internal cross-current section (hatched section)

The transverse flow zone S_m exists above the central baffle in the middle of the tube bundle where the fluid flows perpendicularly to the tube bundle and is as in the equation (Appendix 1).

$$S_m = L_{bc} * \left[L_{bb} + \frac{D_{ctl}}{L_{tp}} (L_{tp} - d) \right] \quad (2)$$

With L_{tp} here denotes the arrangement of the tubes (not triangular generally used).

Figure 4 shows the longitudinal cross-section of a heat exchanger made up of mono-segment baffles. As shown, the number of tubes exposed to cross flow between

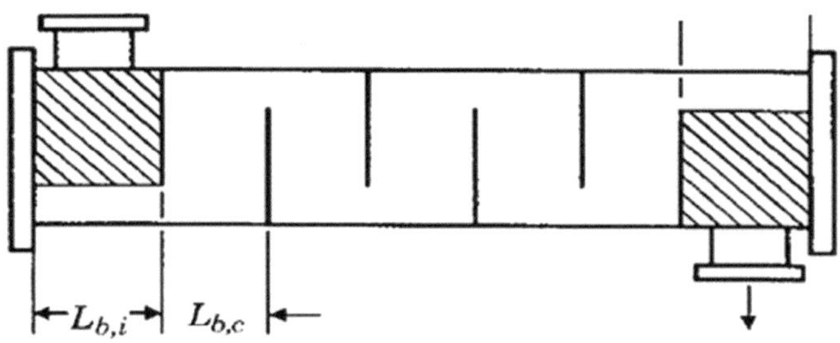

Fig. 3 Crossed current section

the window areas and the flow parallel to that area are needed for the heat transfer calculation. These parameters can be calculated using the equations presented in Appendix 1 [7, 8] (Fig. 5).

- S_{sb} parallel sections

In addition, there is a leakage zone on the plates between the openings of the tube and the outer surfaces, the distance between them being expressed in the form of S_{tb}. Thus, the total leakage area on a baffle element is calculated as in the following equation:

$$S_{tb} = \left\{ \frac{\pi}{4} * \left[(d_e + L_{tb})^2 - d_e^2 \right] \right\} * N_t * (1 - F_w) \tag{3}$$

In the case of unknown values, approximate results are possible with calculations using the standard values provided by TEMA [9]. With the exception of the window

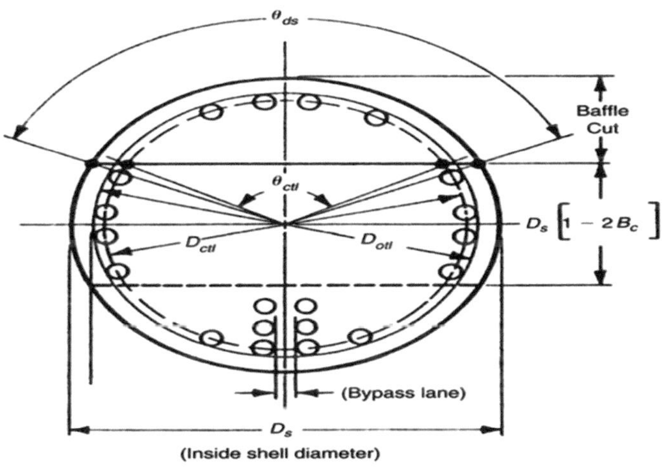

Fig. 4 The current section crossed inside

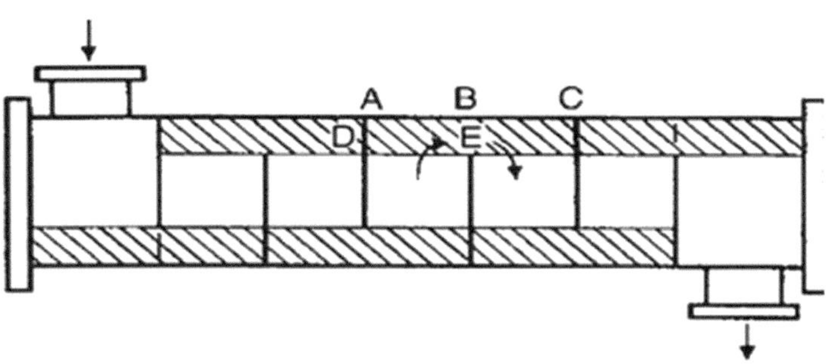

Fig. 5 Parallel sections in a shell and tube heat exchanger

region, the clearance between the outer diameter of the baffle plate and the inner diameter of the calendar forms a leakage flux zone. This area is calculated by the following equation (S_{sb} is the distance between the shell and the baffle) [9].

$$S_{sb} = \frac{D_s * L_{sb} * (2 * \pi - \theta_{ds})}{4} \quad (4)$$

The area S_b between the outside diameter of the tube bundle and the inside diameter of the shell at a central baffle spacing is the area of the bypass flow of the tube bundle and is calculated by the following equation [9].

$$S_b = L_{bc} * \left(D_s - D_{otl} + L_{pl}\right) \quad (5)$$

In this method, in order to achieve the actual values, the heat transfer and shell-side friction coefficients are first calculated for the theoretical ideal cross-flow condition in which the flow is perpendicular to the tube bundle. Then, the coefficients calculated for the ideal case are multiplied by the correction factors determined according to the geometric measurements made in the hull.

2.2 *Calculation of the Shell Side Transfer Coefficient by the Bell-Delaware Method*

To calculate the shell-side transfer coefficient h_s, the Bell-Delaware method and the Taborek's version are used (Appendix 1) [7, 8]:

$$h_s = J_c * J_L * J_r * J_b * J_s * h_i \quad (6)$$

This method takes into consideration various effects that reduce the ideal transfer coefficient h_i on a bank of tubes. Several correction variables are used to adjust for a departure from the ideal flow regime (J_c, J_l, J_r, J_b, J_s) presented in the form of Table 3 which represent the effects and the corresponding correction factors.

Where the ideal transfer coefficient h_i is determined by [10].

Table 3 The shell side transfer coefficient corrective factors h_s

Factors	Influence/impact
J_c	Flow correction factor through baffle window
J_l	Tube/baffle leak correction factor
J_b	Tube bundle flow deviation correction factor
J_r	Factor which depends on the laminar flow regime
J_s	Uneven baffle spacing correction factor

Table 4 Transfer coefficient results

	BEM exchanger
Tube side transfer coefficient h_t (W/m^2 °C)	2786.747
Shell side transfer coefficient h_s (W/m^2 °C)	589.855

$$h_i = \frac{j_i * Cp_{calandre} * G_{calandre} * (\varphi_{calandre})^n}{Pr_{calandre}^{2/3}} \tag{7}$$

For the Colburn J_i factor, two approximation approaches, the Bell approximation (Appendixes 1 and 2) and that of Taborek. Through the adaptation of the Bell-Delaware method to the Taborek version, the basic equation structure is not modified. The adaptation is based on the calculation of characteristic zones and specific numbers of tubes, analogous to the definitions of Taborek's work. The main method is explained in detail previously.

3 Results and Discussion

The Gnielinski correlation for the calculation of the transfer coefficient on the tube side h_t and the Bell-Delaware method for the calculation of the shell side h_s, allow to give the results grouped in Table 4.

These results enables to conclude the global exchange transfer coefficient U for BEM exchanger and to compare with that of study data. The result obtained with the Bell-Delaware method shows a difference from the real values that is rather minor. The calculated value is 473.885 W/m^2. °C while the study data is around 484.74 W/m^2. °C with a deviation of − 2.239% from theory.

The difference obtained maybe due to the geometric parameters whose values are missing and which have therefore been assumed based on the literature, such as the spacing between the L_{bc} baffles and the leakage clearance between the L_{tb} baffles and tubes. Furthermore, the value of the spacing between the L_{bc} baffles (central, inlet and outlet) must be between the minimum $0.2*D_s$ and the maximum 1*Ds (see Appendix 1). Also for the leakage clearance between baffles and tubes L_{tb} must be between 0.397 and 0.794 mm. These values have a great influence on the calculation of the corrective factors of the Bell-Delaware method.

4 Conclusion

In this paper, the BEM shell and tube heat exchanger is modeled by adapting the Bell-Delaware method.

The results are validated by actual LPG plant study data. The model developed for the entire heat exchanger is made up of correlations which take into account all the characteristics of the fluids and the geometry of this equipment. The transfer coefficient on the tube side is calculated by the established Gnielinski correlation and that on the shell side is obtained by adapting the Bell-Delaware method.

In summary, the modeling results show a very good agreement with the real data. In addition, the adapted Bell-Delaware correlation is applicable to other heat exchangers with dual-segment baffles and is an important method for modeling these types.

Appendix 1 Bell-Delaware Method for a Shell and Tube Heat Exchanger

Parameters	Equations
Transfer coefficient h_s	$h_s = J_c{*}J_L{*}J_r{*}J_b{*}J_s{*}h_i$
Ideal transfer coefficient h_i	$h_i = \frac{j_i{*}Cp_{calandre}{*}G_{calandre}{*}(\varphi_{calandre})^n}{Pr_{calandre}^{2/3}}$
Colburn factor j_i	$j_i = a{*}Re_s^b$
Bell's approximations	$a = 1.73\ b = -0.694\ 1 < Res < 100$ $a = 0.717\ b = -0.504\ 100 < Res < 1000$ $a = 0.236\ b = -0.346\ 1000 < Res$
Colburn factor given by Taborek	$j_i = a1{*}\left(\frac{1,33}{\frac{L_{tp}}{de}}\right)^a{*}Re^{a2}$ $a = \frac{a3}{1+0.14{*}Re^{a4}}$ With $a3 = 1.45$ et $a4 = 0.519$ $a1 = 0.321\ a2 = -0.388\ 10000 < Res < 100000$ $a1 = 0.321\ a2 = -0.388\ 1000 < Res < 10000$ $a1 = 0.593\ a2 = -0.477\ 100 < Res < 1000$ $a1 = 1.360\ a2 = -0.657\ 10 < Res < 100$ $a1 = 1.400\ a2 = -0.667\ Res < 10$
Viscosity correction factor	$\varphi_{calandre} = \left(\frac{\mu s}{\mu}\right)^{0.14}$
Reynolds shell side Re	$Re = \frac{G{*}d_e}{\mu}$
Prandt shell side Pr	$Pr = \frac{\mu{*}Cp}{\lambda}$
Mass velocity G	$G = \frac{\dot{m}}{S_m}$

(continued)

(continued)

Parameters	Equations
Passage section S_m	$\mathbf{S_m = L_{bc} * \left[L_{bb} + \frac{D_{ctl}}{L_{tp}}(L_{tp} - d)\right]}$ $L_{bc} = 0.457$ m; $L_{tp} = 0.02381$m $L_{bb} = D_s\text{-}D_{otl}$ $D_{otl} = D_t + D_{ctl}$ $\mathbf{D_{ctl} = \sqrt{\frac{C*Nt}{0.7854}} * L_{tp}}$; C = 0.866 (a triangular step)

Appendix 2 Bell Delaware Method for Calculating Correction Factors J

Parameters	Equations
Factor J_c	$J_c = 0.55 + 0.72*F_c$ $F_c = 1 - 2 * F_w$ $F_w = \frac{\theta_{ctl}}{360} - \frac{\sin\theta_{ctl}}{2*\pi}$ $\theta_{ctl} = 2*\cos^{-1}\left(\frac{D_s}{D_{ctl}}\left[1 - 2*\frac{B_c}{100}\right]\right)$
Factor J_l	$J_l = 0.44*(1 - r_s) + [1 - 0.44*(1 - r_s)]*e^{(-2.2*r_{lm})}$ $r_s = \frac{S_{sb}}{S_{tb}+S_{sb}}$ $S_{sb} = \frac{D_s*L_{sb}*(2*\pi-\theta_{ds})}{4}$ $\theta_{ds} = 2*\cos^{-1}\left[1 - 2*\frac{B_c}{100}\right]$ $S_{tb} = \left\{\frac{\pi}{4}*\left[(d_e + L_{tb})^2 - d_e^2\right]\right\}*N_t*(1 - F_w)$ 0.397 mm $\leq L_{tb} \leq 0.794$ mm $r_s = \frac{S_{sb}+S_{tb}}{S_m}$
Factor J_b	$J_b = \exp^{-C_{bh}*F_{bh}*[1-(1-2*r_{ss})]^{1/3}}$ $C_{bh} = 1.25$ pour Re > 100 $F_{sbp} = \frac{S_b}{S_m}$ With $S_b = L_{bc}*(D_s - D_{otl} + L_{pl})$ $r_{ss} = \frac{N_{ss}}{N_{tcc}}$ $N_{tcc} = \frac{D_s}{L_{pp}}\left[1 - 2*\frac{B_c}{100}\right]$ With $L_{pp} = 0.866*L_{tp}$ (a triangular step)
Factor J_r	$J_r = 1$
Factor J_s	$J_s = 1$

References

1. Bichkar P et al (2018) Study of shell and tube heat exchangers with the effect of types of baffles. In: 2nd international conference on materials manufacturing and design engineering, procedia manufacturing vol 20, pp 195–200
2. Salehi E, Noie Baghban S et al (2000) Analysethermique de l'écoulementcotécalandredanséchangeur de chaleur de tube et calandrel'aide de méthodesexpérimentales et théoriques. Journal international d'ingénierie, vol. N ° 13, p. 13–26, février 2000
3. Palen JW, Taborek J (1969) Solution of shell side flow pressure drop and heat transfer by stream analysis method. Chem Eng Progr Symp Series 65:53–63
4. Bell KJ (1963) Final report of the cooperative research program on shell and tube heat exchangers. University of Delaware, Engineering Experimental Station, Bulletin No. 5
5. Bell KJ (1981) Delaware method for shell side design, heat exchangers: thermal-hydraulic fundamentals and design. Hemisphere/McGraw-Hill, Washington, DC
6. Milcheva I, Heberle F, Brüggemann D (2017) Modeling and simulation of a shell-and-tube heat exchanger for organic rankine cycle systems with double-segmental baffles by adapting the Bell-Delaware method. Appl Therm Eng 126:507–517
7. Uddhage H et al (2016) Design and development of shell and tube heat exchanger by using CFD. IJARIIE, 22395–24396
8. Serna-Gonzalez M et al (2007) Feasible design space for shell and tube heat exchangers using the bell-delaware method. Ind Eng Chem Res 46:143–155
9. Serth RW (2007) Process heat transfer principles and applications. Academic Press, Elsevier
10. Bell KJ (1988) Delaware method for shell-side design. In: Shah RK (ed) Heat transfer equipment design. Delaware method for shell-side design. Hemisphere, New York

Blade Angle Control of a Wind Energy Conversion System

Omessaad Elbeji, Marwa Hannachi, Mouna Benhamed, and Lassaad Sbita

Abstract The PMSG speed uses maximum power point tracking below the rated speed, which corresponds to low and high wind speed, and the maximum energy can be captured from the wind. Pitch angle control is the most common means for adjusting the aerodynamic torque of the wind turbine when wind speed is above rated speed and various controlling variables may be chosen, such as wind speed, generator speed and generator power. As conventional pitch control usually use PI controller, the mathematical model of the system should be known well.

Keywords Blade aerodynamic · Pitch angle · PI controller · Wind turbine

1 Introduction

A wind turbine converts the wind energy captured by its blades into mechanical energy. The ratio between the kinetic energy captured and the mechanical energy produced depends on wind speed, air density and the area swept by the rotor. A moving object is characterized by a kinetic energy that depends on its density (or weight). The denser the air, the greater the energy captured. For example, the density of air is approximately 1.225 kg/m. At normal atmospheric pressure, the temperature is around 15 °C. As air humidity increases, density decreases. Moreover, warm air is less dense than cold air. So at higher altitudes (e.g. on mountains), pressure is lower. This leads to a decrease in air density. These factors, which influence energy production, can differ from one site to another. This has prompted researchers to carry out studies to better select regions where wind turbines can be installed and produce more energy.

O. Elbeji (✉) · M. Hannachi · M. Benhamed · L. Sbita
Research Unit of Energy Processes Environment and Electrical Systems, National Engineering School of Gabes, Omar Ibn El Khattab Street, Zrig, 6029 Gabes, Tunisia
e-mail: bejiomessaad@gmail.com

© The Author(s), under exclusive license to Springer Nature Singapore Pte Ltd. 2024
A. Mellit et al. (eds.), *Proceedings of the 2nd International Conference on Green Energy Conversion System*, Green Energy and Technology,
https://doi.org/10.1007/978-981-97-6148-7_59

1.1 Aerodynamic Stall Systems

In order to be consistent with the grid frequency, certain types of wind turbine must be operated at a fixed rotational speed. The dynamic stall system is a simple and cost-effective solution. It is specific to the shape of the blade (natural limitation). It is based on the principle of the dynamic stall phenomenon. If the wind speed increases (above the nominal value), so does the angle of incidence [1–3]. The operating principle of this wind turbine technology is to produce maximum electrical power equivalent to the turbine's rated output. In fact, if the power is lower than the machine's rated output, the power produced by the wind turbine increases with increasing wind speed. As soon as the power reaches the rated value, it must decrease with increasing wind speed. A wind turbine is therefore characterized by its rated power, a starting wind speed for which the turbine remains power-producing, and a rated speed corresponding to the value of the rated power.

In this technology, the blades are designed in such a way as to guarantee instant lift reduction when the wind speed reaches a value at which power reduction is urgently required. It is therefore impossible to reproduce the rated power again if the wind speed is higher than the rated speed, as the power will decrease too quickly. Another type of power control is the "Active Stall" process, used for high-power machines. This system is based on the same principle as aerodynamic stall. But it features a sub-system that allows the blades to oscillate in response to increasing wind speed. This ensures a gradual reduction in power [1, 2]. The mechanical design of this technique is simple and less costly, since it does not require blade orientation than the technology with blade orientation.

1.2 Orientation Control

This method consists of deflecting the rotor axis horizontally by rotating it around the mast axis, or vertically by tilting the nacelle. This allows the turbine to be positioned slightly away from the direction of the strong wind. If the wind is too strong (above the turbine's limit), the rotor is completely perpendicular to the wind direction. This stops power generation [3–5]. This method requires significant mechanical effort on the mast blades. It is therefore not suitable for large wind turbines.

The above control methods are based on mechanical and aerodynamic means. Whereas the power produced by the wind turbine can be controlled electrically by varying the speed of rotation of the turbine. In fact, the speed of rotation can be regulated according to the desired operating point by means of the machine coupled to a static converter chain in order to control the transfer of power [5, 6]. This control avoids sudden changes in rotor load and compensates for the energy lost through the delay caused by the change in blade position. The conversion of kinetic energy into mechanical energy is based on important physical laws.

2 The Need of a Turbine Protection System

Aerodynamic modeling of the turbine shows that the characteristic of a turbine is highly non-linear, depending essentially on wind speed. On the other hand, the representation of the power versus wind speed characteristic of a wind turbine shows that each turbine is characterized by three characteristic wind speeds; a starting speed from which the turbine produces power, a nominal wind speed from which the power produced by the turbine remains constant and is equal to the nominal speed and a maximum wind speed beyond which the turbine must stop. In this context, the turbine is characterized by four zones. In the first zone, the turbine doesn't produce power because the wind speed is below the starting speed. The power produced by the turbine in the second zone depends on the wind speed picked up by the blades. This is the zone in which the turbine must be controlled to produce maximum wind power. In the third zone, the shaft speed must be kept constant, so the power delivered by the turbine is equal to the rated power. If the turbine reaches zone 3, it must be stopped for protection [6–8].

It is therefore essential to have the turbine checked in the various zones, depending on the operating conditions. For light and medium wind, the control principle is based on controlling the turbine to maximize power output, using techniques to calculate the optimum operating point or algorithms to find the desired point. For strong wind, the aim is to protect the turbine by limiting the power output.

2.1 Turbine Protection Systems

In order to limit the power extracted from the wind to its nominal value, large wind turbines often exploit two essential aerodynamic control techniques. The first technique is based on adjusting the lift of the blades to the wind speed in order to set the power at an approximately constant value. This is how the blade steering system works. The second technique is designed to increase lift losses if wind speed exceeds a certain value, and this is done during blade design. In the case of fixed-speed wind turbines, an aerodynamic stall system is most often used, as it is more economical [7–9]. For variable-speed wind turbines, on the other hand, a blade-steering system is used [1].

2.2 Pitch Actuator Model

The pitch actuator is a mechanical and hydraulic system responsible for orienting the turbine around its longitudinal axis [10]. In strong wind conditions, the turbine's output should be around the nominal value. This is achieved by shifting the lift of the blades through the pitch actuator. This is achieved by shifting the lift of the turbine

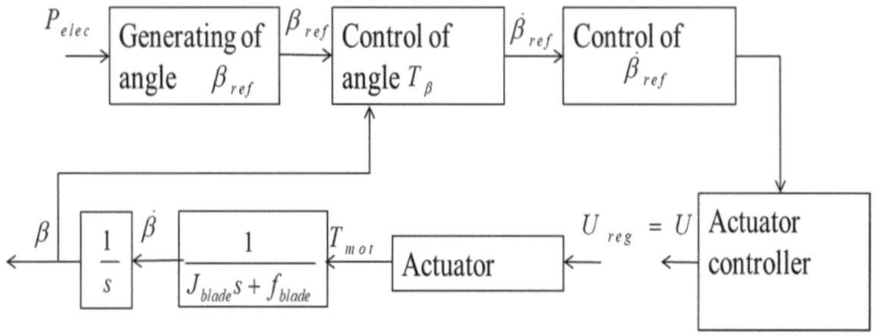

Fig. 1 Block diagram of a blade orientation system

blades through the blade pitching system, as the aerodynamic torque of the turbine changes, and so does the power coefficient of the turbine. The pitch angle is generated by the blade-steering system. The output of a blade steering system is the pitch angle β, and it uses electrical power as input. The block diagram of the blade orientation system can be represented as shown in Fig. 1.

The control process is based on four important subsystems: the reference stall angle generation system (βref), the angle regulation process, the angle speed regulator and the actuator controller.

Generating the reference pitch angle (βref) The reference pitch angle generation process is a very important system for controlling the blade pitch angle. Its role is to determine the reference pitch angle for each wind variation, based on experimental results of the electrical power measured for different values of the angle β. The inverse characteristic, on the other hand, identifies the blade angle required for each power variation. In fact, if the two subsystems; the blade angle speed controller and the actuator controller achieve their objectives in a perfect manner, i.e. quickly in relation to the angle setting dynamics, then we can have U = Ureg, C = Cref, β˙ = β˙ ref. Hence the diagram of a perhaps simplified blade orientation system as shown in Fig. 2.

Modeling the blade steering system essentially involves modeling the blade pitch angle. The equivalent diagram of a closed-loop blade tracking system can be defined as shown in Fig. 3.

The angle can be controlled by a gain or a PI.

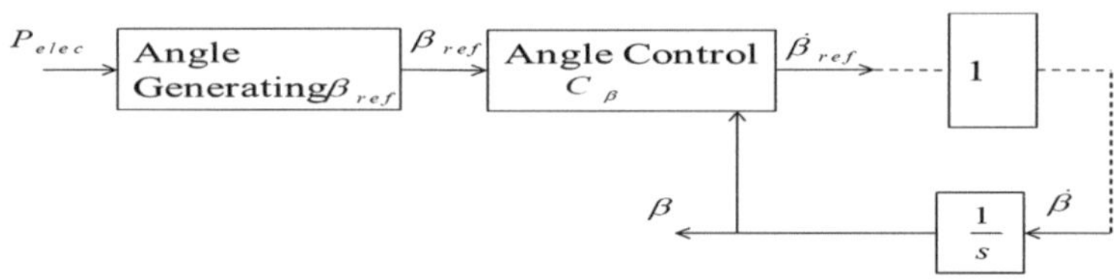

Fig. 2 Simplified block diagram of a blade orientation system

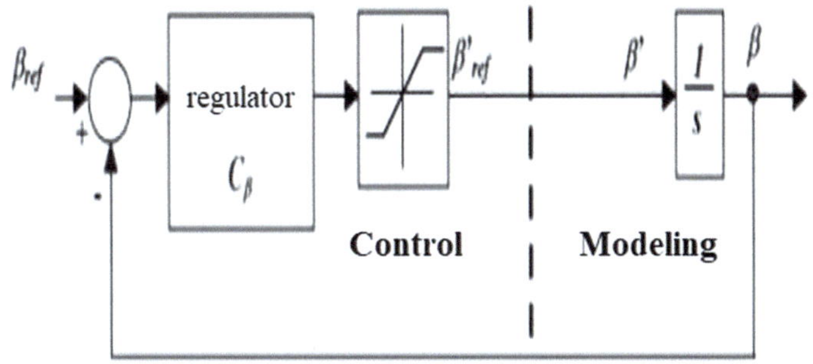

Fig. 3 Modeling of the orientation system

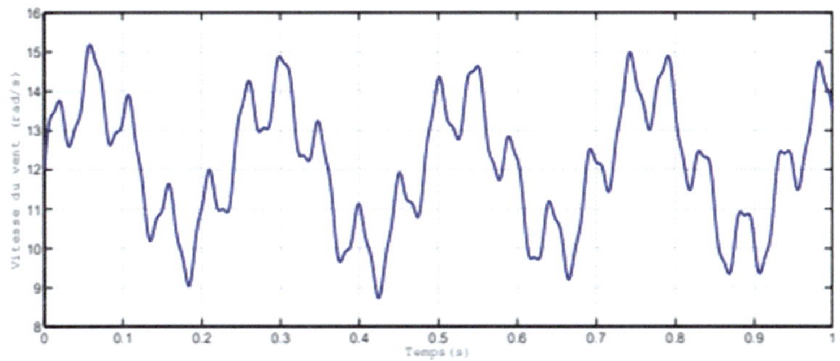

Fig. 4 Wind profile

2.3 Validation of the Stall Angle Control System

In order to validate the set angle control system, a simulation test was carried out with varying wind speed. The simulation results show that the established control system protects the wind turbine in high wind conditions by adjusting the setting angle to the required values (Figs. 4, 5, 6, and 7).

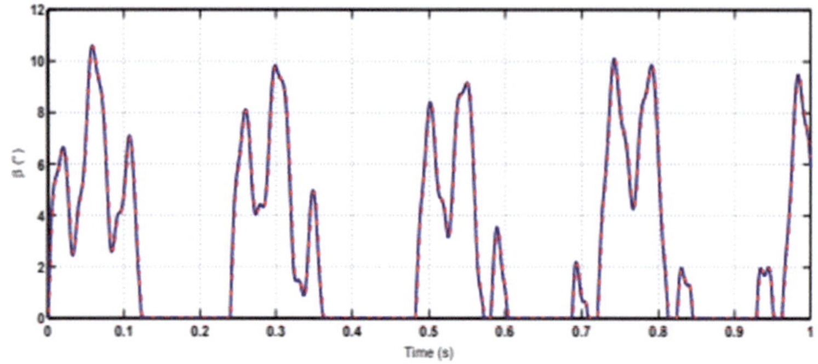

Fig. 5 Pitch angle

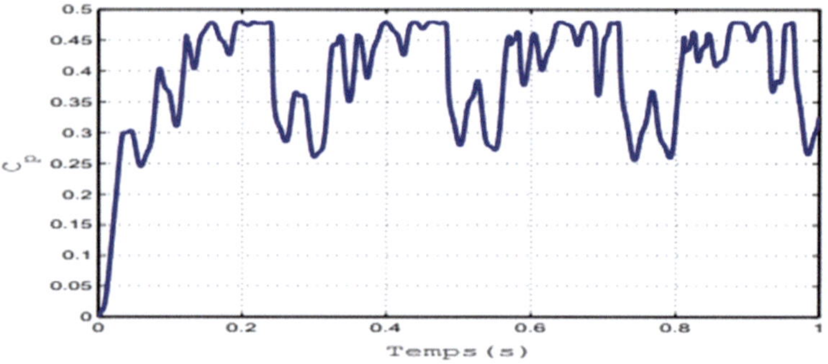

Fig. 6 Coefficient power

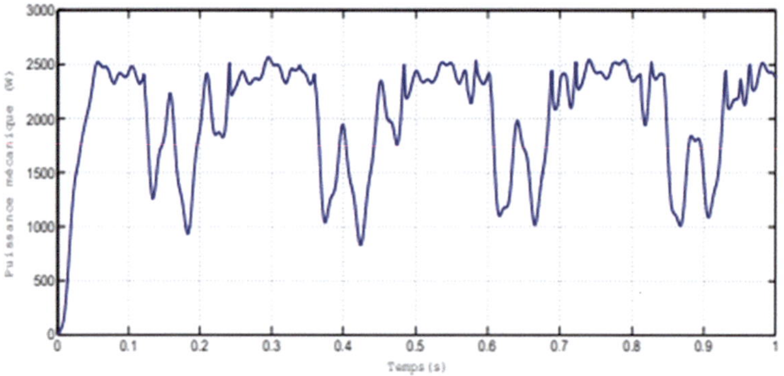

Fig. 7 Mechanical power

3 Conclusion

This paper presents the design of Blade angle Control of a wind energy con troll system. Pitch angle regulation is required in conditions above the rated wind speed when the rotational speed is kept constant. Small changes in pitch angle can have a dramatic effect on the power output. The purpose of the pitch angle control is optimizing the power output of the wind turbine. Below rated wind speed, the pitch setting should be at its optimum value to give maximum power. Also, Preventing input mechanical power to exceed the design limits. Above rated wind speed, pitch angle control provides a very effective means of regulating the aerodynamic power and loads produced by the rotor.

References

1. Ayenew E, Biru G, Mulatu A et al (2023) Modelling and control of wind energy conversion system: performance enhancement. Int J Dyn Control, 1–24
2. Prévost A, Léchappé V, Delpoux R et al (2023) An emulator for static and dynamic performance evaluation of small wind turbines. In: IEEE 32nd international symposium on industrial electronics (ISIE 2023)
3. Ali MN, Soliman M, Ebrahim MA et al (2023) D-decomposition-based multi-objective robust resilient control for blade pitch of wind energy conversion system. Int J Electr Power Energy Syst 146:108781
4. Apata O, Oyedokun DTO (2020) An overview of control techniques for wind turbine systems. Sci Afr 10:e00566
5. Colombo L, Corradini ML, Ippoliti G et al (2020) Pitch angle control of a wind turbine operating above the rated wind speed: a sliding mode control approach. ISA Trans 96:95–102
6. Ganthia BP, Barik SK (2020) Steady-state and dynamic comparative analysis of PI and fuzzy logic controller in stator voltage oriented controlled DFIG fed wind energy conversion system. J Inst Eng (India) Ser B 101(3):273–286
7. Roga S, Wanmali NK, Kisku V et al (2022) Development of pitch angle control algorithm for PMSG based wind energy conversion system. In: 2022 1st international conference on sustainable technology for power and energy systems (STPES). IEEE, pp 1–5
8. Hosseini E, Behzadfar N, Hashemi M et al (2022) Control of pitch angle in wind turbine based on doubly fed induction generator using fuzzy logic method. J Renew Energy Environ 9(2):1–7
9. Castillo OC, Andrade VR, Rivas JJR et al (2023) Comparison of power coefficients in wind turbines considering the tip speed ratio and blade pitch angle. Energies 16(6):2774
10. Hannan MA, Al-Shetwi AQ, Mollik MS et al (2023) Wind energy conversions, controls, and applications: a review for sustainable technologies and directions. Sustainability 15(5):3986

A Methodology for Operational Fault Diagnosis in Electrical Power Transformer: Practical Application

Abdelaziz Lakehal, Lotfi Saidi, Abdelmoumene Hechifa, and Arnaud Nanfak

Abstract Electrical power transformer is critical equipment in power plant and electrical power transmission and distribution. To facilitate the fault diagnosis of this electrical transformer a Bayesian network was developed and used for information fusion. 7 transformer elements were examined. 22 faults were taken care of in this study. 14 information's mainly taken from existing test and measurement equipment have been used and the main lines of their interpretations have been formulated. The main result of this contribution was a useful fault diagnosis manual for handling real problems in electrical power transformer maintenance. This contribution can help and serve as an expert decision support system for maintenance engineer. By detecting, diagnosing and decision making, the availability of electrical power transformer was improved.

Keywords Fault diagnosis · Power transformer · Bayesian network · Decision making · Electrical measurement

A. Lakehal (✉)
Laboratory of Research on Electromechanical and Dependability, University of Souk Ahras, Souk-Ahras, Algeria
e-mail: a.lakehal@univ-soukahras.dz

L. Saidi
Laboratory of Signal Image and Energy Mastery (SIME, LR 13ES03), University of Tunis, ENSIT, 1008 Tunis, Tunisia

A. Hechifa
LGMM Laboratory, Department of Mechanical Engineering, Skikda University, Skikda, Algeria

A. Nanfak
Laboratory of Energy, Materials, Modelling and Methods, National Higher Polytechnic School of Douala, University of Douala, Douala, Cameroon

© The Author(s), under exclusive license to Springer Nature Singapore Pte Ltd. 2024
A. Mellit et al. (eds.), *Proceedings of the 2nd International Conference on Green Energy Conversion System*, Green Energy and Technology,
https://doi.org/10.1007/978-981-97-6148-7_60

1 Introduction

Electrical power transformers are highly critical devices that require prognostics and health management. Therefore, increasing the availability of these devices is one of the main concerns of maintenance managers. However, any predictive maintenance procedure goes through three essential steps: monitoring, diagnosis, decision-making [1].

Electrical measurement analysis is a condition-based maintenance technique for monitoring electrical transformers. It is widely used to assess the state of health of this vital and strategic equipment. The various electrical, thermal, mechanical and environmental constraints that electrical transformers are subjected to cause deviations in electrical measurements compared to standards such as IEEE C57.10.01 and IEC 60137 [2, 3]. Once these values exceed limit thresholds, the transformer becomes undesirable operating conditions and the events that may appear on this equipment are feared.

To evaluate the condition of bushings and the transformer insulation, capacitance and dissipation factor measurements are taken. Aging and breaking down of insulation, or water penetration, increases the amount of energy in the form of heat in the insulation. The level of these losses is measured by the dissipation factor. Winding resistance measurements are made to diagnose winding deterioration. These measurements also make it possible to check the state of the on-load tap-changer. Also by measuring the variation of eddy currents it is possible to detect short-circuits between conductors in parallel as well as local heating.

Dissolved gas analysis represents one of the methods developed to improve the diagnosis and prediction of faults in electrical transformers. Although this method has been the subject of several codes such as: IEEE Std C57.104-2008, IEC Publication 60599 and IEC TC 10 Databases [4–6], it still has weaknesses in the prediction of faults, and the taking of decision. To make this diagnostic technique more reliable, a fusion of information from other sources appears necessary [7].

It should be underlined here that issues related to fault diagnosis and prediction of electrical power transformer has been undertaken already in the last decades. Several investigations have been carried out by researchers to improve and apply these methods in cases of specific solicitations [8]. In recent research works other methods besides traditional methods of diagnostics of electrical transformers have been developed and new precursors have been defined [9]. Although electrical transformers are static equipment, [10] Hong et al. used vibration analysis as technique for monitoring and detecting their faults. Encouraging results occurred in this study, but vibration analysis also has its limitations in predicting faults.

Based on standards or modern diagnostic techniques, uncertainty still exists and decision-making remains a difficult thing. The key to making good decisions and operating transformers in a certain environment is to exploit the information held by experts. This justifies that the majority of current contributions surround themselves around information. A contribution is developed by [11] based on using a statistical notion of causal inference based on information fusion. In this work authors tried give a tool to help in the prediction of electrical behavior arising from common causal dependencies.

The information available on the device especially, that concerning the history of failures and the analysis results represents the essential elements for decision-making. The first contribution of this paper is to made a posteriori analysis method to build a prevention approach. The second one is to reconstitute a malfunction in a logical way, to seek the causes which led to the fault of the transformer element and to define the priorities in terms of corrective maintenance intervention. Through this research we will try to highlight the multi-causality of the fault, identify the most sensitive element of the transformer in order to implement preventive measures to avoid the return of an identical fault and prevent the reproduction of similar or more serious faults.

2 Condition Monitoring and Fault Diagnosis of Electrical Power Transformer

In recent years, several research works on electrical transformers have led to the definition of new diagnostic techniques. The fundamental objective is the detection of the fault and then the diagnosis of its origin. Today, improving the fault diagnosis by using artificial intelligence methods take the advantage regard to the traditional methods. To optimize decision-making, two research paths will be adopted in this paper, one concerning the classic test and measurement on the transformer and the other will be sacrificed to the information fusion from these classical methods.

The international standard IEC 60317 [3] is the reference for insulated bushings for alternating voltages above 1000 V. Aging and breaking down of insulation, or water penetration, increases the amount of energy in the form of heat in the insulation. The dissipation factor measures the level of these losses

Ratio and excitation current measurements are made to assess possible winding deterioration by comparing ratio and measured magnetizing currents, to specifications, to factory measurement results and/or between phases. The deviation criterion of a transformation ratio greater than 0.5% from the theoretical values will be identified as faulty, and as normal if the deviation is lower [3].

Another traditional technique often used for the diagnosis of electrical transformers is the frequency response analysis. The origins of mechanical and electrical faults can be diagnosed by analyzing the frequency response. Mechanical faults can be caused by an external short circuit, for example.

Winding resistance measurements are made to assess possible deterioration of the windings. These measurements also make it possible to check the state of the on-load tap-changer. Resistance measurements are compared between the values of the three phases, or with respect to the values leaving the factory, when available.

Short circuit impedance measurements are made to assess the deterioration and possible displacement of the windings. Also, it is possible to detect short-circuits between conductors and local heating caused mainly by high losses by eddy currents.

Besides the electrical measurements, the state of the insulation is monitored by analysis of the dielectric response. This technique which makes it possible to diagnose the crossings of the active part is based on the evaluation of the water content. Also the evolution of the partial unloading in time is significant in the case of fault in the insulation of the transformers. With this type of measurement it is possible not only to determine the condition of insulating materials but also to predict their failure.

Through the methods mentioned above, information will be collected and analyzed in order to recognize the faulty state of the breast state. This information's will be the input data in the Bayesian model developed later in this paper. Classification of faults is adopted based on the work published by the experts of the company OMICRON [12]. Table 1 summarizes the faults and elements retained in this study.

3 Bayesian Theory

Bayesian theory includes graph theory and probability theory. The Bayesian reasoning is modeled by a network which consists of two elements which are: the structure and the parameters. The structure is defined and optimized by graph theory, while the parameters are given by probability theory. Directed Acyclic Graph represents the structure of the Bayesian network as shown in Fig. 1. The learning of the parameters and the structure allows a better Bayesian modeling. One of the oldest applications of Bayesian inference is decision making [13]. Other researcher used Bayesian reasoning for diagnosis, prediction, reliability and others [14]. Inference using Bayes' theorem makes it possible to update the likelihood of a hypothesis based on a previous estimate of this likelihood and new information.

For example, for diagnostic issues, let events C and F such that C is the cause and F represents the consequence (fault). Bayes' theorem makes it possible to determine the probability of existence of the fault C knowing F, if we know the probabilities of C, B and F knowing C, provided that the probability of F is not equal to 0.

The Bayesian theory is given by:

$$P(C|F) = \frac{P(F|C)P(C)}{P(F)} \tag{1}$$

Inference in Bayesian network is to update probabilities. In this study by using the same model we can make fault diagnosis (causes → symptoms), or fault prediction (symptoms → causes).

The basis of Bayesian computing is probabilistic inference. This Bayesian inference has several advantages over other artificial intelligence methods such as neural

Table 1 Component and fault types

Transformer component Ci	Fault type	Code
Bushings (C1)	Partial breakdown between capacitive graded layers, cracks in resin-bonded insulation	C1F1
	Aging and moisture ingress	C1F2
	Open or compromised measuring tap connection	C1F3
	Partial discharges in insulation	C1F4
CT at bushings (C2)	Current ratio or phase error considering burden, excessive residual magnetism	C2F1
Insulation (C3)	Moisture in solid insulation	C3F1
	Aging, moisture, contamination of insulation fluids	C3F2
	Partial discharges	C3F3
Leads (C4)	Contact problems	C4F1
	Mechanical deformation	C4F2
Tap changer (C5)	Contact problems in tap selector and at diverter switch	C5F1
	Open circuit, short circuit between turns	C5F2
	Contact problems in the off-load tap-changer	C5F3
Windings (C6)	Short-circuits between windings or between turns	C6F1
	Strand-to-strand short-circuits	C6F2
	Open circuits in parallel strands	C6F3
	Short-circuit to ground	C6F4
	Mechanical deformation	C6F5
	Contact problems, open circuits	C6F6
Core (C7)	Mechanical deformation	C7F1
	Floating core ground	C7F2
	Shorted core laminates	C7F3

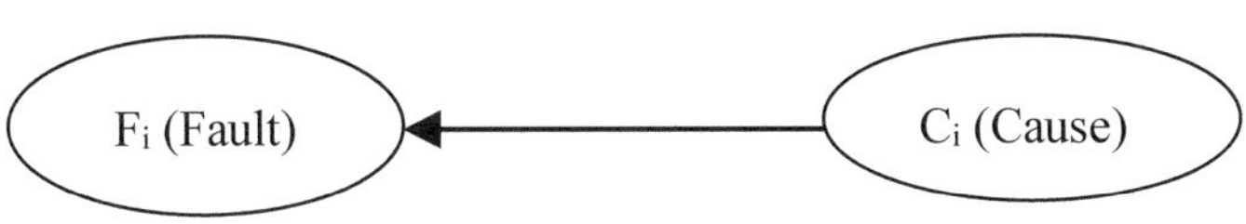

Fig. 1 Simple BN for fault diagnosis

networks and decision trees. Variables are defined as in the other techniques of artificial intelligence while this is not the case for their role. Another advantage is that inference is possible in all directions. For the problem addressed in this paper, it is possible to make inference for diagnosis and prediction from the same model. Bayesian inference is mainly based on the probability tables of the nodes which represent the parameters of the Bayesian network (variables of the model) and the arcs which represent the conditional independences (structure of the network).

The application of Bayesian networks in the field of industrial diagnostics has given promising results [15]. The same applies to the diagnosis and prediction of faults in electrical transformers based on the analysis of gases dissolved in oil [16]. The contribution that Bayesian networks give alongside other artificial intelligence techniques to traditional diagnostic methods is very significant [16].

4 Practical Application and Results

In this section, practical application of the proposed method is made for fault diagnosis of an electrical transformer of a power plant of the Algerian Company of Electricity Production, SONELGAZ/SPE. Analyses were performed on a step-down transformer (15.5/6.6 kV) with a power of 20 MVA (Fig. 2).

The major operational problems in the case study of the electrical power transformer endanger its seven main elements which are: Bushings, CT at bushings, Insulation, Leads, Tap changer, Windings, Core (Table 1). In this application, the choice of these elements is strongly related to the existing test and measurement equipment, but the methodology can be extended to other elements if the related information exists.

Usually, the operational problems of an electrical power transformer are not easily detected and need very developed instrument, and on the basis of the information given by these instruments it's possible to identify the causes of each problem. To achieve a good solution for each problem, its actual causes must be faced. Unfortunately, the diagnosis task is very difficult in the case where some faults occur together. Also, a fault that is not detected in time or detected late can cause the failure of other elements of the electrical transformer.

Fig. 2 The studied electrical power transformer

Another problem is the lack of know-how and experience among the maintenance engineers force them to face any problem by applying temporary measures that postpone and displace it to another service or another job in the power plant. The solution is the use of a computer program and the recording of information from the measuring devices and feedback experience in these programs. The Bayesian network which will be developed in this section represents a decision-making aid tool for the diagnosis of electrical transformer faults.

The task of fault diagnosis is extremely difficult because the problem's symptom–cause relationship is not a unique direct link. It is possible for one fault to have several causes and one cause to result in more than one problem. Also, a cause could be the cause the breakdown of one element of the transformer and at the same time has an impact on other elements.

Now, based on feedback and the experience of operators and maintainers, each cause has a probability and each defect has these causes. The network of Fig. 3 represents the structure of the Bayesian network which makes it possible to diagnose the faults of Insulation. The variables are discrete but there are some cases where it's possible to use continuous variables to model the fault diagnosis of electrical power transformer, for example in the case of dissolved gas analysis variables.

The Conditional Probability Table (CPT) quantifies the causalities in the Bayesian reasoning presented above. Table 2 presents the links between the three faults that can cause transformer insulation breakdown. Each variable in the network of the Fig. 3 can take two states: True (T) and False (F). In this case study the variables are discrete. However, it should be noted that Bayesian networks offer modeling possibilities with continuous variables.

From the CPT of Table 2. It is possible to do some reading as follows:

- If one of the causes C3F1, C3F2 and C3F3 occurs then the Insulation breakdowns.
- If it will not take place for the four causes then the insulation is safety.

To make a diagnosis and make certain decisions based on the Bayesian theory, it is necessary to calculate the probability that an element of the transformer breaks

Fig. 3 BN for insulation fault diagnosis

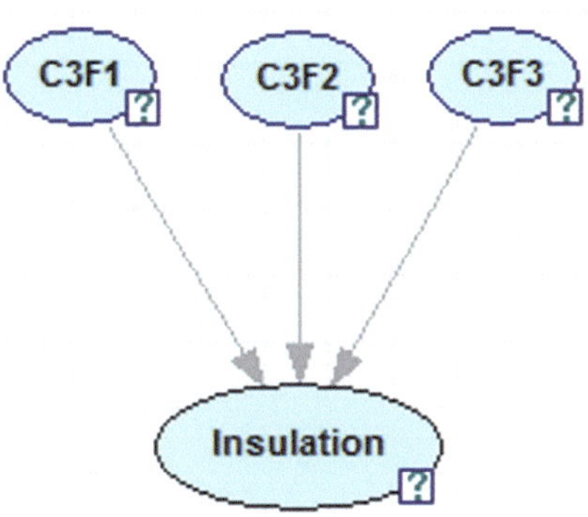

Table 2 Table captions should be placed above the tables

	C3F1	True				True			
	C3F2	True		False		True		False	
	C3F3	T	F	T	F	T	F	T	F
Insulation	True	1	1	1	1	1	1	1	0
	False	0	0	0	0	0	0	0	1

down knowing that it is not broken down immediately. You have to be careful with the terms here, because faulty element means the transformer fails immediately, partially faulty means it fails after a few hours, so not immediately, and acceptable means it does not fail.

In Fig. 4 the causality is strict which means that any new information on the C2F1 variable changes the beliefs of the modeler and the probability that the CT at bushings fault appears is 100% (evidence).

Figures 5, 6, 7, 8 and 9 successively represent the Bayesian networks of faults in Bushings, Leads, Tap changer, Windings and Core of the electrical power transformer.

Referring to the history file of the studied electrical power transformer, the results of the conducted diagnostics, it's possible to define the a priori probabilities for each cause (Table 3). From 45 faults each cause is defined by a probability as given by Table 3. By using the CPT and the Bayesian theorem it is possible to calculate the a posteriori probabilities.

From the results of the inference presented in Table 3, it is possible to define the monitoring operations of the electrical power transformer which, within the framework of systematic preventive maintenance, take place according to a determined periodicity. It is also necessary to increase the frequency of inspections for the

Fig. 4 BN for CT at bushings fault diagnosis

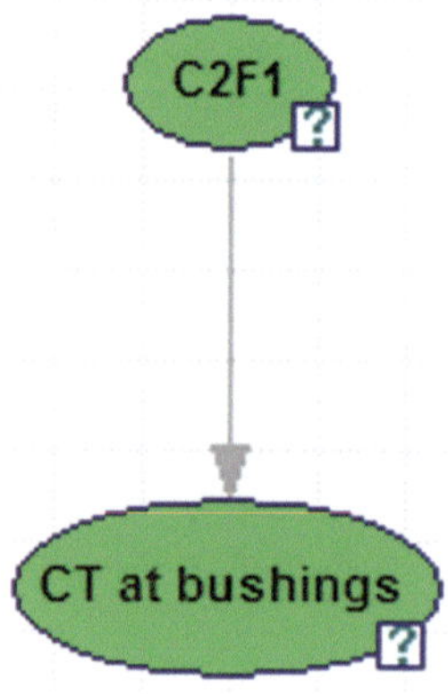

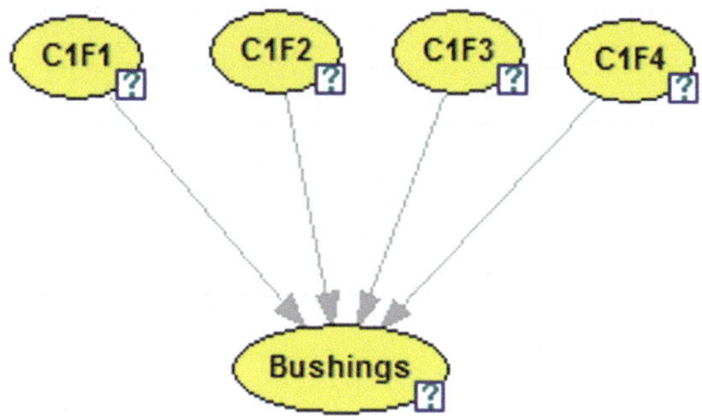

Fig. 5 BN for Bushings fault diagnosis

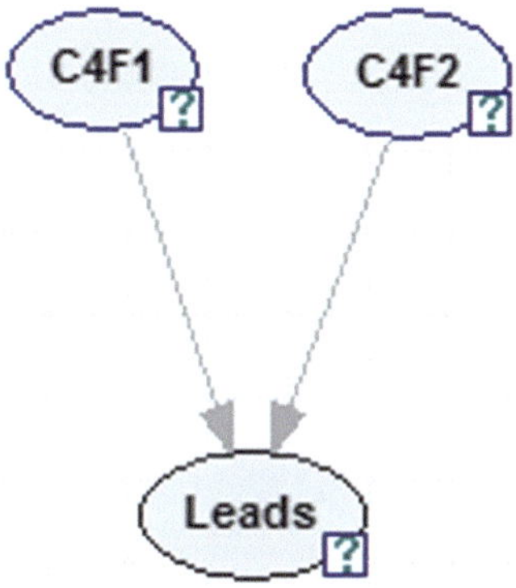

Fig. 6 BN for Leads

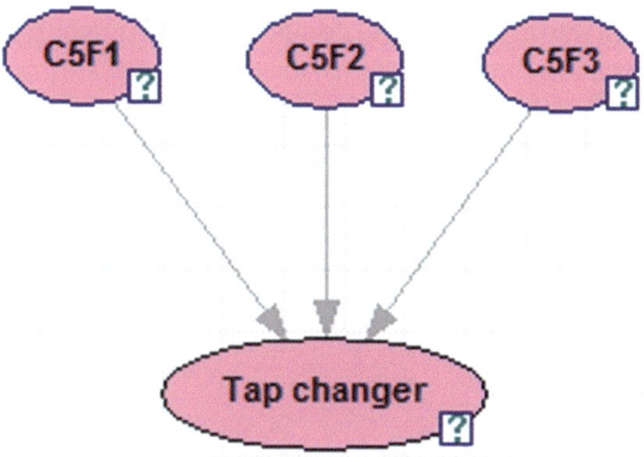

Fig. 7 BN for Tap changer fault diagnosis

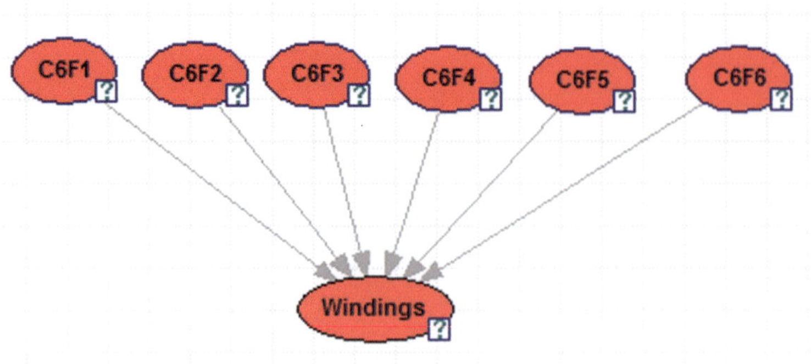

Fig. 8 BN for Windings fault diagnosis

Fig. 9 BN for Core fault diagnosis

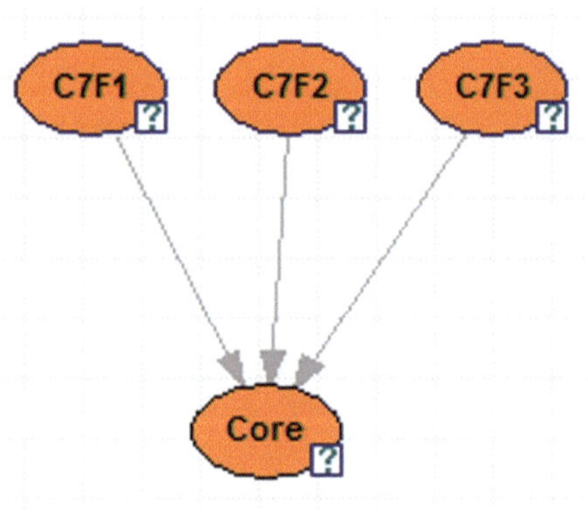

following transformer elements: Bushings (C1), Insulation (C3) and Windings (C6). These interventions correspond to a list of operations defined beforehand which can lead to the dismantling of components and immobilization of equipment. The results shown in Table 3 can be used to anticipate transformer failures. The information used is gathered and merged into the same model.

The results found are specific to the transformer studied in this paper because they depend on the a priori probabilities. Whereas developed Bayesian networks are standard and valid for any other power transformer. In conclusion, if the parameters of the networks change from one study to another, the structure remains the same.

Table 3 The parameters of the developed BN befor and after inference

Transformer component Ci	Code	A priori probabilities	A posteriori probabilities
Bushings (C1)	C1F1	0.0444	0.4755
	C1F2	0.1333	
	C1F3	0.0222	
	C1F4	0.0222	
CT at bushings (C2)	C2F1	0.0222	0.0222
Insulation (C3)	C3F1	0.1333	0.2467
	C3F2	0.1111	
	C3F3	0.0222	
Leads (C4)	C4F1	0.0222	0.0439
	C4F2	0.0222	
Tap changer (C5)	C5F1	0.0222	0.0864
	C5F2	0.0444	
	C5F3	0.0222	
Windings (C6)	C6F1	0.0222	0.2403
	C6F2	0.0667	
	C6F3	0.0222	
	C6F4	0.0444	
	C6F5	0.0888	
	C6F6	0.0222	
Core (C7)	C7F1	0.0444	0.0864
	C7F2	0.0222	
	C7F3	0.0222	

5 Conclusion

Despite the fact that the objective of preventive maintenance is to reduce the transformer failure rate, preventive maintenance is expensive and is not infallible. It therefore becomes legitimate to implement a preventive maintenance policy for the most critical elements of the transformer and then to choose an exclusively corrective policy for the only equipment whose criticality is minor or even zero and this well according to the financial resources. The Bayesian networks presented and used in this paper show the interest and the advantage that can bring in fault diagnosis and the decision making with regard to the actions of electrical power transformer maintenance.

The main objective of a decision statement is to allow a good selection of solutions. In this study a multi-information fusion model was proposed to define the transformer components that have a high priority. Consequently, deciding is knows the cause which will help us come up with a better action of predictive maintenance. Future work will focus on the multisource information fusion. The method will studied the fault prediction of electrical power transformer. The stability, the accuracy and the sensitivity will be examined.

References

1. Elmashtoly AM, Chang CK (2020) Prognostics health management system for power transformer with IEC61850 and internet of things. J Electr Eng Technol 15(2):673–683
2. IEEE Standard for Performance Characteristics and Dimensions for Power Transformer and Reactor Bushings. IEEE Std C57.19.01-2017 (Revision of IEEE Std C57.19.01-2000), pp 1–30, 12 July 2018. https://doi.org/10.1109/IEEESTD.2018.8410922
3. IEC 60137:2017 (2017) Insulated bushings for alternating voltages above 1000 V
4. IEEE Guide for the Interpretation of Gases Generated in Oil-Immersed Transformers, IEEE Std C57.104–2008 (Revision of IEEE Std C57.104–1991) , pp 1–36, 2 Feb. 2009. https://doi.org/10.1109/IEEESTD.2009.4776518.
5. IEC 60599:2022 (2022) Mineral oil-filled electrical equipment in service—guidance on the interpretation of dissolved and free gases analysis
6. Duval M (2001) Interpretation of gas-in-oil analysis using new IEC publication 60599 and IEC TC 10 databases. IEEE Electr Insul Mag 17(2):31–41
7. Zhang Y, Chen HC, Du Y, Chen M, Liang J, Li J, Fan X, Yao X (2021) Power transformer fault diagnosis considering data imbalance and data set fusion. High Voltage 6:543–554
8. Soni R, Mehta B (2021) Review on asset management of power transformer by diagnosing incipient faults and faults identification using various testing methodologies. Eng Fail Anal 128:105634
9. Ma G, Wang Y, Qin W, Zhou H, Yan C, Jiang J, Ju Y (2021) Optical sensors for power transformer monitoring: a review. High Voltage 6(3):367–386
10. Hong K, Huang H, Fu Y, Zhou J (2016) A vibration measurement system for health monitoring of power transformers. Measurement 93:135–147
11. Rodriguez-Rivero J, Ramirez J, Martínez-Murcia FJ, Segovia F, Ortiz A, Salas D, Castillo-Barnes D, Illan IA, Puntonet CG, Jimenez-Mesa C, Leiva FJ, Carillo S, Suckling J, Gorriz JM (2020) Granger causality-based information fusion applied to electrical measurements from power transformers. Inf Fusion 57:59–70
12. OMICRON L2208, mai 2013S.W. www.omicronusa.com
13. Cheng L, Yu T (2018) Dissolved gas analysis principle-based intelligent approaches to fault diagnosis and decision making for large oil-immersed power transformers: a survey. Energies 11(4):913
14. Zhao X, Wang X, Golay MW (2023) Bayesian network–based fault diagnostic system for nuclear power plant assets. Nucl Technol. 1–18
15. Yu W, Zhao C (2019) Online fault diagnosis for industrial processes with bayesian network-based probabilistic ensemble learning strategy. IEEE Trans Autom Sci Eng 16(4):1922–1932. https://doi.org/10.1109/TASE.2019.2915286
16. Lakehal A, Tachi F (2017) Bayesian duval triangle method for fault prediction and assessment of oil immersed transformers. Meas control 50(4):103–109

Blockchain for Machine Learning Protetion Against Attacks in IoT

Walid Dhifallah, Tarek Moulahi, Mounira Tarhouni, and Salah Zidi

Abstract All sectors are adopting open innovations such as intelligent automation with the use of sophisticated technology such as artificial intelligence (AI) to accomplish systematic procedures for Internet of Things (IoT) systems. The practical deployment of AI-enabled smart systems is now limited by a lack of security levels and confidence in detecting IoT system threats. Because Blockchain can decrease AI vulnerabilities and AI can increase Blockchain's performance, these two technologies are complementary. In the literature, there is little research aimed at protecting intrusion detection systems (IDS) based on ML (Machine Learning) against evasion and poisoning attacks adopting, in most cases, statistical schemes based on traditional methods or ML and resulting in additional deployment and runtime costs. Although the power of machine learning tools is insufficient, these techniques are used to protect themselves against attacks. For this reason, we aim, in this article, to propose a new framework called intellig_block for the detection of cyber threats in ML models which will be used to develop IDS. In intellig_block, the execution of the classification technique will be decentralized, by hashing the model file and we integrate this hash as a smart contract. The experimental result shows encouraging results and low execution time and overhead in terms of gas consumed.

Keywords Internet of things (IoT) · Blockchain · Machine learning (ML) · Evasion attack · Poisonous attack

W. Dhifallah (✉) · T. Moulahi · M. Tarhouni · S. Zidi
Laboratoire Hatem Bettaher (IRESCOMATH), University of Gabes, 6029 Gabes, Tunisia
e-mail: dhifallah.walid@gmail.com

T. Moulahi
Department of Information Technology, College of Computer, Qassim University, Braydah, Saudi Arabia

© The Author(s), under exclusive license to Springer Nature Singapore Pte Ltd. 2024
A. Mellit et al. (eds.), *Proceedings of the 2nd International Conference on Green Energy Conversion System*, Green Energy and Technology,
https://doi.org/10.1007/978-981-97-6148-7_61

1 Introduction

In recent years, machine learning (ML)-based technologies have emerged as a plausible alternative for IoT device security. ML is a multifaceted AI approach that can outperform dynamic networks without requiring explicit programming. In health coverage, anomaly detection, fraud prevention, autonomous vehicles, and a variety of other applications, Computer learning approaches may be used to teach the machine to detect different types of assaults and deliver appropriate defensive strategies, which can then be used to make predictions or support decision-making. During the learning sequence of an ML model such as data pre-processing, model training, or during model testing, poisoning or escape attacks may be injected. Many academics suggest AI technologies for IoT systems [1], including ML and Reinforcement Learning (RL). The ML analytical tool is frequently used to analyze vast amounts of data and give helpful information for monitoring, estimating, categorizing, and tracking future IoT activity. However, the growing use of ML technologies has made them appealing targets for attackers seeking to manipulate such techniques for malicious objectives [2]. All ML systems are trained using datasets that are supposed to be representative and reliable for the subject matter in question, allowing for the creation of a genuine system perception of the phenomena of interest. Malicious actors, on the other hand, might wreak havoc on such methods' decision-making algorithms by either aiming the training model or persuading the algorithm to produce the desired result, misclassifying anomalous occurrences. Attackers can degrade overall performance by poisoning and evading data, inducing geared misclassification or improper behavior, and installing backdoors and neural Trojans. The attack types under consideration in this study include poisoning [3] and evasion [4]. The overarching purpose of these models is to minimize the classification's prediction error and maybe mislead the decision support system into intended harmful measured parameters.

Categorization whatever she wants. In the case in point, the attacker may be able to build a dataset of abnormal network-layer protocol behavior and train an anomaly-based intrusion detection system using a labeled attack dataset as the ground truth. As just a consequence, the detection will be unable to identify cyber assaults against this network-layer protocol, putting the fundamental system's future at risk. This assault might be tweaked to have a major influence on the quality of a signature-based intrusion detection system, which oversees identifying malware infiltrating a system or infrastructure.

We identify two significant difficulties from prior research that must be addressed to build a secure solution for the Level of security in IoT applications and centralization.

Beca its many benefits, Blockchain technology is being used to address these concerns. Blockchain is a database system that is protected, diversified, and decentralized Blockchain networks have a dispersed network that connects all nodes, meaning that every exchange and timestamps must be immediately observed, and the absence of an outsider is widely spread. Blockchain technology has come a long way in

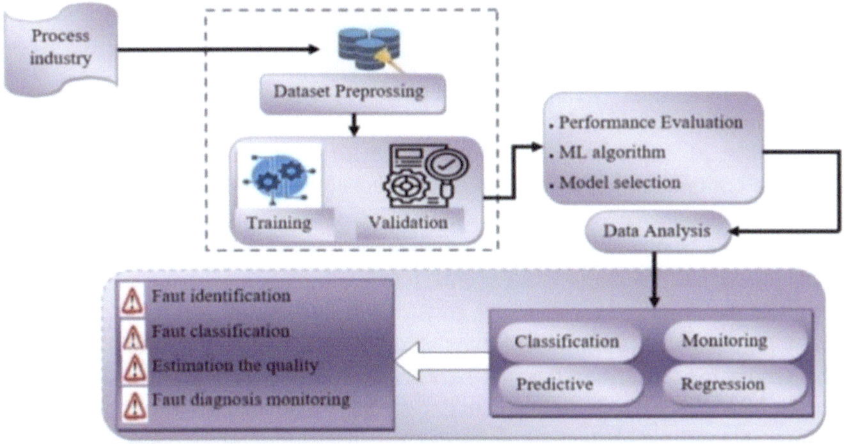

Fig. 1 Process of detecting faults using machine learning

application in a variety of industries, including banking, data security, agriculture, and healthcare in connected devices.

This combination of Cryptocurrencies with AI can help with accuracy, latency, centralization, protection, and confidentiality. Transfers are kept through distributed ledger technology in a database with a numerically signed hashing. As a result, issues like precision, latencies, digital security, and scalability will need to be addressed. These problems are also solved with AI algorithms. With the feature of decentralization, blockchain networks can automatically and quickly validate data, removing a hot spot problem in a virtual machine for big data processing and IoT cryptocurrency support with AI (Fig. 1).

The following are indeed the paper's main factors that contribute:

- Create a new framework to protect machine learning and deep learning against cyber threats by merging them with Blockchain.
- Merge smart contracts and access control to ensure ML security and privacy, then IDS.
- By contrasting the suggested method with an existing one in terms of accuracy, deployment, and execution overhead, you may determine how successful it is.

The paper is structured as follows. Section 2 talks about related work on how Blockchain technology is security in IoT sectors. Introducing our system in Sect. 3. Section 3 describes our proposed scheme. Section 4 The simulation results of the proposed system. Conclusion of this study is in Sect. 5.

2 Related Work

The public study topics of the intelligent decentralized systems on IoT have been invested in and debated by several scholars.

Wright et al. [5] Show that a smart edge for blockchain networks based on Ethereum computation is a reduced, high-precision resource management framework. It allows devices to communicate to outsource calculations to cutting-edge gadgets in a verifiable way in return exchange for money. Swan et al. [6] investigated the benefits of a Blockchain-based approach to Intelligence and human advancement, as well as the architectural approach to Blockchain thinking and its application in the Intelligence blockchain idea. Reyna et al. [7] investigated how Blockchain may enhance the IoT, described the link between Blockchain and IoT, provided investigative obstacles in decentralized IoT systems, and evaluated the most relevant research seeing how blockchain enhances IoT. Banerjee et al. [2] investigated IoT security mechanisms, as well as the paucity of IoT datasets, used between academic and professional organizations. Qian et al. [8] demonstrated high blockchain-based safety control strategies for Connected systems. Nevertheless, there are worries in this study about erroneous traffic control and proof of identity. Rathore et al. [9] Proposed blockchain-based private deep learning techniques to provide reliable data into IoT applications, based on the confluence of the intelligent decentralized systems at the device level.

3 Proposed System Model

The suggested concept is built on the use of blockchain technology. To make blockchain-based apps smarter, ML skills may be utilized. The security of distributed ledgers can be enhanced by employing machine learning. It also provides a chance to improve models by utilizing BT's decentralized design. As illustrated in Fig. 2, we provided an architecture for ML adoption in an intelligent application.

In Fig. 2, we suggest a Blockchain-based architectural scheme that gives security cover to AI-based systems. We utilized smart contracts in this synoptic to safeguard the decision function of ML algorithms against AI assaults. It is divided into three levels: (1) source of data, (2) model training, (3) intelligent secure decentralized model and (4) we be displaying results and evaluating the performance.

Data Source: A system that generates a large amount of data, such as linked devices (IoT sensors, IoT apps...) or the cloud.

Model Learning: In the preceding section, we demonstrated that when learning the ML model, it may be assaulted, resulting in movements.

Intelligent Secure Decentralized Model: To achieve an intelligent secure model, we created an intelligent contract into which the decision function was integrated.

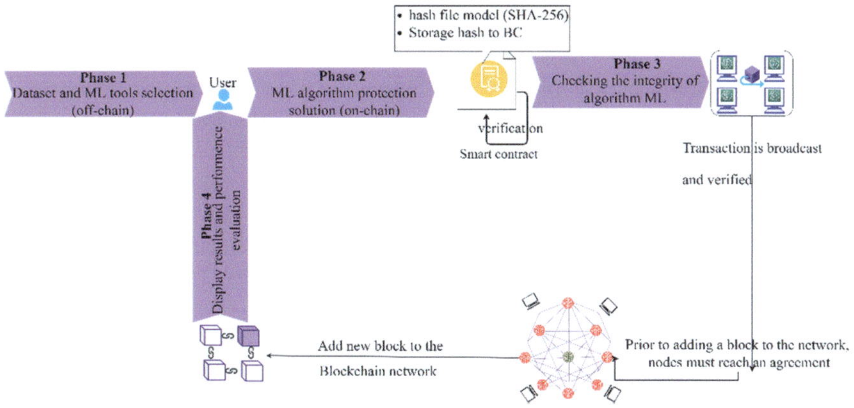

Fig. 2 The design of IoT intelligent block architecture

The landlord utilizes blockchain technology to limit access to authorized individuals and prevent access. The overall security purpose of ML model files is greatly reinforced by this degree of access restriction. To improve security, the model file was kept on the blockchain as encrypted pieces. After a user submits a file upload request, the pieces are encrypted and defragmented to return the user to the original template file.

The user may then use the blockchain to compare the file hash to the calculated hash of the downloaded file to confirm the file's integrity. As a result, the model files' integrity may be established and justified to any third party.

4 Results and Discussion

By merging ML with blockchain, we present intellig_block, a safe ML system based on Blockchain. Intellig_block contributes to the area of machine learning by safeguarding machine learning algorithms from IoT threats. First, a collaborative machine learning paradigm that supports data collection and privacy breaches. Second, to create a safe and dependable model against ML threats, collaborative ML has been applied in a blockchain context (evasion and poisoning). Finally, an intelligent block prototype model was created to test its efficacy in real-world circumstances.

Before running the experiment, the above method made use of underlying programs. The Ganache and MetaMask tools to set up a personal Ethereum Blockchain to test our Solidity contracts.

The blockchain is simulated using Remix IDE (Ethereum) by means of Solidity was used to develop an intelligent contract. To test this hypothesis, the TON_IoT [10] is used. This data is a set of datasets that make up a new generation of IoT and IIoT datasets for evaluating the fidelity and efficiency of various AI-based cybersecurity solutions.

Our smart contracts are implemented using a private blockchain called Ganache. It makes it possible to implement smart contracts, create Dapps, and run tests. Ganache provides ten Ethereum accounts, each with a balance of 100 ether, together with a graphical user interface for tracking all network activity.

Figures 3 and 4 illustrate smart contract compilation and migration to blockchain ganache. The cost of one file of this transaction, as shown in the result of the SVM migration, is 0.00050368 ETH, which is equivalent to £0.65. This conversion was performed on June 03, 2023. Following the transfer of our smart contract, we will build a local virtual server containing the client-side application using the truffle framework. To join our blockchain network, we need to connect to our MetaMask wallet.

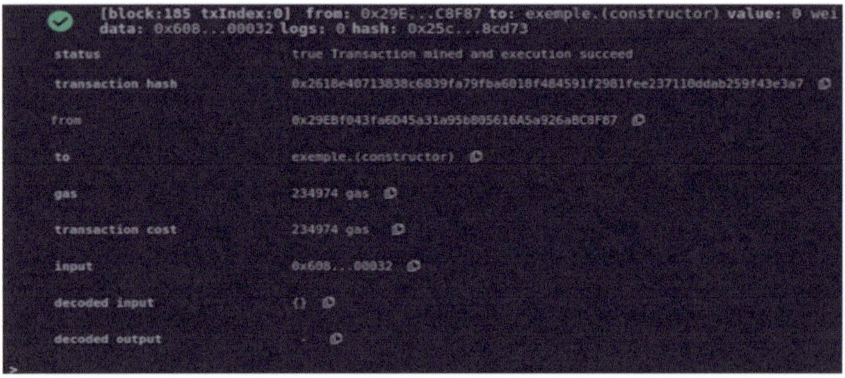

Fig. 3 Journal of the blocks created in a decentralized intelligent system

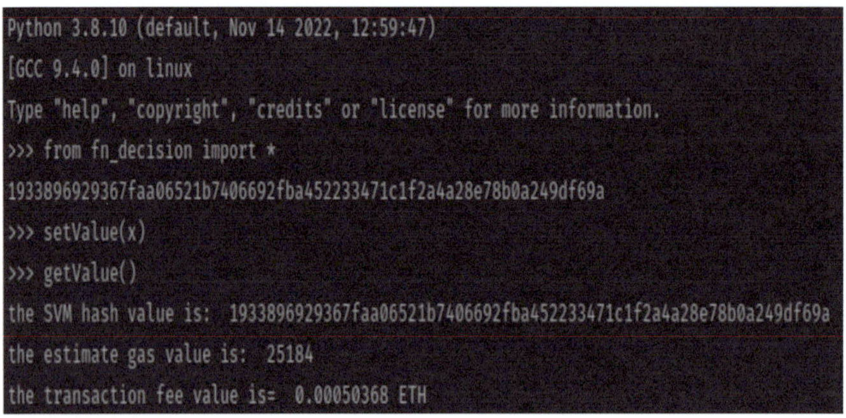

Fig. 4 Smart contract deployment

Table 1 Consumption of gas against the Ml classifier

ML classifier	Gas fee	Gas Used	Gas consumption
SVM	0.00050	22308	**10.616**
Knn	0.00083	41506	**34.449**
DT	0.00053	26506	14.048
RF	0.00053	26506	14.048
MLP	0.00047	25458	11.965
LR	0.00053	47210	25.021
NB	0.00053	22308	11.823

The SVM model file hash has the lowest Gas consumption for storing on the blockchain, whereas the KNN model file hash has the highest Gas consumption

On the Ethereum blockchain, the cost incurred by smart contracts is used to evaluate the proposed platform. These are the units of gas needed to complete a transaction and perform smart contract tasks. Otherwise, it includes both execution cost and transaction cost. The fees for adding smart contract code to the Ethereum blockchain are called transaction costs. It is constrained by the thickness of the smart contract. The fundamental actions performed by the smart contract define its size. While the cost of storing global variables and invoking smart contract methods is the cost of execution. It is also impacted by the calculations performed during the execution of a transaction (Table 1).

We use the following calculation to determine the gas charge for each transaction in our system:

$$Transaction\,Fee = Gas\,Used * Gas\,Price \tag{1}$$

where gas Price represents the number of Gwei required for the transaction and gas Used represents the amount of gas that is used, depending on how much gas is stored and processed for each transaction.

Let's use the add User method as an example, which enables the assignment of roles and accounts to each agent. As shown in Fig. 4, the functions gas Used and gas Price are 84,275 and 20 Gwei, respectively. So,

$$\text{Transaction Fee} = 84,275{*}20 = 1,685,500\text{Gwei} = 0,00000000163\text{ Ether} \tag{2}$$

The results reveal that performing Knn and SVM transactions consume a significant amount of gas Fig. 5. This signifies that the transaction involves a vast quantity of information and uses an IoT of wherewithal in comparison to other methods. The gas used is the amount of gas necessary to accomplish the transaction. The gas charge must cost of doing a transaction. In Table 1 a result, we compute total gas consumption by multiplying the amount of gas used by the cost of gas (Fig. 6).

Table 2 summarizes the findings in a tabular format, and shows the quantity of feedback received, which spans from the ML model to the time spent on each

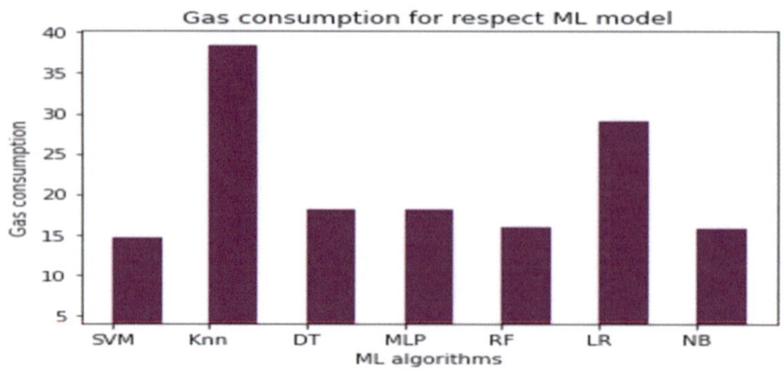

Fig. 5 Gas consumption for respect ML model

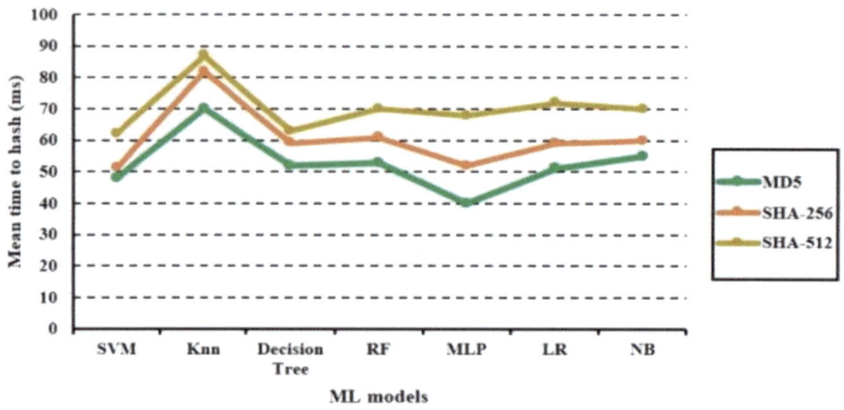

Fig. 6 Secure hash function time velocity ML models

transaction. The average time is the length of time required to accomplish a transfer of funds. that can be seen that each feedback takes approximately the same amount of time. Even while accessing the same service, the IoT might rate the same cloud server many times. The ganache interface is used to calculate the average transaction time. As can be seen from the findings, the Knn takes a long time compared to the other types of input. Fig. 7 shows how the bindings were examined.

Even while accessing the same service, the IoT might rate the same cloud server many times. The ganache interface is used to calculate the average transaction time. As can be seen from the findings, the Knn takes a long time compared to the other types of input. Fig. 7 shows how the bindings were examined. Afterward, we make a comparison between our framework result and the result of work on the same dataset. In intellig_block, the accuracy of MLP is 92.2% and 98% for SVM.

This comparison (Table 3) is based on some criteria deemed important such as blockchain-based, access control, and security.

Table 2 Time transaction Ml algorithm

ML Classifier	Times(s)
SVM	**0.47**
Knn	0.83
DT	0.53
RF	0.53
MLP	**0.47**
LR	0.53
NB	0.53

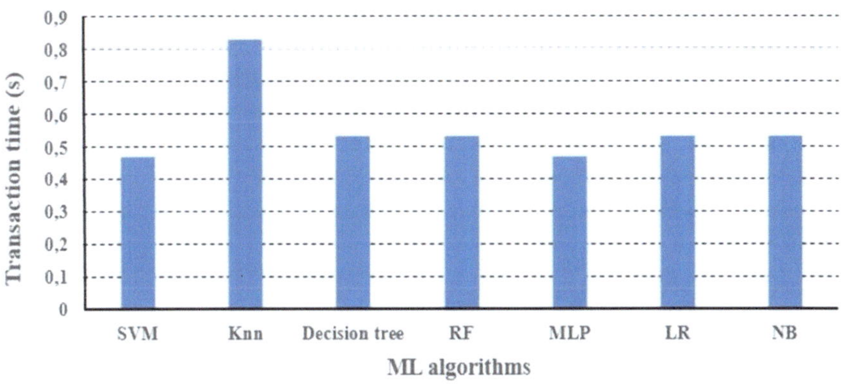

Fig. 7 Average ML model of time transaction/seconds

Table 3 Comparison of the proposed system with related work

Parameter ID	[6]	[7]	[8]	[9]	[10]	Our system
Pr1	√	X	√	X	√	√
Pr2	X	X	√	√	√	√
Pr3	√	X	X	X	√	√
Pr4	√	X	√	X	X	√
Pr5	X	X	X	X	X	√

Pr 1: Blockchain-based; Pr 2: Access Control-based; Pr 3: Security;
Pr 4: Integrity; Pr 5: Multi-agent sys X: Not supported; √: supported

5 Conclusion and Future Work

In this paper, we create a blockchain-based trustworthy service provisioning method to detect assaults on machine learning algorithms throughout this effort. Using the same principle to satisfy an organization's demand for data security, private blockchain would necessitate the authentication of many senior authorities while waiting for agreement. An opportunity verifies the various general agreement for the ML model, considering processing speed, time, and mineral wealth are all important considerations for the development and exploration of data blocks and might be a crucial analysis. The Solidity program code is used to create smart contracts, which are then performed on the Ethereum network. The results of simulations show that our system model can help safeguard IoT devices from malicious ML assaults.

In future work, an intellig_block evaluation trial with a case study can be performed to detect an attack to prove its feasibility and compatibility in IoT.

References

1. Dhifallah W, Tarhouni M, Moulahi T, Zidi S (2021) A Novel Realistic Dataset for Intrusion Detection in IoT based on Machine Learning, In: 2021 International symposium on networks, computers and communications (ISNCC). IEEE, Dubai, United Arab Emirates, pp 1–6. https://doi.org/10.1109/ISNCC52172.2021.9615841
2. Banerjee M, Lee J, Choo K-KR (2018) A blockchain future for internet of things security: a position paper. Digital Commun Network 4(3):149–160. https://doi.org/10.1016/j.dcan.2017.10.006
3. Wattenhofer R (2017) Distributed ledger technology: the science of the blockchain, Second revised Inverted Forest Publishing Aalborg
4. Salah K, Rehman MHU, Nizamuddin N, Al-Fuqaha A (2019) Blockchain for AI: review and open research challenges. IEEE Access 7:10127–10149. https://doi.org/10.1109/ACCESS.2018.2890507
5. Wright K-L, Martinez M, Chadha U, Krishnamachari B (2018) SmartEdge: A Smart Contract for Edge Computing, In: 2018 IEEE International Conference on Internet of Things (iThings) and IEEE Green Computing and Communications (GreenCom) and IEEE Cyber, Physical and Social Computing (CPSCom) and IEEE Smart Data (SmartData), Halifax, NS, Canada: IEEE, pp 1685–1690. https://doi.org/10.1109/Cybermatics_2018.2018.00281
6. Swan M (2015) Blockchain thinking : the brain as a decentralized autonomous corporation [Commentary]. IEEE Technol Soc Mag 34(4):41–52. https://doi.org/10.1109/MTS.2015.2494358
7. Reyna A, Martín C, Chen J, Soler E, Díaz M (2018) On blockchain and its integration with IoT. Challenges and opportunities. Futur Gener Comput Syst 88:173–190. https://doi.org/10.1016/j.future.2018.05.046
8. Qian Y et al (2018) Towards decentralized IoT security enhancement: a blockchain approach. Comput Electr Eng 72:266–273. https://doi.org/10.1016/j.compeleceng.2018.08.021
9. Rathore S, Pan Y, Park JH (2019) BlockDeepNet: a blockchain-based secure deep learning for IoT network. Sustainability 11(14):14. https://doi.org/10.3390/su11143974
10. Moustafa N (2021) A new distributed architecture for evaluating AI-based security systems at the edge: network TON_IoT datasets. Sustain Cities Soc 72:102994. https://doi.org/10.1016/j.scs.2021.102994

Indirect Space Vector Modulation for Matrix Converter

Dalila Chouaya and Mouna Ben Hamed

Abstract Matrix converters (MCs) have become a popular research topic in the last decade due to their potential to provide amplitude and frequency conversion, bidirectional power flow and input displacement factor control, all in a single-stage converter (SSC). These converters are capable of transforming both voltage and frequency simultaneously and can be used in multi-phase systems. However, despite their advantages, matrix converters are not commonly used in commercial applications. There are several reasons for this lack of adoption. study provides a comprehensive analysis of the various aspects of MC topologies, including the principles of bi-directional switch, modulation techniques, input filter design, and protection issues. MC are still highly appealing for certain applications despite their mentioned drawbacks. This is due to the need to avoid energy storage elements like capacitors and inductors, the cost of power semiconductors continuing to decrease, and the MC's ability to be cost competitive while offering bidirectional flow capability and displacement factor control for regeneration requirements. The MC utilizes two types of space vector modulation methods: direct space vector modulation (DSVM) and indirect space vector modulation (ISVM) strategy. This paper focuses exclusively on the examination of the ISVM strategy and its implementation through the use of the MATLAB/Simulink area.

Keywords Matrix converter · Single-stage converter · Direct matrix converter · Indirect matrix converter · Back-to-back converter · Indirect space vector modulation

D. Chouaya (✉) · M. B. Hamed
Processes, Energy, Environment and Electrical Systems (PEESE), National Engineering School of Gabes, University of Gabes, 6029 Gabes, Tunisia
e-mail: dalila.chouaya.eng@gmail.com

© The Author(s), under exclusive license to Springer Nature Singapore Pte Ltd. 2024
A. Mellit et al. (eds.), *Proceedings of the 2nd International Conference on Green Energy Conversion System*, Green Energy and Technology,
https://doi.org/10.1007/978-981-97-6148-7_62

1 Introduction

The MC is a relatively new technological development, although its principles have been known since the 1970s. Recent advances have been made in critical areas of MC operation, such as multi-step commutation techniques. MC has replaced Three phase rectifier and Voltage Source Inverter (VSI) in recent AC-DC-AC conversion developments. MC is an AC-AC converter with superior performance to conventional VSI.

MCs, which offer bi-directional power flow, sinusoidal input/output signals, and controllable input power factor, have been the focus of much attention in recent years. They hold the potential to serve as a suitable replacement for back-to-back converters (B2BCs), but comparative analysis between the two is challenging due to the large number of system characteristics and inherent differences between the two topologies [1].

MCs can be classified into two groups: direct and indirect, also known as dual bridge [2], sparse [3], or two-stage [4] matrix converters. The direct Matrix converter (DMC) (see Fig. 1a) can be configured with a 3 $\times$ 3 array of bi-directional switches, as depicted in Fig. 1b.

The DMC can produce high quality sinusoidal input and output waveforms through the use of suitable modulation techniques, including the Venturini method or space vector modulation (SVM) [5]. The indirect matrix converter (IMC), depicted

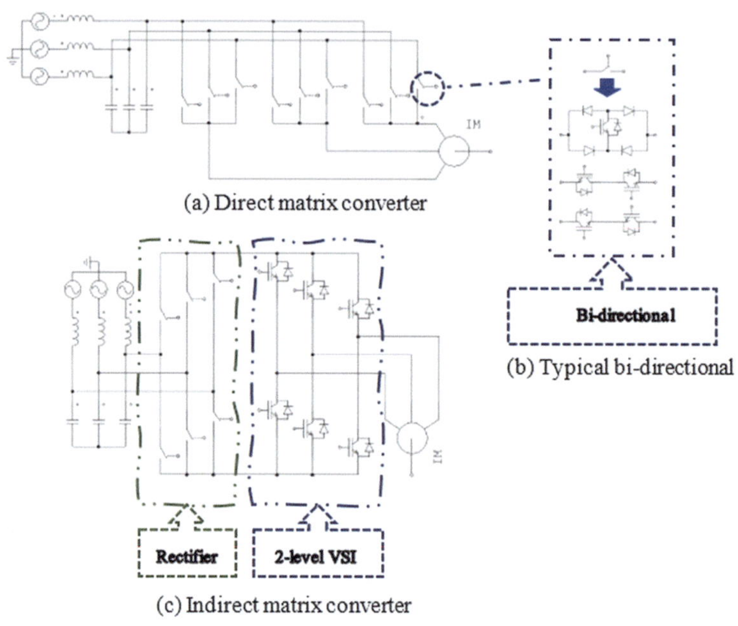

Fig. 1 Matrix converter topologies

in Fig. 1c, comprises a four-quadrant current source rectifier and a 2-level voltage source inverter, and has the capability to generate input and output waveforms of comparable quality to the DMC.

The IMC is sometimes preferred over the DMC because of its simpler and safer switch commutation, the potential to reduce the number of power semiconductor switches, and the ability to build complex converter topologies with multiple input and output ports.

The SVM technique has been effectively utilized in MC to control input power factor, reduce switch commutation, and achieve the highest voltage transfer ratio, with two types, DSVM and ISVM. This paper specifically focuses on the ISVM method of controlling MC by selecting switching functions to achieve the desired output voltages and input currents.

2 Overview of MC

Matrix Converters are a different AC-DC-AC topology without extra passive elements. A 3 X 3 MC is commonly used in power electronic applications due to the three-phase source and load. A schematic diagram of the power circuit in a 3X3 MC is presented in Fig. 2, which displays the bidirectional switch matrix, a potential implementation of the bidirectional switches, and an input filter for smoothing the source currents.

The control stage of the power converter ensures the production of safe control signals for the power switches to achieve the desired output voltage and input current

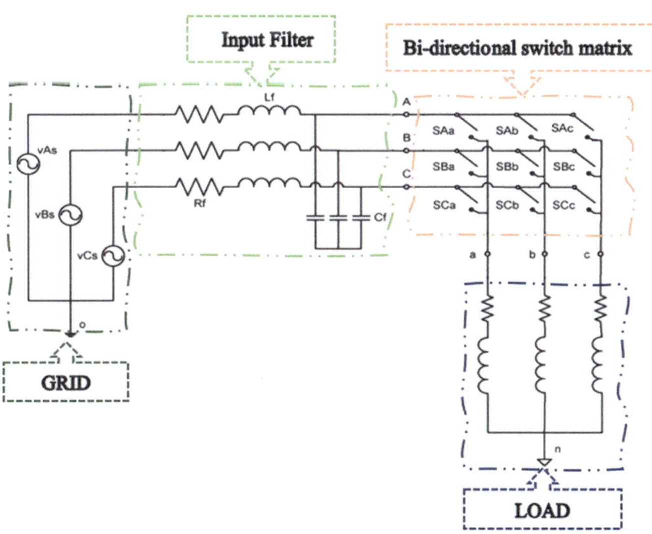

Fig. 2 The schematic diagram of the 3 phase to 3 phase MC topology

waveforms. The control stage of the model will be implemented by employing 9 switching functions for the 9 ideal bidirectional switches.

3 MC Modulation

The Matrix Converters utilize an indirect modulation technique to generate switching pulses by dividing the transfer matrix into inverter and rectifier transfer matrices and calculating the duty ratio for each stage in an alpha–beta environment.

The ISVM algorithm for the MC is akin to the B2B converter, albeit lacking a physical DC link. The ISVM technique partitions the switching stages into an inverter stage and a rectifier stage, joined via a virtual DC link. This results in two transfer functions, one for the inverter and one for the rectifier, as illustrated in Fig. 3.

Where S_{jk} is the switching function of a bi-directional switch, which is defined as:

$$S_{jk}\begin{cases} 1S_{jk}\ closed \\ 0S_{jk}\ opened \end{cases}, j \in \{a, b, c\}, k \in \{A, B, C\} \tag{1}$$

The constraints can be expressed as:

$$S_{ak} + S_{bk} + S_{ck} = 1, k \in \{A, B, C\} \tag{2}$$

The SVM comprises a collection of vectors that are construed as instant space vectors of input and output voltage and current with low Total Harmonic Distortion (THD).

The output voltages V_o and the input currents i_i which can be determined by:

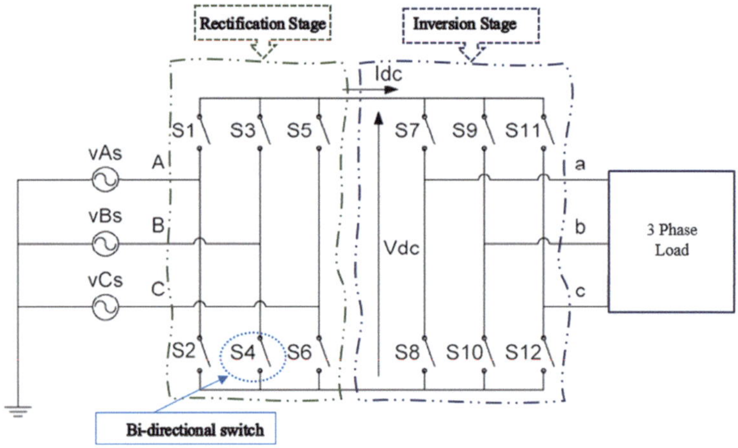

Fig. 3 Matrix converter equivalent circuit

$$V_o = \begin{bmatrix} V_A \\ V_B \\ V_C \end{bmatrix} = \mathrm{T}.\begin{bmatrix} V_a \\ V_b \\ V_c \end{bmatrix} = \mathrm{T}.V_i, i_i = \begin{bmatrix} i_a \\ i_b \\ i_c \end{bmatrix} = \mathrm{T}^T.\begin{bmatrix} i_A \\ i_B \\ i_C \end{bmatrix} = \mathrm{T}^T.i_o \quad (3)$$

The basic idea of the ISVM technique is to decouple the control of the input current and the control of the output voltage.

$$T = \begin{bmatrix} S_{aA} & S_{bA} & S_{cA} \\ S_{aB} & S_{bB} & S_{cB} \\ S_{aC} & S_{bC} & S_{cC} \end{bmatrix} = \begin{bmatrix} S_7 & S_8 \\ S_9 & S_{10} \\ S_{11} & S_{12} \end{bmatrix}.\begin{bmatrix} S_1 & S_3 & S_5 \\ S_2 & S_4 & S_6 \end{bmatrix} = T_{INV}.T_{REC} \quad (4)$$

where T^T is the transpose of the matrix, $V_{A,B,C}$ are the output voltages, $V_{a,b,c}$ are the input voltages, $i_{A,B,C}$ are the output currents and $i_{a,b,c}$ are the input phase currents. T_{INV} and T_{REC} represent respectively the transfer matrices of the inversion stage and rectification stage.

3.1 Rectification Stage

The input current space vector I_{IN} and input voltage space vector V_{IN} are expressed as space vectors using the transformation such as:

$$\begin{cases} I_{IN} = \frac{2}{3}\left[I_a + I_b.e^{j2\pi/3} + I_c.e^{j4\pi/3}\right] \\ V_{IN} = \frac{2}{3}\left[V_a + V_b.e^{j2\pi/3} + V_c.e^{j4\pi/3}\right] \end{cases} \quad (5)$$

The space vector diagram of the rectifier, is divided into six sectors. Based on SVM, the reference vector can be synthesized by two adjacent space vectors (I_γ and I_δ) and the zero current vector "reference vector", I_I^*:

$$I_I^* = d_\gamma I_\gamma + d_\delta I_\delta \quad (6)$$

where d_γ and d_δ are the duty cycles for applying the vector I_γ and I_δ within the switching period. These duty cycles are calculated using the equations given below:

$$\begin{cases} d_\gamma = T_\gamma/T_S = m_C.\sin\left(\frac{\pi}{3} - \theta_C\right) \\ d_\delta = T_\delta/T_S = m_C.\sin(\theta_C) \\ d_{0C} = T_{0C}/T_S = 1 - d_\gamma - d_\delta \end{cases} \quad (7)$$

where m_C is the current modulation index of the rectifier and defines the desired current transfer ratio such as:

$$\begin{cases} m_c = \frac{I_I^*}{I_{DC}} \\ 0 \le m_c \le 1 \end{cases} \quad (8)$$

The current modulation index m_C is often fixed to unity and the voltage modulation index m_V is variable according to a required overall voltage transfer gain.

3.2 Inversion Stage

Since the rectifier stage is usually operated with the condition $m_C = 1$ and unity input displacement angle $\varphi_{IN} = 0$, the V_{DC} become simplified to:

$$V_{DC} = \frac{3}{2}.V_{IN} \tag{9}$$

The output reference vector can be synthesized as below:

$$V_0^* = d_\alpha.V_\alpha + d_\beta.V_\beta \tag{10}$$

The duty cycle of the active vectors is calculated by:

$$\begin{cases} d_\alpha = T_\alpha/T_S = m_V.\sin\left(\frac{\pi}{3} - \theta_V\right) \\ d_\beta = T_\alpha/T_S = m_V.\sin(\theta_V) \\ d_{0V} = T_{0V}/T_S = 1 - d_\alpha - d_\beta \end{cases} \tag{11}$$

where m_V is the voltage modulation index of the inverter and defines the desired voltage transfer ratio such as:

$$\begin{cases} m_v = \frac{\sqrt{3}.V_0^*}{V_{DC}} \\ 0 \leq m_v \leq 1 \end{cases} \tag{12}$$

3.3 Modulation for the Entire Matrix Converter

Using the modulation schemes applied to the rectification and inversion stages. Since both the inverter and the rectifier hexagons contain six sectors, there are 36 combinations or operating mode. Four new vectors, $\left(V_\alpha - I_\gamma\right)$ pair, $(V_\alpha - I_\delta)$ pair, $\left(V_\beta - I_\gamma\right)$ pair and $\left(V_\beta - I_\delta\right)$ pair, are created and duty cycle of new vectors becomes $d_{\alpha\delta}$ and $d_{\beta\delta}$, respectively, as defined below:

$$\begin{cases} d_{\alpha\gamma} = d_\alpha.d_\gamma = m_V.\sin\left(\frac{\pi}{3} - \theta_V\right).\sin\left(\frac{\pi}{3} - \theta_C\right) = T_{\alpha\gamma}/T_S \\ d_{\alpha\delta} = d_\alpha.d_\delta = m_V.\sin\left(\frac{\pi}{3} - \theta_V\right).\sin(\theta_C) = T_{\alpha\delta}/T_S \\ d_{\beta\gamma} = d_\beta.d_\gamma = m_V.\sin(\theta_V).\sin\left(\frac{\pi}{3} - \theta_C\right) = T_{\beta\gamma}/T_S \\ d_{\beta\delta} = d_\beta.d_\delta = m_V.\sin(\theta_V).\sin(\theta_C) = T_{\beta\delta}/T_S \end{cases} \tag{13}$$

During the remaining part of the switching period T_S, zero vector is applied:

$$d_0 = 1 - d_{\alpha\gamma} - d_{\alpha\delta} - d_{\beta\gamma} - d_{\beta\delta} = T_0/T_S \tag{14}$$

4 Simulation and Results (with R-L Load)

The model of the MC shown in Fig. 3 has been simulated using Matlab/Simulink, with the passive R-L load ($R = 100\Omega$ and $L = 40mH$).

In this case, the utilization of R-L load is employed for the MC, and the control strategy adopted is ISVM. The input/output current and voltage characteristics are shown above (Figs. 4 and 5). The input phase voltage and current (V_a, i_a) are shown in Fig. 4a, and the output phase voltage and current (V_A, i_A) are shown in Fig. 4b.

The resultant current and voltage are given in Fig. 5. Here the output voltage as well as current is in sinusoidal in shape. This property is the main attraction of matrix converter.

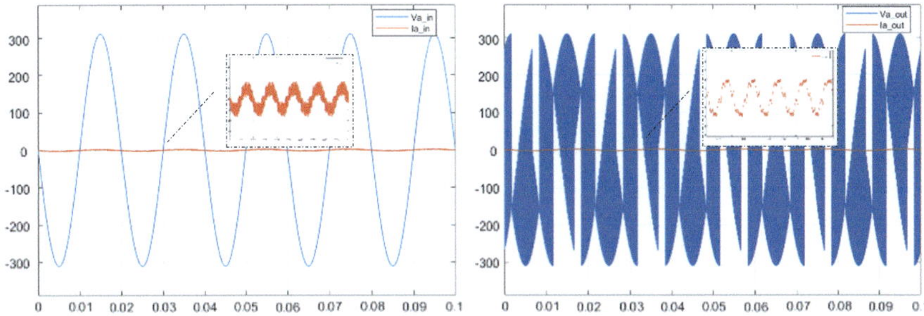

(a) (b)

Fig. 4 **a** The input phase voltage and current per phase (V_a, i_a), **b** The output phase voltage and current per phase (V_A, i_A)

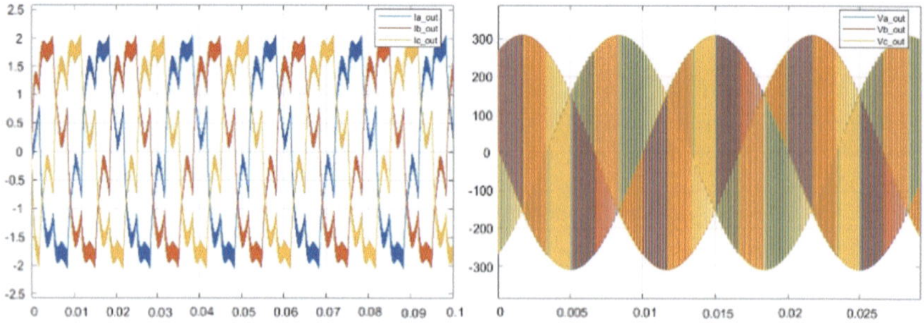

(a) (b)

Fig. 5 **a** The output phase currents $i_{A,B,C}$, **b** The output phase voltages $V_{A,B,C}$

5 Conclusion

The potential of matrix converters to improve technologies related to smart and resilient grids, renewables connection to the grid, electrical drives, E mobility, and sustainable transportation is significant due to advancements in semiconductor technologies, which have enabled higher power densities and increased robustness [6].

The Matrix Converters employ an effective indirect modulation technique to generate switching pulses by dividing the transfer matrix into inverter and rectifier transfer matrices, and calculating the duty ratio at alpha–beta environment.

This paper aims to create a sinusoidal output voltage with a different level than the input voltage using the ISVM technique in MCs. This approach reduces the complexity of space vector modulated multilevel inverters. The ISVM technique employs the inverter stage SVM with rectifier stage SVM to obtain simultaneous output voltage and input current.

Instead of ISVM, various efficient methods, such as DSVM and AV method, can be used to reduce switching pulses in bidirectional switches, thus addressing the main disadvantage of the MC, and can be connected to a motor-based controller to study system parameter variations.

References

1. Do NH, Ho SM, Le QT, Hoang T, Tran TM, Vu P (2022) Comparison between matrix and back-to-back converter in flywheel energy storage systems. Int J Power Electron Drive Syst 13(2):744
2. Saha J, Yadav GNB, Panda SK (2020) A bidirectional matrix-based AC-DC dual-active bridge for modular solid-state-transformers. In: IECON 2020 The 46th Annual Conference of the IEEE Industrial Electronics Society (pp 1136–1141). IEEE

3. Alammari R, Aleem Z, Iqbal A, Winberg S (2019) Matrix converters for electric power conversion: review of topologies and basic control techniques. Int Trans Electr Energy Syst 29(10):e12063
4. Wang H, Zhang Y, Su M, Sun Y, Li X, Zhang G (2018) Control method for the two-stage matrix converter to enhance the linear voltage transfer ratio. IET Power Electron 11(14):2295–2301
5. Shanhu L, Sunpeng C, Xu L, Deng W, Junyu H (2023) A new four-step commutation method for indirect matrix converter with active-voltage vectors being applied. IEEE J Emerg Sel Top Power Electron 11(4):3949–3956
6. Bento A, Paraíso G, Costa P, Zhang L, Geury T, Pinto SF, Silva JF (2021) On the potential contributions of matrix converters for the future grid operation, sustainable transportation and electrical drives innovation. Appl Sci 11(10):4597

ZnO:Al Properties Study for a Transparent Electrode Application

Djouweyda Ferkha and Douniazad Mezdour

Abstract This work deals with the electrical, optical and structural properties of Aluminum (Al) doped Zinc Oxide (ZnO) thin films. The layers were prepared by the sol–gel method and deposited onto glass substrates using the spin coating technique. The effect of Al incorporation on the electrical and optical properties of the samples was investigated by the four-probe method and the UV–Vis spectroscopy while the X-ray diffraction technique was used to investigate their structure. According to the transmittance curves analysis, the films exhibited a high optical transmission up to 90%. It has been found that the apparent optical gap decreases with increasing the doping level. The electrical resistivity was found to be in the range of 10^{-3} Ω. cm for the doped films. All the samples exhibited a hexagonal Wurtzite structure. The best performances were obtained for the 4% Al doped film presenting the highest conductivity and crystallinity combined to a good transparency. These low-cost films could be successfully used as transparent electrodes in Photovoltaic applications.

Keywords ZnO · TCO · Electrical properties · Transmittance

1 Introduction

Transparent conductive materials are an integral part of many optoelectronic devices (Solar cells, organic light emitting diodes, tactile sensors…etc.) [1–3]. Transparent conductive oxides (TCO) are a class of materials combining two properties: High electrical conductivity and transmittance in the visible domain, making them key materials in many applications as transparent electrodes. With a resistivity of about 10^{-4} Ω.cm and a transmittance of 85% in the visible domain, Tin doped Indium Oxide (ITO) is the preferred material [4]. However, its fragility, instability under plasma process, poor mechanical resistance and increasing cost due to its high Indium content

D. Ferkha · D. Mezdour (✉)
Labotatory of Materials Studies, Faculty of Sciences and Technology, University of Jijel, 18000 Jijel, Algeria
e-mail: d_mezdour@univ-jijel.dz

© The Author(s), under exclusive license to Springer Nature Singapore Pte Ltd. 2024

A. Mellit et al. (eds.), *Proceedings of the 2nd International Conference on Green Energy Conversion System*, Green Energy and Technology,
https://doi.org/10.1007/978-981-97-6148-7_63

are all reasons to look for alternative materials. Among known TCOs, Zinc oxide (ZnO) has been extensively studied because of its transparency in the visible and Infrared domain (direct gap energy of 3.37 eV) [5], high electrical conductivity and abundance on the earth at a level of 27 ppm. ZnO is an eco-friendly component with interesting properties for sensing [6] and wave guiding [7] applications, in addition to its easy synthesis and low cost. In this study, electrical and optical properties of pure and Al doped ZnO films are investigated. This aims to understand the key points for improving the performance of transparent oxide electrodes on the optical and electrical levels, allowing their integration in photovoltaic applications.

2 Experimental Procedure

The undoped ZnO solution at a molar concentration of 0.2 mol/l was prepared by dissolving Zinc Acetate dihydrate in Ethanol. After magnetic stirring during 10 mn, an excess of Monoethanolamine (MEA) was added drop wise until reaching a (MEA/ ZnO2 +) ratio of 0.75. The solution was stirred again during 2 h at 60 °C. The obtained solution was homogeneous and transparent. The same procedure was used to obtain doped ZnO: Aluminum Nitrate was added jointly to Zinc Acetate. The thin layers were deposited on glass substrates at a rate of 2000 tr/mn during 5 s. Each layer was dried at 300 °C during 4 mn. All the samples were cured at 180 °C during 180 mn.

The thin layers were characterized by X ray diffraction (XRD) technique to identify the structure and de grain size. The optical transmission in the UV–visible range was used to determine the optical gap. The electrical conductivity was evaluated by the four-probe method.

3 Results and Discussion

3.1 Optical Properties

The use of ZnO layers as transparent electrodes in solar cells and organic light emitting diodes (OLEDs) requires thin conductive layers with high optical transparency. The optical transmission spectra of ZnO thin films grown on glass substrates are shown in Fig. 1. All the layers have a high transparency in the visible domain (300–800 nm). The average transmittance value of the thin films in the visible range is found to be of 85%. The slight variation of the transmittance with increasing the Al concentration is indicative of a good incorporation of aluminum in the ZnO structure. This high transparency is attributed to a low dispersion effect resulting from the structural homogeneity of the layers and the apparent high crystallinity. The obtained transmittance values are comparable to that of ITO.

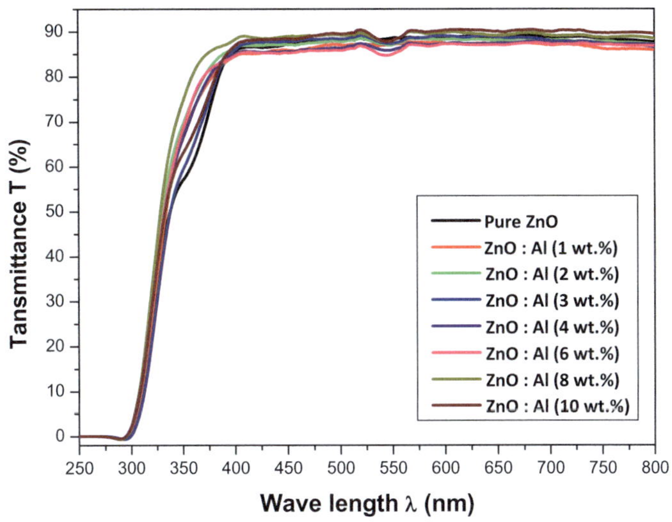

Fig. 1 Transmission spectra T(λ) of Al doped ZnO thin films

The gap energy as a function of the Al concentration was studied using the Tauc's law:

$$(\alpha h\nu)^2 = A\left(h\nu - E_g\right) \quad (1)$$

where α is the absorption coefficient, h the Planck constant, ν the frequency. A is a constant and E_g is the optical gap.

The values of the optical gap are obtained by extrapolating the linear segments of the corresponding curves.

Figure 2 shows the energy dependent plots for thin ZnO layers doped with Al at different concentrations. A slight increase in the gap with the doping rate up to 8% is observed. Indeed, a value of 3.65 eV is registered for the pure Zn0 sample. The gap value reaches 3.72 eV at 8% of Al content. Similar values were reported in previous studies [8]. At 10% of Al, the gap value decreases but remains higher than that of the pure film. It was reported that the optimum impurity doping level for the highest carriers concentration varies within 1–5 at.% for Al doping [9].

This behaviour is in good agreement with the litterature [10, 11]. The increase in the gap with doping is in line with the Burstein-Moss effect [12], as a result of some states close to the conduction band being populated.

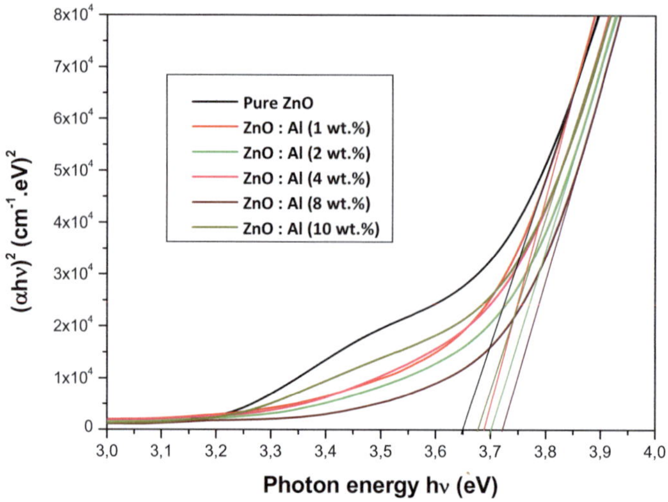

Fig. 2 Determination of the gap energy E_g for the thin layers of ZnO:Al using Tauc's law

3.2 *Electrical Properties*

A JANDEL four probe connected to a Keithley 2400 electrometer was used to measure the sheet resistance $R_\square$ of the samples. The resistivity ρ is expressed by:

$$\rho = R_\square.d \tag{2}$$

where d is the sample thickness.

The mean sheet resistance and resistivity values calculated for each sample are resumed in Table 1.

In the case of the undoped ZnO layer, the charge carriers originate from oxygen gaps and zinc interstitials resulting in a semi conductive behavior. The resistivity of the samples decreases with the increase of the doping rate. The enhancement of conductivity with the increase in the Aluminum concentration may be interpreted by

Table 1 Resistivity values of pure and Al doped ZnO thin films

Sample	Thickness d (nm)	Mean resistance $R_\square$ (Ω/□)	Resistivity ρ (Ω.cm)
Pure ZnO	238	540.76	1.34×10^{-2}
ZnO:Al 1%	231	476.79	1.013×10^{-2}
ZnO:Al 2%	226	445.24	9.81×10^{-3}
ZnO:Al 3%	227	201.82	4.93×10^{-3}
ZnO:Al 4%	229	198.33	4.39×10^{-3}
ZnO:Al 6%	232	219.53	5.09×10^{-3}

the increase in the number of carriers from the Al^{+3} donor ions incorporated in the Zn^{+2} cations substitution or interstitial locations [13]. The lower resistivity value is obtained for the ZnO layer containing 4% of Al. It is one order of magnitude greater than the known value of ITO but is still interesting regarding the layer thickness.

3.3 Structural Properties

Thin layers of ZnO doped at different rates of aluminum were analyzed by the XRD technique. The results are represented in Fig. 3. The obtained spectra show a well crystallized layers having a hexagonal Wurtzite structure along the c axis. Indeed, a main peak is observed at an angle 2θ around 34.5°, except for the ZnO:Al 4% sample exhibiting a preferential orientation according to the (101) plane. The decrease in full width at half height maximum (FWHM) of a XRD peak is usually a sign of an improvement in crystal quality ascribed to an increase in the grain size in the studied sample.

The Scherrer formula estimates the size D of a very small crystal or grain from the width of the diffraction line [14]:

$$D = \frac{0.9.\lambda}{\beta.\cos\theta} \quad (3)$$

where β is the FWHM of the selected diffraction peak in radian, θ is the Bragg angle in ° and λ is the X-ray wavelength ($\lambda = 0.154$ nm).

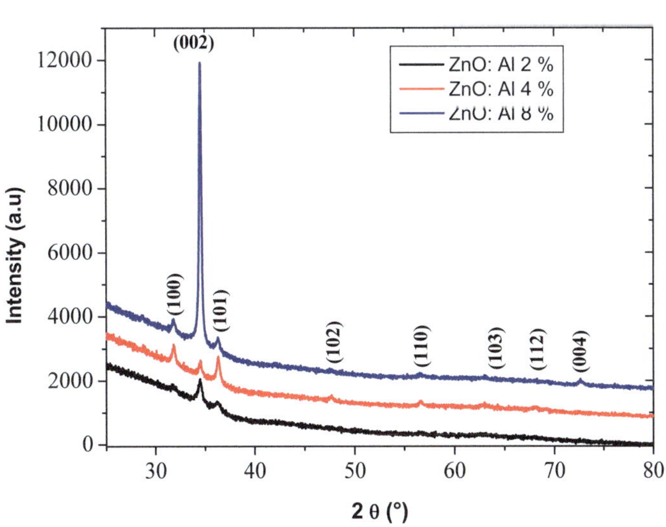

Fig. 3 XRD spectra obtained for thin layers of ZnO doped at different concentrations of Al (2, 4 and 8%)

Table 2 Calculation parameters of the grain size for Al doped ZnO thin films

Sample	(002) Peak location (°)	β (°)	D (nm)
ZnO:Al 2%	34.56	0.58	13.08
ZnO:Al 4%	34.45	0.352	21.56
ZnO:Al 8%	34.53	0.57	13.31

The variation in grain size of ZnO:Al films as a function of the aluminum doping rate is shown in Table 2.

The samples containing 2 and 8% of Al exhibit nearly the same grain size of 13 nm. It is noticed that the thin layer of ZnO:Al 4% is formed of crystallites of larger size compared to the other layers, reaching 21.5 nm. This is characteristic of a better cristallinity and is in good agreement with the electrical properties results.

4 Conclusion

Al doped ZnO thin layers were deposited on glass substrates by the low cost sol-gel technique. The effect of the dopant concentration on the optical, electrical and structural properties of the films has been studied. Optical characterization over a range of 200 to 800 nm showed a high transmittance of the synthesized films comparable to that of ITO. It has been shown that Al doping up to 8% increases the apparent gap. This is attributed to the Burnstein-Moss effect related to high carriers concentration. Good electrical properties were also obtained. The recorded resistivities are one order of magnitude higher than that of the ITO. A hexagonal Wurtzite structure was identified by XRD. A good correlation was found between electrical and structural properties. Indeed, the sample containing 4% of Al presented the lowest resistivity and the highest crystallinity.

These results confirm the suitability of the elaborated materials as transparent electrodes in Photovoltaic applications.

References

1. Wibowo A, Marsudi MA, Ikhlasul Amal M, Ananda MB, Stephanie R, Ardy H, Diguna LJ (2020) ZnO nanostructured materials for emerging solar cell applications. RSC Adv 10:42838–42859
2. Song T-B, Li N (2014) Emerging transparent conducting electrodes for organic light emitting diodes. Electronics 3(1):190–204
3. Li R, Chen G, He M, Tian J, Su B (2017) Patternable transparent and conductive elastomers towards flexible tactile/strain sensors. J Mater Chem C 33:1–5. https://doi.org/10.1039/C7TC02703F
4. MSE supplies home page, https://www.msesupplies.com/pages/material-properties-of-indium-tin-oxide-ito, Last accessed 2023/07/15.

5. Slassi A, Naji S, Benyoussef A, Hamedoun M (2014) On the transparent conducting oxide Al doped ZnO: first principles and Boltzmann equations study. J Alloy Compd 605:118–123
6. Sha R, Basak A, Maity PC, Badhulika S (2022) ZnO nano-structured based devices for chemical and optical sensing applications. Sens Actuators Rep 4:100098
7. Mammeri A, Bouachiba Y, Bouabellou A, Taabouche A, Rahal B, Serrar H, Amara S, Harouni S, Merabti H, Boukentoucha C, Aouati R (2023) Optogeometric and waveguiding properties of multimode ZnO planar waveguide sprayed thin films. Phys B: Condens Matter 663:414965
8. Mallick A, Kole A, Ghosh T, Chaudhuri P, Basak D (2014) Zinc aluminate spinel impurity phase in Al doped ZnO ceramic target and pulsed laser ablated films: Curse or blessing? Sol Energy 108:80–87
9. Mridha S, Basak D (2007) Aluminium doped ZnO films: electrical, optical and photoresponse studies. J Phys D-Appl Phys 40:6902–6907
10. Hafdallah A, Ynineb F, Aida MS, Attaf N (2011) In doped ZnO thin films. J Alloy Compd 509:7267–7270
11. Hammad TM, Salem JK, Harrison RG (2013) Structure, optical properties and synthesis of Co-doped ZnO superstructures. Appl Nanosci 2:133–139
12. Burstein E, Davisson J, Bell E, Turnre W, Lipson H (1954) Infrared photoconductivity due to neutral impurities in germanium. Phys Rev 93:65–68
13. Zhao J-L, Li X-M, Bian J-M, Yu W-D, Gao X-D (2005) Structural, optical and electrical properties of ZnO films grown by pulsed laser deposition (PLD). J Cryst Growth 276(3):507–512
14. Patterson AL (1939) The scherrer formula for X-Ray particle size determination. Phys Rev 56:978–982

Fuzzy Direct Power Control for Enhanced Power Quality Generated by PV System

Sofia Lalouni Belaid and Zoubir Boudries

Abstract Photovoltaic energy is widely used, because it is clean, free and its technology is now well mastered. This paper presents a new fuzzy direct power control (FDPC) technique to control the active power extracted from the photovoltaic array and sent to the grid, to compensate the reactive power and to reduce the total harmonic distortion (THD) of the grid current caused by the connection of non-linear loads. A fuzzy inference system is used to select the switching states of the inverter, and its input includes active/reactive power variations and the position of the line voltage vector. To illustrate the performance of the proposed technique in a grid-connected photovoltaic system, a simulation model was developed using Matlab/Simulink and Sim Power System, under variable irradiation and non-linear load connected to the grid. The simulation results demonstrate the effectiveness of the proposed FDPC technique.

Keywords PV system · Fuzzy MPPT · Fuzzy direct power control (DPC) · Total harmonic distortion (THD)

1 Introduction

Global energy demand has seriously stimulated the use of new energy sources with less impact on the environment [1, 2]. Renewable energy is clean and abundant on earth, for these benefits it has received a lot of attention [3]. Recently, photovoltaic technology is well mastered and developed, which makes its energy widely used [4]. Grid-connected photovoltaic installations have become very popular due to their reliability [5, 6]. Domestic and industrial equipment use electronic circuits whose

S. L. Belaid (✉) · Z. Boudries
Université de Bejaia, Faculté de Technologie, Laboratoire de Technologie Industrielle et de L'Information, 06000 Bejaia, Algérie
e-mail: sofia.lalouni@univ-bejaia.dz

Z. Boudries
e-mail: zoubir.boudries@univ-bejaia.dz

© The Author(s), under exclusive license to Springer Nature Singapore Pte Ltd. 2024

A. Mellit et al. (eds.), *Proceedings of the 2nd International Conference on Green Energy Conversion System*, Green Energy and Technology,
https://doi.org/10.1007/978-981-97-6148-7_64

behavior is non-linear, generating harmonics, thus polluting the network and leading to poor power quality [2]. The traditional simple and low-cost depollution solutions are passive filters [3]. Active filtering is a more interesting solution for the elimination of current harmonics, reactive power compensation and power factor correction [3, 7]. Essentially two parts make up the structure of the active power filter (APF), a power part and a control part [2]. Many references deal with the control of the active power filter in a photovoltaic system [6, 8, 9]. Recently, direct power control (DPC) is widely used [2, 5, 7, 9–11]. It has many advantages such as robustness, speed of response and ease of implementation [9–11]. However, DPC has a major drawback relating to the variable switching frequency which introduces a wide band of harmonics into the grid line current. A new approach, based on fuzzy logic is proposed in this paper, the fuzzy direct power control (FDPC) makes it possible: to minimize the harmonics in the grid current, to inject the photovoltaic power available in the network, to control reactive power and provide unity power factor.

2 Description of the Studied System

Figure 1 presents the studied system. It is composed by a photovoltaic generator connected to the grid via a controlled boost converter and a three-phase inverter. It also includes a non-linear load (uncontrolled rectifier and R, L load). The DC/DC converter performs the MPPT function using a fuzzy logic controller. Meanwhile, to improve the performance of the grid PV system, the inverter is controlled by FDPC.

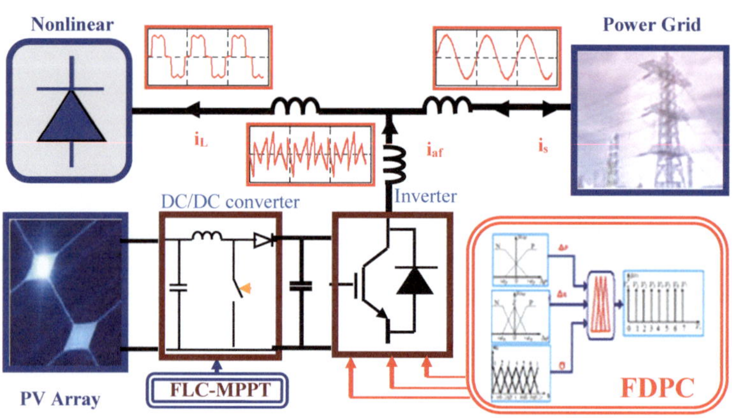

Fig. 1 The global system controlled by fuzzy direct power control FDPC

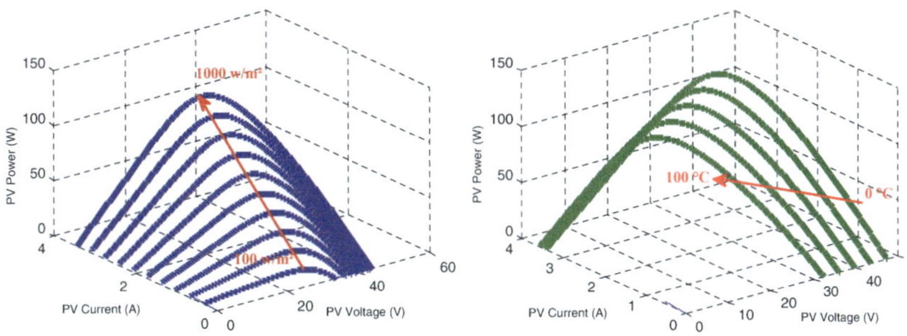

(a) Variable insolation and T=25°C. (b) Variable temperature and insolation G=1000W/m².

Fig. 2 PV panel 3D I-V-P characteristic curves

2.1 Modeling of the PV Array

In the literature [2–5, 7–10] different models of PV arrays are presented. These models vary according to the calculation method, the accuracy and the number of parameters used to simulate its operation. The PV current is calculated by the following equation:

$$I_{pv} = I_{sc}.\{1 - K_1[\exp(K_2 V_{pv}^{K_3}) - 1]\} \tag{1}$$

$$V_{pv,new} = V_{pv} - \beta_{oc}\Delta T_c - R_s\Delta I_{pv} \tag{2}$$

$$I_{pv,new} = I_{pv} + \alpha_{sc}\left(\frac{G}{G_{stc}}\right)\Delta T_c + \left(\frac{G}{G_{stc}} - 1\right)I_{sc,st} \tag{3}$$

where, K_1, K_2, K_3 depend on the PV array parameters. The expression can be found in [4]. The photovoltaic voltage and current vary as the environmental conditions change, using Eqs. (2) and (3).

α_{sc} and β_{oc} are the current and the voltage temperature coefficient respectively, ΔT_c is the temperature variation and R_s is the cell resistance.

Figure 2 shows the 3D I-V-P characteristics of the PV array (110W) simulated using Matlab/Simulink. In order to optimise the efficiency of the PV system, maximum power point tracking (MPPT) based fuzzy logic control (FLC) is used.

2.2 MPPT Based on Fuzzy Logic

The MPPT controller uses the PV voltage and current to computes the PV power; the PV voltage is varied using positive or negative step by a small or large amount in the direction that allows the increase of power P_{pv}. Fuzzy MPPT controller structure

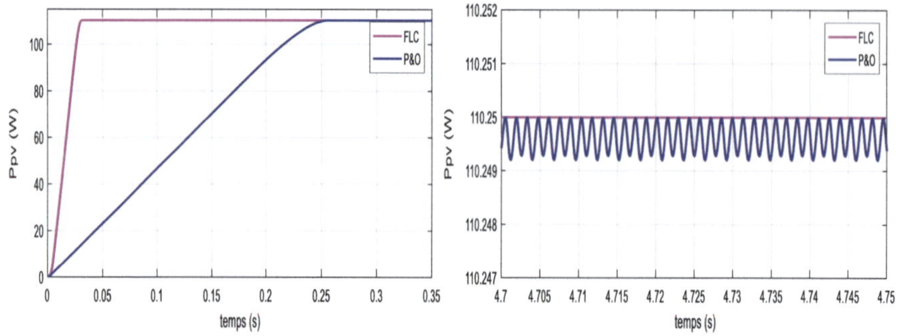

(a)Transient state. (b) Steady state.

Fig. 3 Photovoltaic power waveform

and control rules are given in [4, 5]. The results in Fig. 3 show that the FLC MPPT is fast, reaching the optimum in 0,027 s, whereas the P&O MPPT required 0,25 s and presents oscillations in study state.

2.3 *Active Power Filter (APF) and Fuzzy Direct Power Control (FDPC)*

The structure of the APF essentially consists of two parts, the power and the control part [2, 3]. This last one generates the optimal switching states of the inverter using Fuzzy Direct Power Control (FDPC). The APF aims to compensate harmonic pollution caused by the nonlinear load. Direct power control (DPC) selects the optimum inverter switching states using switching table, which has three inputs: active/reactive power errors and the line voltage vector angular position. DPC's main disadvantage lies with its variable switching frequency, which principally depends on hysteresis controllers. By replacing the hysteresis comparators and the switching table with a fuzzy inference system, this problem can be minimized. Instead of the two states (0, 1) provided by the hysteresis controllers, the FDPC allows to divide the active and reactive power errors into more subsections [5]. In this case, the switching vector will be based on the different levels of the power errors. Table 1 shows the switching table of the proposed control strategy. The active and reactive grid powers are estimated using the following formula:

$$\begin{cases} p_g = i_\alpha.v_\alpha + i_\beta.v_\beta \\ q_g = i_\alpha.v_\beta - i_\beta.v_\alpha \end{cases} \tag{4}$$

The sector position is computed using Eq. (5).

$$(n-2)\frac{\pi}{3} \leq \theta \leq (n-1)\frac{\pi}{3} \tag{5}$$

Table 1 Switching table of FDPC

Θ		θ_1	θ_2	θ_3	θ_4	θ_5	θ_6
Δp	Δq						
P	P	V_3	V_4	V_5	V_6	V_1	V_2
	Z	V_7	V_0	V_7	V_0	V_7	V_0
	N	V_5	V_6	V_1	V_2	V_3	V_4
N	P	V_1	V_2	V_3	V_4	V_5	V_6
	Z	V_0	V_7	V_0	V_7	V_0	V_7
	N	V_6	V_1	V_2	V_3	V_4	V_5

3 Simulation Results and Analysis

The system consists of a 2.2 kW PV generator. MATLAB/Simulink and Sim Power System have been used to model and develop the overall system. The FLC is used to control the MPPT and the inverter is controlled using FDPC. See [10] for the main system parameters used in the simulation. Different irradiance levels are applied as shown in Fig. 4. The result of Fig. 5 shows the PV power extracted using FLC, it provides fast response until irradiance changes. The DC voltage is well controlled to its reference value as shown in Fig. 6. Figure 7 shows the active and reactive power responses. The powers are decoupled and follow their reference values. We first consider the nonlinear load disconnected, then we introduce it at t = 5 s. During the interval [0 5], the PV power is totally injected into the grid, which is represented by the negative sign of grid active power. At t = 5 s, the non-linear load is connected, then, the PV generator supply the load and injects the remaining energy to the grid. During the interval [8 10], the non-linear load consumption will be furnished by the grid because the generated PV power is null. Figure 8 show source current, APF current and load current waveform with their zooms. The APF current and grid current are sinusoidal and in opposition of phase to the source phase voltage (Fig. 9) with a less THD as shown in Fig. 10.

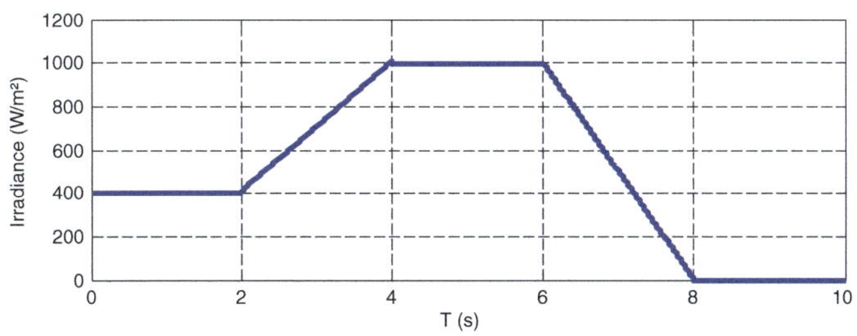

Fig. 4 Solar irradiance

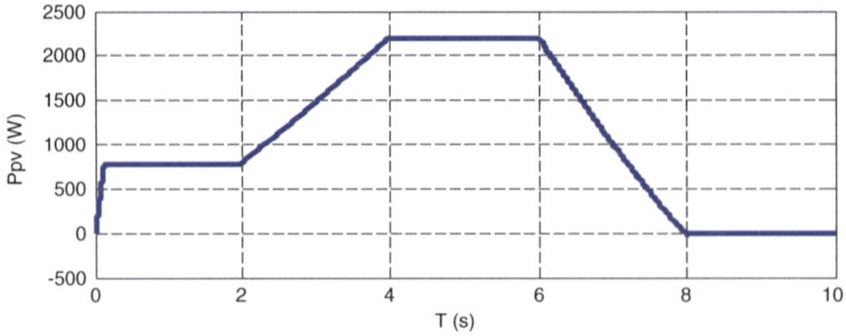

Fig. 5 Generator PV power

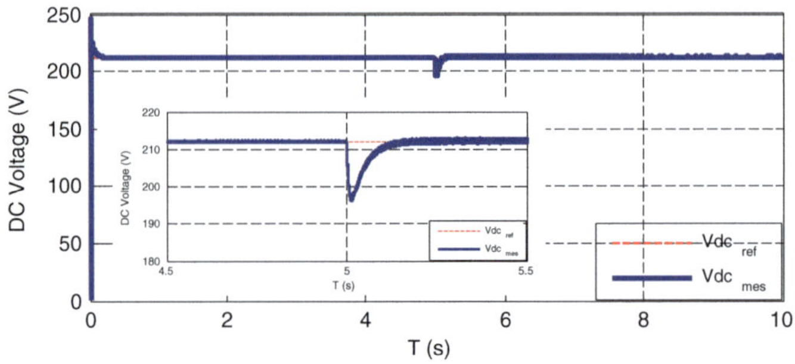

Fig. 6 DC voltage waveform

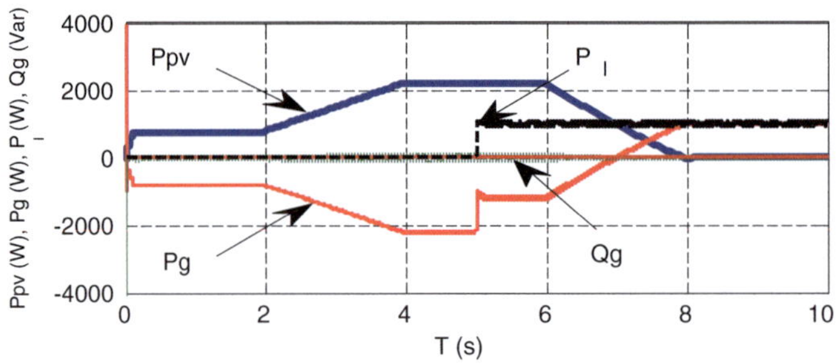

Fig. 7 PV, grid, load powers and reactive power

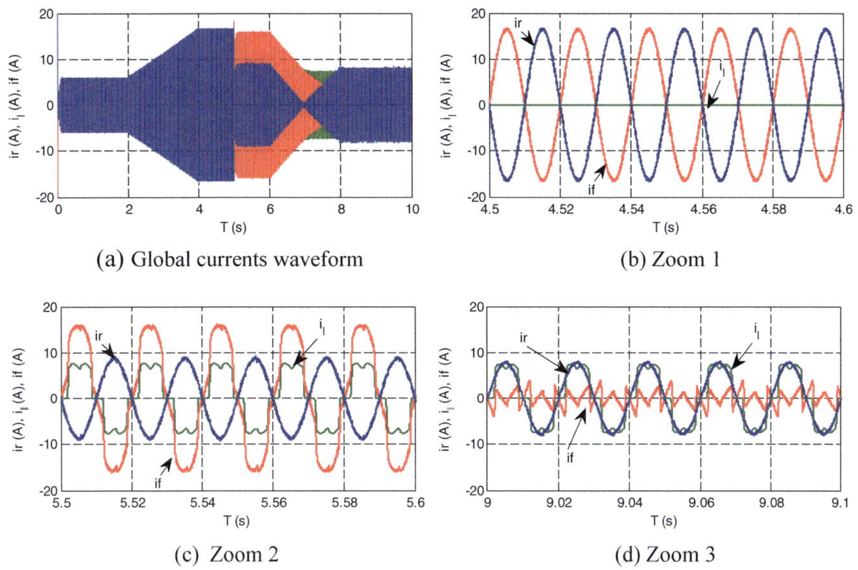

(a) Global currents waveform (b) Zoom 1

(c) Zoom 2 (d) Zoom 3

Fig. 8 Currents waveform of: source, APF and load with zooms

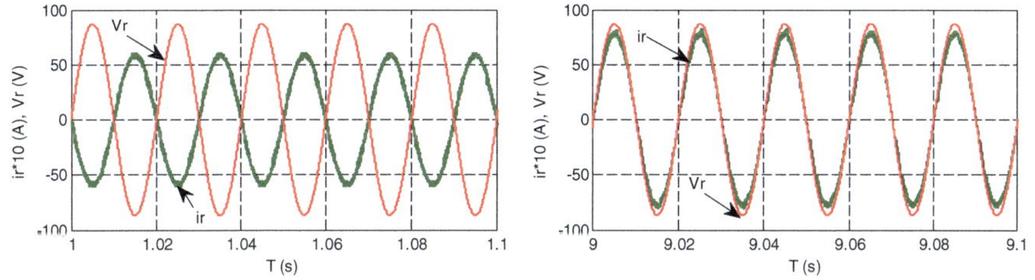

Fig. 9 Source voltage and current (zooms)

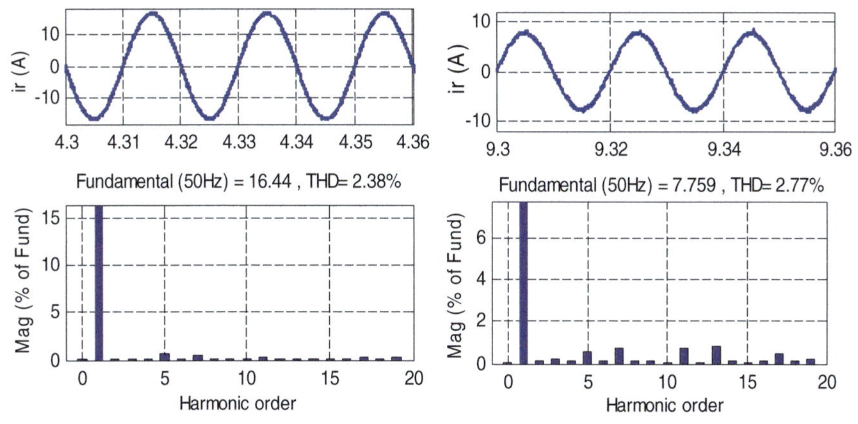

Fig. 10 Grid current THD

4 Conclusion

The effectiveness of FDPC in controlling the grid-connected PV system has been demonstrated in this paper. The studied system feeds a nonlinear load and injects the rest of the generated PV energy into the grid. The harmonic currents caused by the nonlinear load are compensated and the reactive power is well controlled that provide unity power factor. The obtained results, when varying irradiance and load, show that the controlled PV system ensures the functionality of the active power filter by balancing the harmonic current of the nonlinear load. The source currents have a sinusoidal waveform with a less THD, which is below the standard limit.

References

1. Zhang D, Rong C, Ahmad T, Xie H, Zhu H, Li X, Wu T (2022) Review and outlook of global energy use under the impact of COVID-19. Eng Rep 5(3):1–23
2. Bengourina MR, Rahli M, Hassaine L (2017) Direct power control of a grid connected photovoltaic system, associated with an active power filter. Revue des Energies Rnouvelables 20(1):99–109
3. Belaidi R, Haddouche A, Ghribi D, Mghezzi Larafi M (2017) A three-phase grid-connected PV system based on SAPF for power quality improvement. Telkomnika 15(3):1003
4. Deboucha H, Lalouni Belaid S (2017) Improved incremental conductance maximum power point tracking algorithm using fuzzy logic controller for photovoltaic system. Rev Roum Sci Techn–électrotechn Et énerg 62(4):381–387
5. Lalouni Belaid S, Boudries Z, Aoumer A (2019) Direct Power Control Improvement using Fuzzy Switching Table of Grid Connected PV System. In: Belasri A, Beldjilali SA (eds) ICREEC 2019, Springer Proceedings in Energy. Springer, Berlin
6. Tuyen ND, Fujita G (2015) PV-active power filter combination supplies power to nonlinear load and compensates utility current. IEEE Power Energy Technol Syst J 2:1–11
7. Tiwari H, Ghosh A, Ray PK, Subudhi B, Putrus G, Marzband M (2021) Direct Power Control of a Three-phase AC-DC Converter for Grid-connected Solar Photovoltaic System 2021 International Symposium of Asian Control Association on Intelligent Robotics and Industrial Automation, IRIA 2021, pp 125–130
8. Mishra DP, Rout KK, Mishra S, Nivas M, Naidu RK, Salkuti SR (2023) Power quality enhancement of grid-connected PV system. Int J Power Electron Drive Syst (IJPEDS) 14(1):369–377
9. Amrani MN, Dib A (2016) Direct power control for a photovoltaic conversion chain connected to a grid. Rev Roum Sci Techn–Electrotechn et Energ 61(4):378–382
10. Boukezata B, Chaoui A, Gaubert JP, Hachemi M (2016) Power quality improvement by an active power filter in grid-connected photovoltaic systems with optimized direct power control strategy. Electr Power Compon Syst 44(18):2036–2047
11. Trivedi TA, Jadeja R, Bhatt P (2015) A review on direct power control for applications to grid connected PWM converters. Eng Technol Appl Sci Res 5(4):841–849

Energy Security Policy in Tunisia: Discourse and Practice

Haifa Barhoumi

Abstract Increasing energy deficit is not a new challenge for Tunisia. The country is dependent on Algerian gas imports and intent to get its oil supplies from Libya. Its historical and social ties with both countries put it in a privileged position when it needs their support. As to other available options, the Tunisian energy security policy was modified to include renewable energy, and it has not spread. Yet changes at the international and regional levels have challenged the status-quo of Tunisian energy security policy. On the other hand, issues related to energy security have become prominent after 2011 in Tunisia. Within the last decade, populist discourse has challenged the official statistics about national energy resources. In this climate of mistrust, social unrest and rejection of foreign energy companies presence led to the departure of some energy investors while production decreased. This situation has eventually pushed Tunisia to look for new markets for its phosphate exports and new partners. As internal and external factors changed, Tunisia is at a turning point. This paper questions the resilience of Tunisian energy security policy as a discourse and practice after 2011.

Keywords Energy security · Energy policy · Tunisia · Renewable energy

1 Introduction

Experts are still debating the definition of energy security as a concept for it is "polysemic, multi-dimensional, contextual and dynamic in nature." [1, p. 43]. Four key elements stand out: availability, affordability, efficiency and environmental

H. Barhoumi (✉)
Department of Communication, Institut de Presse Et Des Sciences de L'Information (IPSI), Manouba University Campus, Manouba, Tunisia
e-mail: Barhoumi.haifa@gmail.com

© The Author(s), under exclusive license to Springer Nature Singapore Pte Ltd. 2024

A. Mellit et al. (eds.), *Proceedings of the 2nd International Conference on Green Energy Conversion System*, Green Energy and Technology,
https://doi.org/10.1007/978-981-97-6148-7_65

sustainability. In 2000, the UNDP defined energy security as "the availability of energy at all times in various forms, in sufficient quantities, and affordable prices," pointing out that "those conditions must prevail over the long term if energy is to contribute to sustainable development [which requires addressing environmental concerns]." [2, pp. 11–12].

In Tunisia, energy security is not clearly defined. It has been shaped gradually through concerns about energy supply, the need to satisfy an increasing demand in an unstable regional and global context, as well as decision-makers' attempts to reduce consumption through new efficient and sustainable usages. After the ratification of the 1995 Partnership with the European Union, Tunisia chose to increase energy exports. Gradually, its economy has "become largely dependent on its revenues from the energy sector (mines, phosphate and oil) and the tourism sector, while neglecting its manufacturing and agricultural sectors." However, unlike its Maghreb and Gulf neighbors, Tunisia has limited gas and oil reserves [3]. The state opted to emphasize the need to secure energy supply in its subsequent energy policies and legislation, especially when oil prices increased drastically and threatened the sustainability of state economy.

Yet, this choice has proved to be inefficient in the mid and long terms. Unlike the 1990s when the country had an excess in energy production for its smaller number of citizens and rare electrical devices, energy deficit has characterized the Tunisian energy balance since 2000, with a significant increase in demand after the 2011 uprising. For instance, electricity demand increased by 6.2% per year in the 1990s [4, p. 7].

This led to energy dependence in Tunisia. Both oil and gas imports have increased in terms of quantity and cost. For example, Tunisia imported 306.9 thousand tons of crude oil in 2010 (which cost 232.6 million Tunisian dinars) compared to 807.3 thousand tons in 2020 (for the price of 877.1 million Tunisian dinars). Natural gas imports cost 517.9 million Tunisian dinars in 2010 compared to 2129.1 million in 2020 [5, 6]. Gas production decreased within the last decade, as indicated in its annual national statistical reports, while Tunisia has been receiving gas from Algeria "as a royalty on the Algerian Transmed gas pipeline crossing Tunisia to Italy." [3].

Political instability and social unrest after the 2011 Uprising aggravated the situation as Tunisia has become "unable to guarantee the stability of its energy exports." Energy production has fluctuated due to technical issues, employment or environmental protests as well as contractual changes related to energy exploitation. For example, in 2020 Tunisia exported only 3 thousand tons of phosphate compared to almost 690 thousand tons in 2010 due to protests either against foreign energy companies or for employment [5, 6].

The efficiency of the Tunisian energy policy over the last decades is therefore questioned in this paper. As both internal and external factors changed again, Tunisian energy security policy is at a new turning point today.

2 Methodology

Data has been collected from various sources within a timeframe of 70 years, then analyzed from a historical perspective. Constitutions (1956, 2014, 2022), energy related legal documents, official statements and the Official Gazette were used to confirm policy or structural changes, or lack thereof. Earlier data was verified through the summary reports of the First Arab Energy Conference (1979) and Second Arab Energy Conference (1982), in particular Tunisian representatives' statements at these conferences. Documents related to the National Agency for Energy Management (NAEM, aka Agence Nationale pour la Maîtrise de l'Energie) and the Consumer Defense Organization provided details about energy policy implementation mechanisms and results. Samples of energy control awareness campaigns available at the Tunisian Institut de Presse et Sciences de l'Information were reviewed. In terms of energy resources, supply and demand data, relevant yearly reports by the National Institute of Statistics along with international organizations reports about energy in Tunisia were used.

3 Results and Discussion

3.1 Evolution of Energy Policy: From 1970 to 1990s

Tunisia was represented at the First Arab Energy Conference (1979), held to map available "traditional energy sources" in Arab states territories, then at the Second Arab Energy Conference (1982) which focused on "the impact of international developments on the status of energy in the Arab world," and aimed to "clarify energy demand and shape Arab policies to meet this demand" in the short and long terms, while looking for "new and renewable sources" in order to increase their reserves [7, p. 121]. Within thisframework, the Tunisian earliest reports concluded in the 1980s that traditional sources were limited, alternative energies were notwithin immediate reach, while prospective analyses noted future energy demand increase and lack of supply.

Accordingly, the government decided to adopt an energy conservation policy focused on sectorial demand reduction. This policy consisted in monitoring energy demand per sector and recommending available alternatives. In terms of structural reform, a new public institution was created in 1985 to implement this policy: National Agency for Energy (NAE). In practice, they tried to reduce energy demand in the transport, industry and agriculture sectors. For example, they tried to control energy consumption through cars and agricultural machinery technical control, experts' training in industrial use of energy, and awareness campaigns (Ibid). However, while these measures are still applied today, this policy did not have a significant impact. Oil import prices continued to increase after the Gulf

War in the 1990s and energy supply remained a challenge until an energy deficit started by the end of the decade. Some states chose to liberalize the oil market to reduce the cost. Tunisia did not.

3.2 Evolution of Energy Security Policy: From 2000 to 2011

Note After 2000, an efficient and affordable energy policy was needed, as Tunisia had to tackle its energy deficit at a time when oil prices had increased again after the invasion of Iraq. Decision-makers chose to reshape the energy conservation policy by trying to reduce citizens' daily energy demand through new consumption patterns. First, a legislative framework centered on energy security was ratified for the first time through Law n°2004-72, dated 2 August 2004 relating to energy control, later on modified by Law n°2009-7. Second, the existing institutional framework was reformed in tandem to facilitate policy implementation, with the creation of NAEM in 2004. The latter replaced NAE and started a partnership with the Consumer Defense Organization, then cooperated with STEG and other public and private institutions, at local and national levels, to implement a new "National Energy Control Program" over two phases: 2004–2007 and 2008–2011.

This strategy was quite efficient because the government used to have complete control over the media landscape in Tunisia. Between 2004 and 2010, "awareness campaigns" were organized through advertisements and frequent media coverage, as well as events organized by local authorities and reminders posted at public and private institutions alike, while related teaching material was added to school curriculum. Repeated for years through audiovisual and print content, in both public and private spheres, the message was the same. The Tunisian citizen was told: Tunisia does not have enough energy resources. Energy imports have to be reduced because prices increased. The solution is in your hands. Decrease your daily energy consumption.

3.3 Evolution of Energy Security Policy: From 2011 to Today

After 2011, the energy policy implementation was disrupted. With the fall of Ben Ali's political regime and ruling party, and as investigations revealed a web of corruption, some political actors questioned the credibility of official statistics about available national energy resources, including the data published after the independence of the country in 1956. They also questioned the presence of foreign companies and their role in terms of energy exploitation. This rhetoric quickly spread on social media through the hashtag "winou el petrol?" (where's the oil?), leading to protests that blocked energy exports (in Kerkanah for example). This issue was raised again in the 2019 legislative elections by some political candidates who promised the electorate to unveil the "truth" about "hidden" energy reserves once in power [8].

Between 2011 and 2018, violent protests took place in eastern and southern cities that produce energy, leading to foreign investors' departure. For instance, Petrofac left Kerkanah and Tunisia in 2018, after phosphate exports were delayed several times and eventually blocked by citizens. Consequently, Tunisia lost energy revenues for a while, thus exacerbating its economic crisis and energy deficit ([8], pp. 36–37).

In 2017, the government tried to improve the energy security policy by shifting its focus to renewable energies citing their long-term affordability and environmental sustainability. Order n°983 issued by the government in 2017 relating to the creation of a Transnational Energy Fund introduced a distinct funding mechanism for the implementation of renewable energy policy [9, p.28]. Yet, the transformation has been slow, partly due to lack of technology and the relatively high cost of renewable energy forcitizens.

3.4 *Renewable Energy: A Turning Point?*

Even though Tunisia was among the first countries to invest in renewable energies, in 2021, "only 3% of Tunisia's electricity was generated from renewables, including hydroelectric, solar and wind energy." [3]. In September 2022, renewable electricity represented "only 1% of primary energy resources." [10, p. 5]. According to NAEM, "despite all the efforts made for more thantwo decades, the contribution of renewable energies in the Tunisian energy balance is still insignificant. In fact, the public policyfor the development of renewable energy was rather modest until 2009 when it was reinforced with the implementation of moreproactive incentive and regulatory instruments. However, [...] the attitude of public authorities was characterized by some caution,and this prevented the development of some renewable sectors, the production of electricity in particular." [4 , p. 7].

International public aid, like the UN Sustainable Energy for All initiative, and development agencies have been tasked with "[funding] the problem of energy transition" in the Global South. The Tunisian initiatives for a new "energy transition strategy", announced in 2014, 2016 and 2023, were adapted to this rhetoric. Relying on its decades of commitment to environmental sustainability, Tunisia secured funds and investments for several renewable energy projects within the frame of its sustainable development and climate change strategy. For example, on 21 June 2023, the World Bank approved the financing of Tunisia-Italy ELMED project which will connect Tunisia as a "regional hub for renewable energy" to Europe [11].

After the 2011 uprising and in a political landscape that remains unstable, some of the Tunisian policy decision-makers tried to privatize the renewable energy market; however, it was met with opposition and concerns over national interests. In particular, draft law N°2014-06 triggered a controversy and opposition in parliamentary debates. As detailed in the decision of the Temporary Authority to Monitor the Constitutionality of Draft Laws dated 8 October 2014 and relating to the production of electricity from renewable energies by opening electricity projects to private investors, a number of deputies expressed concern over the need to protect natural resources and public institutions, mainly STEG the State power utility company, affirming that the energy sector should remain under state authority. Civil society and the Tunisian General Labor Union mobilized to stop the drafted legislation through the National Constituent Assembly. Then, in 2015, Law 12–2015 and the 2015 Solar Plan were adopted to promote "private sector involvement in energy sector." Today, out of 31 renewable energy projects, "only four are exclusively led by Tunisian companies." [12].

4 Conclusion

For a country that prides itself on its consistent efforts to curb energy demand and to implement renewable energy plans since the 1980s, the objectives of the Tunisian energy policy have been consistent. However, policy implementation was not efficient, otherwise the energy deficit wouldn't have increased in the last 20 years.

The Tunisian energy security policy has been relying on a pragmatic and low-cost approach to face its challenges over the decades. In practice, Tunisia focused on reducing sectorial and public energy demand in terms of quantity. It did not give the priority to investment in renewable energy technology to create its own means for the implementation of this policy. On the other hand, within the last decade policy-makers started to shift towards privatization of renewable energy projects, despite opposition. It is not a long-term solution.

Public opinion is a key variable that can shape the Tunisian energy security policy outcomes, yet the issue has been tackled differently. With energy control responsibility centered on citizens, the 2004–2010 period marked the highest rate of coordinated public awareness campaigns for energy conservation. After 2011, fuel prices increase was used as an incentive to curb demand; however, social unrest and mistrust towards government have been a challenge that froze the energy sector sometimes.

References

1. Rutherford AP (2019) The interplay between energy security and law and policy on green energy development: a socio-legal analysis. PhD thesis, School of Law and Social Justice, The University of Liverpool (United Kingdom)
2. UNDP (2000) World energy assessment energy and challenge of sustainability

3. International Trade Administration (2022) Overview of electrical power systems and renewable energy in Tunisia. Commercial Guide
4. Agence Nationale de la Maitrise de l'Energie (2013) Plan d'Action de Développement des Energies Renouvelables. Tunisie
5. Institut National de la Statistique (2015) Annuaire Statistique de la Tunisie 2010–2014, n°56. Tunisie
6. Institut National de la Statistique (2021) Annuaire Statistique de la Tunisie 2015–2019. N°61. Tunisie
7. Organization of Arab Petroleum Exporting Countries (1982) Summary Report: proceedings of the second Arab energy conference in Qatar from 6 to 11 Mar 1982, vol 1, pp 118–130, Kuwait
8. Barhoumi H (2020) La politique étrangère Tunisienne entre 2011 et 2021. Master's dissertation (not published)
9. UN-ESCWA (2019) Renewable energy: legislations and policies in Arab region
10. Ministère de l'Industrie, des Mines et de l'Energie et Observatoire National de l'Energie et des Mines (2022) Conjoncture Energétique: Rapport mensuel fin septembre 2022. Tunisie
11. World Bank (2023) World Bank approves $268 million project linking Tunisia's energy grids with Europe. https://www.worldbank.org/en/news/press-release. Last accessed 19 Oct 2023
12. Valentin L, Steinfort L (2023) Green colonialism under the guise of energy privatization in Tunisia. Transnational Institute. https://www.tni.org, Last accessed 19 Oct 2023

Enhanced Efficiency of Intermediate Band Solar Cells Through MLP-PSpice Modeling

B. Lakehal, S. Lakehal, M. Mouda, and S. Lakehal

Abstract Intermediate Band Solar Cells (IBSCs) represent a novel and promising advancement in solar cell technology. These cells hold great potential for high conversion efficiencies, with theoretical values reaching 63.2% under black body illumination and 65.1% under the AM1.5 spectrum. Notably, these efficiency levels are comparable to those achieved by optimized triple-junction solar cells, which attain 63.8% and 67.0%, respectively. This paper focuses on the development of an IBSC structure based on ZnTeO and leverages the power of an Artificial Neural Network (ANN) to model and enhance the performance of the IBSC. The use of an ANN brings significant advantages, particularly in reducing the need for an extensive database. By employing a small representative dataset, the ANN-based approach proves to be highly effective in optimizing the IBSC's efficiency. Importantly, the proposed methodology is not limited to IBSCs alone; it can be extended to other types of solar cells as well. This versatility promotes the realization of promising photovoltaic devices that boast both low-cost manufacturing and high efficiency. This research marks a significant step forward in the quest for efficient and affordable solar energy conversion technologies.

Keywords IBSC · I-V characteristics · Hidden layer · Artificial neural network

B. Lakehal · M. Mouda
Health and Safety Institute, Batna2 University, Batna, Algeria
e-mail: b.lakehal@univ-batna2.dz

M. Mouda
Laboratory of Research in Industrial Prevention (LRPI), Batna, Algeria

S. Lakehal (✉)
ESPA Laboratory, Institute of Veterinary Science and Agricultural Sciences, Batna, Algeria
e-mail: b.lakehal@univ-batna2.dz

University of Batna2, Institute of Earth Sciences and University, Batna, Algeria

© The Author(s), under exclusive license to Springer Nature Singapore Pte Ltd. 2024
A. Mellit et al. (eds.), *Proceedings of the 2nd International Conference on Green Energy Conversion System*, Green Energy and Technology,
https://doi.org/10.1007/978-981-97-6148-7_66

1 Introduction

Photovoltaic solar cells with a single junction have demonstrated the ability to achieve power conversion efficiency values close to the maximum theoretical limit [1, 2]. However, their efficiency falls short of the desired level due to energy losses.

Various mechanisms lead to energy losses in solar cells. Optical losses occur because of surface interface reflections in the module, while carrier losses result from recombination losses caused by poor interface or material quality. The most significant losses are inherent to the photovoltaic effect itself. A substantial amount of optical power is lost due to the large volume of infrared light with insufficient energy to elevate an electron into the conduction band (sub-band gap losses). Moreover, high-energy photons can only elevate one electron to the conduction band, wasting excess energy by heating the solar cell (hot-electron or thermalization losses). These fundamental losses directly limit the efficiency to around 40% for commonly used semiconductors and 43.9% for single-junction silicon solar cells, as shown by Shockley and Queisser in 1961 [3, 4]. 3rd generation (3G) concepts aim to harness some of this wasted energy [Green 2006].

Over the past three decades, several intriguing 3G concepts have been explored, such as multi-junction systems like tandem cells, utilizing quantum wells and quantum dots to enhance absorption, employing fluorescent impact ionization to use the kinetic energy of carriers, and exploiting impurity levels and hot electron effects. While these ideas have sparked excitement and interesting discussions, most of them have proven challenging to demonstrate in practice and have often resulted in decreased overall device efficiencies instead of the desired improvements.

The pursuit of a low-cost and highly efficient solar energy conversion device continues to drive global photovoltaic (PV) research. While advancements have improved the power conversion efficiency of tandem and multi-junction solar cells, these devices have become more complex and costlier to manufacture. To address this, researchers have proposed various techniques aimed at increasing solar cell efficiency. These techniques leverage different processes and properties, including photon recycling, impact ionization, angle-dependent selectivity, down-conversion, and up-conversion [5–7].

In recent years, another promising solution to overcome the efficiency limitations of conventional single-junction cells has emerged: intermediate-band solar cells (IBSC) [8]. Additionally, artificial neural networks (ANNs) have proven their success in solving complex problems across various fields, such as pattern recognition, identification, classification, speech, vision, prediction, and control systems [9]. ANNs have shown exceptional capabilities in handling nonlinear, complex, and dynamic tasks with remarkable accuracy. During the design and optimization phases, the researchers utilized the MATLAB interface and implemented the results (optimal architecture, bias, and weights of the network) as a component in the PSpice simulator library. This proposed technique remains applicable to other systems as well, encouraging further experimental efforts to realize the potential of these promising photovoltaic devices [10–13]. This integration of intermediate-band solar cells and artificial neural networks offers a compelling path towards achieving more efficient and cost-effective solar energy conversion technologies.

2 Description of the IBSC Model

The sole proven 3G technology revolves around the utilization of multiple junctions. Single-junction devices reach their peak performance at a specific wavelength equivalent to the band gap; however, their efficiency diminishes at all other wavelengths present in the solar spectrum. In contrast, multi-junction devices incorporate various solar cells with multiple band gaps, precisely tuned to effectively harness the entire spectrum. When incident light reaches these devices, it encounters a wide band gap cell, generating relatively high voltage and efficiently utilizing high-energy photons. Subsequently, lower-energy photons pass through to narrow band gap sub-cells that can absorb transmitted infrared (IR) photons. Predicted maximum efficiencies of 55.9%, 63.8%, and 68.8% are attained for two- (tandem), three-, and four-junction devices, respectively.

In line with the concept of multi-junction systems, material systems based on ZnTe can be employed. Therefore, our focus lies on a novel concept that we believe holds significant practical potential: introducing an intermediate band in ZnTe by incorporating Oxygen, resulting in ZnTeO intermediate band. This intermediate band (ZnTeO) will facilitate the transition of electrons in the valence band with energies below the bandgap energy to the next level, i.e., the conduction band.

Our project delves into the exploration of an extraordinary case involving the ZnTeO intermediate band. The proposed approach aims to profoundly influence carrier transport and recombination properties within the material, ultimately enhancing solar cell conversion efficiency while concurrently reducing manufacturing costs [14].

3 The Neuron Network Training

As we have seen, learning requires a database and a test database, along with determining the number of layers and neurons used in each layer. The proper selection of activation functions and the number of iterations needed to achieve the desired estimated threshold represents the minimum value of the Mean Square Error (MSE) that we aim to obtain. Figure 1 depicts the flowchart that illustrates our structured program in Matlab.

In our specific case, the number of neurons in the output layer of the artificial neural network (ANN) is determined by the number of outputs of the system we are modeling. Since our Intermediate Band Solar Cell (IBSC) has a single output, which is the I-V (Power) curve, we only require one neuron in the output layer. To ensure the accuracy of the ANN model in representing the IBSC's response, we need to

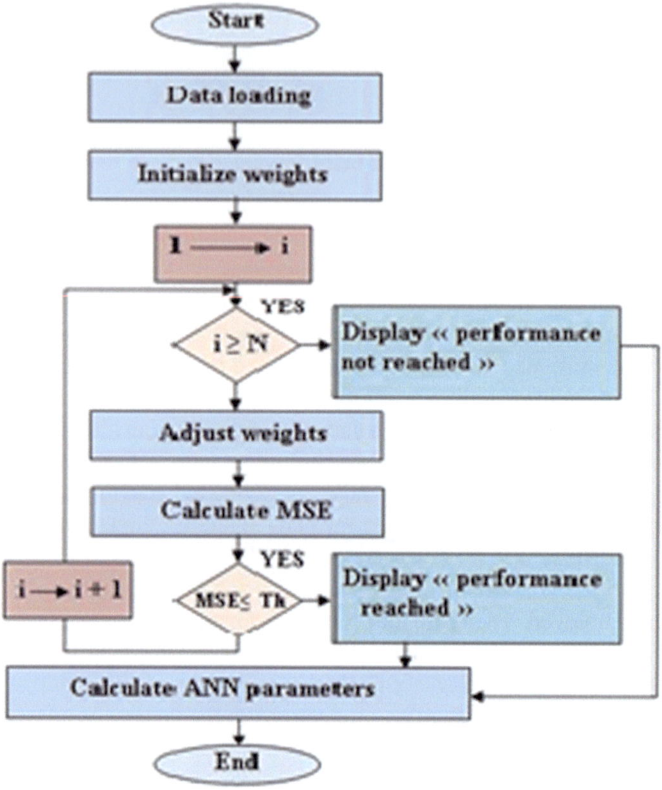

Fig. 1 Flowchart for the training program

determine the optimal number of hidden layers and neurons per layer. In our study, we considered a two-layer architecture with 4 neurons for the first hidden layer and 5 neurons for the second hidden layer. To select the best architecture for the network, we conducted experiments by varying the number of neurons in each layer within certain ranges. Specifically, we tested architectures with 2 to 6 neurons for the first hidden layer and 3 to 6 neurons for the second hidden layer. For each architecture, we observed the variation of the Mean Squared Error (MSE) in the learning and testing phases of the ANN. The MSE provides a measure of how well the model approximates the actual IBSC behavior.

After analyzing the results, Table 1 presents the various architectures tested along with their corresponding MSE values. Among the tested architectures, the third architecture demonstrated an acceptable MSE value, indicating that it is a suitable choice for accurately modeling the IBSC's I-V (Power) curve. This optimized ANN architecture allows us to effectively represent the behavior of the IBSC and pave the way for further advancements in photovoltaic device efficiency.

Table 1 The results for various ANN architectures

	Number of neurons in 1st hidden layer	Number of neurons in 2nd hidden layer	Threshold (Th) for MSE = 10^{-8}
1	2	3	8.1313
2	3	4	4.9679
3	**4**	**5**	**0.01015**
4	5	6	0.09523

The same thing for the choice of the types of activation functions, maximum number iteration and the MSE. The test and observation of the change in error for different values is the solution for the best results. To assess the influence of the threshold Th the number of iterations needed to get MSE < Th. We chose several values for Th, Figs. 2, 3 and 4.

4 Results and Discussion

The model defined for IBSC devices in this Project is applicable for any semiconductor material of interest. In this analysis material parameters for ZnTe are chosen as a prototypical material for IBSC due to recent interest in oxygen-doped ZnTe as an active region for IBSC devices [15]. The chosen sets of device parameters are used in these simulations to investigate the influence of electron transport and carrier recombination.

$W = 1\ \mu m$ with $\alpha I_{C0} = \alpha V_{I0} = 10^3\ cm^{-1}$, $\alpha_{VC} = 10^4\ cm^{-1}$.

The parameters for our device represent the case of a realistic thin-film IBSC where the absorber region thickness is near 1 μm and the intermediate band optical transition is an order of magnitude weaker than the valence- to conduction-band optical transition.

The comparison between the original database and that obtained after learning, using the test database, experimental set was compared with ANN model. Figure 5 represents a comparison of the variation of current as function as output voltage using ANN model at fixed parameters. We can see that proposed model provide a good agreement between them. Indicates that our model accurately expresses the variation and the performance of the IBSC response.

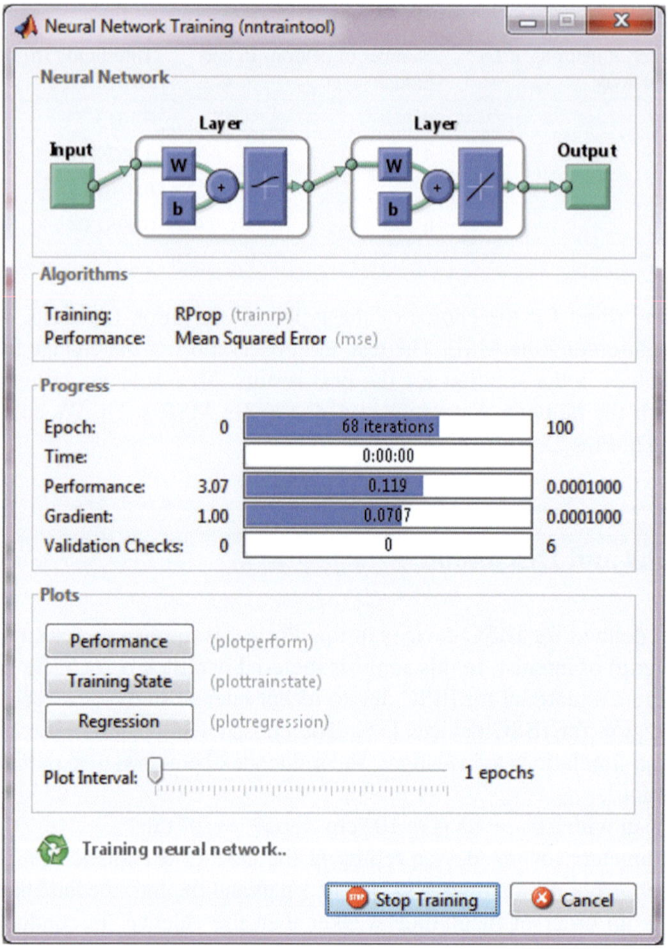

Fig. 2 100 iterations, estimate threshold 10^{-4}

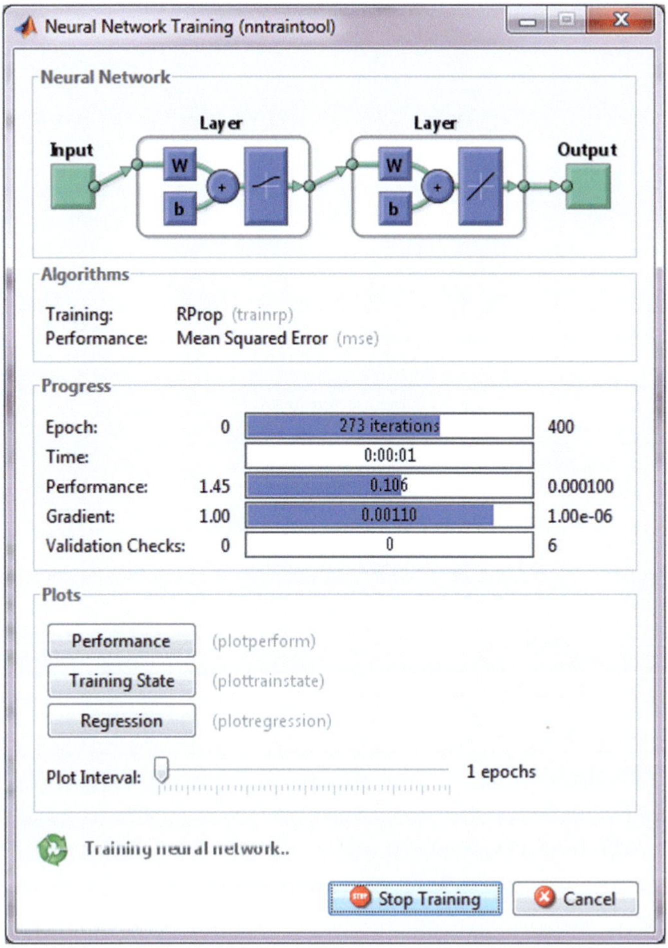

Fig. 3 400 iterations, estimate threshold 10^{-4}

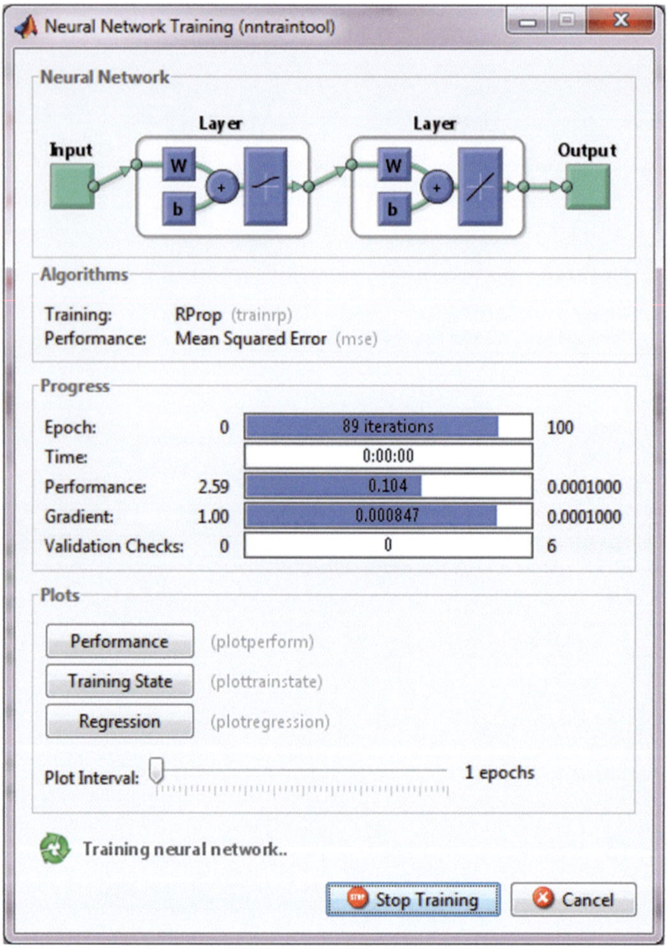

Fig. 4 200 iterations, estimate threshold 10^{-3}

Fig. 5 Comparing the variation of current vs voltage between ANN model an experimental result [4].

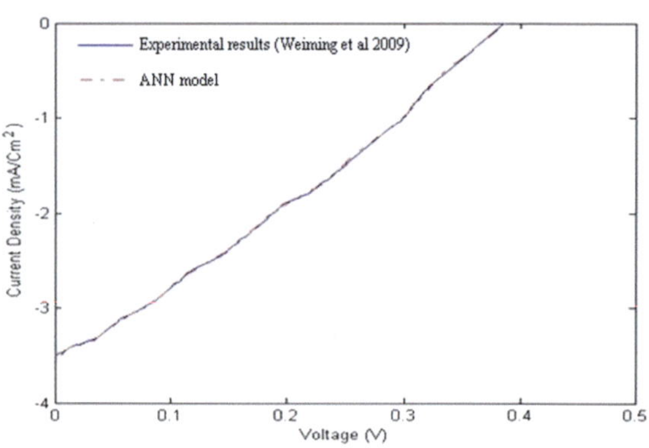

5 Conclusion

This study focuses on the modeling of the current–voltage (I-V) characteristics of intermediate band solar cells (IBSC) based on Zn-Te:O, considering variations in temperature and environmental irradiance. The proposed model aims to understand the impact of different inputs and the overall behavior of the solar cell. The simulation results presented in this work serve as strong evidence of the model's validity. The model employs an approach that involves utilizing an artificial neural network (ANN) for the modeling of intermediate band solar cells. This ANN-based approach helps capture the complex behavior of the IBSC.

To assess the accuracy of the model, it has been thoroughly verified and validated using experimental measurements. The comparison between the model output and the experimental results demonstrates a good agreement, indicating that the model reliably predicts the behavior of the IBSC under varying conditions. Overall, this work highlights the significance of the proposed model in understanding and optimizing the performance of Zn-Te:O-based intermediate band solar cells, thereby advancing the potential for efficient solar energy conversion.

References

1. Martí A, Antolín E, Stanley CR, Farmer CD, López N, Díaz P, Cánovas E, Linares PG, Luque A (2006) Production of photocurrent due to intermediate-to-conduction-band transitions: a demonstration of a key operating principle of the intermediate-band solar cell. Phys Rev Lett 97:247701
2. Ling C, Zhou LQ, Banerjee D, Jia H (2014) Band structures of ZnTe: O alloys with isolated oxygen and with clustered oxygen impurities. J Alloy Compd 584:289–294
3. Shoji Y, Akimoto K, Okada Y (2013) Self-organized InGaAs/GaAs quantum dot arrays for use in high-efficiency intermediate-band solar cells. J Phys D Appl Phys 46:024002
4. Lakehal B, Dibi Z, Lakhdar N, Dendouga A (2015) Electrical equivalent model of intermediate band solar cell using PSpice. Sādhanā 40:1473–1479
5. Martí A, Antolín E, Cánovas E, López N, Linares PG, Luque A, Stanley CR, Farmer CD (2008) Elements of the design and analysis of quantum-dot intermediate band solar cells. Thin Solid Films 516:6716–6722
6. Lin AS, Wang W, Phillips JD (2009) Model for intermediate band solar cells incorporating carrier transport and recombination. J Appl Phys 105:064512
7. Luque A, Marti A (2007) The intermediate band solar cell: progress toward the realization of an attractive concept. Adv Mater 22:160–174
8. Wang W, Lin AS, Phillips JD (2009) Intermediate-band photovoltaic solar cell based on ZnTe:O. J Appl Phys 95:011103
9. Hegedus SS (1997) Current–voltage analysis of a-Si and a-SiGe solar cells including voltage-dependent photocurrent collection. Prog Photovoltaics 5:151–168
10. Wei G, Shiu K-T, Giebink NC, Forrest SR (2007) Thermodynamic limits of quantum photovoltaic cell efficiency. Appl Phys Lett 91:223507
11. Luque A, Martí A, López N, Antolín E, Stanley ECC, Farmer C, Díaz P (2006) Operation of the intermediate band solar cell under non-ideal space charge region conditions and half filling of the intermediate band. J Appl Phys 99:094503
12. Wang W, Lin A, Phillips JD (2008) Electrical Characteristics and Photo-response of ZnO/ZnTe Heterojunction Diodes. J Electron Mater 37:1044–1048

13. Schmeits M, Mani AA (1999) Impurity photovoltaic effect in c-Si solar cells. A numerical study. J Appl Phys 85:2207
14. Lakehal B, Dibi Z, Lakhdar N (2017) New optimized intermediate band (IB) design to improve ZnTeO solar cell performances. J Nano- Electron Phys 09:06005
15. Lakehal B, Dibi Z, Lakhdar N (2017) Performance enhancement of ZnTeO solar cell by optimizing physical and geometrical parameters, International Conference on Green Energy Conversion Systems (GECS), Hammamet, Tunisia, (2017)

Neural Network-Based Super-Twisting Algorithm for Controlling a Floating Offshore Wind Turbine in Low-Wind Region

Ehsan Aslmostafa, Mohamed Assaad Hamida, and Franck Plestan

Abstract This paper studies a hybrid control strategy for enhancing power generation from floating offshore wind turbines (FOWTs) in low-wind conditions. To handle the uncertain dynamics of the FOWT a super-twisting (STW) control strategy is proposed. Later, to improve the performance of the STW control and to reduce the control effort, the STW control and a radial basis function neural network (RBFNN) are integrated to approximate the uncertainties and unmodeled dynamics. Using the OpenFAST simulator, we compare the strategy's effectiveness with and without neural network integration.

Keywords Robust control · Maximum power point tracking · Floating wind turbine · Permanent magnet synchronous generator · Neural network

1 Introduction

In a low-wind area (region 2), maximizing power output from wind turbines is challenging. Methods like tip speed ratio (TSR) control, which monitors wind and rotational speed to optimize torque are used. These keep turbines near peak power coefficient, extracting maximum energy [1]. In this study, we utilize a permanent magnet synchronous generator (PMSG) for power generation which offers an exceptional torque-to-inertia ratio that makes the PMSG adaptable to environments with low wind profiles [2].

In Region 2, various FOWT control methods are implemented. In [3], a novel backstepping-based speed control to achieve MPPT is presented which utilizes the

E. Aslmostafa (✉) · M. Assaad Hamida · F. Plestan
Nantes Université, Ecole Centrale Nantes, CNRS, LS2N, Nantes 44000, France
e-mail: ehsan.aslmostafa-jarchelou@ec-nantes.fr

M. Assaad Hamida
e-mail: mohamed.hamida@ec-nantes.fr

F. Plestan
e-mail: franck.plestan@ec-nantes.fr

© The Author(s), under exclusive license to Springer Nature Singapore Pte Ltd. 2024
A. Mellit et al. (eds.), *Proceedings of the 2nd International Conference on Green Energy Conversion System*, Green Energy and Technology,
https://doi.org/10.1007/978-981-97-6148-7_67

reluctant part of the electromagnetic torque via the maximum torque per ampere (MTPA) strategy. In a comparative study, [4] compares classical PI controllers with backstepping control for wind energy conversion systems, illustrating the latter's superior performance in transient modes. In practical scenarios, such as FOWTs with uncertain dynamics while taking the OpenFAST platform, backstepping control might not be a good choice as it is designed based on the system model. Hence, there is a demand for a controller that can manage unmodeled dynamics and the uncertain part of the system. Sliding mode control (SMC), as a robust control strategy used in FOWTs to handle uncertainties, demonstrates strong tracking performance, and robustness. However, it may face chattering issues, increasing damage to turbine components. Our paper proposes nonlinear control strategies based on super-twisting (STW) algorithms to overcome the limitations of the SMC control methods. The STW algorithms provide robustness against uncertainties and reduction of chattering [5]. In such a control strategy, very little information on the system model is required and this reduces the tuning effort and guarantees a high level of performance. Nonetheless, little literature has focused on the development of such control strategies on FOWT in a low-wind region. In [6], the efficacy of a second-order sliding mode control algorithm (SOSMC) is demonstrated. The study develops a PMSG-based wind emulator to evaluate the performance of integrating SOSMC with an STW algorithm in a wind energy conversion system. To further enhance the performance of the robust controller, an approximation of the uncertain part can be utilized. In this study, we propose a hybrid STW control combined with an artificial neural network (ANN) approach. This hybrid approach aims to achieve improved robustness and stabilization in a FOWT by approximating the unknown parts. In [7], the authors put forward an STW controller for wind turbines, leveraging artificial intelligence (AI) capabilities specifically for a doubly fed induction generator (DFIG). A robust integral SMC intended to manage the active and reactive powers of a turbine along with a sturdy MPPT algorithm influenced by an ANN controller was proposed by [8]. To the best of the authors' knowledge, no prior research has explored the implementation of an NN-based STW control for low-wind conditions (Region 2) in a FOWT, considering the OpenFAST realistic platform.

2 System Modeling and Description

The 5MW OC3 FOWT model used in this study has been proposed by the National Renewable Energy Laboratory (NREL) and simulated through the OpenFAST platform [9]. A more realistic model evolved through OpenFAST cannot be described as a simple mathematical model for control design due to its complexity as a result, a reduced model must be used as detailed in the sequel. The aerodynamic power (1), captured by the wind turbine's rotor blades, and the aerodynamic torque (2) of the rotor, which is extracted from a wind turbine, is given as

$$P_a = \frac{1}{2}\rho\pi R^2 C_p(\lambda, \beta)v^3 \tag{1}$$

$$T_a = \frac{1}{2}\frac{C_p(\lambda, \beta)}{\lambda}\rho\pi R^3 v^2 \tag{2}$$

where R is the length of each blade plus hub radius, v is the wind speed, ρ is the density of air and C_p is the power coefficient which is highly nonlinear and depends on blade aerodynamic design and wind turbine operating conditions. Let λ^* symbolize the tip speed ratio (TSR) that optimizes the power coefficient C_p. Subsequently, the turbine should operate at speed ω^* which can be calculated as follows:

$$\omega^* = \frac{\lambda^*}{R}V - k\dot{\varphi} \tag{3}$$

with λ^* as the optimal TSR and k as a positive constant. The concept involves balancing rotor speed and platform pitching. When the platform pitches forward, the rotor speed increases to counteract it. This convergence is achieved through control mechanisms and supported by reference [10].

2.1 Floating Offshore Wind Turbine Modeling

The drive train dynamics of the wind turbine reads as

$$\dot{\omega} = \frac{1}{J}\left(T_a - n_g T_g\right), \quad \omega_g = n_g\omega \tag{4}$$

with ω the rotor speed, ω_g the generator speed, T_a and T_g the aerodynamic and generator torque, respectively, n_g the gearbox ratio between the high-speed shaft and low-speed shaft, and J the overall inertial. Due to the complex dynamics of a FOWT, it is necessary to develop a control design based on a nonlinear uncertain system. Here, the FOWT dynamics is considered a class of nonlinear systems represented by [10].

$$\dot{\mathbf{X}} = \mathbf{F}(\mathbf{X}, \mathrm{t}) + \mathbf{G}(\mathrm{t})\mathrm{U} \tag{5}$$

where $\mathbf{X}$ represents the state vector and is defined as $\mathbf{X} = \left[\varphi \ \dot{\varphi} \ \omega\right]^T$, such that φ, $\dot{\varphi}$, and ω denote the platform pitch, platform pitch rate, and rotor speed, respectively. The control input is the generator torque T_g (details are provided later). Notice that functions $\mathbf{F}$ and $\mathbf{G}$ are bounded because the wind speed and rotor speed are limited and cannot be formally expressed; it is why it is necessary to design a robust control strategy as described in the next section.

2.2 PMSG Modeling

The selection of PMSG is primarily based on its superior efficiency compared to traditional generators. In order to apply the proposed controls, we take into account the mathematical electrical model of the surface-mounted PMSG dynamics, as presented in the (d, q)-reference frame, according to [11]:

$$\begin{aligned} \dot{i}_d &= -\frac{R_s}{L_d} i_d + p\frac{L_q}{L_d}\omega i_q + \frac{1}{L_d} v_d \\ \dot{i}_q &= -\frac{R_s}{L_q} i_q - p\frac{L_d}{L_q}\omega i_d - p\frac{1}{L_q}\phi_f \omega + \frac{1}{L_q} v_q \end{aligned} \tag{6}$$

with i_d and i_q the currents along the d and q-axes respectively, ω the rotor speed and v_d and v_q the voltages along the d and q-axes respectively, and are used as the control inputs for the taken system (5) since the turbine is coupled to a PMSG. The parameters R_s are the stator resistance, L_d the inductance along the $d-$axis, L_q along the q-axis respectively, p the number of pole pairs, and ϕ_f the permanent magnet flux. Notice that the electromagnetic torque by assuming symmetric magnetic characteristics along the d-axis and the q-axis ($L_d = L_q$) reads as $T_{em} = \frac{3}{2} p\phi_f i_q$.

Remark 1 To achieve an operation that maximizes electromagnetic torque, it is sufficient to set the direct stator current to zero. Hence, we select the current reference $i_d^* = 0$. By controlling the current i_d to a reference value i_d^*, which is zero, we ensure the non-existence of d-axis stator current, thereby eliminating any reluctance torque.

2.3 Model of a FOWT Equipped with a PMSG

Based on a model presented in (5), the model of a FOWT equipped by a PMSG in a low-wind region can be rearranged as follows

$$\dot{\mathrm{x}} = \mathrm{f}(\mathrm{x}, \mathrm{t}) + \mathrm{g}(\mathrm{t})\mathrm{u} \tag{7}$$

Here, $\mathbf{x} = \left[\varphi\ \dot{\varphi}\ \omega\ i_d\ i_q\right]^T$ and also, $\mathbf{f}$ depends on perturbations like waves and wind, as well as electrical parameters like resistance and inductance that can vary. The term $\mathbf{g}$ depends on inductance and has constant positive values. Since the perturbation and rotor speed is assumed to be bounded, the uncertain functions f and g are also bounded. Besides, the control input is $u = [v_d\ \ v_q]^T$. It should be noted that the generator torque is controlled through v_q.

3 Control Design

When the wind speed is operating in Region 2, the main objective is to maximize the power output while reducing the platform oscillation by controlling the generator torque as a function of the rotor speed. To achieve this goal, the stator voltages (v_d and v_q) are taken as the control inputs of the control system to control the state variables. To design a robust control scheme to achieve the control objectives, we take the model of a FOWT equipped with a PMSG as (7) and to further enhance the performance of the robust controller, an approximation of the uncertain part can be utilized. In this work, an RBF neural network is integrated into the STW control. In other words, an NN-based STW with the aim of achieving improved robustness and stabilization for generator torque control in a FOWT, by approximating the unknown function is presented. Implementing this approach enables us to study the system's overall behavior in a more realistic situation (OpenFAST platform). This will lead to fewer errors and some improved performance while having less fatigue which will be shown in the simulation result section.

Lemma 1 *Based on* [12], *if we consider the following nonlinear system:*

$$\dot{\sigma} = f(\sigma, t) + u + \delta(t) \tag{8}$$

such that $f(\sigma, t)$ is considered a known nonlinear function, u is the control input, and $\delta(t)$ is an external disturbance. Also, σ is called the sliding variable in [13]. *Under the assumption that there exist positive constants δ_M and Δ, ensuring that the perturbation $\delta(\sigma, t)$ and its derivative $\dot{\delta}(\sigma, t)$ are bounded, i.e., $|\delta(\sigma, t)| < \delta_M$, and $|\dot{\delta}(\sigma, t)| < \Delta$ and supposing that $f(\sigma, t)$ is well-known, an STW approach can be defined as*

$$\begin{aligned} u &= -f(\sigma, t) - k_1|\sigma|^{\frac{1}{2}}\,\text{sign}(\sigma) + v \\ \dot{v} &= -k_2\,\text{sign}(\sigma). \end{aligned} \tag{9}$$

If the gains k_1 and k_2 are set according to [14], *the sliding variable σ and its first-time derivative $\dot{\sigma}$ converge to 0 in the sense of finite time stability.*

In contrast to the system adopted in Lemma 1, the term $f(\sigma, t)$ is is not well-known in most of the practical systems (such as FOWT) and it is not easy to design the proposed controller in Eq. (9) due to unknown model uncertainties. In addition, an assumption on the boundedness of function $f(\sigma, t)$ can not be a general condition, especially from a stability point of view. In the sequel, a neural network is proposed to approximate the function $f(\sigma, t)$.

3.1 Radial Basis Function Neural Network

The RBFNN is employed to estimate $f(x,t)$ by leveraging the universal approximation attribute [15]. We designate the input vector as $x = [\mathbf{X}_1, \mathbf{X}_2, \cdots, \mathbf{X}_n]^T$ and the radial basis vector of the RBF network as $h(x) = [h_1(x), h_2(x), \cdots h_m(x)]^T$, where each h_j represents a Gaussian basis function and read as

$$h_j = \exp\left(-\frac{|x - c_j|^2}{2b_j^2}\right), \quad j = 1, 2, \cdots m \tag{10}$$

Here, j symbolizes the $j-$th node in the hidden layer, $c_j = \left[c_{j1}, c_{j2}, \cdots c_{jn}\right]^T$ denotes the center vector of the j−th node, and $j = 1, 2 \cdots n$, while b_j implies the width of the Gaussian basis function. In the taken RBFNN, the weight from the input layer to the hidden layer is set as 1.0, and the weight vector transitioning from the hidden layer to the output layer is denoted as $\hat{W} = \left[\hat{w}_1, \hat{w}_2, \cdots, \hat{w}_m\right]^T$ where $\hat{W}$ signifies the vector of weights and it is updated using the following adaptive law

$$\dot{\hat{w}}_j = \gamma_j \sigma h_j(x) \tag{11}$$

such that γ_j is the learning rate of the adaptive law for the weights. Moreover, σ is the defined sliding variable for a system based on the states x. The output of the RBFNN with ideal approximation is $y(x) = {W^*}^T h(x)$ where W^* is the weight vector of the network with ideal values. However, an approximation error ϵ must be considered which is bounded as $|\epsilon| \leq \epsilon_N$. Therefore, the output estimated signal is obtained as

$$\hat{y}(x) = \hat{W}^T h(x) \tag{12}$$

such that

$$y(x) - \hat{y}(x) = \tilde{W}^T h(x) + \epsilon \tag{13}$$

where $\tilde{W} = W^* - \hat{W}$.

3.2 Neural Network-Based Super Twisting Controller

In this part, we propose an STW controller in conjunction with an RBFNN. Taking into account the system represented in Eq. (8) and under the assumption that $f(\sigma, t)$ is not in hand, the estimation of the unknown function is feasible. The sliding variable σ acts as the network's input. Following this, the output of the network is determined as per the succeeding equation:

$$\hat{f}(\sigma, t) = \hat{W}^T h(\sigma) \tag{14}$$

such that $\hat{W}$ is updating protocols of weights obtained in the sense of Lyapunov stability as

$$\dot{\hat{W}}_j = \gamma_j \sigma h_j(\sigma) \tag{15}$$

where γ_j is the learning rate of the adaptive law for the weights. Then, NN-based STW control based on (8) can be written as

$$\begin{aligned} u &= u_{nn} + u_{stw} \\ u &= -\hat{f}(\sigma, t) - k_1|\sigma|^{\frac{1}{2}} \operatorname{sign}(\sigma) + v \\ \dot{v} &= -k_2 \operatorname{sign}(\sigma). \end{aligned} \tag{16}$$

If the gains k_1 and k_2 are adjusted as proposed in [14], and the bound of the approximation error ϵ is taken into consideration, the sliding variable σ and its first derivative $\dot{\sigma}$ converge toward 0. Subsequently, the nonlinear system, under the influence of STW and the NN approximation algorithm, attains stability. This results in the sliding variable σ and its first derivative $\dot{\sigma}$ being guided toward the origin asymptotically. It is important to note that the setting of gains, considering the approximation error, necessitates high gains, a condition that could be restrictive. Nevertheless, this paper's primary focus is to illustrate the advantages and efficiency of employing the NN-based strategy in conjunction with STW control on the FOWT. In the sequel, we are going to design q and d reference voltages based on the above discussion.

3.3 Quadrature-Axis Voltage Design

Take (7), the control inputs based on our previous discussions can be taken as the generator voltage, i.e., v_q and v_d, and the output of the system is taken as $y_1 = \omega - \omega^*$ and $y_2 = i_d - i_d^*$. The objective is to design controllers to ensure that both (ω) and (i_d) track their corresponding reference values. To design an STW based on an RBFNN, firstly, the rotor rotational speed error is taken as

$$\sigma_\omega = \omega - \omega^* \tag{17}$$

To design v_q, by defining speed error as (17) the relative degree equals two as the control input v_q appears in the second derivative to the system output y_1. To apply the control input (16), the following auxiliary sliding variable with the convergence rate $\alpha > 0$ is considered

$$s_1 = \dot{\sigma}_\omega + \alpha \sigma_\omega. \tag{18}$$

From Eq. (4), according to [10], we can rewrite it as $\dot{\omega} = a_1(\cdot) + b_1(\cdot) i_q$ and the function $a_1(\cdot)$ is unknown and bounded and $b_1(\cdot)$ is an unknown positive constant function, then we can take $b_1(\cdot) = b_{1n}(\cdot) + \delta b_1(\cdot)$, where b_{1n} is a known constant and $\delta b_1(\cdot)$ is bounded perturbation such that $b_{1n}(\cdot) > |\delta b_1(\cdot)|$. Then from Eqs. (4)–(7), by doing some simplification, and taking $b_{1n} = -(3p\phi i_q/2J)$ the time derivative of (18) is as

$$\begin{aligned}\dot{s}_1 &= \ddot{\sigma}_\omega + \alpha\dot{\sigma}_\omega \\ &= \underbrace{\dot{a}_1 + \alpha a_1 + \delta b_1 \dot{i}_q + \alpha\delta b_1 i_q}_{\text{unknown and bounded}} + b_{1n}\dot{i}_q + \alpha b_{1n} i_q + \ddot{\omega}^* + \alpha\dot{\omega}^* \\ &= \mathrm{A}(\mathrm{x}, t) + Bv_q + b_{1n}(-\frac{R_s}{L_q} i_q - p\frac{L_d}{L_q}\omega i_d - p\frac{1}{L_q}\phi_f\omega + \alpha i_q) + \ddot{\omega}^* + \alpha\dot{\omega}^*\end{aligned} \tag{19}$$

and the term A is unknown and assumed to be bounded and the term $\mathrm{B} = b_{1n}/L_q$. The first and second derivative of reference rotor speed ω^* is assumed to be available. Now, based on discussions on Lemma 1 and the proposed controller in (16) to compensate for the effect of unknown function A, the $q-$axis voltage is designed as $v_q = u_{eq1} + u_{stw1} + u_{nn1}$ and the equivalent control is

$$u_{eq1} = -b_{1n}\left(-\frac{R_s}{L_q} i_q - p\frac{L_d}{L_q}\omega i_d - p\frac{1}{L_q}\phi_f\omega + \alpha i_q\right) - \ddot{\omega}^* - \alpha\dot{\omega}^*. \tag{20}$$

Finally, the designed control is as follows and this guarantees $s_1 = \dot{s}_1 = 0$.

$$v_q = \frac{1}{B}\big(u_{eq} - \hat{A} - k_1|s_1|^{\frac{1}{2}}\,\mathrm{sign}(s_1) - k_2\int \mathrm{sign}(s_1)\,dt\big). \tag{21}$$

3.4 Direct-Axis Voltage Design

In this part, the objective is to force the current i_d to track the reference $i_d^* = 0$. To ensure this feature, the following current tracking error is considered as

$$\sigma_d = i_d - i_d^* \tag{22}$$

where i_d^* is taken to be zero as discussed in Remark. 1. Now, take (6), the dynamic of the d-axis current by considering the parameters variation can be rewritten in an uncertain form as

$$\dot{\sigma}_d = a_2(\cdot) + b_2(\cdot)v_d \tag{23}$$

Notice that, in (6), each parameter is composed of a known nominal part and unknown uncertainty one (e.g., the resistance R_s can be written as $R_s = R_{sn} + \delta R_s$, and R_{sn} being its nominal value and δR_s the associated uncertainty). Then, one gets $a_2(\cdot) = a_{2n}(\cdot) + \delta a_2(\cdot)$ and $b_2(\cdot) = b_{2n}(\cdot) + \delta b_2(\cdot)$ with a_{2n} and b_{2n} the nominal part reading as

$$a_{2n} = -\frac{R_s}{L_d} i_d + p\frac{L_q}{L_d}\omega_g i_q, \quad b_{2n} = \frac{1}{L_d}. \tag{24}$$

The designed control input for the d-axis read as

$$v_d = u_{eq2} + u_{stw2} + u_{nn2} \tag{25}$$

and the equivalent control is

$$u_{eq2} = -\frac{a_{2n}}{b_{2n}} \tag{26}$$

Finally, the designed control is as and it guarantees $\sigma_d = \dot{\sigma}_d = 0$.

$$v_d = \frac{1}{b_{2n}}\Big(u_{eq2} - \delta\hat{a}_2 - \delta\hat{b}_2 - k_3|\sigma_d|^{\frac{1}{2}}\,\text{sign}(\sigma_d) - k_4\int \text{sign}(\sigma_d)\,dt\Big). \tag{27}$$

Remark 2 Given that the unknown functions in our model may be subjected to external disturbances such as wind and wave forces, it is reasonable to consider the first and second derivatives of these unknown functions to be bounded. This assumption acknowledges the impact of real-world conditions on our model, lending greater practical relevance to our findings.

4 Simulations and Analysis

4.1 Simulation Set-up

A 5MW NREL-designed OC3 FOWT is simulated using OpenFAST software. The mechanical parameter of the wind turbine and the installed PMSG are acquired from [16]. The control strategy is executed via Matlab/SIMULINK, connected to OpenFAST, employing a 24 DOF model. Simulations run for 600 s using ODE1 (Euler) with 0.0125 s fixed-step. STW controller gains $k_1 = 5$ and $k_2 = 0.01$; neural network hidden layer contains 27 neurons; Gaussian basis function parameters are $c_j = 4$ and $b_j = 1$. The neural network's initial weights, chosen randomly between $[-2, 2]$, use a learning rate $\gamma = 0.001$. Controllers operate under an irregular wave (3.25 m height, 9.7 s peak spectral period) and stochastic wind (9 m/s mean, 10% turbulence). The wind and wave profiles are shown in Fig. 1. In this section, the STW control with constant gains will be compared in detail to an NN-based STW control.

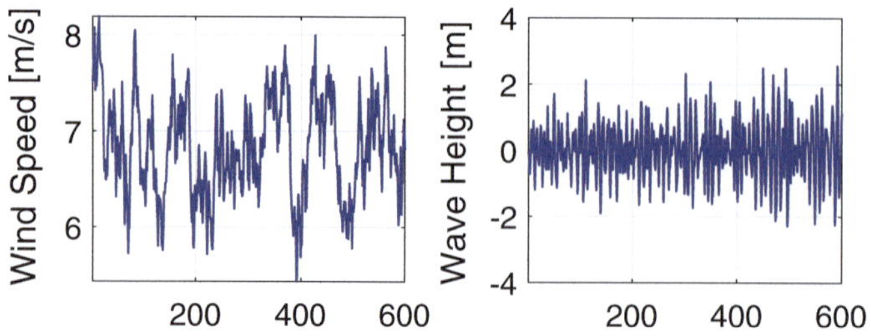

Fig. 1 Wind and wave profiles versus time

4.2 Results and Analysis

Here, we compare NN-based control to the constant gain strategy (without NN integration) [14], also various criteria are considered to show the merits and demerits of each control. The damage equivalent loads (DEL) index is defined for evaluating the possible fatigue loads on the wind turbine structure in different sections. Here, the DEL is computed for the main structural sections of the FOWT such as the tower base (TB), blade root (BR), fair-lead force (FF), and anchor force (AF) of the three mooring lines.

In Fig. 2, the evolution of the main variables for FOWT has been presented in order to see the performances of the controllers in completely similar environmental conditions. The optimal power P^* is calculated based on (1) and as seen in Fig. 2 (upper one), both implemented control present a satisfying result in achieving the optimal power tracking. Also in Fig. 2 (bottom one), the performance of the proposed controls in achieving the tracking of the desired rotor speed (w^*) which is obtained by the TSR-algorithm is illustrated. Note that all of the indexes have been computed for [200 600] (s). For a more in-depth analysis of the results, Tables 1 and 2 are presented, where they display the corresponding improvement percentages. According to Table 1, the mean value of the generated power for both controls is the same, however, there is a reduction in the rotor speed tracking error by 3.33%. This reduction leads to decreased fatigue on the FOWT, a point which will be further explored in the subsequently given criterion. In the context of another performance measure, considering roll, pitch, yaw, and their respective rate changes, the NN-based STW strategy exhibits superior efficacy. This is particularly noticeable in the value of roll and roll rate, which have witnessed improvements by 2.19% and 7.14% respectively. Referencing Table 2, it is evident that the NN-based STW method exhibits comparatively superior performance in managing the Damage Equivalent Load (DEL) of the FOWT. Particularly in the TB side-to-side parameter, an improvement of 5.33% is observed. Similarly, in the cases of FF2 and AF2, enhancements of 3.48% and 3.57% respectively are seen. Contrarily, in the context of certain parameters such as FF3 and AF3, the STW method with constant gains outperforms the NN-based STW approach.

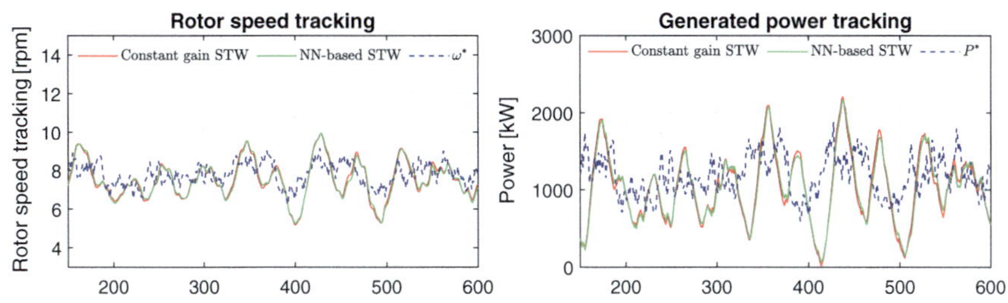

Fig. 2 Tracking of the generated power and the rotor speed versus time

Table 1 Performance indicator parameters

Parameters	STW	NN-based STW	Improvement (%)
Total generated power	1.0331	1.0334	0.02
Rotor speed error	0.2726	0.2638	3.33
Roll	0.1069	0.1046	2.19
Pitch	1.5637	1.5621	0.10
Yaw	0.0908	0.0903	0.55
Roll rate	0.0120	0.0112	7.14
Pitch rate	0.1398	0.1395	0.21
Yaw rate	0.0503	0.0501	0.39

Table 2 DEL values of structure loads

Parameters	STW	NN-based STW	Improvement (%)
TB side-to-side	1032.18	979.877	5.33
TB fore-aft	14699.6	14682.9	0.11
TB torsional	112.878	110.737	1.93
BR edge-wise	2527.74	2527.90	≈ 0
BR flap-wise	654.012	656.618	− 0.39
FF1	24884.7	24841.5	0.17
FF2	18155.7	17545.0	3.48
FF3	16692.5	17040.3	− 2.08
AF1	24658.8	24619.1	0.16
AF2	17984.1	17362.8	3.57
AF3	16475.8	16853.4	− 2.29

5 Conclusion

This research investigated a hybrid control strategy for improving power generation from a FOWT in a low-wind condition. This was accomplished by integrating an STW control strategy with an RBFNN. The STW control addressed the complex dynamics of the FOWT, while the RBFNN aimed to approximate uncertainties and unmodeled dynamics which can reduce the dependency on precise system modeling. The effectiveness of this approach was confirmed through the OpenFAST simulator, which demonstrated an MPPT power output, reduced pitch motion, and decreased fatigue.

References

1. Kumar D, Chatterjee K (2016) A review of conventional and advanced MPPT algorithms for wind energy systems. Renew Sustain Energy Rev 55:957–970; 2(5):99–110
2. Pan L, Shao C (2020) Wind energy conversion systems analysis of PMSG on offshore wind turbine using improved SMC and extended state observer. Renew Energy 161:149–161
3. Hamida M, Taleb M, Marie A, Zhang C, Testelin P, Plestan F (2022) A novel control strategy of variable speed floating wind turbine in region II. In: OCEANS 2022. IEEE, Chennai, pp 1–6
4. Benakcha M, Benalia L, Ammar A, Bourek A (2019) Wind energy conversion system based on dual stator induction generator controlled by nonlinear backstepping and PI controllers. Int J Syst Assur Eng Manag 10(4):499–509
5. Levant A (2003) Higher-order sliding modes, differentiation and output feedback control. Int J Control 76(9–10):924–941
6. Matraji I, NI-Durra A, Errouissi R (2018) Design and experimental validation of enhanced adaptive second-order SME for PMSG-based wind energy conversion system. Int J Electr Power Energy Syst 103:21–30
7. Sami I, Ullah S, Amin SU, Al-Durra A, Ullah N, Ro J-S (2022) Convergence enhancement of super-twisting sliding mode control using artificial neural network for DFIG-based wind energy conversion systems. IEEE Access 10:97625–97641
8. Chojaa H, Derouich A, Chehaidia SE, Zamzoum O, Taoussi M, Elouatouat H (2021) Integral sliding mode control for DFIG based WECS with MPPT based on artificial neural network under a real wind profile. Energy Rep 7:4809–4824
9. Jonkman J, Butterfield S, Musial W, Scott G (2009) Definition of a 5-mw reference wind turbine for offshore system development. Tech. rep., National Renewable Energy Lab.(NREL), Golden, CO, USA
10. Zhang C, Plestan F (2020) Power and motion control of a floating wind turbine: an original approach based on adaptive second order sliding mode control. In: IFAC world congress, Berlin, Germany
11. Hong C-M, Chen C-H, Tu C-S (2013) Maximum power point tracking-based control algorithm for PMSG wind generation system without mechanical sensors. Energy Convers Manage 69:58–67
12. Feng Z, Fei J (2018) Super-twisting sliding mode control for micro gyroscope based on RBF neural network. IEEE Access 6:64993–65001
13. Shtessel Y, Edwards C, Fridman L, Levant A (2014) Sliding mode control and observation. Springer
14. Chalanga A, Plestan F (2017) Finite time stabilization of an uncertain chain of integrators by integral sliding mode approach. IFAC-PapersOnLine 50(1):9613–9618

15. Ren X, Lewis FL, Zhang J (2009) Neural network compensation control for mechanical systems with disturbances. Automatica 45(5):1221–1226
16. Zhang C, Plestan F (2021) Adaptive sliding mode control of floating offshore wind turbine equipped by permanent magnet synchronous generator. Wind Energy

The Removal of Heavy Metals Effluent from Wastewater by Nanoparticle Based on Polyoxometalates

Nacéra Zabat and Nawel Nedjah

Abstract In this study, a lacunar Polyoxometalate (POM) of Dawson type $(\alpha_2P_2W_{17}O_{61})^{10-}$ was prepared and tested in the complexation of (Ni^{2+}) in aqueous solution. The stability of the formed complex $(\alpha_2P_2W_{17}O_{61}Ni)^{8-}$ according to the reactional medium pH and the time were studied. The stoichiometry and the constant of stability of this formed complex were also studied. From where it was noted that this compound has rather high constant of stability, it was calculated to be $\beta = 1.57 \times 10^6$. Thus this compound has interesting proprieties for fixing heavy metals such us nickel existing in industrial waste water.

Keywords Polyoxometalates · Nanoparticle · Waters pollution · Nickel · Spectroscopic methods

1 Introduction

The importance of the use of water in industry and by the urban population carries out to think more and more on the means necessary and objective to carry out the good quality of this essential element. To this end and since many years, the work were realized implementing increasingly elaborate materials allowing bettering determine the whole of the encountered problems; and to reach thresholds of very low detection such as they are imposed by the legislation according to the nature of water to analyze. The proportioning of metals in traces knew these last years the use of

N. Zabat (✉)
Laboratory of Organic Synthesis-Modeling and Optimization of Chemical Processes, Department of Process Engineering, Faculty of Engineering, Badji Mokhtar-Annaba University, P.O. Box 12, 23000 Annaba, Algeria
e-mail: zabatnassira@yahoo.fr; nassira.zabat@univ-annaba.dz

N. Nedjah
Physical Metallurgy and Materials Properties Laboratory, National Higher School of Technology and Engineering ENSTI, Badji Mokhtar University, Annaba, Algeria

© The Author(s), under exclusive license to Springer Nature Singapore Pte Ltd. 2024
A. Mellit et al. (eds.), *Proceedings of the 2nd International Conference on Green Energy Conversion System*, Green Energy and Technology,
https://doi.org/10.1007/978-981-97-6148-7_68

sophisticated analytical methods requiring expensive and not very accessible equipment. In fact, complexation is a method, which remains largely used in the metal traces proportioning; it is based on the use of the ligands whose synthesis does not cease growing. The lacunars Polyoxometalates (POMs) of Dawson type [1] are materials which lend themselves perfectly to reactions of addition leading to mixed compounds [2, 3]. This property can be exploited to fix heavy metals to be present in aqueous medium, at end of qualitative analyses and quantitative. These Polyoxometalates (POMs) have remarkable electrochemical and electrocatalytic properties [4–6]. These compounds also showed very interesting applications in analytical chemistry [7], homogeneous and heterogeneous catalysis [8, 9], in science of the materials [10], in water treatment [11], and even in field of electrochemical sensors [12]. In this context a lacunar Polyoxometalate was synthesized $(\alpha_2P_2W_{17}O_{61})^{10-}$ and used as a ligand for complexation of nickel (Ni^{2+}) in aqueous solution. This heavy metal is considerate as dangerous and toxic element can be present in the industrial liquid effluents. In this objective a spectroscopy characterization by IR and UV–Visible of the formed complex $(\alpha_2P_2W_{17}O_{61}Ni)^{8-}$ has realized. The spectrophotometric study makes it possible to determine the optimal conditions supporting the formation of this complex namely:

- pH of the reactional medium;
- Stability according to time of complexation;
- Stoichiometry or ratio ligand /metal of the complexation reaction;
- Constant of stability of the formed complex.

2 Experimental Method

2.1 Synthesis of the Complex

The Polyoxometalate $(\alpha_2P_2W_{17}O_{61}Ni)^{8-}$ was prepared by addition of nickel nitrate on the lacunar compound $(\alpha_2P_2W_{17}O_{61})^{10-}$ according to method described in the literature [13]. The lacunar compound $(\alpha_2P_2W_{17}O_{61})^{10-}$ was synthesized by stereospecific alkaline degradation of saturated species $\alpha P_2W_{18}O_{62}$ [14], this species is obtained by polycondensation of tungstate ions in acid medium, in presence of a heteroatom in the occurrence phosphorus. Degradation related to the elimination of an octahedral $WO_4{}^{2-}$ located in site (a). The various structures of these heteropolyanions are presented on Fig. 1.

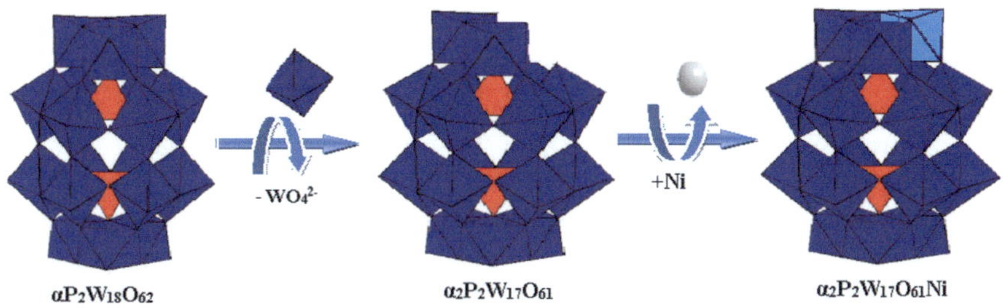

Fig. 1 Different structures of Polyoxometalates

2.2 *Characterization of the Polyoxometalates by Infra-Red Spectroscopy (I.R)*

For better appreciating the purity of these prepared compounds (HPA), the characterization by the IRTF was carried out. The spectra IR were recorded in solid phase in KBr (1 mg of sample for 300 mg of KBr) on a spectrophotometer (Shimadzu japan). The recorded spectra are presented in Fig. 2.

In the spectrum of the lacunar compounds $(\alpha_2P_2W_{17}O_{61})^{10-}$, it is notices a band PO located at 1050 cm^{-1}. Delcheff [15] showed that the existence of such a band represented an important disturbance in a direction of tetrahedron PO_4^-, Direction of the gap in this case. This band disappears for the complex $(\alpha_2P_2W_{17}O_{61}Ni)^{8-}$. The area of the bands PO then becomes similar to that observed on the spectrum of the saturated species $(\alpha P_2W_{18}O_{62})^{6-}$. The addition of nickel seems to take place in the vicinity of the gap.

Fig. 2 Infra-red spectra of different Polyoxometalates

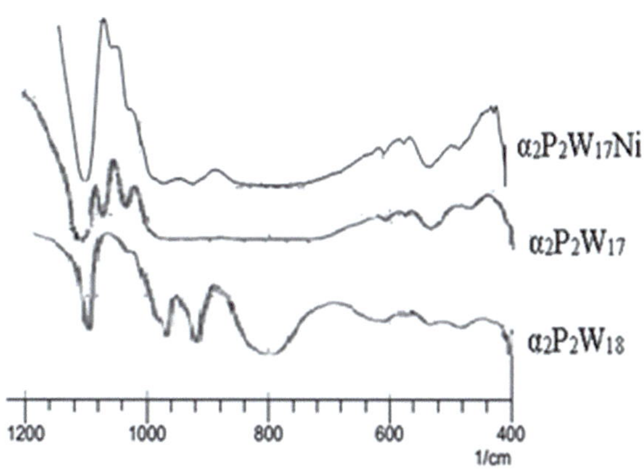

Table 1 Variation of absorbance complex $(\alpha_2P_2W_{17}Ni)^{8-}$ according pH

pH		3	4	5	6	8	9	10	11
Ni^{2+}	ABS	0,269	0,241	0,255	0,269	0,813	0,984	0,333	0,493
	λ(nm)	215	302	323	216	321	308	228	287

3 Results and Discussion

3.1 *Influence of pH on the Formation of the Complex*

This study was carried out with a different pH (acid and basic) by making a UV–Visible sweeping. It was proceeded with a equimolaire mixture of POM/Metal. The solutions were prepared as follows:

- Quantity of buffer solution;
- A same quantity of solution ligand (POM) at a concentration of 5×10^{-4} M;
- A same quantity of metal at identic concentration of ligand;
- The reference solution consists of:
- Quantity of corresponding buffer solution and a double quantity of ligand solution.

The results obtained, are grouped on Table 1.

According to the values obtained in Table 1, it is notices that the degree of formation of the complex is deducted from the value of the recorded maximum absorbance. The formation of the complex is very weak in acid medium. On the other hand, medium basic is more favorable to their formation. This may be due to the possibility of dimerization of the Polyoxometalate in an acid medium [16]. The optimal pH for the formed complex $(\alpha_2P_2W_{17}O_{61}Ni)^{8-}$ is pH = 9 at a maximum wavelength = λ 308 nm.

3.2 *Influence of Time on the Formation of the Complex*

To determine the stabilization time for the formed complex $(\alpha_2P_2W_{17}O_{61}Ni)^{8-}$, variation of the absorbance of this complex according to the time was followed. The same solutions and the optimum pH founded were kept. The results obtained on Fig. 3 reveal that the formed complex is stable in a rather broad time interval (one hour). A similar result has been observed in the literature [17].

3.2.1 Determination of the Stoichiometry Reaction

To determine the stoichiometry of the complexation reaction or the report of the ligand–metal concentrations (C_L/C_M), the calculation methods were used: method of the series isomolaires and method of saturation [18].

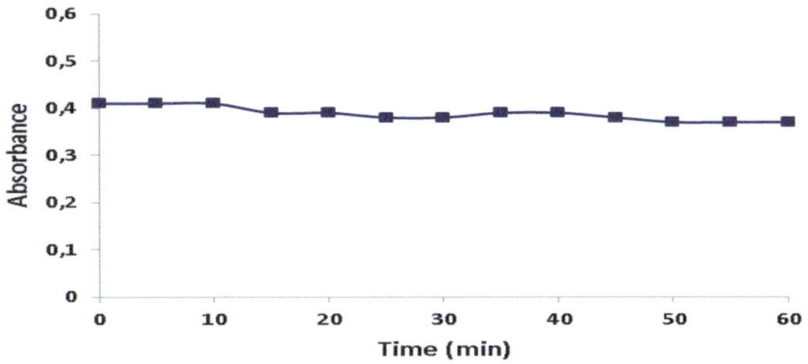

Fig. 3 Variation of the absorbance of the formed complex $(\alpha_2P_2W_{17}O_{61}Ni)^{8-}$ Versus time

Fig. 4 Variation of the absorbance of $(\alpha_2P_2W_{17}O_{61}Ni)^{8-}$ versus C_L/C_M by isomolaire method

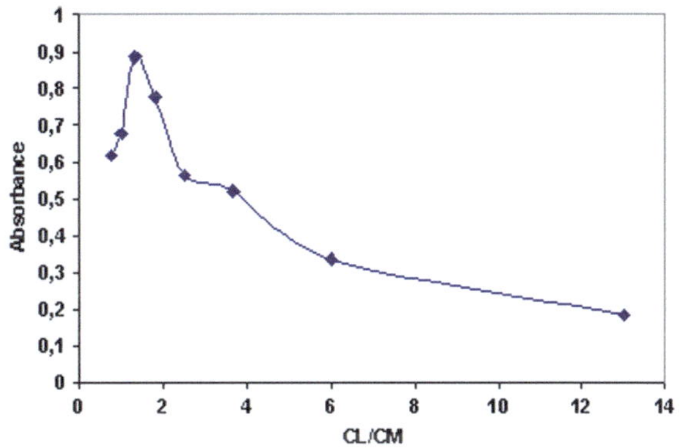

3.3 *Method of the Isomolaires Series*

In this method, the volume of the buffer solution (pH $-$ 9) is maintained constant whereas the volume of the solutions of ligand and nickel at the same concentration (5.10^{-4}M) are mixed in variable proportions in order to keep constant total volume. Figure 4 reproduces the variation of the absorptance of the complex according to the report of concentration (C_L/C_M).[19]

According to the principle of the method of the series isomolaires, the stoichiometry deduced from report C_L/C_M leading to the maximum value of the absorbance. Figure 4 shows that the stoichiometry of the formed complex is equal to 1.

3.4 *Method of Saturation*

This method consists in varying the volume of the ligand and the buffer solutions, while the volume of the metal is maintained constant; in manner the total volume is kept constant. The values of this method become source data in calculation of the

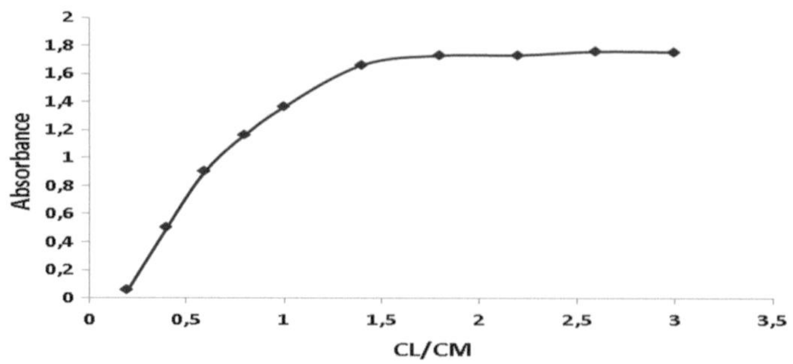

Fig. 5 Variation of the absorbance of $(\alpha_2P_2W_{17}O_{61}Ni)^{8-}$ versus C_L/C_M by saturation method

constant of stability by the method of balance displacement. The variation of the absorbance of the solutions according to report C_L/C_M is represented on Fig. 5.

According to Fig. 5 it is noted that the curves of saturation of the formed complex present two parts: the first part corresponds to an increase in absorbance and the second presents a stage of saturation. The abscissa of intersection of the tangents of the two parts gives the ratio ligand–metal (C_L/C_M) corresponding to the stoichiometry of the reaction of complexation. This figure shows that the stoichiometry of the formed complex $(\alpha_2P_2W_{17}O_{61}Ni)^{8-}$ is equal to one ($C_L/C_M = 1$). This result is concordant with work established by Ciabrini and Contant [20], they find a stoichiometric ratio of 1:1, for Polyoxometalate $(\alpha_2P_2W_{17})$ complexed with cerium (III and IV). These results confirm once again the previously results obtained by the series isomolaires method.

3.4.1 Determination of the Constant of Stability

The stoichiometry (n) and the constant of stability (β) of the formed complex are determined by linearization of the results obtained by the method of saturation according to the following equation:

$$\mathrm{Log}\,(\mathrm{Ai}/\mathrm{Amax} - \mathrm{Ai}) = \mathrm{nLog}C_{\mathrm{Li}} + \mathrm{Log}\,\beta \tag{1}$$

Ai absorbance of the complexe corresponding to the increasing part of the curve of saturation.
Amax maximum absorbance corresponding to the stage of the curve of saturation.
n stoechiometric report (C_L/C_M).
C_{Li} concentration of the ligand corresponding to each value of Ai.
β constant of stability.

By tracing Log (Ai/Amax—Ai) according to $\mathrm{Log}C_{\mathrm{Li}}$ for four points of the ascending part of the curve of saturation, Figure 6 is obtained.

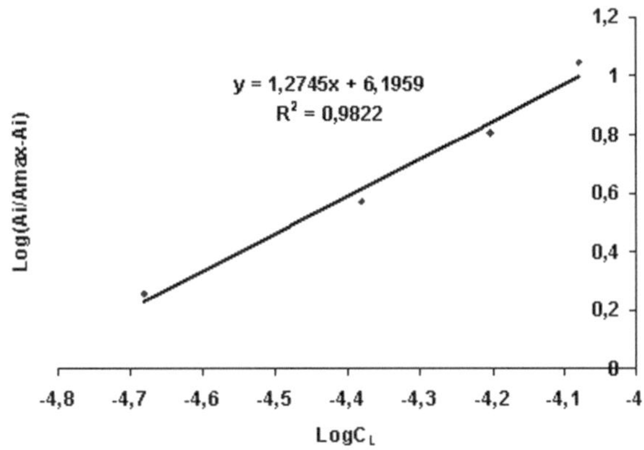

Fig. 6 Variation of $LogC_{Li}$ according to Log (Ai/Amax—Ai) for the complex$(\alpha_2P_2W_{17}Ni)^{8-}$

The equation of the linear regression in Fig. 6 shows that the stoichiometry or ligand–metal ratio (n = 1.2745) and (Logβ = 6.1959).Therefore the value of the stability constant is very high ($\beta = 1{,}57 \times 10^6$). A similar result has been observed in the literature [21].

4 Conclusion

This work concerned the study of the chelating properties of the Dawson type Polyoxometalate $(\alpha_2P_2W_{17}O_{61})^{10-}$ with respect to the heavy metal Ni^{2+} present in aqueous medium. The synthesized complex was characterized by various spectroscopic techniques UV Visible and Infra-red (IR). The characterization by spectroscopy IR, showed that the metal element (Ni^{2+}) tends to fill the gap of lacunar compound $(\alpha_2P_2W_{17}O_{61})^{10-}$, thus leading to a new saturated specie $(\alpha_2P_2W_{17}O_{61}Ni)^{8-}$.

The study of the stability of the formed complex by UV -Visible spectrophotometry revealed the following results:

- Optimal pH of the formed complex $(\alpha_2P_2W_{17}O_{61}Ni)^{8-}$ is pH = 9 (basic medium).
- The time of stability for the formed complex reach one hour.
- The stoichiometry of the complexation reaction or the report C_L/C_M is equal to 1.

In addition the constant of stability of the formed complex $(\alpha_2P_2W_{17}O_{61}Ni)^{8-}$ is enough raised $\beta = 1{,}57 \times 10^6$ indicating the possibility of use of this compound, in particular in aqueous medium without risk of decomposition. According to these results, it was noted that, the lacunar Polyoxometalate prepared $(\alpha_2P_2W_{17}O_{61})^{10-}$ has interesting properties of complexation in aqueous solution. Of this fact they can be used in the field of water treatment to fix toxic metals; it can be also proposed in chemical analysis and used as catalyst in homogenous catalysis.

Acknowledgements The authors thank the Ministry of Higher Education and Scientific Research, Algeria for the financial support through the PRFU project (N°: A16N01UN230120180001).

References

1. Dawson B (1953) The structure of the 9 (18)-heteropoly anion in potassium 9 (18)-tungstophosphate, K6 (P2W18O62) 14H2O. Acta Crystallogr A 6(2):113–126
2. Contant R, Abbessi M, Thouvenot R, Hervé G (2004) Dawson Type Heteropolyanions. 3.Syntheses and ^{31}P, ^{51}V and ^{183}W NMR Structural Investigation of Octadeca (molybdo−tungsto−vanado) diphosphates Related to the [H2P2W12O48] 12-Anion. Inorgan Chem 43(12):3597–3604
3. Belghiche R, Contant R, Lu YW, Keita B, Abbessi M, Nadjo L, Mahuteau J (2002) Synthesis and caracterization of Fe-or Cu-substituted molybdenum-enriched tungstodiphosphates Eur. J Inorg Chem 6:1410
4. Keita B, Girard F, Nadjo L, Contant R, Belghiche R, Abbessi M (2001) Chemistry, Cyclic voltammetric evidence of facilitation of the reduction of nitrite by the presence of molybdenum in Fe-or Cu-substituted heteroplytungstates. J. Electroanalytical. 508:70–80
5. Keita B, Mbomekalle I-M, Nadjo L, De Oliveira P, Ranjbari A, Contant R (2005) Vanadium-substituted Dawson-type polyoxometalates as versatile electrocatalysts. C.R. Chimie. 8:1057–1066
6. Gkika E, Troupes A, Hiskia A, Papaconstantinou E (2005) Photocatalytic reduction of chromium and oxidation of organics by polyoxometalates. Appl Catal B 62:28–34
7. Liu S, Tang Z, Wang E, Dong S (1999) Fabrication and characterization of heteropolyanion Langmuir-Blodget. Thin Solid Films 339:277–283
8. Kozhevnikov IV (1997) $PMO_{12-n}V_nO^{(3+n)-}{}_{40}$ heteropolyanions as catalysts for aerobic oxidation. J Mol Catal A Chem 117:151–158
9. Stobbe-Kreemers AW, Van Der Zon M, Makkee M, Scholten JJF (1996) Palladium salts of heteropolyanions as catalysts in heterogeneous Wacker oxidation of 1-butene. J Mol Catal A Chem 107:247–253
10. Clemente-Juan JM, Clemente-León M, Coronado E, Galán-Mascarós JR, Giménez-Saiz C, Gómez-García CJ (1998) Magnetic clusters and conducting molecular materials from polyoxometalates. Comptes Rendus de l'Acad des Sci Series IIC-Chem 1(5–6):305–317
11. Zabat N, Samar ME, Abbessi M (2010) Complexation de l'aluminium par des hétéropolyanions lacunaires: récupération par membrane liquide émulsionnée. Revue Synthèse 21:75–85
12. Zabat N, Abbessi M, Kherrat R, Jaffrezic-Renault N, Zougar S, Grid A (2008) Elaboration of Impedancemetric sensor for cobalt detection by polymeric membrane incorporating a heteropolyanion of Dawson type. Mater Sci Eng C 28:1006
13. Keita B, Lu YW, Nadjo L, Contant R, Abbessi M, Canny J, Richet M (1999) Electrochemical and catalytic behaviour of Dawson-type complexes derived from $[(1),2,3\text{–}P_2Mo_2\,W_{15}\,O_{61}]^{10-}$and first transition metal ions. Electroanal Chem 477:146–157
14. AbbessiM (1989) Synthèse et propriétés d'hétéropolyanions vanado-molybdo-tungsto-phosphoriques en vue d'une application en catalyse d'oxydation PhD thèses, Université Paris VI, France
15. Rocchiccioli-Deltcheff C, Thouvenot R (1979) Vibrational studies of heteropolyanions related to a-$P_2W_{18}O_{62}{}^{6-}$ I-infrared evidence of the structure ofa1 and α2-$P_2W_{17}O_{61}{}^{10-}$. Spectrosc Lett 12:127–138
16. Keita B, Girard F, Nadjo L, Contant R, Canny J, Richet M (1999) Metal ion complexes derived from thea1 isomer of $(P_2W_{17}O_{61})^{10-}$:comparison with the corresponding α_2 species. Electroanal Chem 478:76
17. Contant R, Ciabrini JP (1981) Stereospecific preparations of new n-molybdo-(18-n)-tungsto-2-phosphates and related "defect" compounds (n= 2, 4 or 5). J Inorgan Nucl Chem 43(7):1525–1528
18. Hamlaoui ML, Vlassenko K, Messadi D (1990) Constante de stabilité des complexes de quelques métaux de transition avec l'hétéropolyanion non saturé $[P_2W_{17}O_{61}]^{10-}$. CR Acad Sci Paris SérieII. 311:795–798

19. Lu Y-W, Keita B, Nadjo L (2004) Rational approach of the stoichiometries of lanthanide complexes with α_2-$[P_2W_{17}O_{61}]^{10-}$heteropolytungstate in aqueous solution. Polyhedron 23:1579
20. Ciabrini JP, Contant R (1993) Mixed heteropolyanions Synthesis and formation constants of Cerium (III) and Cerium (IV) complexes with lacunary tungstophosphates. J Chem Res 391:2720–2744
21. Contant R, Abbessi M, Canny J, Richet M, Keita B, Belhouari A, Nadjo L (2000) Synthesis, characterization and electrochemistry of complexes derived from [(1), 2, 3–P2Mo2W15O61] 10–and first transition metal ions. Eur J Inorgan Chem 2000(3):567–574

Modeling of an Ejector Compression Refrigerating Machine Performance Operating with Low GWP Working Fluids

I. Ouaz, I. Saad, S. Maalej, and M. C. Zaghdoudi

Abstract A model is developed to assess the thermodynamic performance of an Ejector-Compression Refrigerating Machine (ECRM). Four Low GWP refrigerants with zero Ozone Depletion Potential (ODP) are considered: R1234yf, R152a, R290, and R717. The thermodynamic performance of the ECRM is evaluated in critical and non-critical conditions. In critical conditions, the simulation tests show that the critical pressure and critical temperature increase with the boiler temperature; and the entrainment ratio falls as the generator temperature augments. In non-critical conditions, the results reveal that there exists an optimum generator temperature and a motive ratio for which the entrainment ratio is maximum. The best entrainment ratio is achieved with R290 with the highest generator temperature and lowest motive and compression ratios, and the lowest entrainment is obtained with R717 with the highest generator temperature, motive ratio, and compression ratio.

Keywords Refrigeration · Ejector · Low GWP · Thermodynamic performance · Entrainment ratio

1 Introduction

Theoretical and experimental studies have been carried out to choose the most efficient refrigerant for a given ejector geometry. Dorantes and Lallemand [1] developed a model to evaluate the thermodynamic performance of an ECRM according to the nature of the working fluid. They found that R141b, R123, RC318, and R142b lead to the best entrainment ratio; however, R22 yields the lowest one. Al-Khalidy [2] has theoretically studied the effect of using R717, R11, R113, and R114 on the efficiency of an ECRM. It has shown that the refrigerant R113 is the best suitable fluid. Sun

I. Ouaz · I. Saad · S. Maalej · M. C. Zaghdoudi (✉)
Laboratoire Matériaux, Mesures et Applications (MMA, LR11ES25), University of Carthage, Institut National des Sciences Appliquées et de Technologie (INSAT), B.P. N° 676, Centre Urbain Nord, 1080 Tunis, Tunisia
e-mail: chaker.zaghdoudi@insat.rnu.tn

© The Author(s), under exclusive license to Springer Nature Singapore Pte Ltd. 2024
A. Mellit et al. (eds.), *Proceedings of the 2nd International Conference on Green Energy Conversion System*, Green Energy and Technology,
https://doi.org/10.1007/978-981-97-6148-7_69

[3] studied eleven refrigerants, including water, CFCs (R11, R12, R113), HCFCs (R21, R123, R142b), HFCs (R134a, R152a), a cyclic organic compound refrigerant (RC318), and an azeotrope (R500). They showed that the ECRM operating with steam allows for the lowest thermodynamic performance and that the system operating with R152a gives the best entrainment ratio. Selvaraju and Mani [4] studied the ECRM performance operating with different refrigerants: R134a, R152a, R290, R600a, and R717. They have shown that R134a exhibits the highest entrainment ratio in nominal conditions. Boumaraf and Lallemand [5] have studied the performance of an ECRM in a nominal regime as a function of source temperatures and various working fluids among them R22, R290, R134a, R152a, RC318, and R124. They demonstrated that R290 is adapted to the ECRM operation at low generator temperatures despite a lower entrainment ratio than that obtained with R124 and R152a for higher generator temperatures.

Most of the considered working fluids are not eco-friendly and their GWP is high. Recently, a new generation of refrigerants has been proposed to overcome environmental issues [6–9]. The present work aims to quantify the thermodynamic performance of an ECRM operating with zero-ODP eco-friendly refrigerants having low GWP values. From the previous literature study, four refrigerants are identified: R1234yf (GWP = 4, A2L2 safety group), R152a (GWP = 124, A2L2 safety group), R290 (GWP = 20, A3 safety group), and R717 (GWP = 0, B2 safety group). The effects of the generator temperature and the motive pressure ratio on the entrainment ratio are highlighted. The results of this study will be used to optimize the choice of the working refrigerant that will be considered in our future tests on a solar ECRM which was tested previously with R141b [10].

2 Analysis of the Ejector Cooling Cycle

In an ECRM, the primary flow gets in the ejector over a convergent-divergent nozzle, where the flow changes from subsonic to supersonic and generates a low-pressure region at the nozzle exit. This drives the secondary flow originating from the evaporator (Fig. 1). The primary and secondary flow do not mix directly at the nozzle exit, and the mixing happens in the constant constant-area section of the ejector (Fig. 2). Then, the mixture is subjected to transverse schoch waves and flows to the diffuser and the mixture pressure increases to reach the pressure of the condenser (Fig. 2). Finally, a part of the condensate is pumped into the generator and the other part is sent to the evaporator through an expansion valve (Fig. 1).

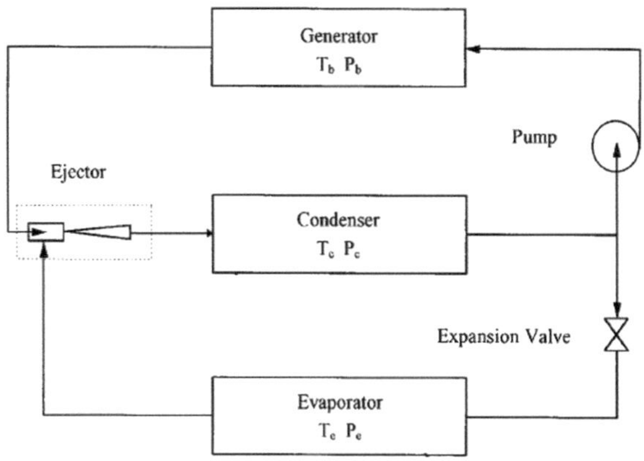

Fig. 1 Main components of the ECRM

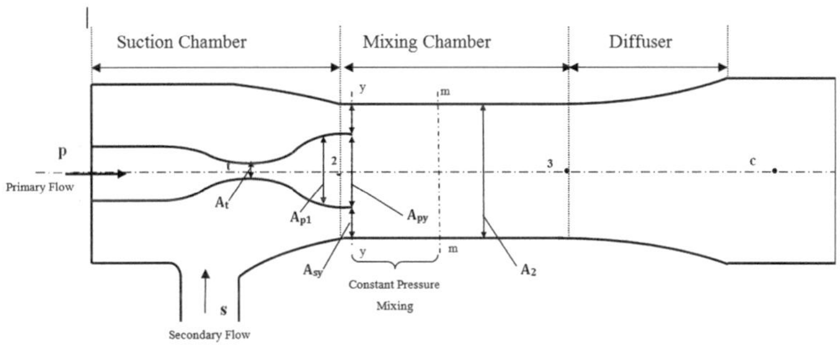

Fig. 2 Geometrical characteristics of the ejector

3 Mathematical Model Formulation

A theoretical model is developed by considering perfect gas law, mass, momentum, and energy balance equations as well as the equations established for isentropic transformations [10]. Hence, the following equations are considered

$$\frac{P}{\rho} = rT \tag{1}$$

$$h = c_p T \tag{2}$$

$$\frac{P}{\rho^{\gamma}} = \text{constant} \tag{3}$$

$$\frac{P}{T^{\frac{\gamma}{\gamma-1}}} = \text{constant} \tag{4}$$

$$c^2 = \frac{\gamma P}{\rho} \tag{5}$$

P, ρ, and T are the pressure, density, and temperature of the fluid, respectively, c_p is the specific heat at constant pressure, γ is the ratio of the specific heat at constant pressure and volume, c is the sonic velocity, and r is the gas constant.

The energy balance equations, the isentropic relations, and the isentropic efficiency, η_p, for the primary flow, are considered to derive the expressions of the temperature, pressure, and velocity at the neck of the primary nozzle (section A_t) according to [10]

$$\frac{T_t}{T_g} = \frac{1}{1 + \frac{\gamma-1}{2} M_t^2} \tag{6}$$

$$P_t = P_g \left[\left(1 - \frac{1}{\eta_p} \right) + \frac{1}{\eta_p} \left(\frac{1}{1 + \frac{\gamma-1}{2} M_t^2} \right) \right]^{\frac{\gamma}{\gamma-1}} \tag{7}$$

$$V_t = \sqrt{\frac{\gamma r T_g M_t^2}{1 + \frac{\gamma-1}{2} M_t^2}} \tag{8}$$

$$M_t = \frac{V_t}{\sqrt{r T_t}} \tag{9}$$

P_g and T_g are the generator pressure and temperature, respectively. In these equations, the Mach number at the primary nozzle throat is $M_t = 1$ since the flow in this location is sonic.

The mass flow rate of the primary flow at the primary nozzle throat depends on the cross-section of the primary nozzle throat, A_t, according to

$$\dot{m}_p = \frac{P_t}{r\, T_t} A_t \; V_t \tag{10}$$

The expressions for the temperature, the pressure, the speed, and the Mach number at the primary nozzle exit (section A_{p1}) are derived from the energy balance equations according to [10]

$$T_{p1} = \frac{T_g}{1 + \frac{\gamma-1}{2} M_{p1}^2} \tag{11}$$

$$P_{p1} = P_g \left[\left(1 - \frac{1}{\eta_{p1}} \right) + \frac{1}{\eta_{p1}} \left(\frac{1}{1 + \frac{\gamma-1}{2} M_{p1}^2} \right) \right]^{\frac{\gamma}{\gamma-1}} \tag{12}$$

$$V_{p1} = \sqrt{\frac{\gamma r T_g M_{p1}^2}{1 + \frac{\gamma - 1}{2} M_{p1}^2}} \tag{13}$$

The mass flow rate at the primary nozzle exit is

$$\dot{m}_p = \frac{P_{p1}}{r\, T_{p1}} A_{p1}\ V_{p1} \tag{14}$$

From Eqs. (10) to (14), we can deduce the expression of A_{p1} as a function of A_t and M_{p1} according to

$$\frac{A_{p1}}{A_t} = \frac{1}{M_{p1}} \left[\left(\frac{2}{\gamma + 1} \right) \left(1 + \frac{\gamma - 1}{2} M_{p1}^2 \right) \right]^{\frac{\gamma + 1}{2(\gamma - 1)}} \tag{15}$$

The Mach number, M_{p1}, is calculated iteratively so that the equality of Eqs. (10) and (14) is ensured.

The flow characteristics at the secondary nozzle exit (section A_{sy}) are obtained from the energy balance, the isentropic relations, and the isentropic efficiency, η_s according to [10]

$$T_{sy} = \frac{T_e}{1 + \frac{\gamma - 1}{2} M_{sy}^2} \tag{16}$$

$$P_{sy} = P_e \left[\left(1 - \frac{1}{\eta_s} \right) + \frac{1}{\eta_s} \left(\frac{1}{1 + \frac{\gamma - 1}{2} M_{sy}^2} \right) \right]^{\frac{\gamma}{\gamma - 1}} \tag{17}$$

$$V_{sy} = \sqrt{\frac{\gamma r T_e M_{sy}^2}{1 + \frac{\gamma - 1}{2} M_{sy}^2}} \tag{18}$$

$$M_{sy} = \frac{V_{sy}}{\sqrt{\gamma r T_{sy}}} \tag{19}$$

P_e and T_e are the evaporator pressure and temperature, respectively. η_s and M_{sy} are the isentropic efficiency and the Mach number of the secondary flow, respectively.

The secondary flow mass flow rate is expressed as

$$\dot{m}_s = \frac{P_{sy}}{r\, T_{sy}} A_{sy}\ V_{sy} \tag{20}$$

A_{sy} is the cross-section of the secondary flow at the level of the primary nozzle outlet (Fig. 2).

The pressure, P_{py}, the temperature, T_{py}, the Mach number, M_{py}, and the velocity, V_{py} are at the inlet of the mixing chamber are obtained according to [10]

$$T_{py} = \frac{T_g}{1 + \frac{\gamma-1}{2} M_{py}^2} \tag{21}$$

$$P_{py} = P_g \frac{1}{\eta_{py}} \left[\left(\eta_{py} - 1 \right) + \left(\frac{1}{1 + \frac{\gamma-1}{2} M_{py}^2} \right) \right]^{\frac{\gamma}{\gamma-1}} \tag{22}$$

$$V_{py} = M_{py} \sqrt{\gamma r T_{py}} \tag{23}$$

$$M_{py} = \sqrt{\frac{2}{\gamma - 1} \left[\left[\frac{1}{\left[\left[\frac{\eta_{py}}{\eta_{p1}} \left[\frac{P_{py}}{P_{p1}} \right]^{\frac{\gamma-1}{\gamma}} \left[\left(\eta_{p1} - 1 \right) + \left(\frac{1}{1+\frac{\gamma-1}{2} M_{p1}^2} \right) \right] \right] - \left(\eta_{py} - 1 \right) \right]} \right] - 1 \right]} \tag{24}$$

The cross-section of the primary flow at the inlet of the mixing chamber, A_{py}, is determined from

$$\dot{m}_p = \frac{P_{py}}{r\, T_{py}} A_{py}\, V_{py} \tag{25}$$

A_{sy} is deduced from the following relation

$$A_2 = A_{py} + A_{sy} \tag{26}$$

The velocity, the temperature, and the Mach number at section, A_2, before the shock wave are expressed as [10]

$$V_2 = \frac{\dot{m}_p\, V_{py} + \dot{m}_s V_{sy}}{\left(\dot{m}_p + \dot{m}_s \right)} \tag{27}$$

$$T_2 = \frac{\left[\dot{m}_p \left(c_p T_{py} + \frac{V_{py}^2}{2} \right) \right] + \left[\dot{m}_s \left(c_p T_{sy} + \frac{V_{sy}^2}{2} \right) \right]}{\left(\dot{m}_p + \dot{m}_s \right) \mathrm{c}_p} - \frac{V_2^2}{2\, \mathrm{c}_p} \tag{28}$$

$$M_2 = \frac{V_2}{\sqrt{\gamma\, \mathrm{r}\, \mathrm{T}_2}} \tag{29}$$

After the shock wave, the flow becomes subsonic, and the pressure, the Mach number, the temperature, and the velocity, are

$$P_3 = P_2 \frac{2}{\gamma + 1} \left[\left(\gamma\, M_2^2 \right) - \frac{\gamma - 1}{2} \right] \tag{30}$$

$$M_3 = \sqrt{\frac{M_2^2 + \frac{2}{\gamma-1}}{\frac{2}{\gamma-1} M_2^2 - 1}} \tag{31}$$

$$T_3 = T_2 \frac{1 + \frac{\gamma-1}{2} M_2^2}{1 + \frac{\gamma-1}{2} M_3^2} \tag{32}$$

$$V_3 = M_3 \sqrt{\gamma \; r \; T_3} \tag{33}$$

The pressure, P_4, at the diffuser exit is determined from the application of the energy conservation principle for which the velocity at the condenser is supposed to be zero according to

$$P_4 = P_3 \left(1 + \frac{\gamma - 1}{2} M_3^2\right)^{\left(\frac{\gamma}{\gamma-1}\right)} \tag{34}$$

The entrainment ratio is calculated according to

$$\mathrm{U} = \frac{\dot{m}_s}{\dot{m}_p} \tag{35}$$

4 Computation Methodology

A computer simulation program is developed. The thermodynamic properties are calculated by using the NIST REFPROP program and the calculating program is written in MATLAB language. Hereafter is the detailed calculation procedure:

(1) Fixing the values of T_g, P_g, T_e, P_e, P_c.
(2) Fixing the values of γ, η_p, η_s, and molecular mass $\overline{M}$ of the working fluid.
(3) Calculating the specific heat $cp = \gamma r/(\gamma - 1)$ and the gaz constant $r = R/\overline{M}$ where R is perfect gas constant ($R = 8.314 \text{ J mol}^{-1} \text{ K}^{-1}$).
(4) Entering the geometrical characteristics: A_t, A_{p1}, A_2.
(5) Calculating the flow characteristics at the primary throat nozzle (T_t, P_t, V_t, and $\dot{m}_p$) assuming that the fluid achieved sonic velocity at the throat.
(6) Calculating the flow characteristics downstream of the primary throat nozzle.

    6.1. Fixing an arbitrary Mach number M_{p1}.
    6.2. Calculating T_{p1}, P_{p1}, and V_{p1}.
    6.3. Varying the value of the Mach number M_{p1} to obtain equality between the calculated value of the section ratio A_{p1}/A_t obtained with Eq. (15) and that fixed by the ejector geometry.

(7) Calculating the flow characteristics of the secondary nozzle.

7.1. Fixing an arbitrary value of M_{sy}.
7.2. Calculating T_{sy}, P_{sy}, and V_{sy}.
7.3. Varying the value of the Mach number M_{sy} to obtain equality between the condenser pressure calculated by Eq. (34) and that fixed by the operating conditions.

(8) Calculating the flow characteristics of the mixing chamber.

8.1. Calculating $P_{py} = P_{sy}$, M_{py}, V_{py}, T_{py}, and A_{py}.
8.2. Calculating A_{sy} and $\dot{m}_s$.

(9) Calculating the flow characteristics of the mixing chamber at section A2: $P_2 = P_{sy}$; V_2, T_2, M_2, and $\dot{m}_2 = \dot{m}_s + \dot{m}_p$.
(10) Calculating the flow characteristics of the mixing chamber between Sects. 2 and 3: $A_3 = A_2$, P_3, M_3, T_3, and V_3.
(11) Calculate the pressure P_4, which must be equal to the condensation pressure (P_c).
(12) Calculation of the entrainment ratio according to Eq. (35).

5 Results and Discussion

In this work, the geometrical characteristics of the ejector are $d_t = 2.8$ mm (throat diameter of the primary nozzle), $d_p = 4.5$ mm (exit primary nozzle diameter), $d_m = 6$ mm (mixing chamber diameter). They correspond to those of our experimental set-up [10]. The variations of the entrainment ratio with P_c ($T_g = 90$ °C and $T_e = 5$ °C) show that there is a critical pressure P_c* beyond which the entrainment ratio starts to decrease (Fig. 3). The value of P_c* varies from one fluid to another. The P_c* values are equal to 0.93 MPa (T_c* = 40 °C), 1.241 MPa (T_c* = 48 °C), 1.53 MPa (T_c* = 45 °C) and 1.555 MPa (T_c* = 40 °C) for R152a, R123yf, R290 and R717 respectively. It should be noted that the entrainment rate U obtained with R290 under critical conditions ($P_c < P_c$*) is the highest and is worth 0.340, whereas R717 has the lowest entrainment rate, i.e. approximately half that of R290 (U = 0.176). Given these results, it can be concluded that R290 is the optimal fluid under these conditions. Indeed, R290 allows the highest entrainment ratio but also a high critical pressure corresponding to the highest critical temperature that allows the machine to operate for high condensation temperatures under critical (nominal) conditions (primary and secondary sonic flows).

The variations in the entrainment ratio as a function of the generator temperature, T_g, are shown in Fig. 4. In these simulations, we considered a condensation temperature $T_c = 35$ °C and an evaporation temperature $T_e = 5$ °C. It can be seen that the curves present maxima, i.e. there is an optimum generator temperature for which the entrainment ratio is maximum. Thus, the optimal values of T_g are equal to 75 °C for R1234yf, R152a, and R717 while for R290, the optimal value of T_g is 80 °C.

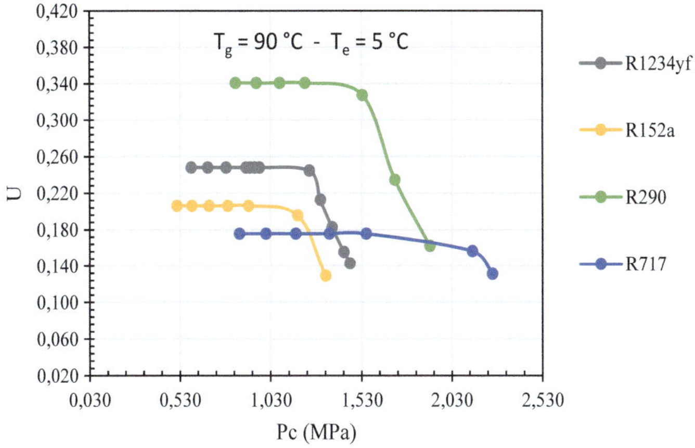

Fig. 3 Variations of U versus Pc ($T_g = 90$ °C, $T_e = 5$ °C)

The optimal entrainment ratios, U_{opt}, for R1234yf, R152a, and R717 are respectively 0.429, 0.401, and 0.372 while for R290, $U_{opt} = 0.5$. It should be noted that for energy saving, the ECRM can be operated with R290 at a generator temperature comparable to other fluids, i.e. 75 °C and the entrainment ratio will then be 0.451, which remains the highest. It should also be noted that below 75 °C, all the fluids provide very similar entrainment ratios, whereas, for generator temperatures above T_{gopt} (non-energy efficient operation), the best performance is obtained with R290 then follows R1234yf then R152a, and finally R717. The variations in the entrainment ratio as a function of the motive ratio $\xi = P_g/P_c$ are shown in Fig. 5 as a function of T_g. This performance indicator allows for optimization of the choice of the working fluid with the operating pressures when it is known that this parameter will define the thickness of the various components of the installation and consequently its cost. The curves also show maxima for optimal ξ ratios. The ξ_{opt} values are 2.572, 2.538, 2.652, and 2.747 for R290, R1234yf, R152a, and R717 respectively. Thus, it can be seen that R290 and R1234yf have optimum motive ratio values, which are comparable and remain slightly lower than those of the other fluids (R152a and R717).

The presence of an optimum T_g temperature is explained as follows. When T_g increases, the mass flow rates of the secondary (subsonic) and primary (supersonic) fluids also increase, but the increase in the flow rate of the secondary fluid is greater than that of the primary fluid, which leads to an increase in U. When T_g increases further, the secondary flow approaches the sonic flow or even becomes sonic and its flow rate is then maximum and becomes constant while the mass flow rate of the primary flow continues to increase leading to the decrease of the entrainment ratio.

The variations of the entrainment ratio U* calculated under critical conditions as a function of the critical motor ratio $\xi^* = P_g/P_c^*$ are presented in Fig. 6. It can be seen that these variations are located on the same curve whatever the fluid, indicating that the variations of U* as a function of ξ^* do not depend on the nature of the fluid. Otherwise, for a given critical motor ratio, all fluids provide the same performance

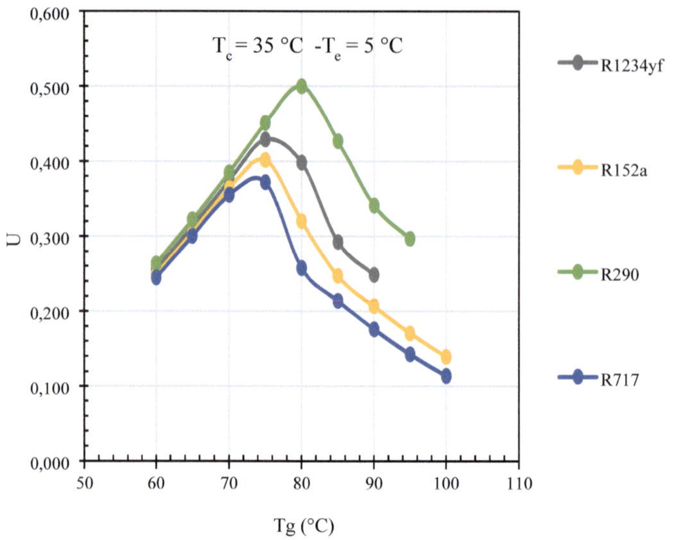

Fig. 4 Variations of U versus T_g

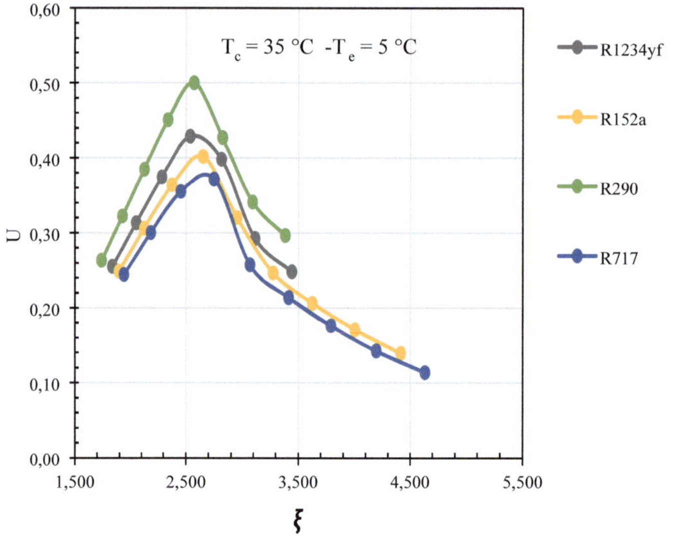

Fig. 5 Variations of U versus ξ

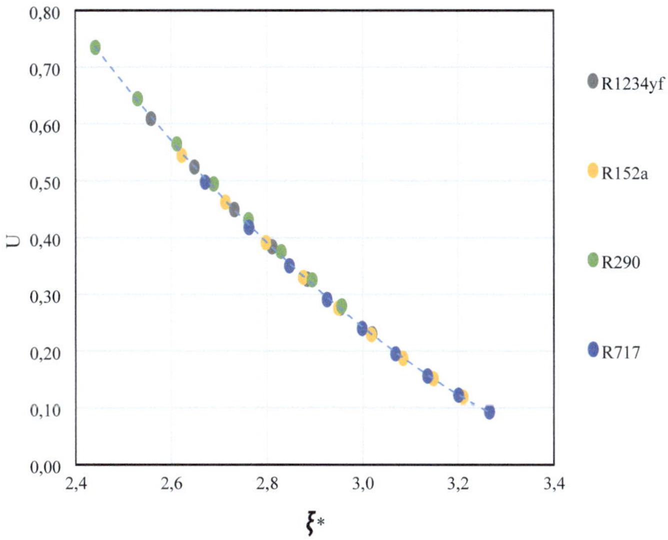

Fig. 6 Variations of U versus ξ*

(same entrainment ratio). This result is of major practical importance. Indeed, if we know the equation governing the variations of U* as a function of ξ* which is independent of the fluid, we can determine U* and P_c* by fixing the couple (T_g, ξ*). We then know that if $P_c < P_c$*, the entrainment ratio is constant and equal to U*. We can then vary T_g by fixing the critical motive ratio ξ* to determine the temperature T_{gopt} allowing us to obtain the optimal value of U. We can then predict the optimal conditions for ECRM operation.

The variations of U as a function of the evaporation temperature, T_e, are presented in Fig. 7. The entrainment ratio increases with T_e with superior performance for R290 whose entrainment ratio reaches 0.6 for T_e = 20 °C. It should be noted that operating at high evaporating temperatures is possible when coupling ECRM and a conventional vapor compression machine is considered. Under these conditions, the cold is produced at the level of the vapor compression machine whose condenser is cooled by the evaporator of the ECRM, which makes it possible to increase the COP of both machines.

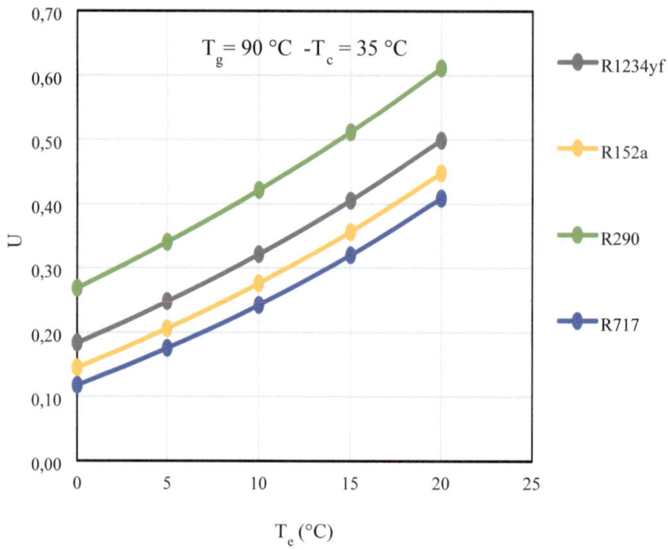

Fig. 7 Variations of U versus T_e

6 Conclusion

In this study, a model is developed to determine the thermodynamic performances of an ECRM operating with zero ODP and low GWP refrigerants (R1234yf, R152a, R290, and R717). The results indicate that the best entrainment ratio is obtained with R290, especially for generator temperatures higher than 75 °C, and for lower generator temperatures, the considered working fluids performed almost identically. It is also demonstrated that there exists an optimum generator for which the entrainment ratio is maximum. Also, exists an optimum motive ratio for which the entrainment ratio is maximum. In addition, it is shown that the critical entrainment ratio U* variations with the critical motive ratio ξ* are independent of the working and it decreases when ξ* augments. Finally, the entrainment ratio increases with the evaporator temperature and it is the highest for R290.

References

1. Dorantes R, Lallemand A (1995) Influence de la nature des fluides, purs ou en mélanges non-azéotropiques, sur les performances d'une machine de climatisation à éjecto-compresseur. Rev Int Froid 18(1):21–30
2. Al-Khalidy N (1998) An experimental study of an ejector cycle refrigeration machine operating on R113. Int J Refrig 21(8):617–625
3. Sun DW (1999) Comparative study of the performance of an ejector refrigeration cycle operating with various refrigerants. Energy Convers Manage 40:873–884

4. Selvaraju A, Mani A (2004) Analysis of an ejector with environment-friendly refrigerants. Appl Therm Eng 24:827–838
5. Boumaraf L, Lallemand A (2005) Comportement d'un éjecteur dans des conditions de fonctionnement nominales et non nominales. In : 12èmes Journées internationales de thermique, Tanger, Maroc pp 279-282
6. Gagan J, Smierciew K, Lukaszuk M, Butrymowicz D (2018) Investigations of thermal performance of ejection refrigeration system driven by low grade heat. Appl Therm Eng 130:1121–1138
7. Aidoun Z, Ameur K, Falsafioon M, Badache M (2019) Current advances in ejector modeling, experimentation and applications for refrigeration and heat pumps. Part 1: Single-phase ejectors. Inventions 4(15):73
8. Fan C, Yan G, Yu J (2019) Thermodynamic analysis of modified solar assisted ejector-compression heat pump cycle with zeotropic mixture R290/R600a. Appl Therm Eng 150:42–49
9. Wang X, Yan Y, Li B, Hao X, Gao N, Chen G (2020) Prospect of solar-driven ejector-compression hybrid refrigeration system with low GWP refrigerants in Summer of Gangzhou and Beijing. Int J Refrig 117:230–236
10. Ouaz I, Maalej S, Saad I, Zaghdoudi MC (2021) Thermodynamic performances of solar-powered ejector refrigerating machine. Sol Energy 224:765–786

Treatment of Real Textile Wastewater by Adsorption onto Naturel Tunisian Clay

Mohamed Ismail, Tesnim Dhiss, and Hedi Ben Amor

Abstract This study focused on the removal of organic matter from real textile effluents by natural clay. The raw clay used in this work was collected from the region of El-Hamma in Tunisia. The clays were characterized by means of Fourier-transform infrared spectroscopy (FTIR) and BET surface area analysis. The different physical and chemical parameters of the textile discharges (pH, Electrical conductivity, SS, COD) were evaluated. Experimental results showed that the maximum COD removal depends on the adsorbent dose and contact time. The adsorption kinetics is well described by the pseudo-first-order model. The equilibrium adsorption data were analysed by Langmuir and Freundlich isotherm models.

Keywords Adsorption · COD · Clay · Textile effluent · Isotherms

1 Introduction

Textile is one of the important industries in Tunisia. These industries generate large amount of wastewater which contains dyes and auxiliary chemicals such as salts, surfactants, detergents, oils, suspended and dissolved solids [1, 2]. Thus, the textile wastewater contains strong color and high concentration of organic matter (COD) [3]. The release of untreated colored wastewaters into receiving water causes damage to the environment as they may significantly affect photosynthetic activity in aquatic life due to reduced light penetration [4]. Furthermore, dyes used in the textile industry may be toxic to aquatic organisms and some of these dyes are suspected carcinogens. Therefore, there is a considerable need to treat these effluents prior to discharge. There are several methods to treat textile wastewater using conventional wastewater treatment processes like adsorption, coagulation-flocculation, ozonation, oxidation,

M. Ismail (✉) · T. Dhiss · H. B. Amor
Research Laboratory of Process, Energy, Environment and Electrical Systems (PEESE), School National of Engineers of Gabes, University of Gabes, 6072 Gabes, Tunisia
e-mail: ismail.issatg@gmail.com; mohamed.ismail@issatgb.u-gabes.tn

© The Author(s), under exclusive license to Springer Nature Singapore Pte Ltd. 2024
A. Mellit et al. (eds.), *Proceedings of the 2nd International Conference on Green Energy Conversion System*, Green Energy and Technology,
https://doi.org/10.1007/978-981-97-6148-7_70

nanofiltration membrane, photocatalysis and electrochemical methods. The adsorption process is one of the most efficient methods in the removal of contaminants from wastewater. Adsorbent such as activated carbon is very suitable for reducing the organic substance (COD/BOD) and color. However, commercially available activated carbon is becoming too expensive. Many studies have been undertaken to investigate the use of low-cost adsorbents [5]. The application of natural clays in the wastewater treatment has attracted more consideration due to their low coast, nontoxicity, abundance in nature and great cation exchange capacity.

The purpose of the present study is to evaluate the potential of Tunisian natural clay as a low-cost adsorbent for the COD reduction in a real dyeing wastewater from textile industry operating in Tunisia. Adsorption studies were carried out under various parameters such as contact time and adsorbent dose. The adsorption kinetic data was tested by pseudo-first order and pseudo-second-order models. The equilibrium data were analyzed using Langmuir and Freundlich isotherm models.

2 Materials and Methods

2.1 Materials

2.1.1 Sampling

Textile wastewater (TWW) samples were collected from textile industry located at Monastir (Tunisia). The wastewater is coming from the dyeing, washing, and bleaching baths, which is currently treated with coagulation/flocculation. But the obtained treated water quality is not quietly into compliance with the Tunisian standard of wastewater discharge to public sewers. Table 1 shows the basic quality parameters of the wastewater sample i.e. COD, pH, temperature, suspended solids, turbidity and conductivity, which were analyzed in our laboratory.

Table 1 Physico-chemical analysis of textile wastewater (final effluent-after coagulation/flocculation treatment)

Parameters	Unit	value
pH	–	6.15
Temperature	°C	27–30
COD	mgO_2/L	835
Turbidity	NTU	119
Suspended solid	mg/L	142
Conductivity	mS/cm	3.84
λmax	nm	665

Table 2 Chemical composition of natural clay

	SiO_2	Al_2O_3	Fe_2O_3	SO_3	MgO	CaO	CO_2
Percentage%	52.07	17.94	6.29	3.07	1.77	0.27	1.49

2.1.2 Preparation of Raw Clay

The raw clays used in this work were collected from El-Hamma region in Tunisia. The following process was used to purify the raw clays: a clay dispersion was placed in a graduated cylinder to extract the fine fraction of the fraction with a size more than 2 mm during the static sedimentation of the particles in suspension using Stokes equation [5]. Principal characteristics of the clay are summarized in Table 2.

2.1.3 Characterization of the Clays

The chemical properties of the clays were highlighted by a Fourier-transform infrared spectrometer (Nicolet Instruments, Nexus 870, Madison, WI, USA). The FTIR spectra were carried out in the range of 400–4000 cm^{-1}. The BET surface areas were evaluated using a Quanta Chrome Nova 3000e automated surface area analyzer (American Quanta Chrome Instruments Company, USA).

2.2 Adsorption Procedure

The sample effluent was set for batch adsorption experiments on purified clay to determine various equilibrium and kinetic parameters. Batch sorption tests were conducted in 250 ml conical flasks using an average shaker's speed of 500 rpm. After the flasks has been shaken for different time periods varying between 1 and 90 min, the treated wastewater sample was filtered through 0.45 μm filter paper and final equilibrium COD was analyzed using a spectrophotometer (HACHDR/2000) at the wavelength of 620 nm.

The amount of adsorption at equilibrium, qe, was calculated from Eq. (1):

$$q_e = (C_0 - C_e)\frac{V}{m} \tag{1}$$

where C_0 and C_e are the liquid-phase concentrations of adsorbate at initial and at equilibrium, respectively. V (mL) is the volume of the solution and m (g) is the mass of dry adsorbent used. The reduction percentage at time t can be calculated as follows:

$$\% = \frac{(C_0 - C_t)}{C_0} \times 100 \tag{2}$$

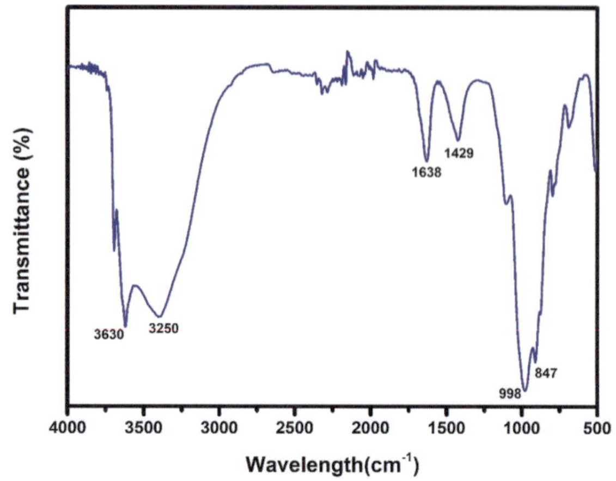

Fig. 1 FTIR spectra of El-Hamma clay

3 Results and Discussions

3.1 Characterization of Clay

3.1.1 FTIR Spectroscopy Analysis

The FTIR spectra of El-Hamma clays (Fig. 1) shows the different adsorption bands characteristic of clays recorded between 4000 and 400 cm^{-1}. The O–H stretching vibrations of Al–OH, Al, and adsorbed H_2O molecules and O–H bending vibration of the adsorbed water molecule were observed at 3630 and 3250 and 1638 cm^{-1} [7, 8]. The peak at 1429 cm^{-1} relates to the C–O stretching vibration of carbonate [9]. The peak at 980 cm^{-1} was assigned to the Si−O stretching vibration of bentonite tetrahedral sheets [10].

3.1.2 BET Surface Area Analysis

The Brunauer–Emmett–Teller (BET) method was applied to clays to determine the textural parameters such as specific surface area (S_{BET}) and pore volume (V_P) and pore size. The results of N_2 adsorption isotherms are given in Fig. 2 and Table 3.

3.2 Effect of Adsorbent Dose

In order to determine the effect of the adsorbent dose on COD reduction, 100 mL of sample was treated with different doses of adsorbent ranging between 0 and 0.6 g. Figure 3 represents the effect of various adsorbent doses onto COD using clay. The

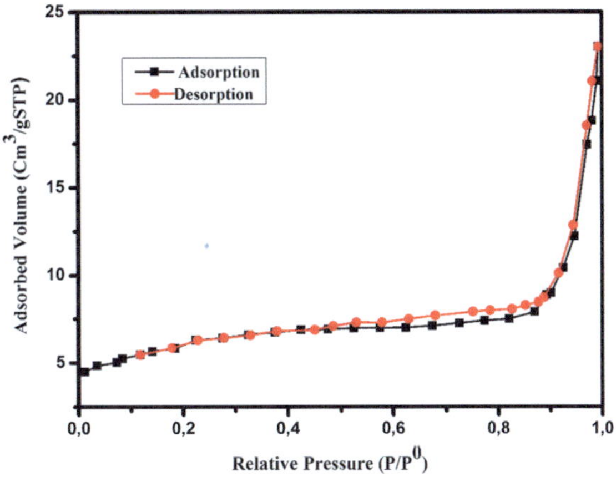

Fig. 2 Adsorption–desorption N_2 profile of Clay

Table 3 Textural properties of natural clay

Adsorbent	SBET (m^2/g)	Pore Volume (cm^3/g)	Pore size(Å)
El-Hamma clay	23.18	0.0595	10.32

percentage removal of COD increases with increasing adsorbent dose up to 0.3 g, there-after, any increase did not exert any significant increase in the removal of COD. This is due to the greater availability of the exchangeable sites or surface area at higher concentration of the adsorbent.

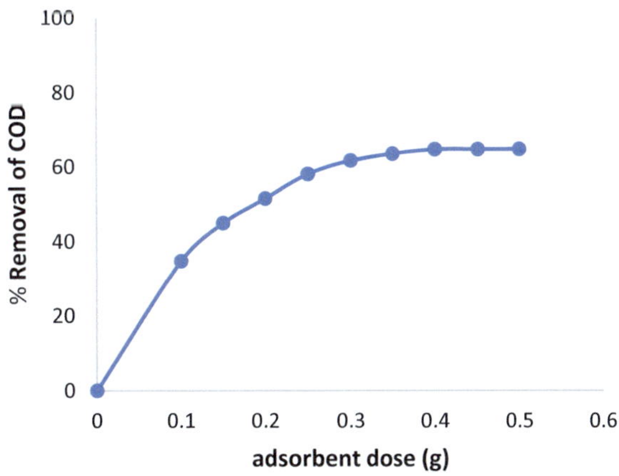

Fig. 3 Effect of adsorbent dose on the adsorption of COD (T = 25 °C, V = 100 mL)

3.3 Adsorption Kinetics

Figure 4 shows the effect of contact time on percentage of reduction of COD. It is found that the rate of removal was rapid during the initial 60 min and thereafter, no significant change in the rate of removal was observed. The initial fast adsorption may be attributed to large, uncovered surface area of clay at the beginning for the adsorption of the solute molecules. With further increasing time, the adsorption rate becomes slower, which is probably due to the diffusion of the solute molecules into the pores of the sorbent.

The adsorption kinetics analysis is a very important aspect on the study of the removal of pollutants from aqueous solution. Lagergren's pseudo-first-order model is the oldest known one describing the kinetic adsorption based on the adsorption capacity. It is expressed by the following equation:

$$Ln\frac{(q_e - q)}{q_e} = -K_1.t \quad (3)$$

where q and qe (mg/g) are the adsorption capacities at time t (min) and at equilibrium, respectively, and k_1 (min^{-1}) is the first order rate constant. Integration by applying the initial conditions $q_t = 0$ at $t = 0$ and $q = q_t$ at $t = t$.

The application of this models gave Fig. 5.

The concordance of the results obtained with the LAGERGREN equation confirms that it is a pseudo first order.

The values of the Kads adsorption rate constant for the two doses of clay (0.2 and 0.4 g) are calculated from these straight lines and are collated in Table 4:

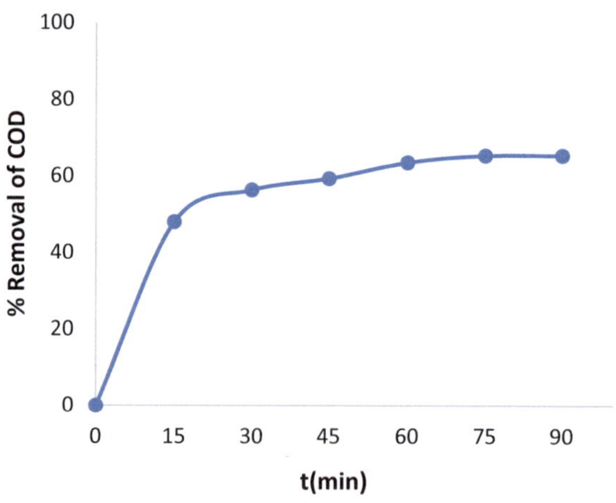

Fig. 4 Effect of contact time on the adsorption of COD (T = 25 °C, m = 0.3 g/100 mL)

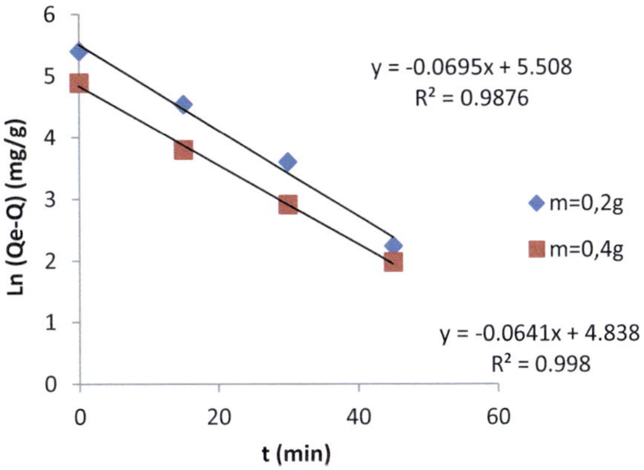

Fig. 5 Pseudo-first-order kinetics of the adsorption

Table 4 LAGERGREN constants for the different doses of clay

Dose of clay (g)	0,2	0,4
K_{ads} (min^{-1})	0,0695	0,0641

3.4 *Adsorption Isotherms*

The adsorption isotherms were analyzed by the Langmuir and Freundlich models and the results were presented in Fig. 6a, b.

The Langmuir isotherm model assumes that adsorption takes place at specific homogeneous sites on the surface. It is described by the following equation:

$$\frac{C_e}{q_e} = \frac{1}{q_m} C_e + \frac{1}{K_L q_m} \tag{4}$$

where Ce is the equilibrium concentration of the adsorbate, qe (mg/g) is the amount of adsorbate adsorbed per unit mass of adsorbent, qm (mg/g) and KL (L/mg) are Langmuir isotherm constants related to maximum adsorption capacity and rate of adsorption, respectively.

The Freundlich model is an empirical equation used to describe heterogeneous systems and is expressed by the following equation:

$$Lnq_e = \frac{1}{n} LnC_e + LnK_F \tag{5}$$

where K_F is the adsorption capacity and 1/n is the adsorption intensity.

The slope and the intercept of each linear plot in Fig. 6a, b are used to calculate Langmuir and Freundlich parameters which are regrouped in Table 5 along with associated correlations.

Fig. 6 **a** Langmuir isotherm **b** Freundlich isotherm

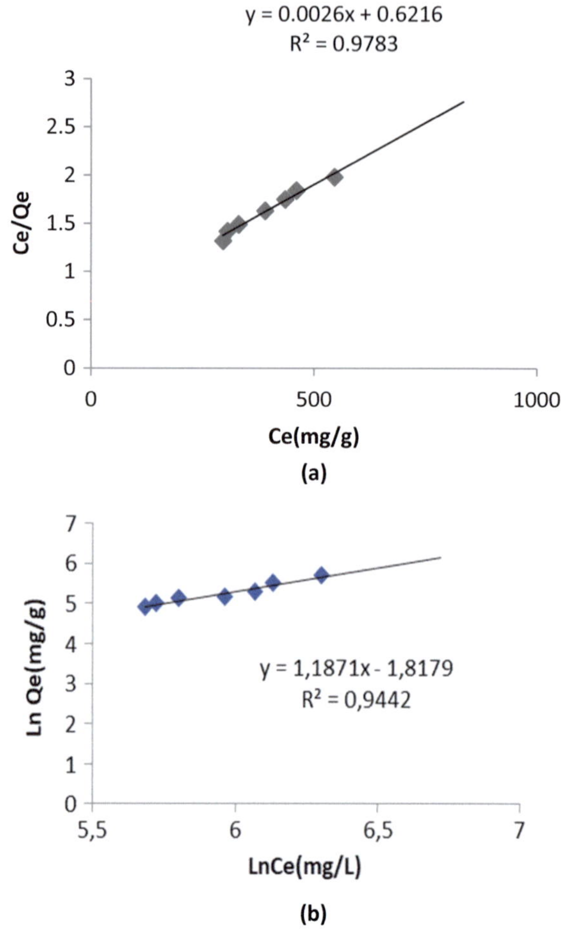

Table 5 Langmuir and Freundlich isotherms constants and correlations coefficients

Langmuir isotherm			Freundlich isotherm		
q_m, (mg/g)	K_L(L/g)	R^2	K_F (mg/g)	1/n	R^2
384.07	0.0041	0.978	0.162	0.84	0.944

The R^2 values were above 0.96 for the Freundlich and the Langmuir isotherms, indicating a very good fit by both models. The Freundlich constant ($1/n < 1$) revealed that the adsorption is favorable.

4 Conclusion

In this study, the ability of using natural clay from the south of Tunisia as adsorbents has been explored, with the aim of eliminating organic compounds from real effluent textile which are evaluated by the COD. The equilibrium data were well described by both the Langmuir and the Freundlich isotherm models with a maximum adsorption capacity of 384.07 mg/g. Pseudo-first kinetic model describe well the adsorption.

References

1. Babu BR, Parande AK, Raghus S, Prem Kumar T (2007) Cotton textile processing: Waste generation and effluent treatment. J Cotton Sci 11:141–153
2. Vandevivere PC, Bianchi R, Verstraete W (1998) Treatment and reuse of wastewater from the textile industry: review of emerging technologies. J Chem Technol Biotechnol 72:289–302
3. Ponnusami V, Srivastava SN (2009) Studies on application of teak leaf powders for the removal of color from synthetic and industrial effluents. J Hazard Mat 169:1159–1162
4. Liakou S, Pavlou S, Lyberatos G (1997) Ozonation of azo dyes. Water Sci Technol 35:279–286
5. Banat IM, Nigam P, Singh D, Marchant R (1996) Microbial decolourization of textile dye containing effluents. Biores Technol 58:217–227
6. Gee JW, Bauder GW, Particle-size analysis, In: Methods of soil analysis. Part 1. Physical and mineralogical methods, Part 1. 2n., A. Klute, Ed., pp 383–411
7. Sravanthi K, Ayodhya D, Swamy PY (2019) Green synthesis, characterization and catalytic activity of 4-nitrophenol reduction and formation of Benzimidazoles using bentonite supported zero valent iron nanoparticles. Mater Sci Energy Technol 2:298–307
8. Mousavi SE, Younesi H, Bahramifar N, Tamunaidu P, Karimi-Maleh H (2022) A novel route to the synthesis of α-Fe2O3@ C@ SiO2/TiO2 nanocomposite from the metal-organic framework as a photocatalyst for water treatment. Chemosphere 297:133992
9. Barakan S, Aghazadeh V (2019) Separation and characterisation of montmorillonite from a low-grade natural bentonite: using a non- destructive method. Micro Nano Lett. 14:688–693
10. Annan E, Nyankson E, Agyei-Tuffour B, Armah SK, Nkrumah-Buandoh G, Hodasi JA, Oteng-Peprah M (2021) Synthesis and characterization of modified kaolin-bentonite composites for enhanced fluoride removal from drinking water. Adv Mater Sci Eng 1:6679422

SAPF for Power Quality Improvement and Compensation of Harmonics Generated by Compact Fluorescents Lamps Connected to PV System

Alazrak Achwak, Hajjej Mohamed, and Sbita Lassaad

Abstract The research focuses on the solar generator's components, which are connected to a DC bus that supplies power to the load. To optimize energy harvesting, a maximum power point tracking (MPPT) system is incorporated into the inverter control. The MPPT algorithm enables the power converters to constantly identify the highest power output from the photovoltaic (PV) array, accounting for changes in solar irradiance and temperature. The inverter efficiently converts DC power to AC power while integrating the MPPT function to regulate the system's voltage and current. Additionally, the system is connected to an active filter, which helps reduce load harmonics.

Keywords PV system · Power quality · Harmonic compensation · Shunt active power filter · SVM

1 Introduction

In power systems, the significance of power quality issues has grown substantially in recent years, primarily due to the widespread adoption of semiconductor devices and nonlinear loads in various applications. Among the various power quality concerns, harmonic currents have emerged as a major problem. To address these issues, Shunt Active Power Filters (SAPFs) have become a common choice for managing current harmonics, reactive power compensation, and neutral current compensation in distribution systems. SAPFs are typically connected at the Point of Common Coupling (PCC) and have proven to be more effective in mitigating current

A. Achwak (✉) · H. Mohamed · S. Lassaad
Research Laboratory of Process, Energies, Environment and Electrical Systems, National Engineering School of Gabes, University of Gabes, Zrig Eddakhlania 6029, Tunisia
e-mail: achwaklazrakg@gmail.com

S. Lassaad
e-mail: lassaad.sbita@enig.rnu.tn

© The Author(s), under exclusive license to Springer Nature Singapore Pte Ltd. 2024
A. Mellit et al. (eds.), *Proceedings of the 2nd International Conference on Green Energy Conversion System*, Green Energy and Technology,
https://doi.org/10.1007/978-981-97-6148-7_71

harmonics compared to passive power filters (PPFs), which have limitations such as fixed compensation, resonance effects, and specific sizing [1].

This study presents a control system for a photovoltaic (PV) system that is connected to a nonlinear load. The main components of the system consist of solar arrays connected to an inverter through a DC bus [2, 3]. To address the rapid changes in solar irradiance and temperature, the control system incorporates a maximum power point tracking (MPPT) technique in the inverter's operation. The objective is to operate the PV generator at its maximum power point (MPP), which is the point of maximum power output.

Various algorithms can be employed for MPPT, and in this work, the Perturb & Observe (P&O) method is utilized [4–6]. The P&O algorithm utilizes an error signal as input and provides the duty ratio of the DC/DC converter (or its variation) as output, aiding in finding the MPP. Additionally, the system includes bidirectional DC–DC converters to connect a battery to the DC bus, allowing energy storage and bidirectional power flow. This setup enhances the system's flexibility and efficiency by enabling energy storage and sup-ply as needed. The SAPF is connected in parallel in the load side to compensate the Hight harmonic currents presented in the load. The full equivalent model of the studied system is presented in Fig. 1.

This paper is organized as follows: Sect. 2 describes the PVG with the MPPT algorithms. Section 3 present a review of the SVM control technique of the DC-AC inverter. A three-phase four-wire Shunt Active Power Filter SAPF with four legs and a split capacitor topology is presented in detail in Sect. 4. Also, it presents a review of instantaneous power theory and the hysteresis bloc control for the SAPF. The simulation results and harmonic mitigation are discussed in the next section (Sect. 5). Finally, Sect. 6 concludes this work.

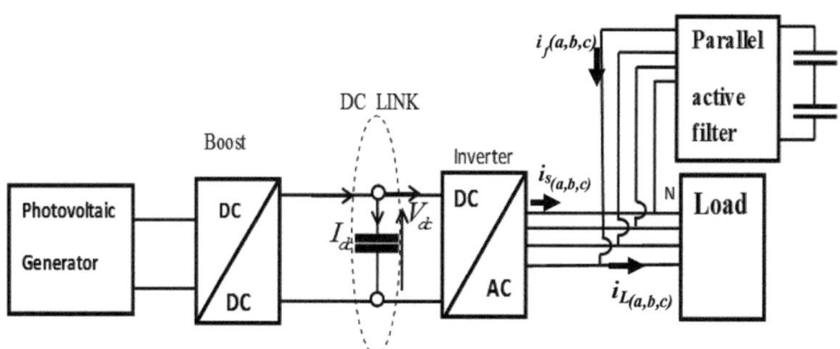

Fig. 1 Equivalent model of PV system with active parallel filter

2 PVG Model with MPPT of DC/DC Converter

PV cells are organized into larger units called modules or panels, and these modules are further combined to form PV arrays, also known as the PV generator. By arranging the cells in series and parallel configurations, the desired output voltage and current can be achieved. Figure 2 illustrates the equivalent circuit representation of solar cells arranged in N_s series and N_p parallel configurations.

The PV generator's power production can be accurately predicted by a simplified algebraic model, which is based on the current–voltage relationship defined in Eq. (1).

$$I_{pv} = N_p I_{ph} - N_p I_s \left\{ \exp\left[\frac{1}{A.V_{th}} \left(\frac{V_{pv}}{N_s} + \frac{R_s.I_{pv}}{N_p} \right) \right] - 1 \right\} - \frac{N_p}{R_p} \left(\frac{V_{pv}}{N_s} + \frac{R_s.I_{pv}}{N_p} \right) \quad (1)$$

where

I_{pv}	PV array output currents, in A;
V_{pv}	PV array output voltage, in V;
I_{ph}	Solar cell photocurrent in A;
I_s	Solar cell diode reverse saturation current;
A	Solar cell diode P–N junction ideality factor;
R_s and R_p	series and parallel resistances.

2.1 DC/DC Converter

The DC-DC boost converter serves as a bridge between the PV array and the inverter, connecting the output DC voltage of the PV modules to the DC link voltage at the input of the voltage source inverter.

This converter plays a crucial role in ensuring that the PV system's characteristics align with the equivalent impedance of the inverter's DC bus. By continuously adjusting the duty cycle of the DC-DC converter, it can effectively track and adapt to the changing atmospheric conditions and load. Consequently, the PV system can

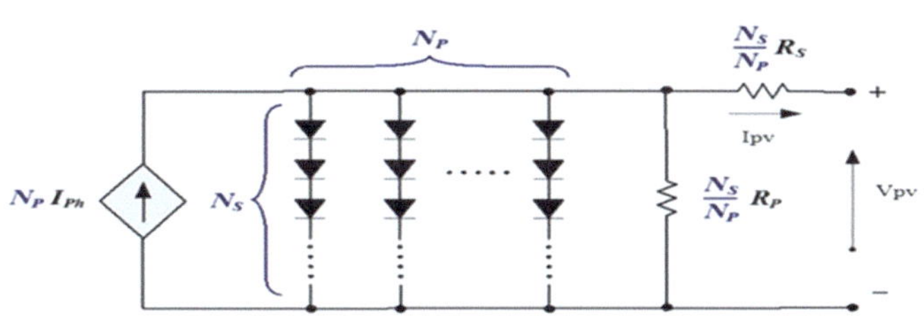

Fig. 2 Equivalent circuit of the PV generator

be operated near its Maximum Power Point (MPP) at any given moment, optimizing power generation and enhancing overall efficiency.

2.2 Perturb and Observe Algorithm

The P&O control principle involves introducing a minor disturbance in the voltage V_{PV}, leading to a corresponding change in power. Figure 3 illustrates that an increase in voltage results in increased power when the operating point is to the left of the maximum power point (PPM), and vice versa, a decrease in power occurs when the operating point is to the right of the PPM. Similarly, the same reasoning applies when there is a decrease in voltage.

Therefore, when a voltage disturbance results in increased power, the disturbance direction is maintained to move closer to the PPM. Conversely, if the power decreases due to the disturbance, the direction of the disturbance is reversed to guide the operating point towards convergence with the PPM.

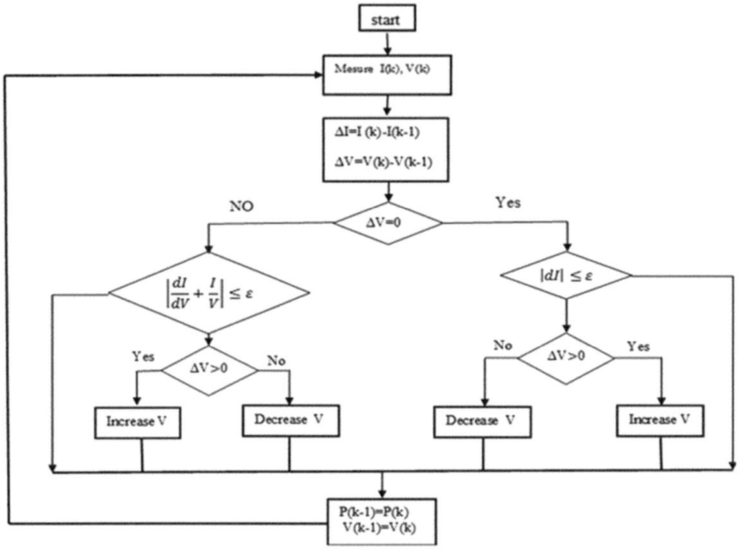

Fig. 3 P&O algorithm

3 Space Vector Modulation (SVM)

The PWM gate signals for the converter control the switching of a 2-level three-phase voltage source inverter. The inverter consists of six switches distributed across three legs, responsible for regulating the phase voltage and, consequently, the current of the induction generator.

By representing the "ON" and "OFF" states of the upper switches as "1" and "0," respectively, for each leg, we can derive up to eight distinct states for the inverter outputs. These eight states correspond to different voltage combinations, which are summarized in Table 1 along with the resulting phase voltage in both (abc) and (αβ) reference frames. These eight inverter output voltages can be visualized as eight voltage vectors ranging from [0, 0, 0] to [1,1,1], as shown in Fig. 4.

Then we obtained;

$$\begin{cases} V_{AN} = (2L_1 - L_2 - L_3).\dfrac{V_{dc}}{2} \\ V_{BN} = (-L_1 + 2L_2 - L_3).\dfrac{V_{dc}}{2} \\ V_{CN} = (-L_1 - L_2 + 2L_3).\dfrac{V_{dc}}{2} \end{cases} \tag{2}$$

The converter study is carried out in a fixed reference frame with two axes (α, β) by the following passage:

Table 1 Space vector state

L1	L2	L3	V_{AN}	V_{BN}	V_{CN}	V_α	V_β
1	1	1	0	0	0	0	0
1	0	0	$2V_{dc}/3$	$-V_{dc}/3$	$-V_{dc}/3$	$2V_{dc}/3$	0
1	1	0	$V_{dc}/3$	$V_{dc}/3$	$-2V_{dc}/3$	$\sqrt{(1/6)}.V_{dc}$	$\sqrt{(1/2)}.V_{dc}$
0	1	0	$-V_{dc}/3$	$2V_{dc}/3$	$-V_{dc}/3$	$\sqrt{(1/6)}.V_{dc}$	$\sqrt{(1/2)}.V_{dc}$
0	1	0	$-2V_{dc}/3$	$V_{dc}/3$	$V_{dc}/3$	$-\sqrt{(2/3)}.V_{dc}$	0
0	0	1	$-V_{dc}/3$	$-V_{dc}/3$	$2V_{dc}/3$	$\sqrt{(1/6)}.V_{dc}$	$-\sqrt{(1/2)}.V_{dc}$
1	0	1	$V_{dc}/3$	$-2V_{dc}/3$	$V_{dc}/3$	$\sqrt{(1/6)}.V_{dc}$	$-\sqrt{(1/3)}.V_{dc}$
0	0	0	0	0	0	0	0

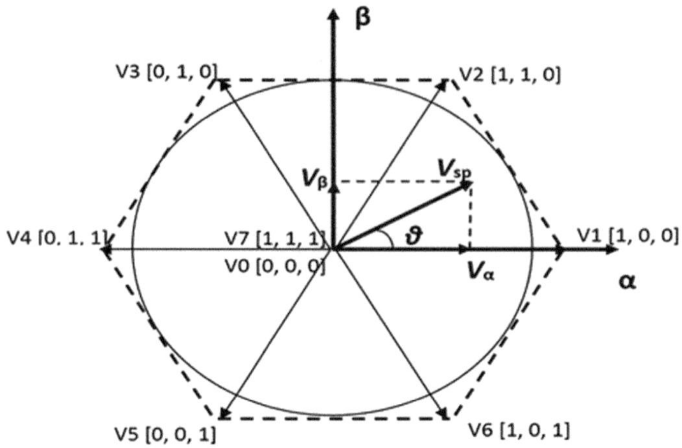

Fig. 4 Eight inverter voltage space vectors

$$\begin{bmatrix} V_\alpha \\ V_\beta \end{bmatrix} = \begin{bmatrix} \cos(0) & \cos\left(0-\frac{2\Pi}{3}\right) & \cos\left(0+\frac{2\Pi}{3}\right) \\ -\sin(0) & -\sin\left(0-\frac{2\Pi}{3}\right) & -\sin\left(0+\frac{2\Pi}{3}\right) \\ \frac{1}{\sqrt{2}} & \frac{1}{\sqrt{2}} & \frac{1}{\sqrt{2}} \end{bmatrix} \cdot \begin{bmatrix} V_{AN} \\ V_{BN} \\ V_{CN} \end{bmatrix} \quad (3)$$

Then:

$$\begin{cases} V_\alpha = \sqrt{2/3}\left[V_{AN}\frac{1}{2}V_{BN} - \frac{1}{2}V_{CN}\right] \\ V_\beta = \sqrt{2/3}\left[\frac{\sqrt{3}}{2}V_{BN} - \frac{\sqrt{3}}{2}V_{CN}\right] \end{cases} \quad (4)$$

4 Three Phase Shunt Active Power Filter Presentation

Parallel Static Compensators, also known as Shunt Active Power Filters (SAPF), are highly sophisticated electrical devices primarily utilized within power systems to address challenges arising from harmonic distortions and reactive power. The fundamental principle behind SAPFs involves the utilization of cutting-edge power electronics and sophisticated control algorithms, which enable them to analyze the current waveform of the electrical network. When these devices identify the presence of harmonic currents or an excess of reactive power, they promptly generate counteracting currents with opposite phases, effectively neutralizing these undesirable effects. This real-time injection of compensating currents ensures that the system's voltage stability is maintained, the power factor is enhanced, and electrical power is utilized more efficiently. With their remarkable ability to swiftly respond to load variations and swiftly rectify power quality issues, SAPFs play a crucial role as valuable assets within contemporary power grids, ultimately leading to improved reliability and minimized energy losses. The block diagram of the SAPF connected to the power PV circuit and the nonlinear load are already illustrated in Fig. 1; when harmonics, current imbalances $i_{L_{(a,b,c)}}$, or other harmful distortions are detected in the electrical system (in the load side), the parallel active filter acts by injecting compensating currents $i_{f_{(a,b,c)}}$ with phases opposite to those of the disturbances. These compensating currents effectively cancel out the undesirable effects, there by contributing to improving the quality of electrical main energy $i_{S_{(a,b,c)}}$.

From the literature There are many topologies of Active Power Filter APF [7, 8], The SAPF presenting in this manuscript is a Three phase four wire SAPF with four legs and a split capacitor topology [9], because it's robustness and ability to compensate all currents levels, also for the neutral line compensation i_{nL}. Figure 5 present the detailed circuit diagram of the studied inverter.

As for those researchers are focus on the power part of the SAPF, other researchers are focus on the control part [10], in which, in this paper we use the Instantaneous Reactive Power Theory (known also as PQ Theory) introduced in 1983, by Akagi et al. [11], to calculate the compensation currents $i_{c_{(a,b,c)}}*$. The PQ theory employs the Clarke transformation to convert the three-phase voltages and currents forms (a-b-c) coordinates to (α-β-0) coordinates, allowing for the calculation of instantaneous power components based on this transformation.

Fig. 5 Circuit diagram of the three phase four wire SAPF

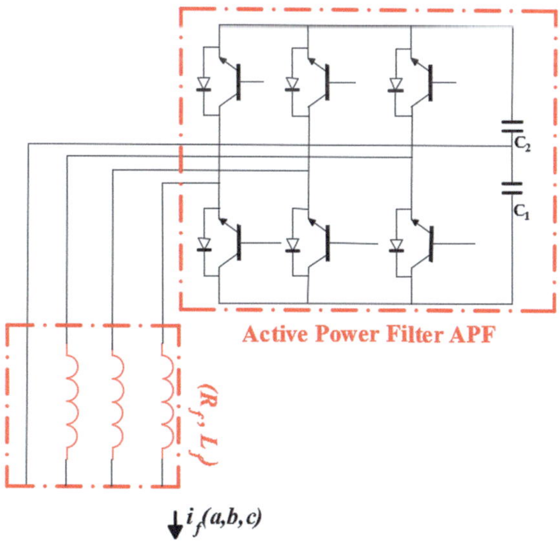

One of the significant advantages of the PQ theory lies in its capability to segregate the zero sequence components, which have no impact on the α and β components. Consequently, these zero sequence components can be disregarded in balanced systems.

Also in this paper, a simple fixed-band hysteresis control bock is used to control switched Power Device of the SAPF ($F_{(a,b,c)}$ and $\overline{F}_{(a,b,c)}$) because of its robustness and simplicity of implementation.

5 Results and Discussions

We model and simulate the block diagram of the equivalent model of PV generator system with the SAPF presented in Fig. 1 under Matlab/Simulink software. The nonlinear load contains 75 General Electric lamps model distributed over 3 AC phases (25 lamps in parallel for each phase) [12]. This load is powered up with the PV generator from Softech 1STH-215P panel model. Also, a 3 phase SAPF is modeled to compensate harmonic generated by the nonlinear load. The main characteristics of the proposed system are illustrated in Table 2.

Figure 6a illustrates the different variations of irradiance, and the MPPT (Maximum Power Point Tracking) based on the Perturb and Observe (P&O) method promptly tracks the prospective maximum power point within a short duration. The P&O MPPT algorithm exhibits robustness against rapid changes in atmospheric conditions, resulting in minimal oscillations around the Maximum Power Point (MPP) and achieving high efficiency of up to 92%. Figure 6b, c displays the output voltage of the PVG while in Fig. 6d we present the DC-link voltage across the capacitor is managed constant almost during the irradiance change with a drop voltage lower than 25 V; the recovery time is about 0.02 s. Thus, this result confirms the efficiency of the DC-link voltage control system.

The AC power output section of this work also work in best conduction, the load currents waveforms $i_{L_{(a,b,c)}}$ of the CFL are presented in Fig. 7a, in which we observe that the load signals is shifted to the source currents waveforms, the current is almost not sinusoidal (very Hight THD). Thus, means that the load is a source of pollution to the main power. Also, and after connected the SAPF at t = 0.02 s we can notice that the source current waveforms $i_{S_{(a,b,c)}}$ remain sinusoidal, and in phase with the voltage source. Meanwhile, we present the compensation currents $i_{C_{(a,b,c)}}$.

Table 2 PV_G and SAPF system parameters

Parameters	Value
Source Voltage and frequency (V_S, F_S)	230 V, 50 Hz
Filter inductance (R_f, L_f)	1 mΩ, 0.1 μH
DC Link Capacitor (C_1, C_2)	2200 μF
Reference voltage V_{ref}	600 V

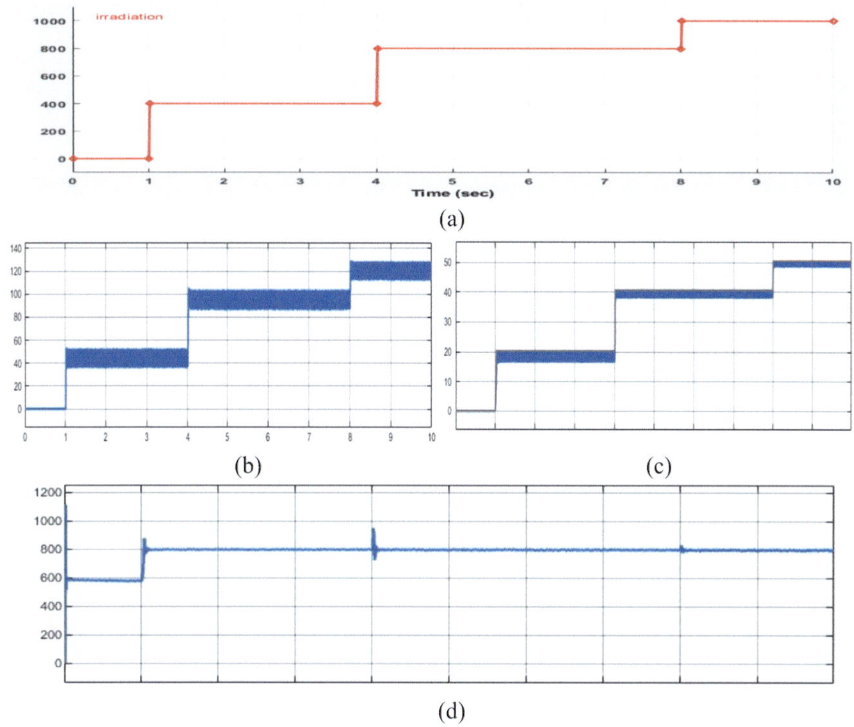

Fig. 6 **a** Variation of Irradiance (Irr (W/m2)), **b** PV array output currents (I_{PV} (A)), **c** PV array output voltage (V_{PV} (V)), and **d** dc-bus voltage (V_{dc} (V))

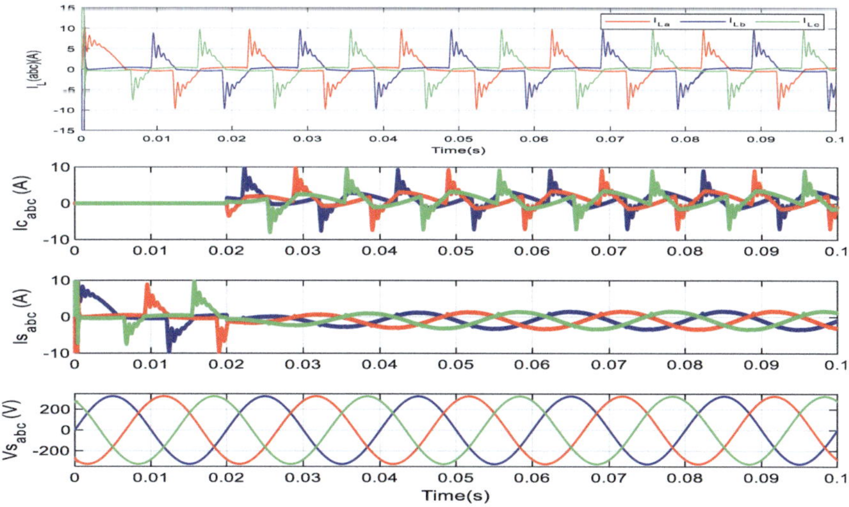

Fig. 7 Source currents waveforms at different point before and after compensation

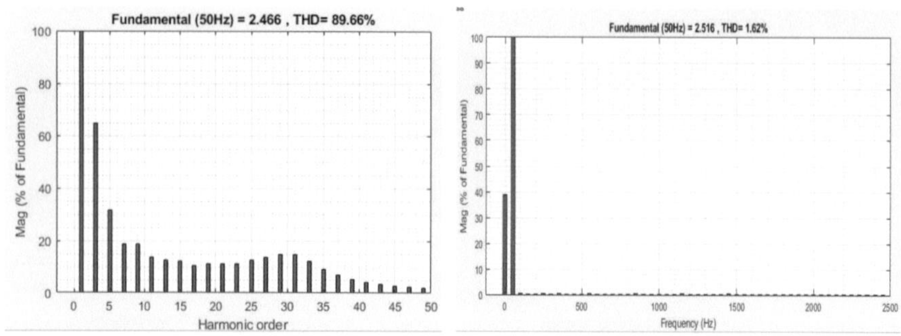

(a) (b)

Fig. 8 FFT analysis of source current ($i_{s_{(a,b,c)}}$(A)): **a** without SAPF, **b** with SAPF

It can be seen from Fig. 8a, b that the THD has been decreased considerably of source and load voltage from 89 to 1.6%. This shows the efficiency of this PQ method which is based to control the SAPF. The founded THD value of the source current reached is less than 5% imposed by the international standards [13].

6 Conclusions

This paper presented a three phase four wire SAPF for harmonic compensation generated by nonlinear load consist of Compact Fluorescents Lamps. Powered up with a PV system. The global system with active power main the high quality of the current and voltage of the AC load.

References

1. Akagi H (2005) Active harmonic filters. Proc IEEE 93(12):2128–2141
2. Mehiri A, Hamid A, Almazrouei S (2017) The effect of shading with different PV array configurations on the grid-connected PV system. Int Renew Sustain Energy Conf IRSEC 2017:1–6. https://doi.org/10.1109/IRSEC.2017.8477420
3. Yadav AP, Thirumaliah S, Haritha G, Scholar PG (2012) Comparison of MPPT Algorithms for DC-DC Converters Based PV Systems. Int J Adv Res Elect Electron Instrument Eng 1(1):18–23
4. Kumar M, Ansari F, Jha AK (2011) Maximum power point tracking using perturbation and observation as well as incremental conductance algorithm. Int J Res Eng Appl Sci 1(4):19–30
5. Colin H, Duvauchelle C, Moine G, Tanguy Y, Gaiddon B, Tran-Quoc T (2010) Raccordement des installations photovoltaïques au réseau public de distribution électrique à basse tension: cadre réglementaire, impacts et préconisations, Projet ESPRIT, Avril
6. Hanen Abbes HA (2013) Etude comparative de cinq algorithmes de commande MPPT pour un système photovoltaïque. J Renew Energ 17(3):435–445
7. Nagi Reddy B et al (2019) Performance and dynamic analysis of single switch AC-DC buck-boost buck converter. Int J Innov Technol Explor Eng 8(4):307–313

8. Ramachandra Y et al (2019) Analysis of recent developments in brushless DC motors controlling techniques. Int J Innov Technol Explor Eng 8(5):522–526
9. Singh B et al (1998) Harmonic elimination, reactive power compensation and load balancing in three-phase, four wire electric distribution systems supplying non-linear loads. Int J Elect Power Syst Res 44(2):93–100
10. Dahono PA (2009) New hysteresis current controller for single-phase full-bridge inverters. IET Power Electron 2(5):585–594
11. Akagi H, Kanazawa Y, Nabae A (1983) Generalized theory of the instantaneous reactive power in three-phase circuits, IPEC'83—Int power electronics conf., Tokyo, Japan, pp 1375–1386
12. Mohamed H, Mohamed N and Lassaad S (2021) Experimental investigations of a CFLs currents harmonics injection into an electrical network grid, 2021 4th International symposium on advanced electrical and communication technologies (ISAECT), pp 01–05, https://doi.org/10.1109/ISAECT53699.2021.9668434
13. IEEE Recommended practices and requirements for harmonic control in electric power systems, IEEE Std 519–2014

Effects of the Porosity Distribution on Liquid–Water Transport in the Cathode of a PEMFC Fuel Cell

Faycel Khemili and Mustapha Najjari

Abstract The liquid water generation in the Gas Diffusion Layer (GDL) of a Proton Exchange Membrane Fuel Cell (PEMFC) increases the resistance to oxygen flow toward the catalyst layer. Water flooding inside the GDL can affect the PEMFC performance especially at higher current densities. Therefore, a good understanding of the effect of liquid water amount in the GDL is crucial to water management and, subsequently, to the performance of the fuel cell. The purpose of the present study is to investigate the effect of the microstructure characteristics of the GDL on the water flooding and liquid water distribution inside the GDL. A one-dimensional theoretical model has been developed. Results indicate that the porosity gradient has a significant effect on the liquid water saturation and the performance of the PEM fuel cell.

Keywords Membrane fuel cell · Gas diffusion layer · Water flooding

1 Introduction

In the Proton Exchange Membrane Fuel cell (PEMFC), the porous Gas Diffusion Layer (GDL) holds a pivotal position within the cathode assembly. It serves as a structural support for the polymer electrolyte membrane (PEM) and a conduit for electron transfer through its solid matrix. Moreover, the GDL assumes the dispersion of reactant gases across the catalyst layer and the expulsion of resultant liquid water from electrochemical reactions within the catalyst layer.

F. Khemili
College of Science, University of Hafar Al Batin, Hafar Al Batin, Kingdom of Saudi Arabia

National Engineering School of Monastir, LESTE, University of Monastir, Monasir, Tunisia

M. Najjari (✉)
National Engineering School of Gabes, University of Gabes, LR18ES34, Gabes, Tunisia
e-mail: najjari98@gmail.com

© The Author(s), under exclusive license to Springer Nature Singapore Pte Ltd. 2024
A. Mellit et al. (eds.), *Proceedings of the 2nd International Conference on Green Energy Conversion System*, Green Energy and Technology,
https://doi.org/10.1007/978-981-97-6148-7_72

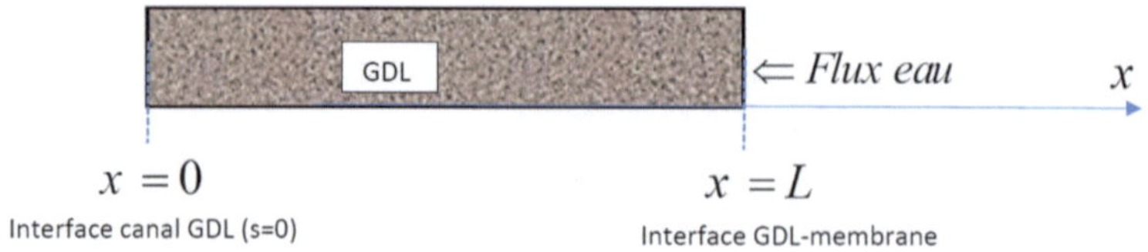

Fig. 1 Physical model

However, excessive accumulation of liquid water within the porous structure of the GDL can trigger a phenomenon known as flooding. This inundation clogs the GDL's pores, reduces the PEMFC efficiency, particularly under high current densities. In response to this challenge, a novel GDL design incorporating a microporous layer (MPL) has been introduced. This innovative configuration is anticipated to enhance the concurrent transport of gases and water, as proposed by notable researchers [1–4].

In the literature, many research works have been devoted to study the mechanisms of the capillary transport within the porous gas diffusion layer (GDL) [5–11].

The present study introduces a novel one-dimensional steady-state mathematical model, based on capillary pressure– liquid saturation interconnection within the GDL of the PEMFC. We investigate specially how the porosity spatial distribution (microstructure) influences the liquid water distribution within the GDL.

2 Theoretical

2.1 Physical Model

A basic theorical model describing the capillary liquid flow has been developed to predict the liquid water saturation distribution inside the cathode gas diffusion layer. A schematic representation of the PEMFC cathode considered in this study is shown in Fig. 1.

2.2 Governing Equations

The continuity equation of liquid water in the cathode GDL can be written as:

$$\frac{d}{dx}(\rho_l u_l) = 0 \tag{1}$$

where u_l is the liquid water velocity and ρ_l is the liquid water density.

The liquid water velocity is given by Darcy's law:

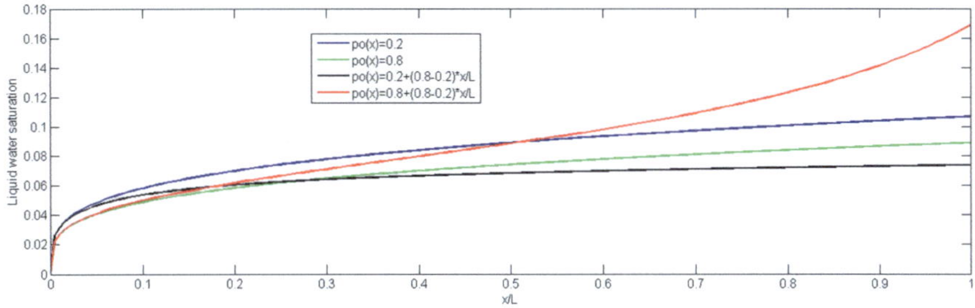

Fig. 2 Liquid water saturation profile across the porous GDL

$$u_l - -\frac{Kk_{rl}}{\mu_l}\left(\frac{dp_l}{dx}\right);$$

K is the absolute permeability of the GDL, k_{rl} is the relative permeability of the liquid phase, μ_l and p_l are respectively the dynamic viscosity and the pressure of the liquid phase. The relative permeability increases with liquid water saturation s. k_{rl} is expressed as $k_{rl} = s^3$. . The liquid saturation s is defined by the ratio of the liquid volume and the void space volume (Fig. 2).

The difference between the gas pressure and liquid phase pressure is known as capillary pression:

$$p_c = p_g - p_l = \frac{2\sigma\cos(\theta)}{r}$$

where σ is the surface tension at the liquid–gas interface, θ is the contact angle of liquid water wetting the pore wall, and r the capillary radius of diffusion pore.

The integration of the Eq. (1) over the thickness of the cathode leads to:

$$-\rho_l\frac{Kk_{rl}}{\mu_l}\frac{dp_l}{dx} = N_l^{CC} \tag{2}$$

where N_l^{CC} is the net water flux across the membrane from anode to cathode characterized by a net water transport coefficient, α: $N_l^{CC} = -(1+2\alpha)\frac{I}{2F}M_{H_2O}$

The capillary pressure between the two phases, liquid, and gas, can be modeled by the following empirical equation [12]:

$$p_c = (\frac{\varepsilon}{K})^{1/2}\sigma\cos(\theta)J(s) \tag{3}$$

where K is the absolute permeability of the GDL and ε is the GDL porosity.

The term $J(s)$ is the Leverett function, which represents the dimensionless capillary pressure as a function of liquid saturation. For the hydrophilic ($\theta \prec 90°$) and hydrophobic ($\theta \succ 90°$) GDL, the Leverett functions are given by the following relations [13]:

$$J(s) = \begin{cases} 1.417s - 2.120s^2 + 1.263s^3 for\theta \succ 90^\circ \\ 1.417(1 - s) - 2.120(1 - s)^2 + 1.263(1 - s)^3 for\theta \prec 90^\circ \end{cases}$$

We assume that the porosity is a function of position in the form:

$$\varepsilon(x) = \varepsilon_0 + (\varepsilon_1 - \varepsilon_0)\frac{x}{L}$$

where ε_0 and ε_1 are, respectively, the porosities at the interfaces x = 0 and L.

After mathematical development, we obtain an ordinary differential equation:

$$\begin{cases} \frac{d\chi}{dz} = f(z, \chi(z)) \\ \chi(0) = 0 \end{cases} \tag{4}$$

where:

$$f(z, \chi(z)) = \frac{4\left(\frac{-(1+2\alpha)\frac{I}{2F}M_{H_2O}}{\frac{\rho_l^{H_2O}}{\mu_l}K^{1/2}\sigma\cos(\theta)} - \frac{\rho_l^{H_2O}\frac{K\chi^{\frac{3}{4}}}{\mu_l}g}{\sigma\cos(\theta)\frac{K^{1/2}}{\mu_l}}\right)}{(\varepsilon(z))^{1/2}\left(1.417 - 4.240\chi^{0.25} + 3.789\chi^{0.5}\right)} - \frac{4\left(1.417\chi - 2.120\chi^{1.25} + 1.263\chi^{1.5}\right)\frac{d}{dz}\left((\varepsilon(z))^{1/2}\right)}{(\varepsilon(z))^{1/2}\left(1.417 - 4.240\chi^{0.25} + 3.789\chi^{0.5}\right)}$$

and $\chi = s^4$.

The Tafel equation [14] is used to describe the electrochemical reaction along the membrane/cathode interface, namely,

$$\eta_c = \frac{RT}{\alpha_c F}\ln\left(\frac{C_{ref}}{I_0}\frac{I}{((1 - s)C_O)_{x=L}}\right) \tag{5}$$

I_0 is the exchange current density, α_c is the cathode transfer coefficient, F is the Faraday constant, R is the universal gas constant, T is the absolute temperature, I is the local current density and C is the molar concentration of the oxygen.

3 Results and Discussions

Equation (4) provide mathematical equations to estimate the distribution of liquid water saturation inside the GDL. The saturation profiles across the GDL thickness s(x), for different assigned parameters (ε_0 and ε_1, the porosities at the interfaces x = 0 and L.) have been calculated with MATLAB software. Equation (5) simulates the output current density against the voltage (polarization curve), analysing the response of the fuel cell.

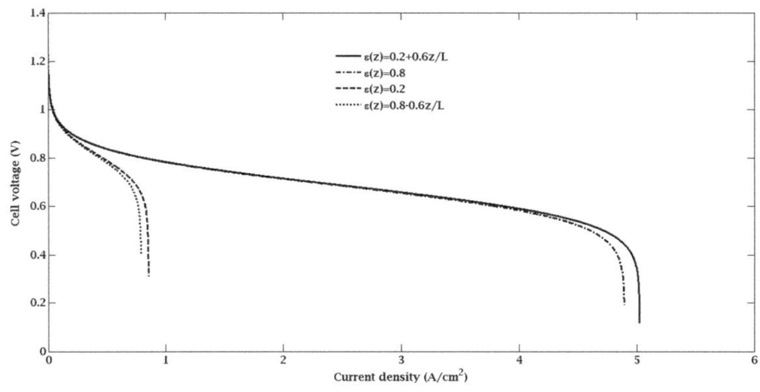

Fig. 3 Fuel cell performance for different GDL porosity

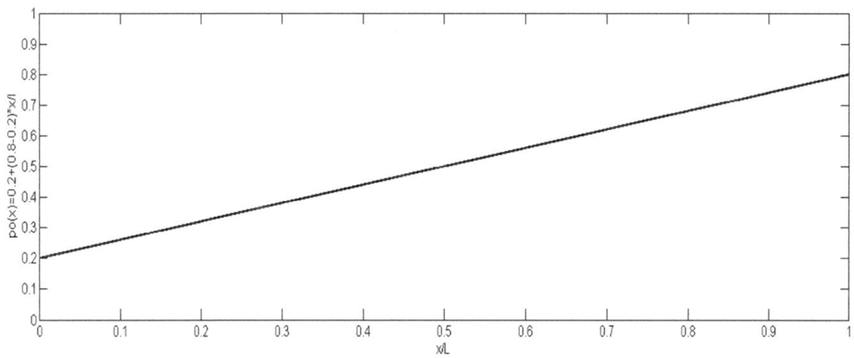

Fig. 4 Positive porosity gradient

It is noted that for a positive porosity gradient the saturation at the membrane cathode interface is weaker than for those for a negative porosity gradient and also for the case of uniform porosities (significant and low porosities). Liquid water evacuation from the GDL is improved when using a positive porosity gradient medium.

It can be seen from Fig. 3 that the performance of a PEM fuel cell has been enhanced for a negative porosity gradient (Fig. 4).

References

1. Tang H, Wang SH, Pan M, Yuan R (2007) Porosity-graded micro-porous layers for polymer electrolyte membrane fuel cells. J Power Sources 166(1):41–46
2. Wang X, Zhang H, Zhang J, Xu H, Zhu X, Chen J, Yi BL (2006) A bi-functional micro-porous layer with composite carbon black for PEM fuel cells. J Power Sources 162(1):474–479

3. Wang XL, Zhang HM, Zhang JL, Xu HF, Tian ZQ, Chen J, Zhong HX, Liang YM, Yi BL (2006) Micro-porous layer with composite carbon black for PEM fuel cells. Electrochim Acta 51(23):4909–4915
4. Weber AZ, Newman J (2005) Effects of microporous layers in polymer electrolyte fuel cells. J Electrochem Soc 152(4):677–688
5. Khemili F, Najjari M, Ben Khadher N, Ben NS (2009) Two-dimensional modelling of Transport phenomena by (CVFE) method in the porous cathode of a PEM fuel cell. J Porous Med 12(12):1181–1193
6. Najjari M, Khemili F, Nasrallah SB (2008) The effects of the cathode flooding on the transient responses of a PEM fuel cell. Renew Energy 33(8):1824–1831
7. Khemili F, Najjari M (2022) Analytical solution for liquid water saturation and oxygen flowin the porous cathode of a proton exchange membrane fuel cell: effect of surface wettability. J Porous Med 25(7):1
8. Najjari M, Khemili F, Nasrallah SB (2013) The effects of the gravity on transient responses and cathode flooding in a proton exchange membrane fuel cell. Int J Hydrog Energy 38(8):3330–3337
9. Khemili F, Najjari M, Nasrallah SB (2012) Transient thermal model for proton exchange membrane fuel cells. J Fuel Cell Sci Technol 9(2):1
10. Khemili F, Najjari M, Nasrallah SB (2012) A comprehensive model for evaluating the performance of proton exchange membrane fuel cells. Int J Heat Technol 30(1):1–9
11. Khemili F, Najjari M, Nasrallah SB (2011) Effects of the temperature on two-phase transport phenomena and electrochemical responses of a proton exchange membrane fuel cells (PEMFC). J Porous Med 14(9):1
12. Kumbur EC, Sharp KV, Mench MM (2007) On the effectiveness of Leverett approach for describing the water transport in fuel cell diffusion media. J Power Sources 168(2):356–368
13. Wanyuan S, Eru K, Nobuyuki O (2008) Effect of capillary pressure on liquid water removal in the cathode gas diffusion layer of a polymer electrolyte fuel cell. J Power Sources 182(1):112–118
14. Wang ZH, Wang CY, Chen KS (2001) Two-phase flow and transport in the air cathode of proton exchange membrane fuel cells. J Power Sources 94(1):40–50

PIC Based Bipolar and Unipolar SPWM for Pure Sine Wave Single-Phase Inverter

Aymen Chaaira, Habib Kraiem, Rabiaa Gamoudi, and Lassaad Sbita

Abstract This paper presents a detailed comparative study of bipolar and unipolar Sinusoidal Pulse Width Modulation (SPWM) techniques in DC-AC inverters, focusing on their efficacy in reducing harmonic distortions, which are detrimental to power system performance. Both SPWM strategies are implemented using the cost-effective and high-performance PIC18F2431 microcontroller, tasked with controlling a single-phase inverter to produce a pure sine wave output. An in-depth analysis of Total Harmonic Distortion (THD) levels at various modulation index reveals that while both SPWM methods significantly curtail harmonics, the unipolar SPWM inverter exhibits a more uniform THD reduction across the entire modulation range. Conversely, the bipolar SPWM inverter, though starting with higher THD levels, demonstrates a drastic reduction as modulation index increase, especially when enhanced by the application of an LC low-pass filter. This filter is instrumental in further refining the inverter's output, achieving impressively low THD values below 1% at full modulation, and ensuring a high-quality sinusoidal AC waveform. The comparative results accentuate the nuanced performance differences between the two SPWM strategies and underscore the critical role of LC filters in optimizing inverter output for a variety of power applications.

Keywords Bipolar SPWM · DC-AC power inverter · PIC18F2431 · THD · Unipolar SPWM

A. Chaaira (✉) · R. Gamoudi · L. Sbita
Processes, Energy, Environment and Electrical Systems Laboratory, University of Gabes, Gabes, Tunisia
e-mail: aymen.chaaira@gabes.r-iset.tn

L. Sbita
e-mail: lassaad.sbita@enig.rnu.tn

H. Kraiem
Department of Electrical Engineering, College of Engineering, Northern Border University, Arar 73222, Saudi Arabia
e-mail: ALhabeeb.Kareem@nbu.edu.sa

© The Author(s), under exclusive license to Springer Nature Singapore Pte Ltd. 2024
A. Mellit et al. (eds.), *Proceedings of the 2nd International Conference on Green Energy Conversion System*, Green Energy and Technology,
https://doi.org/10.1007/978-981-97-6148-7_73

1 Introduction

DC-AC inverters are used in a variety of high-voltage applications such as power distribution systems, solar systems, and energy transfer improvements. However, they produce harmonic distortion throughout the commutation process. These enormous voltage surges are induced by nonlinear loads in the energy system. All power grids, electrical equipment, and communication networks are impacted. Choosing the right modulation strategy is critical in dealing with this problem. To eliminate harmonic distortion, this study applies SPWM. In fact, SPWM has gotten a lot of attention and has been explored especially for high-power converters. It can be used with a low operating frequency to obtain higher levels of spectral efficiency, as measured by lower THD [1–7].

Because the SPWM are difficult to execute in real time, numerous methods are employed in practice to save sinusoidal wave form reference in a look-up table [4, 5].

In this paper, the resulting SPWM control signal is implemented in low-cost high-performance PIC18F2431 microcontroller. It operates a single-phase pure sine wave inverter. Then, the high order harmonics content is ameliorated by filtering the inverter output.

2 Bipolar and Unipolar SPWM Control

The concepts of Bipolar and Unipolar SPWM represent two pivotal control strategies in power inverter. Both methods aim to modulate the output of an inverter to closely emulate a sine wave, which is essential for converting direct current (DC) to a smooth alternating current (AC). However, they differ in their operational principles, advantages, and suitability for various applications.

Bipolar SPWM involves alternating the output voltage of the inverter between the positive and negative values of the DC supply voltage as illustrated in Fig. 1a. This results in an output waveform that oscillates above and below the zero-voltage line, closely mimicking a true sine wave. The primary advantage of Bipolar SPWM lies in its ability to produce a high-quality output with lower harmonic distortion. This makes it particularly suitable for applications where power quality is critical, such as in sensitive medical equipment, precision laboratory instruments, or audio systems where signal integrity is paramount [1–3, 6–8].

Unipolar SPWM, in contrast, switches the output voltage between a single polarity (positive or negative) and zero as illustrated in Fig. 1b. Unlike Bipolar SPWM, the waveform in Unipolar SPWM does not cross the zero-voltage line. It either stays above zero, switching between the positive peak and zero, or below zero, switching between the negative peak and zero. The major advantage of Unipolar SPWM is the reduced stress on the switching devices, as the voltage does not swing between the positive and negative extremes. This reduced stress can lead to increased efficiency and potentially longer component lifespan [1–9].

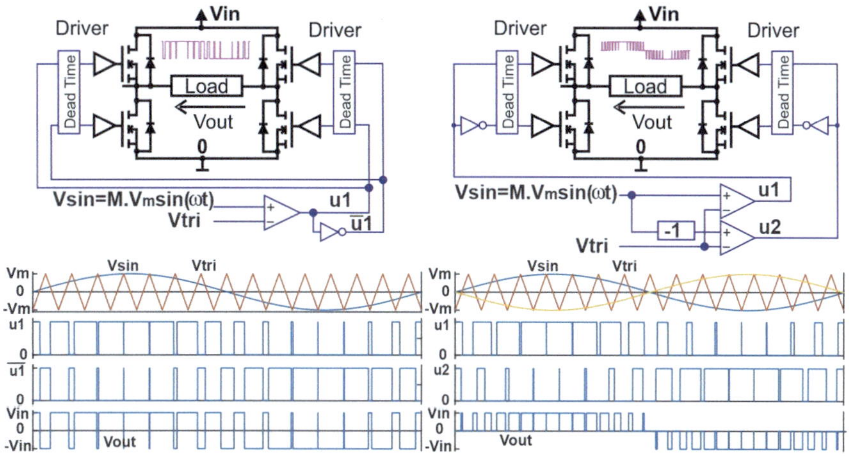

(a) : Bipolar SPWM inverter (b) : Unipolar SPWM inverter

Fig. 1 Bipolar and unipolar SPWM

3 Simulations Results and Discussion

The simulation performed using ISIS software is presented in Fig. 2. It is composed of:

- The PIC18F2431 microcontroller is adept at generating SPWM signals. It operates using an 8 MHz crystal, which is effectively quadrupled to 32 MHz by the microcontroller's internal Phase Locked Loop (PLL). This acceleration allows the microcontroller to execute operations at a rate of 8 Million Instructions Per Second (MIPS).
- The L6203 full-bridge driver is employed, notable for its:
- Compatibility with supply voltages reaching up to 48 V.
- Capability to handle total RMS currents up to 4A.
- Low on-state resistance, typically just 0.3 Ω at 25 °C.

The spectrum inverter's output voltage is characterized by a fundamental sinusoidal frequency of 50 Hz, along with harmonics starting at a minimum frequency of 10 kHz. To refine this output into a clean sinusoidal voltage, we introduce a low-pass filter with a cutoff frequency strategically set at 1 kHz, which is high above the fundamental frequency yet below the harmonic frequencies. The filter design, featuring an inductance of 2.5 mH and a capacitance of 10 μF, effectively smooths the output while preserving the desired 50 Hz signal.

The block diagram of the bipolar and unipolar SPWM control is presented in Fig. 3.

The block diagram of L6203 circuit is presented in Fig. 4.

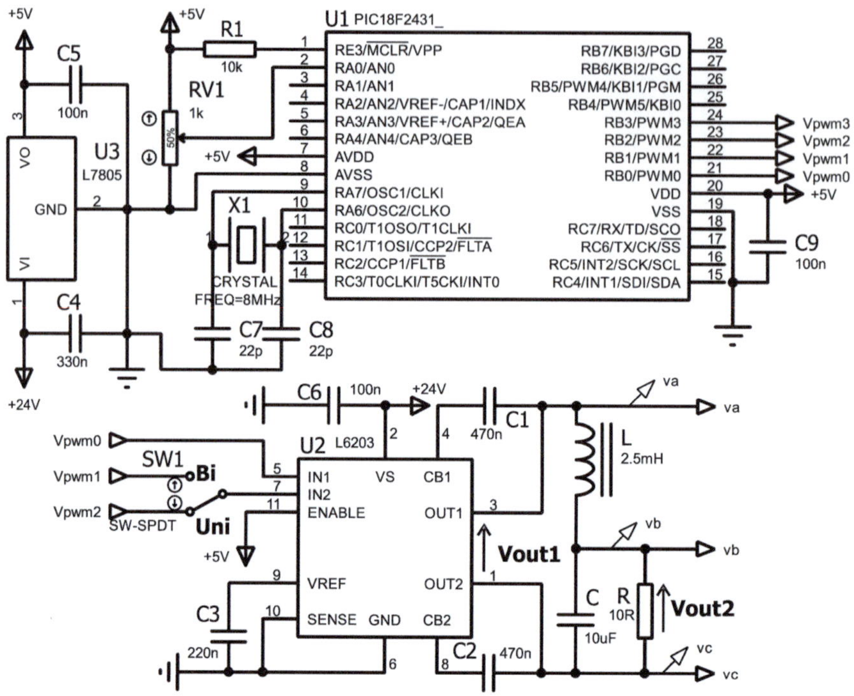

Fig. 2 Electronic circuit of the sinewave single phase inverter under ISIS

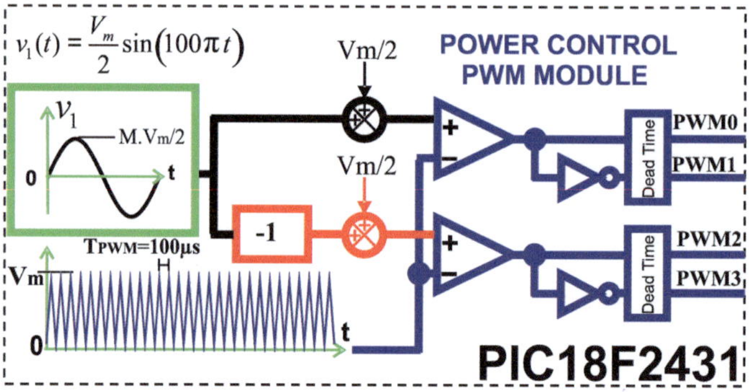

Fig. 3 Block diagram of implementation of SPWM

Figures 5, 6, 7 and 8 demonstrates the inverter's output voltage waveforms and frequency spectra at various modulation index, both before and after the integration of an LC filter. It is evident that the post-filter waveforms are more closely resemble a pure sinusoidal shape, indicative of significant attenuation of higher-frequency components. This attenuation becomes more pronounced as the modulation index rises, suggesting that the filter's ability to smooth the waveform improves

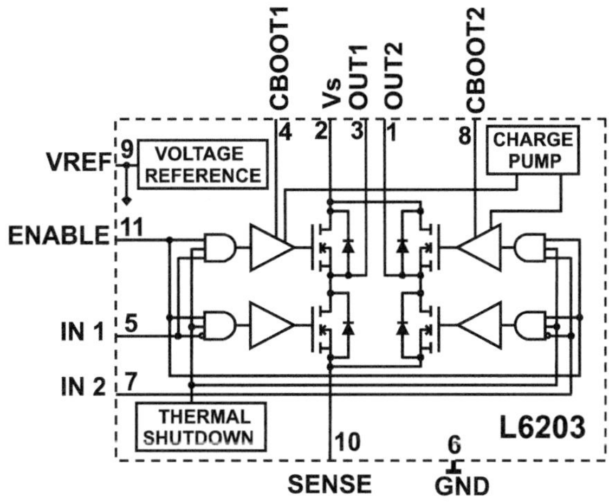

Fig. 4 Block diagram of L6203

with increased modulation. Such a trend not only substantiates the efficacy of the LC filter in enhancing the power quality by reducing harmonic distortion but also highlights the filter's role in maintaining waveform integrity across a range of operational conditions. This behavior underscores the critical impact of the filter in achieving a cleaner AC output, which is particularly beneficial for sensitive electronic devices that require high-quality power inputs.

Integrating observations from Figs 9 and 10, the Total Harmonic Distortion (THD) levels of the inverter output (Vout1) before filter and (Vout2) after filter at different modulation indices. For Vout1, the THD starts at extremely high values at lower modulation index, which is typical for bipolar or unipolar SPWM inverters without filtering and decreases as the modulation index increases. However, once the LC filter is applied, the THD values for Vout2 are significantly lower, indicating effective harmonic suppression. These observations underscore the LC filter's critical role in reducing harmonic distortion, thereby enhancing the quality of the power output from the inverter. The consistent decrease in THD with the LC filter across the modulation spectrum demonstrates the filter's efficiency in producing a more sinusoidal output, which is desirable for sensitive electronic devices and applications requiring precise power quality.

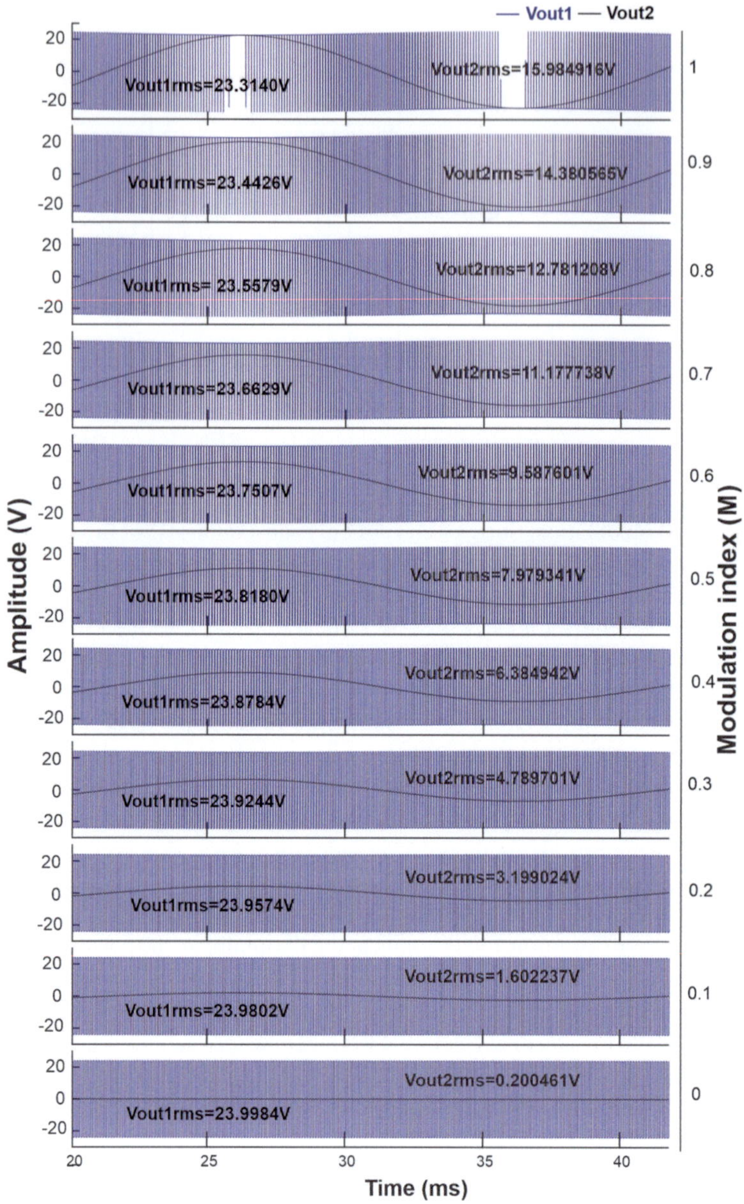

Fig. 5 Bipolar SPWM inverter output voltage waveforms at various modulation index before and after LC filter application

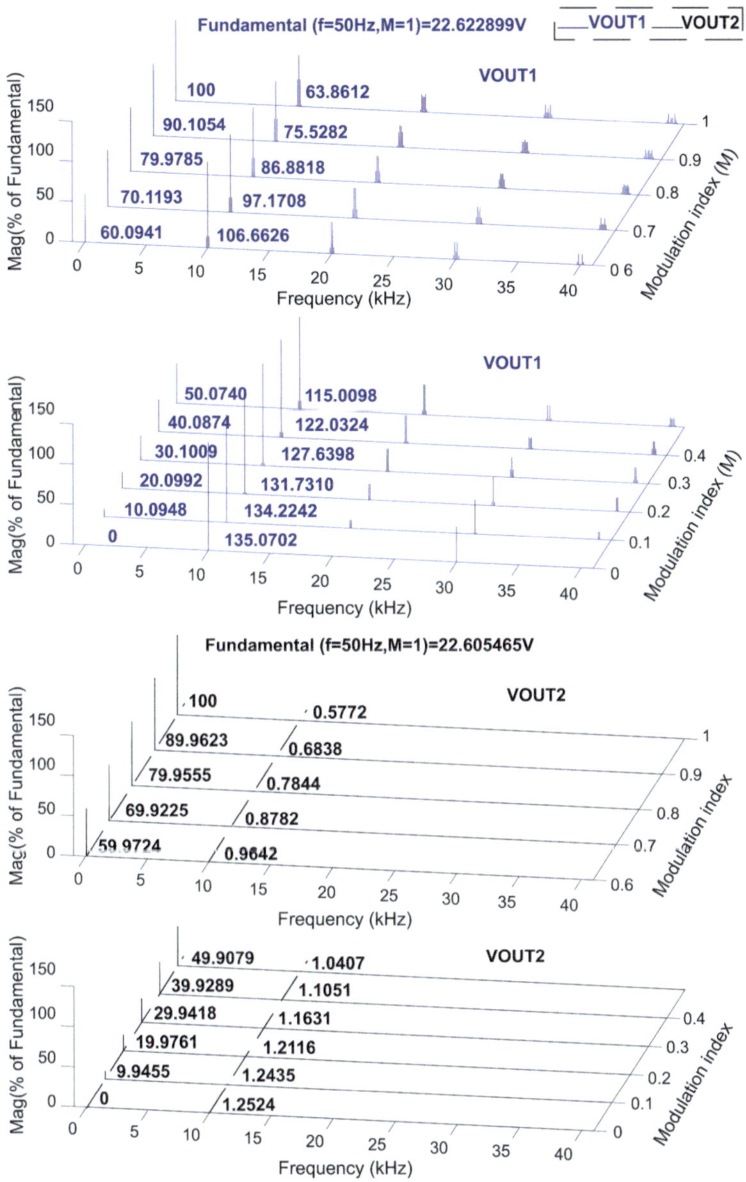

Fig. 6 Bipolar SPWM inverter output harmonics at various modulation index before and after LC filter application

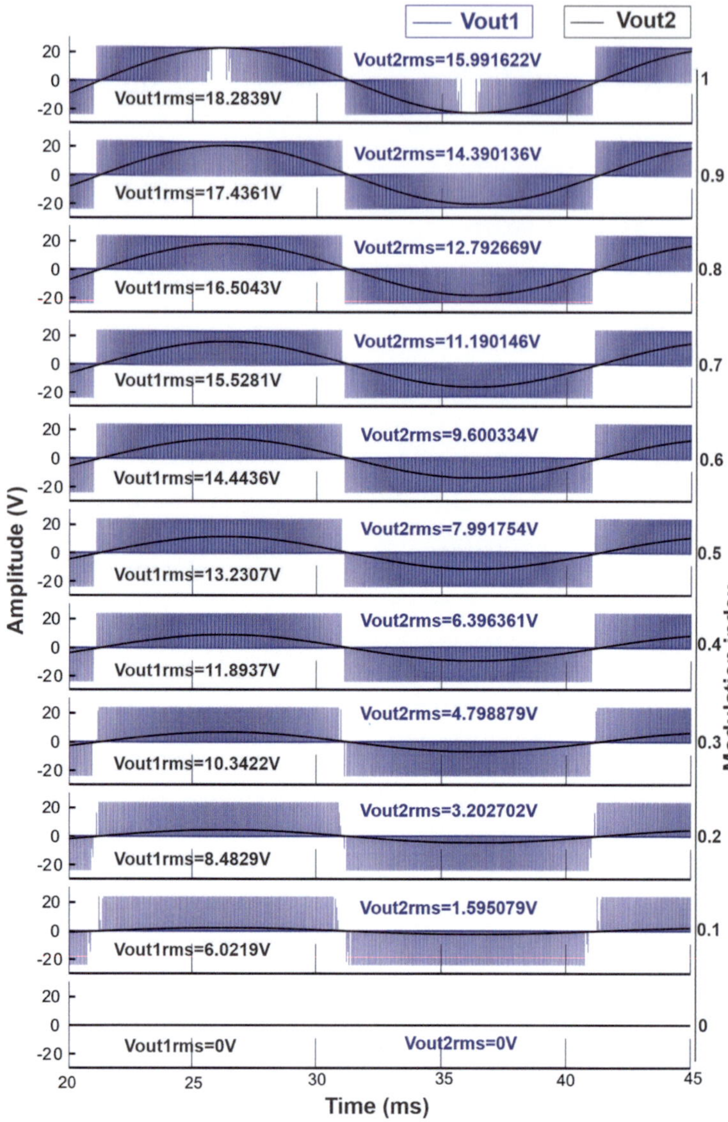

Fig. 7 Unipolar SPWM inverter output voltage waveforms at various modulation index before and after LC filter application

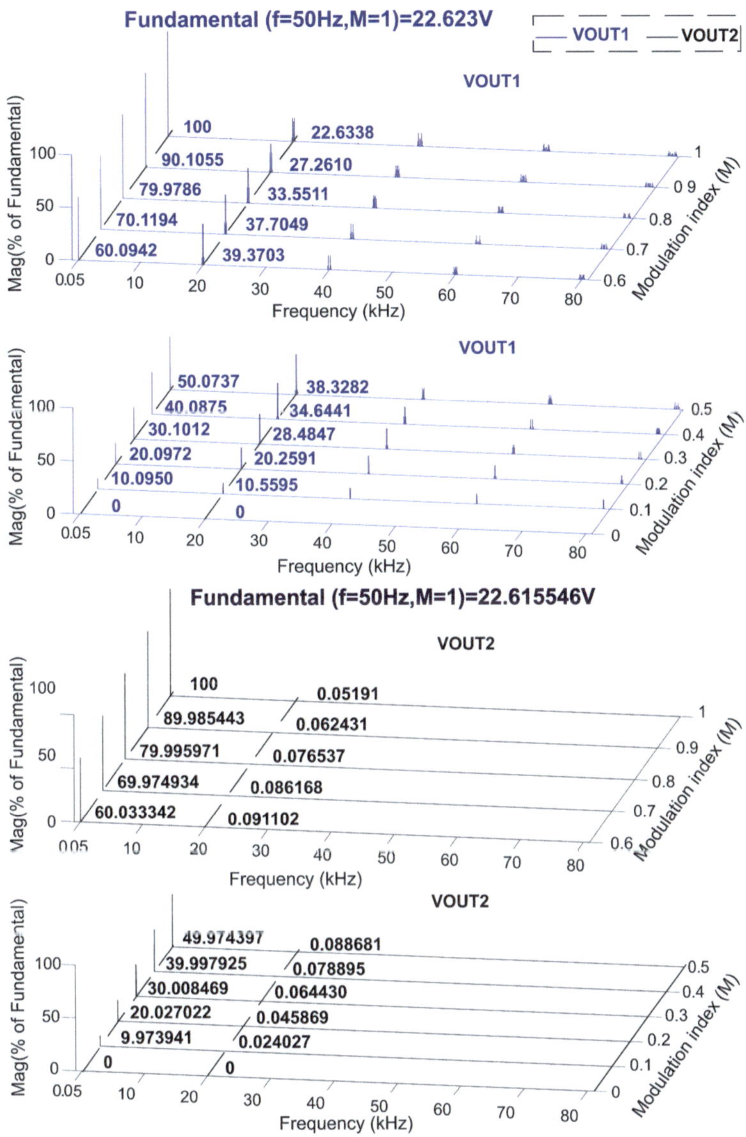

Fig. 8 Unipolar SPWM inverter output harmonics at various modulation index before and after LC filter application

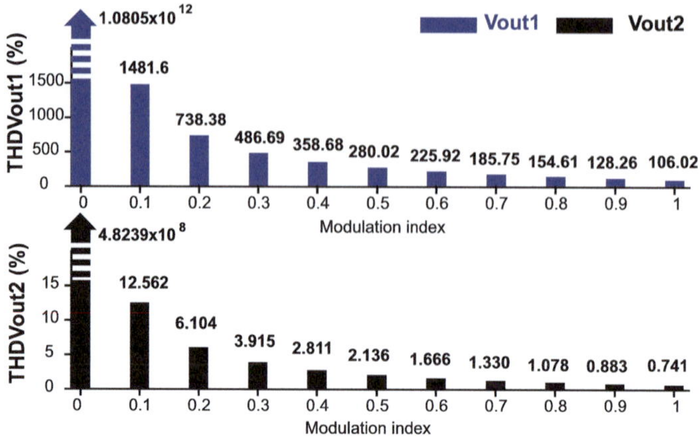

Fig. 9 Bipolar SPWM inverter THD at various modulation index before and after LC filter application

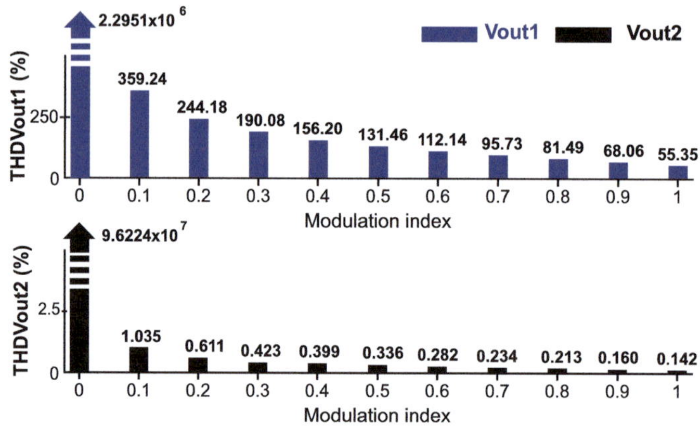

Fig. 10 Unipolar SPWM inverter THD at various modulation index before and after LC filter application

4 Conclusion

This paper presents a comparative study between the bipolar and unipolar Sinusoidal Pulse Width Modulation (SPWM) inverters, focusing on their Total Harmonic Distortion (THD) performance across various modulation index. The bipolar inverter shows a higher initial THD at lower modulation indices, which suggests a greater harmonic content in its output. With an increase in the modulation index. Both inverter types demonstrate a significant reduction in THD. However, the reduction in the unipolar inverter is more uniform across the modulation range, while the bipolar inverter shows a more pronounced decrease as it approaches higher modulation index. The introduction of an LC filter is highly effective in both configurations, particularly for the bipolar inverter, where it significantly reduces the initially higher THD levels. At full modulation, both inverters achieve impressively low THD percentages below 1%, with the unipolar inverter marginally outperforming at 0.142% compared to the bipolar inverter's 0.741%, underscoring the nuanced efficiencies of the filtering process in different SPWM strategies. The study indicates that while both bipolar and unipolar inverters benefit from the LC filter implementation, the initial THD and the rate of THD reduction as the modulation index increases vary. The unipolar inverter displays a more consistent THD reduction, suggesting potential efficiency advantages in applications requiring a broad modulation range. Conversely, the bipolar inverter, starting with higher THD values, achieves THD levels comparable to the unipolar inverter after filtering, highlighting the LC filter's vital role in enhancing the inverter's output quality.

References

1. Gouse Basha S, Venkatesan M (2020) A review on topologies for transformer-less grid-connected PV inverter. In: Sengodan T, Murugappan M, Misra S (eds) Advances in electrical and computer technologies. Springer Singapore, Singapore, pp 1151–1165. https://doi.org/10.1007/978-981-15-5558-9_98
2. Hannan S, Aslam S, Ghayur M (2018) Design and real-time implementation of SPWM based inverter. 2018 international conference on engineering and emerging technologies (ICEET). IEEE, Lahore, Pakistan, pp 1–6. https://doi.org/10.1109/ICEET1.2018.8338637
3. Birbir Y, Yurtbasi K, Kanburoglu V (2019) Design of a single-phase SPWM inverter application with PIC micro controller. Eng Sci Technol Int J 22:592–599. https://doi.org/10.1016/j.jestch.2018.11.014
4. Ahmed N, Khan ZR (2021) Microcontroller based pure sine wave inverter. 2021 IEEE international conference in power engineering application (ICPEA). IEEE, Malaysia, pp 173–177. https://doi.org/10.1109/ICPEA51500.2021.9417841
5. Islam SMM, Sharif GM (2009) Microcontroller based sinusoidal PWM inverter for photovoltaic application. 2009 1st international conference on the developments in renewable energy technology (ICDRET). IEEE, Dhaka, pp 1–4. https://doi.org/10.1109/ICDRET.2009.5454227
6. Xiao W (2021) Power electronics step-by-step: design, modeling, simulation, and control. McGraw Hill, New York
7. Albu RD, Gordan CE (2023) Single-phase SPWM inverter ordered with microcontroller. 2023 17th international conference on engineering of modern electric systems (EMES). IEEE, Oradea, Romania, pp 1–5. https://doi.org/10.1109/EMES58375.2023.10171665
8. Ruan X, Wang X, Pan D, Yang D, Li W, Bao C (2018) Design of LCL filter. Control techniques for LCL-type grid-connected inverters. Springer Singapore, Singapore, pp 31–61. https://doi.org/10.1007/978-981-10-4277-5_2
9. Qazalbash AA, Amin A, Manan A, Khalid M (2009) Design and implementation of microcontroller based PWM technique for sine wave inverter. 2009 international conference on power engineering, Energy and electrical drives. IEEE, Lisbon, Portugal, pp 163–167. https://doi.org/10.1109/POWERENG.2009.4915171

Battery Management Optimization with Supervision Stage: Application for Cathodic Protection System

Arezki Fekik, Lyes Abbassen, Mohamed Lamine Hamida, Benabderrahamane Nabila, Hathat Lydia, and Hakim Denoun

Abstract This study presents an innovative battery management strategy for cathodic protection systems fueled by photovoltaic sources. Addressing the corrosion of vital structures such as pipelines, the integration of photovoltaic generators offers a dynamic solution. The system encompasses a 640Ah Lithium-Ion battery, a supervisory control module, and three operational switches ("Charging", "Direct", "Discharging"). Through derived equations, the battery's charging and discharging behaviors are meticulously examined based on photovoltaic and battery power interactions. Simulation results validate the effectiveness of the Maximum Power Point Tracking (MPPT) control mechanism and underscore the supervisory module's significance in protecting battery health and fortifying cathodic protection. This research underscores the pivotal role of efficient battery management in prolonging battery life and ensuring steadfast pipeline safeguarding. By harmonizing photovoltaic energy, supervisory control, and battery management, this approach presents a robust and sustainable avenue for energy optimization within cathodic protection systems.

Keywords Lithium-ion battery · Battery management · Supervision · Control command · State of charge · Hybrid system · Photovoltaic power · Optimization

---

A. Fekik (✉) · B. Nabila · H. Lydia
Department of Electrical Engineering, Faculty of Sciences and Applied Sciences, Bouira University, Bouira, Algeria
e-mail: A.fekik@univ-bouira.dz

L. Abbassen
SONATRACH, Direction Central Recherche and Development, Boumerdes, Algeria

M. L. Hamida · H. Denoun
Electrical Engineering Advanced Technology Laboratory (LATAGE), University Mouloud Mammeri of Tizi-Ouzou, Tizi-Ouzou, Algeria

© The Author(s), under exclusive license to Springer Nature Singapore Pte Ltd. 2024
A. Mellit et al. (eds.), *Proceedings of the 2nd International Conference on Green Energy Conversion System*, Green Energy and Technology,
https://doi.org/10.1007/978-981-97-6148-7_74

1 Introduction

Various metallic structures such as gas pipelines, water pipelines, etc., are often susceptible to deterioration caused by corrosion, which is defined as the destruction of metal through chemical or electrochemical reactions. One solution involves protecting the metal by making it contact another metal, creating a cell and a current flowing between the two [1, 2]. A generator is used to accelerate this process, direct the current in the desired direction, and maintain its intensity. This protection process is known as cathodic protection. Implementing this method using a photovoltaic generator offers several advantages, including its ability to operate effectively under all outdoor conditions, especially in remote locations where installations are often situated. This approach proves to be the most advantageous and truly suitable. The solar energy input is periodic due to its alternation between day and night, but it's also random as it depends on meteorological conditions. This variability necessitates the use of an energy storage system such as fuel cells, supercapacitors, and batteries, which are typically employed in photovoltaic systems and will also be used in our system [3–5]. One of the electrical characteristics commonly used to describe a battery is its Capacity. Battery capacity represents the amount of current it can supply over time. It's expressed in Ampere-hours or Watt-hours. Efficient battery management in hybrid power systems is crucial to ensure optimal utilization of stored energy. In this article, we introduce an innovative approach based on a 640Ah Lithium-Ion battery, incorporating a control command and a supervision stage to ensure both secure and efficient battery behavior, aimed at achieving optimal cathodic protection for pipelines.

2 System Modelling

The proposed system shown in Fig. 1 consists of a Photovoltaic panel, a Boost converter, a battery, and the protection system, which is modeled by a resistor:

2.1 PV Modelling

In the ideal case, the PV cell is represented by a current generator I_{ph} induced by the light flux, connected in parallel with a diode representing the PN junction of the semiconductor. However, in reality, there are several models of electrical circuits for a photovoltaic cell [6].

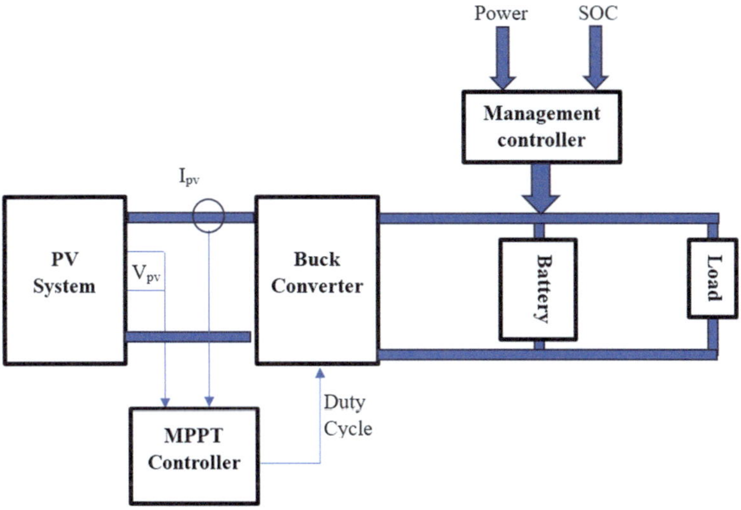

Fig. 1 Global system

Generally, this circuit introduces parasitic series resistance r_s and parallel resistance r_p in addition to the ideal circuit to account for dissipative phenomena within the cell, as shown in Fig. 2.

The series resistance r_s represents the contact and connection resistance, while another parallel resistance (the shunt resistance r_p) represents the leakage current [7]. Therefore, the mathematical model for the current–voltage characteristic of a PV cell can be described using Kirchhoff's law at the nodes through the following equation:

$$I_{pv} = I_{ph} - I\left[\exp\left(\frac{\beta}{n}(V + Ir_s)\right) - 1\right] - \frac{(V + Ir_s)}{r_{sh}} \text{ with } \beta = \frac{Q}{kT} \quad (1)$$

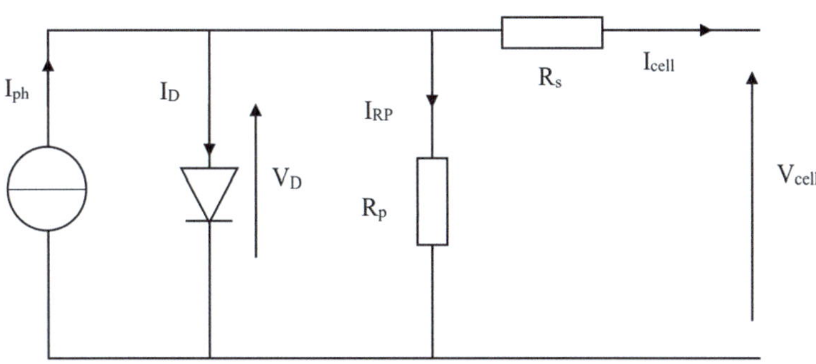

Fig. 2 Electrical model of a photovoltaic cell

where:

I_{ph}: The photocurrent generated by photons.

I: The saturation current in Amperes.

I_{pv}: The current delivered by the PV cell.

Q: Electron charge (1.602 × 10^− 19 C).

K: Boltzmann constant (1.381 × 10^− 23 J/K).

T: Effective temperature of the cell in Kelvin.

r_s: Series resistance characterizing various contact and connection resistances, and the resistivity of collector grids in ohms (Ω).

r_{sh}: Shunt or parallel resistance that model junction leakage currents, depending on how it is implemented in ohms (Ω).

2.2 Buck Converter Modeling

The operating principle of the Buck converter relies on the periodic switching of the power switch K to create state transitions that enable the storage of energy in an inductor and its release to a load. When the switch is closed (from 0 to DT), the diode is blocked, and current flows through the inductor, storing energy. When the switch is open (from DT to T), the diode becomes conductive, and current flows through it to the load.

However, it should be noted that this device adheres to the principle of energy conservation, so if $V_{in} > V_{out}$, then $I_{in} < I_{out}$ [8].

There are thus two possible topologies resulting from a given position of the switch. By applying Kirchhoff's laws to each topology, we can derive the equations that define it. It's important to note that the following modeling is done for continuous conduction mode.

- From 0 to DT: The switch K is closed, and the diode is reverse-biased, as shown in Fig. 3.

Applying Kirchhoff's voltage law to this circuit yields the following equations:

$$V_{in}(t) = V_L(t) + V_{out}(t)\text{with}V_k(t) = 0v \quad (2)$$

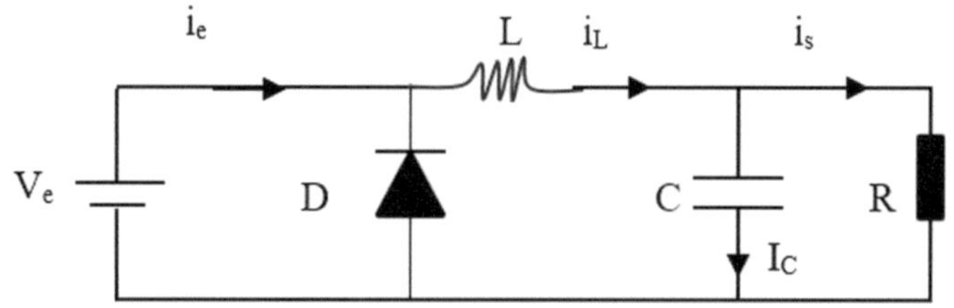

Fig. 3 Equivalent circuit of the Buck converter with switch K closed

This implies:

$$V_L = V_{in}(t) - V_{out}(t) \tag{3}$$

So

$$L\frac{di_L(t)}{dt} = V_{in}(t) - V_{out}(t) \tag{4}$$

$$i_L(t) = \frac{V_{in} - V_{out}}{L}t + I_{Lmin} \tag{5}$$

We know that for t = 0, we will have a minimum current I_L and for t = DT, we will have a maximum current I_L, so we can deduce that:

$$i_L(DT) = I_{Lmax} = \frac{V_{in} - V_{out}}{L}DT + I_{Lmin} \tag{6}$$

So, the peak-to-peak current ripple Δi_L can be determined by the following formula:

$$\Delta i_L = I_{Lmax} - I_{Lmin} = \frac{V_{in} - V_{out}}{L}DT\,\text{Avec}\; T = \frac{1}{f_s} \tag{7}$$

- From DT to T: The switch K is open, and the diode is conducting, as shown in Fig. 4.

Applying Kirchhoff's voltage law to this circuit yields the following equations

$$V_L(t) = -V_{out}(t)\text{with } V_{out}(t) = V_c(t)\text{et } V_d(t) = V_c(t) \tag{8}$$

$$-L\frac{di_L(t)}{dt} = V_{out}(t) \tag{9}$$

$$i_L(t) = \frac{V_{out}}{L}t + I_{Lmax} \tag{10}$$

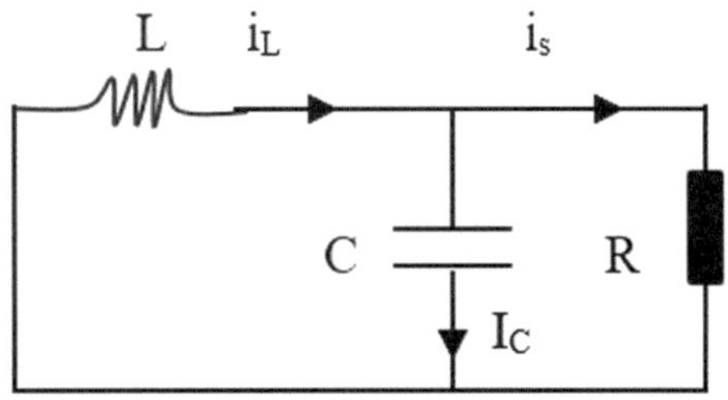

Fig. 4 Equivalent circuit of the Buck converter with switch K open

We know that for $t = DT$, we will have a maximum current I_L and for $t = T$, we will have a minimum current I_L, so we can deduce that:

$$i_L(T) = I_{Lmin} = \frac{-V_{out}}{L}(T - DT) + I_{Lmax} \tag{11}$$

So, the peak-to-peak current ripple Δi_L can be determined by the following formula:

$$\Delta i_L = I_{Lmax} - I_{Lmin} = \frac{V_{out}}{L}(\mathrm{T} - DT) with T = \frac{1}{f_s} \tag{12}$$

In steady-state, the energy stored in each component of the Buck converter remains constant at the beginning and end of each switching cycle. Consequently, the current I_L flowing through the inductance also remains constant at the beginning and end of each switching cycle. This implies that the current in the inductance does not change significantly during a switching period and remains practically stable [9]. This allows us, based on Eqs. (7) and (12), to obtain:

$$\frac{V_{out}}{L}(\mathrm{T} - DT) = \frac{V_{in} - V_{out}}{L} DT \rightarrow \frac{V_{out}}{V_{in}} = D \tag{13}$$

2.3 Management of Battery Charging and Discharging

This module comprises a battery with a capacity of 640 Ah, featuring a programmable state of charge, along with a logical supervisory control that enables the regulation of battery behavior in critical situations using three switches: "Charging", "Direct", and "Discharging".

We can express the charging behavior of the battery (P_{load}) using equations:

When P_{pv} is greater than P_{load} (battery charging):

$$P_{Load} = P_{pv} + P_{bat} \tag{14}$$

When P_{pv} is less than P_{load} (battery discharging):

$$P_{Load} = P_{pv} - P_{bat} \tag{15}$$

where:

P_{load}: Battery charging power (positive for charging, negative for discharging).

P_{pv}: Photovoltaic power (solar panels).

P_{bat}: Battery power (positive for discharging, negative for charging, depending on the charging/discharging conditions in the table and the "Charging", "Direct", "Discharging" switches).

3 Simulation Results

In order to verify the robustness of the MPPT control, we implemented the (P&O) algorithm in the impedance adapter section of the Simulink block diagram of the PC system. Variations in the illumination was carried out using "Signal Builder" blocks n Simulink, as illustrated in Fig. 5.

The current of the Buck adaptation stage follows this variation in the same direction, while delivering a continuous voltage of 48 V with a duty cycle around the same value (60%). This is because the illumination has a significant effect on the current delivered by the PV generator to the load, as shown in Fig. 6. These results validate the operation of the MPPT control.

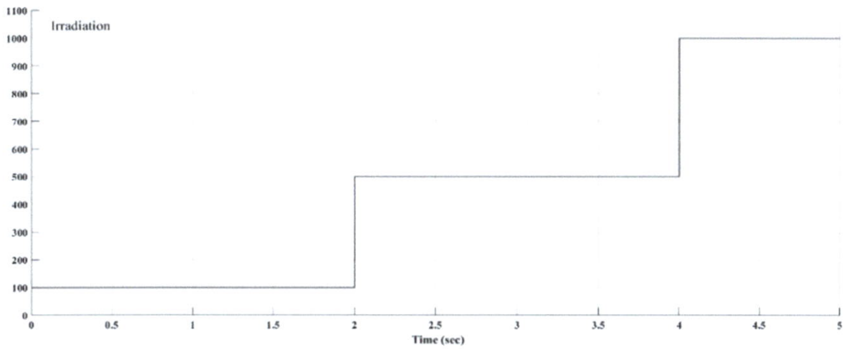

Fig. 5 The signal generated by the Signal Builder block to vary the irradiation

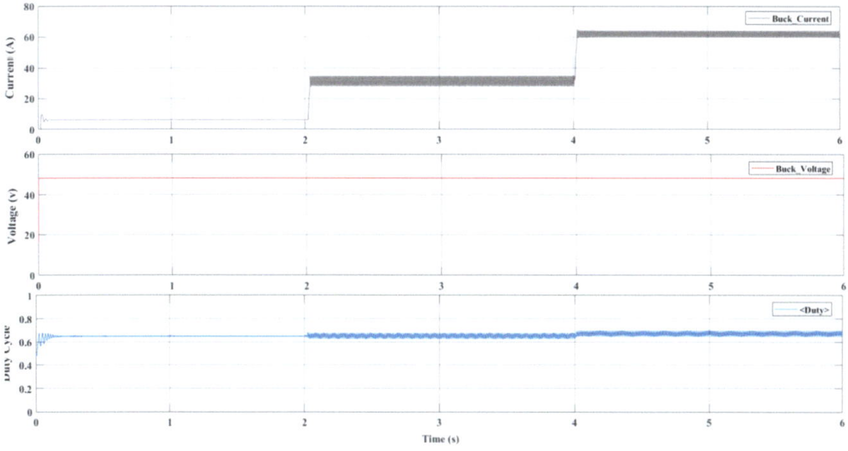

Fig. 6 Effect of irradiation on the current and voltage at the output of the DC–DC Buck converter relative to MPPT control

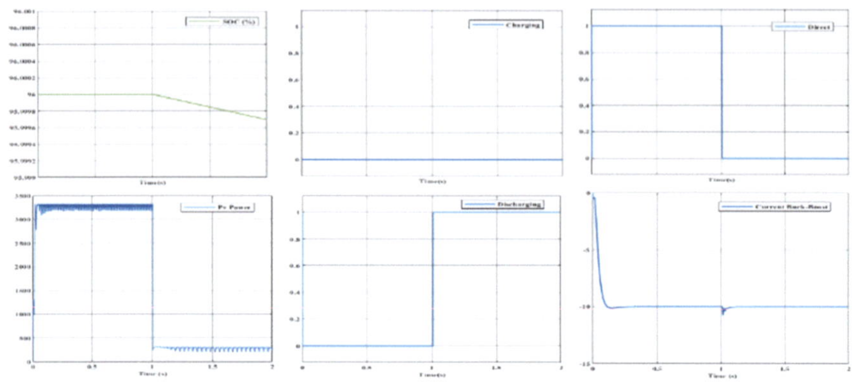

Fig. 7 State of Charge (SOC) greater than 95%

Figures 7 and 8 display the battery parameters and the performance of the proposed supervision stage in ensuring battery protection and cathodic protection system efficacy. The results have been presented through graphs illustrating the battery's state of charge (SOC), as well as the corresponding charge and discharge characteristics. These data reveal the impact of battery management on charge and discharge currents based on SOC and specific charge parameters. Under normal operation, the battery charges when the power delivered by the PV generator is sufficient. It interrupts the charging cycle during periods of insufficient production to provide the necessary energy supplement, thus fulfilling the energy demand of the load. It's important to note that the charging current (negative) may vary based on the battery's SOC, charger capacity, and specific charge parameters.

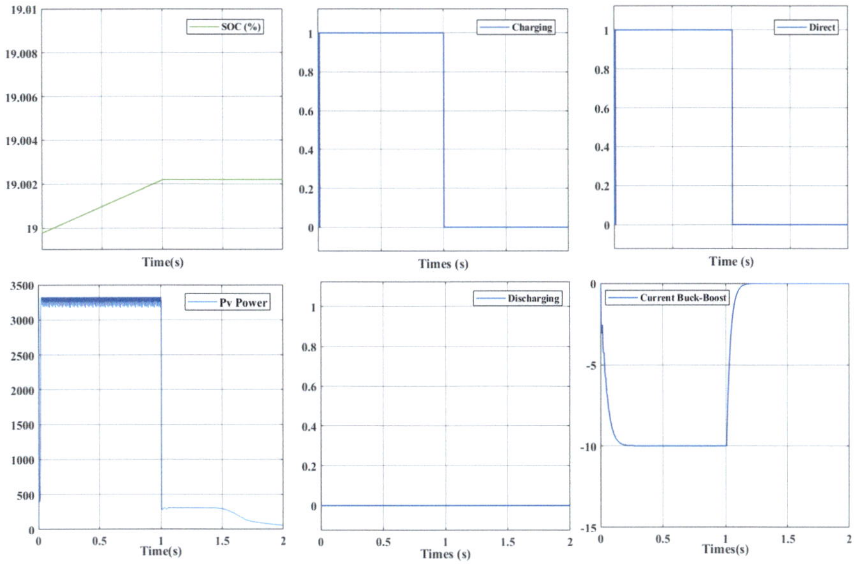

Fig. 8 SOC below 20%

4 Conclusion

This article has presented an innovative approach to optimize battery management in the context of a cathodic protection system for pipelines powered by a photovoltaic source. The utilization of a 640 Ah Lithium-Ion battery in conjunction with a supervisory stage and control mechanism has been demonstrated as an effective method to ensure both secure and efficient battery behavior in critical scenarios. Various battery SOC states have been examined, illustrating how the battery management reacts according to different charge levels. The supervisory stage has played a vital role in safeguarding the battery against overcharging and excessive discharging, thereby ensuring an extended battery lifespan. This promising approach to battery management in cathodic protection systems powered by photovoltaic energy showcases how the synergy between the battery, supervisory stage, and control mechanism can ensure dependable performance, prolonged battery life, and effective pipeline protection. Consequently, this contributes to a more sustainable and cost-effective utilization of energy resources.

References

1. Assad H, Ganjoo R, Sharma PK (2023) Corrosion inhibitors: introduction, classification and selection criteria. In: Grafted biopolymers as corrosion inhibitors: safety, sustainability, and efficiency, pp 145–169

2. Davis JR (2000) Corrosion control by cathodic and anodic protection. In: Corrosion: understanding the basics, pp 407–427
3. Fekik A, Azar AT, Hameed IA, Hamida ML, Amara K, Denoun H, Kamal NA (2023) Enhancing photovoltaic efficiency with the optimized steepest gradient method and serial multi-cellular converters. Electronics 12(10):2283
4. Hamida ML, Fekik A, Denoun H, Ardjal A, Azar AT (2023) Sliding mode controller for a single-phase six-level flying capacitor inverter—application for a standalone PV system. In: Power electronics converters and their control for renewable energy applications, pp 301–325
5. Othman S, Barbot JP, Sbita L, Alali MA, Ghanes M, Fekik A (2023) Hybrid renewable energy system based on Petri nets control for a single-phase flying capacitor multilevel DC–AC converter. In: Power electronics converters and their control for renewable energy applications, pp 145–166
6. Seck A, Thiam M, Sarr M, Ka I, Wade M (2022) Comparative study of photovoltaic models using simulation and experimental studies. Int J Renew Energy Res 12:1705–1711
7. Aouchiche N (2020) Conception d'une commande MPPT optimale à base d'intelligence artificielle d'un système photovoltaïque, Doctoral dissertation, Bourgogne Franche-Comté
8. Leng Z, Liu Q (2017) A simple model predictive control for Buck converter operating in CCM. In: 2017 IEEE international symposium on predictive control of electrical drives and power electronics (PRECEDE). IEEE, pp 19–24
9. Fang JS, Tsai SH, Yan JJ, Chen PL, Guo SM (2021) Realization of DC–DC buck converter based on hybrid H_2 model following control. IEEE Trans Industr Electron 69(2):1782–1790

Speed and Current Control of a Brushless DC Motor Powered by a PEM Fuel Cell Using an Anti-Wind-Up PI

Yamina Jouili, Radhia Garraoui, Mouna Ben Hamed, and Lassaad Sbita

Abstract The principal technology investigated here is a "Polymer-Electrolyte Membrane (PEMFC)" fuel cell stack, which is used to power a DC motor without a brush (BLDC). In order to regulate the current and speed of brushless DC motors, this work presents a back-calculation method specifically made for anti-wind-up controllers. The performance of the proposed controller is analyzed via simulated operations at varying operational settings, such as with varying reference speeds. The proposed controller is compared to the standard proportional-integral (PI) controller in terms of overshoot, errors, rise times, and settling times. Matlab/Simulink is used to model the system, allowing us to see how the suggested brushless DC motor reacts to the proposed controller.

Keywords PEM fuel cell · BLDC motor · PI controller · Anti-windup PI-back calculation controller

1 Introduction

The Polymer-Electrolyte Membrane Fuel Cell (PEMFC) is a fascinating device that could change how we drive innovation and sustainability in the fields of energy conversion and propulsion systems. This first part sets the stage for the next part, which talks about PEMFC uses in the context of brushless DC motors. This shows how these two new technologies work well together. PEM fuel cells have recently gotten a lot of attention as a possible source of energy because they are very efficient, don't give off much pollution, and don't hurt the earth much [1, 2]. They use the power made when hydrogen is burned to make up for the pollution caused by internal

Y. Jouili (✉) · R. Garraoui · M. B. Hamed · L. Sbita
Department Electric-Automatic, University of Gabes, Omar Ibn Elkatab Road, 100190 Gabes, Tunisia
e-mail: jouiliyamina@gmail.com

PEESE (Processes, Energetics, Environments Electrical Systems) Research Laboratory LR18ES34, University of Gabes, 10587 Gabes, Tunisia

© The Author(s), under exclusive license to Springer Nature Singapore Pte Ltd. 2024
A. Mellit et al. (eds.), *Proceedings of the 2nd International Conference on Green Energy Conversion System*, Green Energy and Technology,
https://doi.org/10.1007/978-981-97-6148-7_75

combustion engines. On the other hand, brushless DC motors are popular in areas as different as aerospace and robotics because of how precise and flexible they are. Because they can be used for a lot of different things and work well, they are very useful. Combining PEMFCs and brushless DC motors is a big step forward in the search for energy sources that are good for the environment. Due to their high energy density and quick response time, PEMFCs can be used to power brushless DC motors in a way that is both efficient and good for the environment. This partnership can be used in electric cars, renewable energy systems, and many other places where reliable power distribution and precise regulation are important. Here, we talk about PEMFC technology, its perks, and how it can be used to power brushless DC motors. Together, these two technologies hold the promise of a cleaner, more efficient, and more sustainable future.

Control devices frequently use PI controllers to regulate mechanical as well as electrical speed and current. These controllers can experience integrator wind-up, which causes delayed and oscillatory responses due to error accumulation. Anti-wind-up approaches target this issue to enhance the system's functioning and reliability.

This paper is organized as follows: Sect. 2 details Modeling and describing a PEM fuel cell and Sect. 3 describes a mathematical model commonly describing the electrical and mechanical behavior of a BLDC motor. Section 4 defines The PI anti-wind-up control system. Section 5 is presenting the simulation results. The conclusion of the paper is outlined in Sect. 6.

2 Modeling and Describing a PEM Fuel Cell

2.1 The PEMFC Operating Principle

The fuel and oxidizer in PEM fuel cells are hydrogen (H2) and air (O2). Catalytically splitting hydrogen into protons (H +) and electrons (e−) at the anode. An electric current is created by protons passing through the electrolyte membrane and electrons flowing through an external circuit.

Air oxygen interacts with protons and electrons at the cathode to make water. Another catalyst assists this process. Only water is created by this electrochemical process, making PEM fuel cells ecologically neutral.

2.2 Modeling

In this presentation, a rudimentary model of the proton-exchange membrane fuel cell is depicted, focusing exclusively on its voltage response. The determination of the voltage response in a Proton Exchange Membrane Fuel Cell (PEMFC) involves the deduction of the reversible thermodynamic voltage from the voltage losses incurred due to reaction activation, gas diffusion, and ohmic resistances.

The Eq. (1) can be utilized for the purpose of calculating the voltage output [2–4].

$$V_{cell} = E_{Nest} - V_a - V_c - V_{ohm} \tag{1}$$

The voltage drops represented by terms like () occur when the first term is the thermodynamic potential.

As a result, we can determine the voltage response of the fuel cell stack by

$$\mathrm{V_{fc}} = \mathrm{N}_b \cdot \mathrm{V_{cel}} \tag{2}$$

N denotes the quantity of fuel cells contained within the stack.

In [1, 2], the voltage drops of the PEM membrane fuel cell model are described mathematically. A typical elementary cell has a nominal voltage of 0.8 V and a maximum current density of 1 A/cm2. The total efficiency and effectiveness of the PEFMFC system is dependent on these voltage drops. High power output is possible thanks to the nominal voltage of 0.8 V and the maximum current density of 1 A/cm2.

The characteristics of the Proton Exchange Membrane (PEM) fuel cell are illustrated in Fig. 1. The shape of the curve is contingent upon voltage losses. The cell voltage initially begins at its highest value and subsequently reduces due to the impact of a drop in voltage [10].

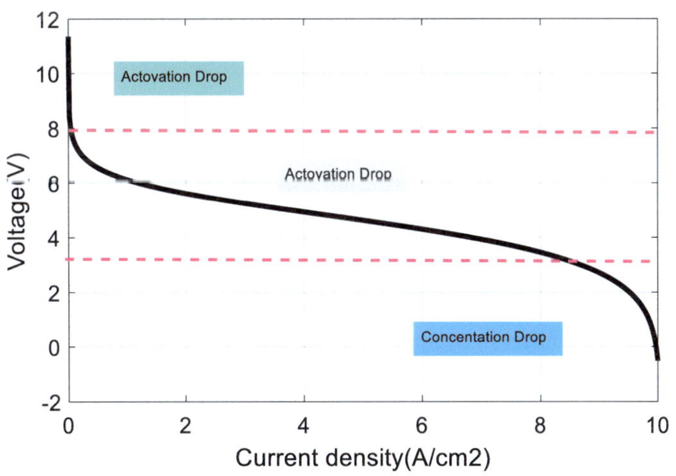

Fig. 1 The polarization characteristic of the Proton Exchange Membrane Fuel Cell (PEMFC)

3 BLDC Motor

A set of equations in a mathematical model commonly describe the electrical and mechanical behavior of a BLDC motor. It is possible to define the voltages at the BLDC motor's output using the following matrices:

$$\begin{bmatrix} Vo_a \\ Vo_b \\ Vo_c \end{bmatrix} = \begin{bmatrix} r\ 0\ 0 \\ 0\ r\ 0 \\ 0\ 0\ r \end{bmatrix} \cdot \begin{bmatrix} io_a \\ io_b \\ io_c \end{bmatrix} + \begin{bmatrix} Li - Mi & 0 & 0 \\ 0 & Li - Mi & 0 \\ 0 & 0 & Li - Mi \end{bmatrix} \cdot \frac{d}{dt} \cdot \begin{bmatrix} io_a \\ io_b \\ io_c \end{bmatrix} + \begin{bmatrix} ea \\ eb \\ ec \end{bmatrix} \tag{3}$$

The variables re, L_i, and M_i denote the stator resistance, self-inductance, and mutual inductance of the winding, respectively. The symbols $io_{(abc)}$ and $Vo_{(abc)}$ are used to denote the currents and voltages in each phase of the BLDC motor, respectively.

The motion of a mechanical system can be described by the following expression:

$$T_{em} = \frac{1}{J}\frac{d\omega_m}{dt} + \beta.\omega_m(t) + TL(t) \tag{4}$$

where T_L is the torque that the load exerts, J is the moment of inertia, and β is the coefficient of static friction.

4 The PI Anti-WIND-UP Regulator

The PI anti-wind-up control system integrates the proportional and integral components while incorporating an anti-wind-up mechanism. This mechanism prevents excessive error accumulation in the integral term, which has the potential to induce instability or other undesirable effects in the control structure.

The anti-wind-up mechanism incorporated inside a proportional-integral (PI) controller serves to restrict the buildup of errors by modifying the integral term in response to the saturation of the control signal. Furthermore, the PI Anti-Wind-Up regulator exhibits a high level of versatility, making it readily deployable across a range of control systems. Its extensive adoption is observed in diverse industrial sectors, including robotics, process control, and automotive applications.

5 Simulation Results

The simulation work is based on the control scheme mentioned in Fig. 2 which contains two regulation loops, an external regulation loop for the speed and an internal regulation loop for the current, these two regulation loops are based on the conventional PI and the PI anti wind up controllers.

5.1 Gas Pressure at Standard Conditions

In this scenario, the proton exchange membrane fuel cell (PEMFC) generator works at a consistent voltage value (24.4 V), while maintaining a hydrogen partial pressure of 1.486 atm and an oxygen partial pressure of 0.98 atm as mentioned in Fig. 3a–c illustrate the representation of the output power as well as the voltage output signals originating from the Proton Exchange Membrane Fuel Cell (PEMFC) and the DC/DC boost converter.

Additionally, the duty cycle signal is also displayed by its respective curve in the Fig. 3d. The presented curves illustrate the performance characteristics of the Perturb and Observe (P&O) technique employed for the purpose of monitoring and sustaining the optimal operational efficiency of the Proton Exchange Membrane Fuel Cell (PEMFC) stack.

Figure 4 shows the torque and back EMF of a BLDC motor curves. The PI anti-wind up with a back-calculation mechanism is used to regulate the velocity and the current of the BLDC. The motor curve torque showed acceptable ripples. This shows

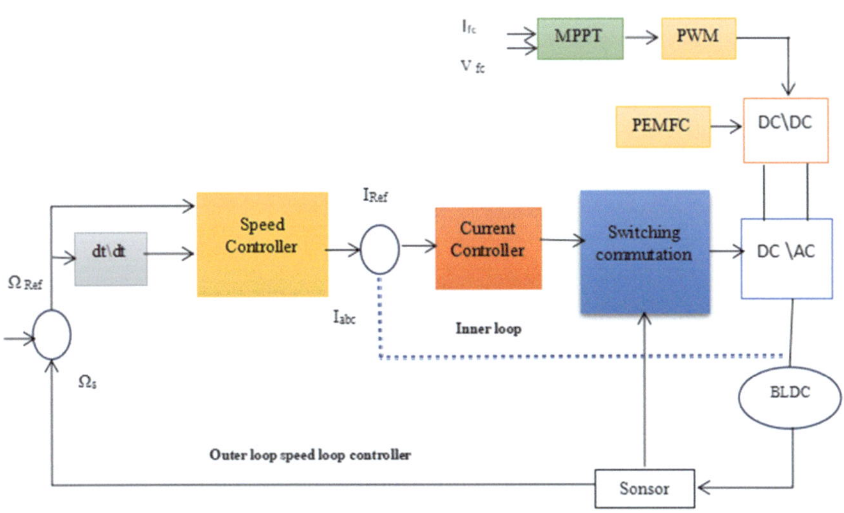

Fig. 2 Cascaded control structure of BLDCM driver

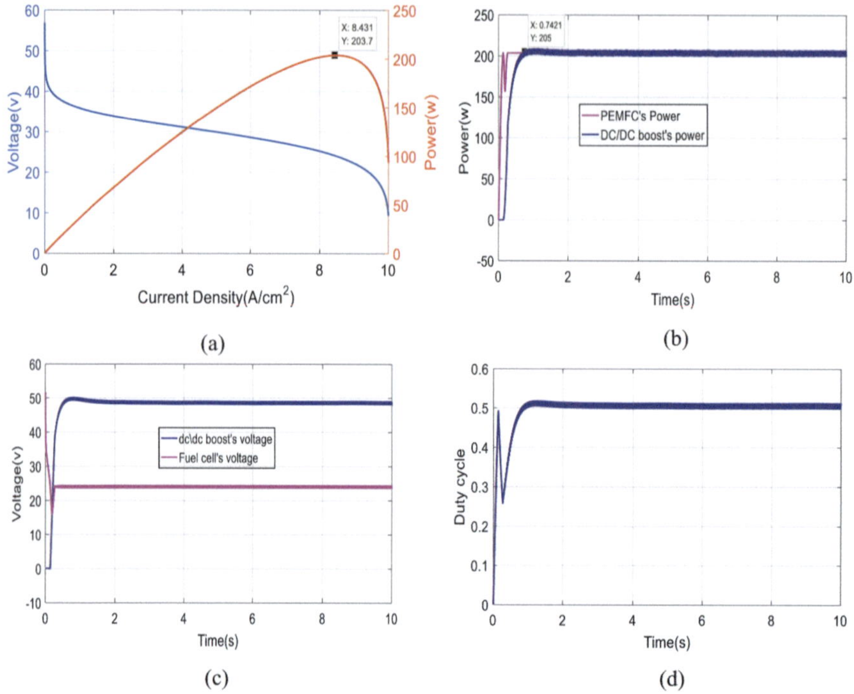

Fig. 3 **a** PEM fuel cell generator characteristics, **b** DC link power and PEMFC generator power, **c** DC boost and fuel cell generator output voltage and **d** duty cycle response under constant conditions

that the BLDC motor can continue to operate consistently under a range of torque loads and speeds.

The simulated speed responses of a BLDC motor with both conventional PI controller and PI anti-wind-up controller are shown in Fig. 5. A load torque of 0.5 Nm is used in these observations. The recommended controller enhances the reliability of the BLDC motor. The recommended controller: PI anti-wind-up promptly returns the system to its intended value, reducing undershoot, reducing rise time, and accelerating steady-state response in various operating conditions.

5.2 *Variable-Speed Scenario Outcomes*

The speed reference pace shifts from 900 tr/min to 980 tr/min at $t = 4$ s, and then from 980 tr/min to 920 tr/min at $t = 7$ s. By simulating changes in the reference speed mentioned in Fig. 6.

The effectiveness of the suggested controller was tested. So, the rise time (R.T.), the setting time (S.T.), steady-state error (S.S.E.), overshoot (OVER), and undershoot (UNDER), As part of the control performance traits, peak time (P.T.) and peak

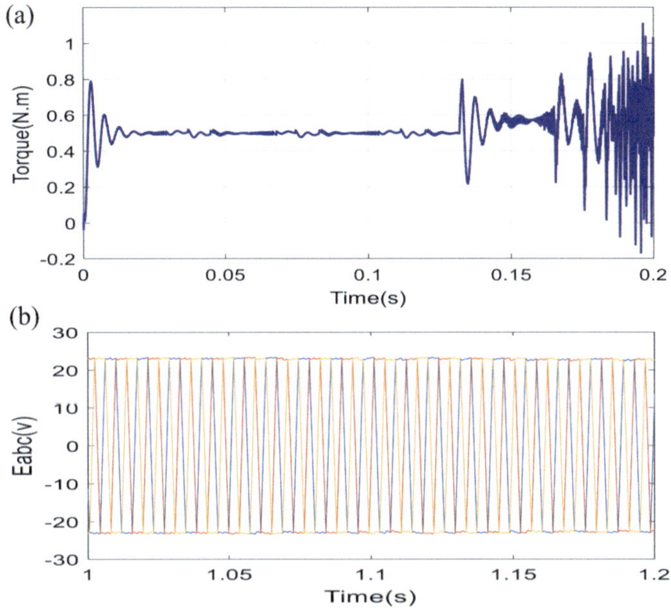

Fig. 4 **a** Torque curves of BLDC motor and **b** stator back EMF_{abc} curves of BLDC motor

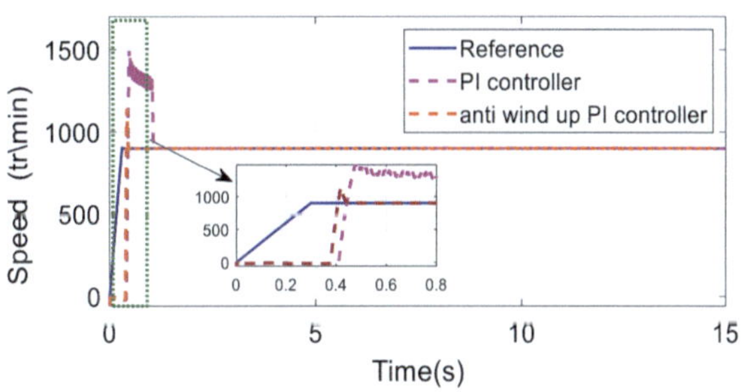

Fig. 5 Standard operating-condition speed response from a brushless DC motor

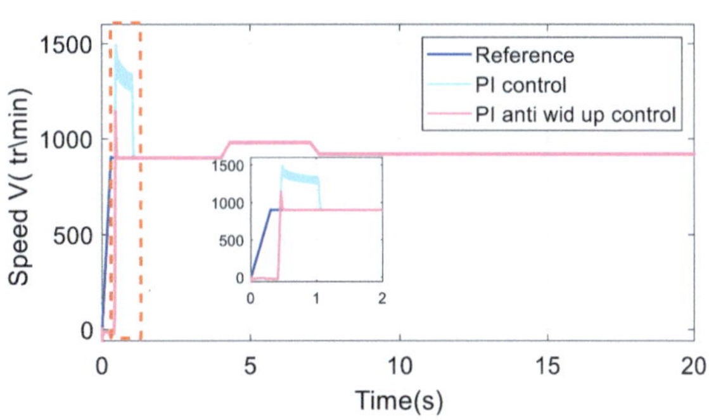

Fig. 6 A speed response for the BLDC motor with variable condition

Table 1 Ccontrol performance metrics for BLDC speed

Controller	R. T(s)	Peak	P. T(s)	OVER (%)	UNDER (%)	S. T	S.S.E(rad/s)
PI	0.437	1490.63	0.476	65.625	0	1.052	0
PI Anti Win up	0.437	1142.22	0.454	26.888	0	0.47	0

value (P.V.) are calculated and looked at the Table 1, it presents a control numerical performance comparisons between both controllers. This table shows that the recommended controller exhibits better settling time, overshoot, and steady-state error than the PI controller.

6 Conclusion

To summarize, this paper presents a new proposition for a PI controller that aims to address the issue of wind-up in the back-calculation procedure for the speed and current controllers of a BLDC motor, which is operated by a Proton Exchange Membrane (PEM) fuel cell. The controller that has been suggested utilizes sophisticated control algorithms in order to improve the overall performance and efficiency of the system. Extensive simulations have provided empirical evidence that the proportional-integral (PI) controller efficiently mitigates the issues related to wind-up, thereby substantially enhancing the dynamic response and stability of the motor. This enhances the dependability and agility of the Brushless DC (BLDC) motor, specifically in situations when the power supply is a Proton Exchange Membrane (PEM) fuel cell.

In general, the proposed proportional-integral (PI) controller demonstrates a significant progress in the regulation of brushless DC (BLDC) motors, particularly within the realm of sustainable energy alternatives. The effective implementation and proved benefits of integrating Brushless Direct Current (BLDC) motors with Proton Exchange Membrane (PEM) fuel cells present promising opportunities for the development of more efficient and ecologically sustainable power production systems (Table 2 and 3).

Table 2 BLDC parameters

Name	Value	Units
viscous damping	0.0003035	N.m.s
Inertia	0.000062	Kg.m2
Pole pairs	4	–
Static friction	0	N.m
Phase resistance	0.28	Ω
Phase inductance	6e–4	H
Constant Torque	0.33423	N.m./Apeak

Table 3 Fuel cell model parameters

Parameters		
Symbol	Value	Units
T	298.15	(Kelvin)
A	162	(cm2)
B	0.1	(V)
R_c	0.0003	(Ω)
L	175*10^– 6	(cm)
Ψ	23	–
J_{max}	0.062	(A/cm)
N	10	Cell
ε_1	0.95312	–
ε_2	– 0.00312	–
ε_3	– 7.7*10^-5	–
ε_4	1.87*10^-4	–
Ph_2	1.486	(atm)
Po_2	0.98	(atm)

References

1. Yamina JM, Garraoui R, Mouna BH (2020) Pem fuel cell with conventional MPPT. In: 2020 17th international multi-conference on systems, signals & devices (SSD). IEEE, pp 249–255
2. Jouili Y, Youssef MAB, Hamed B, Sbita L (2021) Brushless DC motor fed by PEM fuel cell stack for mini UAV's. In: 2021 12th international renewable energy congress (IREC). IEEE, pp 1–6
3. Derbeli M, Barambones O, Silaa MY, Napole C (2020) Real-time implementation of a new MPPT control method for a DC-DC boost converter used in a PEM fuel cell power system. Actuators MDPI 9(4):105
4. Abdalla SA, Abdullah SS, Kassem AM (2022) Performance enhancement and power management strategy of an autonomous hybrid fuel cell/wind power system based on adaptive neuro fuzzy inference system. Ain Shams Eng J 13(4):101655

5. Jouili Y, Garraoui R, Hamed MB, Sbita L (2022) Anti-windup self-adaptive PI-Fuzzy speed controller for brushless DC motor. In: 2022 IEEE international conference on electrical sciences and technologies in Maghreb (CISTEM), vol 4. IEEE, pp 1–6
6. Xie W, Wang JS, Wang HB (2019) PI controller of speed regulation of brushless DC motor based on particle swarm optimization algorithm with improved inertia weights. Math Problems Eng 2019(1):2671792
7. Tariq M, Bhattacharya TK, Varshney N, Rajapan D (2016) Fast response Antiwindup PI speed controller of brushless DC motor drive: modeling, simulation and implementation on DSP. J Electr Syst Inf TechnolOgy 3(1):1–13

Validating the Techno-Economic and Environmental Sustainability of Solar PV Technology in Agriculture Using RETScreen Experts to Assess Its Viability

Khadija Rajah, Ons Ghriss, Ali Fguiri, and Mohamed-Razak Jeday

Abstract The use of clean energy has now become a major concern for every society. Their relevance is strongly influenced by the economic impact of the facilities and their environmental, social and economic benefits compared to conventional energy sources, particularly fossil fuels (coal, gas, oil, etc.) As the global demand for sustainable energy solutions continues to rise, the integration of renewable energy sources into various sectors becomes increasingly important. This feasibility study focuses on the integration of photovoltaic (PV) systems in the greenhouse structures of Zina Fresh, a company specialized in production and export of cherry tomatoes. To achieve this, RETScreen Expert software is used to validate the techno-economic and environmental sustainability of installing a grid-connected solar photovoltaic system in the company. Climatic data from the National Aeronautics and Space Administration (NASA) for the various locations and some financial parameters were used for the assessment. From the result, all the selected locations are recommended for solar photovoltaic project from all the available indices. The annual electrical potential of the installation amounts to 430 MWh for the first year, the best payback period of 6.7 years and the maximum GHG emission reduction of 184 tons of CO2 which is equivalent to 16.9 ha of forest absorbing carbon.

Keywords ETScreen expert · Solar photovoltaic · Feasibility analysis · Greenhouse gas emission · Payback period

K. Rajah · O. Ghriss (✉) · A. Fguiri · M.-R. Jeday
Research Laboratory "Processes, Energetics Environment and Electrical Systems", National Engineering School of Gabes (ENIG), Gabes University, Omar Ibn Kattab ZRIG, 6029 Gabes, Tunisia
e-mail: onsgh@hotmail.fr

© The Author(s), under exclusive license to Springer Nature Singapore Pte Ltd. 2024

A. Mellit et al. (eds.), *Proceedings of the 2nd International Conference on Green Energy Conversion System*, Green Energy and Technology,
https://doi.org/10.1007/978-981-97-6148-7_76

1 Introduction

The global agricultural sector stands at a crossroads, facing a formidable challenge to meet the growing demand for food, fiber, and bioenergy while simultaneously addressing the pressing issues of climate change and resource scarcity [1]. In this context, the integration of renewable energy sources into agricultural practices has emerged as a promising solution to enhance both productivity and sustainability. Among these renewable energy technologies, solar photovoltaic (PV) systems have garnered significant attention for their potential to revolutionize the agricultural landscape. Solar PV technology offers a unique set of advantages that can empower farmers and agribusinesses to optimize their operations [2]. By harnessing the power of sunlight, these systems can generate clean and reliable electricity, reducing the dependence on fossil fuels and grid electricity. This not only lowers energy costs but also mitigates greenhouse gas emissions, thereby contributing to climate change mitigation efforts. Moreover, solar PV technology presents an opportunity for energy diversification and resilience, particularly in remote or off-grid agricultural regions. Despite the potential benefits, the adoption of solar PV technology in agriculture is not without challenges. The viability of such systems depends on various factors, including their techno-economic feasibility and their environmental impact. Accurate assessment of these aspects is crucial for farmers, policymakers, and investors to make informed decisions regarding the implementation of solar PV technology in agricultural settings. In this scientific article, we embark on a comprehensive journey to validate the techno-economic and environmental sustainability of solar PV technology in agriculture. To achieve this, we employ the powerful and widely recognized RETScreen Expert software as our primary analytical tool [3]. RETScreen Expert is renowned for its capability to conduct rigorous and in-depth assessments of renewable energy projects, providing insights into their feasibility, performance, and impact. Our study will delve into the specific nuances of solar PV deployment in agriculture, taking into consideration factors such as geographical location, climate conditions, crop types, and energy demand. We will assess the economic viability by examining the initial investment costs, ongoing operational expenses, and potential revenue streams associated with solar PV installations in agricultural contexts. Furthermore, we will evaluate the environmental sustainability, quantifying the reductions in greenhouse gas emissions and resource conservation attributed to the adoption of this technology. By conducting this research, we aim to contribute valuable insights to the ongoing discourse surrounding sustainable agriculture and renewable energy integration. Our findings will offer a practical guide for stakeholders in the agricultural sector, enabling them to make informed decisions that balance economic prosperity with environmental responsibility. Ultimately, we aspire to promote the widespread adoption of solar PV technology in agriculture as a pivotal step toward a more sustainable and resilient future for our planet and its food-producing ecosystems.

Table 1 Annual energy balance

Energy balance (2022)													
Month	1	2	3	4	5	6	7	8	9	10	11	12	Total
MWh	82	76	103	97	87	124	46	17	68	42	65	83	902

2 Materials and Methods

2.1 *Greenhouse Description*

Zina Fresh is a geothermal company that specializes in producing and exporting cherry tomatoes. He studied greenhouse is located in Chenchou El-Hamma Gabes Tunisie (latitude 33, 9°, longitude 9, 8°). The energy consumption of the greenhouse amounts to 902 MWh for the year 2022. The energy consumed during the year of this greenhouse are detailed in Table 1 [4].

2.2 *Simulation with RETScreen Expert Software*

The RETScreen Expert tool was used to evaluate the feasibility of implementing a 247.5 kWp solar power plant, the module will be bifacial monocrystalline technology from JA SOLAR, JAM72S30 525–550/MR/1500 V. The main characteristics of the modules are as follows (Table 2):

The annual electrical potential amounts to 488 MWh for the first year. The PV installation provides 54% of the company's total need. With the passage of time, the degradation of photovoltaic modules will cause a loss of the order of 0.5% of annual production.

Table 2 Properties of the inverter and the solar PV system [5]

Parameter	Specification
Slope	10°
Azimuth	0°
Power capacity	247.5 kWp
Type of solar panel	Mono-Si
Model	JA SOLAR, JAM72S30 525–550/MR/1500 V
Number of models	450
Efficiency	21,3%
Nominal operating cell temperature	43 °C
Temperature coefficient	COV—0,275%/°C, Isc + 0,045%/°C

3 Results and Discussion

3.1 *Financial Analysis of Solar PV System*

The financial analysis model of the "RETScreen Expert" software makes it possible to determine the viability criteria of the project studied in particular the net present value (NPV), the internal rate of return (IRR) and the return on equity (RCP). The results in the Fig. 1 show the economic viability of the solar PV project.

The IRR and the break-even point of equity payback was 6.7 years while the simple payback was 8.7 years. It was seen that the PV system's simple payback was far less than the system's life expectancy. The PV system had an IRR-assets of 17.4%. Therefore, the project will be attractive to investors.

3.2 *Carbon Analysis*

For the carbon analysis, the GHG emission factor which is part of the RETScreen Expert data base was used in the assessment. During the carbon analysis, the RETScreen Expert tool calculates the amount of GHG emission reduction by means of using solar PV as a power generation source. Figure 2 shows the results of the emission analysis.

For the carbon analysis, the GHG emission factor which is part of the RETScreen Expert data base was used in the assessment. During the carbon analysis, the RETScreen Expert tool calculates the amount of GHG emission reduction by means of using solar PV as a power generation source. The proposed case will allow an annual average reduction of GHG emissions of the order of 184 teqCO2, the equivalent of approximately 33,7 unused automobiles and trucks. It shows that the solar system contributes to the reduction of emissions and therefore support the United Nations Framework Convention on Climate Change (UNFCCC) on low carbon development strategies.

4 Conclusion

In summary, this study carried out by the "RETScreen Expert" software allowed us to calculate annual energy gains, GHG emissions and financial analysis criteria (NPV, IRR and RCP). Indeed, the technical–economic analysis and environmental impact of proposed renewable energy solutions shows that all Purely renewable solutions can reduce consumption costs and emissions GHG emissions for these structures.

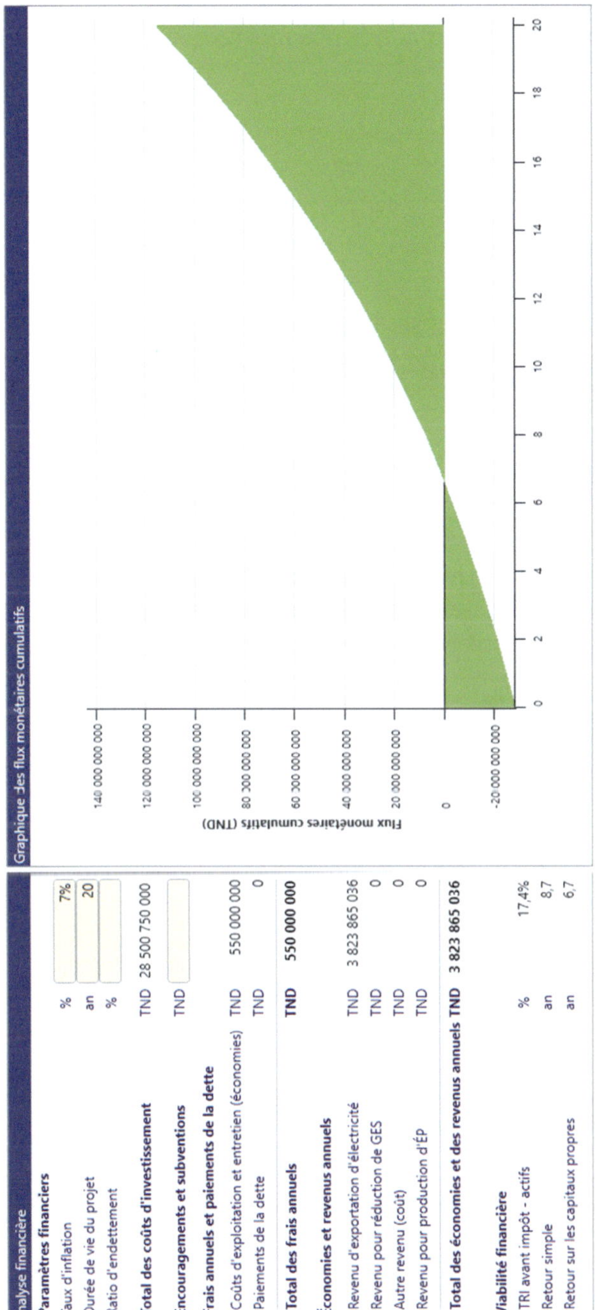

Fig. 1 Cumulative cash flow of the project

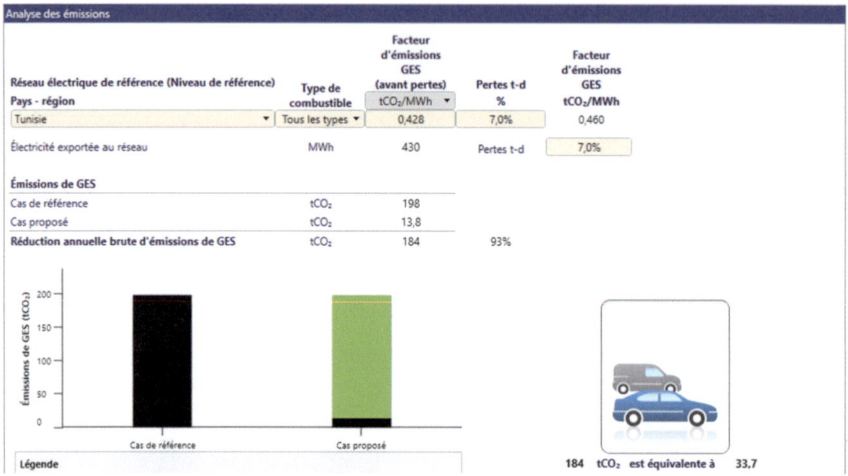

Fig. 2 The GHG emission reduction potential of solar PV project

- The annual electrical potential amounts to 488 MWh for the first year. The PV installation provides 54% of the company's total need.
- The IRR and the break-even point of equity payback was 6.7 years while the simple payback was 8.7 years
- The proposed case will allow an annual average reduction of GHG emissions of the order of 184 teqCO2, the equivalent of approximately 33, 7 unused automobiles and trucks.

References

1. Nations U (2015) Transforming our world: the 2030 agenda for sustainable development. https://doi.org/10.1201/b20466-7
2. Mi P, Zhang J, Han Y, Guo X (2021) Study on energy efficiency and economic performance of district heating system of energy saving reconstruction with photovoltaic thermal heat pump. Energy Convers Manag 247:114677
3. Yimen N, Dagbasi M (2019) Multi-attribute decision-making: applying a modified Brown-Gibson model and RETScreen software to the optimal location process of utility-scale photovoltaic plants. Processes 7(8):505
4. Hoberg V, Peterson G (2015) Multi-criteria decision-making for comparing energy choices–a Hoberg course brief
5. Technical data, Bombeatec company specializing in high-power solar pumping solutions Spain

MIX
Papier aus verantwortungsvollen Quellen
Paper from responsible sources
FSC® C105338

If you have any concerns about our products,
you can contact us on
ProductSafety@springernature.com

In case Publisher is established outside the EU,
the EU authorized representative is:
Springer Nature Customer Service Center GmbH
Europaplatz 3, 69115 Heidelberg, Germany

Printed by Libri Plureos GmbH
in Hamburg, Germany